**Association for
Computing Machinery**

Advancing Computing as a Science & Profession

e-Energy'13

Proceedings of the Fourth ACM International Conference on

Future Energy Systems

Sponsored by:

ACM SIGCOMM

Supported by:

CITRIS

**Association for
Computing Machinery**

Advancing Computing as a Science & Profession

The Association for Computing Machinery
2 Penn Plaza, Suite 701
New York, New York 10121-0701

Notice to Past Authors of ACM-Published Articles
ACM intends to create a complete electronic archive of all articles and/or other material previously published by ACM. If you have written a work that has been previously published by ACM in any journal or conference proceedings prior to 1978, or any SIG Newsletter at any time, and you do NOT want this work to appear in the ACM Digital Library, please inform permissions@acm.org, stating the title of the work, the author(s), and where and when published.

ISBN: 978-1-4503-2052-8 (Digital)

ISBN: 978-1-4503-2283-6 (Print)

Additional copies may be ordered prepaid from:

ACM Order Department
PO Box 30777
New York, NY 10087-0777, USA

Phone: 1-800-342-6626 (USA and Canada)
+1-212-626-0500 (Global)
Fax: +1-212-944-1318
E-mail: acmhelp@acm.org
Hours of Operation: 8:30 am – 4:30 pm ET

Printed in the USA

General Chairs' Welcome

Welcome to e-Energy 2013. This is an important year for e-Energy, which has become the flagship ACM Sigcomm conference on Energy Systems thanks to the support of the Sigcomm Executive Committee. It aims to be the premier venue for researchers working in the broad areas of Computing and Communication for Smart Energy Systems (including the Smart Grid), and in Energy-efficient Computing and Communication Systems.

The conference is held in the beautiful (LEED Platinum) Brower Center at Berkeley in California, not far from the UC Berkeley campus, the Lawrence Berkeley National Laboratory, and the Silicon Valley.

Our objective was to provide you a very high quality one-track conference dealing with research issues at the frontier of both ICT and Energy. Thanks to the dedicated efforts of the organizing committee and the technical program committee, co-chaired by S. Keshav and J. Kurose, we are excited to offer you an excellent program composed of 22 full paper presentations, 18 posters, one panel and two keynotes from researchers coming from universities and industries in North America, Europe and Asia.

By bringing together researchers in a high-quality single-track conference with significant opportunities for individual and small-group interaction, e-Energy 2013 serves as a major forum for presentations and discussions that will shape the future of this area.

David Culler
e-Energy 2013 Co-Chair
University of California Berkeley

Catherine Rosenberg
e-Energy 2013 Co-Chair
University of Waterloo

Message from the Technical Program Chairs

Welcome to the *ACM e-Energy 2013 Conference* being held in Berkeley, California! As Co-Chairs of the Technical Program Committee (TPC), we are delighted to introduce this year's very strong technical program. The program committee accepted 22 full papers selected from 76 submissions in addition to 16 posters and 2 demos. These represent a substantial and diverse body of work that showcases the rapid maturation of work in our field. This year's program includes papers ranging from building energy management to picogrids and data center energy management to false data injection in grid estimation systems to energy storage systems and electric vehicle charging.

ACM e-Energy is truly a global conference - both submitted and accepted papers, posters and demos have authors hailing from 20 countries. Interestingly, the final program has only two papers with affiliations to US institutions. We were fortunate to work with an outstanding technical program committee of 29 members from around the world that brought a diverse range of expertise to the table. Members of the PC were drawn from 11 countries on 4 continents, with substantial representation from Europe as well as the US. The hallmark of a high quality conference is a thoughtful and careful review process that provides valuable feedback to all authors, and *ACM e-Energy* definitely met that high bar. We are extremely grateful for the hard work, and the very thorough and insightful paper reviews, provided by TPC members.

ACM e-Energy used a two-round review process, with a first round of three reviews for each paper. Papers that did not receive a high-confidence review in the first round or received reviews with high variance were assigned additional reviews in a second round of reviewing. After the second round, we initiated online discussions on all submitted papers. The thoughtful asynchronous discussion of papers in advance of the TPC meeting allowed us to discuss 38 papers in depth during an online, day-long TPC meeting that crafted the final program.

We believe that the collective efforts of the TPC has created an extremely strong and technically vibrant program that will greatly interest and inspire all attendees while setting a high standard for subsequent editions of the conference.

S. Keshav
2013 ACM e-Energy TPC Co-Chair
University of Waterloo

Jim Kurose
2013 ACM e-Energy TPC Co-Chair
University of Massachusetts Amherst

Table of Contents

Session: Data Center Energy Management

Session: Energy Efficient Networking and Network Inference

Session: Keynote #3

Session: Smart Homes and Buildings

Session: Poster Papers

e-Energy 2013 Conference Organization

General Chairs: David Culler *(University of California, Berkeley, USA)*
Catherine Rosenberg *(University of Waterloo, Canada)*

Program Chairs: S. Keshav *(University of Waterloo, Canada)*
Jim Kurose *(University of Massachusetts, USA)*

Proceedings Chair: Anirban Mahanti *(NICTA, Australia)*

Local Arrangements Chair: Barath Raghavan *(Google, USA)*

Publicity Chair: Vincenzo Mancuso *(IMDEA Networks, Spain)*

Treasurer & Registration Chair: Sarvapali Ramchurn *(University of Southampton, UK)*

Web Chair: Omid Ardakanian *(University of Waterloo, Canada)*

Steering Committee: Hermann de Meer *(University of Passau, Germany)*
David Hutchison *(Lancaster University, United Kingdom)*
Henning Schulzrinne *(Columbia University, USA)*
Karin Anna Hummel *(ETH Zurich, Switzerland)*
Marco Ajmone Marsan *(Politecnico di Torino, Italy)*
Antonio Fernandez Anta *(IMDEA Networks, Spain)*

Program Committee: Anil Aswani *(UC Berkeley, USA)*
Suman Banerjee *(University of Wisconsin, USA)*
A. J. Brush *(Microsoft Research, USA)*
Sid Chau *(Masdar Institute, UAE)*
Florin Ciucu *(DT Labs, Germany)*
Jon Crowcroft *(University of Cambridge, UK)*
Hermann De Meer *(University of Passau, Germany)*
Nada Golmie *(NIST, USA)*
Carla Gomes *(Cornell University, USA)*
Daniel Gmach *(HP Labs, USA)*
Shiv Kalyanaraman *(IBM Research, India)*
Daniel Kofman *(LINCS, France)*
Jean-Yves Le Boudec *(EPFL, Switzerland)*
Steve Low *(Caltech, USA)*
Anirban Mahanti *(NICTA, Australia)*
Marco Marsan *(Politecnico Turino, Italy)*
Klara Nahrstedt *(UIUC, USA)*
Jean-Marc Pierson *(IRIT, France)*
Mary-Anne Piette *(LBNL, USA)*
Sarvapalli Ramchurn *(University of Southampton, UK)*

Ultra-Large Scale Control and Communications in Energy Delivery Systems

Dr. Jeffrey D. Taft
CISCO

Scaling Distributed Energy Storage for Grid Peak Reduction

Aditya Mishra, David Irwin, Prashant Shenoy, and Ting Zhu‡
University of Massachusetts Amherst   ‡Binghamton University
{adityam,irwin,shenoy}@cs.umass.edu   ‡tzhu@binghamton.edu

ABSTRACT

Reducing peak demand is an important part of ongoing smart grid research efforts. To reduce peak demand, utilities are introducing variable rate electricity prices. Recent efforts have shown how variable rate pricing can incentivize consumers to use energy storage to cut their electricity bill, by storing energy during inexpensive off-peak periods and using it during expensive peak periods. Unfortunately, variable rate pricing provides only a weak incentive for distributed energy storage and does not promote its adoption at scale. In this paper, we present the storage adoption cycle to describe the issues with incentivizing energy storage using variable rates. We then propose a simple way to address the issues: augment variable rate pricing with a surcharge based on a consumer's peak demand.

The surcharge encourages consumers to flatten their demand, rather shift as much demand as possible to the low-price period. We present PeakCharge, which includes a new peak-aware charging algorithm to optimize the use of energy storage in the presence of a peak demand surcharge, and use a closed-loop simulator to quantify its ability to flatten grid demand as the use of energy storage scales. We show that our system i) reduces upfront capital costs since it requires significantly less storage capacity per consumer than prior approaches, ii) increases energy storage's ROI, since the surcharge mitigates free riding and maintains the incentive to use energy storage at scale, and iii) uses aggregate storage capacity within 18% of an optimal centralized system.

Categories and Subject Descriptors

J.7 [**Computer Applications**]: Computers in Other Systems—*Command and control*

Keywords

Energy, Battery, Electricity, Grid, Peak shaving

1. INTRODUCTION

As is now well-known, a significant fraction of the electric grid's capital and operational expenses (CapEx and OpEx)

result from satisfying its peak power demands. For example, recent work estimates that 10%-18% of North American CapEx, in terms of energy generation capacity, is idle and wasted over 99% of the year [8]. Similarly, peak demand also influences OpEx, by i) requiring utilities to operate high cost and inefficient "peaking" generators to meet demand [1], ii) contributing to higher transmission charges, which are set based on peak demand, and iii) forcing utilities to offset supply shortages by purchasing electricity in the wholesale market at inopportune times, i.e., when it is most expensive. Thus, reducing peak demand and its impact on CapEx and OpEx is an important part of ongoing smart grid research efforts. One way to reduce peak demand that has received significant attention in the research community is leveraging energy storage to shift some demand from peak to off-peak periods. To shift demand, prior work proposes to store energy during off-peak periods, which increases off-peak demand, and use it during peak periods, which then decreases peak demand [4, 7, 11, 13, 22, 23, 24].

To implement the approach, utilities may either i) install large-scale *centralized* energy storage systems at strategic points in the grid, such as at power plants and substations [13], and directly control when they store and release energy, or ii) incentivize consumers to install and control their own small-scale energy storage systems *distributed* at buildings throughout the grid. Prior research has focused largely on the latter case, since the increasing adoption of variable rate pricing plans by utilities [6, 16, 21] provides an incentive [4, 7, 11, 22, 23, 24], and endowing buildings with energy storage has additional value-added benefits, e.g., providing power during outages and conditioning power to increase its quality. Since variable rate pricing plans charge higher rates during periods of peak demand, consumers that store energy during off-peak periods—when prices are low— and use it during peak periods—when prices are high—are able to lower their electricity bill. While many energy storage technologies exist, including pumped water storage, flywheels, and compressed air, batteries are currently the most viable option for storing energy at building-scale.

Prior research analyzes the potential savings for residential [4, 7, 23, 24] and industrial [11, 22] consumers to install batteries. The focus is largely on cost-benefit analyses using existing pricing plans, which vary electricity's price per kilowatt-hour (kWh). Unfortunately, for the reasons below, these plans *provide only a weak incentive for distributed energy storage and do not promote its adoption at large scales.*
Large Upfront Capital Costs. Since today's pricing plans typically exhibit low prices during off-peak nighttime

periods and high prices during peak daytime periods, they incentivize consumers to shift *all of their demand* to the off-peak period. Of course, the cost of batteries limits the amount of storage capacity available to shift demand. In our prior work on SmartCharge, we show that for a residential home with near the average U.S. electricity usage, ~24kWh of capacity[1] maximizes the return-on-investment (ROI) when taking into account battery costs [15]. Given typical battery lifetimes, we estimate the *annual* amortized cost to maintain 24kWh of energy storage to be $1416 [15]. Since the annual electricity bill for an average U.S. home is $1419 [5], battery costs effectively prevent (at current price levels) a positive ROI using this much energy storage.

Rebound Peaks and Grid Instability. Current pricing plans incentivize *all consumers* to charge their batteries during off-peak, low-price periods. Thus, at large scales, simultaneous battery charging during off-peak periods will trigger rebound peaks if prices do not change to reflect the resulting increases in off-peak demand. Our prior work shows that if prices do not change and 100% of consumers install 24kWh of energy storage, then the peak demand period will migrate to the (previously) off-peak period and actually *increase*, rather than decrease, peak demand by nearly 120% [15]. Note that most variable rate pricing plans in use are Time-of-Use (TOU) plans with rates that do not react quickly to changes in demand, but instead are manually reset by utilities on an infrequent basis, e.g., monthly or seasonally [16].

Uncertain Return-on-Investment. One way to prevent rebound peaks is to alter electricity rates in real-time as peak and off-peak demand changes. Although not widespread, some utilities are experimenting with real-time pricing (RTP) plans for residential consumers, where rates vary dynamically each hour based on demand [6, 21]. Unfortunately, consumers only benefit from energy storage by exploiting the difference between peak and off-peak prices. With RTP plans, as peak demand declines and off-peak demand rises due to the increasing use of energy storage, the difference between the peak and off-peak price narrows, reducing energy storage's benefits [24]. In the extreme, if grid demand is near flat then the price of electricity will be similar at all times [4, 15, 24]. Once the peak/off-peak price differential is not large enough to compensate for the conversion losses from storing energy in batteries, there is no benefit to using energy storage. Our prior work estimates that grid demand would be near flat once just 22% of consumers install 24kWh of energy storage [15], which is consistent with related work [4, 24]. Consumers are unlikely to invest in energy storage with such uncertain future long-term benefits.

Socialized Benefits and Free Riders. For residential consumers, the annual cost to install and maintain battery-based energy storage is much higher—around 10X for average consumers in the U.S.—than the annual savings on an electric bill using current battery costs, electricity rates, and pricing plans [15]. However, prior work does not consider the grid-wide reductions in generation costs from lowering the grid's aggregate peak demand. Unfortunately, with existing pricing plans, these cost savings are distributed (or socialized) across all consumers, since they manifest themselves as cheaper electricity rates. Thus, variable rate pricing plans provide a weak, non-optimal incentive for energy storage. Strengthening the incentive requires eliminating free riders

to ensure that the consumers that invest in energy storage reap its full benefits, especially given the large capital costs.

The problems above arise from the interaction between current pricing plans and battery charging algorithms that minimize cost. We argue that solving these problems requires re-designing both pricing plans and charging algorithms to explicitly encourage energy storage adoption. In particular, any charging algorithm should prevent grid instability regardless of the pricing plan, similar to how TCP prevents Internet congestion even though it does not maximize end-user bandwidth. Likewise, pricing plans should sustain, not eliminate, the incentive to use energy storage as capacity scales. Finally, the charging algorithm and pricing plan should work together to ensure a stable grid, while also maximizing energy storage's ROI at scale.

1.1 Contributions

Ideally, energy storage distributed at buildings throughout the grid would behave like centrally-controlled energy storage of equal capacity. That is, the "right" fraction of buildings would i) charge their batteries whenever grid demand is below average and ii) discharge their batteries whenever grid demand is above average, such that aggregate grid demand remains flat and constant at the average. Of course, ensuring the behavior of any self-organizing distributed system emulates that of an equivalently-sized centralized system is challenging. In this case, determining when and how many batteries should charge requires explicit feedback from the grid and coordination among all buildings, which does not scale. This paper targets an alternative approach: designing a charging algorithm and pricing plan where individual consumers (partially) flatten their own demand. As we discuss, our distributed approach does not require global coordination between consumers and the utility, and addresses each of the issues with scaling distributed energy storage.

The main drawback to incentivizing consumers to flatten their own demand is that it may require more aggregate energy storage capacity to flatten grid demand than the minimum required using a centralized approach. Since batteries are expensive, minimizing overall storage capacity and distributing it as widely as possible among consumers is critical to reducing per-consumer capital costs and increasing ROI. Our hypothesis is that, when consumers' peak demand is well-aligned, a charging algorithm and pricing plan that flattens each consumer's demand uses aggregate storage capacity near the optimal centralized approach. In evaluating our hypothesis we make the following contributions.

Scalable Design. We describe the *storage adoption cycle* that arises as energy storage scales. We show that existing charging algorithms and pricing plans cannot simultaneously minimize an electric bill and ensure grid stability at scale. In particular, preventing rebound peaks requires some (explicit or implicit) feedback from the grid to signal algorithms to rate-limit charging as demand rises. To resolve the cycle, we propose augmenting variable rate plans with a peak demand surcharge, and then modifying charging algorithms to account for it. Our system, called PeakCharge, is a complete redesign of our SmartCharge system [15] that optimizes a consumer's electricity costs in the presence of a peak demand surcharge.

Closed-loop Experimentation. We implement a closed-loop simulator, which replays traces of real household demand, using a representative generator dispatch stack,

---

[1]Operated at a maximum of 45% depth-of-discharge.

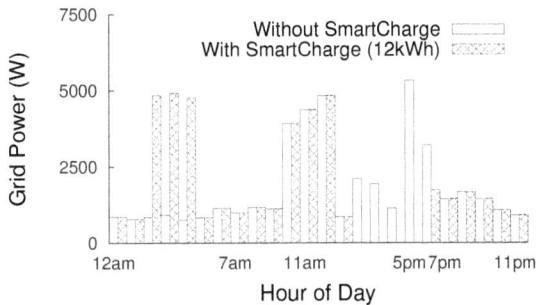

Figure 1: Prior switch-based architectures do not significantly lower an individual building's peak demand. Figure from [15].

which specifies the cost to generate electricity as demand rises, to dynamically compute electricity rates based on demand. Our simulator is closed-loop since our charging algorithm reacts to the rates, which in-turn alters demand and then changes the rates. In contrast, prior work has evaluated energy storage using only open-loop simulations, where consumer behavior does not affect prices. Using our simulator, we experimentally verify the undesirable behavior of existing charging algorithms and pricing plans at scale.

Grid- and Consumer-scale Evaluation. We evaluate both the grid- and consumer-scale effects of PeakCharge, comparing it with prior "greedy" approaches that store as much energy as possible during low-price periods. Our analysis shows that, when compared with these systems, PeakCharge i) reduces upfront capital costs since it requires significantly less storage capacity per consumer and ii) increases ROI, since a peak surcharge mitigates free riding and maintains energy storage's incentive at large scales, while requiring aggregate storage capacity within 18% of optimal.

2. OVERVIEW AND APPROACH

Our work leverages the use of battery-based energy storage systems to reduce electricity costs. We assume an intelligent battery-based energy storage system that is capable of determining when, and how much, to charge and discharge batteries based on variable electricity rates over time to minimize electricity costs. To be cost-effective, these systems must i) limit energy storage capacity due to battery costs, which, amortized over their lifetime, are currently $100-$200 per year per kWh of usable capacity for the VRLA/AGM lead acid variety widely used in stationary energy storage systems, and ii) account for the ~20% conversion loss from storing energy in batteries. Note that, since a lead-acid battery's lifetime is a function of its depth-of-discharge (DOD), a 24kWh battery operated at 50% DOD has only 12kWh of usable capacity. As in past work, we consider both the savings from batteries and their cost (20% energy loss and capital cost) when considering a system's ROI.

2.1 PeakCharge Architecture

Previously proposed architectures for leveraging energy storage [15] use a programmatic power transfer switch, which allows them to toggle a building's power supply between the grid and a battery. Thus, in addition to a charging algorithm that decides when and how much to charge batteries, the system also decides when to toggle the building's power supply between the grid and the battery, based on

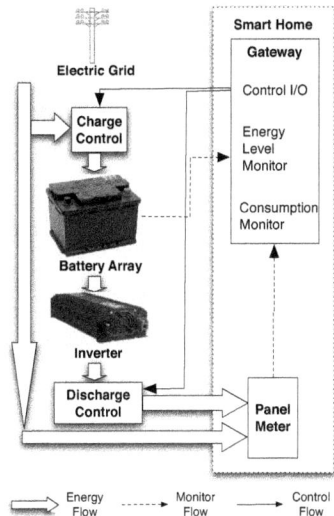

Figure 2: PeakCharge architecture, which includes a battery capable of programmatically controlling the rate of discharge wired in parallel with the grid.

expectations of future prices and demand. Of course, when batteries supply power, the building's load dictates the rate of discharge due to Kirchhoff's laws. Although not programmatic, such switches are common in commercial standby UPS systems, which automatically switch to battery power when grid voltage falls below a preset threshold. The coarse switching architecture works well in previous systems, since they connect to the grid and charge batteries during lengthy low-price periods at night before switching to battery power during lengthy high-price periods during the day.

In contrast, we assume a system architecture that is capable of controlling a battery's rate of discharge independent of the building's load. For example, if a building is consuming 1kW of power, the system is able to control the fraction of the 1kW the battery supplies, with the grid supplying the remainder. Thus, the system may choose to satisfy 1kW of demand using 500W via the battery and 500W via the grid, or using 200W via the battery and 800W via the grid. Controlling the rate of discharge is necessary for PeakCharge's approach, which encourages buildings to flatten their demand rather than simply shift large amounts of demand from daytime to nighttime. As Figure 1 demonstrates, for individual buildings, the simple switching architecture does not significantly reduce (or flatten) an individual building's peak demand. The figure (from [15]) illustrates how, due to off-peak battery charging, our prior switch-based SmartCharge system simply shifts the original peak demand to the off-peak period to minimize electricity costs.

There are two primary ways to control a battery's rate of discharge. A simple approach is to install multiple switches capable of switching separate fractions of a building's load between grid and battery power. For example, the system may be able to individually switch each circuit. In this case, the system controls the rate of discharge by monitoring the load on each circuit and switching some subset of circuits to the battery to achieve a specific rate of discharge. An alternative, cleaner approach depicted in Figure 2 is to connect the battery in parallel to the grid and use a discharge controller to programmatically limit the rate of discharge. These controllers use pulse-width modulation (PWM) to control the charge or discharge rate by connecting and disconnecting the battery at rapid frequencies. Unfortunately, controllers capable of programmatically setting the rate of discharge are not widely available, since their primary purpose today is in testing equipment [26]. However, programmatic control may become more widespread in the future, since recent work beyond our own also requires this capability [14, 25]. We assume this latter method is available to control the discharge rate in PeakCharge.

Finally, both our work and prior work derives from the fact that the marginal cost for a utility to generate each additional watt of power increases non-linearly as utilities activate additional generators to satisfy increasing demand. Utilities maintain a dispatch stack of generators: as grid demand rises utilities activate, or "dispatch," additional generators in ascending order of their marginal cost. Figure 3 shows the demand-cost function we use to compute generation costs based on demand in our closed-loop simulator, and demonstrates the non-linear relationship between cost and demand. To derive our function, we scaled real demand-cost data from the Southeastern U.S. from a 2008 report [9] by the Federal Energy Regulatory Commission (FERC) to match the peak demand in our traces, discussed in Section 5, while also ensuring a median electricity cost of 10¢/kWh, which is near the average cost of electricity in the U.S. We then fitted an exponential function to this scaled data for use in our simulations.

2.2 The Storage Adoption Cycle

Figure 4 depicts the *storage adoption cycle* that arises from the use of distributed energy storage at large scales to minimize electricity costs in the presence of variable rate electricity prices. At the top of the figure, variable demand for power first causes the price of electricity over time to change based on the demand-cost function from Figure 3. Variable pricing, in turn, incentivizes consumers to adopt energy storage to reduce their costs by shifting demand to low price periods. However, as more consumers shift demand using energy storage, the difference between the grid's peak and off-peak demand narrows resulting in a flatter grid demand profile. As a result, prices also flatten to reflect the new demand distribution. Unfortunately, flat prices eliminate the incentive to use energy storage, which causes demand to vary again and the cycle to repeat. Of course, our depiction is idealistic. It assumes, first, a high enough price differential to warrant energy storage, which is not the case today, as outlined in Section 1. Second, grid/utilities may not explicitly want to incentivize energy storage. In practice, completing each step would take a long time, potentially requiring significant regulatory changes and large capital investments. Further, the phenomenon depicted in Figure 4, and described below is only for illustrative purposes to demonstrate potential trends as the use of energy storage scales.

The storage adoption cycle may also cause grid instability if prices do not react fast enough to changes in demand. To demonstrate the potential for instability, we ran a simple experiment using our trace-driven closed-loop simulator to show how grid power demand could experience significant oscillations even if utilities alter prices each day based on the previous day's demand. Day-ahead planning is common, since consumers require some pricing feedback to adjust their behavior and utilities require some advance notice

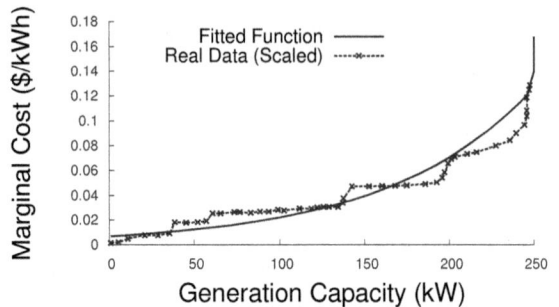

Figure 3: The model we use in our simulator of the marginal cost to generate electricity as demand increases. The fitted function we use is based on scaled data from a recent FERC report [9].

to activate generators. The simulator, which we discuss in Section 5, takes traces of demand as input; in this case, we use demand traces we collected of 194. For our experiment, 49 of the 194 homes have usable energy storage capacity that is 50% of their average daily demand, where each home uses a "greedy" charging algorithm similar to prior work [15], which minimizes electricity costs by charging as much as possible during the lowest price periods. Our simulator is closed-loop, since it computes the next day's prices each hour using Figure 3's demand-cost function and the previous day's demand.

Figure 5 shows how the peak demand periods change dramatically each day. On the first day, everyone charges during the low-price period at night (12am-6am), which increases demand during that period and, hence, also increases the price of electricity during that time on the second day. As a result, on the second day the lowest-price period shifts to the morning (6am-12pm), which is the low-demand period from the previous day, and causes peak demand to shift dramatically from the nighttime 12am-6am period to the morning 6am-12pm period. Since generators require lead time to activate, utilities carefully plan generator dispatch schedules each day based on the previous day's demand. If demand were to change dramatically each day, as in this scenario, the grid would be incapable of balancing supply and demand. This simple example highlights how naive battery charging-discharging algorithms can potentially cause grid instability for some of the existing pricing plans, like [6], [16], under certain conditions. Note that, in this paper, we do not evaluate the cost-benefit tradeoff between the excessive distributed storage (caused by our approach) and the potential affects of the storage adoption cycle.

2.3 An Optimal Approach

Before describing PeakCharge's charging algorithm, we first define and consider an optimal centralized battery charging scheme. Ideally, to minimize generation costs based on the demand-cost function from Figure 3, the optimal approach would shift aggregate grid demand such that it was the same—equal to average demand—all the time. If we assume a centrally controlled battery array, then an optimal algorithm simply charges and discharges batteries whenever grid demand is below or above average, respectively, such that demand is always equal to the average. With this algorithm, the minimum energy capacity necessary to flatten demand is equal to the maximum capacity ever required

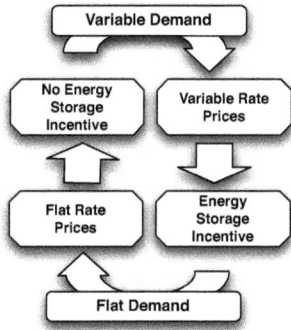

Figure 4: Illustration of the storage adoption cycle.

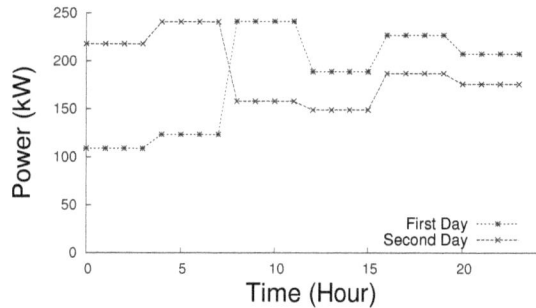

Figure 5: Load oscillations in our simulated microgrid, in presence of day-ahead real time pricing.

to charge or discharge the batteries to sustain the average. Of course, as mentioned in Section 1, a centralized system has drawbacks compared to the distributed approach, which provides value-added benefits to consumers.

As an example, Figure 6 depicts a grid demand profile from the first day of our trace of 194 homes, as well as the average demand for the day. In this case, the maximum capacity required to charge or discharge the battery occurs between hour 16 and 23 and equals 392kWh (equivalent to the area between the instantaneous demand and the average demand from hour 16 to 23). With the optimal approach this 392kWh of storage would reduce generation costs by 23% based on our demand-cost function in Figure 3. If this storage capacity were distributed evenly among all 194 homes, then each home would need only 2.02kWh of usable energy storage (or 4.5kWh of rated capacity used at 45% depth-of-discharge to maximize lifetime). This capacity is over 5X less than the 24kWh of rated capacity each home requires to maximize energy storage's ROI based on our previous SmartCharge work [15], which uses existing variable rate pricing plans and a "greedy" charging algorithm. Qualitatively, this result holds for any greedy battery charging-discharging approach for price saving, and not just SmartCharge. Since battery costs scale linearly with capacity, maintaining 5X less capacity decreases costs by 5X (from $1416 amortized per year to maintain 24kWh to $266 per year to maintain 4.78kWh). The example demonstrates how minimizing capacity, and distributing it as widely as possible among consumers reduces the ROI per consumer of energy storage.

3. SCALABLE DESIGN

The storage adoption cycle discourages distributed energy storage from scaling to a large fraction (> 20%) of consumers. Unfortunately, variable rate electricity prices incentivize consumers with energy storage to use greedy battery charging algorithms, which charge batteries as much as possible during the lowest price periods to minimize electricity costs. At large scales, the use of greedy charging algorithms results in either i) large rebound peaks (if prices do not react to changing demand), ii) grid instability (if prices react slowly to changing demand as in Figure 5), or iii) no benefit to the consumer (if prices react quickly to changing demand by flattening). None of these outcomes is desirable. Variable rate pricing is effective at reducing peak demand today only because electricity's price elasticity of demand is typically low, i.e., consumers do not react strongly to

changes in electricity's price. As a result, only a small fraction of consumer demand shifts to low price periods. In contrast, large-scale distributed energy storage makes electricity's price completely elastic with demand, causing a large fraction of demand to shift to the lowest price period.

Properly incentivizing distributed energy storage at scale requires rethinking electricity pricing plans. Our premise is that augmenting existing variable rate pricing plans with a peak demand surcharge (or penalty) is a simple and effective way of addressing the storage adoption cycle. A peak demand surcharge bills consumers X/kWh based on their peak demand over an N minute interval within some billing period M. Typical values are $N = 1$ hour and $M = 1$ day; for example, in this case, the consumer in Figure 1 would incur an additional charge for using ∼5kWh during their peak hour of that day. Utilities already use such a peak demand surcharge for large, primarily industrial, consumers. Put simply, a large peak demand surcharge incentivizes consumers to flatten their own demand to minimize their peak, rather than simply shift as much demand to the lowest price period. Of course, if consumers flatten their own demand, then grid demand will also flatten. We discuss below, how penalizing peak usage addresses the problems from Section 1. While the incentive to flatten due to a peak surcharge disappears once a home reaches peak demand, we design our proposed peak-aware charging algorithm to avoid peaks when possible. Our results indicate that the algorithm is successful most of the time.

3.1 Benefits of a Peak Demand Surcharge

First, flattening a consumer's demand takes significantly less energy storage capacity than shifting all of it to the lowest price period. For example, Figure 7 shows that, while 12kWh of usable energy storage is only capable of shifting a fraction of demand to the low price period, it is more than enough to completely flatten the original demand from Figure 1. As a result, the approach encourages distributing aggregate storage capacity widely across consumers, requiring less storage capacity per consumer, and resulting in lower upfront capital costs and higher per-consumer ROI. In effect, to flatten grid demand, the approach incentivizes a large number of consumers to install a small amount of energy storage (and make a small investment), rather than incentivizing a small number of consumers to install a large amount of energy storage (and make a large investment). Second, the approach prevents rebound peaks and grid instability, since consumers are (partially) flattening, rather than shifting, their demand. Third, the approach maintains the incentive to use energy storage as capacity scales, since con-

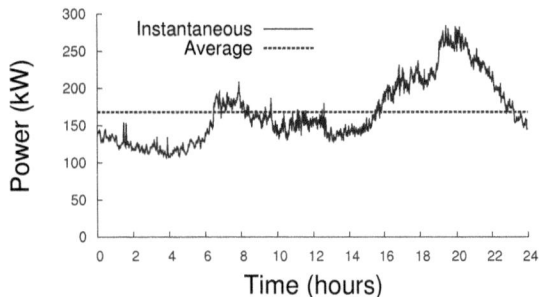

Figure 6: Instantaneous and average grid demand for 194 homes in our trace.

sumers always benefit from not paying an additional peak demand surcharge, regardless of other consumers' behavior. Finally, utilities can mitigate free riding by altering the peak demand surcharge, in addition to the electricity rate, as generation costs change, since only the set of consumers with energy storage are able to automatically optimize for peak demand. Thus, a higher peak demand surcharge and lower rates will penalize consumers with energy storage less than consumers without it.

Of course, utilities must use a peak demand surcharge in conjunction with existing variable rate schemes, since only charging based on peak demand would encourage *more* energy use. For example, with only a peak demand surcharge, if a residential consumer runs a dryer that causes their peak load to be 7kW over an hour, the consumer has *no incentive to ever reduce their demand below 7kW*. Since the home in Figures 1 and 7 has an average electricity usage of ∼1kW, billing solely based on peak demand would allow the consumer to use 7X more electricity at no extra cost. Thus, utilities must balance the size of the peak demand surcharge with the electricity rates to encourage flattening without incentivizing consumers to use significantly more electricity. We examine how the size of the peak demand surcharge affects our battery charging algorithm in Section 5.

While this paper focuses primarily on how a peak demand surcharge addresses the storage adoption cycle, it also has other benefits. For instance, homes without energy storage could reduce their electricity costs using automated load scheduling techniques that flatten demand, e.g., via Smart-Cap [3] or nPlug [10]. Consumers have little monetary incentive to use these techniques today, since most deferrable loads, e.g., refrigerators, air conditioners, heaters, dehumidifiers, are unable to defer their usage (by up to 12 hours) to low-price nighttime periods without causing significant harm, e.g., spoiled food or an uncomfortable environment. In addition, as recent work shows, flattening demand using a battery preserves privacy [14, 25], since it removes power variations that Non-Intrusive Load Monitoring (NILM) algorithms use to identify appliance usage and behavioral patterns. Unfortunately, with existing variable rate plans consumers with a battery must choose to either use it to reduce their electricity bill or preserve privacy, but not both. A peak demand surcharge could enable consumers to minimize their electricity bill *and* preserve their privacy.

3.2 Drawbacks of a Peak Demand Surcharge

The primary drawback to encouraging consumers to flatten their own demand is that, in aggregate, it may require consumers to install more energy storage capacity than nec-

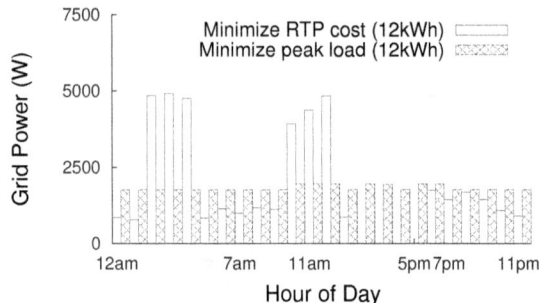

Figure 7: While 12kWh of energy storage is capable of shifting only a fraction of demand to the low price period, it is more than enough to completely flatten the demand from Figure 1.

essary to flatten grid demand. To understand why, consider a simple grid with only two homes, where each day the first home uses 1kW from 12am-12pm and 2kW from 12pm-12am, while the second home uses 2kW from 12am-12pm and 1kW from 12pm-12am. In this case, to flatten their own demand, each home requires 6kWh of energy storage for a total of 12kWh of capacity, which the homes would charge at a rate of 500W/hour when usage is 1kW and discharge at a rate of 500W/hour when usage is 2kW. However, in aggregate, the two homes' demand is already flat—using exactly 3kW all the time—without any energy storage. Thus, in this case, energy storage is not necessary. The waste occurs because the peak periods of the two homes are not aligned with each other; if their peak periods were exactly aligned then they would each require 6kWh to flatten demand (and 12kWh would be the minimum capacity necessary to flatten aggregate demand). In general, the aggregate energy storage capacity necessary to flatten grid demand by flattening each home's demand will diverge more from the optimal amount the more the peak and off-peak periods of the homes become less aligned. Of course, in practice, homes in the grid exhibit peak demand at similar times, which naturally reduces the divergence from optimal. We quantify this divergence using our closed-loop simulator and demand traces in Section 5.

As we discuss, augmenting existing variable rate plans with a peak demand surcharge requires rethinking the greedy charging algorithms used in prior work. Below, we present PeakCharge's peak-aware charging algorithm, which minimizes a consumer's electricity bill in the presence of a peak-demand surcharge.

4. PEAK-AWARE CHARGING

Our initial approach to designing PeakCharge's battery charging algorithm was to simply modify the algorithm from our prior work on SmartCharge [15]. SmartCharge uses a linear program (LP) that executes at the beginning of each day and takes as input next-day electricity prices, which are typically well-known, and expected demand each hour to determine how to charge and discharge a battery throughout the day. The LP is optimal, i.e., minimizes costs, if future demand is known. For completeness, we include a description of this LP and its constraints in the Appendix. Since future demand is not known, we use machine learning, specifically a support vector machine with a polynomial kernel, to predict next-day demand over the five multi-hour periods each day in Ontario's TOU pricing plan [16]. On average, our predictions were within 6% of demand for a representative home, and SmartCharge's LP achieved cost savings within 10-15% of an oracle with perfect future knowledge. As a result, for PeakCharge, we initially added constraints to SmartCharge's LP to account for the new peak demand surcharge. As with SmartCharge, given perfect future knowledge, the LP is optimal at minimizing costs. Unfortunately, our experimental results were far from optimal: the PeakCharge variant *did not* result in flatter consumer demand and *did not* minimize electricity costs. In this case, rates were based on the Ontario TOU scheme, and the peak demand surcharge was applied to the peak hour of each day.

We found the reason for the poor results to be that, with a peak-demand surcharge, the LP is highly sensitive to the prediction of the peak demand hour each day. Unfortunately, predicting next-day demand at the granularity of an hour is much less accurate than over the multi-hour periods used in SmartCharge. Further, ensuring high prediction accuracy for the peak hour is more difficult that ensuring a high average accuracy. In contrast, SmartCharge, which only optimized for variable electricity prices, is much less sensitive to prediction accuracy. While there are corner cases the LP is able to optimize for with accurate predictions, in general, it will always charge a battery as much as possible during the lowest-rate nighttime periods. Thus, simply charging the battery at its maximum rate overnight, regardless of the predictions of next-day demand, accounts for the vast majority of the savings in SmartCharge and other systems. One implication of a sensitivity to demand prediction accuracy is that optimizing for a peak demand surcharge becomes more difficult the longer the time interval M the peak is evaluated over, since predictions predictions are generally less accurate over longer time horizons. For instance, many utilities charge industrial consumers a surcharge for their peak demand hour within an entire month, requiring them to accurately predict demand (including the peak) each hour of the month to determine when to charge and discharge the battery using our LP above.

4.1 Optimizing for the Peak

Based on our experiences with the LP, rather than retrofit a greedy algorithm originally designed to minimize costs for variable electricity rates, to account for a peak demand surcharge, we instead began by designing an algorithm to minimize costs for a peak demand surcharge in the absence of variable electricity rates. Our starting point is the same algorithm we outlined in Section 2.3 for flattening grid demand, but applied to individual buildings. The system selects a target average power and then simply charges and discharges batteries whenever demand is below or above the target, respectively, such that demand is always equal to the average. Note that rather than run our algorithm once per day (at the beginning of the day) using predictions of next-day demand, as with SmartCharge, this algorithm naturally operates in an online manner, adjusting the charging and discharging of the battery in real time based on changing demand. This peak-centric algorithm works well as long as i) the target average is near the actual average power, and ii) the storage capacity is large enough to flatten demand.

If the target average is too small, then the approach will not store enough energy to reduce the peak by its maximum amount; if the target is too large, then it will store more

energy than necessary throughout the day. However, importantly, while the algorithm is sensitive to a prediction of average power, it does not require shorter time-scale predictions of future demand, e.g., hourly day-ahead predictions, as in the LP approach. Average power predictions over long time periods tend to be much more accurate than demand predictions over short time-scales far into the future. In fact, when predicting average power, the longer the time-scale, generally the more accurate the prediction [19], e.g., the average power of a home each year tends to vary less than each day. In addition, accurate predictions of average power over long periods, e.g., a day or month, do not require sophisticated methods [19, 20]. In this paper, to predict average demand over an interval, we simply use the average demand over the previous interval.

If the available storage capacity is too small, then the approach may discharge batteries when demand is only slightly above average, causing there to be little energy left for the highest peaks. In this case, short time-scale predictions of future demand are necessary to optimize use of the available storage capacity, i.e., save stored energy for the highest peaks each day or month. Thus, with a peak demand surcharge, the less storage capacity a consumer has, the more fine-grained and accurate the predictions required to minimize cost. However, importantly, as discussed earlier, flattening consumer demand requires much less energy storage capacity than shifting it to take advantage of variable rates.

4.2 Optimizing for Peaks and Variable Rates

Our peak-centric algorithm focuses only on flattening demand. As a result, it minimizes a consumer's electricity costs when their bill is based solely on a peak demand surcharge in absence of variable rates charged per kWh of energy use. Given our basic algorithm, we must modify it to optimize for cost in the presence of both a peak demand surcharge and variable rates. With a high peak demand surcharge the algorithm should behave like the peak-centric algorithm, and with a low peak demand surcharge the algorithm should behave greedily, i.e., by charging at its maximum rate during low-price periods. To understand the decision of whether to behave greedily or peak-centric, consider the inequality below, which compares the benefit of greedily taking advantage of variable rates versus the cost of raising peak demand. In this case, we consider only two rate periods: high and low, where C_{high} is the cost per kWh during the high rate period and C_{low} is the cost per kWh during the low rate period. In addition, T is the length of the low-price period, P is the cost per kWh of usage during the peak hour each day, e_{loss} is the energy conversion loss as a percentage stored energy (typically 80% in practice), and X_{max} is the maximum charging rate of the battery.

$$X_{max}e_{loss}C_{high}T - X_{max}C_{low}T > X_{max}P \qquad (1)$$

The left side of the inequality is the maximum monetary benefit of greedily charging the battery at its maximum rate during the low-price period and then discharging it during the high-price period, while the right side is the cost of the peak demand surcharge from charging the battery at its maximum rate. If the inequality holds then the benefit of charging greedily during the lowest-price period is greater than the cost, signaling that a consumer should act greedily. If not, then a consumer may benefit from acting peak-centric by charging (or discharging) at less than the maximum rate during low-price periods. Unfortunately, determining exactly how much less to charge (or discharge) than the maximum is challenging, requiring the same accurate short time-scale, e.g., hourly, predictions of future demand that SmartCharge's LP requires. As a result, we adopt a heuristic approach using four simple cases, as outlined below, based on whether the electricity rate is high or low and the demand is above or below average.

- If the electricity rate is **low** and demand is **below average**, then greedily charge at the maximum rate if (1) holds, else charge at a rate to sustain the target average demand.

- If the electricity rate is **low** and demand is **above average**, then greedily charge at the maximum rate if (1) holds, else discharge at a rate to sustain the target average demand.

- If the electricity rate is **high** and demand is **below average**, then greedily discharge at the full rate (bounded by the building's demand) if (1) holds, else do nothing.

- If the electricity rate is **high** and demand is **above average**, then greedily discharge at the full rate (bounded by the building's demand) if (1) holds, else discharge at a rate to sustain the target average demand.

Rather than add more cases, to extend the approach to multiple rate periods, we simply divide each period into two bins, based on whether its price is higher and lower than average, and compute C_{high} and C_{low} by taking the average of the cost per period (weighted by the length of the period) in each respective bin. Based on the cases, if the inequality holds then the algorithm simply acts greedily by charging at the maximum rate when the electricity price is low, while discharging at the maximum rate (bounded by the building's demand) when the electricity price is high. In contrast, if the inequality does not hold, then the algorithm simply toggles to using the peak-centric algorithm, with one exception. If the electricity rate is high and demand is below average, it balances the objective of the greedy algorithm, i.e., to discharge, and the peak-centric algorithm, i.e., to charge, by doing nothing. Note that, in the extreme, since variable rates are based on the grid's demand, as grid demand flattens the rates will equal each other and the algorithm will become entirely peak-centric.

Using our peak-aware algorithm above, when the peak demand surcharge is high relative to the electricity rates, the algorithm above charges and discharges the battery to hit the expected average demand; in contrast, when it is low, the algorithm devolves to a greedily charges the battery at the maximum rate during the lowest price periods.

4.3 Summary

Our *peak-aware* algorithm above optimizes for a peak demand surcharge by using inequality (1) to determine when to act greedily and when to optimize for the peak. In the next section, we compare the peak-aware algorithm with an online *greedy* algorithm that is conceptually similar to our previous LP-based approach, but operates in an online manner by charging at the maximum rate during the lowest-price periods at night and discharging during the highest price periods during the day. Since battery capacity is typically much lower than each day's energy usage, this simple variant performs similarly to our previous LP-based approach.

As an additional point of comparison, we also experiment

with a variant of the greedy algorithm with an additional congestion parameter, which limits the maximum charging rate of the battery by a factor P_{limit}, which is between 0 and 1. Enforcing a limit on the battery charging rate is a simple way to ensure grid stability and prevent rebound peaks, even using greedy charging. Of course, in practice, this parameter requires feedback and enforcement from a utility, which could either directly disseminate P_{limit} to consumers or allow them to indirectly infer it, e.g., by using subtle changes in line voltage as a signal of grid demand as in nPlug [10]. We show that while our *congestion-aware greedy variant* prevents rebound peaks and grid instability, without a peak demand surcharge, it reduces the savings (and ROI) of energy storage for consumers.

5. EVALUATION

To evaluate the charging algorithms from the previous section, we built a closed-loop simulator that takes as input traces of building energy usage. The simulator is closed-loop, since it determines the price of electricity each hour of each day based on the demand from i) the same hour on the previous day and ii) the demand-cost function in Figure 3; we call this Day-Ahead Real-Time (or DART) pricing, since each day's prices are known at the beginning of the day. In addition to DART, our simulator also supports open-loop TOU pricing, where prices do not change based on demand. As in our prior work [15], we use TOU prices based on Ontario's rates [16]; specifically, 6.3¢ per kWh from 11pm to 6am (off-peak period), 11.8¢ per kWh between 6am to 10am and 4pm to 11pm (peak periods), and 9.9¢ per kWh from 10am to 4pm (mid-peak period).

In addition to the rate plans above, the simulator also supports a peak demand surcharge in $/kW of peak usage. The surcharge applies to the highest average demand over a 30 minute sliding window across each day. Our default surcharge in the experiments below, unless otherwise noted, is $3/kW. This surcharge is high relative to the rates, i.e., inequality (1) does not hold, although with DART pricing rates may rise (since they vary every day based on demand). As in our prior work [15], we use a maximum charge rate of C/4 for the usable storage capacity, i.e., the battery charges to full capacity in 4 hours, which translates to a C/8 rate for a battery used at 45% DOD. We use power demand traces from 194 homes, which have an in-panel energy meter to record usage each minute, for ten consecutive days. While our traces are not at utility scale, i.e., with tens of thousands of residential homes as well as commercial and industrial buildings, they are sufficient to verify the trends in using energy storage at scale and to explore the behavior of our algorithm. However, the benefits of storage at scale will certainly vary based on the characteristics of each grid's (and building's) demand profile.

As with our previous work [2], we plan to make our traces available for download from our Smart* data repository located at http://smart.cs.umass.edu. Finally, we experiment with the algorithms from the previous section—greedy, peak-aware, and congestion-aware greedy—using DART (closed-loop) and TOU (open-loop) rate plans, examining both the grid-scale and consumer-scale effects.

5.1 Grid-scale Effects

We first examine the effect of rebound peaks when consumers use energy storage at large scales. Figure 8 shows the

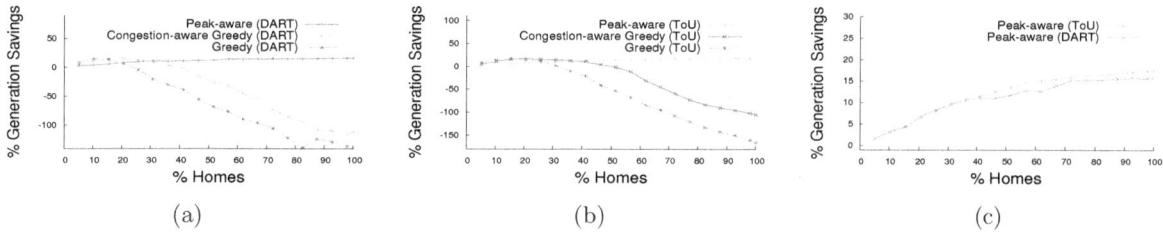

Figure 8: Generation cost savings compared to using no energy storage for both closed-loop DART (a) and open-loop TOU (b) pricing plans. Zoom-in of generation cost savings for peak-aware algorithm (c).

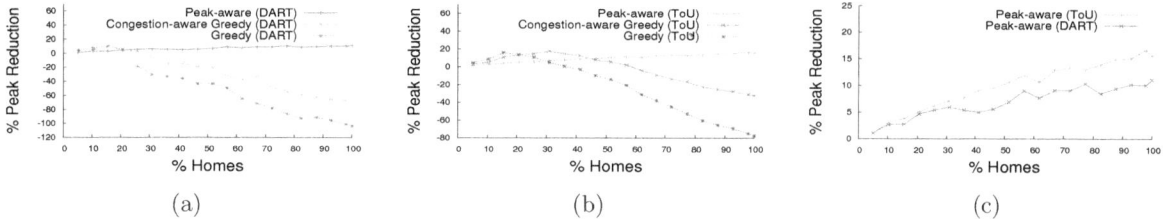

Figure 9: Peak reduction as a percentage compared to using no energy storage for both closed-loop DART (a) and open-loop TOU (b) pricing plans. Zoom-in of peak reduction for peak-aware algorithm (c)

savings in generation costs across the entire grid compared to no energy storage, as the percentage of homes using energy storage scales up with both DART (a) and TOU (b) pricing. In this case, each home has usable energy storage capacity that is 50% of their average daily demand. The graph shows that as the number of homes using energy storage scales up, the greedy algorithm, akin to our SmartCharge algorithm, increases generation costs, i.e., the savings are negative, due to simultaneous battery charging and large rebound peaks after 20% of homes use energy storage (with DART).

TOU pricing scales slightly better than DART because the closed-loop pricing results in very low rates when consumers are not charging, which reflect in the next day's prices. Since TOU rates do not change, it does not suffer from oscillations in peak demand or prices. The congestion-aware greedy variant (with a limit on the charging rate of 60% the maximum rate) in both cases is more scalable, but still results in rebound peaks once enough homes adopt energy storage (35% for DART and 45% for TOU). In contrast, the peak-aware algorithm steadily decreases the grid's generation costs as more homes use energy storage, signaling that homes and the grid are successfully flattening demand. The congestion-aware greedy variant demonstrates an important point: consumers could prevent rebound peaks by rate-limiting their charging, but they would reduce their cost savings.

Figure 8(c) zooms in on the Peak-aware(DART), Peak-aware(ToU) lines from Figure 8(a) and (b), respectively. It demonstrates the steadily increasing, rather than decreasing, cost savings as more homes use energy storage. Notice maximum savings from the three variants is similar: this reflects that in each case the aggregate energy storage is sufficient to flatten demand. The difference with the peak-aware algorithm is that it distributes this capacity across 100% of consumers, while the other algorithms distribute it across a much smaller set of consumers. Figure 9 shows the corresponding reduction in peak demand for the same experiment, which, as expected, shows similar trends as the generation cost savings. Namely, the grid's peak demand steadily decreases, rather than increases, as more homes use

energy storage with the peak-aware algorithm. As an example of this decrease in peak demand, Figure 10(a) and (b) show the time-series of power usage of an example day in our trace both with and without energy storage (for DART and TOU pricing).

We also run a similar experiment, but rather than vary the percentage of homes using energy storage we vary the amount of energy storage each home has as a fraction of its demand. In this case, 100% of homes have energy storage. Figures 11(a) and (b) show the generation cost savings and peak reduction, respectively. In both cases, the results show that each home only needs energy storage capacity that is a small fraction of its average demand. In Figure 11(a), generation cost savings stop increasing once homes have energy storage capacity that is 20% of their average demand. Similarly, Figure 11(b) shows the grid's peak not decreasing further at the same 20% threshold. At the 20% threshold, the aggregate storage capacity across homes is near the optimal storage capacity required to flatten demand (from Section 2.3). In contrast, using our previous work on SmartCharge without a peak demand surcharge, a representative home required 50% of their average demand in storage capacity to maximize their ROI [15].

The previous experiments used the same, relatively high, peak demand surcharge of $3/kW, such that inequality (1) does not hold and our peak-aware algorithm focuses on flattening demand. Figure 12 demonstrates the percentage peak reduction across the 194 homes in our traces as we vary the peak demand surcharge for both TOU and DART pricing plans. In this case, all homes use energy storage with usable capacity that is 50% of their average demand. As expected, for low surcharge values the homes are greedy resulting in large rebound peaks—larger than the original peaks—from simultaneous battery charging during low-price periods. However, once the peak demand surcharge passes the threshold defined by inequality (1) the homes switch to flattening their demand. In this case, for high peak demand surcharges the algorithm reduces the peak 10-15%. The threshold near $0.60/kW represents the tipping point

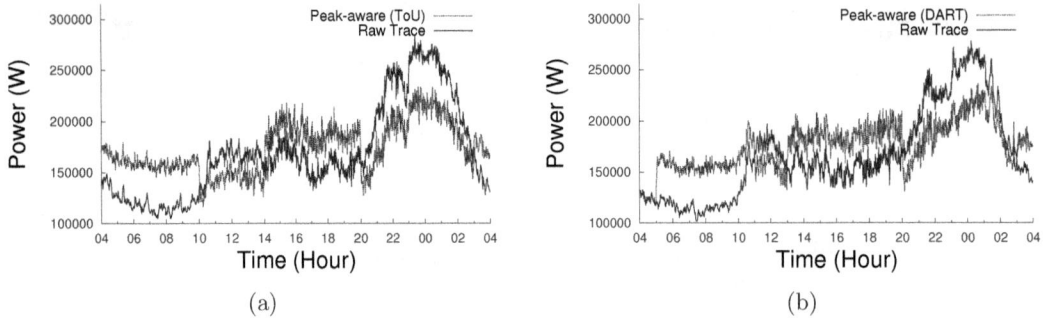

Figure 10: Time series of aggregate grid demand for TOU (a) and DART (b) pricing for both without energy storage and using our peak-aware algorithm with each home having energy storage.

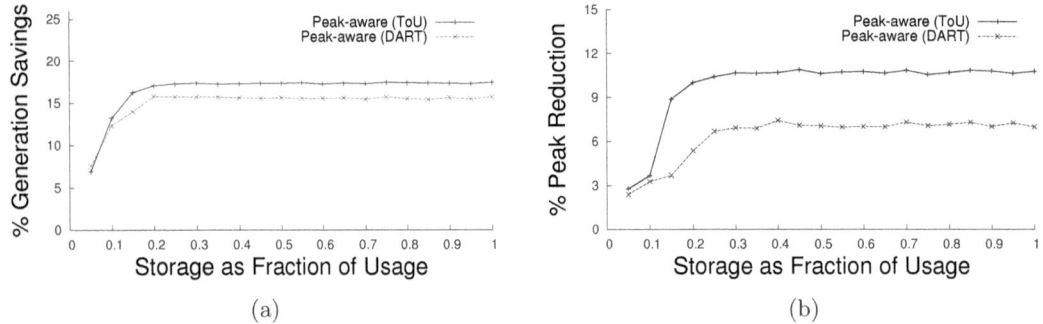

Figure 11: Cost savings (a) and grid peak reduction (b) as a function of each home's energy storage capacity.

where the benefit of charging the battery at its maximum rate during the lowest-price hour is not worth the cost of the peak demand surcharge.

Lastly, we revisit the analysis of the optimal minimum amount of energy storage capacity necessary to flatten grid demand in a centrally controlled system (from Section 2.3). For that trace, the minimum capacity required was 393kWh. For the same trace, our peak-aware algorithm requires 481kWh of aggregate capacity across the grid to maximize its peak reduction and generation savings from flattening demand. Figure 10 shows the high-level trend of aggregate demand achieved by the peak-aware algorithm relative to raw consumption. As the figure demonstrates, the peak-aware algorithm is not able to perfectly flatten demand due to inaccuracy in choosing the target average, which may result in high peaks if a home's battery is empty during a period of high demand. While predicting average demand over a long period is more accurate than predicting hourly demand over the same period, it is not perfect due to changing consumer behavior. Despite the inaccuracy, the approach comes within 18% of the optimal centralized approach.

5.2 Consumer-scale Effects

With distributed energy storage in use large scales, the peak reductions and cost savings for individual consumers mirror the reductions and savings in the grid. To show this, we use the TOU pricing scheme, which, as shown above, performs similarly to DART. We take a representative home with near the average demand for a home in our trace and look at its individual peak reduction and cost savings (as both a percentage and in dollars) as function of the home's usable energy storage capacity for our peak-aware algorithm. Figure 13 shows the results. As in the grid, the peak-aware algorithm reaches its maximum peak reduction

(Figure 13(a)) when usable storage capacity is only 20% of average demand (rather than 50% with a greedy approach like SmartCharge [15]). Further, for less than 2X the energy storage capacity, the percentage cost savings in the electric bill (Figure 13(b)) is in the same 10-15% range as we found in SmartCharge. Thus, the ROI for the consumer is much higher, since consumers achieve similar savings using much less energy storage capacity, which overwhelmingly dominates the cost of the system. Finally, Figure 13 shows the average per day dollar savings, which mirrors the trend in the overall percentage savings.

The results above clearly demonstrate the benefits, as the use of energy storage scales, of incentivizing consumers to flatten their own demand using a peak demand surcharge. For the similar cost savings per consumer using energy storage, the approach requires much less storage capacity, resulting in a higher ROI. Further, this ROI does not diminish as more consumers use energy storage, since the utility is able to control each consumer's incentivize to flatten demand using the peak demand surcharge independent of the electricity rates. This also mitigates free riding, since consumers without energy storage are less able to control their peak demand and benefit from reducing their peak. By incentivizing all consumers to use energy storage, the approach encourages distributing the aggregate energy storage necessary to flatten grid demand across a wide set of consumers, which all share in the savings. In prior work, with only variable electricity rates, the consumers not using energy storage also benefit from lower overall generation costs (and electricity rates) as storage capacity increases, which, in turn, diminishes the savings for the consumers that use energy storage.

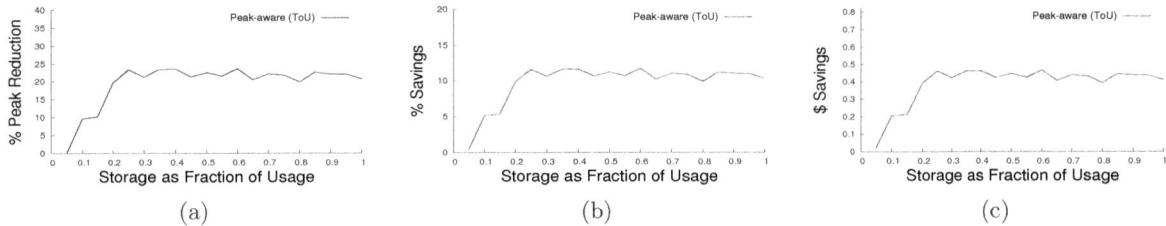

Figure 13: Percentage peak reduction (a), percentage cost savings (b), and dollar cost savings (c) for an individual home using our peak-aware algorithm as the home's energy storage capacity varies using our peak-aware algorithm under a peak-demand surcharge.

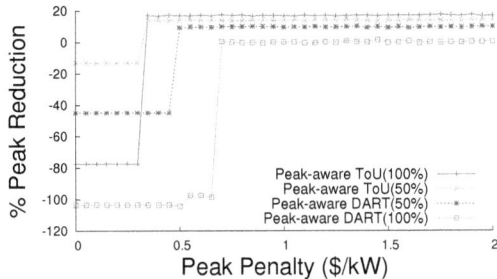

Figure 12: Increasing the peak demand surcharge prevents rebound peaks in the grid by incentivizing consumers to flatten their demand.

6. RELATED WORK

As mentioned in Section 1, numerous researchers have studied the use of energy storage at homes and buildings to shift demand and cut electricity bills under emerging variable rate electricity pricing plans. Daryanian et al. [7] was the first to propose this form of energy arbitrage. This work, as well as work by van de ven et al. [23], study the problem from a theoretical standpoint, e.g., assuming certain demand distributions, without evaluating their solutions on real data. More recently, our own work on SmartCharge [15], as well as work by Carpenter et al. [4], study a similar problem in a realistic setting taking into account battery inefficiencies, stochastic demand in residential settings, and existing variable rate pricing plans in Ontario and Illinois. Both papers mention the problems with scaling distributed energy storage to many consumers, but neither i) explores the full implications of large scale adoption, including the decreasing ROI for consumers with storage as adoption scales nor ii) proposes or evaluates a solution to the problem. Johnson et al. [12] formulate the peak shaving problem based on a pricing plan where customers are billed for their peak usage. The authors present an optimal offline algorithm, and a competitive online algorithm for solving the problem. However, they do not focus on or evaluate the proposed algorithms at large scales, as the use of energy storage increases.

While the data sets in these papers are different, they both show that ~20% of homes using energy storage maximizes the grid's peak reduction. After this point, rebound peaks and simultaneous battery charging begin to reduce energy storage's benefits, ultimately leading to a higher peak usage than without energy storage if prices do not react to demand. In earlier work, Vytelingum et al. [24] shows formally that under variable electricity rate pricing plans there is a Nash equilibrium that maximizes social welfare, e.g., cost savings, once only 38% of U.K. households use energy storage (based

on a U.K. data set). Although slightly higher than the ~20% of homes found above, the paper's trend is the same: beyond a certain point with existing variable rate electricity prices the benefits of consumers installing energy storage begin to decrease. We argue that, due to the high cost of batteries, when designing incentivizes for distributed energy storage, the goal should be to encourage the distribution of aggregate capacity as widely as possible among consumers.

While the work above focuses on residential settings, prior work has also looked at similar problems from the perspective of industrial consumers, particularly data centers [11, 22], but has not examined the impact of storage at scale. Prior work also highlights the effect of variable rate pricing on grid stability [17, 18], showing that real-time pricing has the potential to create an unstable closed feedback loop. We show this experimentally in Figure 5 in the presence of large-scale energy storage. Finally, we know of no work that proposes and evaluates using a peak demand surcharge to maintain a stable grid and prevent rebound peaks by incentivizing consumers to flatten their own demand.

7. CONCLUSION

This paper examines the effects of using energy storage distributed at buildings and homes throughout the grid to flatten grid demand. In particular, we show that as more consumers adopt energy storage, a number of problems arise that impact grid stability and generation costs. As a result, we propose to augment traditional variable rate electricity pricing plans with a substantial peak demand surcharge, which incentivizes consumers to flatten their demand rather than shift it all to a low-price period. Utilities already use peak demand surcharges for large industrial consumers; we argue that, to incentivize distributed energy storage, they may want to broaden their use to other consumers.

We then design PeakCharge, which includes an online algorithm to minimize electricity costs in the presence of variable rates and the peak demand surcharge. Using a closed-loop simulator, we show that our algorithm is effective at both i) maintaining the incentives for consumers to use energy storage at large scales and ii) ensuring grid stability. Further, our results indicate the aggregate energy storage capacity to flatten grid demand by incentivizing consumers to flatten their own demand is within 18% of the minimum, optimal capacity to flatten grid demand in a centralized system. Since flattening a home's demand requires over 2X less storage capacity per consumer to maximize consumer savings, it significantly reduces the ROI of energy storage, which is dominated by battery costs.

13

References

[1] Electric Generator Dispatch Depends on System Demand and the Relative Cost of Operation. In *U.S. Energy Information Administration: Today in Energy*, August 17th 2012.

[2] S. Barker, A. Mishra, D. Irwin, E. Cecchet, P. Shenoy, and J. Albrecht. Smart*: An Open Data Set and Tools for Enabling Research in Sustainable Homes. In *SustKDD*, August 2012.

[3] S. Barker, A. Mishra, D. Irwin, P. Shenoy, and J. Albrecht. Smartcap: Flattening peak electricity demand in smart homes. In *PerCom*, March 2012.

[4] T. Carpenter, S. Singla, P. Azimzadeh, and S. Keshav. The Impact of Electricity Pricing Schemes on Storage Adoption in Ontario. In *e-Energy*, May 2012.

[5] D. Cauchon. Usatoday, household Electricity Bills Skyrocket. http://www.usatoday.com/money/industries/energy/story/2011-12-13/electric-bills/51840042/1, December 13th 2011.

[6] CNT energy. Dynamic Pricing and Smart Grid. http://www.cntenergy.org/pricing/, 2011.

[7] B. Daryanian, R. Bohn, and R. Tabors. Optimal Demand-side Response to Electricity Spot Prices for Storage-type Customers. *TPS*, 4(3), August 1989.

[8] A. Faruqui, R. Hledik, and J. Tsoukalis. The Power of Dynamic Pricing. *Electricity Journal*, 22(3):42–56, February 2009.

[9] State of the Markets Report 2008. Technical report, Federal Energy Regulatory Commission, August 2009.

[10] T. Ganu, D. Seetharam, V. Arya, R. Kunnath, J. Hazra, S. Husain, L. DeSilva, and S. Kalyanaraman. nPlug: A Smart Plug for Alleviating Peak Loads. In *e-Energy*, May 2012.

[11] S. Govindan, A. Sivasubramaniam, and B. Urgaonkar. Benefits and Limitations of Tapping into Stored Energy for Datacenters. In *ISCA*, June 2011.

[12] M. P. Johnson, A. Bar-Noy, O. Liu, and Y. Feng. Energy Peak Shaving with Local Storage. *Sustainable Computing: Informatics and Systems*, 1(3), 2011.

[13] I. Koutsopoulos, V. Hatzi, and L. Tassiulas. Optimal Energy Storage Control Policies for the Smart Power Grid. In *SmartGridComm*, September 2011.

[14] S. McLaughlin, P. McDaniel, and W. Aiello. Protecting Consumer Privacy from Electric Load Monitoring. In *CCS*, October 2011.

[15] A. Mishra, D. Irwin, P. Shenoy, J. Kurose, and T. Zhu. SmartCharge: Cutting the Electricity Bill in Smart Homes with Energy Storage. In *e-Energy*, May 2012.

[16] Ontario Energy Board: Electricity Prices. http://www.ontarioenergyboard.ca/OEB/Consumers/Electricity/Electricity+Prices, 2012.

[17] M. Roozbehani, M. A. Dahleh, and S. K. Mitter. On the stability of wholesale electricity markets under real-time pricing. In *CDC*, December 2010.

[18] M. Roozbehani, M. Rinehart, and M. A. Dahleh. Real-time pricing with ex-post adjustments in competitive electricity markets: Volatility and efficiency analysis. In *Allerton*, September 2011.

[19] N. Sharma, J. Gummeson, D. Irwin, and P. Shenoy. Cloudy computing: Leveraging weather forecasts in energy harvesting sensor systems. In *SECON*, June 2010.

[20] N. Sharma, P. Sharma, D. Irwin, and P. Shenoy. Predicting Solar Generation from Weather Forecasts Using Machine Learning. In *SmartGridComm*, October 2011.

[21] Smart Grid Information Clearinghouse: Legislation and Regulation. http://www.sgiclearinghouse.org/Legislation?q=node/1705, 2011.

[22] R. Urgaonkar, B. Urgaonkar, M. Neely, and A. Sivasubramaniam. Optimal Power Cost Management Using Stored Energy in Data Centers. In *SIGMETRICS*, March 2011.

[23] P. van de ven, N. Hegde, L. Massoulie, and T. Salonidis. Optimal Control of Residential Energy Storage Under Price Fluctuations. In *ENERGY*, May 2011.

[24] P. Vytelingum, T. Voice, S. Ramchurn, A. Rogers, and N. Jennings. Agent-based Micro-storage Management for the smart grid. In *AAMAS*, May 2010.

[25] W. Yang, N. Li, Y. Qi, W. Qardaji, S. McLaughlin, and P. McDaniel. Minimizing Private Data Disclosures in the Smart Grid. In *CCS*, October 2012.

[26] Zivan. Battery Discharge at Constant Current: Technical Features and User Manual. http://www.zivanusa.com/pdf/SBM%20Gb.pdf, 2012.

Parameter	Definition
T	Time in T discrete intervals 1 to T
I	Length of each of the discrete time intervals
s_i	Power charged in interval i
d_i	Power discharged in interval i
e	Battery efficiency, $0 \le e \le 1$
p_i	Grid power demand in interval i
c_i	Power cost per kWh in interval i
m_i	Charge for electricity in interval i
C	Battery capacity in kWh
l_i	Aggregate grid demand in interval i.
L	**Peak grid demand over all intervals**
c'	**Peak demand surcharge**

Table 1: Parameter definitions for linear program.

APPENDIX

Below is the modification of the LP from SmartCharge to minimize an electricity bill using energy storage in the presence of a peak demand surcharge, given future knowledge of next-day prices and next-day demand each hour. Table 1 defines the optimization's parameters. The formal objective is to minimize $\sum_{i=1}^{T} m_i$ each day, given constraints below. The first five constraints are present in SmartCharge's original LP: constraints (1) and (2) ensure positive energy is charged and discharged from the battery, constraint (3) bounds the battery's charging rate, constraint (4) preserves conservation of energy (including energy conversion efficiency), and constraint (5) bounds the battery's capacity. The final three constraints (in bold) are necessary to optimize for a peak demand surcharge: constraint (6) computes the bill based on variable rate prices and the peak demand surcharge, constraint (7) represents grid's demand in the ith interval, and constraint (8) is the size of the peak demand surcharge.

$$s_i \ge 0, \forall i \in [1, T] \tag{2}$$
$$s_i \le C/4, \forall i \in [1, T] \tag{3}$$
$$d_i \ge 0, \forall i \in [1, T] \tag{4}$$
$$\sum_{t=0}^{i} d_t \le e * \sum_{t=0}^{i} s_t, \forall i \in [1, T] \tag{5}$$
$$(\sum_{t=0}^{i} s_t - \sum_{t=0}^{i} d_t/e) * I \le C, \forall i \in [1, T] \tag{6}$$
$$\mathbf{m_i} = (\mathbf{p_i} + \mathbf{s_i} - \mathbf{d_i}) * \mathbf{I} * \mathbf{c_i} + \mathbf{L} * \mathbf{c'}, \forall \mathbf{i} \in [\mathbf{1}, \mathbf{T}] \tag{7}$$
$$\mathbf{l_i} = \mathbf{p_i} + \mathbf{s_i} - \mathbf{d_i}, \forall \mathbf{i} \in [\mathbf{1}, \mathbf{T}] \tag{8}$$
$$\mathbf{l_i} \le \mathbf{L}, \forall \mathbf{i} \in [\mathbf{1}, \mathbf{T}] \tag{9}$$

Acknowledgements. We thank our shepherd, Chi-Kin Chau, and reviewers for their comments. This research was supported by NSF grants CNS-1143655, CNS-0916577, CNS-0855128, CNS-0834243, CNS-0845349, CNS-1217791.

Impact of Storage on the Efficiency and Prices in Real-Time Electricity Markets

Nicolas Gast
IC-LCA2
EPFL, Switzerland
nicolas.gast@epfl.ch

Jean-Yves Le Boudec
IC-LCA2
EPFL, Switzerland
jean-yves.leboudec@epfl.ch

Alexandre Proutiere
KTH, Sweden
and INRIA, France
alepro@kth.se

Dan-Cristian Tomozei
IC-LCA2
EPFL, Switzerland
dan-cristian.tomozei@epfl.ch

ABSTRACT

We study the effect of energy-storage systems in dynamic real-time electricity markets. We consider that demand and renewable generation are stochastic, that real-time production is affected by ramping constraints, and that market players seek to selfishly maximize their profit. We distinguish three scenarios, depending on the owner of the storage system: (A) the supplier, (B) the consumer, or (C) a stand-alone player. In all cases, we show the existence of a competitive equilibrium when players are price-takers (they do not affect market prices). We further establish that under the equilibrium price process, players' selfish responses coincide with the social welfare-maximizing policy computed by a (hypothetical) social planner. We show that with storage the resulting price process is smoother than without.

We determine empirically the storage parameters that maximize the players' revenue in the market. In the case of consumer-owned storage, or a stand-alone storage operator (scenarios B and C), we find that they do not match socially optimal parameters. We conclude that consumers and the stand-alone storage operator (but not suppliers) have an incentive to under-dimension their storage system. In addition, we determine the scaling laws of optimal storage parameters as a function of the volatility of demand and renewables. We show, in particular, that the optimal storage energy capacity scales as the volatility to the fourth power.

Categories and Subject Descriptors

C.4 [**Performance of Systems**]: Modeling techniques.; G.3 [**Probability and Statistics**]: Stochastic processes.

Keywords

Electricity Pricing; Energy Storage System; Market Efficiency; Energy Economics.

1. INTRODUCTION

The process of liberalizing electricity markets is underway worldwide. It amounts to replacing tightly regulated monopolies with lightly regulated competitive markets [28]. Electricity production is managed through scheduling decisions. In a first stage, producers commit to an energy generation schedule determined through forecasts for the demand and generation of renewable energy for the following day. In a second stage, decisions are made in real-time to compensate for forecast errors in load and production. The price volatility in real-time electricity markets[1] raises the question of the efficiency of these markets: Does the selfish behavior of the various actors lead to a socially acceptable situation?

Electricity markets are highly complex dynamical systems. They incorporate renewable energy sources, such as wind and solar, that are highly volatile; loads that are mostly inelastic; and generation units, that are subject to friction and real-time constraints. To avoid blackouts and due to physical constraints, real-time scheduling of energy is critical.

A model of an electricity market that takes into account these dynamical aspects is proposed in [7, 8]. The authors study the competitive equilibriums in a real-time electricity market where demand is stochastic and energy generation is subject to ramping constraints. They show that if all actors are price-takers (they do not affect market prices), then there exists a competitive equilibrium that is *efficient*: more specifically, the selfish behavior of actors leads indeed to a socially optimal scheduling of generation. However, they also show that the prices that guarantee such an equilibrium exhibit considerable volatility: they oscillate between 0 and a "choke-up" price, and do not concentrate around the marginal production cost. This model has been extended to incorporate network constraints [25, 27], or the presence of renewables [18]; see also [26] for a survey.

Our motivation in this paper is to understand the role of storage in compensating volatility in real-time electricity markets. This is highly relevant for markets with a high penetration of renewables. Exploitation of renewable energy is encouraged in many countries as a means to reduce $CO2$ emissions. However, renewable energy sources, such as wind and solar photovoltaic, are not dispatchable. A side effect of their high penetration is the increase in the volatility of electricity generation and thus of prices, according to [18]. Therefore, in order to compensate for their volatility, a high penetration of renewables needs to be supported by mechanisms such as storage systems (batteries or pump-hydro) or fast-ramping generators (essentially gas-fired turbines). Storage can be operated by an energy producer, a consumer, or by a stand-alone storage operator. In the last case, the

---

[1]the peak to mean ratio can be as high as several thousands, *e.g.* the prices observed in California in 2000-2001 [16].

storage owner needs to generate revenue, hence energy is stored when market prices are low and is provided to the grid when prices are high. We are interested, in particular, in understanding whether the market efficiency results of [26] continue to hold; e.g., is it socially optimal to have stand-alone storage operators that react to real-time prices?

Contributions. We extend the wholesale real-time market model of [26] to incorporate a storage system with losses due to the charge/discharge cycles. We consider three scenarios depending on the owner of the storage: (A) the supplier, (B) the consumer, or (C) a stand-alone real-time storage operator. We show that in all three cases, the market is efficient: there exists a price process such that the selfish behaviors of the players coincide with a socially optimal use of the storage and scheduling of the generation. When the storage energy capacity is large, this price process becomes smooth. Moreover, irrespective of the considered scenario, *the same* decisions concerning bought/sold energy, real-time production, and storage system operation lead to social optimality. These decisions are enforced via *the same* incentive (pricing) scheme.

We show numerically that when the storage belongs either to the consumer or to a stand-alone real-time storage operator, the storage energy capacity that maximizes the storage owner welfare is strictly smaller than the socially optimal storage energy capacity. Consequently, even though the market is efficient when the storage parameters are fixed, consumers and stand-alone storage operators still have an incentive to under-dimension their storage systems.

Finally, we study the effect of the volatility and of the ramping capabilities of generators on the optimal storage parameters. We show that when the volatility is σ, the optimal energy capacity of the storage system scales as σ^4, and its maximum charging/discharging power scales as σ^2. When the ramping capability of the generators is ζ, the optimal storage energy capacity scales as $1/\zeta^3$, and its maximum charging/discharging power scales as $1/\zeta$. We conclude that a linear increase in the ramping capability of fast-ramping generators (such as gas turbines) entails a cubic decrease in the required energy capacity. In view of the high cost of storage capacity, a paradoxical situation arises: to accommodate a large deployment of renewables and to compensate for the resulting generation volatility, there is an incentive to deploy conventional fast-ramping generators (that have high carbon dioxide emissions) rather than to invest in storage.

Road-Map. The rest of the paper is structured as follows. We first describe the market and storage model in Section 3. We study the control problem from a social point of view in Section 4. Section 5 contains the main theoretical results: we show the existence of competitive equilibria in all three scenarios and prove that they are socially optimal. We study the incentives for actors to install storage devices in Section 6. We investigate the relations between volatility, ramping capabilities and optimal storage energy capacity in Section 7. Finally, we conclude in Section 8.

2. RELATED WORK

A large part of the work related to the economic aspects of storage systems investigates optimal energy storage strategies for profit maximization in the electricity market, assuming that storage owner are price-takers [5, 12, 17]. For a wind-farm owner, the authors of [9] study optimal storage strategies in a day-ahead market. Authors often assume that prices are known. The uncertainties due to the price variability and forecasting are also studied in [1, 4].

The economic viability for storage owners is also an important question. On the one hand, it is shown in [24] that there is a strong economic case for storage installations in the New York City region. A similar analysis is conducted for the PJM interconnection (US east-coast) in [20] for which the authors show a moderate storage capacity is viable. On the other hand, according to [22], "*storage is not viable from a system perspective until extremely large levels of wind power are seen on the system*" in Ireland – a country that envisioned 80% of the electrical consumption generated by wind power plants. Moreover, it is suggested in [15] that pumped-storage hydro-plant operators need to change their business model (currently electricity price arbitrage) because the potential diminishes with the increased penetration of renewables: Inexpensive energy is generated during the noon consumption peak by PV installations.

From a social planner's perspective, it is shown [21] that the undesired effects of the volatility of renewables can be mitigated via the use of energy storage, with a manageable increase in energy costs, based on a study in the UK. At a European scale, the authors of [2, 14, 23] show how to use model predictive control to update day-ahead production schedule and mitigate energy curtailments. The use of storage can also compensate for forecast uncertainties. Generation scheduling policies that minimize energy losses and the use of fast-ramping generators are developed in [3, 11].

The question of the efficiency of the control by prices of storage devices is raised in a few papers. Using traces of the real market bids data of the French day-ahead market in 2009, the authors of [13] evaluate the storage operation by a public or a private operator by simulation. They show that the public operation leads to higher social welfare and lower cost of supply than the private operation. Using traces from the Swiss market, authors of [6] also show that the optimal control for a price-taker storage owner are socially optimal. However, these two papers only consider a single-run of simulation and do not provide theoretical guarantee.

The authors of [19] go one step further and obtain theoretical guarantees on a model of a purely static setting: loads are predictable on-peak and off-peak periods, and the price depends linearly on the load. They show that for the same three cases of ownership structure as ours (storage belongs to producers, to consumers or to independent actors), the selfish behavior of price-taker agents lead to a socially optimal use of storage. However, when agents influence prices, storage tends to be underused when owned by producers or independent actors or overused if owned by consumers. The reason is that the higher the use of storage is, the smoother the price is. Our model differs greatly from [19] as it incorporates the dynamical aspects of demand and generation.

3. MARKET AND STORAGE MODEL

In this section we first present a model of a two-stage (day-ahead and real-time) electricity market consisting of two main actors, a consumer and a supplier. Both can have access to an energy storage system in the real-time stage of the market, depending on the scenario. We then describe the storage system, and we analyze three scenarios, depending on the owner of the storage system: (A) the supplier, (B) the consumer, or (C) a stand-alone player.

3.1 Two-stage Electricity Markets

We consider an electricity market with two stages: a day-ahead stage and a real-time stage. The two main actors or players in this market are a *supplier* who produces electricity, and a *consumer* who buys electricity and serves a fairly large number of end-users.

In the *day-ahead* market, each day, players forecast a demand profile $d^{da}(t)$ and plan generation $g^{da}(t)$ for the next day. The price of electricity is set by market mechanisms to $p^{da}(t)$, and the consumer agrees to buy an amount of electricity $g^{da}(t)$ at this price. As often in markets where demand is inelastic and unpredictable, the consumer purchases an extra amount of goods over the predicted demand as a precautionary measure. Hence we assume that $g^{da}(t) = d^{da}(t) + r^{da}$, where r^{da} is called the *fixed reserve*. In addition to the conventional energy sources, the supplier uses renewable energy sources (*e.g.*, wind, solar) that provide a volatile production. The forecast of generation of renewables constitutes a fraction of the planned generation $g^{da}(t)$, and is complemented with conventional energy.

The *real-time* market deals with situations where the actual demand cannot be met, due to the volatility of the demand and of the production of renewable energy. The actual demand at time t is $D^a(t) = d^{da}(t) + D(t)$, where $D(t)$ is referred to as the *real-time demand*, and can be either positive or negative. The renewable forecast error (the difference between the actual and the forecasted production of renewable energy) is denoted by $\Gamma(t)$. The supplier reacts in real-time to unpredictable changes in demand and renewable generation, and she produces an amount of energy $G(t)$ from additional and usually expensive sources. The total amount of electricity produced at time t is then $G^a(t) = g^{da}(t) + \Gamma(t) + G(t)$, and the corresponding real-time reserve is defined as $R(t) = G^a(t) - D^a(t)$. The energy produced in real-time is sold at a price $P(t)$ at time t. In [8], the authors characterize the competitive equilibria arising in the real-time market – they analyze the amount of real-time energy bought by the consumer, the amount of real-time energy $G(t)$ produced by the supplier, and the price $P(t)$, resulting from the strategic behavior of the market actors. In this paper, we are interested in studying the effect of the existence of storage on the real-time market. We make the same assumptions as in [8], *i.e.*:

(A1) Volatility. We model the forecast errors made in the demand and in the generation of renewable energy, $Z(t) = \Gamma(t) - D(t) + r^{da}$, as a driftless Brownian motion with volatility σ, *i.e.*, the variance of $Z(t)$ is $\sigma^2 t$.

(A2) Prices. We assume that the market actors are price takers: the actors strategically define their actions assuming that the prices are exogenous, and that they cannot influence or manipulate these prices.

(A3) Information. We suppose that the market actors share the same information: Before making a decision at time t, they have access to the exogenous data up to time t (*i.e.*, $(D(\tau), \Gamma(\tau), P(\tau), \tau \leq t)$ and to all the past decisions of the other players up to time t. Mathematically, this assumption means that there exists a filtration $\{\mathcal{F}_t, t \geq 0\}$ such that all processes are $\mathcal{F}_t$-adapted.

3.2 Storage Model

We consider a storage system characterized by (i) the maximum amount of energy $B_{\max}$ it can store, *i.e.*, its energy capacity; (ii) the maximum speed $C_{\max}$ at which it can

be charged, referred to as the maximum charging power; (iii) the maximum instantaneous amount of energy $D_{\max}$ we can extract from it, *i.e.*, its maximum discharging power. We model the efficiency of a charge-discharge cycle as follows: only a fraction $\eta \in (0, 1]$ of the injected energy is stored, and the stored energy can be extracted without losses.

More precisely, let $B(t)$ denote the stored energy at time t, and let $u(t)$ be the instantaneous energy extracted from the storage. $u(t) < 0$ means that we are currently storing energy. The storage level evolves as follows:

$$\frac{\partial B}{\partial t} = -u(t)(\mathbb{1}_{\{u(t)>0\}} + \eta \mathbb{1}_{\{u(t)<0\}}). \qquad (1)$$

The storage control process $u = (u(t), t \geq 0)$ satisfies the following constraints: at any time t,

$$-C_{\max} \leq u(t) \leq D_{\max}, \qquad (2)$$
$$u(t) \geq 0 \text{ if } B(t) = B_{\max}, \text{ and } u(t) \leq 0 \text{ if } B(t) = 0. \qquad (3)$$

We write $u \in X_B$ if (2)-(3) are satisfied. Note that (3), together with (1), is equivalent to $0 \leq B(t) \leq B_{\max}$ for all t.

3.3 Scenario A: Storage at the Supplier

We first focus on the case where the supplier controls the storage system. The storage system is used only in the real-time stage of the market. At this stage, deterministic processes describing day-ahead demand, generation, and prices $(d^{da}(t), g^{da}(t), p^{da}(t), t \geq 0)$ are known from the day-ahead market. The real-time market model for this scenario is represented in Figure 1. Next, we describe the strategic decisions that the actors may take, and introduce the notion of dynamic competitive equilibrium.

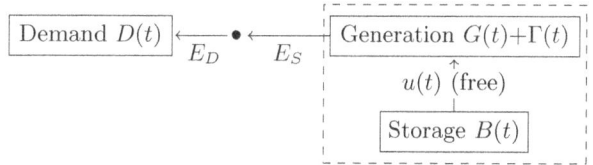

Figure 1: Real-time market model in Scenario A: storage is at the supplier.

Consumer. The strategic decisions taken over time by the consumer are represented by a process E_D, where $E_D(t)$ is the energy bought on the real-time market at time t. We denote by X_D the set of all possible processes E_D. We denote by v the consumer utility per unit of satisfied demand. When the demand exceeds the acquired energy, *i.e.*, if $E_D(t) + g^{da}(t) < D^a(t)$, the consumer bears the cost of a blackout, and suffers from a loss of utility c^{bo} per unit of unsatisfied demand. Recall that the price of electricity in the real-time market is $P(t)$ at time t, and hence the payoff[2] $U_D(t)$ of the consumer at time t is:

$$U_D(t) := v \min(D^a(t), E_D(t) + g^{da}(t))$$
$$- c^{bo}(D^a(t) - E_D(t) - g^{da}(t))^+ - (P(t)E_D(t) + p^{da}(t)g^{da}(t)).$$

This payoff can be decomposed as $U_D(t) = \bar{U}_D(t) + W_D(t)$, where

$$\bar{U}_D(t) = v(d^{da}(t) + D(t)) - p^{da}(t)g^{da}(t),$$
$$W_D(t) = -(v + c^{bo})(E_D(t) - D(t) + r^{da})^- - P(t)E_D(t).$$

---

[2]Throughout the paper, we use the standard notations: $(a)^+ = \max(a, 0)$ and $(a)^- = \max(-a, 0)$.

17

The term $\bar{U}_D(t)$ only contains quantities that cannot be controlled by the consumer in the real-time market. Thus, the quantity of interest is the second term $W_D(t)$. By abuse of notation, we refer to this second term (rather than $U_D(t)$) as the consumer's payoff.

The consumer's objective is to maximize her welfare defined as her long-run discounted expected payoff:

$$\max_{E_D \in X_D} \mathcal{W}_D := \mathbb{E} \int_0^\infty e^{-\gamma t} W_D(t) \, dt.$$

The expectation is taken over the random process $Z(t)$.

Supplier. The supplier controls three quantities: the real-time energy generation $G(t)$, the storage system via the storage control process $u(t)$, and the energy sold on the real-time market $E_S(t)$. We say that (E_S, G, u) satisfies the constraints of the supplier and we write $(E_S, G, u) \in X_S$ if

- the following ramping constraints for real-time generation are satisfied: $\forall t' > t$, $\quad \zeta^- \le \frac{G(t') - G(t)}{t' - t} \le \zeta^+$, where $\zeta^- < 0$ and $\zeta^+ > 0$ are called the ramping capabilities;
- the storage constraints are satisfied: $u \in X_B$;
- for all $t \ge 0$, the supplier sells at most $E_S(t) \le \Gamma(t) + G(t) + u(t)$ on the real-time market.

We denote by c the marginal cost of real-time energy generation. We assume that the energy generated by renewable sources is free. The total generation cost of the conventional energy sources planned in the day-ahead market is equal to $c^{da}(t)$. The supplier's payoff at time t is given by:

$$U_S(t) = P(t)E_S(t) + p^{da}(t)g^{da}(t) - cG(t) - c^{da}(t).$$

As for the consumer, we write $U_S(t) = \bar{U}_S(t) + W_S(t)$, where

$$\bar{U}_S(t) = p^{da}(t)g^{da}(t) - c^{da}(t),$$
$$W_S(t) = P(t)E_S(t) - cG(t).$$

The term $\bar{U}_S(t)$ contains only quantities that cannot be controlled by the supplier in real-time. Thus, we focus on the second term, $W_S(t)$. Again, by abuse of notation we refer to $W_S(t)$ (rather than $U_S(t)$) as the supplier's payoff. Observe that the supplier's utility is increasing in E_S. Hence, setting $E_S(t) = \Gamma(t) + G(t) + u(t)$ maximizes her payoff in X_S.

The objective of the supplier is to maximize her welfare:

$$\max_{(E_S, G, u) \in X_S} \mathcal{W}_S := \mathbb{E} \int_0^\infty e^{-\gamma t} W_S(t) \, dt.$$

Dynamic Competitive Equilibrium. As in Cho and Meyn [8], we introduce the notion of dynamic competitive equilibrium. When the supplier owns and controls the storage, a dynamic competitive equilibrium is defined as follows:

DEFINITION 1. *(Dynamic competitive equilibrium, storage at the supplier) A dynamic competitive equilibrium is a set of price and control processes $(P^e, E_D^e, E_S^e, G^e, u^e)$ satisfying*

$$E_D^e \in \arg\max_{E_D \in X_D} \mathcal{W}_D, \tag{4}$$

$$(E_S^e, G^e, u^e) \in \arg\max_{(E_S, G, u) \in X_S} \mathcal{W}_S, \tag{5}$$

$$E_D^e = E_S^e. \tag{6}$$

In the above definition, (4) means that E_D^e constitutes an optimal control from the consumer's perspective. Similarly, (5) states that (E_S^e, G^e, u^e) is optimal from the supplier's perspective. Finally, (6) is the market constraint. Note that in (4), the consumer is not subject to the supplier's constraints and vice-versa for (5). See [8] for a discussion.

3.4 Scenario B: Storage at the Consumer

When the consumer has control of the storage, her strategic decisions are represented by the process pair (E_D, u), where $u(t)$ is the storage control process, *i.e.*, the amount of power discharged at time t from the storage.

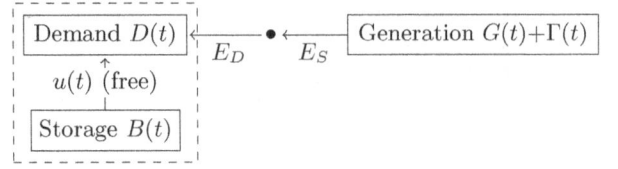

Figure 2: Market model in scenario B: storage is at the consumer.

By abuse of notation we still write the constraints of the consumer as $(E_D, u) \in X_D$. The energy bought is complemented using the storage system, and thus the consumer's payoff becomes

$$W_D(t) = -(v + c^{bo})(E_D(t) + u(t) - D(t) + r^{da})^- - P(t)E_D(t).$$

The supplier controls the energy sold E_S and the real-time generation G. We say that they satisfy the supplier constraints and we write $(E_S, G) \in X_S$ if and only if G satisfies the ramping constraints and if for all $t \ge 0$, $0 \le E_S(t) \le G(t) + \Gamma(t)$. The supplier's payoff remains the same as in the case she controls the storage: $W_S(t) = P(t)E_S(t) - cG(t)$.

In this scenario, the definition of a dynamic competitive equilibrium is similar to that presented in the case the supplier owns the storage: the only difference is that (4) and (5) are replaced respectively by $(E_D^e, u^e) \in \arg\max_{E_D, u \in X_D} \mathcal{W}_D$ and $(E_S^e, G^e) \in \arg\max_{(E_S, G) \in X_S} \mathcal{W}_S$. Note that here the expression of $\mathcal{W}_D$ is modified compared to (4).

3.5 Scenario C: Stand-Alone Storage Operator

In this scenario, the storage is owned by a third player, the *stand-alone storage operator* who seeks to maximize her profit via arbitrage on the real-time market: buying energy at low prices and selling at high prices.

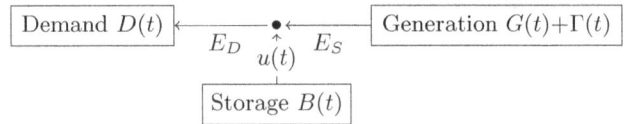

Figure 3: Market model in Scenario C: stand-alone storage operator.

The only control action of the storage operator is $u(t)$, the power discharged from the storage system at time t. Her constraints depend on the storage system parameters and are summarized writing $u \in X_B$. The payoff of the storage operator at time t is $W_B(t) = P(t)u(t)$.

The control, constraints, and payoff of the consumer are the same as in Section 3.3. In particular, her payoff is

$$W_D(t) = -(v + c^{bo})(E_D(t) - D(t) + r^{da})^- - P(t)E_D(t).$$

The control, constraints, and payoff of the supplier are the same as in Section 3.4. Again, the supplier's payoff remains equal to $W_S(t) = P(t)E_S(t) - cG(t)$.

The definition of a dynamic competitive equilibrium is again similar to the one presented in the case where the supplier owns the storage. The difference is that (5) is replaced by $(E_S^e, G^e) \in \arg\max_{(E_S,G)\in X_S} \mathcal{W}_S$, and (6) by $E_S(t) + u(t) = E_D(t)$. In addition a competitive equilibrium maximizes the welfare of the storage operator: $u^e \in \arg\max_{u\in X_B} \mathcal{W}_B$.

4. THE SOCIAL PLANNER'S PROBLEM

In this section we assume that the system is controlled by a single entity, a hypothetical social planner, who decides what is bought and sold, the generation and storage control processes, and whose objective is to maximize the system welfare over all feasible controls. After defining the social planner's optimal control problem, we characterize its solution, *i.e.*, the socially optimal controls. This analysis will be used later in the paper to state whether the market is efficient, or the strategic behavior of the various market actors has a negative impact in terms of social efficiency.

4.1 Social Welfare

A time t, the total system payoff $U(t)$ is either defined as $U(t) = U_D(t) + U_S(t)$ when the storage is controlled by the supplier (or the consumer) or as $U(t) = U_D(t) + U_S(t) + U_B(t)$ in case the storage is controlled by a stand-alone market actor. It can be easily checked that the total payoff does not depend on the market actor controlling the storage. Let us, for instance, assume that the storage is controlled by the supplier. As previously, we decompose the total payoff into a controllable part $\bar{W}(t)$ and an uncontrollable part $\bar{U}(t)$. We have, for any $t \geq 0$:

$$\bar{W}(t) = -(v + c^{bo})(E_D(t) - D(t) + r^{da})^- \\ + P(t)(E_S(t) - E_D(t)) - cG(t). \quad (7)$$

When the market constraint $E_S = E_D$ is further imposed, the controllable part of the total payoff does not depend on the price $P(t)$ and is an increasing function of $E_S(t)$. Hence, the socially optimal controls satisfy: $E_S(t) = E_D(t) = \Gamma(t) + G(t) + u(t)$. These controls must also satisfy the generation and storage constraints, *i.e.*, $(\Gamma + G + u, G, u) \in X_S$. Recall that the reserve process R is defined as the difference of the generated energy and of the demand:

$$R(t) := \Gamma(t) + G(t) - D(t) + r^{da}.$$

By abusing the notation, we write $(R, u) \in X_S$ if and only if $(\Gamma + G + u, G, u) \in X_S$. The controllable part of the total payoff can be further reduced and written as

$$W(t) = -(v + c^{bo})(R(t) + u(t))^- - cR(t). \quad (8)$$

Compared with $\bar{W}(t)$, defined in (7), we substracted the uncontrollable term $c(\Gamma(t) - D(t) + r^{da})$ to obtain $W(t)$.

A social planner want to maximize the social welfare $\mathcal{W}$:

$$\mathcal{W} := \mathbb{E}\int_0^\infty e^{-\gamma t}[-(v + c^{bo})(R(t) + u(t))^- - cR(t)]dt.$$

over all couple of processes $(R, u) \in X_S$. The solution of this problem is denoted by $(R^\star, u^\star)$, and the corresponding optimal controls at the various market actors are $E_D^\star = E_S^\star = R^\star + u^\star + D - r^{da}$, $G^\star = R^\star + D - \Gamma - r^{da}$, and $u^\star$.

4.2 Socially Optimal Controls

The following theorem characterizes the socially optimal controls. These controls define, for any given reserve r and storage state b, the way real-time generation evolves, *i.e.*,

$g = \frac{\partial G}{\partial t}$, and the storage control process u. All proofs are presented in Appendix.

THEOREM 1. *The socially optimal controls satisfy:* (*Generation*) *For each b, there exists a threshold $\phi(b)$ such that the socially optimal generation control $g^\star$ is ζ^+ if $r < \phi(b)$ and ζ^- if $r > \phi(b)$. ϕ is a nonincreasing function.* (*Storage*) *The optimal storage control $u^\star$ satisfies:*

$$u^\star = \begin{cases} \max(-r, -C_{\max}) & \text{if } r \geq 0 \text{ and } b < B_{\max}, \\ \min(-r, D_{\max}) & \text{if } r < 0 \text{ and } b > 0, \\ 0 & \text{otherwise}. \end{cases} \quad (9)$$

The value function of the social planner problem is:

$$V(r,b) = \sup_{(R,u)\in X_S} \mathbb{E}_{(r,b)} \int_0^\infty e^{-\gamma t} \\ \times \left[-(v + c^{bo})(R(t) + u(t))^- - cR(t)\right] dt, \quad (10)$$

where (r, b) is the initial condition: $R(0) = r$ and $B(0) = b$, and $\mathbb{E}_{(r,b)}[\cdot] := \mathbb{E}[\cdot | R(0) = r, B(0) = b]$.

In the proof of the above theorem, we establish that $\frac{\partial V}{\partial r}$ is well defined (almost everywhere), and that the threshold function ϕ characterizing the optimal generation control is $\phi(b) = \sup\{r : \frac{\partial V}{\partial r}(r, b) \geq 0\}$. The optimal storage control $u^\star$ has a simple interpretation: When the reserve is positive, the supplier first serves the demand and puts the remaining energy in the storage while ensuring that storage constraints are satisfied (e.g. the charging power cannot exceed $C_{\max}$). If the reserve is negative, the supplier can serve only part of the demand using the generated energy and has to extract energy from the storage to serve the remaining demand (if this is at all possible).

Under the socially optimal controls, we denote by $B^\star$ the storage level process. The system dynamics are characterized by $(R^\star, B^\star)$ and admit a steady-state whose stationary distribution is denoted by π. Under the assumption described in Section 3, $R^\star$ is a Brownian motion with variance σ and whose average drift varies over time when the generation control $g^\star$ switches values (e.g. from ζ^+ to ζ^-).

4.3 Numerical Example and Energy Units

To illustrate the socially optimal controls defined in Theorem 1, we plot the threshold function $b \mapsto \phi(b)$ in Figure 4(a) for different values of the storage energy capacity $B_{\max}$. We have the storage level b on the x-axis and the reserve r on the y-axis. Using the same representation, a sample-path of $(R^\star, B^\star)$ is plotted in Figure 4(b). The sample path starts at $B_{\max}(0) = 2.5$ u.e. and $R(0) = 1$ u.p.. The vector field is its drift. It corresponds to the optimal controls $(u^\star, g^\star)$. The dashed line is the function $b \mapsto \phi(b)$.

In all figures, the storage energy capacity $B_{\max}$ is expressed in *units of energy* (u.e.) and the maximum charging/discharging powers $C_{\max}$ and $D_{\max}$ are expressed in *units of power* (u.p.). For a variability σ and a ramping capability ζ, we choose that one unit of energy is equal to σ^4/ζ^3 and one unit of power corresponds to σ^2/ζ. We refer to Section 7 for the reasons for these scalings.

For example, let us consider the scenario envisioned for the UK in 2020 [3, 11] where wind power is used to cover 20% of the total electricity consumption – this corresponds to 26GW of peak power. At the scale of the country, the variability of the demand is small compared to the variability of the wind generation. This variability is due to uncertain

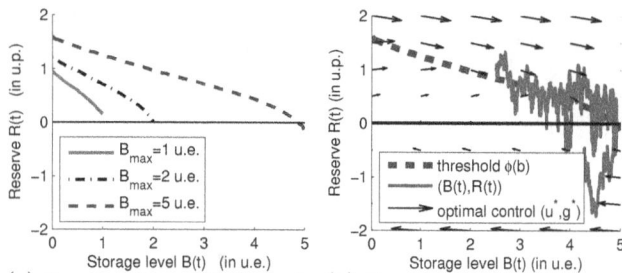

(a) Function $b \mapsto \phi(b)$ for various values of the storage energy capacity $B_{\max}$.

(b) Sample of a trajectory of the optimal reserve and storage processes. $B_{\max} = 5\,\mathrm{u.e.}$

Figure 4: Illustration of the optimal control law for the social planer's problem.

forecast. Using the same data as [11], the square of the volatility of the wind generation is $\sigma^2 = 0.6\,(GW)^2/h$. This means that if the ramping capability $\zeta = 1\mathrm{GW/h}$, one unit of energy corresponds to 360MWh and one unit of power to 600MW. For a ramping capability of $\zeta = 2\mathrm{GW/h}$, one unit of energy corresponds to 45MWh and one unit of power to 300MW.

Throughout the paper, all numerical values are obtained by discretizing in time and space the original model to obtain a discrete Markov decision process. This control problem is then solved using brute-force dynamic programming to obtain the function $b \mapsto \phi(b)$. The simulation of a sample-path (Figures 4(b) and 5) is obtained using ϕ in a discrete-time continuous-space model and the steady-state performance indicator (Figures 6 to 11(b)) are computed by computing the stationary measure of the discrete-time discrete-space model. Unless otherwise specified, the parameters used for the simulations are $\sigma=1$, $\zeta^+=1$, $\zeta^-=3$, $B_{\max}=1\,\mathrm{u.e.}$, $C_{\max}=D_{\max}=3\,\mathrm{u.p.}$, $\eta=1$, $\gamma=0.01$, $c=1$ and $(v + c^{bo})=5$.

5. DYNAMIC COMPETITIVE EQUILIBRIA AND MARKET EFFICIENCY

Under a dynamic competitive equilibrium, the price process is such that the decisions taken over time by the various market actors maximize their respective welfares. In this section, we first prove that for all scenarios, the market is efficient in the sense that any dynamic competitive equilibrium maximizes the social welfare. We then prove the existence of such an equilibrium: we provide explicit expressions for the equilibrium price process and for the strategic controls used by the various market actors in equilibrium. We conclude the section by showing numerically the effect on prices of the presence of storage: in absence of storage, prices are volatile and can only take two values, 0 and the "choke-up" price $v + c^{bo}$ [8], whereas with storage, prices are smoother and oscillate around the marginal production cost c as the storage energy capacity grows large.

5.1 Market Efficiency

We first assume that dynamic competitive equilibria exist, and we prove that these equilibria are efficient in the sense that the corresponding controls maximize the system social welfare. This result is often referred to as the *social welfare theorem* in economics. A similar result has been established in [26] without storage. We show that the market remains efficient even in presence of storage.

THEOREM 2. *(Social Welfare Theorem) Assume that a dynamic competitive equilibrium exists. Then:*
(i) any competitive equilibrium maximizes the social welfare;
(ii) conversely, for any control processes (E_D^e, E_S^e, G^e, u^e) maximizing the social welfare, there exists a price process P^e such that $(P^e, E_D^e, E_S^e, G^e, u^e)$ is a competitive equilibrium.

5.2 Equilibria: Existence and Properties

If dynamic competitive equilibria exist, we know that they are socially efficient (even in presence of storage). We show that competitive equilibria indeed exist and characterize the corresponding price and control processes. More precisely, we identify a price process P^* that can lead to an equilibrium (in fact, as it turns out, P^* is the only price process leading to an equilibrium), and show that if (E_D^*, E_S^*, G^*, u^*) are socially optimal controls, $(P^*, E_D^*, E_S^*, G^*, u^*)$ is a competitive equilibrium

Let (R^*, u^*) be the reserve-storage control process maximizing the social welfare (starting at $R^*(0) = r$ and $B^*(0) = b$). Denote by B^* the corresponding storage level process, and define the price process P^* as:

$$P^*(t) = \begin{cases} 0 & \text{if } R^*(t)+u^*(t) > 0, \\ \eta \frac{\partial V}{\partial b}(R^*(t), B^*(t)), & \text{if } R^*(t)+u^*(t) = 0, R^*(t) > 0, \\ \frac{\partial V}{\partial b}(R^*(t), B^*(t)), & \text{if } R^*(t)+u^*(t) = 0, R^*(t) \leq 0, \\ v + c^{bo}, & \text{if } R^*(t)+u^*(t) < 0. \end{cases}$$
(11)

THEOREM 3. *Let P^*, G^* and u^* be defined as above. Then*

(i) $(P^*, E_D^* = \Gamma + G^* + u^*, E_S^* = \Gamma + G^* + u^*, G^*, u^*)$ is a competitive equilibrium when the storage is at the supplier.

(ii) $(P^*, E_D^* = \Gamma + G^*, E_S^* = \Gamma + G^*, G^*, u^*)$ is a competitive equilibrium when the storage is at the consumer.

(iii) $(P^*, E_D^* = \Gamma + G^* + u^*, E_S^* = \Gamma + G^*, G^*, u^*)$ is a competitive equilibrium when there is a stand-alone storage owner.

In particular, this theorem implies that at the equilibrium, the price, generation and storage control processes are the same in all three scenarios. Moreover these controls maximize the social welfare.

5.3 Equilibrium Price Distribution: the Impact of Storage

At the equilibrium, the price process (11) is a function of the optimal reserve and storage control processes. We plot in Figure 5 the evolution over time of the prices $P^*(t)$ and of the storage level $B^*(t)$. We fix the maximum charging and discharging powers to $C_{\max}=D_{\max}=3\,\mathrm{u.p.}$ and we compare the results for four values of energy capacity of storage: $B_{\max}=0\,\mathrm{u.e.}$ (*i.e.* no storage), $1\,\mathrm{u.e.}$, $3\,\mathrm{u.e.}$ and $10\,\mathrm{u.e.}$, where $1\,\mathrm{u.p.} = \sigma^2/\zeta$ and $1\,\mathrm{u.e.}=\sigma^4/\zeta^3$ (see §4.3). We use the same random seed so that $D(t)-\Gamma(t)$ is the same in all four cases.

As shown in [26], when there is no storage the prices oscillate between 0 and the "choke-up" price $(v + c^{bo})$ (Figure 5(a)). When $B_{\max} > 0$, the prices always remain between 0 and $(v+c^{bo})$, but they are smoother. When $B_{\max}$ is large and $\eta = 1$ (Figure 5(c)), the prices are almost constant and close to the marginal cost of production $c = 1$. When $B_{\max}$ is large and $\eta = 0.8$ (Figure 5(d)), the prices oscillate around $c = 1$, in this case, between 0.88 and $1.1 = 0.88/\eta$.

The optimal reserve and storage level process (R^*, B^*) is stationary. We compute numerically the distribution of

(a) Without storage   (b) $B_{\max} = 2$ u.e., $\eta = 1$.

(c) $B_{\max} = 10$ u.e., $\eta = 1$ (d) $B_{\max} = 10$ u.e., $\eta = 0.8$

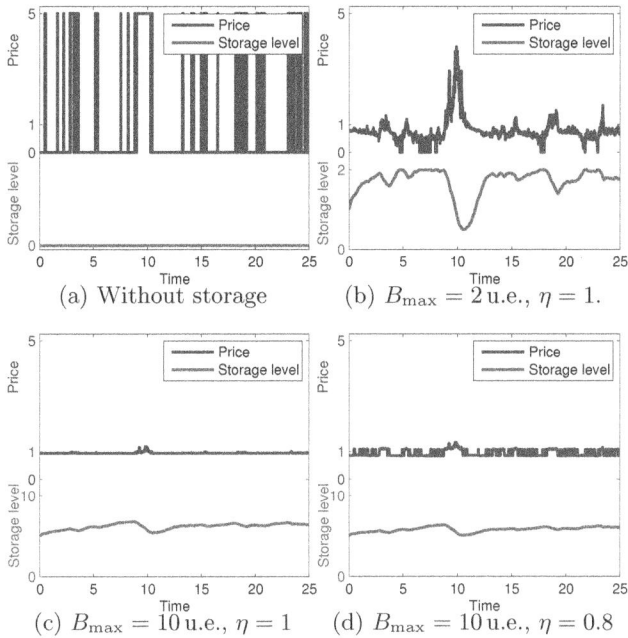

Figure 5: Evolution of prices and storage level over time for various storage energy capacities $B_{\max}$. For $B_{\max} = 10$ u.e., we compare $\eta = 0.8$ and $\eta = 1$.

prices in the steady-state regime of the process (R^*, B^*). The results are reported in Figure 6. Again, we observe that when there is no storage, the price takes only the two values 0 and $(v + c^{bo})$. When $\eta = 1$, the price tends to concentrate around the marginal production cost $c = 1$ as $B_{\max}$ increases. For $\eta = 0.8$, the price signals whether the storage owner should charge (price≈0.88) or discharge (price≈1.1).

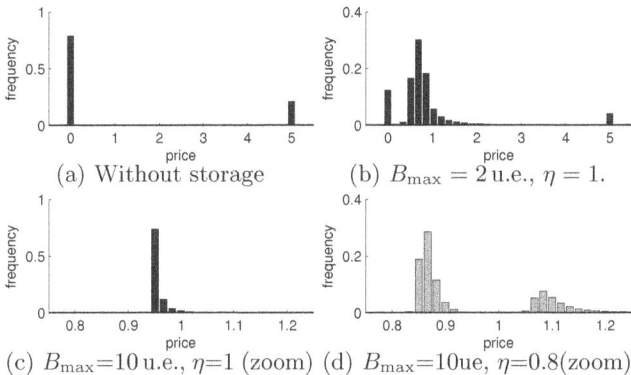

(a) Without storage   (b) $B_{\max} = 2$ u.e., $\eta = 1$.

(c) $B_{\max} = 10$ u.e., $\eta = 1$ (zoom)   (d) $B_{\max} = 10$ ue, $\eta = 0.8$ (zoom)

Figure 6: Steady-state distribution of prices for various storage energy capacities $B_{\max}$. For $B_{\max} = 10$ u.e., we zoom on $c = 1$ to compare $\eta = 0.8$ and $\eta = 1$.

6. STRATEGIC INVESTMENT IN STORAGE

In this section, we study numerically the welfare of the different actors as a function of the energy capacity of storage and of the maximum charging/discharging powers. We first define and compute the socially optimal energy capacity (§6.1). We then show that when the supplier owns the storage, her optimal energy capacity is the socially optimal

one. However, when the storage controlled by the consumer or by an independent actor, their optimal energy capacity is strictly lower. Finally, we show in §6.3 that the same results hold for the maximum charging/discharging powers.

6.1 Socially Optimal Energy Capacity

The total payoff, given by (8), is composed of two terms: the benefit for the consumer minus the cost of blackout, equal to $-(v + c^{bo})(R(t) + u(t))^-$, and minus the cost of producing the energy $-cG(t)$. Thus, if the initial reserve and storage process (r, b) is distributed as the stationary distribution π of the optimal reserve and storage processes, the expected social welfare is

$$- \mathbb{E}_{(r,b)\sim\pi} \int_0^\infty e^{-\gamma t} \left[(v + c^{bo})(R^* + u^*)^- + cR^* \right] dt$$

$$= \frac{1}{\gamma} \mathbb{E}_{(r,b)\sim\pi} \left[-(v + c^{bo})(R^* + u^*)^- - cR^* \right]. \quad (12)$$

The social welfare increases as the energy capacity ($B_{\max}$) or maximum charging/discharging power ($C_{\max}$ and $D_{\max}$) of the storage system increases. Thus, if we neglect the cost of installing additional energy capacity, the greater the storage system is, the greater the social welfare is.

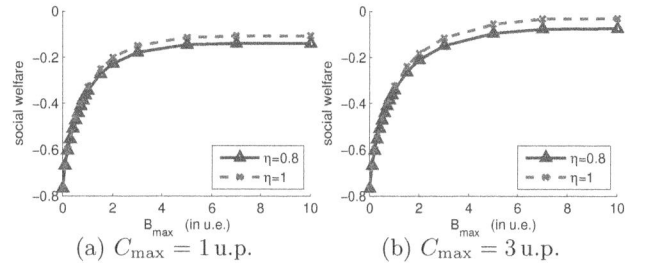

(a) $C_{\max} = 1$ u.p.   (b) $C_{\max} = 3$ u.p.

Figure 7: Expected social welfare in the stationary regime as a function of the energy capacity $B_{\max}$.

In practice, however, storage capacity is expensive. Thus, installing additional storage capacity is worthwhile only as long as the resulting welfare gain is important. In Figure 7, we plot the average social welfare in a stationary regime, given by (12), as function of the storage capacity $B_{\max}$ for two values of the maximum charging/discharging power, $C_{\max} = D_{\max} = 1$ u.p. and $C_{\max} = D_{\max} = 3$ u.p., and for two values of the storage efficiency, $\eta = 0.8$ and $\eta = 1$. We observe that in all cases, the gain in welfare is important for low storage capacities and saturates rapidly. For example, when $C_{\max} = 1$ u.p., Figure 7(a), the saturation occurs for $B_{\max} \approx 4$ u.e.. For $C_{\max} = 3$ u.p., the saturation occurs for $B_{\max} \approx 7$ u.e.. We call these values the socially optimal storage capacities.

6.2 Storage Operator's Revenue

6.2.1 Storage at the Supplier

Let us first assume that the storage belongs to the supplier. As the market is efficient, the reserve process R and storage control u are equal to the socially optimal reserve processes R^* and u^*. Thus, at time t, the supplier sells $R^*(t) + u^*(t)$ at prices $P^*(t)$ and has a cost of production of $-cR^*(t)$. Her instantaneous payoff is $P^*(t)(R^*(t) + u^*(t)) - cR^*(t)$. The equilibrium price $P^*(t)$ is equal to

0 when $R^*(t) + u^*(t) > 0$ and to $(v + c^{bo})$ when $R^*(t) + u^*(t) < 0$. This shows that the supplier's payoff equals $-(v + c^{bo})(R^*(t) + u^*(t))^- - cR^*(t)$, which is exactly the total payoff. This shows that if the storage belongs to the supplier, her welfare increases as the energy capacity increases in the same proportion as the social welfare. Hence, the optimal energy capacity from a supplier-owned storage perspective is the same as the socially optimal capacity.

6.2.2 *Storage at the Consumer and at a Stand-Alone Storage Operator*

When the the consumer owns the storage, her payoff at time t is equal to $-(v + c^{bo})(R^*(t) + u^*(t))^- - P^*(t)R^*(t)$. The price $P^*(t)$ is equal to 0 when $R^*(t) + u^*(t) > 0$ and to $v + c^{bo}$ when $R^*(t) + u^*(t) < 0$. Thus we can write

$$P^*(t)R^*(t) = P^*(t)(R^*(t) + u^*(t)) - P^*(t)u^*(t)$$
$$= -(v + c^{bo})(R^*(t) + u^*(t))^- - P^*(t)u^*(t).$$

This shows that the consumer's payoff is equal to $P^*(t)u^*(t)$. If the storage is owned by a stand-alone storage operator, her instantaneous payoff will be exactly the same: at time t, a quantity $u^*(t)$ is sold at price $P^*(t)$. Thus, the average welfare of the consumer is equal to that of a stand-alone storage operator:

$$\mathbb{E}_{(r,b)\sim\pi} \int_0^\infty dt e^{-\gamma t} P^*(t)u^*(t) = \frac{1}{\gamma}\mathbb{E}_{(r,b)\sim\pi}[P^\star \cdot u^\star]. \quad (13)$$

In Figure 8, we plot the average welfare of a stand-alone storage operator as a function of the energy capacity $B_{\max}$. The system parameters are the same as those used to compute the social welfare in Figure 7. When $C_{\max} = D_{\max} = 1$, the expected welfare saturates for $B_{\max} = 2$ u.e. and diminishes slightly afterward. When $C_{\max} = D_{\max} = 3$ u.p., the expected welfare is maximal for $B_{\max} \approx 1.5$ u.e. and decreases sharply afterward. It diminishes almost to zero for $\eta = 0.8$ when $B_{\max}$ goes to infinity. In both cases, the expected welfare is maximal for a finite energy capacity and this capacity is much lower than the socially optimal capacity. This means that the consumers and the stand-alone storage operators have an incentive to undersize their storage.

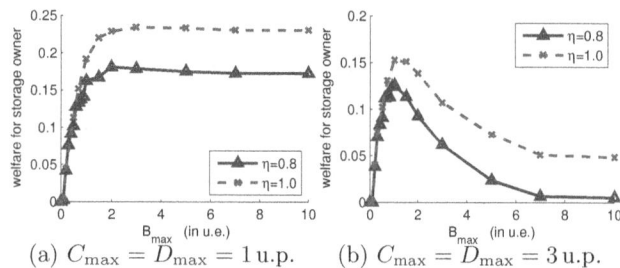

(a) $C_{\max} = D_{\max} = 1$ u.p. (b) $C_{\max} = D_{\max} = 3$ u.p.

Figure 8: Expected welfare of the consumer when she owns the storage or of the stand-alone storage operator as a function of the energy capacity $B_{\max}$.

This result seems paradoxical, as it implies that even if we neglect the cost of storage, a storage owner would make less money with a larger energy capacity. The explanation comes from the price-taking assumption (A2). When the energy capacity grows, the price variations diminish. As a stand-alone storage owner gains only from buying at low price

and selling at high price, her gain diminishes as the prices variability decreases. This situation is radically different when the supplier owns the storage. In this case, a higher storage leads to lower losses and therefore diminishes the production cost. This results in higher gain for the supplier even with a large storage.

6.3 Optimal Maximum Charging/Discharging Powers

As we just observed, when the storage belongs to the consumer or to a third party, the storage owner has an incentive to undersize her energy capacity, compared to a social planner. Figure 9 shows that we have a similar phenomenon regarding the maximum charging/discharging power.

In Figure 9(a), we plot the expected social welfare as a function of the maximum charging/discharging power $C_{\max} = D_{\max}$. This curve is similar to Figure 7 and the optimal charging/discharging powers are $C_{\max} = D_{\max} \approx 1$ u.p.. In Figure 9(b), we plot the expected welfare of a stand-alone storage operator. As for the energy capacity, it has a maximum for a relatively small value $C_{\max} = D_{\max} \approx 1$ u.p. and then decreases quickly. We plot these curves for $B_{\max} = 5$ u.e. but their shapes are similar for other values of $B_{\max}$, even small values like $B_{\max} = 0.5$ u.e..

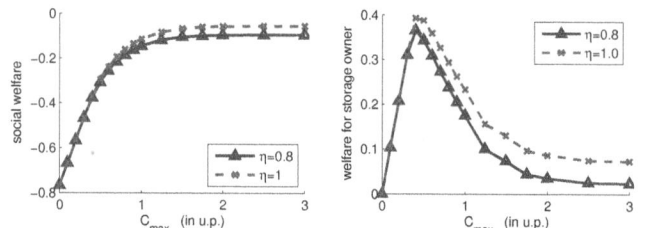

(a) Social welfare or welfare of the supplier when she owns the storage (scenario A). (b) Welfare for the customer in scenario A or for a stand-alone storage operator.

Figure 9: Welfare of the players as a function of maximum charging/discharging power $C_{\max} = D_{\max}$.

7. VOLATILITY AND STORAGE SIZING

We now study the effect of σ (the volatility of the demand and renewable generation process) on the way the storage should be dimensioned. We show that if the ramping capability of the generators remains unchanged, then the storage energy capacity needed to mitigate this volatility scales as σ^4, and that the required charging/discharging powers scale as σ^2. If the volatility is fixed and the ramping capability ζ varies, then the storage capacity needed to mitigate the volatility scales as $1/\zeta^3$, and the maximum charging/discharging powers should scale as $1/\zeta$. In particular, this implies that to maintain a fixed optimal energy capacity, ζ should scale as $\sigma^{4/3}$.

This supports the idea that increasing the amount of power coming from renewables could lead to an increase in carbon dioxide emissions. Doubling the amount of renewables would result in multiplying the volatility by 2. In this case, it is probably easier and less expensive to install generators with $2^{4/3} \approx 2.5$ times larger ramping capability rather than to multiply the energy capacity by 16. However, these fast-ramping generators (*e.g.* gas turbines) tend to emit more carbon dioxide than conventional generators.

7.1 Scaling Laws for the Storage Capacities

Let us recall that σ is the volatility[3] of the difference of demand and of the renewable generation process and that ζ^- and ζ^+ are the ramping capabilities of the real-time generators. The next theorem shows that if the storage capacity B_{max} is scaled as σ^4/ζ^3 and the maximum charging/discharging powers C_{max} and D_{max} are scaled as σ^2/ζ, then the social welfare scales as σ^2/ζ.

In Section 6, we defined the socially optimal storage parameters. They correspond to the knee of the curve of Figures 7 and 9(a) beyond which installing new storage capacity leads to a negligible increase of the social welfare. Therefore, this theorem implies that the optimal storage energy capacity needed to accommodate the volatility scales as σ^4/ζ^3, and the maximum charging/discharging powers scale as σ^2/ζ:

$$B_{max} = \Theta(\sigma^4/\zeta^3) \quad \text{and} \quad C_{max} = \Theta(\sigma^2/\zeta).$$

THEOREM 4. *Let R, B be the socially optimal reserve and storage level processes when the system parameters are*

$$(\sigma, \zeta^+, \zeta^-, B_{max}, C_{max}, D_{max}, \gamma).$$

Then, when the system parameters become

$$(x\sigma, y\zeta^+, y\zeta^-, \frac{x^4}{y^3}B_{max}, \frac{x^2}{y}C_{max}, \frac{x^2}{y}D_{max}, \frac{x^2}{y^2}\gamma),$$

the corresponding socially optimal processes $R^{x,y}, B^{x,y}$ are

$$R^{x,y}(t) = \frac{x^2}{y}R\left(\frac{y^2}{x^2}t\right) \quad \text{and} \quad B^{x,y}(t) = \frac{x^4}{y^3}B\left(\frac{y^2}{x^2}t\right).$$

Moreover, let $\mathcal{W}$ be the social welfare for the initial parameters. The social welfare for the rescaled parameters is $(x^2/y)\mathcal{W}$.

The proof of this theorem consists in verifying that the proposed scaling works. It is detailed in [10].

7.2 Practical Implications

Theorem 4 implies that if ζ and σ are multiplied by the same factor x, then the optimal storage parameters increase linearly in x. Figure 10 illustrates this fact: we plot the optimal social welfare as a function of the available B_{max} for four values of x. Each time the remaining parameters are the same, and we choose $C_{max} = D_{max} = 1000$ to release the charge/discharge constraints of the storage. Figure 10 shows that for $x = 1$ (respectively 2, 4, 6), the optimal energy capacity is approximately 8 (respectively 15, 30 and 50), that is to say linear in x (as expected).

For a fixed ramping constraint ζ, Theorem 4 states that the optimal energy capacity increases as the scaling factor of σ to the fourth power. In Figure 11(a) we scale ζ by the a fixed constant $y = 3$, we scale σ by a various factors $x = 1, 2, 3, 4$, and we plot the resulting optimal social welfare. For the four values of x, the respective optimal values of the energy capacity are approximately $7, 100, 400, 1000$ u.e., which correspond to roughly $\Theta(x^4)$, as expected.

For a fixed scaling of the volatility σ, by Theorem 4 the optimal storage size decreases as the cube of the scaling factor of the ramping constraint ζ. This is illustrated in

---

[3]Note that the volatility corresponds to the standard deviation and not to the variance: as $D-\Gamma$ is a Brownian motion, this means that $D(t)-\Gamma(t) - D(0)+\Gamma(0)$ has a variance $\sigma^2 t$.

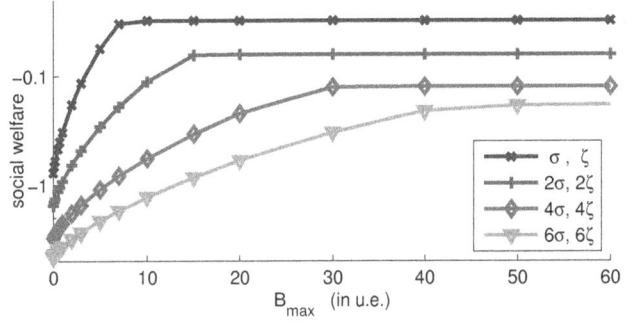

Figure 10: Social welfare as a function of B_{max} when σ and ζ are scaled by the same factor. B_{max} is expressed in unit of energy with 1 u.e. $= \sigma^4/\zeta^3$. The y-axis is in log-scale.

Figure 11(b), where we consider a fixed σ, and we scale ζ by factors $y = 1, 2, 3, 4$. We plot the optimal social welfare, and we observe the decreasing values of optimal energy capacity $10, 1.5, 0.5, 0.3$ ($\approx \Theta(y^{-3})$).

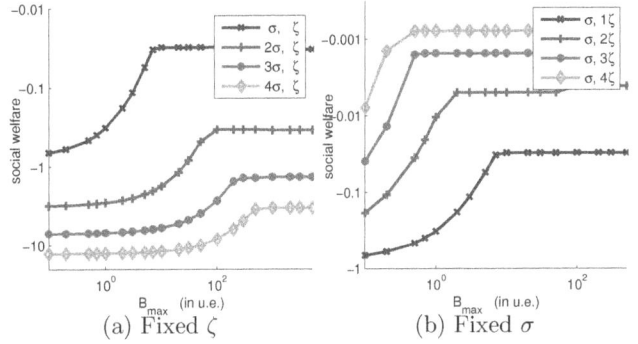

(a) Fixed ζ (b) Fixed σ

Figure 11: Social welfare as a function of the energy capacity B_{max} when one of σ or ζ is fixed. The plot is log-log scale. B_{max} is expressed in unit of energy with 1 u.e. $= \sigma^4/\zeta^3$.

8. CONCLUSION

We have shown that under the price-taking assumption, electricity markets remain efficient when we introduce storage capabilities in the system: they lead to an efficient allocation of generation and storage control and smooth the prices. However, we have shown that there is no incentive for consumers or third-party actors to install large storage devices, despite the fact that the required storage capacity needed to accomodate real-time flucuations explodes with high variability.

There are still many open questions. A first question concerns the case of oligopolies: Does the market remain efficient if a small number of players can influence prices? Another question is whether the market efficiency results also hold in a physical network as described in [26]. More specifically, how are these results affected by the placement of storage devices in such a network? Finally, the lack of incentives for actors to install large storage capacity raises the question of designing political incentives to encourage the development of storage systems.

APPENDIX

A. PROOF OF THEOREM 1

To derive socially optimal controls, we use the following structural properties for the value function V:

LEMMA 1. (i) $(r, b) \mapsto V(r, b)$ is concave.
(ii) V is sub-additive: for all $r^- \leq r^+$ and $b^- \leq b^+$,

$$V(r^-, b^-) + V(r^+, b^+) \leq V(r^-, b^+) + V(r^+, b^-).$$

(iii) For all $b^- \leq b^+$, and r:

$$0 \leq V(r, b^+) - V(r, b^-) \leq (b^+ - b^-)(v + c^{bo}).$$

The proof of the lemma is presented at the end of this section. We first prove the optimality of $g^\star$. We start from a reserve $R(0) = r$ and storage $B(0) = b$. The storage control is fixed and optimal. Due to the concavity of V, $\frac{\partial V}{\partial r}$ is well defined almost everywhere. Let $\delta > 0$. We denote by $d(\delta) = \int_0^\delta Z(t) dt$, and by $b' = B(\delta)$. Consider two generation controls g_1 and g_2 different over $[0, \delta]$, but equal to the optimal control after δ. Denote by V_1 and V_2 the expected social welfare obtained using g_1 and g_2, respectively. Define $\xi_1 = \frac{1}{\delta} \int_0^\delta g_1(t) dt$ and $\xi_2 = \frac{1}{\delta} \int_0^\delta g_2(t) dt$. We can show that:

$$\begin{aligned}
V_1 - V_2 &= e^{-\gamma \delta} \mathbb{E}[V(r + \xi_1 \delta - d(\delta), b') \\
&\quad - V(r + \xi_2 \delta - d(\delta), b')] + O(\delta^2) \\
&= \delta(\xi_1 - \xi_2) \frac{\partial V}{\partial r}(r, b) + O(\delta^2).
\end{aligned}$$

We deduce that the optimal generation control $g^\star$ is such that $g^\star = \zeta^+$ if $\frac{\partial V}{\partial r} > 0$ and $g^\star = \zeta^-$ if $\frac{\partial V}{\partial r} < 0$. Now since V is concave, the threshold function $\phi(b) = \sup\{r : \frac{\partial V}{\partial r} \geq 0\}$ is well defined, and for $r < \phi(b)$ (resp. $r > \phi(b)$), the optimal generation control is $g^\star = \zeta^+$ (resp. ζ^-). The fact that ϕ is nonincreasing is directly deduced from (ii) in Lemma 1.

Next we prove the optimality of $u^\star$. We establish that $u^\star$ is optimal when starting from a reserve $r = R(0)$ over a small time interval $[0, \delta]$. The generation control is fixed, and we denote by $r' = R(\delta)$. Let $b = B(0)$. The proof consists in analyzing several cases depending on the initial reserve value r. Due to space limitation, we provide the analysis of one of the cases. The other cases are treated in [10].

Assume that $r \in (0, C_{\max})$ and $\delta > 0$ is such that $C_{\max} \geq R(s) \geq 0$ for all $s \in [0, \delta]$ (with high probability). We have $u^\star(s) = -R(s)$ for all $s \in [0, \delta]$. If this control is used, the expected social welfare is denoted by V_1. Now assume that instead we use the control $u \in [-C_{\max}, D_{\max}]$, in which case, the expected welfare is V_2. At time δ, under control $u^\star$, the state of the storage is $b + c_1$, and under u, it is $b + c_2$. We have $c_1 = \int_0^\delta \eta R(s) ds$, and

$$\begin{aligned}
c_2 &= -\int_0^\delta (\eta u(s) 1_{u(s)<0} + u(s) 1_{u(s) \geq 0}) ds \\
&\leq -\int_0^\delta \eta(u(s) 1_{u(s) \leq -R(s)} + u(s) 1_{-R(s) < u(s) < 0}) ds \\
&\leq -\int_0^\delta \eta(u(s) 1_{u(s) \leq -R(s)} - R(s) 1_{-R(s) < u(s) < 0}) ds.
\end{aligned}$$

We deduce that: $c_2 - c_1 \leq \eta \int_0^\delta (R(s) + u(s))^- ds$. Now observe that under control u, if $R(s) + u(s) < 0$ we have a blackout, and hence:

$$\begin{aligned}
V_1 - V_2 &= (v + c^{bo}) \int_0^\delta ds e^{-\gamma s} (R(s) + u(s))^- \\
&\quad + e^{-\gamma \delta} \mathbb{E}[V(r', b + c_1) - V(r', b + c_2)].
\end{aligned}$$

From (iii) in Lemma 1,

$$V_1 - V_2 \geq (1 - \eta)(v + c^{bo}) \int_0^\delta ds (R(s) + u(s))^- \geq 0.$$

Proof of Lemma 1. (i) and (ii) are straightforward and details can be found in [10]. Regarding (iii), let $b^- \leq b^+$ and u_- and u_+ be optimal controls starting from b^- and b^+. Let B_+ be the storage level process starting at b^+ under control u_+. We introduce the control u (and the corresponding storage level process B that starts from b^-):

$$u(t) = \begin{cases} u_+(t) & \text{if } B(t) > 0 \text{ or } u_+(t) < 0 \\ 0 & \text{otherwise} \end{cases}$$

By abuse of notation, we denote $V(u)$ the value function of the control u. As u_- is optimal starting from b^- and u is a valid control starting from b^-, we have:

$$\begin{aligned}
V(b^+) - V(b^-) &= V(u_+) - V(u) \\
&= \mathbb{E}\left[\int_0^\infty e^{-\gamma t}(v + c^{bo})[(R(t) + u(t))^- - (R(t) + u_+(t))^-] dt\right] \\
&\leq \mathbb{E}\left[\int_0^\infty (v + c^{bo})(u_+(t) - u(t)) dt\right] \leq (v + c^{bo})(b^+ - b^-).
\end{aligned}$$

The first inequality is obtained remarking that $x \mapsto -(R + x)^-$ is 1-Lipschitz. The last inequality is obtained by combining the facts that $b^+ - B_+(T) = \int_0^T f(u_+(t)) dt$, $b^- - B(T) = \int_0^T f(u(t)) dt$, where $f(x) = x(1_{x>0} + \eta 1_{x \leq 0})$, and that by construction $B(T) \leq B_+(T)$ for any T – the inequality is deduced by letting $T \to \infty$.

B. PROOF OF THEOREM 2

To simplify notations, we define $E = (E_D, E_S, G, u)$ and we write $E \in X_C$ if $E_D \in X_D$ and $(E_S, G, u) \in X_S$. To emphasize the dependence of the social welfare on the process E, we use the notations $\mathcal{W}(E)$. Similarly we denote by $\mathcal{W}_D(E_D, P)$ and $\mathcal{W}_S(E_S, G, u, P)$ the welfares of the consumer and the supplier. Moreover, we define the inner product of two stochastic processes F_1 and F_2 as: $\langle F_1, F_2 \rangle := \mathbb{E} \int_0^\infty e^{-\gamma t} F_1(t) F_2(t) dt$. Using this notation, the consumer's welfare becomes $\mathcal{W}_D := \langle W_D, \mathbf{1} \rangle$, where $\mathbf{1}$ is the constant process, always equal to 1. The proof is detailed for scenario A. It applies *mutatis mutandis* to the two other scenarios.

To prove the result, we proceed as in [26] and we interpret the social welfare optimization problem as the problem of maximizing $\mathcal{W}_D(E_D, P) + \mathcal{W}_S(E_S, G, u)$ subject to the constraints $E \in X_C$ and $E_D = E_S$. The last constraint is relaxed, and the price process P serves as corresponding Lagrange multipliers. The Lagrangian is:

$$\begin{aligned}
\mathcal{L}(E, P) &= -\mathcal{W}(E) + \langle P, E_D - E_S \rangle \\
&= -\mathcal{W}_D(E_D, P) - \mathcal{W}_S(E_D, G, u, P).
\end{aligned}$$

The dual function h is defined as: $h(P) = \inf_{E \in X_C} \mathcal{L}(E, P)$. Let $E \in X_C$. If $E_D = E_S$, weak duality holds: $h(P) \leq \mathcal{L}(E, P) = -\mathcal{W}(E)$. We further establish that for $E^e \in X_C$ such that $E_D^e = E_S^e$, (E^e, P) is a competitive equilibrium if and only if $h(P) = -\mathcal{W}(E)$.

1. Assume that (E^e, P) is a competitive equilibrium. Then:

$$h(P) = \inf_{E \in X_C} \mathcal{L}(E, P) = - \sup_{E_D \in X_D} \mathcal{W}_D(E_D, P)$$
$$- \sup_{(E_S, G, u) \in X_S} \mathcal{W}_S(E_S, G, u, P)$$
$$= -\mathcal{W}_D(E_D^e, P) - \mathcal{W}_S(E_S^e, G^e, u^e, P)$$
$$= \mathcal{L}(E^e, P) = -\mathcal{W}(E).$$

2. Conversely, assume that under (E^e, P), $h(P) = -\mathcal{W}(E)$. Since $(E_D^e, E_S^e) \in X_T$, we deduce that: $h(P) = \mathcal{L}(E^e, P)$, which in turn implies that:

$$- \sup_{E_D \in X_D} \mathcal{W}_D(E_D, P) - \sup_{(E_S, G, u) \in X_S} \mathcal{W}_S(E_S, G, u, P)$$
$$= -\mathcal{W}_D(E_D^e, P) - \mathcal{W}_S(E_S^e, G^e, u^e, P).$$

We conclude that E_D^e maximizes the welfare from the consumer's perspective, and that (E_S^e, G^e, u^e) maximizes supplier's welfare under price process P.

From the above analysis, we deduce that any competitive equilibrium maximizes the social welfare. Now let (E^e, P^e) be a competitive equilibrium, and let $E \in X_C$ be socially optimal. Since both E and E^e maximize the social welfare, we have: $-\mathcal{W}(E) = -\mathcal{W}(E^e) = h(P^e)$. This implies that (E, P^e) is a competitive equilibrium.

C. PROOF OF THEOREM 3

The proof is detailed for scenario A. The two other scenarios B and C can be treated analogously.

The proof of the theorem consists in showing that under the price process $P^\star$, the controls maximizing the social welfare are also optimal from the consumer's and supplier's perspectives. We first prove the following result.

LEMMA 2. We have: $\frac{\partial V}{\partial r} = \mathbb{E} \int_0^\infty e^{-\gamma t}(P^\star(t) - c)dt$.

PROOF. Let $\epsilon > 0$ (typically small). We first prove that we can choose $\delta > 0$ uniformly w.r.t. (r, b) such that:

$$V(r, b) - V(r + \epsilon, b) \geq \epsilon \mathbb{E} \int_0^\delta e^{-\gamma t}(-(1 + \epsilon)P^\star(t) + c)dt$$
$$+ e^{-\gamma \delta} \mathbb{E}\left[V(R^\star(\delta), B^\star(\delta)) - V(R^\star(\delta) + \epsilon, B^\star(\delta))\right]. \quad (14)$$

Let (R_1, B_1) (resp. (R_2, B_2)) be the socially optimal reserve and storage level processes when starting at (r, b) (resp. $(r + \epsilon, b)$). Observe that the controls leading to the processes $(R_1 + \epsilon, B_2)$ are suboptimal when starting from $(r + \epsilon, b)$. We deduce that: $V(r, b) - V(r + \epsilon, b) \geq G(\delta) + H(\delta)$, where

$$G(\delta) = \mathbb{E} \int_0^\delta e^{-\gamma t}(W(R_1(t), B_1(t)) - W(R_2(t), B_2(t)))dt,$$
$$H(\delta) = e^{-\gamma \delta} \mathbb{E}\left[V(R_1(\delta), B_1(\delta)) - V(R_1(\delta) + \epsilon, B_2(\delta))\right].$$

We consider several cases depending on the initial reserve value r. Due to space limitations, we provide the analysis of one of the cases – the other cases are treated in details in [10]. Assume that $0 < r < C_{\max}$ and for all $t \leq \delta$, $R_1(t), R_2(t) \in [0, C_{\max}]$. In this case, for all $t \leq \delta$, $u_1(t) = -R_1(t)$ and $u_2(t) = -R_2(t)$. Note that $R_1(t) + \epsilon$ is suboptimal when starting from $(r + \epsilon, b)$, and so we can assume that $u_2(t) = -R_1(t) - \epsilon$. This leads to $B_2(\delta) = B_1(\delta) + \epsilon\delta\eta$ (because

$u_1(t), u_2(t) < 0$). We have: $G(\delta) \geq \epsilon \mathbb{E} \int_0^\delta e^{-\gamma t} c\,dt$, and

$$H(\delta) = e^{-\gamma \delta}\Big(\mathbb{E}\left[V(R_1(\delta), B_1(\delta)) - V(R_1(\delta) + \epsilon, B_1(\delta))\right]$$
$$+ \mathbb{E}[V(R_1(\delta) + \epsilon, B_1(\delta)) - V(R_1(\delta) + \epsilon, B_1(\delta) + \epsilon\delta\eta)]\Big),$$
$$\geq e^{-\gamma \delta}\Big(\mathbb{E}\left[V(R_1(\delta), B_1(\delta)) - V(R_1(\delta) + \epsilon, B_1(\delta))\right]$$
$$+ \mathbb{E}[V(R_1(\delta), B_1(\delta)) - V(R_1(\delta), B_1(\delta) + \epsilon\delta\eta)]\Big),$$

where the above inequality comes from (ii) in Lemma 1. Now we can select $\delta > 0$ small enough and uniformly w.r.t. $(r, b) \in [0, C_{\max}] \times [0, B_{\max}]$ such that:

$$e^{-\gamma \delta}\mathbb{E}[V(R_1(\delta), B_1(\delta)) - V(R_1(\delta), B_1(\delta) + \epsilon\delta\eta)]$$
$$\geq -\epsilon(1 + \epsilon)\mathbb{E} \int_0^\delta e^{-\gamma t} \eta \frac{\partial V}{\partial b}(R_1(t), B_1(t))dt.$$

We deduce that:

$$G(\delta) + H(\delta) \geq \epsilon \mathbb{E} \int_0^\delta e^{-\gamma t}(-(1 + \epsilon)P^\star(t) + c)dt$$
$$+ e^{-\gamma \delta}\mathbb{E}\left[V(R_1(\delta), B_1(\delta)) - V(R_1(\delta) + \epsilon, B_1(\delta))\right].$$

Using a similar analysis for the other cases (see [10]), Equation (14) follows. Iterating (14), we obtain that for any $k > 0$

$$V(r, b) - V(r + \epsilon, b) \geq \epsilon \mathbb{E} \int_0^{k\delta} e^{-\gamma t}(-(1 + \epsilon)P^\star(t) + c)dt$$
$$+ e^{-\gamma k\delta}\mathbb{E}\left[V(R^\star(\delta), B^\star(\delta)) - V(R^\star(\delta) + \epsilon, B^\star(\delta))\right].$$

Letting $k \to \infty$, and then $\epsilon \to 0$, we conclude that:

$$\frac{\partial V}{\partial r} \leq \mathbb{E} \int_0^\infty e^{-\gamma t}(P^\star(t) - c)dt.$$

The converse inequality is obtained by similar arguments. □

We now prove that $E_D^\star = \Gamma + G^\star + u^\star$ optimizes the consumer's welfare under price $P^\star$. The consumer's welfare is

$$\mathbb{E} \int_0^\infty e^{-\gamma t} - (v + c^{bo})(E_D(t) - D(t) + r^{da})^- - P^\star(t)E_D(t)dt.$$

As there is no ramping constraints in the consumer's optimization (i.e., $E_D \in X_D$), an optimal strategy is myopic in the sense that for any $t \geq 0$, an optimal strategy $E_D^\star(t)$ satisfies $E_D^\star(t) \in \arg\max_e\{-(v + c^{bo})(e - D(t) + r^{da})^- - P^\star(t)e\}$.

Recall that $R^\star = G^\star + \Gamma^\star - D + r^{da}$. We consider three cases:
1. When $P(t) = 0$ (i.e. $R^\star(t) + u^\star(t) > 0$) any $e \geq D(t) - r^{da}$ maximizes the consumer's payoff. Thus $E_D^\star(t) = \Gamma(t) + G^\star(t) + u^\star(t) > D(t) - r^{da}$ is optimal for the consumer.
2. When $0 < P^\star(t) < (v + c^{bo})$ (i.e. $R^\star(t) + u^\star(t) = 0$), her payoff is maximized for $E_D^\star(t) = D(t) - r^{da} = \Gamma(t) + G^\star(t) + u^\star(t)$.
3. When $P(t) = v + c^{bo}$, her payoff is maximized for any $E_D^\star(t) \leq D(t) - r^{da}$. Thus, $E_D^\star(t) = \Gamma(t) + G^\star(t) + u^\star(t) < D(t) - r^{da}$ is optimal for the consumer in that case.

It remains to show that $(R^\star, u^\star)$ maximizes the welfare of the supplier. By assumption (A1-A3), the real time price $P(t)$ and and the generation of renewable energy $\Gamma(t)$ are uncontrollable by the players. Moreover, as $E_S(t) \leq \Gamma(t) + u(t) + G(t)$ and the payoffs of the supplier is increasing in E_S, a supplier will chose $E_S(t) = \Gamma(t) + u(t) + G(t)$. Thus, a direct computation shows that the welfare of the supplier can be solely expressed as a function of the reserve process R

and the control process u. Her payoff can be written (up to uncontrollable parts) as $W_S(t) = P^\star(t)(R(t) + u(t)) - cR(t)$.

We define by X_R the set of possible generation controls: $(R, u) \in X_S$ iff $R \in X_R$ and $u \in X_B$. This decomposition is possible because the constraints on generation and storage are not coupled. Observe that the optimal control problem that the supplier solves is equivalent to:

$$\max_{(R,u) \in X_S} \mathbb{E} \int_0^\infty e^{-\gamma t}(P^\star(t)(R(t) + u(t)) - cR(t))dt.$$

The corresponding value function $V_S(r', b)$ is:

$$V_S(r', b) = \sup_{(R,u) \in X_S} \mathbb{E}_{(r',b)} \int_0^\infty e^{-\gamma t} \\ [P^\star(t)(R(t) + u(t)) - cR(t)]dt.$$

We have $V_S(r', b) = V_R(r') + V_B(b)$ where

$$V_R(r') = \sup_{R \in X_R} \mathbb{E}_{r'} \int_0^\infty e^{-\gamma t}(P^\star(t) - c)R(t)dt$$

$$V_B(b) = \sup_{u \in X_B} \mathbb{E}_b \int_0^\infty e^{-\gamma t}P^\star(t)u(t)dt.$$

Optimality of $R^\star$: From the above expression, we simply deduce that: $\frac{\partial V_R}{\partial r'} = \mathbb{E}_{r'} \int_0^\infty e^{-\gamma t}(P^\star(t) - c)dt$. From Lemma 2, we conclude that $\frac{\partial V_S}{\partial r} = \frac{\partial V}{\partial r}$, and hence $R^\star$ also maximizes the supplier's welfare.

Optimality of $u^\star$: The proof is similar to that of the social optimality of $u^\star$. It relies on the observation that $V_B(b') - V_B(b) \le (v + c^{bo})(b' - b)$. See [10] for details.

D. REFERENCES

[1] ABGOTTSPON, H., BUCHER, M., AND ANDERSSON, G. Stochastic dynamic programming for unified shortand medium-term planning of hydro power considering market products. In *International Conference on Probabilistic Methods Applied to Power Systems(PMAPS), Istanbul, Turkey* (2012).

[2] ARNOLD, M., AND ANDERSSON, G. Model predictive control of energy storage including uncertain forecasts. In *Power Systems Computation Conference (PSCC), Stockholm, Sweden* (2011).

[3] BEJAN, A., GIBBENS, R., AND KELLY, F. Statistical aspects of storage systems modelling in energy networks. *46th Annual Conference on Information Sciences and Systems* (2012).

[4] CASTRONUOVO, E., AND LOPES, J. On the optimization of the daily operation of a wind-hydro power plant. *Power Systems, IEEE Transactions on 19*, 3 (2004), 1599–1606.

[5] CASTRONUOVO, E. D., AND CAS LOPES, J. A. P. Optimal operation and hydro storage sizing of a wind-hydro power plant. *International Journal of Electrical Power & Energy Systems 26*, 10 (2004), 771 – 778.

[6] CHATZIVASILEIADIS, S., BUCHER, M., ARNOLD, M., KRAUSE, T., AND ANDERSSON, G. Incentives for optimal integration of fluctuating power generation.

[7] CHEN, M., CHO, I., AND MEYN, S. Reliability by design in distributed power transmission networks. *Automatica 42*, 8 (2006), 1267–1281.

[8] CHO, I., AND MEYN, S. Efficiency and marginal cost pricing in dynamic competitive markets with friction. *Theoretical Economics 5*, 2 (2010), 215–239.

[9] GARCIA-GONZALEZ, J., DE LA MUELA, R., SANTOS, L., AND GONZÁLEZ, A. Stochastic joint optimization of wind generation and pumped-storage units in an electricity market. *IEEE Transactions on Power Systems 23*, 2 (2008), 460–468.

[10] GAST, N., LE BOUDEC, J., PROUTIÈRE, A., AND TOMOZEI, D. Impact of storage on the efficiency and prices in

real-time electricity markets. Tech. rep., EPFL. http://infoscience.epfl.ch/record/183149, 2013.

[11] GAST, N., TOMOZEI, D., AND LE BOUDEC, J. Optimal storage policies with wind forecast uncertainties. *Greenmetrics* (2012).

[12] GRAVES, F., JENKIN, T., AND MURPHY, D. Opportunities for electricity storage in deregulating markets. *The Electricity Journal 12*, 8 (1999), 46–56.

[13] HE, X., DELARUE, E. ABD D'HAESELEER, W., AND GLACHANT, J.-M. Coupling electricity storage with electricity markets: a welfare analysis in the french market. *TME working paper - Energy and Environment* (2012).

[14] HEUSSEN, K., KOCH, S., ULBIG, A., AND ANDERSSON, G. Unified system-level modeling of intermittent renewable energy sources and energy storage for power system operation. *Systems Journal, IEEE 6*, 1 (2012), 140 –151.

[15] HILDMANN, M., ULBIG, A., AND ANDERSSON, G. Electricity grid in-feed from renewable sources: A risk for pumped-storage hydro plants? In *8th International Conference on the European Energy Market (EEM)* (2011), pp. 185 –190.

[16] JOSKOW, P., AND KAHN, E. A quantitative analysis of pricing behavior in california's wholesale electricity market during summer 2000. Tech. rep., National Bureau of Economic Research, 2001.

[17] KORPAAS, M., HOLEN, A. T., AND HILDRUM, R. Operation and sizing of energy storage for wind power plants in a market system. *International Journal of Electrical Power & Energy Systems 25*, 8 (2003), 599 – 606.

[18] MEYN, S., NEGRETE-PINCETIC, M., WANG, G., KOWLI, A., AND SHAFIEEPOORFARD, E. The value of volatile resources in electricity markets. In *CDC* (2010), IEEE, pp. 1029–1036.

[19] SIOSHANSI, R. Welfare impacts of electricity storage and the implications of ownership structure. *Energy Journal 31*, 2 (2010), 173.

[20] SIOSHANSI, R., DENHOLM, P., JENKIN, T., AND WEISS, J. Estimating the value of electricity storage in PJM: Arbitrage and some welfare effects. *Energy Economics 31*, 2 (2009), 269–277.

[21] STRBAC, G., SHAKOOR, A., BLACK, M., PUDJIANTO, D., AND BOPP, T. Impact of wind generation on the operation and development of the UK electricity systems. *Electric Power Systems Research 77*, 9 (2007), 1214 – 1227.

[22] TUOHY, A., AND O'MALLEY, M. Impact of pumped storage on power systems with increasing wind penetration. In *Power & Energy Society General Meeting.* (2009), IEEE, pp. 1–8.

[23] ULBIG, A., AND ANDERSSON, G. On operational flexibility in power systems. In *IEEE PES General Meeting, San Diego, USA* (2012).

[24] WALAWALKAR, R., APT, J., AND MANCINI, R. Economics of electric energy storage for energy arbitrage and regulation in new york. *Energy Policy 35*, 4 (2007), 2558–2568.

[25] WANG, G., KOWLI, A., NEGRETE-PINCETIC, M., SHAFIEEPOORFARD, E., AND MEYN, S. A control theorist's perspective on dynamic competitive equilibria in electricity markets. In *Proc. 18th World Congress of the International Federation of Automatic Control (IFAC)* (2011).

[26] WANG, G., NEGRETE-PINCETIC, M., KOWLI, A., SHAFIEEPOORFARD, E., MEYN, S., AND SHANBHAG, U. Dynamic competitive equilibria in electricity markets. In *Control and Optimization Theory for Electric Smart Grids.* Springer, 2011.

[27] WANG, G., NEGRETE-PINCETIC, M., KOWLI, A., SHAFIEEPOORFARD, E., MEYN, S., AND SHANBHAG, U. Real-time prices in an entropic grid. In *Innovative Smart Grid Technologies (ISGT), IEEE PES* (2012), IEEE, pp. 1–8.

[28] WILSON, R. Architecture of power markets. *Econometrica 70*, 4 (2003), 1299–1340.

DC Picogrids: A Case for Local Energy Storage for Uninterrupted Power to DC Appliances

Sunil Kumar Ghai, Zainul Charbiwala,
Swarna Mylavarapu, Deva P. Seetharam
IBM Research, India

Rajesh Kunnath
Radio Studio
Chennai, India

ABSTRACT

An increasing number of appliances now operate on DC and providing uninterrupted power supply (UPS) to them through outages requires two conversions: first from an energy store, typically a DC battery, to AC mains and then from AC mains to the DC input required by the appliance. The energy storage and DC-to-AC inversion are usually centrally located and tied to existing AC distribution lines to amortize costs and battery capacity. In this paper, we argue that adding energy storage locally to each DC appliance and managing it intelligently can lead to higher efficiency and lower average cost. We term this topology a *DC picogrid* as it mimics a scaled down independent microgrid. Our contribution is the design and evaluation of a smart picogrid controller that a) identifies the power source and b) decides on battery charging or discharging based on the power source. As we expect DC picogrids to co-exist with AC UPSes, we must ensure that the DC picogrid does not draw power from the UPS's battery but charges from the macrogrid when available. To accomplish this, we exploit the fact that AC distribution from the macrogrid exhibits sufficiently distinct characteristics compared to an AC UPS or a diesel generator. Our picogrid controller uses a Hidden Markov Model for state estimation that uses temporally correlated fluctuations in line voltage and frequency for discrimination. We show through data from four settings that the controller can identify its supply source with over 90% accuracy, and that efficiency recovered from conversion losses could result in 30% reduction in energy consumption.

Categories and Subject Descriptors

I.5.4 [**Applications**]: Signal processing

General Terms

Design, Experimentation

Keywords

DC Picogrid, Power Source Identification, Conversion Losses

1. INTRODUCTION

The proliferation of consumer electronics and communication technologies coupled with the rise of data centers, all of which natively operate on direct current (DC), is beginning to shift the balance of energy loads into a DC regime [6]. By some estimates, our residential DC appliances already constitute 15% of global residential electricity consumption and this quantity is set to double by 2022 [5]. This trend is further fueled by the rapidly declining costs of semiconductor based electronics and a general rise in the global ownership levels of gadgets [5].

DC is being used in a diverse set of other applications as well, from LED lighting (and electronic ballasts in fluorescent lighting), to variable frequency/speed drives in ventilation systems, to public transportation in many parts of the world, to the high-voltage DC lines used for long distance power transmission and for coupling of separate alternating current (AC) systems [11]. Furthermore, most renewable energy sources produce native DC power.

As generation, transmission and distribution has traditionally been AC, DC appliances are preceded by an AC-to-DC conversion generally in the form of an external power supply. External power supplies have recently been the focus of energy efficiency measures and every supply (<250W) must now meet a stringent international marking protocol [9]. In reality, however, efficiencies above 90% are rarely achieved [23]. The key concern, though, is for DC appliances that require uninterrupted power. Data centers fall in this category, as does DC lighting in homes and offices and home appliances (audio/video, network equipment, etc) in outage prone geographies. Uninterrupted power supply (UPS) is provided through an intermediate energy store, typically a battery bank, during supply outages. The battery bank is inherently DC and tying it to the AC distribution network entails a separate DC-to-AC conversion step, called inversion. The conversion process and associated losses are depicted in Figure 2.

The EMerge Alliance is a growing consortium that proposes a DC distribution system to eliminate conversion losses. DC appliances are powered directly from a battery bank and the battery bank is tied to the larger macrogrid through a battery charger. If DC appliances require voltages that are incompatible with the DC distribution voltage, a high efficiency switch mode DC-to-DC converter can be employed [28]. A number of installations around the world have successfully evaluated DC distribution for data centers [27, 22, 4] but this requires separate cabling infrastructure for high voltage DC. The Emerge Alliance also recommends short runs

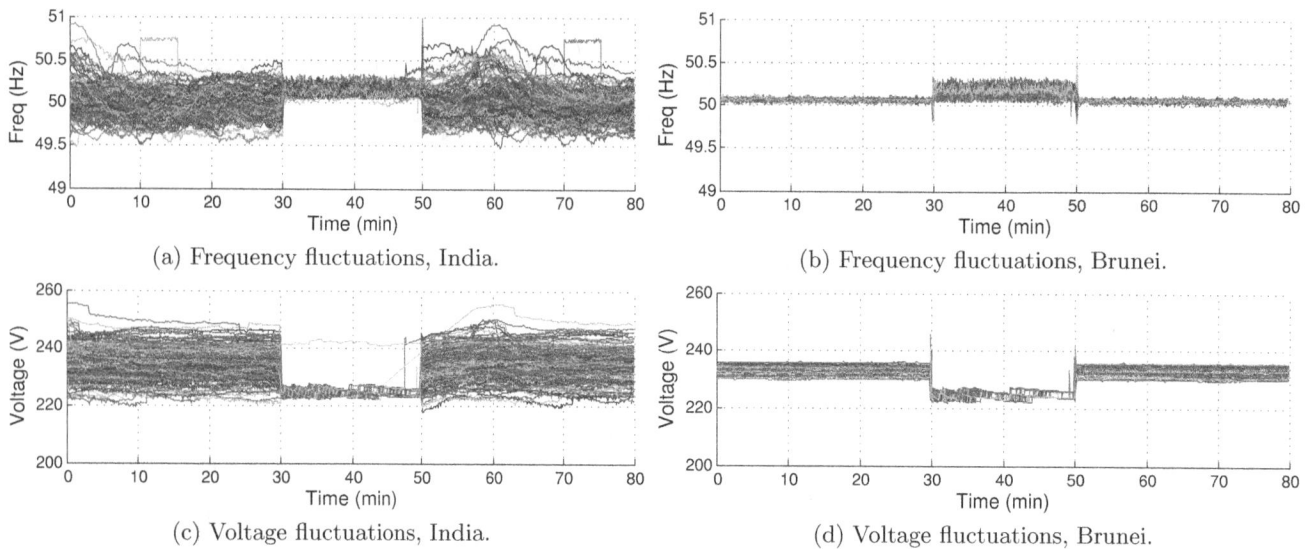

(a) Frequency fluctuations, India.

(b) Frequency fluctuations, Brunei.

(c) Voltage fluctuations, India.

(d) Voltage fluctuations, Brunei.

Figure 1: Frequency and voltage fluctuations before, during and after 200 emulated outages observed at the output of an AC UPS connected to the macrogrid in Bangalore, India and Bandar Seri Begawan, Brunei. Each line represent one outage instance. The middle section represents the F/V sensed during an outage when the UPS is in inversion mode. Data collected from a Numeric HPH1400 inverter.

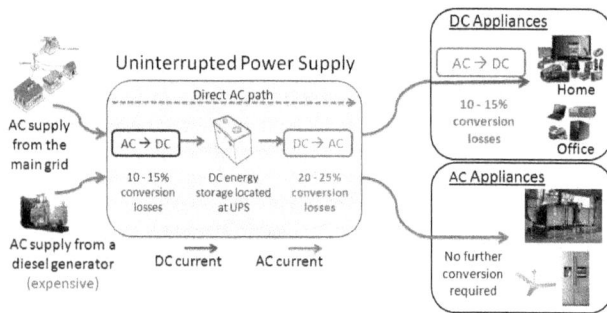

Figure 2: Multiple conversion losses when providing uninterrupted power to DC appliances

of low voltage DC, but this incurs significant resistive losses that increase quadratically with the power delivered to DC appliances. The requirement of a dual distribution network is especially discouraging for home users. Additionally, if home users install an AC UPS for AC loads that require uninterrupted power, separate wiring is needed to connect to the DC distribution system to ensure that DC loads do not draw power from the AC UPS battery during outages, as this would defeat the very purpose of DC distribution.

should be connected to the same cabling as the uninterruptible AC loads, but it should determine whether an outage exists at the input of the AC UPS and thus whether to draw

This paper attempts to make the bold case of placing energy storage, a battery, at (or within) each DC appliance. By managing the battery energy intelligently, we show that DC appliances can be powered more efficiently at lower average cost. We term this topology a *DC picogrid*, a miniature version of a smart microgrid that uses technology to optimize production, distribution, and consumption and that can operate independently by managing its own expenses

and production capacity. The DC picogrid within an appliance would present an AC input to the macrogrid, but would use an AC-to-DC battery charger instead of an external power supply. This topology has the innate advantage of needing no additional wiring while still eliminating the two unnecessary conversion losses that accrue with AC UPSes powering DC appliances. Imagine Figure 2 with the DC appliances connected to the grid only through the direct AC path. Another advantage is the ability of consumers to transition organically – each appliance can be replaced independently of others. While adding a battery increases the bulk and price of each appliance, we believe that due to systemic efficiencies gained from such an architecture, the total volume and cost to the consumer can actually reduce. A primary reason for the cost reduction is the exclusion of the complicated inversion step in DC picogrids. A full analysis of the cost involved is deferred until Section ??. It should be noted that compared to an AC UPS or a DC distribution, our system precludes sharing battery capacity across appliances. We argue that this issue may not manifest in many scenarios as users may require the DC appliances (servers, lighting and routers, for example) to be powered simultaneously through an outage.

As we expect that DC appliances will be wired up with other AC loads that need to be provided uninterrupted power using an AC UPS, the DC picogrid must have some way of distinguishing its source of supply. The key contribution of this paper is the design and evaluation of a smart picogrid controller that a) identifies the power source and b) decides on battery charging or discharging based on the power source. The smart picogrid controller may also schedule battery charging in the future based on time of use pricing, load patterns, renewables source prediction and outage forecasts. In this paper, however, we chiefly focus on the problem of power supply source identification.

Figure 1 shows the line frequency and voltage fluctuations observed at the output of an AC UPS (inverter) in two locations. One in Bangalore, India at a corporate office and one in Bandar Seri Begawan, Brunei at an academic institute. The plots show the frequency and voltage synchronized in time over 200 emulated outages (by disconnecting the input supply of the inverter) of 20 min each, with 30 min of grid supply on each side of the outage. While the quality of the grid between the two locations is vastly different, note that the outages are visually perceptible because of changes in characteristics between grid supply and inverter generated supply. This observation forms the basis of our source identification method. Furthermore, we observe that the frequency and voltage fluctuations are strongly temporally correlated during an outage. In order to exploit this, we train a Hidden Markov Model (HMM) that performs inferences on whether the source is connected to the grid or the backup by discriminating frequency and voltage (F/V) characteristics. The HMM employed in this work contains two hidden states, one for direct grid connectivity and the other for a backup source that turns on during an outage. Section 3 explains the power systems theory behind the observed changes in characteristics. Note, also, that the characteristics are visible when the backup system is an AC diesel generator (see Figure 16b and 16f).

The HMM based source identification requires a training phase. Empirically, we have found that training the HMM on one just outage allows it to gather sufficient statistics for successful disambiguation. Intuitively, this is because the distributions of F/V during backup remain consistent over long periods. Details on the methodology and formulation are presented in Section 4. Evaluation of the source identification system and measurements corresponding to conversion losses are detailed in Section 5. The next section outlines prior work that precedes ours and, in some ways, has inspired it.

2. RELATED WORK

Owing to the increasing prevalence of DC appliances and advancements in semiconductor technologies, the concept of subgrids is attracting considerable attention [18, 17, 25]. A milli, micro or nano grid [18] essentially refers to a distributed energy system [1] where a variety of small, modular power-generating technologies are combined with load management and storage systems to improve the quality and reliability of electric supply. Unlike the historic "top-down" approach taken by conventional grids, distributed energy systems follow a "bottom up" approach in terms of controls and distribution. These systems are typically located at end-consumer sites where the energy generated is locally consumed.

A microgrid integrates local (distributed) generation with local storage, provides both AC and DC output voltage and can operate in both grid-tied and in islanded modes [21]. H. Kakigano et al. [15] propose a house level microgrid where each house has a co-generation system and shares the power among the houses. Nanogrids are small scale microgrids. A nanogrid has at least one load, usually outputs low voltage DC, may have control and can interpolate with other nano or microgrids through gateways [21]. Kinn[16] proposes a smart nanogrid for a domestic electrical system that operates at or below 50V DC. The inherent architecture of micro or nanogrids is immune to conversion losses. However, it requires a DC distribution network. Design issues have been considered for such networks in both residential and com-

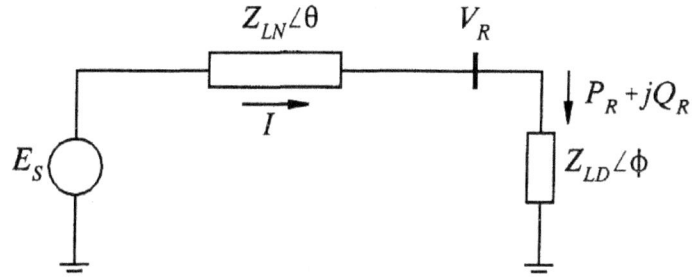

Figure 3: A two-bus power system

mercial buildings [14, 24]. While microgrids are expected to grow rapidly in developing nations in the future[12], significant rewiring in building infrastructure and challenges such as voltage control, power flow control, load sharing, protection and stability limit their wider adaptability. The proposed picogrid architecture does not require any change in the existing AC distribution network and gains efficiency without installing a DC distribution network.

Johnston et. al. [13] consider picogrid as an appliance level power distribution network, with ultra low power demand such as laptops, smart phones, tablets, sensor networks, USB 2.0/3.0 devices etc. Wang et al. [29] regard picogrid as a network of electronics devices in a home or building, whereas, Boroyevich et al. [3] view electrical system in a plug-in hybrid electric vehicle (PHEV) a picogrid. None of the previous work on picogrids analyze the potential of coupling DC appliances with an AC distribution network and present the controller architecture.

Central battery banks in subgrids are used to mitigate the unreliability of supply due to intermittent nature of renewables sources in islanded mode. However, in grid-connected mode, the goal is to prevent propagation of load fluctuations to the grid. Zhou et al. [30] evaluates a dynamic energy management scheme with battery and capacitor banks to address these issues. Smart battery charging schemes [10, 20] implement a demand response technique for laptops and smart phones and schedule the battery charging based on user's preferences and load on the grid. Mishra et al. [19] evaluate benefits of central battery arrays at home and proposes a system that optimally schedules battery charging during hours when electricity prices are low. It is concluded that if such a system is widely deployed, not only does it benefit end users but also reduces peak demand. Local sensing of line voltage or frequency at the appliance [7] has been used to infer the aggregate demand or power imbalance at higher levels in the AC distribution network, but identification of AC source has not been considered. Srinivasan et. al. [26] provide an architecture for switching off an AC electrical appliance when local power supply generator is used but do not evaluate local storage.

3. POWER SYSTEMS BACKGROUND

This section uses power systems theory to explain how fluctuations in the line voltage and frequency can serve as good indicators of determining the power source.

3.1 Voltage fluctuations with varying demand

Figure 3 shows a simple "power system" wherein a load is connected to a generator using a transmission line. $\tilde{E}_S$ is the generator voltage, $\tilde{V}_R$ is the load voltage, $\tilde{Z}_{LN}$ is the

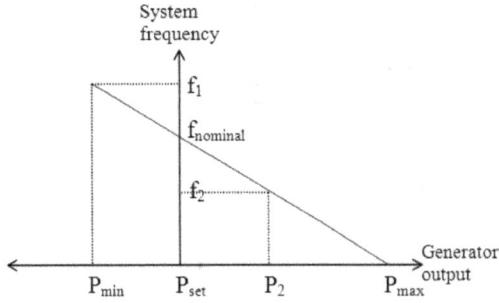

Figure 4: Load-frequency control characteristics

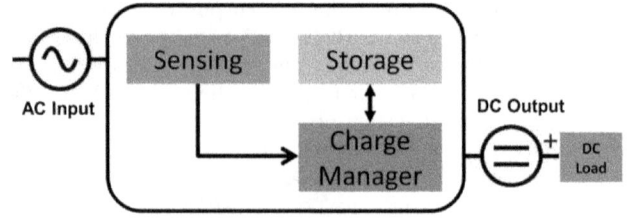

Figure 5: A depiction of DC picogrid controller architecture. It includes a sensing module, a charge manager and a local storage unit. The controller sits between AC supply and DC load

transmission line impedance, and $\tilde{Z}_{LD}$ is the load impedance (all quantities are vectors). The current flowing through the line and load, $\tilde{I}$ is given by

$$\tilde{I} = \frac{\tilde{E}_s}{\tilde{Z}_{LN} + \tilde{Z}_{LD}}$$

$$\text{where} \tilde{Z}_{LN} = Z_{LN} \angle \theta = Z_{LN} \cos\theta + jZ_{LN} \sin\theta$$

$$\tilde{Z}_{LD} = Z_{LD} \angle \phi = Z_{LD} \cos\phi + jZ_{LD} \sin\phi$$

Here θ is phase angle between reactive and resistive components of the line impedance while ϕ is the phase angle between the load current and voltage. Now the magnitude of current I is given by

$$I = \frac{E_S}{\sqrt{(Z_{LN}\cos\theta + Z_{LD}\cos\phi)^2 + (Z_{LN}\sin\theta + Z_{LD}\sin\phi)^2}}$$

Therefore the magnitude of load voltage V_R is:

$$V_R = Z_{LD} \times I$$

$$= \frac{Z_{LD} \times E_S}{\sqrt{(Z_{LN}\cos\theta + Z_{LD}\cos\phi)^2 + (Z_{LN}\sin\theta + Z_{LD}\sin\phi)^2}} \quad (1)$$

Since the source voltage E_S and transmission line impedance $Z_{LN}\angle\theta$ are generally constant, the load voltage V_R is essentially a function of the magnitude of load impedance Z_{LD} and the power factor $\cos\phi$. To minimize reactive power consumption, appliances are usually designed to have high power factor (0.9 to 1). Thus from Eq. (1) we see that the load voltage V_R is dominated by the magnitude of load impedance Z_{LD}. As the load increases (i.e. impedance decreases), the load voltage V_R decreases and *vice versa*. Therefore, continuous demand shifts on the grid generate voltage fluctuations. The same theory applies for a household inverter or a building level diesel generator; however, since only local load variations effect the voltage, aggregate demand varies much lesser compared to the larger macrogrid which keeps the voltage at the household level relatively stable.

3.2 Frequency fluctuations with varying demand

Conventionally, the grid frequency is regarded as an indicator of imbalance between generation and demand. During imbalance, the output of each generator is automatically adjusted to meet the demand. This changes the system frequency according to the *load-frequency* characteristics of the generators as shown in Figure 4. The plot shows that when the generation is higher than P_{set} (the generation needed to support a fixed load), the frequency drops. On the other

hand, if it is less than P_{set}, the frequency shoots up. The output frequency in inverter is decided by the manufacturer and is oblivious to the connected loads. For diesel generators, as mentioned earlier, since only local changes in demand have an impact, frequency remains relatively stable.

4. DC PICOGRID CONTROLLER

Figure 5 depicts the architecture of a DC picogrid controller which consists of three main components, (i) sensing, (ii) charge manager, and (iii) local storage. The controller takes an AC input, provides a DC output, and mediates between AC supply and DC appliance. The following sections explain the functionality of each component.

4.1 Sensing for Source Identification

The module senses line voltage and frequency at regular intervals to determine the AC power source in real time. As explained in the previous section, frequency and voltage exhibit continuous fluctuations in unstable grids where there is a high mismatch between generation and demand. In contrast, the output frequency band in local household inverters is relatively oblivious to connected load or demand variations. Similarly, due to negligible changes in demand in a home or a building, as compared to grid, local generators do not demonstrate much change in the sensed parameters. Considering that the parameters are widely dispersed in grid as compared to the local source, we evaluate maximum likelihood based algorithms to estimate the source. Moreover, in many parts of the world that suffer from high shortage of electricity, utility companies resort to load shedding. These power cuts occur sporadically throughout the day. Some utility companies do publish advance notice of planned outages. To capture temporal patterns, we propose the application of statistical tools to model time series data, the Hidden Markov Model (HMM).

The HMM employed contains two hidden states, one for direct grid connectivity and the other for a backup source that turns on during an outage (the model extends to multiple different back systems as well). The HMM observes samples of frequency and voltage sensed at the input of the DC picogrid's battery charger to infer what state the supply source is in.

We assume that time is discretized in Δ increments, such that $t = \Delta k$. The state is denoted by $s(k) = \{S_0, S_1, ..., S_N\} \forall k \in \{1...n\}$. N is the number of states and n is the sequence length. The observations are the sensed frequency $f(k)$ and voltage $v(k)$. Voltage observations are discretized into m_v equi-spaced bins between v_{min} and v_{max}. Frequency ob-

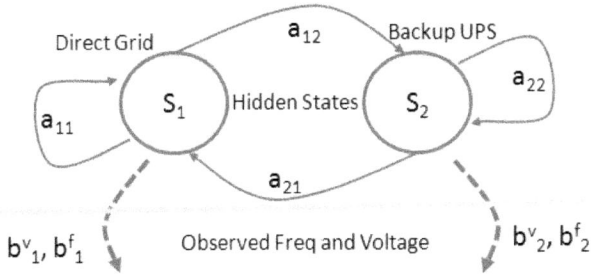

Figure 6: A pictorial representation of the two state Hidden Markov Model.

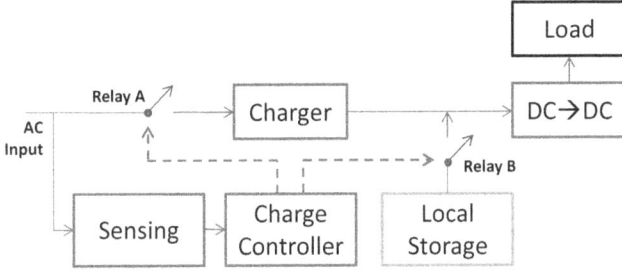

Figure 7: Charge Controller - Two relays that can be controlled to power the appliance based on the source

AC Source	Battery Charged?	Relay A	Relay B
Grid	No	Yes	Yes
Grid	Yes	Yes	Yes/No
Inverter	No	Yes	No
Inverter	Yes	No	Yes
None	No	No Energy	
None	Yes	Yes/No	Yes

Table 1: Relay states based on AC source and battery charge status

servations are discretized into m_f equi-spaced bins between f_{min} and f_{max}. We use two independent 2-state Hidden Markov Models, one for frequency and another for voltage and combine the outputs for final inference.

Transition probability matrix:

$$\begin{aligned} A &= \{a_{ij}\} \ \forall i,j \in \{1,2\} \\ a_{ij} &= \Pr[s(k+1) = S_j | s(k) = S_i] \\ a_{ij} &\geq 0 \\ \sum_j a_{ij} &= 1 \end{aligned}$$

Observation symbols:

$$\begin{aligned} V &= \{v_1, v_2, ..., v_{m_v}\} \\ F &= \{f_1, f_2, ..., f_{m_f}\} \end{aligned}$$

Emission Matrix for Voltage Observations:

$$\begin{aligned} B^v &= \{b_j^v(l)\} \ \forall j \in \{1,2\}, 1 \leq l \leq m_v \\ b_j^v(l) &= \Pr[v = v_l | s = S_j] \\ b_j^v(l) &\geq 0 \\ \sum_l b_j^v(l) &= 1 \end{aligned}$$

Emission Matrix for Frequency Observations:

$$\begin{aligned} B^f &= \{b_j^f(l)\} \ \forall j \in \{1,2\}, 1 \leq l \leq m_f \\ b_j^f(l) &= \Pr[f = f_l | s = S_j] \\ b_j^f(l) &\geq 0 \\ \sum_l b_j^f(l) &= 1 \end{aligned}$$

Initial State Probability Matrix:

$$\begin{aligned} \pi &= \{\pi_i\} \ \forall i \in \{1,2\} \\ \pi_i &= \Pr[s(1) = S_i] \\ \pi_i &\geq 0 \\ \sum_i \pi_i &= 1 \end{aligned}$$

The above parameters of the HMM model are estimated in a short training phase using the forward procedure[2]. The training phase requires that the user indicate to the DC picogrid controller when an outage occurs. We find that training the model with just one outage is sufficient to gather statistics about the emission probabilities. Good estimates of the transition probabilities are feasible using historical outage data. One could imagine the training to be explicit using, say, a button on the picogrid controller, or implicit, by the user turning off the appliance for a predetermined time during an outage.

4.2 Charge Management

Once the source has been sensed, the charge controller acts as a coordinator between three components, (i) AC power source, (ii) local storage, and (iii) connected DC appliance. Based on the identified source, the controller takes two key actions – first, it toggles input power source for the connected load between AC supply and internal storage unit, second, it controls discharging and charging of local battery.

Figure 7 shows one basic design where the controller drives the system based on two relays. The grid availability, battery state and local supply collectively decide (Table 1) the state of the relays, and hence power source for the connected load. The picogrid controller would choose to use grid power whenever available to serve the load and to keep the local energy storage topped up. When the power source switches to the AC UPS during an outage, the picogrid controller would serve the DC load for as long as possible from the local energy store. When local storage is depleted, the picogrid controller would start to draw power from the AC UPS, but would not start recharging the battery until grid power is available again.

Battery charging can be optimized based on multiple parameters as shown in Figure 8. The controller may learn usage patterns of the appliance, perform outage prediction to anticipate an outage, take user input and time of usage pricing into consideration while scheduling battery charging. Similar techniques have been used extensively [10, 20, 19] in the past to reduce energy costs for end users and peak demand for utility companies. Such systems essentially time-shift the demand using central or distributed battery banks. While none of the existing work considers AC power source

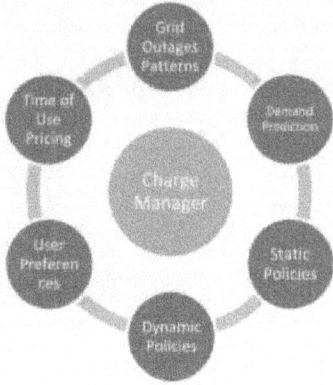

Figure 8: The charge controller toggles the flow of output supply through AC mains or internal storage. It optimizes charging of the local storage based on several parameters.

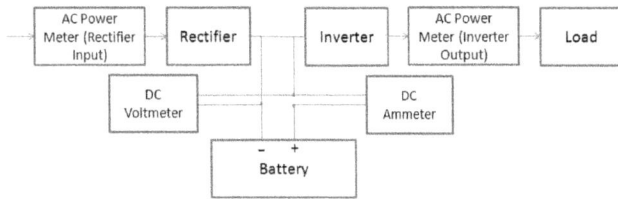

Figure 9: Experimental setup to calculate conversion losses in central storage with 150AH lead acid batteries

as a variable to reduce conversion losses, we believe inclusion of source information in optimization models may be straightforward. Therefore, in our evaluation we focus on source determination and cost benefit analysis of the proposed system, as explained in the next section.

5. EVALUATION

In this section, we present the experimental evaluation of the proposed system. We provide empirical evidence to compare losses in central and appliance level battery backup architectures. We then evaluate the performance of existing machine learning algorithms to identify the AC power source.

5.1 Conversion Losses

We evaluate conversion losses in two systems, i) lead acid batteries connected to a central UPS, and (ii) an AC-to-DC adapter connected to a 12V LED light. Controlled experiments were performed to calculate losses at each conversion step. Plug level in-line energy meters were used to measure AC power flow and DC multimeters were used to measure DC power. The AC power meters sense the line voltage, frequency and power every ten seconds and upload the data to a central server over a WiFi network.

5.1.1 Conversion Losses in Central UPS

Two 150AH 12V (CRTT150AH) Amaron batteries were connected in series to a HPH1400 NUMERIC 1400VA UPS as shown in Figure 9. The batteries were fully charged before the experiment and a 1000W room heater was used as a load to discharge them. This serves as a reasonable best case scenario for the characterization experiment as the room

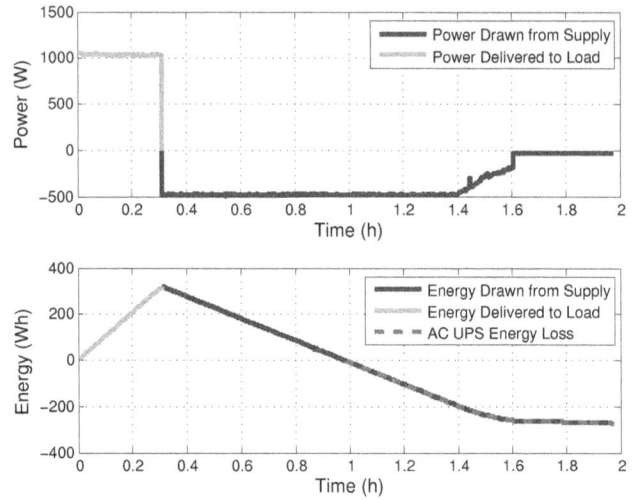

Figure 10: Power and energy characterization of HPH1400 AC UPS through a discharge and charge cycle.

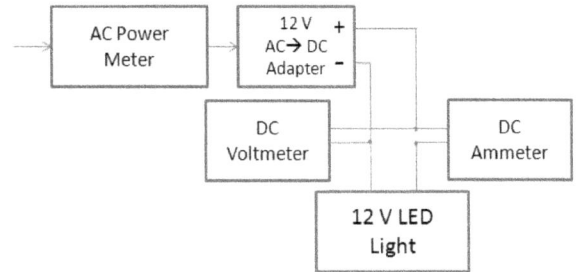

Figure 11: Experimental setup to calculate conversion loss in a AC/DC adapter

heater provides a power factor very close to 1 since it is a mostly resistive load, is a constant power load and also loads the UPS around its optimal efficiency point.

As a central storage system goes through rectification, battery round trip and inverter losses before energy reaches the load during an outage, aggregated efficiency of the UPS was calculated by measuring AC input before rectifier and AC output after inverter. The fully charged batteries were discharged for around 20 minutes and then charged back until they stopped drawing a high amount of charging current. The result from one such experiment is shown in Figure 10. Positive power and energy accumulation indicates power delivered to the load. Negative power and energy declination indicates power required to recharge the battery bank. Dashed line indicates the energy overhead for the specific discharge-charge cycle. The net UPS efficiency observed was 54.89% (shown in Figure 15 as UPS).

5.1.2 AC-to-DC Adapter Conversion Loss

DC loads use an external AC-to-DC adapter or an internal rectifier, which further adds up to the conversion losses. The efficiency of a 12V/3A AC-to-DC adapter was empirically estimated using the experimental setup as shown in Figure 11. A 12V/12W LED light was connected to the adapter. Figure 12 shows conversion losses in the adapter with average efficiency value of 82.09% (Figure 15 (AC Adapter)). Therefore, a DC load connected to a central UPS and a

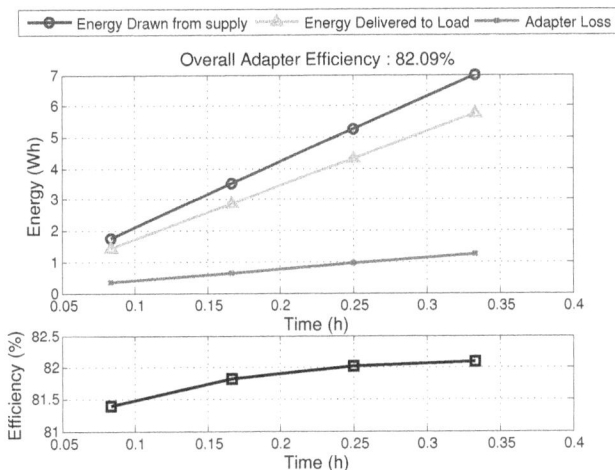

Figure 12: Conversion losses in a AC-DC converter (12 V/3A output)

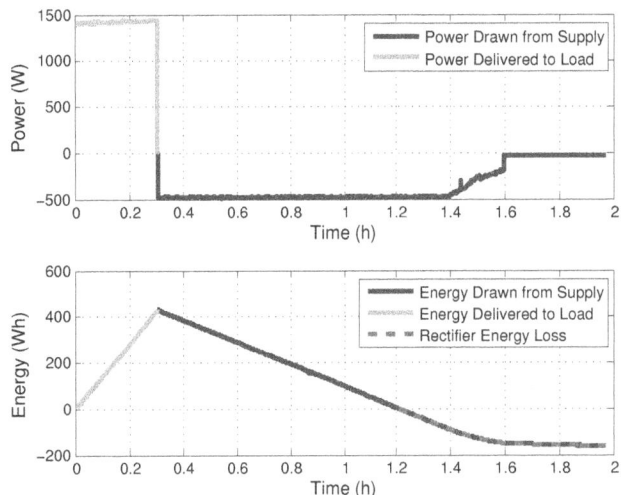

Figure 13: Experimental setup to calculate conversion loss in a DC Picogrid.

AC-to-DC adapter sustains 45.07% of energy loss during an outage (shown in Figure 15 as Central Storage).

5.1.3 Conversion Losses in DC Picogrid

The experimental setup to estimate the losses in a DC picogrid is shown in Figure 13. Since the DC picogrid architecture for a DC load consist of only rectifier and battery, losses until the two steps were calculated by measuring AC input before rectifier and DC output from the battery, giving efficiency of 74.18% (shown in Figure 15 as DC Picogrid). The detailed characterization is shown in Figure 14. The increase in efficiency is visible, compared to Figure 10, by noting the overall energy overhead required to recharge the batteries after the same discharge. The difference between the plots is that Figure 10 also includes loss in efficiency due to the inversion process while Figure 14 does not require the inversion process. We have considered a DC load that uses the battery voltage directly. In many applications, however, a DC-to-DC conversion may be needed. We show that significant conversion losses occur at each step and with two less conversions in appliance level local storage, the collective efficiency gains could be around 30% for DC appliances depending upon the inverter efficiency

5.2 AC Power Source Identification

In this section, we evaluate the performance of existing machine learning algorithms to identify AC power source. Frequency and voltage grid data was collected in two cities in India: Bangalore and Chennai and one city in Brunei:

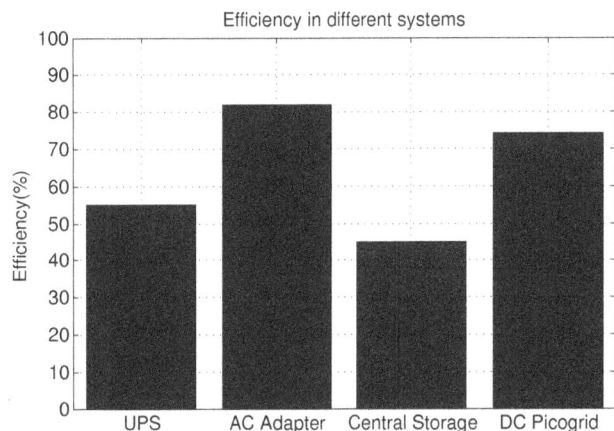

Figure 14: Power and energy characterization for a DC picogrid based on the rectifier on the HPH1400 through a discharge and charge cycle.

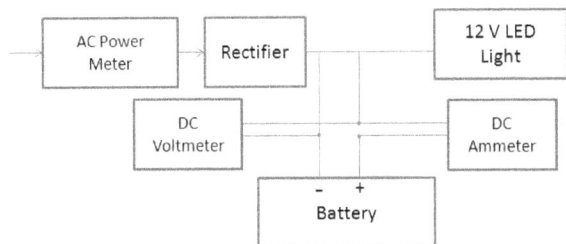

Figure 15: Comparison of system efficiency in central storage and DC picogrid architectures

Bandar Seri Begawan. Inverter output data was collected from two different manufacturers and diesel generator data was collected in an apartment complex in Bangalore for evaluation. The data collected in the Indian cities represents an unstable grid, which usually has inadequate supply, whereas, data collected in Brunei represents a stable grid with surplus capacity. In each of the cities, a few locations were instrumented with plug level energy monitors to report the line voltage and frequency every ten seconds. One of the instrumented homes in Bangalore had a building level diesel generator installed, which powers the building during an outage. Outages were recorded in this instance by a separate plug level meter connected to a non-backed up circuit. We use six weeks (December 2012 to mid-January 2013) of voltage and frequency time series data in our experiments and crop them to extract the relevant pieces.

5.2.1 Ground Truth

Inverters are equipped with a switch that allows them to be bypassed when the grid is online. Therefore, AC output of an inverter signifies grid characteristics when the grid power is available. However, as AC power is generated using local battery arrays during an outage, the properties of generated AC waveforms are controlled by the DC to AC inverter.

Over 200 outages were emulated using an HPH 1400 Numeric inverter (connected with 150AH batteries). For each emulated outage, grid input to the inverter was cut-off for 20 minutes every hour. Figure 1a/1c compares frequency and voltage values observed at inverter output with grid data in Bangalore, where each time series represents an emulated outage. Each time series has three parts, (i) first 30 minutes correspond to grid data before an outage, (ii) next 20 minutes correspond to inverter generating output during an outage, and (iii) last 30 minutes correspond to grid data after the outage. Similarly, Figure 1b/1d associate inverter data with the Bruneian national grid data respectively. As is clear in all the figures, high fluctuations in frequency and voltages are observed in the Indian grid, whereas, the Bruneian grid shows much higher stability. Figure 16 compares distribution of frequency and voltage time series between grid and inverter.

5.2.2 Inference Performance

To evaluate the performance of our HMM based inference system, we compared to an unsupervised k-means clustering approach for disambiguation and a k-nearest neighbor approach that included training. The final inference accuracies with the false positive and false negative rates are listed in Table 2. Results from the different settings show that training using 20% of the data set delivers a marginal improvement over training using just one outage. The primary reason for this is the consistency of F/V characteristics observed from the AC UPS output.

Figure 17 illustrates the performance of the HMM 1-Outage method compared to k-means. We observe that while unsupervised learning methods do not perform well in general, both HMM and KNN are able to distinguish the two states accurately. False positives occur when the system incorrectly estimates that AC power is being fed from a backup source such as an AC UPS or a diesel generator. False negatives occur when the system incorrectly estimates that the main grid is available. From Table 1, it is possible to assess the implication of these errors. False positives might make the picogrid controller switch in the local energy storage even though grid power is available. False negatives might result in charging the local energy storage from the central UPS battery bank. Both these settings reduce the efficiency of the overall picogrid system, but neither are detrimental to its operation.

It might be noted that results from the Bruneian data is especially good due to the clean distinction in the frequency and voltage characteristics. Furthermore, the KNN classifier performs almost as well as the HMM based classifier. In general, however, we expect the HMM based system to outperform the KNN classifier, especially when there are changes in the F/V distributions that can be explained through temporal correlations.

6. COST BENEFIT ANALYSIS

The major expense of a storage system is its battery array. Eliminating conversion losses not only increases system efficiency, also lowers the battery capacity requirements as compared to central storage. In this section, we estimate

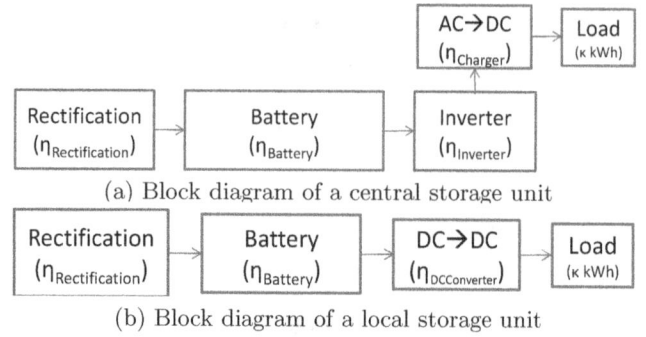

(a) Block diagram of a central storage unit

(b) Block diagram of a local storage unit

Figure 18: Block diagrams for both central and local storage units

yearly battery and electronics cost for both central and local storage architecture. The following system parameters are defined as shown in Figure 18:

$\eta_{Rectifier}$: Rectification (AC → DC) Efficiency
$\eta_{Battery}$: Battery Round Trip Efficiency
$\eta_{Inverter}$: Inverter (DC → AC) Efficiency
$\eta_{Charger}$: External power supply (AC → DC) Efficiency
$\eta_{DCConverter}$: DC → DC Conversion Efficiency
e : Energy price ($) per unit of kWh
$MTBF$: Electronics mean time between failure (MTBF)
κ : Outage demand of a geography (kWh/year). Energy delivered by the system over a year.
$C_{Central-Equipment}$: Equipment cost for central storage
$C_{Local-Equipment}$: Equipment cost for local storage

The lifetime of a battery is defined by a number of parameters, including depth of discharge, number of cycles, environmental conditions etc. Total energy delivered by a battery over its lifetime (in kWh) can be defined as:

$$L = C \cdot DOD \cdot N(DOD)$$

where,
C : Battery capacity (kWh at rated nominal voltage)
DOD : Depth of discharge in each cycle
N : Number of cycles as a $f(DOD)$
$C_{\$/kWh}$: Battery cost in terms of capacity ($/kWh)

We calculate the cost of three main components in each architecture, (i) battery, (ii) conversion losses, and (iii) equipment cost.

6.1 Battery Cost ($/year)

The per unit energy cost of the battery over its lifetime is $(C_{\$/kWh} * C)/(L)$. As κ is the amount of energy delivered to the load, energy supplied by the battery over a year in the central storage architecture will be $\kappa/\eta_{Inverter} * \eta_{Charger}$. Therefore, battery cost for central storage architecture is:

$$Cost_{Battery-Central} = \frac{C_{\$/kWh} * C}{L} \cdot \frac{\kappa}{\eta_{Inverter} * \eta_{Charger}}$$

(2)

Similarly, amount of energy given by the battery in local storage will be $\kappa/\eta_{DCConverter}$, therefore, battery cost

$$Cost_{Battery-Local} = \frac{C_{\$/kWh} * C}{L} \cdot \frac{\kappa}{\eta_{DCConverter}}$$

(3)

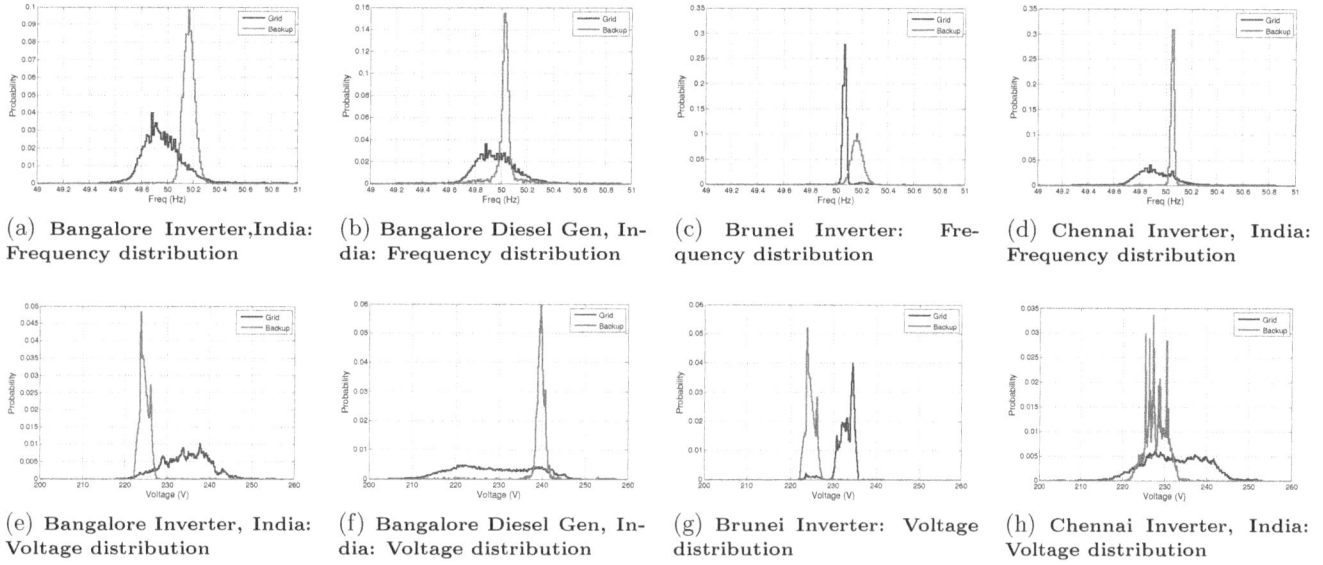

(a) Bangalore Inverter,India: Frequency distribution

(b) Bangalore Diesel Gen, India: Frequency distribution

(c) Brunei Inverter: Frequency distribution

(d) Chennai Inverter, India: Frequency distribution

(e) Bangalore Inverter, India: Voltage distribution

(f) Bangalore Diesel Gen, India: Voltage distribution

(g) Brunei Inverter: Voltage distribution

(h) Chennai Inverter, India: Voltage distribution

Figure 16: Comparison of frequency and voltage time series distribution between national grid, inverter and a diesel generator across data sets from India and Brunei.

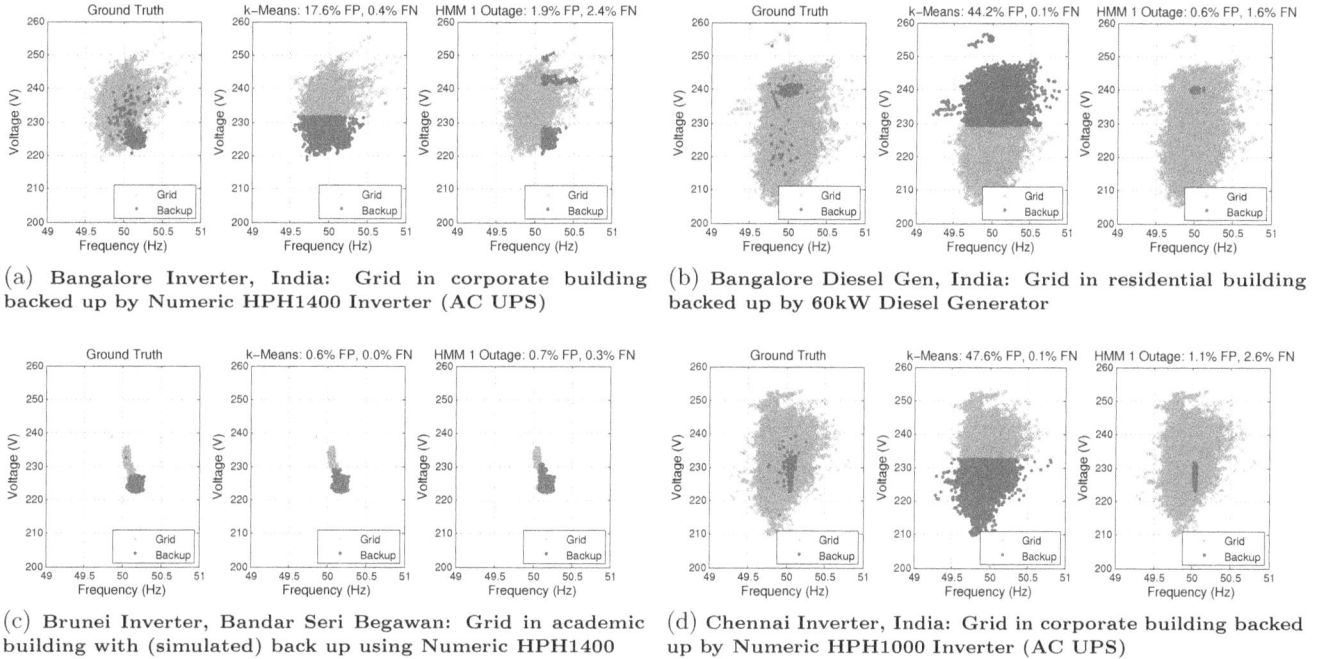

(a) Bangalore Inverter, India: Grid in corporate building backed up by Numeric HPH1400 Inverter (AC UPS)

(b) Bangalore Diesel Gen, India: Grid in residential building backed up by 60kW Diesel Generator

(c) Brunei Inverter, Bandar Seri Begawan: Grid in academic building with (simulated) back up using Numeric HPH1400

(d) Chennai Inverter, India: Grid in corporate building backed up by Numeric HPH1000 Inverter (AC UPS)

Figure 17: Performance comparison of k-Means clustering and HMM based source identification trained on 1-outage, across data sets from India and Brunei in different settings.

FP / FN / Acc (%)	Bangalore Inverter	Bangalore Diesel Gen	Brunei Inverter	Chennai Inverter
k-Means	17.6 / 0.4 / **82.0**	44.2 / 0.0 / **55.8**	0.0 / 0.6 / **99.4**	47.2 / 0.0 / **52.8**
HMM, Unsupervised	3.7 / 0.6 / **95.7**	22.7 / 0.0 / **77.3**	0.9 / 0.1 / **99.0**	16.6 / 0.1 / **83.3**
K-NN, trained on $\frac{1}{5}$th of data	3.4 / 0.9 / **95.6**	0.5 / 2.5 / **97.0**	0.1 / 0.1 / **99.8**	3.4 / 1.7 / **94.9**
HMM, trained on $\frac{1}{5}$th of data	1.4 / 1.9 / **96.7**	0.7 / 2.2 / **97.1**	0.1 / 0.0 / **99.9**	1.8 / 1.5 / **96.7**
K-NN , trained on 1 outage	5.1 / 0.9 / **94.1**	2.2 / 1.2 / **96.6**	0.0 / 0.7 / **99.3**	10.1 / 0.9 / **89.0**
HMM, trained on 1 outage	1.9 / 2.4 / **95.7**	0.6 / 1.6 / **97.8**	0.7 / 0.3 / **99.0**	1.1 / 2.6 / **96.3**
Outages / Data set size (hr)	200 / 194.2	29 / 667.4	200 / 185.8	29 / 613.7

Table 2: Performance of source identification methods on different data sets.

6.2 Conversion Loss Cost ($/year)

If κ is the amount of energy delivered to the load, the input energy required before the rectifier in the central storage architecture will be $\kappa/(\eta_{Rectifier} * \eta_{Battery} * \eta_{Inverter} * \eta_{Charger})$. The cost of energy loss is:

$$e \cdot \kappa \cdot \left(\frac{1}{\eta_{Rectifier} * \eta_{Battery} * \eta_{Inverter} * \eta_{Charger}} - 1 \right) \quad (4)$$

Similarly, the cost of energy loss in local storage

$$e \cdot \kappa \cdot \left(\frac{1}{\eta_{Rectifier} * \eta_{Battery} * \eta_{DCConverter}} - 1 \right) \quad (5)$$

6.3 Equipment cost ($/year)

To calculate the cost of electronics equipment, we estimate their life period through Mean Time Between Failure (MTBF). MTBF is indicative of electronic economic life. It is a calculated number based on statistically determined reliability numbers assigned to each component used in a design, and depends on the stress and thermal environment[1]. Equipment costs per year for central and local storage are defined as Equations 6 and 7 respectively.

$$\frac{C_{Central-Equipment}}{MTBF} \quad (6)$$

$$\frac{C_{Local-Equipment}}{MTBF} \quad (7)$$

Therefore, total cost of central storage system (Equation 2 + Equation 4 + Equation 6)

$$Cost_{Central} = \quad (8)$$

$$\frac{C_{\$/kWh} * C}{L} \cdot \frac{\kappa}{\eta_{Inverter} * \eta_{Charger}} +$$
$$e \cdot \kappa \cdot \left(\frac{1}{\eta_{Rectifier} * \eta_{Battery} * \eta_{Inverter} * \eta_{Charger}} - 1 \right) +$$
$$\frac{C_{Central-Equipment}}{MTBF}$$

Total cost for local storage (Equation 3 + Equation 5 + Equation 7)

$$Cost_{Local} = \quad (9)$$

$$\frac{C_{\$/kWh} * C}{L} \cdot \frac{\kappa}{\eta_{DCConverter}} +$$
$$e \cdot \kappa \cdot \left(\frac{1}{\eta_{Rectifier} * \eta_{Battery} * \eta_{DCConverter}} - 1 \right) +$$
$$\frac{C_{Local-Equipment}}{MTBF}$$

Based on the prices of batteries and other equipment used in the experiments to compare conversion losses, we now evaluate the cost of both the architectures and aim to find

[1]National Energy Renewable Laboratory, http://www.nrel.gov/docs/fy06osti/38771.pdf

Parameter(P)	Value	C/P	Value
$\eta_{Rectifier}$	0.75	$MTBF$	5 yrs
$\eta_{Battery}$	0.75	$C_{Central-Equip}$	$200
$\eta_{Inverter}$	0.75	$C_{Local-Equip}$	$350
$\eta_{Charger}$	0.85	C	1.8kWh
$\eta_{DCConverter}$	0.9	DOD	0.3
e ($/kWh)	$0.10	N & $C_{\$/kWh}$	1000/$120

Table 3: Values and price of different parameters and components used for calculations

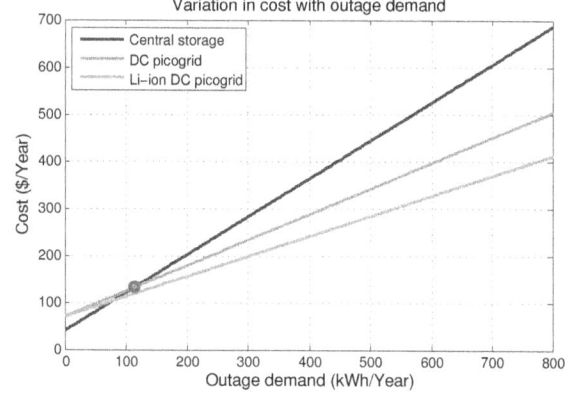

Figure 19: Variation in the cost of central vs DC picogrid storage architecture with outage demand.

the value of κ for which local storage becomes more economical than central storage. NUMERIC 1400 VA UPS inverter used in the experiments costs around $200 and the 36W AC-to-DC adapter costs $6. Adding the cost of sensing and controller hardware in the adapter (approximately) and making it equivalent to 1.4KVA, we take the equipment cost of local storage unit to be $350. As DC →DC conversions are more efficient than AC →DC or DC→AC conversions, values for $\eta_{DCConverter}$ is higher than other conversion parameters. All the experiments were done using lead acid batteries; DOD is taken as 30 percent with 1000 cycles. $MTBF$ has been taken as 5 years as per the study done by National Energy Renewable Laboratory mentioned earlier. We plot Equations 8 and 9 in Figure 19, which shows that for $\kappa > 114$ kWh/yr a DC picogrid with local storage is more cost effective than an AC UPS with central energy storage. Figure 19 also illustrates the potential saving when Li-ion battery chemistry is used. Li-ion is an especially promising technology for DC picogrids as they have higher energy density leading to smaller physical dimensions.

7. DISCUSSION

Appliance level storage exists in many mobile DC appliances such as laptops, mobile phones, cameras etc. The need for on-board storage in the past has mostly surfaced from mobility requirements, rather than from efficiency issues. However, with the emergence of DC appliances and industry standards for DC distribution networks, such as the Emerge Alliance, DC appliances are becoming widely deployed. A report from the US Department of Energy [8] evaluates benefits of moving from AC appliances to DC in residential places with a DC distribution network. The report considers cooling, lighting, space heating, clothes washing and dish-washing loads and predicts savings of about 33%. Battery powered LED lights; consumer electronics with in-built storage such as televisions are already available in the market.

A major difference between central and appliance level storage currently is that of battery chemistry. Existing central storage scheme installs lead acid batteries because of low cost and easy availability. The installations are designed to run minimal set of loads (such as Fans, Lights, a TV etc.) for a specific period of time. Whereas, appliance level batteries are of Li-Ion chemistry because of its high energy density which fits with mobility requirements.

Inverters are most efficient when they operate at peak power output and the rated efficiency mentions this value. However, most inverters operate at lesser load factors, which further decreases the system efficiency. There is a capacity loss and consequent energy losses in the battery due to asymmetrical charge-discharge regimes. Since the discharge current is always much higher that the charge current, the capacity during discharge is lesser. This accounts for storage losses and further decreases the efficiency.

Low power electronic equipment such as mobiles, routers, switches, tablets, laptops have a switched mode supply which draws current only during peaks of the voltage waveform. This results in sharp current pulses. As the number of low-power electronics goods proliferate, the peaks will be pronounced more and more since all of them draw current in synchronicity around the voltage peaks. This increases the apparent power. In case of inverters driving these loads, this will result in higher conduction and switching losses. Also, such inverters need to have higher surge handling specifications to service low average loads. Higher peak currents at the output would also result in higher battery conversion losses

8. CONCLUSION

This paper provides a fresh perspective on the problem of eliminating conversion losses for uninterrupted operation of DC appliances: DC picogrids. We propose that energy storage be co-located with each DC appliance and design and evaluate a smart picogrid controller that can identify the source of power supply. By disconnecting the charger from the AC mains during outages, we can disconnect the input from a pre-existing AC UPS. This enables using the existing AC cabling infrastructure while accruing the benefits of a DC power distribution system.

We show that power supply source identification can use the distinction in characteristics in line frequency and voltage between the grid and the back up source and employ a two-state HMM for inference. The HMM is able to exploit temporal correlations in the data, but we find that a k-NN approach works satisfactorily on our data as well. The performance of our algorithms was evaluated on four data sets from different parts of the world and they are able to achieve over 90% accuracy in identifying the correct source. A cost benefit analysis shows that the DC picogrid can reduce costs to the consumer by eliminating the complex electronics embedded in the inversion process. A rough measurement of the conversion losses for commercially available inverters and battery chargers illustrates that gains of 30% are easily obtainable. We admit that these gains are hard to generalize as

newer inverter and charger technologies may tip the balance of the equations in either direction.

9. ACKNOWLEDGMENTS

This material is based upon work supported by the UBD|IBM Centre in Brunei Darussalam. Any opinions, findings, conclusions or recommendations expressed in this material are those of the author(s) and do not necessarily reflect the views of the UBD|IBM Centre. The authors would like the thank Prof. Liyanage Chandratilak De Silva and Teck Sion Wong at the University of Brunei Darussalam, Brunei Darussalam for access to the Bandar Seri Begawan power grid data.

10. REFERENCES

[1] ALANNE, K., AND SAARI, A. Distributed energy generation and sustainable development. *Renewable and Sustainable Energy Reviews 10*, 6 (2006), 539 – 558.

[2] BAUM, L. E., PETRIE, T., SOULES, G., AND WEISS, N. A maximization technique occurring in the statistical analysis of probabilistic functions of markov chains. *The annals of mathematical statistics 41*, 1 (1970), 164–171.

[3] BOROYEVICH, D., CVETKOVIC, I., DONG, D., BURGOS, R., WANG, F., AND LEE, F. Future electronic power distribution systems a contemplative view. In *Optimization of Electrical and Electronic Equipment (OPTIM), 2010 12th International Conference on* (2010), IEEE, pp. 1369–1380.

[4] CARLSSON, U., FLODIN, M., AKERLUND, J., AND ERICSSON, A. Powering the internet-broadband equipment in all facilities-the need for a 300 v dc powering and universal current option. In *Telecommunications Energy Conference, 2003. INTELEC'03. The 25th International* (2003), IEEE, pp. 164–169.

[5] ELLIS, M., AND JOLLANDS, N. *Gadgets and Gigawatts: Policies for Energy Efficient Electronics*. OECD/IEA, 2009.

[6] FAIRLEY, P. Dc versus ac: The second war of currents has already begun [in my view]. *Power and Energy Magazine, IEEE 10*, 6 (2012), 104–103.

[7] GANU, T., SEETHARAM, D., ARYA, V., KUNNATH, R., HAZRA, J., HUSAIN, S., DE SILVA, L., AND KALYANARAMAN, S. nplug: a smart plug for alleviating peak loads. In *Future Energy Systems: Where Energy, Computing and Communication Meet (e-Energy), 2012 Third International Conference on* (2012), IEEE, pp. 1–10.

[8] GARBESI, K. Catalog of dc appliances and power systems.

[9] HAWLEY, J., AND ECONOMOU, M. Trends in energy efficiency regulations and initiatives for consumer external power supplies. In *Sustainable Systems and Technology (ISSST), 2010 IEEE International Symposium on* (2010), IEEE, pp. 1–6.

[10] HILD, S., LEAVEY, S., GRÄF, C., AND SORAZU, B. Smart charging technologies for portable electronic devices. *CoRR abs/1209.5931* (2012).

[11] HIROSE, K. Dc power demonstrations in japan. In *Power Electronics and ECCE Asia (ICPE & ECCE),*

2011 IEEE 8th International Conference on (2011), IEEE, pp. 242–247.

[12] HTTP://WWW.PIKERESEARCH.COM/NEWSROOM/REMOTE-MICROGRIDS-WILL-HELP-MEET-SOARING-ENERGY-DEMAND-IN-THE-DEVELOPING WORLD. Remote microgrids will help meet soaring energy demand in the developing world.

[13] JOHNSTON, J., COUNSELL, J., BANKS, G., DIRECTOR, A., AND STEWART, M. Beyond power over ethernet: the development of digital energy networks for buildings.

[14] KAKIGANO, H., MIURA, Y., AND ISE, T. Configuration and control of a dc microgrid for residential houses. In *Transmission & Distribution Conference & Exposition: Asia and Pacific, 2009* (2009), IEEE, pp. 1–4.

[15] KAKIGANO, H., MIURA, Y., ISE, T., MOMOSE, T., AND HAYAKAWA, H. Fundamental characteristics of dc microgrid for residential houses with cogeneration system in each house. In *Power and Energy Society General Meeting - Conversion and Delivery of Electrical Energy in the 21st Century, 2008 IEEE* (july 2008), pp. 1 –8.

[16] KINN, M. Proposed components for the design of a smart nano-grid for a domestic electrical system that operates at below 50v dc. In *Innovative Smart Grid Technologies (ISGT Europe), 2011 2nd IEEE PES International Conference and Exhibition on* (2011), IEEE, pp. 1–7.

[17] LASSETER, R., AKHIL, A., MARNAY, C., STEPHENS, J., DAGLE, J., GUTTROMSON, R., MELIOPOULOUS, A., YINGER, R., AND ETO, J. The certs microgrid concept. *White paper for Transmission Reliability Program, Office of Power Technologies, US Department of Energy* (2002).

[18] MARNAY, C. Future roles of milli-, micro-, and nano-grids.

[19] MISHRA, A., IRWIN, D., SHENOY, P., KUROSE, J., AND ZHU, T. Smartcharge: cutting the electricity bill in smart homes with energy storage. In *Proceedings of the 3rd International Conference on Future Energy Systems: Where Energy, Computing and Communication Meet* (New York, NY, USA, 2012), e-Energy '12, ACM, pp. 29:1–29:10.

[20] MURTHY, N., TANEJA, J., BOJANCZYK, K., AUSLANDER, D., AND CULLER, D. Energy-agile laptops: Demand response of mobile plug loads using sensor/actuator networks.

[21] NORDMAN, B. Evolving our electricity systems from the bottom up.

[22] PRATT, A., KUMAR, P., AND ALDRIDGE, T. Evaluation of 400v dc distribution in telco and data centers to improve energy efficiency. In *Telecommunications Energy Conference, 2007. INTELEC 2007. 29th International* (2007), IEEE, pp. 32–39.

[23] RASMUSSEN, N., AND SPITAELS, J. A quantitative comparison of high efficiency ac vs. dc power distribution for data centers. *White Paper# 127 of APC Inc* (2007).

[24] SALOMONSSON, D., AND SANNINO, A. Low-voltage dc distribution system for commercial power systems with sensitive electronic loads. *Power Delivery, IEEE Transactions on 22*, 3 (july 2007), 1620 –1627.

[25] SAVAGE, P., NORDHAUS, R., AND JAMIESON, S. Dc microgrids: Benefits and barriers. *published for Renewable Energy and International Law (REIL) project, Yale School of Forestry and Environmental Studies* (2010), 0–9.

[26] SRINIVASAN, B., SINGH, R., AND BUSSA, N. Switching off an ac electrical appliance when local power supply generator is used, Nov. 5 2009. WO Patent WO/2009/133,494.

[27] SYMANSKI, D., AND WATKINS, C. 380vdc data center at duke energy. *Emerging Technology Summit, Nov 9* (2010).

[28] WALKER, G., AND SERNIA, P. Cascaded dc-dc converter connection of photovoltaic modules. *Power Electronics, IEEE Transactions on 19*, 4 (2004), 1130–1139.

[29] WANG, X., AND YI, P. Security framework for wireless communications in smart distribution grid. *Smart Grid, IEEE Transactions on 2*, 4 (2011), 809–818.

[30] ZHOU, H., BHATTACHARYA, T., TRAN, D., SIEW, T., AND KHAMBADKONE, A. Composite energy storage system involving battery and ultracapacitor with dynamic energy management in microgrid applications. *Power Electronics, IEEE Transactions on 26*, 3 (2011), 923–930.

Hidden Costs of Power Cuts and Battery Backups

Deva P. Seetharam
IBM Research, India

Ankit Agrawal
IBM Research, India

Tanuja Ganu
IBM Research, India

Jagabondhu Hazra
IBM Research, India

Venkat Rajaraman
Solarsis India

Rajesh Kunnath
Radio Studio, India

ABSTRACT

Many developing countries suffer from intense electricity deficits. For instance, the Indian electricity sector, despite having the world's fifth largest installed capacity, suffers from severe energy and peak power shortages. In February 2013, these shortages were 8.4% (7.5 GWh) and 7.9% (12.3 GW) respectively. To manage these deficits, many Indian electricity suppliers induce several hours of power cuts per day that impact a large number of their customers. Many customers use lead-acid battery backups with inverters and/or diesel generators to power their essential loads during those power cuts. The battery backups exacerbate the deficits by wasting energy in losses (conversion and storage) and by increasing the load (by immediately charging the batteries) when the grid is available. The customers also end up incurring additional costs due to aforementioned losses and due to limited lifetimes of batteries and inverters. In this paper, we discuss the issues with power cuts and backups in detail and illustrate their impact through measurements and simulation results.

Categories and Subject Descriptors

H.4 [**Information Systems Applications**]: Miscellaneous; D.2.8 [**Software Engineering**]: Metrics—*complexity measures, performance measures*

General Terms

Electricity storage, inverters, battery backups, uninterrupted power supply, UPS, losses, power grids

Keywords

Electricity storage; inverters; battery backups; losses; power grids

1. INTRODUCTION

Many developing countries suffer from intense electricity deficits. For instance, the Indian electricity sector, despite

having the world's fifth largest installed capacity [7], suffers from severe energy and peak power shortages. In February 2013, these shortages[1] were 8.4% (7.5 GWh) and 7.9% (12.3 GW) respectively [7]. Specific regions could suffer from larger shortages. For example, Tamil Nadu, a southern Indian state, reported 18.4% (1.2 GWh) and 13.5% (1.5 GW) energy and peak power shortages for the month of February 2013 [7]. In fact, India has long struggled with electricity deficit and as a consequence, millions of consumers have been suffering with inadequate power supply [31, 35].

To manage the deficits, many Indian electricity suppliers induce several hours of power cuts per day that affect large percentage of their customers. Many of those customers use a backup power source (such as a Diesel Generator (DG) and/or an inverter with a battery) to power their essential loads (such as lights and fans) during those power cuts [37]. Diesel generators do not depend on grid for their operations. However, they require periodic fuel refills and can generate fumes. Moreover, due to local fuel prices, electricity generated by them could be expensive. In India, cost of one kWh generated by a DG can be three times more expensive than a kWh bought from a electricity retailer [38]. This price difference could get much larger as the diesel subsidies are being gradually removed [10].

Inverter backups charge their batteries when there is mains supply and power the loads using the batteries when there is a power cut. Compared to DGs, the inverter backups can be simpler (no fuel refills) and cleaner (no fumes) to operate. But, they aggravate the power deficits by wasting energy in losses (conversion and storage) and by increasing the load (by charging the batteries) on the power grids whenever the mains supply is available. The customers also end up incurring additional costs due to aforementioned losses and due to limited lifetimes of inverters and batteries [37].

While inverter backups are used by residential consumers and small offices, Uninterruptible Power Supplies (UPS) are commonly used as power backups in commercial environments [37]. Inverter backups and UPS contain the same set of components, and they are similar except for the changeover (from mains to battery or vice versa) delay: the former takes about 500 milliseconds to a second and the latter takes only about 20 milliseconds[2]. UPS are typically used to protect computers, data centers, telecommunication equipment or other electrical equipment where an unexpected power

---

[1]The shortages include the load from battery backups as well. Load-specific consumption details are not available.
[2]There is also a special kind of UPS called the online UPS where the batteries are always connected to the inverter, so that no power transfer switches are necessary.

disruption could cause serious business disruption or data loss. Some consumers, especially the commercial ones, use a combination of DGs and battery backups to handle power cuts.

The cumulative impact of battery backups on grids and consumers could be aggravated as more Indian consumers are able to afford inverters due to their increased purchasing power. Initially, they were used to fulfill only primary demands, such as powering a few fans and lights; however, now they are also being used to run desktops, air conditioners and the many other household appliances during power cuts. More importantly, the Indian power inverter market is expected to grow until 2016-17 [13]. This growth can aggravate supply-demand gap unless the inverters are lossless. Moreover, there exists a potential for economic disparity with the higher-income groups being able to afford backups with higher capacity and worsening the demand-supply gap increasing the frequency of power cuts in economically backward areas.

Since the impact of these inverter backups can increase non-linearly with the number of inverters in use, it is important to have a framework for analyzing their impact. Using detailed measurements and models of inverter backups, we analyze their impact on electricity suppliers and consumers. We also demonstrate the large-scale effects using a multi-agent simulation with load/generation profiles, multiple levels of inverter penetration, different battery capacities, duration of power cuts, etc.

The rest of the paper is organized as follows: Section 2 presents the technical details of inverters and batteries. An approach for quantifying the impact of battery backups is presented in Section 3. Simulation setup and results are presented in Section 4. Section 5 reviews the relevant existing body of work and finally, section 6 concludes with a discussion about future work.

2. BACKGROUND

2.1 Inverters

Figure 1 presents a simplified block diagram of a single-phase inverter. An inverter, using the charge stored in the attached battery, can give constant AC voltage (either 220V AC or 110V AC depending on the line voltage of the particular country) at its output socket when the AC mains power supply is not available. Inverters meant to handle power cuts usually have a built-in charger and the focus of this paper is on such inverters.

The inverter functioning can be explained under two situations [8]:
1. *When the AC mains power supply is available:* the AC mains sensor will sense the availability of mains supply and will activate a relay that will directly pass the line supply to the output socket so that the loads can be driven by the mains. The line supply is also connected to the battery charging section where it is converted to a DC voltage (usually, 12V DC or 24V DC) for charging the attached battery.
2. *When the AC mains power supply is not available:* the AC mains sensor will sense the unavailability of mains supply and cutoff the output socket from mains. Simultaneously, it will activate relays that connect the change-over section to inverter section where DC to AC conversion takes place. There are multiple realizations of inverter circuitry based on cost considerations. Simple designs use an oscillator to generate the mains cycle with a push-pull transistor

circuitry connected to a transformer while more sophisticated ones employ a Pulse Width Modulation (PWM) based approach using digital signal processors driving a full bridge with filters at the output for precise sine wave generation and control.

The majority of inverters used in India produce quasi sine-wave outputs. These inverters produce a distinct sound when fans are operated and consumers are able to easily distinguish between mains and inverter operation. However, decreasing cost of computing and processors is making the quasi sine wave inverters obsolete and they are gradually being replaced with their pure sine wave counterparts.

Manufacturers of residential inverters (500V, 600VA, 800VA and 1000VA) are under constant pressure to decrease prices and this comes at the cost of performance and flexibility. Ease of use dictates that users need not bother with settings after the installation. Efficiency is often compromised since the user is not aware of the operating costs and benefits. Further, in order to decrease cost of the overall solution which includes the battery cost (the most significant contributor to cost), inverter charging current is optimized for a narrow range of battery capacities. As a result, using a higher-capacity battery would not be beneficial because charging it will take more time and may not get fully charged, if there are frequent power cuts.

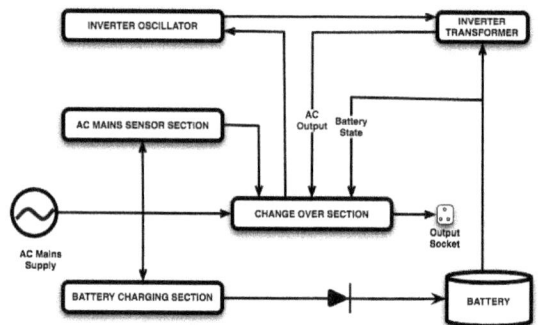

Figure 1: Simplified block diagram of a single-phase inverter

2.2 Batteries

Lead-Acid batteries are commonly used with inverters. The technology is affordable, proven and easy to manufacture, despite poor volumetric efficiency and relatively low depth of discharge. Most residential inverter systems use a single battery, even though it is not uncommon for multiple batteries (usually connected together in series) being used in installations that require high capacity backups.

Lead Acid batteries are characterized by voltage and capacity. The capacity of a battery is usually specified for of 20-hour[3] discharge cycle. Discharging the battery faster (at a higher current) will reduce available capacity. For instance, the Panasonic LC-XA12100P has 100Ah capacity when discharged in 20 hours and 55Ah capacity, when discharged in an hour [27]. This relationship between discharge rate and available capacity is given by well known Peukert's Law [20] which relates battery capacity to discharge rate:

$$C_p = I^k.t \qquad (1)$$

where C_p is the amp-hour capacity at 1 A discharge rate, I

---

[3]Known as C20 rating which means the battery is designed to provide 1/20th of the stated Ah capacity for 20 hours.

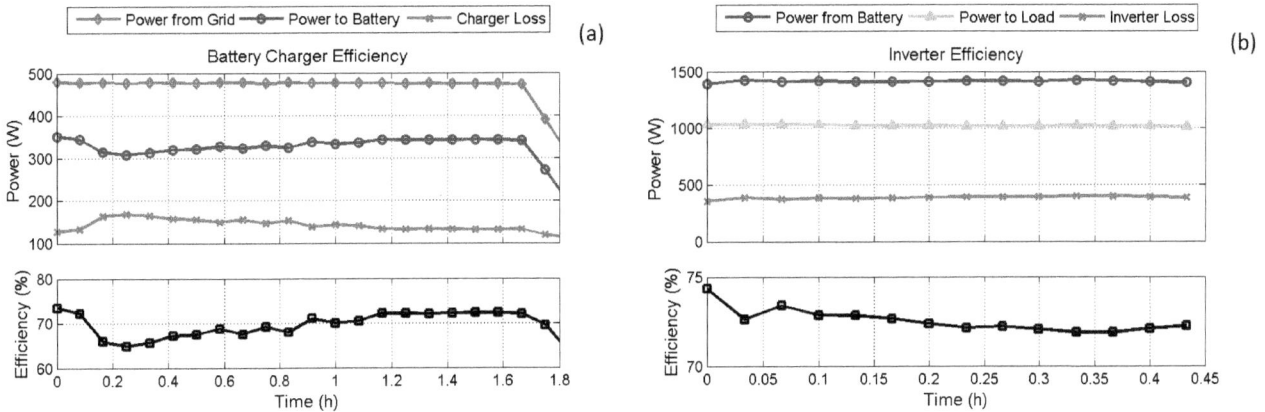

Figure 2: (a) Losses incurred by an inverter during charging, (b) Losses incurred by an inverter during discharging (powering a single-coil 1040W room heater.)

is the discharge current in Amperes, t is the discharge time in hours and k is the Peukert coefficient (typically 1.1 to 1.3).

Another important factor that must be considered is the depth of discharge as it has a non-linear effect on the life of batteries. For example, the Industrial Energy IEL 12150 battery, when discharged 30% (depth of discharge) of its capacity every time, can withstand 1200 cycles; whereas, at a depth of discharge of 50%, the life of battery falls by more than half to about 450 cycles [26].

3. IMPACT ANALYSIS

The negative impact of battery backups can be analyzed in terms of various direct costs incurred by electricity suppliers and consumers. These costs are interconnected in complex ways. For example, if the schedule of power cut is known in advance, consumers can preplan and complete their activities before or after those hours. However, from the point of view of suppliers, this may not be beneficial since the energy consumption is not reduced as much as it time shifted. As a result, unscheduled power cuts would be more beneficial for the suppliers and less convenient for the consumers; on the other hand, scheduled power cuts would be more convenient to the consumers and less effective for the suppliers.

In the following sections, each of the costs is explained in detail and is expressed as mathematical expressions. Since the parameters of the equations arise from the interplay of various socioeconomic and technological factors, it may not be straightforward to obtain their values. Nevertheless, these equations can be used to get a perspective of the costs using estimated (for instance, from data collected through well-designed customer surveys) parameter values.

3.1 Impact on the Electricity Supplier

3.1.1 Energy Loss

Every unit of electricity that passes through a battery backup suffers various losses. To illustrate these losses, we measured the charging and discharging losses of a 1400 VA inverter (Numeric Digital HPH 1400) with a 150 Ah battery (Amaron). As illustrated in Figure 2, the rectification and inversion efficiencies are about 70%.

As a result, an electricity supplier must provide additional energy to compensate for such losses.

$$LE = \sum_{i=1}^{N} e_i \left(\frac{1 - \eta_i}{\eta_i} \right) \qquad (2)$$

where LE is the total loss incurred by the supplier during a power cut, N is the total number of inverters, e_i is the energy served by the ith inverter and η_i is the total efficiency from mains to load through the inverter.

The total efficiency η_i can be expressed as:

$$\eta_i = \eta_r \eta_c (1 - l_s) \qquad (3)$$

where η_r is the efficiency of rectification process (AC to DC conversion) used for charging the battery, η_c is the efficiency of inversion process (DC to AC) used while powering the loads from batteries, and l_s is the electrochemical loss (in percentage) incurred in charge-discharge cycles.

Typically in developing countries, the volume of inverters sold is highest in the sub-1kVA range. These inverters are used for powering a few lights, fans and a TV. For such inverters, η_r and η_c are around 80%, and l_s is around 10%. It is important to note the efficiency depends on the delay between charging and discharging. If this delay is significant (in the order of months), batteries could leak the charge and the efficiency could reduce.

Since the inverters are usually neither monitored nor metered, it may not be possible to get the exact inverter specific data, therefore equation (2) can be rewritten as the following to estimate the energy lost during a power cut:

$$LE = E \times \frac{\delta}{100} \times \frac{IF_a}{100} \times \frac{IP}{100} \times \left(\frac{1 - \eta_a}{\eta_a} \right) \qquad (4)$$

where E is the baseline energy demand during the power cut time, δ is the % energy deficit with respect to the baseline demand, IF_a is the % consumer load that is served by inverters, averaged over all consumers with the inverters, IP is the % inverter penetration, and η_a is the inverter efficiency from mains to load through the inverter, averaged over inverter efficiencies η_i (defined in (3)) for all the consumers.

In addition to the above mentioned losses, energy is also lost in the form of tare losses [6] in operating the inverters even when they are not serving any load. These tare losses can accumulate to significant energy wastage when there are a large number of inverters without appropriate loss control mechanisms.

3.1.2 Rebound Effect

Since the battery chargers do not follow any managed charging schedules, they could start recharging the batteries as soon as the mains supply becomes available. Such simultaneous charging from multiple backups could impose additional load on the grid. The additional load from battery chargers and deferred loads on the grid during the hours that follow a power cut can be expressed as:

$$AE = \frac{E \times \frac{\delta}{100} \times \frac{IF_a}{100} \times \frac{IP}{100}}{\eta_c \eta_r (1 - l_s)} + E_d \qquad (5)$$

where E, δ, IF_a and IP are same as defined in equation (4) and E_d is the total amount of deferred energy consumption from other time-shiftable loads such as washing machines, dish washers,etc.

There are a few important factors to consider about this rebound effect:

Scheduling power cuts - If the power cut ended just before the peak hours, the backups will aggravate the peak load by charging during the peak hours. Moreover, the distribution infrastructure that is already operating close to its capacity will be further taxed by this additional load. On the other hand, if the power was cut during the peak hours, the backups could alleviate peak loads. However, the power must be cut through the peak hours to avoid any rebounding during those hours or additional load must be cut to handle deferred consumption, if the power can't be cut for the entire duration of peak hours. It must be noted that these battery backups could be beneficial if there is an overall energy surplus but there is a power shortage during peak hours.
Differential energy prices - Since the price of electricity in bulk markets[4] varies with time, the economic impact of rebound effect must be carefully considered. Without power cuts, consumers would have used even their non-essential loads during those hours and cost of supplying power for those loads could have been high (it is possible that powering the secondary loads may not have been an option due to deficits). But, that cost savings must be carefully compared with additional costs incurred when the rebound happens. This cost difference can be expressed as follows:

$$NB = \sum_{i=1}^{n} OE_i \times p_{t1} - \sum_{j=1}^{m} AE_j \times p_{t2} \qquad (6)$$

where NB is the net benefit to the utility company, OE_i is the baseline energy demand i.e. the original energy (for both primary and secondary loads) that would have been consumed if there were no power cuts at the ith time slot, AE_j is the rebounded energy consumption as defined in (5) at the jth time slot. p_t1 and p_t2 are price per kWh of energy during the power cut and rebound hours respectively. n and m are the total number of hours of power cut and hours of rebound respectively.

---

[4]Indian bulk power markets use time-varying Availability Based Tariff (ABT)[4] as the electricity pricing mechanism.

3.1.3 Peak-to-Average Ratio of Inverter Current Consumption

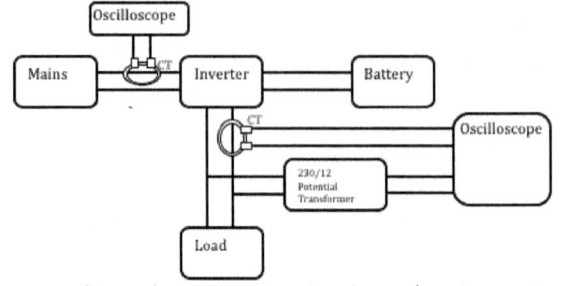

Figure 3: Setup for measuring the charge/discharge characteristics of a single-phase inverter (Numeric HPS1000, 1kVA inverter) with a battery (Amaron Tubular 150AH, 12V).

Most inverters employ switched mode charging since it offers better efficiency than linear charging. Switched mode charging results in a non-sinusoidal current waveform with high peak to average current. The high peak to average ratio implies that the apparent power requirement is high and this places high instantaneous demand on the electrical infrastructure. To illustrate this behavior, we studied an inverter using the measurement setup presented in Figure 3 to measure the charging/discharging characteristics and the corresponding inefficiencies. As shown in Figure 4, peak current (2.52A) and peak-to-average ratio (3.6) for the current can be quite high. When multiple inverters start charging immediately after a power cut, the cumulative current drawn from the grid can be quite high.

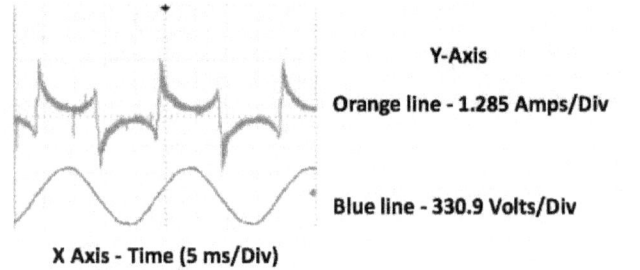

Y-Axis

Orange line - 1.285 Amps/Div

Blue line - 330.9 Volts/Div

X Axis - Time (5 ms/Div)

Figure 4: High peak-to-average inverter charging current. (Blue line represents supply voltage and the orange line is the charging current.)

3.2 Impact on the Consumers

3.2.1 Higher Electricity Cost

Consumers end up paying higher price per unit of power due to the conversion and storage losses. The additional cost incurred by the consumer i, AC_i, during a power cut can be expressed as:

$$AC_i = E_i \times IF_i \times p_u \times \left[\frac{1}{\eta_c \eta_r (1 - l_s)} - 1 \right] \qquad (7)$$

where E_i is the original electricity demand of the ith consumer during the power cut, IF_i is *Inverter load factor* - the fraction of original demand (E_i) served by the inverter i and p_u is the price for one unit of electricity.

3.2.2 Inconvenience Cost

The inconvenience cost experienced by consumers due to the power cuts depends upon various factors. It varies for every consumer and depends upon the factors like time of day and type of appliances affected due to power cut etc. We model the inconvenience cost by using the categorization of appliances and utility functions suggested in [22].

Let $q_{i,a}(t)$ is the power demanded by i^{th} consumer at time slot t for an appliance $a \in A_i$, where A_i is the set of appliances used by consumer i. Let $q_{i,a}$ denote the vector of power demands over all possible time slots. Let $U_{i,a}(q_{i,a})$ is a utility function that provides the utility obtained by user i for consuming electricity as $q_{i,a}$.

Due to the power cuts, the consumption vector is altered depending upon whether the appliance a can run on inverter or not. Let $q_{i,a}^{cut}$ denotes the new consumption vector obtained due to the power cuts in a given day. The total inconvenience cost, IC_i, incurred by consumer i can be computed as:

$$IC_i = \sum_{a \in A_i} U_{i,a}(q_{i,a}) - U_{i,a}(q_{i,a}^{cut}) \qquad (8)$$

where the utility functions $U_{i,a}(q_{i,a})$ vary depending upon the appliance types - temperature controlling appliances (like AC, Refrigerator), deferrable appliances (like washing machine, dish washer etc.), essential loads (like lighting) and interactive appliances (like TV, entertainment devices)- and could be used as suggested in [22].

3.2.3 Battery Life

In general, the battery life (the number of charge/discharge cycles) reduces with the increase in Depth of Discharge (DoD). The DoD in turn depends on the duration of power cuts and load levels during those hours. The depth of discharge during a power cut for consumer i can be approximated as:

$$DOD_i = \left(\frac{D \times w_i}{v_i \times A_i} \right) \qquad (9)$$

where D is duration of the power cut in hours, w_i is the amount of power supplied by i^{th} inverter during that power cut, A_i is the total capacity of the battery in ampere hours, and v_i is the nominal voltage of the battery in volts.

Since deep discharges can reduce the life of batteries, inverters prevent deep discharges by shutting down the inverter at a pre-determined voltage. The open terminal voltage as a reference for the State of Charge (SoC) is reliable only if many factors (such as battery age, ambient temperature, etc) are considered and if the battery is allowed to rest for a few hours before the measurement [5].

After sustained use, the lead plates eventually weaken and are no longer able to store energy [33]. Used lead-acid batteries (ULABs) are either discarded or recycled. Because of the toxic materials within these used batteries, the Basel Convention has included ULABs on its list of materials classified as hazardous waste [2]. However, since the environmental protection laws in developing countries are not strict, these batteries usually do not get disposed properly. Unregulated recycling industries and informal methods of extracting lead can cause high levels of environmental contamination.

3.2.4 Available Capacity

As explained above, the available capacity of a battery is determined by the charging and discharging rates. Typi-

cally, in backup systems, batteries are sized such that they can be fully recharged in five hours. This ensures the batteries would be fully recharged, if two consecutive power cuts are separated by a gap of at least five hours.

Inverter systems have an asymmetrical charge-discharge characteristic. Discharge currents can be very high and sometimes can be more than 10 times the charge current. Moreover, since the battery voltage (12 V/ 24 V) can be much smaller (1/20th or 1/10th) than the line voltage (230 V) required by loads, the discharge current must be correspondingly larger (20 times or 10 times) than the actual current drawn by the loads to maintain the same power rating (VA). Further, while charge currents are constant, discharge currents vary dynamically depending on load conditions.

The available capacity as a function of discharge current can be computed using Peukert's Law. However, this model doesn't include the recovery effect (when discharged intermittently, batteries can recover during idle periods) [20], an important feature of backup systems where batteries get charged and discharged to various depths as per the availability of mains supply and the energy requirement of loads. The Kinetic Battery Model (KiBaM) model [24] is more accurate since it accounts for (a) decrease in capacity with increasing charge or discharge rates, (b) recovery in charge, (c) increase of voltage with charging current and state of charge, and (d) decrease in voltage with discharging current and state of charge. The KiBaM uses a chemical ki-

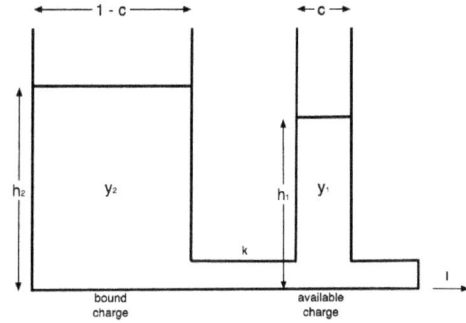

Figure 5: Two-well model used by KiBaM)

netics process as its basis. In the model, as illustrated in Figure 5, the battery charge is distributed over two wells: the available-charge and the bound-charge wells. The available charge well supplies electrons directly to the load, whereas the bound-charge well supplies electrons only to the available-charge well. The rate at which charge flows between the wells depends on the difference in heights of the two wells, and on a parameter k. The parameter c gives the fraction of the total charge in the battery that is part of the available-charge well.

Computing Available Charge

The KiBaM model calculates available capacity as follows:

$$
\begin{aligned}
y_1 =\,& y_{1,0}.e^{-k't} + \frac{(y_0 k'c - I)(1 - e^{-k't})}{k'} \\
& - \frac{Ic(k't - 1 + e^{-k't})}{k'} \\
y_2 =\,& y_{2,0}.e^{-k't} + y_0(1 - c)(1 - e^{-k't}) \\
& - \frac{I(1 - c)(k't - 1 + e^{-k't})}{k'}
\end{aligned}
\qquad (10)
$$

where k' is defined as $k' = \frac{k}{c.(1-c)}$, and $y_{1,0}$ and $y_{2,0}$ are the amount of available and bound charge, respectively, at $t = 0$. And, $y_0 = y_{1,0} + y_{2,0}$.

Computing Terminal Voltage
The battery is modeled as a voltage source in series with an internal resistance. The level of the voltage varies with the depth of discharge. The voltage is given by:

$$V = E - IR_0 \qquad (11)$$

where I is the discharge current and R_0 is the internal resistance. E is the internal voltage, which is given by:

$$E = E_0 + AX + \frac{CX}{D - X} \qquad (12)$$

where E_0 is the internal battery voltage of the fully charged battery, A is a parameter reflecting the initial linear variation of the internal battery voltage with the state of charge, C and D are parameters reflecting the decrease of the battery voltage when the battery is progressively discharged, and X is the normalized charge removed from the battery. These parameters can be obtained from discharge data.

3.2.5 Electrical Noise

Most low-cost inverters are quasi sine wave (or modified sine wave) type inverters since they are cheaper to manufacture than the ones that supply pure sine wave voltage output. The output of the inverter (measured using the setup in Figure 3) is shown in Figure 6 and as shown, it only approximates sine wave output. During discharging, these inverters can have their capacities reduced when driving electronic loads since these loads demand current at the peak of the waveform. Since the waveform is more like a step, the current behavior is also more impulse-like generating higher harmonics. This can be a problem when powering sensitive electronic equipment or Hi-Fi audio. Although

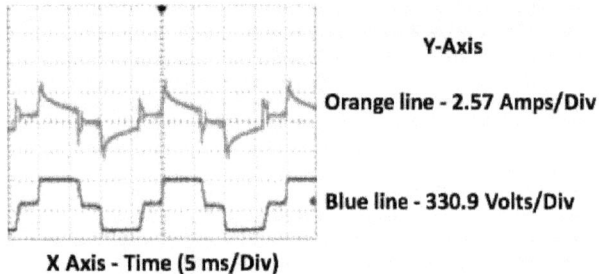

Y-Axis

Orange line - 2.57 Amps/Div

Blue line - 330.9 Volts/Div

X Axis - Time (5 ms/Div)

Figure 6: A quasi sine wave inverter powering a number of consumer electronics appliances. (Blue line represents the inverter output voltage and the orange line is the current drawn by the loads.)

these quasi-sine wave inverters are being phased out by the large manufacturers, we have noticed that some of the cheap unbranded inverters still use this design.

4. SIMULATION RESULTS

In this section, we present the overall large-scale effects of the power cuts and battery backups observed in simulation. Firstly, using a multi-agent simulation system, we study the positive feedback effect of battery backups. Next, we analyze how the durations of power cuts and their frequency affect the battery life. Finally, we discuss the changes in grid stability as a function of inverter penetration using a power system simulator.

4.1 Feedback spiral of electricity deficit and battery backups

In this section, we discuss the positive feedback loop that exists between the energy deficit and battery backups. We study this aggregate effect using a multi-agent simulation.

4.1.1 Simulation Setup

Consumer segments: In simulation, we create a population of 1000 consumer agents and divide them uniformly into three consumer groups. Each group is assigned a set of configurable parameters: maximum total load, primary load, deferrable load, battery capacity and inverter/ battery efficiencies (rectification, inversion, and storage). The groups are assigned maximum total loads of 1 KW, 2 KW, and 3 KW. We categorize the consumer demand into three different categories: 1. Primary load - it includes the essential load like lighting, fans and some interactive appliances which can be run on inverter, 2. Deferrable load- it includes the loads like washing machine, water pumps etc. which can be time-shifted, 3.Non-deferrable secondary loads - it includes loads like air conditioner, refrigerator which are not time-shiftable and can be expensive to run on inverter. We assume primary loads to be 25% of total loads. The secondary load of a consumer is defined as difference of the respective total load and the primary load, and deferrable load is assumed to be 30% of secondary load.Based on the percentage of inverter penetration, a subset of consumer agents are assigned inverters with operating parameters that match the primary load requirements of the corresponding consumer agents.

Demand Modeling: The original consumer demand of electricity at every time step (1 minute) is computed using a parameter: time of use probability, that determines the amount of electricity a customer would be using at a specific time of the day. We use the residential demand pattern as a bimodal distribution with one peak occurring in the morning and the other in the evening as specified in [29]. At every time step, if there is no power cut for a given consumer, its original demand is added to the total load on the grid. If the consumer has an inverter and if the attached battery needs recharging, that charging requirement is added to the grid load. Similarly, if there are any deferred loads due to the previous power cuts, those are also added to the load served by the grid. This total demand that is the sum of original, charging and deferred loads is referred to as *actual demand*. In case of a power cut, if a consumer agent has an inverter with a sufficiently charged battery, that consumer's entire primary load would be served by the inverter.

Supply Capacity Modeling: The capacity of the electricity supplier at each hour is a configurable property. For generating different scenarios with different peak power deficits and energy deficits, different supply capacities are allocated to hours across a day.

Shortage calculation: The shortage for every hour is computed as the difference between total demand (Sum of original, charging and deferred loads. The latter two can be zero, if there were no recent power cuts.) and the available capacity at that hour.

Power cuts: The decision to cut power is determined based on the computed shortage for that hour. Based on the percentage of shortage, a subset of consumers are randomly (over uniform probability distribution) selected from each of the consumer groups. This ensures required power cut spread uniformly over different consumer groups.

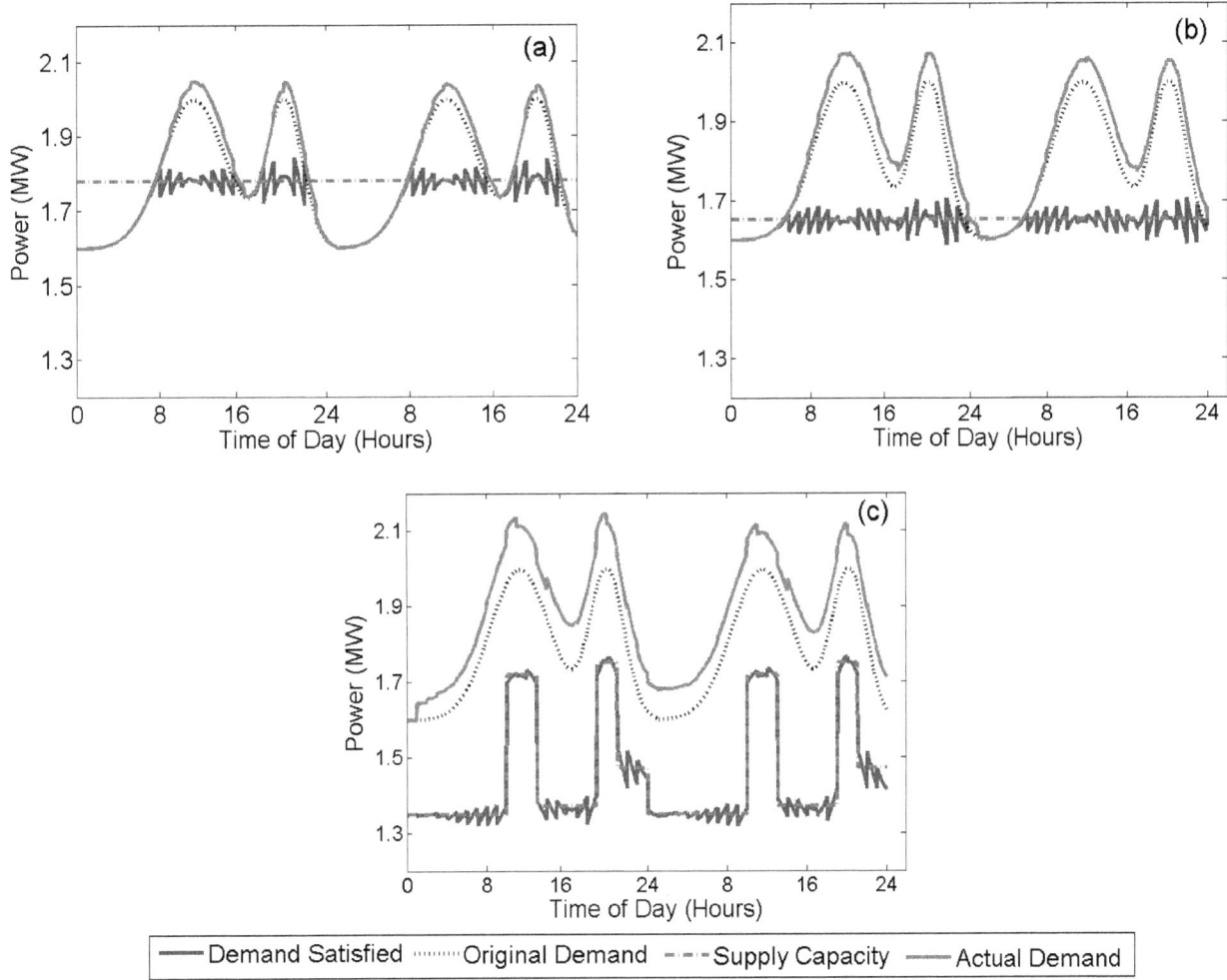

Figure 7: Multi-agent simulation of 1000 consumers with 20% inverter penetration for three different scenarios - (a)Scenario 1: No energy deficit, Peak power deficit - 10%, (b)Scenario 2: Energy deficit - 8%, Peak power deficit - 15%, (c)Scenario 3: Energy deficit - 18% Peak power deficit - 13% (Real-world Scenario of Feb, 2013 in Tamilnadu state, India [7])

Inverter	Scenario 1		Scenario 2		Scenario 3	
Penetration (%)	Avg. Power Cut Hours	Losses (%)	Avg. Power Cut Hours	Losses (%)	Avg. Power Cut Hours	Losses (%)
0	1.025	0	2.427	0	5.825	0
10	1.090	0.096	2.577	0.209	6.091	0.456
20	1.139	0.182	2.685	0.411	6.334	0.934
30	1.174	0.274	2.779	0.631	6.577	1.445
40	1.217	0.372	2.893	0.869	6.806	1.982

Table 1: The effect of different % inverter penetrations on average number of hours of power cut per consumer and energy deficit (due to inverter losses) considering three different scenarios.

Losses calculation: We take into consideration three types of losses: rectification loss, inversion loss, and storage loss. We calculate these losses using rectification, inversion and storage efficiencies that are usually around 80%, 80% and 90% respectively. For every unit of energy served by an inverter, effectively $1/.8$ (=1.25) units will be drawn by inverter from battery. This leads to 25% of inversion loss. Similarly for every one unit of usable energy to be drawn from battery, $1/.9$ (=1.11) units of energy will need to be stored. So, this amounts to total storage losses of $.11*100/.8$ = 13.75%. On top of these, there will be additional $(.25*100)/(.8*0.9)$ =34.77% of rectification losses in for every unit of usable energy drawn, to effectively put back one unit of charge in the battery. These three losses together sum up to 76.52%.

4.1.2 Evaluation

To study the effect of power cuts and increasing inverter penetrations in different energy deficit scenarios, we run the simulation for multiple combinations of peak power deficit and energy deficit. For each scenario, we run the simulation for inverter penetration values of 0,10,20,30, and 40%.
We consider following scenarios:
Scenario 1: No energy deficit, Peak power deficit - 10%

Scenario 2: Energy deficit - 8%, Peak power deficit - 15%
Scenario 3: Energy deficit - 18% Peak power deficit - 13%
(Real-world Scenario of Feb, 2013 in Tamil Nadu state, India [7]).

Figure 7 represents the supply and demand situation for a multi-agent simulation of 1000 consumers with 20% inverter penetration considering three aforementioned scenarios. The dotted black line represents original demand in the absence of power cuts, based on ToU probabilities. The green dotted line specifies the supply capacity. If the original demand is greater than the supply capacity at that time, the power cuts are initiated and the thick blue line shows the demand satisfied. As a consequence of prior power cuts, some of the deferrable loads and inverter charging loads get accumulated and add to the original demand. This actual demand is represented by the red line. Even for a moderate (20%) level of inverter penetration, for the scenarios with and energy deficits, Figure 7 shows that the accumulated demand (due to earlier power cuts) is significantly higher than the original demand. This aggravates deficit, causing additional power cuts.

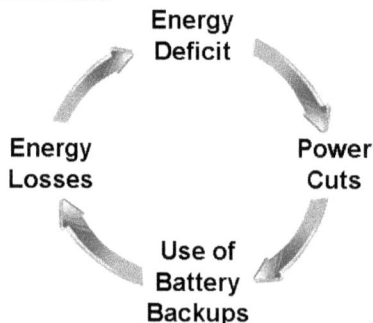

Figure 8: Feedback loop: energy deficit and battery backups

We further evaluate the effect of different inverter penetration values on this feedback loop by observing two parameters: *average power cut duration per consumer per day* and *% of energy losses with respect to the original demand*.

As shown in Table 1, the average power cut duration increases with increase in inverter penetration. This is primarily because of the unscheduled charging of inverters. When power comes back after a power cut, all the inverters start charging simultaneously, aggravating deficits leading to additional power cuts. Moreover, the energy losses increase drastically with every 10% increase in inverter penetration. Additionally, as shown in Table 1, impact of increased inverter penetration becomes more significant with increase in energy deficit. We found that in absence of energy deficit even at high levels of inverter penetrations, losses incurred are less than 0.4%, which may not be significant. Therefore, in absence of energy deficit even if peak power deficit exists, inverters can be used to alleviate peak loads. However, loss percentages go up to 2% when 18% energy deficit pre-exists. So, in the case of pre-existing energy deficit, increase in inverter penetrations adds to the deficit, leading to larger number of power cuts.

Increased duration of power cuts and energy losses together cause a spiral effect as shown in Figure 8. With the increase in average power cut duration, the load getting served by inverter increases. Increased inverter loads cause more energy losses. As supply capacity remains same, more losses results in greater energy deficit that leads to even more power cuts. This spiral effect becomes more prominent in presence of high energy deficit that exists in India,

adding to the seriousness of the matter in the long run. Also, additional power cuts can potentially increase inverter penetration [13], thereby worsening this effect. An important social consequence could be the subset of population that can afford inverters may end up depriving energy for the others who cannot afford them.

4.2 Impact of power cuts on battery life

In this section we analyze how the duration of power cuts and the frequency (consequent interval between two power cuts) affect the life of batteries.

4.2.1 Power cut duration and discharge current

As explained above, a battery can go through a certain number of charge-discharge cycles (called cycle life) before it loses its capacity to store energy. The cycle life depends on the Depth of Discharge (DoD) per cycle. The DoD in turn depends on the duration of power cuts and/or the load levels during those power cuts. To illustrate this effect, we studied a model (KIBAM [24]) battery with 200 Ah capacity. As illustrated in Figure 9, we varied the duration of power cuts with three different constant (for the duration of power cuts) discharge currents. As shown, the cycle life decreases rapidly with the duration and/or the current levels.

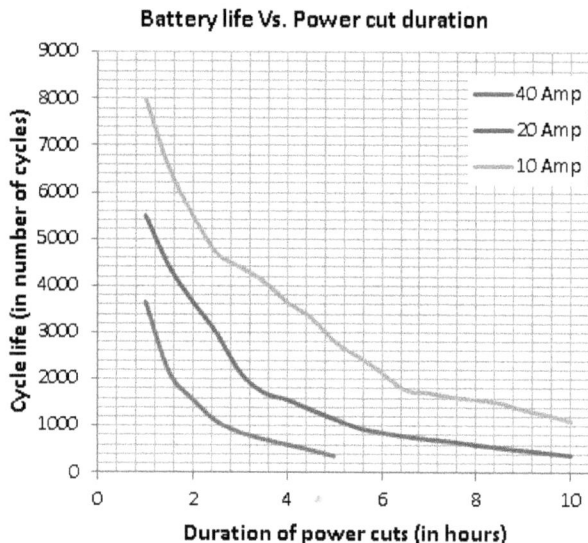

Figure 9: Relationship between duration of power cuts (with different discharge currents) and battery life.

4.2.2 Recovery duration

The time interval between two consecutive power cuts determines the amount of time a battery gets to recover and recharge itself. This recovery actually adds to the total available capacity [24]. To understand this effect, we studied a model (KIBAM [24]) battery with 100 Ah capacity and explored the impact of recovery duration with constant charge (20 Amps - corresponding to C/5) and discharge (40A[5] currents for different durations (50 - 70 mins) of power cuts. As illustrated in Figure 10, the battery life increases rapidly with increase in recovery duration and conversely, decreases with decrease in recovery interval.

---

[5] As explained above, the AC current supplied to loads could be much smaller as the voltage needs to be stepped up to 230 V from the battery terminal voltage.

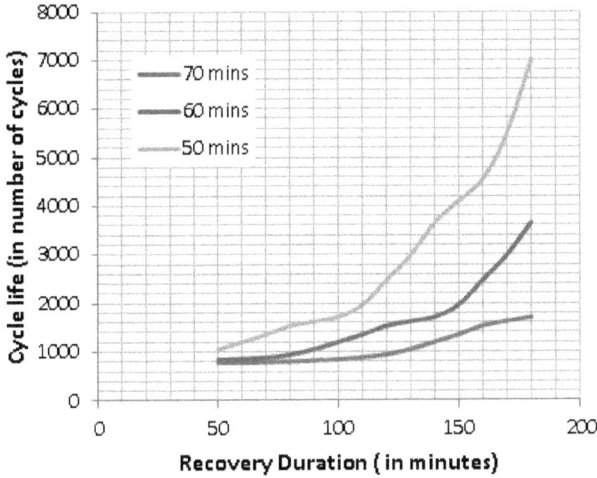

Figure 10: Relationship between recovery interval (with different power cut durations) and battery life

4.3 Impact of Inverters on the Grid Stability

Increase in inverter penetration leads to decreased voltage stability and frequency stability. The former happens due to reactive power shortage caused by battery charging systems that are not power factor corrected and the latter happens because of the supply-demand imbalance caused by aforementioned energy losses. In this section, we study these stability issues in simulation using a benchmark IEEE 300 bus system which is the largest openly available system [18]. This system includes 300 substations, 304 transmission lines, 199 load buses, 107 transformers and 69 generators. This system could be a good representative of Northern Region Electricity Broad (NREB, India) system having 390 buses. For simulation purpose, inverters were assumed to be placed at all the load buses. Inverter loads were assumed to be composed of two components: battery charging load and inverter losses. Battery charging load was assumed to be proportional to the overall load on that bus. We simulated four types of inverters having efficiency of 60%, 70%, 80%, and 90% respectively. Simulations were made for the worst case scenario when all the inverters get charged simultaneously just after when power got restored following a power cut.

Real and reactive power equations with inverter load are represented as follows:

$$P_k = P_g - (P_d + P_{inv}) \qquad (13)$$
$$Q_k = Q_g - (Q_d + Q_{inv}) \qquad (14)$$

$$P_k = \sum_{j=1}^{N} |V_k||V_j| (G_{kj}cos(\theta_k - \theta_j) + B_{kj}sin(\theta_k - \theta_j)) \qquad (15)$$

$$Q_k = \sum_{j=1}^{N} |V_k||V_j| (G_{kj}sin(\theta_k - \theta_j) - B_{kj}cos(\theta_k - \theta_j)) \qquad (16)$$

where P_k and Q_k are active and reactive power injections, respectively at node k.
P_{inv} is inverter load, P_g and Q_g are active and reactive power injections, respectively. V and θ are voltage magni-

tude and voltage phase angle respectively. G_{kj} and B_{kj} are the conductance and susceptance, respectively for the line between nodes k and j.

These set of equations are solved iteratively using standard fast decoupled load flow method which calculates voltage stability, frequency stability and network losses in the system.

4.3.1 Voltage instability

In general voltage stability is determined from PV (Active power-Voltage) characteristics [36] of the system. At the "knee" of the PV curve, voltage drops rapidly with an increase in MW transfer due to inverter penetration. After certain percentage of inverter penetration, voltage falls rapidly and finally, system becomes unstable. This effect is characterized using an index called voltage stability index (VSI) [30] as defined bellow:

$$V_{si} = \frac{\partial P_i / \partial \delta_i}{\sum_{j=1 j \neq i}^{n} B_{ij} V_j} \qquad (17)$$

Where, n is number of buses in the system, P_i is real power injection at bus i, V_i is magnitude of voltage at bus i, δ_j is phase angle of voltage at bus j and B_{ij} is element of network admittance matrix. During normal condition VSI value is close to 1 and with the increase in stress VSI value reduces and grid collapse occurs at VSI value of 0.5. Figure 11 illustrates how VSI changes as a function of penetration of inverter load and inverter efficiency.

Figure 11 shows that inverters with 60-90% efficiency, VSI gradually decreases until 20% penetration and beyond 20% VSI starts decreasing drastically and the grid becomes unstable at 25% penetration. This happens because of non-linear effects caused by inverter penetration. Similarly, for inverters with 80-90 % efficiency, VSI gradually decreases until 25% of penetration and the grid becomes unstable at 30% of penetration. It shows that the grid can sustain higher level of inverter penetration, if inverters are efficient.

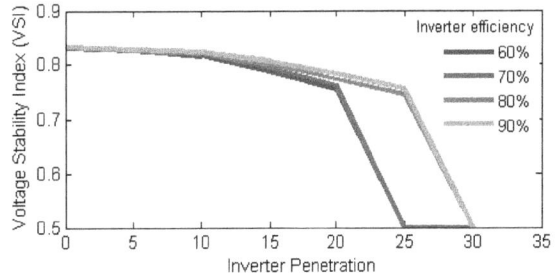

Figure 11: Voltage stability with penetration of inverters

4.3.2 Frequency stability

We also studied the effect of inverter penetration on grid frequency. As inverters create additional load on the grid, and generators can not increase their out put immediately, a power mismatch happens in the grid. This power mismatch is responsible for frequency dip in the system. Normally, for a 50Hz system, tolerable frequency range is between 49.5 Hz and 50.5 Hz.

Figure 12 shows how grid frequency changes with inverter penetration and inverter loss. Figure 12 shows system frequency gradually decreases until 15-20% of inverter penetration and thereafter frequency dips very fast. This happens because initially generators meet the extra load using stored kinetic energy in the flywheel and beyond certain point generators fall short of sufficient kinetic energy and frequency dips very sharply. Simulation results show that the grid can safely sustain only 21-27% of inverter penetration depending on inverter efficiency.

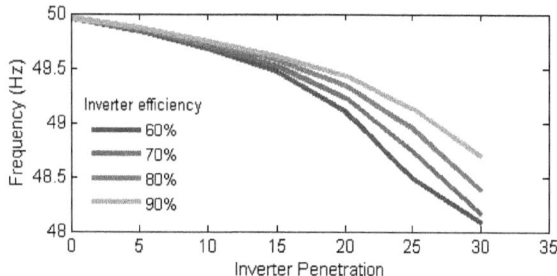

Figure 12: Frequency stability with penetration of inverters

4.3.3 Network loss

We also studied how network loss varies with inverter penetration. This is an important aspect because inverter penetration can magnify this loss considerably. As network loss is proportional to $I^2 \times R$, cumulative network loss highly depends on peak to average current. As discussed in previous section, inverter penetration increases peak to average ratio which effectively increases loss in the network. Figure 13 presents how network loss varies with inverter penetration. It can be seen from the figure that beyond 15% penetration, network loss increases almost exponentially.

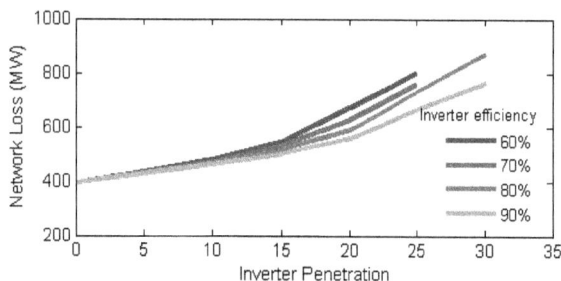

Figure 13: Network loss with penetration of inverters

5. RELATED WORK

Since ability to accommodate all types of distributed storage options and renewable energy sources is one of main characteristics of smart grid [23], there has been a great interest in understanding the impact of various energy storage technologies on the power grid. The existing body of work can be broadly classified into four categories: understanding the impact of power cuts, Vehicle to Grid (V2G), understanding the impacts of Electric Vehicles (EVs) on the power grids and grid-wide storage systems.

5.1 Impact of Power Cuts

Fisher-Vanden [12] present the impact of power cuts on the productivity of industrial consumers and on the environment in the context of developing countries like China.

Wartsila corporation conducted a study that examines the severity of power outages in these cities and the different ways consumers deal with these outages. The study quantifies the costs incurred by consumers who invest in power back-up mechanisms - both initial as well as operating costs [37].

Our focus is different: we study the systemic effects of power cuts and of battery backups.

5.2 Vehicle-to-grid (V2G)

Vehicle-to-grid (V2G) describes a system in which plug-in electric vehicles communicate with the electricity supplier to provide demand response services by either injecting some of the energy stored in their batteries into the grid or by throttling their charging rate. Kempton et al [21] present systems and processes required for implementing V2G. Tomic et al [32] discuss how battery-electric vehicles can provide regulation services to a specific electricity market and evaluated the economic potential of two utility-owned fleets of battery-electric vehicles to provide regulation services in four US regional markets.

The focus of V2G literature is different from ours in the following ways: their main objective is to analyze how V2G power could be sold to high-value, short-duration power markets such as regulation, spinning reserves and peak load when there is an explicit request for such services from the grid operator. On the other hand, we focus on uncoordinated inverter backups that recharge themselves whenever they require and whenever the mains supply is available.

5.3 Grid-wide Energy Storage

Future electrical grids are expected to include significant dedicated energy storage elements at different locations: they may be installed at generating stations to smoothen variations in the intermittent sources, in the transmission networks to even out peak transmission loads, or at substations and feeders in distribution networks to absorb variations in electrical loads [11, 19]. Ardakanian et al [1] study the effect of storage on the loading of neighborhood pole-top transformers. Using the techniques developed for sizing buffers and communication links in communication networks, the authors compute the potential gains from transformer upgrade deferral due to the addition of storage.

Similar to V2G storage, the charging and discharging of these dedicated storage elements will be coordinated by electricity suppliers. As a result, these grid-wide storage technologies will not introduce the unintended consequences like the inertia backups.

5.4 Impact of Electric Vehicles

Several researchers have analyzed the impacts of electric-drive vehicles (such as plug-in hybrid electric vehicles (PHEVs) and battery electric vehicles (BEVs)) charging on the power grids. Putrus et al [28] analyze the impact of EVs on power distribution networks using a typical distribution network that is serving a demand profile that usually occurs in Europe. Clement et al [9] analyze the potential impact of PHEVs on the Belgian distribution networks. Hadley et al [16] analyze the potential impacts of PHEVs on electricity demand, supply, generation structure, prices, and associated emission levels in 2020 and 2030 in 13 regions specified by the North American Electric Reliability Corporation (NERC). Hadley et al [15] conduct an analysis of what the grid impact may be in 2018 with one million PHEVs added to a region of the USA that includes South Carolina, North Carolina, and much of Virginia.

Impact of electric vehicles is primarily dependent on the local driving patterns. On the other hand, the impact of inverter backups is mainly a function of grid deficit and involves a positive feedback loop. In other words, more consumers invest in inverters when the power cuts increase; power cuts become more frequent and/or longer with increase in energy and power deficits; the inverters worsen the energy deficit through conversion losses and power deficit by charging whenever the mains supply is available. Moreover, as the number of power cuts increase, these backups will be forced into a higher number of charge/discharge cycles despite their limited durability and higher cost per kWh of electric energy. As a result, the consumers also stand to lose.

To summarize, the existing body of work has not analyzed the impact of power cuts and battery backups in detail as presented in this paper.

6. DISCUSSIONS AND FUTURE WORK

The developing countries such as India are likely to struggle with power shortages for the next several years unless electricity generation (central or distributed) capacity is increased drastically to match the exploding demand [17]. The power deficit could worsen when electricity grid is expanded to the unserved communities in these countries [6]. As a result, power cuts and battery backups will continue to be an integral part of these geographies. The impact analysis presented in this paper could be used to assess the various costs incurred by the society and electricity suppliers in these situations. It is important to note that many of these costs will remain relevant even if high-efficiency ($> 95\%$) inverters and low-loss ($< 5\%$) batteries become available because rebound effects experienced by the suppliers and inconvenience costs to consumers cannot be eliminated by those advancements.

The costs incurred by the different stake holders could increase exponentially with the duration of power cuts and the number of backups employed. Given that, when the inverter penetration becomes significant, the impact analysis could be used as a basis for designing solutions and setting policy mandates. A few specific recommendations could be:

- Electricity suppliers could manage the charging schedules of batteries through centralized technologies such as Demand Response programs [34] or through decentralized technologies such as nPlugs that can sense the grid conditions locally without any explicit communication from the suppliers[14].

- Electricity suppliers could introduce time-of-use pricing mechanisms to discourage consumers from charging the batteries during peak hours. Without such a differential pricing structure, the consumers and inverter manufacturers have no economic incentives for maintaining grid-friendly charging schedules.

- Consumers can demand more details about and controls over charging/discharging schedules of their battery backup so that they can optimize its operational characteristics. For instance, a small LCD display could provide various details such as state of charge, average discharging rates, etc; and a control user interface (even simple buttons and knobs) could be provided so that the consumers can control charging schedules, maximum discharge rates, etc.

- Government agencies such as BEE [3] could enforce strict efficiency standards for inverters and batteries.

- Governments, instead of attempting to curb the use of inverters [25], can mandate that a certain percentage of every inverter's capacity must supplied by renewable energy sources at the site.

We are considering several future extensions to our work. Firstly, we plan to study a much larger number of backup system configurations: inverters (capacities, conversion efficiencies, etc), chargers (charging rates, charging patterns, etc) and batteries (capacity, charge/discharge characteristics, etc). Secondly, while measuring the inconvenience cost of consumers, we are attempting to include richer models of electrical appliances such as the one proposed by Li et al [22]. Thirdly, do a cost-benefit analysis between a government providing subsidies to encourage local energy generation and incurring various costs in such backups. Finally, we are hoping to establish a framework for optimizing the charging/discharging schedules for a given level of inverter penetration and power/energy deficits. However, optimizing for a large number of inverters with different load characteristics may not be straightforward since the optimizer would require fine-grained real-time performance characteristics of the backup systems. Moreover, these characteristics could change with ambient temperature, number of charge/discharge cycles, power quality, etc.

7. ACKNOWLEDGMENTS

We are grateful to Sunil K. Ghai and Zainul Charbiwala for providing valuable comments and for sharing the inverter charging/discharging efficiency data. We wish to thank Hari S. Gupta and anonymous reviewers for their constructive feedback that helped us to improve this paper.

8. REFERENCES

[1] O. Ardakanian, C. Rosenberg, and S. Keshav. On the Impact of Storage in Residential Power Distribution Systems. In *ACM Greenmetrics Worskshop at Sigmetrics*, June 2012.

[2] The Basel Convention on the Control of Transboundary Movements of Hazardous Wastes and Their Disposal. http://www.basel.int/text/17Jun2010-conv-e.pdf.

[3] Bureau of Energy Efficiency, Ministry of Power, Government of India. http://www.beeindia.in.

[4] B. Bhushan. ABC of ABT. http://www.nldc.in/docs/abcabt.pdf.

[5] I. Buchmann. *Batteries in a Portable World: A Handbook on Rechargeable Batteries for Non-Engineers*. Cadex Electronics Inc, 2011.

[6] What are AC solar panels? http://www.motherearthnews.com/ask-our-experts/AC-solar-panels-zb0z09zblon.aspx#axzz2QAtzbKMn.

---

[6]In India, 400 million (approximately 30% of the population) people do not have access to the power grid [17].

[7] Central Electricity Authority, Ministry of Power, Government of India. Monthly Power Supply Position. `http://www.cea.nic.in/monthly_power_sup.html`, February 2013.

[8] How an Inverter works. `http://www.circuitstoday.com/how-an-inverter-works`, August 2008.

[9] K. Clement, E. Haesen, and J. Driesen. Coordinated charging of multiple plug-in hybrid electric vehicles in residential distribution grids. In *Power Systems Conference and Exposition, 2009. PSCE '09. IEEE/PES*, pages 1 –7, march 2009.

[10] Diesel prices to be hiked by 40-50 paise every month: Oil Minister M Veerappa Moily. `http://articles.economictimes.indiatimes.com/2013-02-01/news/36684521_1_diesel-prices-litre-prices-in-small-doses`, 2013.

[11] EPRI-DOE. *Handbook of Energy Storage for Transmission and Distribution Applications*. Electric Power Research Institute, 2003.

[12] K. Fisher-Vanden, E. T. Mansur, and Q. J. Wang. Costly Blackouts? Measuring Productivity and Environmental Effects of Electricity Shortages. Working Paper 17741, National Bureau of Economic Research, January 2012.

[13] Frost and Sullivan. Indian Power Inverter Market Veering Toward Double-Digit Growth Rates Until 2017. `http://www.frost.com/prod/servlet/press-release.pag?docid=223228220&gon11032=PSMI1`, February 2011.

[14] T. Ganu, D. P. Seetharam, V. Arya, R. Kunnath, J. Hazra, S. A. Husain, L. C. D. Silva, and S. Kalyanaraman. nPlug: A Smart Plug for Alleviating Peak Loads. In *Third International Conference on Future Energy Systems, e-Energy*, May 2012.

[15] S. W. Hadley. Impact of Plug-in Hybrid Vehicles on the Electric Grid. `http://www.ornl.info/info/ornlreview/v40_2_07/2007_plug-in_paper.pdf`, October 2006.

[16] S. W. Hadley and A. Tsvetkova. Potential Impacts of Plug-in Hybrid Electric Vehicles on Regional Power Generation. `http://www.ornl.gov/info/ornlreview/v41_1_08/regional_phev_analysis.pdf`, January 2008.

[17] IEA. World Energy Outlook. `http://www.iea.org/Textbase/npsum/weo2010sum.pdf`, 2010.

[18] IEEE 300 Bus Power Flow Test Case. `http://www.ee.washington.edu/research/pstca/pf300/pg_tca300bus.htm`.

[19] S.-I. Inage. Prospects for Large-Scale Energy Storage in Decarbonised Power Grids, 2009.

[20] M. R. Jongerden and B. R. Haverkort. Which battery model to use? *IET Software*, 3(6):445–457, 2009.

[21] W. Kempton and J. Tomic. Vehicle-to-grid power implementation: From stabilizing the grid to supporting large-scale renewable energy. *Journal of Power Sources*, 144(1):280 – 294, 2005.

[22] N. Li, L. Chen, and S. H. Low. Optimal demand response based on utility maximization in power networks. In *IEEE Power and Energy Society General Meeting (PESGM)*, 2011.

[23] T. Logenthiran and D. Srinivasan. Intelligent management of distributed storage elements in a smart grid. In *Power Electronics and Drive Systems (PEDS), 2011 IEEE Ninth International Conference on*, pages 855 –860, dec. 2011.

[24] J. F. Manwell and J. G. McGowan. Lead acid battery storage model for hybrid energy systems. *Solar Energy*, 50:399–405, 1993.

[25] India Today. To beat power crisis, Kerala puts curb on use of inverters and induction cookers. `http://indiatoday.intoday.in/story/kerala-power-crisis-inverters-induction-cookers/1/228892.html`.

[26] Industrial Energy. IEL12150 data sheet. `http://www.industrialenergy.com.sg/iel-series/`.

[27] Panasonic. LC-XA1200p data sheet. `http://www.panasonic.com/industrial/includes/pdf/Panasonic_VRLA_LC-XA12100P.pdf`.

[28] G. Putrus, P. Suwanapingkarl, D. Johnston, E. Bentley, and M. Narayana. Impact of electric vehicles on power distribution networks. In *Vehicle Power and Propulsion Conference, 2009. VPPC '09. IEEE*, pages 827–831, sept. 2009.

[29] D. P. Seetharam, T. Ganu, and B. Jayanta. Sepia : A Self-Organizing Electricity Pricing System. In *IEEE PES Innovative Smart Grid Technologies Asia*, May 2012.

[30] A. K. Sinha and D. Haźarika. Comparative study of voltage stability indices in a power system. *Electrical Power and Energy Systems*, 22(8):589–596, 2000.

[31] Soutik Biswas. Ten Interesting Things about India Power. `http://www.bbc.co.uk/news/world-asia-india-19063241`, July 2012.

[32] J. Tomic and W. Kempton. Using fleets of electric-drive vehicles for grid support. *Journal of Power Sources*, 168(2):459 – 468, 2007.

[33] Used Lead Acid Batteries: Factsheet. `http://www.environment.gov.au/settlements/chemicals/hazardous-waste/publications/lead-acid-fs.html`.

[34] U.S. Department of Energy. Benefits of demand response in electricity markets and recommendations for achieving them, 2006. `http://eetd.lbl.gov/ea/emp/reports/congress-1252d.pdf`.

[35] Vikas Bajaj. India Struggles to Deliver Enough Power. `http://www.nytimes.com/2012/04/20/business/global/india-struggles-to-deliver-enough-electricity-for-growth.html?ref=asia`, April 2012.

[36] Voltage Stability Using PV Curves. `http://www.powerworld.com/files/S06PVCurves.pdf`.

[37] Wartsila. Real Cost of Power. `http://www.wartsila.com/en_IN/media/reports/rcop`, July 2009.

[38] Unleashing the Potential of Renewable Energy in India. `http://siteresources.worldbank.org/INDIAEXTN/Resources/Reports-Publications/Unleashing_potential_of_Renewable_Energy_in_India.pdf`.

An Ensemble Model for Day-ahead Electricity Demand Time Series Forecasting

Wen Shen, Vahan Babushkin, Zeyar Aung, and Wei Lee Woon
Masdar Institute of Science and Technology
PO Box 54224, Abu Dhabi
United Arab Emirates
{wshen, vbabushkin, zaung, wwoon}@masdar.ac.ae

ABSTRACT

In this work, we try to solve the problem of day-ahead prediction of electricity demand using an ensemble forecasting model. Based on the Pattern Sequence Similarity (PSF) algorithm, we implemented five forecasting models using different clustering techniques: K-means model (as in original PSF), Self-Organizing Map model, Hierarchical Clustering model, K-medoids model, and Fuzzy C-means model. By incorporating these five models, we then proposed an ensemble model, named Pattern Forecasting Ensemble Model (PFEM), with iterative prediction procedure. We evaluated its performance on three real-world electricity demand datasets and compared it with those of the five forecasting models individually. Experimental results show that PFEM outperforms all those five individual models in terms of Mean Error Relative and Mean Absolute Error.

Categories and Subject Descriptors

I.5 [**Pattern Recognition**]: Models, Clustering

General Terms

Design, Experimentation

Keywords

Time Series Forecasting, Ensemble Model, Clustering

1. INTRODUCTION

In the electrical power industry, it is essential for decision makers to accurately predict the future values of variables such as electricity demand or price. This process of forecasting or prediction is called *Time Series Forecasting* or *Time Series Prediction*.

To forecast electricity demand and/or price, various forecasting techniques have been studied in the literature. These techniques include the wavelet transform models [1], the ARIMA models [2], the GARCH models [3], the Artificial Neural Network models [4] the hybrid model (basically a combination of Artificial Neural Networks and Fuzzy Logic) [5], the nearest neighbors methodology [6], the Support Vector Machines framework [7] and the Least-Square Support Vector Machine models [8].

Martínez-Álvarez *et al.* [9] proposed a Label-Based Forecasting (LBF) Algorithm using K-means clustering to predict electricity pricing time series. Using the mean squared error as a metric, they demonstrated that the LBF method outperforms several other methods including Naïve Bayes, Neural Networks, ARIMA, Weighted Nearest Neighbors and the Mixed Models. Based on the LBF algorithm, they later developed the Pattern Sequence-based Forecasting (PSF) algorithm [10], which predicts the future evolution of a time series based on the similarity of pattern sequences. They reported that PSF produced a significant improvement in the prediction of energy time series compared to several well-known techniques including its predecessor LBF algorithm.

In this work, we propose a Pattern Forecasting Ensemble Model (PFEM). Five PSF-style forecasting models are deployed, namely the K-means model (PSF itself), Self-Organizing Map model, Hierarchical Clustering model, K-medoids model, and Fuzzy C-means model. Each model is first applied separately, producing its respective forecasted values; we then perform a weighted combination of those values in an iterative manner in order to realize a better energy time series forecasting model.

We evaluated the performance of PFEM on three publicly available electricity demand datasets from the New York Independent System Operator (NYISO) [11], the Australia's National Electricity Market (ANEM) [12], and the Ontario's Independent Electricity System Operator (IESO) [13]. It was observed that the PFEM was able to provide more accurate and reliable forecasts than the five individual models, including PSF.

The rest of the paper is organized as follows. In Section 2, we briefly introduce the general principles of the PSF algorithm and give a general description of the five clustering methods which are the basis for our five individual forecasting models respectively. In Section 3, we present our proposed Pattern Forecasting Ensemble Model (PFEM). In Section 4, we evaluate the performance of PFEM and compare its performance with five individual models based on K-means (i.e., same as original PSF), Self-Organizing Map, Hierarchical Clustering, *K*-meloids, and Fuzzy C-means. In Section 5, we reach the conclusion and discuss the future work.

2. BACKGROUND

In this section, we present a short description of each of the five clustering methods used in PFEM, namely: K-means, Self-Organizing Map, Hierarchical Clustering, K-medoids, and Fuzzy C-means. Each of these methods can be used to generate a separate forecasting model, as described later in Section 3. However, in the following subsection, we begin by describing the basic concept of the PSF algorithm, upon which the proposed PFEM algorithm is based.

2.1 PSF Algorithm

The Pattern Sequence-based Forecasting (PSF) algorithm [10] is the basis of the proposed prediction algorithm, where the basic set of steps taken are repeated for each of the five individual forecasting models which comprise the PFEM algorithm. The general idea of the PSF algorithm can be described as follows:

1. Firstly, a clustering method is used to divide all of the 24-hour segments in the training dataset into a set of similar categories (K-means was the method of choice for clustering in PSF). This allows us to assign a label for each day in the training set, producing an associated label sequence. This sequence as well as the associated cluster centers are stored for use in the subsequent steps of this process.

2. To predict the demand for a given day, the category label for the days leading up to the day in question are determined based on similarity to the cluster centers previously obtained. This results in a label sequence which forms a kind of "fingerprint".

3. The sequence of labels generated during the training phase is searched for occurrences of this fingerprint, and matching instances are collected.

4. The demand profiles for the days immediately following each of these matches are extracted. The prediction is then generated by taking the average of all of these profiles.

2.2 Clustering Methods

2.2.1 K-means Clustering

K-means clustering [14] is a simple unsupervised learning algorithm which partitions N observations into k disjoint subsets C_j containing N_j data points. The K-means clustering algorithm aims to find the minimum value (or the local minima in most cases) of an objective function. The objective function J is described as follows:

$$J = \sum_{j=1}^{K} \sum_{n \in C_j} d(x_n, \mu_j) \qquad (1)$$

where μ_j is the cluster centroid for points in C_j and $d(x_n, \mu_j)$ points from their respective cluster centers, is a chosen distance measure between a data point x_n and the cluster center μ_j. In most cases, the Euclidean distance is used as a metric. The algorithm will terminate when no new partitions occur.

K-means is a greedy algorithm and as such its performance closely relies on the choice of the parameter k and the appropriate selection of the initial cluster centers [15].

2.2.2 Self-Organizing Map

Self-Organizing Map (SOM) [16] is an unsupervised learning artificial neural network which uses a neighborhood function to preserve the topological properties of the input space and maps high-dimensional data onto a 2-dimensional grid.

SOMs can be used to produce low dimensional representations of the data that preserve similarities between points in the data. Due to SOM's ability to preserve topological properties and good visualization features, they perform well for the prediction of non-linear time series [17].

2.2.3 Hierarchical Clustering

Hierarchical Clustering is a widely used data analysis tool which seeks to build a binary tree of the data that successively merges similar groups of points. Among all clustering techniques known in the literature, Hierarchical Clustering offers great versatility since it does not require a pre-specified number of clusters [18]. Instead, it only requires a measure of similarity between groups of data points.

In our work, we will use agglomerative Hierarchical Clustering. The algorithm can be described as follows: given a dataset $D = (x_1, x_2, \ldots, x_n)$ of n points, it first calculates a distance matrix M with all of the pairwise distances between points. Then it conducts the following process recursively until D has only a single data point:

1. Choose the two points x_i, x_j from D such that the distance between the two points is the minimum.

$$i, j = \arg \min_{i,j} d(x_i, x_j) \qquad (2)$$

where $i \neq j$.

2. Cluster x_i, x_j to form a new point c. The new point could be the mean of the two points or some other metric.

3. Remove x_i, x_j from D and insert c into D. Recalculate the pairwise distance matrix M.

2.2.4 K-medoids Clustering

The K-medoids algorithm [19] is an adaptation of the K-means algorithm in which a representative item, or a medoid is chosen for each cluster at each iteration. Medoids for each cluster are calculated by finding object i within the cluster that minimizes $\sum_{j \in C_i} d(i, j)$, where C_i is the cluster containing object i and $d(i, j)$ is the distance between objects i and j.

There are two advantages of using existing objects as the centers of the clusters. Firstly, a medoid object serves to usefully describe the cluster. Secondly, there is no need for repeated calculation of distances at each iteration, since the K-medoids algorithm can simply look up distances from a distance matrix.

The K-medoids algorithm can be described as follows [19]:

1. Choose k data points at random to be the initial cluster medoids.

2. Assign each data point to the cluster associated with the closet medoid.

3. Recalculate the positions of the k medoids.

4. Repeat Step 2 and Step 3 until the medoids become fixed.

2.2.5 Fuzzy C-means Clustering

Fuzzy C-means clustering [20] groups data by assigning a membership value linking each data point to each cluster center. This value is a reflection of the distance between the cluster center and the data point. Let $X = \{x_1, x_2, x_3, \ldots, x_n\}$ be the set of data points and $V = \{v_1, v_2, v_3, \ldots, v_c\}$ be the set of cluster centers. The Fuzzy C-means clustering algorithm is described as follows [20]:

1. Randomly select c cluster centers.

2. Calculate the fuzzy membership u_{ij} of ith data point to j^{th} cluster center using:

$$u_{ij} = \frac{1}{\sum_{k=1}^{c} \left(\frac{d_{ij}}{d_{ik}}\right)^{\left(\frac{2}{m-1}\right)}} \quad (3)$$

where n is the number of data points, d_{ij} is the distance between i^{th} data and j^{th} center, m is the fuzziness index and greater than or equal to 1.

3. Compute the fuzzy centers v_j using:

$$v_j = \frac{\sum_{i=1}^{n} x_i (u_{ij})^m}{\sum_{i=1}^{n} (u_{ij})^m}, \ \forall j = 1, 2, \ldots, c \quad (4)$$

4. Repeat Step 2 and 3 until the minimum value of objective function J is achieved or $d(U^{(k+1)}, U^{(k)}) < \beta$. β is the termination threshold between [0,1], $U = (u_{ij})_{n \times c}$ is the fuzzy membership matrix and J is defined by the following formula:

$$J(U, V) = \sum_{i=1}^{n} \sum_{j=1}^{c} (u_{ij})^m (x_i, v_j) \quad (5)$$

where $d(x_i, v_j)$ is the distance between i^{th} data point and j^{th} cluster center.

Unlike K-means where data points exclusively belong to one cluster, in the Fuzzy C-means algorithm data points are assigned memberships to each cluster center and as such can belong to more than one cluster at a time. Therefore, Fuzzy C-means clustering produces better results for overlapped dataset compared with K-means clustering.

3. PATTERN FORECASTING ENSEMBLE MODEL (PFEM)

The Pattern Forecasting Ensemble Model (PFEM) consists of four phases: data preprocessing, applying individual clustering methods, building individual forecasting models (based on corresponding clustering results), and iterative ensembling. Figure 1 shows the general idea behind PFEM. Details about the four phases are given below.

3.1 Data Preprocessing

The same feature selection and data normalization methodologies used for the original PSF model [10] were adopted for this study. This will facilitate direct comparison of PSF with the proposed method.

3.2 Applying Individual Clustering Methods

In this phase, five individual clustering methods: K-means, Self-Organizing Map, Hierarchical Clustering, K-medoids and Fuzzy C-means are applied on the preprocessed data. In each clustering exercise, cluster labels (which are discrete

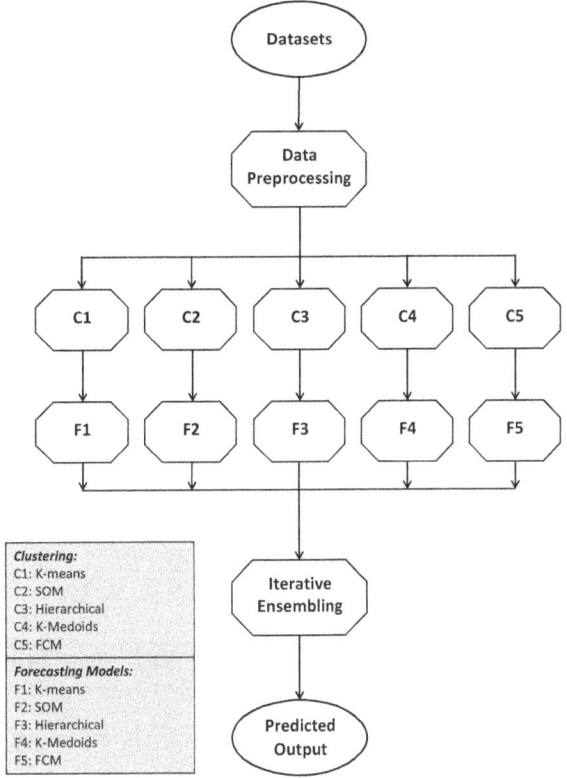

Figure 1: The general framework of Pattern Forecasting Ensemble Model.

values like 1, 2, 3) are assigned to all the real-number values in the preprocessed time-series data. Different clustering methods usually produce different cluster outputs form the same input data. Thus, the sets of labels assigned to the preprocessed data by the five methods will apparently be different from each other.

For each individual clustering method, several parameters need to be configured. The optimal parameter configurations for each method are determined by employing both statistical analyses (using well-defined indexes) and empirical analyses (based on the accuracy of the subsequent forecasting results). These parameters are briefly discussed below.

3.2.1 Number of Clusters

For K-means, K-medoids, Fuzzy C-means and SOM, the numbers of clusters are required to be specified in advance. For this reason, statistical analyses, namely the Silhouette index [21], the Dunn index [22] and the Davies-Bouldin index [23] are used to determine in how many groups the original dataset has to be split. For Hierarchical Clustering, it is not necessary to specify how many clusters as a priori. This is because the end result is a dendrogram where the leaves are data points and interior nodes represent a cluster made up of all children of the nodes. With the dendrogram, one can choose various number of clusters to use by cutting the tree at some particular height. However, we also employ the three indexes to determine the optimal number of clusters for Hierarchical Clustering for our convenience.

3.2.2 The Distance Function

In K-means, K-medoids, Fuzzy C-means and SOM, we use the Euclidean distance function to calculate similarities. In Hierarchical Clustering, the Mahalanobis distance [24] is utilized. This is because the respective indexes give more accurate indications than that with Euclidean distance according to our experimental analysis.

3.2.3 Window Size

The window size (the length of the sequence label) is selected through 12 folds cross validation as the K-means (original PSF) model does as described in [10].

3.2.4 Parameters for Fuzzy C-means Clustering

With lower value of the termination threshold β, we can get the better result. But it may require more iterations and thus more time to terminate. Therefore, a compromise between the accuracy and the efficiency is needed. In our experiments, we set the termination threshold $\beta = 10^{-5}$.

The fuzziness parameter m significantly effects the fuzziness of the clustering partition and hence affects the prediction results [25]. As the fuzziness m approaches 1 from above 0, the partition becomes hard ($u_{ik} \in \{0, 1\}$) and v_i are ordinary means of the clustering. As $m \to \infty$, the partition becomes completely fuzzy ($u_{ik} = 1/c$) and the cluster means are all equal to the mean of the dataset X.

3.3 Building Individual Forecasting Models

By changing the underlying clustering algorithm, different label sequences are produced. In this way, five individual predicting models are constructed using the same basic procedure used for the PSF algorithm [10] (see Section 2.1 for details). The mechanisms of all these five models are exactly the same except for the source label sequences they use. In this way, K-means model (same as original PSF), Self-Organizing Map model, Hierarchical Clustering model, K-medoids model, and Fuzzy C-means model are built separately. The five models produce their respective prediction results, which are subsequently used as the inputs in the next Iterative Ensembling phase.

3.4 Iterative Ensembling

In the iterative ensembling phase, the ensemble model is constructed by using the five individual forecasting models produced in the previous phase. The idea of iterative ensembling is to create a new model derived from a linear combination of several models with coefficients (weights) to minimize the forecasting error rates.

In the training stage, the actual values and the predicted values produced by the individual models are employed to select the weights which give the minimum prediction error rates. Then the weights are re-normalized. In the testing phase, the newly produced weights are used to predict the value. After prediction, this sample is incorporated into the training dataset. The weights are re-calculated, and new samples are learned iteratively in the same way. The idea of our iterative ensembling is inspired by AdaBoost [26].

The formal process is described as following:

1. Initialize the vector of observation weights $\mathbf{w}^{(0)} = (w_1^{(0)}, w_2^{(0)}, \ldots, w_n^{(0)})$, where n is the number of participating forecasting models for ensemble learning ($n = 5$ in our case):

$$w_i^{(0)} = 1/n \quad \forall i = 1, 2, \ldots, n. \tag{6}$$

2. For a training dataset with M days, for $m = 1$ to M:

$$\mathbf{P}^{(m)} = \sum_{i=1}^{n} w_i^{(m-1)} \mathbf{P}_i^{(m)} \tag{7}$$

where $\mathbf{P}^{(m)} = (P_1^{(m)}, P_2^{(m)}, \ldots, P_{24}^{(m)})$ is the vector of combined predicted values for 24 hours in day m, and $\mathbf{P}_i^{(m)} = (P_{i1}^{(m)}, P_{i2}^{(m)}, \ldots, P_{i24}^{(m)})$ is the vector of predicted values for 24 hours in day m generated by the individual forecasting model i.

(a) Define the prediction error rate for the iteration m:

$$err^{(m)} = \frac{1}{24} \sum_{h=1}^{24} \frac{\mid P_h^{(m)} - A_h^{(m)} \mid}{\bar{A}^{(m)}} \tag{8}$$

where $\mathbf{A}^{(m)} = (A_1^{(m)}, A_2^{(m)}, \ldots, A_{24}^{(m)})$ is the vector of actual values for 24 hours in day m and $\bar{A}^{(m)}$ is the average of actual values for 24 hours in that day.

(b) Calculate new weights for the iteration m:

$$\mathbf{w}^{(m)} = \operatorname*{argmin}_{w_i^{(m-1)}, i=1, \ldots, n} err^{(m)}, \tag{9}$$

such that $0 \leq w_i^{(m)} \leq 1, \sum_{i=1}^{n} w_i^{(m)} = 1$

3. Calculate the initial weights for testing:

$$\mathbf{w} = \frac{1}{m} \sum_{m=1}^{M} \mathbf{w}^{(m)} \tag{10}$$

Re-normalize $\mathbf{w}$.

4. Produce the predicted values using Equation (7) with the weights obtained in Step 3.

5. Add testing sample to the training dataset, increase M by 1, recalculate the initial weights for testing using Step 2, 3 and 4 for the next testing sample.

4. EXPERIMENTAL ANALYSIS

4.1 Performance Metrics

Several metrics are used to evaluate the performance of the PFEM approach:

- Mean Error Relative(MER):

$$MER = 100 \times \frac{1}{N} \sum_{h=1}^{N} \frac{|\hat{x}_h - x_h|}{\bar{x}} \tag{11}$$

where $\hat{x}_h$ are predicted and actual demand at hour h, respectively, $\bar{x}$ is the mean demand of the day and N is the number of predicted hours.

- Mean Absolute Error(MAE):

$$MAE = \frac{1}{N} \sum_{h=1}^{N} |\hat{x}_h - x_h| \tag{12}$$

- Mean Absolute Percentage Error(MAPE):

$$MAPE = \frac{100\%}{n} \sum_{i=1}^{n} \left| \frac{x_i - \hat{x}_i}{x_i} \right| \qquad (13)$$

where n is the number of the samples. x_i is the actual value and $\hat{x}_i$ is the forecast value.

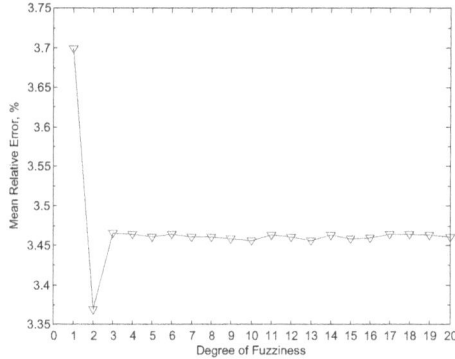

Figure 2: Correlation between mean relative error and degree of fuzziness for NYISO electricity demand data.

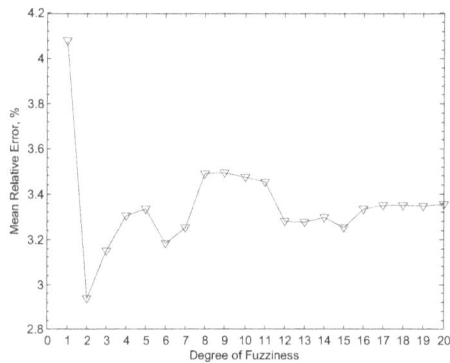

Figure 3: Correlation between mean relative error and degree of fuzziness for ANEM electricity demand data.

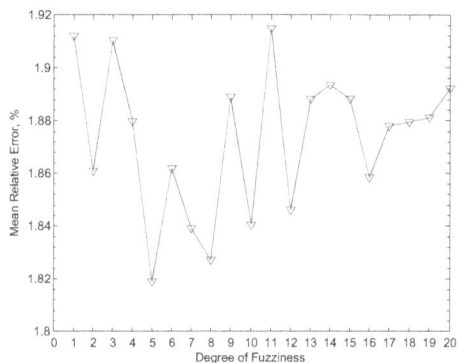

Figure 4: Correlation between mean relative error and degree of fuzziness for IESO electricity demand data.

4.2 Experimental Settings

In order to evaluate the performance of the PFEM, three real-world electricity demand datasets-the New York Independent System Operator (NYISO), the Australian Energy Market Operator (ANEM) and the Ontario's Independent Electricity System Operator (IESO) from 2007-2011 are employed.

For NYISO dataset, historical demand for Capital area was considered because the Capital district is a representative sample in terms of electricity consumption in the New York state. As for AEMO, load data for the New South Wales area was used since New South Wales is the most populated Australian state with high electricity demand.

To construct the Pattern Forecasting Ensemble Model for predicting the electricity demand of a day in 2009, Data from 2007 was used to train the individual clustering classifiers which were subsequently used to generate predictions for the data in 2008 and 2009. Then the predicted values and the actual time series of the data in 2008 were utilized to calculate the initial weights for testing following Step 2 described in section 3.4. This process is categorized as the training phase of the PFEM. With the preliminarily predicted time series in 2009 and the initial weights computed in the previous step, the final prediction of the electricity demand time series in 2009 could be produced by using methods shown in Step 3, 4 and 5 in section 3.4. For comparison, we also employed the five PSF-based individual learning models to forecast the 2009 time series. The 2007 and 2008 datasets were used for training and the 2009 dataset for testing. Likewise, we also performed the forecasting experiments for the 2010 dataset using the 2008 and 2009 datasets for training.

4.2.1 Number of Clusters

In order to find optimal numbers of clusters, we analyze the data for 3 years. For all the three datasets the training set is chosen from Jan 1, 2007 to 31, Dec 2007.

The optimal numbers of clusters are chosen based on the majority voting method of Silhouette, Dunn and Davies-Bouldin indices. The maximum values of Silhouette and Dunn indices indicate the optimal numbers of clusters when the best possible clustering is obtained, that is, the distance among elements in the same cluster is minimized and the distance among elements from different clusters is maximized (for Davies-Bouldin index the optimal number of clusters corresponds the minimum value). If all three indices give different results, the optimal numbers of clusters are chosen based on the sub-optimal numbers of clusters. The numbers of clusters are calculated for 10 times and the most frequent values are selected. Table 1 shows the numbers of clustering for different markets in the year 2007 with the three indices. Similarly, the numbers of clusterings are chosen for the three datasets in the year 2007, 2008, 2007-2008 and 2008-2009, which can be seen in Table 2.

4.2.2 Window Size

The dataset for 1 year (2007) was normalized (in order to eliminate the noise introduced by the presence of global trend) and divided into 12 parts, each part consisting of a one-month long subset of given data for one year (2007). Then, starting from the first month and rotating through all twelve months in the year, one of the months was used for testing and the remaining 11 months for training. For each day of the testing set, given the known optimal number

Market	K-means				SOM				Hierarchical				k-medoid				Fuzzy C-means			
	Silh.	DU	DB	sel.	Silh.	DU	DB	sel.	Silh.	DU	DB	sel.	Silh.	DU	DB	sel.	Silh.	DU	DB	sel.
NYISO	3	3	4	**3**	4	4	4	**4**	2	6	5	**5**	6	8	6	**6**	3	3	4	**3**
ANEM	5	5	4	**5**	2	4	4	**4**	4	4	4	**4**	4	6	5	**6**	2	4	3	**3**
IESO	3	4	3	**3**	3	4	3	**3**	3	3	3	**3**	4	4	3	**4**	3	4	4	**4**

Table 1: Numbers of clusters for different markets evaluated with Silhouette (denoted as Silh.), Dunn (DU) and Davies-Bouldin (DB) indexes for five clustering methods(2007). "sel." indicates the selected optimal numbers of clusters.

Method	2007			2008			2007-2008			2008-2009		
	NYISO	ANEM	IESO	NYISO	ANEM	IESO	NYISO	ANEM	IESO	NYISO	ANEM	IESO
K-means	3	5	3	3	4	3	5	5	3	6	5	3
SOM	4	4	3	6	3	8	7	6	4	4	5	6
Hierarchical	4	4	3	5	2	4	5	4	4	5	5	5
K-medoids	6	5	4	6	5	3	6	5	5	7	6	4
Fuzzy c-means	3	3	4	6	5	4	6	3	4	3	4	4

Table 2: Optimal numbers of clusters, when the 5 clustering methods achieve the best clustering, for NYISO, ANEM and IESO markets (2007, 2008, 2007-2008 and 2008-2009)

Method	W=1	W=2	W=3	W=4	W=5	W=6	W=7	W=8	W=9	W=10	W=11	W=12	W=13	W=14	Selected
K-means	3.0940	2.9106	**2.7445**	2.7570	2.8066	2.8485	2.8835	2.9108	2.9733	2.8843	2.9120	2.8740	3.0363	2.8942	3
SOM	3.4371	**3.4217**	3.5281	3.5693	3.7037	3.7310	3.8234	3.7682	3.8117	3.7612	3.8120	3.8012	3.8411	3.8697	2
Hierarchical	3.5894	3.5067	3.4721	**3.4541**	3.6008	3.6149	3.6478	3.6845	3.7076	3.7423	3.8077	3.8259	3.8264	3.8273	4
K-medoids	3.3078	3.2503	**3.2092**	3.3177	3.3370	3.3669	3.5249	3.5572	3.4905	3.4773	3.5413	3.4736	3.5832	3.4611	3
Fuzzy C-means	3.4852	3.4788	**3.4526**	3.4602	3.5494	3.5669	3.6486	3.6482	3.6210	3.6671	3.6289	3.6653	3.6792	3.6947	3

Table 3: MER-based selection of optimal window sizes for NYISO Demand Dataset(2007) through 12 folds cross-validation

Method	W=1	W=2	W=3	W=4	W=5	W=6	W=7	W=8	W=9	W=10	W=11	W=12	W=13	W=14	Selected
K-means	3.0940	2.9106	**2.7445**	2.7570	2.8066	2.8485	2.8835	2.9108	2.9733	2.8843	2.9120	2.8740	3.0363	2.8942	3
SOM	3.0176	2.9963	**2.8940**	2.9266	2.9685	2.8958	2.8962	2.9515	2.9874	3.0121	3.0818	3.1324	3.1456	3.0827	3
Hierarchical	3.1001	2.8099	2.7063	2.6131	**2.4989**	2.5378	2.5636	2.5896	2.6228	2.6762	2.7008	2.7121	2.7157	2.7223	5
K-medoids	3.0831	2.8597	**2.7627**	2.8017	2.8297	2.9381	2.8089	3.0244	2.8452	2.9224	2.9302	2.9858	2.9351	3.0995	3
Fuzzy C-means	3.1135	3.0842	2.9290	2.8923	2.8718	2.8538	**2.8304**	2.8512	2.8827	2.8912	2.9262	2.9769	3.0288	3.0519	7

Table 4: MER-based selection of optimal window sizes for ANEM Demand Dataset(2007) through 12 folds cross-validation.

Method	W=1	W=2	W=3	W=4	W=5	W=6	W=7	W=8	W=9	W=10	W=11	W=12	W=13	W=14	Selected
K-means	2.6414	2.5821	2.5561	2.5948	2.5621	**2.5560**	2.5716	2.5964	2.5946	2.5935	2.5687	2.5887	2.5710	2.5966	6
SOM	2.6650	2.4498	**2.4414**	2.4440	2.4705	2.4976	2.5375	2.5569	2.5904	2.6308	2.6372	2.6392	2.6562	2.6322	3
Hierarchical	2.5956	2.4754	**2.4404**	2.4580	2.4588	2.4755	2.4734	2.4769	2.4624	2.4597	2.4608	2.4700	2.4811	2.4902	3
K-medoids	2.6218	2.4190	2.5062	**2.3822**	2.4567	2.4788	2.5751	2.4740	2.5218	2.5317	2.5806	2.5735	2.5629	2.6450	4
Fuzzy C-means	2.5261	2.3559	**2.3413**	2.3476	2.3746	2.3938	2.3932	2.4416	2.4662	2.5103	2.5100	2.5031	2.5102	2.5003	3

Table 5: MER-based selection of optimal window sizes for IESO Demand Dataset(2007) through 12-fold cross-validation.

Method	2007			2008			2007-2008			2008-2009		
	NYISO	ANEM	IESO	NYISO	ANEM	IESO	NYISO	ANEM	IESO	NYISO	ANEM	IESO
K-means	3	3	6	5	5	5	4	5	8	5	5	7
SOM	2	3	3	2	9	7	3	5	4	4	8	4
Hierarchical	4	5	3	5	6	4	4	5	3	5	5	5
K-medoids	3	3	4	2	9	4	2	12	5	3	6	5
Fuzzy c-means	3	7	3	2	9	7	2	9	4	5	6	6

Table 6: The selected window sizes of NYISO, ANEM and IESO Demand Datasets (2007, 2008, 2007-2008, 2008-2009) through 12 folds cross-validation

of clusters and proposed sizes of window, predictions were made and the respective MER calculated. The MER for given month is obtained by averaging across all days in the testing set. Iterating through the different window sizes, the correlation and dependence between size of window and the MER is evident. Finally, from obtained MER and size of

Market	Error	K-means		SOM		Hierarchical		K-medoids		Fuzzy C-means		PFEM	
		Err	σ	Err	σ	Err	σ	Err	σ	Err	σ	Err	σ
NYISO	MER	3.11	0.41	3.06	0.44	2.92	0.31	2.97	0.37	3.18	0.43	2.76	0.35
	MAE	39.16	6.88	38.5	7.38	36.79	5.92	37.27	6.21	39.89	6.61	34.78	6.31
	MAPE	3.18	0.42	3.12	0.44	2.99	0.32	3.03	0.38	3.26	0.43	2.82	0.36
ANEM	MER	2.98	0.862	3.18	0.76	2.76	0.91	2.79	0.73	2.58	0.67	2.55	0.80
	MAE	259.66	85.92	283.23	76.98	244.83	85.89	249.25	71.88	229.28	65.22	228.35	79.00
	MAPE	2.96	0.902	3.25	0.81	2.78	0.95	2.86	0.77	2.63	0.71	2.61	0.84
IESO	MER	2.42	0.36	2.67	0.36	2.5	0.41	2.34	0.29	2.31	0.27	2.23	0.25
	MAE	384.02	59.27	422.61	49.22	394.26	58.62	371.21	45.23	364.71	39.31	354.99	46.92
	MAPE	2.49	0.38	2.74	0.36	2.58	0.43	2.41	0.31	2.37	0.30	2.30	0.27

Table 7: Summary performance results of models tested on demand data of NYISO, ANEM and IESO markets for 2009.

window values, the optimal size of window that corresponds to minimal mean relative error is selected.

Tables 3, 4 and 5 present the prediction errors with respect to MER in relation to the size of window and the clustering technique used for the NYISO, ANEM and IESO electricity demand datasets respectively. The window sizes selected for each of the three datasets are shown in the last column of each table. Correspondingly, we selected the optimal sizes of window for 2007, 2008, 2007-2008, 2008-2009 datasets using the same method. The results are shown in Table 6.

4.2.3 Determining Fuzziness Parameter

In order to find the optimal values of the fuzziness parameters that minimize the mean relative error we employed the same cross validation technique that was used to determine the optimal size of window.

In the case of NYISO and ANEM datasets, as fuzziness increases, the error rates first drop to the bottom and then increase to a certain level. Then there will be fluctuations when the fuzziness is greater that 2. The is because the degree of data overlap in this real-world dataset is relatively low. The optimal fuzziness parameter is 2 with corresponding MER values 3.37% and 2.93% as shown in Figure 2 and Figure 3.

For IESO dataset, the optimal fuzziness parameter is 5, when MRE of 1.82% is achieved as shown in Figure 4.

4.3 Experimental Results

Table 7 and Table 11 show the annual mean and the annual standard deviation of MER, MAE and MAPE for each learning model based on the testing results of three datasets in 2009 and 2010, respectively.

4.3.1 NYISO Dataset

Table 8 shows the Mean Error Relative (MER) and Mean Absolute Error (MAE) obtained for all five individual models based on the PSF algorithms and the proposed PFEM in the year 2009. Among all the five individual models, Hierarchical Clustering gives the forecasting results with the lowest MER (2.92%) and MAE(36.79MW). The PFEM outperforms all other models in terms of both MER(2.76%) and MAE(34.78MW).

Figure 5 demonstrates the best results for a day's electricity demand forecasting with respect to MER in the year 2009. Figure 6 gives the worst prediction results for a day in the same year. Table 12 summarizes the performance of all 5 PSF-style algorithms and the PFEM algorithm for 2010. In this case, the lowest mean annual MER (2.77%) and MEA (36.57MW) are achieved with K-medoids model. Again, PFEM outperforms all 5 individuals models with MER (0.74%) and MAE (36.27MW). Figure 11 illustrates

the best prediction electricity demand and Figure 12 gives the worst prediction results for a day in the same year.

4.3.2 ANEM Dataset

As for the Australian electricity demand time series, our ensemble model also produces the best prediction results with the MER of 2.55% and a MAE of 228.35MW.

The Fuzzy C-means defeats all other models except the PFEM. Table 9 presents the details about the performance of all models for the year 2009 dataset. Figure 7 and 8 illustrate the best and the worst forecasting results for the ANEM dataset respectively. The lowest MER it can be obtained is 0.744% while the highest error rate is 7.051%.

For 2010 the Hierarchical Clustering based model beats all other individual models with mean annual MER 2.45% and MAE 214.61MW, only PFEM achieves MER and MAE of 2.39% and 210.91MW. The best and the worst day-ahead demand forecasting results for 2010 are represented on Figure 13 and Figure 14 correspondingly.

4.3.3 IESO Dataset

With respect to the IESO dataset, Fuzzy C-means based PSF outperforms all other individual models with a MER of 2.31% and a MAE of 364.71MW. The PFEM gives a slightly better results than the Fuzzy C-means based model. Figure 9 and 10 present the best and worst forecasting of the PFEM in a day of the year 2009. Figure 15 and Figure 16 illustrate the best and worst prediction curves respectively. It shows that the PFEM is able to predict the day-ahead electricity load time series very accurately in the best case.

Considering the results for 2010, the Hierarchical Clustering based model outperforms all other 4 individual models in terms of MER (2.18%) and MEA (354.15MW). The PFEM outperforms all models with MER = 2.10% and MAE = 345.30MW. The best and worst day-ahead electricity demand prediction results for 2010 are represented on Figure 15 and Figure 16. Noticeably, that the best and worst forecasting results for 2009 and 2010 are achieved with MER of 0.81% and 0.59% for best and 4.05% and 6.26% for worst predictions correspondingly.

5. CONCLUSION AND FUTURE WORKS

In this work, a novel forecasting model for electricity demand time series is proposed. Named the "Pattern Forecasting Ensemble Model" (PFEM), the new method is based on the pre-existing PSF algorithm, but uses a combination of five separate clustering models: the K-means model (PSF itself), the SOM model, the Hierarchical Clustering based model, the K-means model and the Fuzzy C-means model. The optimal values of parameters such as k, c and the win-

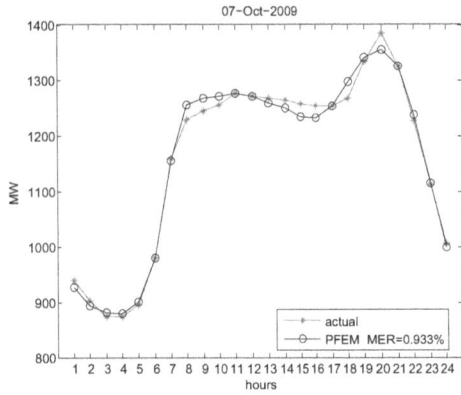

Figure 5: Best prediction of PFEM for NYISO dataset(2009).

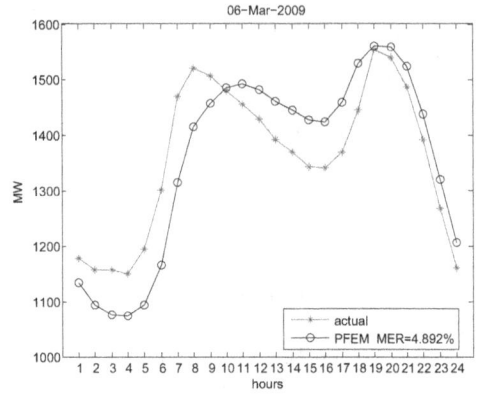

Figure 6: Worst prediction of PFEM for NYISO dataset(2009).

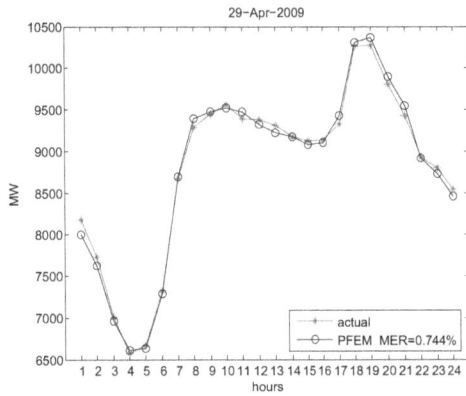

Figure 7: Best prediction of PFEM for ANEM dataset(2009).

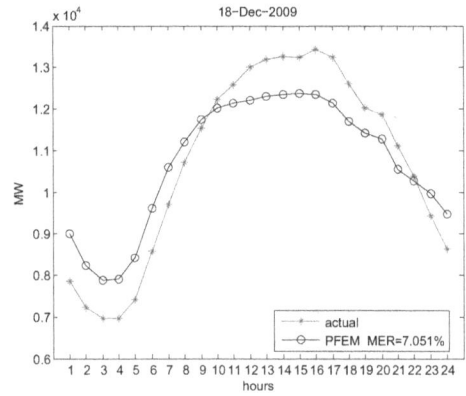

Figure 8: Worst prediction of PFEM for ANEM dataset(2009).

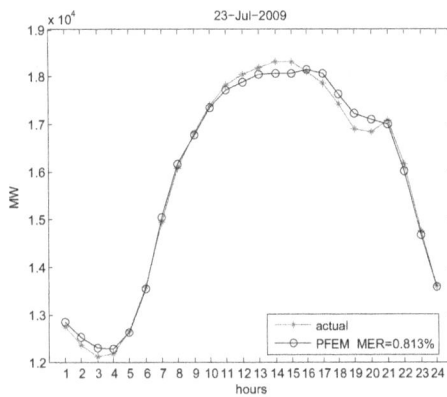

Figure 9: Best prediction of PFEM for IESO dataset(2009).

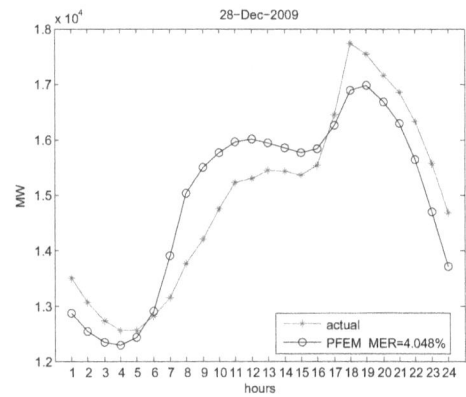

Figure 10: Worst prediction of PFEM for IESO dataset(2009).

dow size were also determined through an empirical approach. We then evaluate the performance of the PFEM and that of the other five individual models on three real-world electricity demand datasets. Experimental results indicate that our proposed approach gives superior results compared with all the other five individual models in terms of both MER and MAE. For future work, we intend to explore non-linear combinations of these individual models and conduct a quantitative evaluation of this approach along with a comparison to the related linearly combined ensemble models.

6. REFERENCES

[1] A. J. Conejo, M. A. Plazas, R. Espinola, and A. B. Molina, "Day-ahead electricity price forecasting using the wavelet transform and ARIMA models," *IEEE Transactions on Power Systems*, vol. 20, no. 2, pp. 1035–1042, 2005.

[2] G. Janacek, "Time series analysis forecasting and control," *Journal of Time Series Analysis*, vol. 31, no. 4, p. 303, 2010.

[3] R. Garcia, J. Contreras, M. Van Akkeren, and J. Garcia, "A GARCH forecasting model to predict day-ahead electricity prices," *IEEE Transactions on Power Systems*, vol. 20, no. 2, pp. 867–874, 2005.

[4] B. Neupane, K. S. Perera, Z. Aung, and W. L. Woon, "Artificial neural network-based electricity price forecasting for smart grid deployment," in *Proceedings of the 2012 IEEE International Conference on Computer Systems and Industrial Informatics (ICCSII'12)*, pp. 1–6, IEEE, 2012.

[5] N. Amjady, "Day-ahead price forecasting of electricity markets by a new fuzzy neural network," *IEEE Transactions on Power Systems*, vol. 21, no. 2, pp. 887–896, 2006.

[6] A. T. Lora, J. M. R. Santos, A. G. Expósito, J. L. M. Ramos, and J. C. R. Santos, "Electricity market price forecasting based on weighted nearest neighbors techniques," *IEEE Transactions on Power Systems*, vol. 22, no. 3, pp. 1294–1301, 2007.

[7] J. H. Zhao, Z. Y. Dong, X. Li, and K. P. Wong, "A framework for electricity price spike analysis with advanced data mining methods," *IEEE Transactions on Power Systems*, vol. 22, no. 1, pp. 376–385, 2007.

[8] Z. Aung, M. Toukhy, J. Williams, A. Sanchez, and S. Herrero, "Towards accurate electricity load forecasting in smart grids," in *Proceedings of the 4th International Conference on Advances in Databases, Knowledge, and Data Applications (DBKDA'12)*, pp. 51–57, IARIA, 2012.

[9] F. Martínez-Álvarez, A. Troncoso, J. C. Riquelme, and J. S. Aguilar-Ruiz, "LBF: A labeled-based forecasting algorithm and its application to electricity price time series," in *Proceedings of the 8th IEEE International Conference on Data Mining (ICDM'08)*, pp. 453–461, IEEE, 2008.

[10] F. Martínez-Álvarez, A. Troncoso, J. C. Riquelme, and J. S. Aguilar Ruiz, "Energy time series forecasting based on pattern sequence similarity," *IEEE Transactions on Knowledge and Data Engineering*, vol. 23, no. 8, pp. 1230–1243, 2011.

[11] "New York Independent System Operator (NYISO)." http://www.nyiso.com/public/index.jsp.

[12] "Australia's National Electricity Market (ANEM)." http://www.aemo.com.au/Electricity.

[13] "Independent Electricity System Operator (IESO)." http://www.ieso.ca/default.asp.

[14] J. MacQueen et al., "Some methods for classification and analysis of multivariate observations," in *Proceedings of the 5th Berkeley Symposium on Mathematical Statistics and Probability*, vol. 1, pp. 281–297, University of California, Berkeley, 1967.

[15] K. Hammouda and F. Karray, "A comparative study of data clustering techniques," tech. rep., Pattern Analysis and Machine Intelligence Research Group, University of Waterloo, 2000.

[16] J. Vesanto and E. Alhoniemi, "Clustering of the self-organizing map," *IEEE Transactions on Neural Networks*, vol. 11, no. 3, pp. 586–600, 2000.

[17] A. Chitra and S. Uma, "An ensemble model of multiple classifiers for time series prediction," *International Journal of Computer Theory and Engineering*, vol. 2, no. 3, pp. 454–458, 2010.

[18] P. Rodrigues, J. Gama, and J. P. Pedroso, "LBF: Hierarchical time-series clustering for data streams," in *Proceedings of the 1st International Workshop on Knowledge Discovery in Data Streams (IWKDDS'04)*, pp. 22–31, ACM, 2004.

[19] A. Reynolds, G. Richards, and V. Rayward-Smith, "The application of k-medoids and PAM to the clustering of rules," pp. 173–178, Springer, 2004.

[20] R. L. Cannon, J. V. Dave, and J. C. Bezdek, "Efficient implementation of the fuzzy c-means clustering algorithms," *IEEE Transactions on Pattern Analysis and Machine Intelligence*, vol. 8, no. 2, pp. 248–255, 1986.

[21] P. J. Rousseeuw and L. Kaufman, *Finding Groups in Data: An Introduction to Cluster Analysis*. John Wiley & Sons, 1990.

[22] J. C. Dunn, "Well-separated clusters and optimal fuzzy partitions," *Journal of Cybernetics*, vol. 4, no. 1, pp. 95–104, 1974.

[23] D. L. Davies and D. W. Bouldin, "A cluster separation measure," *IEEE Transactions on Pattern Analysis and Machine Intelligence*, vol. 1, no. 2, pp. 224–227, 1979.

[24] R. De Maesschalck, D. Jouan-Rimbaud, and D. L. Massart, "The Mahalanobis distance," *Chemometrics and Intelligent Laboratory Systems*, vol. 50, no. 1, pp. 1–18, 2000.

[25] D. Dembele and P. Kastner, "Fuzzy C-means method for clustering microarray data," *Bioinformatics*, vol. 19, no. 8, pp. 973–980, 2003.

[26] J. Zhu, S. Rosset, H. Zou, and T. Hastie, "Multi-class AdaBoost," *Statistics and Its Interface*, vol. 2, no. 3, pp. 349–360, 2009.

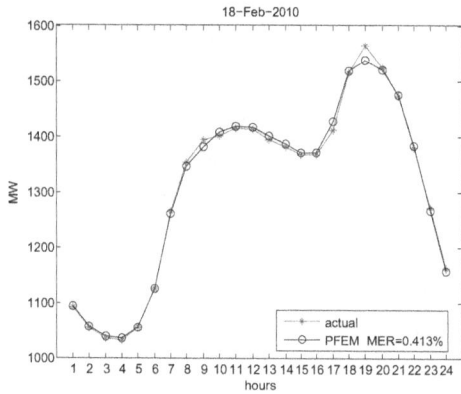

Figure 11: Best prediction of PFEM
for NYISO dataset(2010).

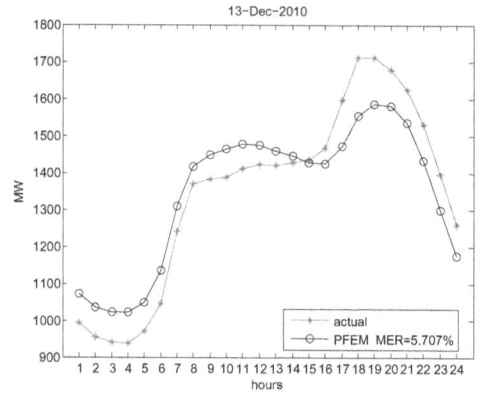

Figure 12: Worst prediction of PFEM
for NYISO dataset(2010).

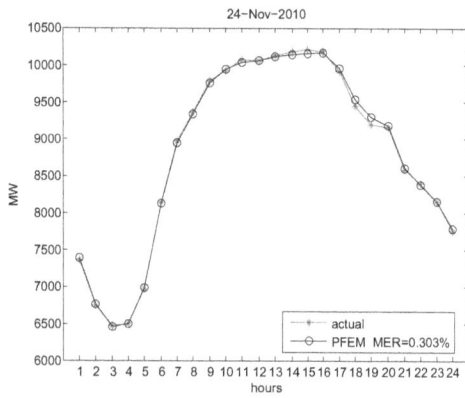

Figure 13: Best prediction of PFEM
for ANEM dataset(2010).

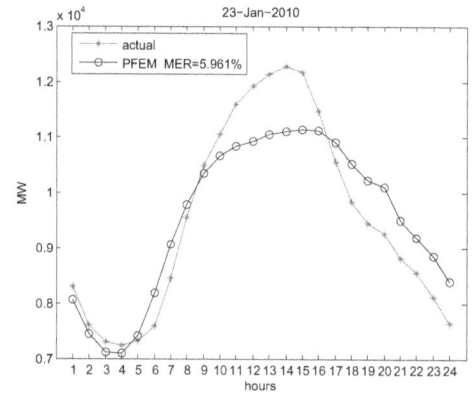

Figure 14: Worst prediction of PFEM
for ANEM dataset(2010).

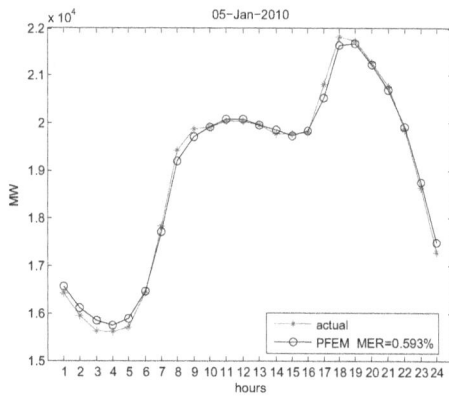

Figure 15: Best prediction of PFEM
for IESO dataset(2010).

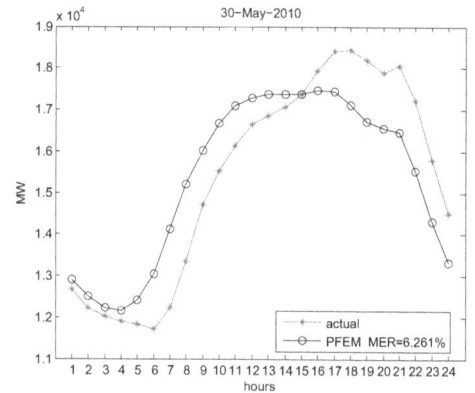

Figure 16: Worst prediction of PFEM
for IESO dataset(2010).

Month	K-means		SOM		Hierarchical		K-medoids		Fuzzy C-means		PFEM	
	MER(%)	MAE(MW)	MER(%)	MAE(MW)	MER(%)	MAE(MW)	MER(%)	MAE(MW)	MER(%)	MAE(MW)	MER(%)	MAE(MW)
Jan.	3.09	43.45	2.89	40.57	2.61	36.72	2.76	38.83	2.85	40.17	2.55	35.85
Feb.	2.89	39.02	2.93	39.72	2.62	35.39	2.75	37.37	2.83	38.32	2.46	33.33
Mar.	3.38	41.43	3.26	39.94	3.24	39.79	3.19	39.39	3.69	45.22	3.07	37.65
Apr.	3.41	38.09	3.38	37.84	3.05	34.37	3.65	41.54	3.64	40.75	3.13	35.02
May.	3.44	37.94	3.03	33.36	2.76	30.32	2.69	28.36	3.32	36.53	2.56	28.62
Jun.	3.15	39.09	3.12	38.69	3.09	38.15	2.86	35.36	2.99	37.07	2.72	33.84
Jul.	3.04	40.89	2.85	37.57	3.02	39.81	2.98	39.08	2.98	39.17	2.89	38.27
Aug.	3.59	51.99	3.61	52.29	3.41	49.65	3.55	51.36	3.64	52.76	3.31	48.19
Sept.	2.47	29.98	2.37	28.75	3.02	36.81	2.37	28.84	2.57	31.32	2.29	27.83
Oct.	2.26	26.17	2.27	26.24	2.33	27.78	2.87	33.11	2.57	29.89	2.22	25.52
Nov.	3.06	35.02	3.24	37.15	2.74	31.17	2.91	33.16	3.36	38.38	2.74	31.41
Dec.	3.53	46.89	3.75	49.91	3.14	41.59	3.08	40.84	3.72	49.17	3.16	41.89
Mean	3.11	39.16	3.06	38.50	2.92	36.79	2.97	37.27	3.18	39.89	2.76	34.78

Table 8: Performance of all models tested on NYISO demand data for 2009.

Month	K-means		SOM		Hierarchial		K-medoids		Fuzzy C-means		PFEM	
	MER(%)	MAE(MW)	MER(%)	MAE(MW)	MER(%)	MAE(MW)	MER(%)	MAE(MW)	MER(%)	MAE(MW)	MER(%)	MAE(MW)
Jan.	5.92	468.94	5.15	482.18	4.97	457.47	4.42	410.97	4.09	376.24	4.52	420.75
Feb.	3.93	369.69	3.68	347.89	3.29	310.16	3.27	309.89	2.94	277.27	3.13	295.26
Mar.	1.88	162.09	2.67	230.52	2.37	196.76	2.12	181.62	2.19	170.89	1.86	160.05
Apr.	2.62	210.94	2.76	223.38	2.09	169.46	2.18	177.11	2.07	169.42	1.97	159.63
May.	2.48	208.27	2.37	205.69	1.86	162.91	2.32	200.77	2.04	177.88	1.87	163.81
Jun.	2.32	220.06	2.68	246.29	2.02	192.58	2.25	212.23	1.94	184.12	1.96	186.31
Jul.	2.38	224.77	2.61	246.67	2.08	198.16	2.24	212.59	2.14	202.69	2.21	208.52
Aug.	2.61	231.65	2.97	264.59	2.27	201.07	2.58	230.34	2.28	201.73	2.18	193.78
Sept.	2.24	187.95	2.83	238.16	2.29	185.64	2.28	192.23	2.28	191.59	2.03	170.49
Oct.	2.96	248.99	3.17	266.39	3.19	261.27	2.83	236.04	2.54	214.06	2.52	211.71
Nov.	3.23	296.75	3.72	338.18	3.29	300.62	3.52	320.97	3.25	297.08	3.24	295.43
Dec.	3.25	285.82	3.59	308.92	3.44	301.88	3.57	306.34	3.28	288.41	3.11	274.51
Mean	2.98	259.66	3.18	283.23	2.76	244.83	2.79	249.25	2.58	229.28	2.55	228.35

Table 9: Performance of all models tested on ANEM demand data for 2009.

Month	K-means		SOM		Hierarchial		K-medoids		Fuzzy C-means		PFEM	
	MER(%)	MAE(MW)	MER(%)	MAE(MW)	MER(%)	MAE(MW)	MER(%)	MAE(MW)	MER(%)	MAE(MW)	MER(%)	MAE(MW)
Jan.	1.94	347.33	1.96	357.19	1.63	295.57	1.95	355.46	1.95	346.89	1.81	329.19
Feb.	2.31	399.73	2.36	406.96	1.91	331.03	2.04	354.55	2.68	347.96	2.18	376.81
Mar.	2.27	369.71	2.75	449.66	3.05	495.62	2.68	439.05	2.29	374.83	2.27	370.83
Apr.	2.54	376.76	3.78	450.44	2.46	367.02	2.43	362.47	2.29	341.58	2.27	337.68
May.	2.24	310.57	2.74	383.07	2.83	400.38	2.38	334.71	2.68	381.81	2.09	294.88
Jun.	2.89	444.41	2.62	406.57	2.74	420.28	2.46	376.94	2.45	376.13	2.49	382.84
Jul.	2.17	326.39	2.55	387.35	2.43	366.49	2.25	340.94	2.07	313.55	2.03	301.91
Aug.	2.93	482.93	2.96	494.02	2.85	473.09	2.75	454.78	2.79	461.18	2.75	456.64
Sept.	2.63	400.62	2.32	351.71	2.49	364.35	2.23	337.52	2.19	331.82	2.21	336.37
Oct.	1.96	285.77	2.96	445.78	2.34	353.89	2.16	325.12	2.22	334.91	2.01	302.12
Nov.	2.84	438.99	3.28	504.81	2.82	434.97	2.83	433.89	2.63	402.67	2.51	387.04
Dec.	2.53	425.14	2.58	433.88	2.57	428.49	2.02	340.67	2.17	363.35	2.28	383.68
Mean	2.42	384.02	2.67	422.61	2.50	394.26	2.34	371.21	2.31	364.71	2.23	354.99

Table 10: Performance of all models tested on IESO demand data for 2009.

Market	Error	K-means		SOM		Hierarchal		K-medoids		Fuzzy C-means		PFEM	
		Err	σ	Err	σ	Err	σ	Err	σ	Err	σ	Err	σ
NYISO	MER	3.09	0.61	3.11	0.55	2.97	0.54	2.77	0.54	3.07	0.56	2.74	0.57
	MAE	40.72	11.45	40.87	10.86	40.87	10.86	36.57	10.38	40.35	10.84	36.27	11.11
	MAPE	3.16	0.49	3.12	0.55	2.99	0.58	2.79	0.57	3.26	0.63	2.78	0.58
ANEM	MER	2.78	0.56	2.89	0.39	2.45	0.71	2.64	0.52	2.73	0.51	2.39	0.46
	MAE	243.88	52.78	254.17	38.63	214.61	60.99	232.11	47.29	239.93	46.91	210.92	44.11
	MAPE	2.83	0.58	2.96	0.41	2.48	0.52	2.69	0.55	2.78	0.52	2.44	0.48
IESO	MER	2.29	0.48	2.52	0.47	2.19	0.45	2.28	0.46	2.59	0.53	2.10	0.44
	MAE	372.48	91.21	410.51	89.67	354.15	83.38	368.67	86.72	422.13	102.99	345.30	87.04
	MAPE	2.35	0.52	2.59	0.51	2.24	0.48	2.34	0.49	2.66	0.56	2.18	0.48

Table 11: Summary performance results of models tested on demand data of NYISO, ANEM and IESO markets for 2010.

Month	K-means MER(%)	K-means MAE(MW)	SOM MER(%)	SOM MAE(MW)	Hierarchical MER(%)	Hierarchical MAE(MW)	K-medoids MER(%)	K-medoids MAE(MW)	Fuzzy C-means MER(%)	Fuzzy C-means MAE(MW)	PFEM MER(%)	PFEM MAE(MW)
Jan.	2.68	36.77	2.89	39.63	2.61	35.63	2.61	35.83	2.73	37.32	2.46	33.95
Feb.	2.53	33.14	2.33	30.87	2.04	27.18	2.03	26.89	2.28	30.28	1.96	25.89
Mar.	2.52	29.63	2.57	30.14	2.61	30.65	2.52	29.76	2.64	31.06	2.42	28.40
Apr.	2.93	31.87	3.01	32.72	2.63	28.56	2.28	24.93	3.12	33.88	2.42	26.26
May.	3.35	42.02	3.41	42.64	3.23	40.45	2.99	37.83	3.43	42.85	2.94	37.30
Jun.	3.06	41.34	2.97	40.06	3.26	44.17	2.97	40.39	2.92	39.55	2.89	39.10
Jul.	3.98	62.22	3.95	62.41	3.93	54.28	3.31	47.94	4.05	63.71	3.44	50.46
Sept.	3.83	50.44	3.68	48.68	3.71	48.93	3.49	45.96	3.67	52.98	3.27	42.65
Oct.	2.32	26.04	2.45	27.88	2.33	26.42	2.27	25.62	3.81	50.29	2.08	23.51
Nov.	2.71	32.35	2.83	33.97	2.82	33.71	2.67	32.02	2.48	27.84	2.57	30.88
Dec.	3.27	44.62	3.28	44.87	2.73	37.26	2.38	32.37	3.16	43.07	2.59	35.32
Mean	3.09	40.72	3.11	40.87	2.97	39.06	2.77	36.57	3.01	40.35	2.74	36.27

Table 12: Performance of all models tested on NYISO demand data for 2010.

Month	K-means MER(%)	K-means MAE(MW)	SOM MER(%)	SOM MAE(MW)	Hierarchical MER(%)	Hierarchical MAE(MW)	K-medoids MER(%)	K-medoids MAE(MW)	Fuzzy C-means MER(%)	Fuzzy C-means MAE(MW)	PFEM MER(%)	PFEM MAE(MW)
Jan.	4.41	400.34	3.84	351.47	4.15	378.93	3.91	356.08	3.93	359.73	3.63	335.61
Feb.	2.73	254.43	3.32	309.16	2.55	240.69	3.14	290.54	2.62	244.52	2.64	246.29
Mar.	2.54	221.64	2.77	239.93	2.32	200.71	2.43	211.67	2.13	185.98	2.15	186.79
Apr.	2.55	204.82	2.84	229.27	2.68	216.33	2.58	207.24	2.62	211.01	2.35	188.32
May.	2.48	217.79	2.74	239.92	1.96	171.71	2.35	205.07	2.93	258.05	2.06	181.64
Jun.	2.42	227.29	2.34	221.69	1.74	166.11	2.14	203.51	2.35	222.64	2.02	192.97
Jul.	2.28	217.94	2.51	240.11	1.54	150.04	1.97	190.69	2.47	236.19	1.98	188.91
Aug.	2.71	247.68	2.81	257.91	1.91	176.93	2.37	218.62	2.72	249.59	2.27	209.07
Sep.	2.75	228.26	2.58	217.47	2.00	170.51	2.39	202.92	2.64	222.79	2.02	171.29
Oct.	3.31	271.44	3.06	251.00	2.86	234.09	2.95	240.93	3.48	285.75	2.67	218.72
Nov.	2.63	218.76	3.05	254.64	2.72	226.68	2.58	214.96	2.41	200.01	2.36	197.03
Dec.	2.67	216.29	2.89	237.65	2.98	242.67	2.97	243.05	2.52	202.94	2.64	214.38
Mean	2.78	243.88	2.89	254.17	2.45	214.61	2.64	232.10	2.73	239.93	2.39	210.91

Table 13: Performance of all models tested on ANEM demand data for 2010.

Month	K-means MER(%)	K-means MAE(MW)	SOM MER(%)	SOM MAE(MW)	Hierarchical MER(%)	Hierarchical MAE(MW)	K-medoids MER(%)	K-medoids MAE(MW)	Fuzzy C-means MER(%)	Fuzzy C-means MAE(MW)	PFEM MER(%)	PFEM MAE(MW)
Jan.	2.09	364.46	2.16	376.11	1.61	281.01	1.85	321.80	2.13	370.98	1.53	304.54
Feb.	1.85	318.37	1.79	309.31	1.68	289.51	1.66	287.60	1.81	312.67	1.61	275.35
Mar.	2.06	323.03	2.49	391.73	2.28	357.84	2.26	352.95	2.47	387.77	2.12	330.70
Apr.	1.80	256.00	2.79	404.61	2.02	291.81	1.95	279.72	2.75	396.48	1.82	259.58
May.	2.57	397.83	2.92	456.09	2.59	397.54	2.48	383.41	2.82	443.48	2.37	370.40
Jun.	2.76	439.81	2.46	398.17	2.38	367.08	2.41	386.29	2.79	451.16	2.25	358.47
Jul.	3.09	548.15	3.42	612.35	2.96	525.82	3.23	564.81	3.74	671.03	3.09	549.91
Aug.	2.77	478.41	3.02	533.66	2.76	480.63	2.81	484.82	3.16	562.05	2.63	462.74
Sep.	2.88	441.52	2.75	420.97	2.58	395.93	2.64	403.06	2.71	415.73	2.38	361.03
Oct.	1.67	242.83	1.98	292.05	1.85	270.93	1.83	266.62	2.25	329.57	1.66	243.50
Nov.	2.09	326.84	2.42	378.94	1.85	290.25	2.33	365.46	2.51	391.24	2.06	322.79
Dec.	1.97	332.63	2.08	352.31	1.79	301.48	1.94	327.58	1.98	333.53	1.85	304.69
Mean	2.32	372.48	2.52	410.51	2.18	354.15	2.28	368.670	2.59	422.14	2.10	345.30

Table 14: Performance of all models tested on IESO demand data for 2010.

A Cloud-Based Consumer-Centric Architecture for Energy Data Analytics

Rayman Preet Singh, S. Keshav, and Tim Brecht
{rmmathar, keshav, brecht}@uwaterloo.ca
School of Computer Science, University of Waterloo,
Waterloo, Ontario, Canada

ABSTRACT

With the advent of utility-owned smart meters and smart appliances, the amount of data generated and collected about consumer energy consumption has rapidly increased. Energy usage data is of immense practical use for consumers for audits, analytics, and automation. Currently, utility companies collect, use, share, and discard usage data at their discretion, with no input from consumers. In many cases, consumers do not even have access to their own data. Moreover, consumers do not have the ability to extract actionable intelligence from their usage data using analytic algorithms of their own choosing: at best they are limited to the analysis chosen for them by their utility. We address these issues by designing and implementing a cloud-based architecture that provides consumers with fast access and fine-grained control over their usage data, as well as the ability to analyse this data with algorithms of their choosing, including third party applications that analyse that data in a privacy preserving fashion. We explain why a cloud-based solution is required, describe our prototype implementation, and report on some example applications we have implemented that demonstrate personal data ownership, control, and analytics.

Categories and Subject Descriptors

C.0 [**Computer Systems Organization**]: General—*System architectures*; D.2.11 [**Software**]: Software Architectures

Keywords

Home energy, data privacy, data analytics, third party applications, system architecture

1. INTRODUCTION

Utilities around the world are deploying "smart meters" to record and report energy consumption readings to utility central offices. This enables different prices to be charged for electricity based on the time of day and eliminates the cost

of a monthly visit by a meter reader. The time series of meter readings, originally meant only for customer billing, has unanticipated uses. On the one hand, customers who have access to their usage data can get a real-time, fine-grained view into their electricity consumption patterns. When suitably analysed, this can reveal potential cost savings and customized guidance on the benefits from energy conservation measures, such as installing insulation, solar panels, or purchasing energy-efficient products. On the other hand, this same data stream can reveal private information about the customer, for example, when they are home and when they are not, the appliances they own, and even, in some cases, which TV channel or movie they are watching [30, 39]!

Unlike traditional utility-centric approaches to data management in the smart grid, we instead take a consumer-centric approach [37]. We believe that consumers would like to:

- have control over their own data while outsourcing data storage and persistence to an infrastructure provider

- get a single view into data collected from multiple sources

- give access to their data to analytic algorithms of their choice, but without giving up data privacy

These goals are not achieved by any existing solution. Today, many utilities do not even provide consumers with access to their own usage data. Even the utilities that give consumers access to data, such as those participating in the Green Button initiative [6], or those that provide rudimentary analytics, do not allow consumers to use analytic algorithms of their own choosing. Finally, no current system gives consumers fine-grained control over who can access the data, and the granularity and period of time at which it can be accessed.

Building on the rich infrastructure of modern clouds, we have designed and implemented cloud-based personal data and execution containers that persistently store data and offer an environment for the execution of arbitrary analytic algorithms. Consumers can use these containers to grant fine-grained access to their data to third parties. These containers also allow secure and private control of home appliances from any Internet-enabled device.

The key contributions of our work are:

- The design of a system that allows consumers to own and control access to their energy usage data and have it analysed using algorithms of their choice

- A proof-of-concept implementation of our architecture on modern cloud computing platforms

- An evaluation of the system architecture with respect to data access and control

The remainder of the paper is organized as follows. Section 2 describes related research. Section 3 outlines the goals and requirements of our system and Section 4 explains the rationale for our architecture. Section 5 presents a detailed description of the architecture followed by a description of our prototype implementation in Section 6. We evaluate our architecture in Section 7, discuss practical implications in Section 8 and limitations in Section 9. Our conclusions are presented in Section 10.

2. RELATED WORK

We group related work into the following categories: Personal Data, Energy Data, Energy Data Privacy and Systems Architectures.

Personal Data: Researchers have proposed ecosystems built around an individual's data, such as health records, smart meter data, data concerning banking, taxation and shopping [41]. McAuley et al. [38] introduce the concept of *dataware* that defines the processes of obtaining, accessing and using an individual's data. Haddadi et al. [31] report that the ethical and legal consequences of gathering individuals' data are not yet fully determined, but it is understood that the individual co-owns any data concerning them. We focus on applying the concept of *privacy-preserving dataware* to an individual's energy data and investigate the goals, architecture, and mechanisms needed to implement such a system.

Energy Data: Currently, various utility and software companies provide consumers with access to their energy data through web portals. Examples include initiatives like Green Button [6], analytics providers like Opower [10], utility companies such as Waterloo North Hydro [14], San Diego Gas and Electric [13], and software projects like Google Powermeter [5] and Microsoft Hohm [8] (both now defunct). While these portals allow residential and commercial consumers to download data about their energy consumption (or *energy data*), the consumer is responsible for its long term storage and use. In many cases, the data is only available for a limited time (e.g., three months [14]) and hence such portals do not provide consumers with a durable storage solution. Secondly, the data analytics available to consumers is at the utilities' discretion. As a result they are deprived of potentially better analytics through third-party applications. We focus on building a platform to circumvent these problems.

Energy Data Privacy: Analysing smart meter data in a privacy-preserving fashion has been the focus of much research. Most work has focused on applying obfuscation, aggregation and homomorphic encryption to energy data [18,29,35,36,47,49]. Other work develops cryptographic protocols for achieving the same goal [39,44]. Shi et al. [51] propose a cryptosystem where an *aggregator* can compute the sum of multiple energy values from their ciphertexts, without access to the individual energy values because they have been encrypted under different keys. Rajagopalan et al. [43] propose a framework to quantify the privacy and usefulness of energy data and propose a model to control the tradeoff between them. This motivates us to build a system which

employs this work for privacy-preserving application development for energy data.

System architectures: Previous work has focussed on energy data management for commercial buildings and office spaces, while aiming to achieve extensibility, scalability, and/or performance [11, 19, 25, 48]. Such systems are designed for users with expertise in the understanding of energy data (e.g., people specializing in building operations or energy managers). Residential consumers are unlikely to possess such levels of expertise, thus necessitating privacy-preserving third party applications (e.g., energy data analytics). To our knowledge our consumer-centric approach has been overlooked by existing work. Secondly, existing systems do not allow fine-grained access control over data streams, which is essential for privacy-preserving sharing of data.

Many other consumer-centric solutions [24,28,40,42,52,57] target various forms of personal data (e.g., healthcare, energy, mobile sensors, photos, videos). They provide data consolidation and ownership by aggregating data, but require exposing data to third parties thereby putting privacy at risk. Other work addresses data transformation before releasing it to third parties [20, 33, 34], consequently gaining privacy at the cost of inhibiting applications that require access to raw data. Càceres et al. [23,50] found that consumers' interests are best served by hosting their data on *virtual individual servers* in the cloud. We extend this approach to enable the in-depth analysis of consumer data such as time-series consumption data, by third party applications while preserving the privacy of the consumer.

3. GOALS AND REQUIREMENTS

Our main goal is to design a system that allows consumers to aggregate their data from multiple sources, control how that data is accessed and shared, and to allow them to quickly and easily access that data from any device, at anytime, from anywhere. These top-level goals translate into the following subgoals.

Consolidation: To allow a single view into multiple data streams and cross-correlation between different time series, the system should automatically consolidate energy usage data from multiple sources.

Durability: To allow analysis of usage history, a consumer's energy data should be always available, irrespective of its time of origin.

Portability: To prevent lock-in to a single provider, data and computation should be portable to different cloud providers.

Privacy: To preserve privacy, the system should allow a consumer to determine which other entities can access the data and at what level of granularity.

Flexibility: The system should allow consumers a free choice of analytic algorithms.

Integrity: The system should ensure that a consumer's energy data has not been tampered with by a third party.

Scalability: The system should scale to large numbers of consumers and large quantities of time series data.

Extensibility: It should be possible to add more data sources and analytic algorithms to the system.

Good Performance: Data analysis times and access latencies should be minimized.

Universal Access: Consumers should be able to get real-time access to their data on their Internet-enabled mobile devices.

4. DESIGN RATIONALE

We now describe the high-level rationale for our design by considering and eliminating several alternative approaches. The essential elements of any system that stores and analyses energy usage data are a data store, denoted D, and an application runtime that is the locus of execution of analytic algorithms, denoted AR.

The simplest possible system is one where the consumer stores data in their own home and uses a home-based computer for running data analytics. This case is shown as Case I in Figure 1. This solution provides portability, privacy, flexibility, and a certain amount of scalability, and extensibility. However, it requires consumers to be responsible for data collection and consolidation, and ensuring data durability (unfortunately very few consumers routinely backup their data). It also assumes that users' home computers are powerful enough to run sophisticated analytic algorithms over large data sets, which may not necessarily be the case, especially with the increasing proliferation of tablet devices. Moreover, if the home computer is behind a firewall, the solution does not provide good performance or universal access. For these reasons we do not believe this simple solution meets our design goals and the desires of consumers.

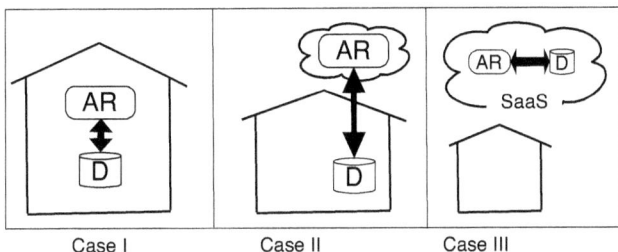

Figure 1: Solutions for energy data management. D: energy consumption data, AR: application runtime.

Consumers can avoid placing the computational burden on their home machines while, to some extent, preserving data privacy by storing data locally and sending data to analytic algorithms running in the cloud. This is shown as Case II in Figure 1. To preserve data privacy, data may be encrypted in a way that allows operations on ciphertext [51], or randomized values may be added to the data (i.e., dithering) to obfuscate details of consumer behaviour. This approach, however, allows a limited types of data computations and suffers from many of the same problems as the prior solution: the need for consumers to manage consolidation and durability, and the potentially poor performance.

Yet another approach would be to place both the data and the application runtime in the cloud using the "Software as a Service" (SaaS) approach. This is shown as Case III in Figure 1. Here, the SaaS provider would provide data consolidation and durability, freeing the consumer from these responsibilities. This approach, typified by the Microsoft Hohm [8] and Google Powermeter [5] approaches (both defunct), fails to provide privacy, extensibility, and flexibility, but does provide good performance and universal access.

Learning from the pros and cons of these three solutions, our goal is to provide a design that supports privacy, flexibility, and extensibility of data storage in the home combined with the consolidation, durability, good performance and universal access that can be obtained from a cloud-based solution. Specifically, we propose that a consumer shall have access to a "virtual home" or **VHome** that provides both data storage and an application runtime. Critically, the data access policies for data stores in the VHome are controlled not by the cloud provider, who would only be providing "Infrastructure as a Service" (IaaS), but by the consumer. As we demonstrate in Section 7 this solution meets all the design goals presented in Section 3.

By keeping the data and application runtime resident in the cloud, our solution allows VHome providers to support data consolidation and durability. Cloud-based data storage also allows low-latency universal access to the data, and relieves consumers of data consolidation and warehousing tasks. However, because consumers own their VHomes, they do not lose privacy or flexibility. We have engineered our solution for scalability, extensibility, and portability, as discussed later in the paper, thus meeting all of the design goals.

In the remainder of the paper, we present the details of our design, describe some applications we have implemented and evaluate whether or not our approach is successful in meeting our goals.

5. ARCHITECTURE

This section describes the architecture of our system. Details of our prototype implementation can be found in Section 6.

Figure 2 shows an overview of our system. It has four main components, from left to right, (a) the home-resident *gateway* (labelled Gateway), (b) the *virtual home* (labelled VHome) hosted by an SaaS provider in an IaaS cloud, (c) *cloud-based applications* (labelled CBA) also hosted in the cloud by other SaaS providers, and (d) *User interfaces* (labelled Remote UIs) for access to the gateway and the VHome from Internet-based devices. We now discuss each component.

5.1 Gateway

The gateway is a home-resident and consumer-controlled architectural element that provides two main services. First, it collects home energy production and usage data and uploads it over a secure connection to the cloud-based virtual home. Second, it provides an interface to allow the home owner to control devices in the home from Internet-connected devices.

The gateway interacts with smart appliances and monitors that are already network-capable and also, using add-on hardware such as Internet-controlled power strips, with legacy devices. Communication typically uses one or more types of channels such as USB, Zigbee, Ethernet, WiFi, RPL [56], or ZWave [17]. Usage data is uploaded from the home to the virtual home over a secure communication channel. This assures data durability and relieves the consumer from the need for data warehousing.

The gateway authenticates remote users and accepts control commands from them. These control commands either configure the gateway, request data uploads, or request that actions be taken by appliances and devices. A gateway may

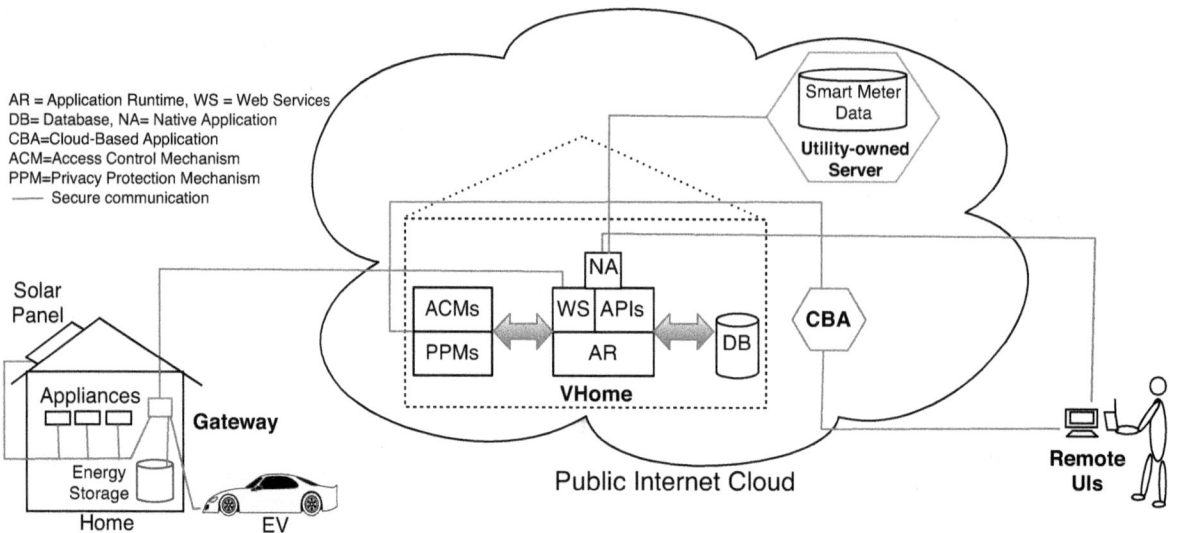

Figure 2: System Overview.

5.2 VHome

A virtual home or VHome is a virtualized execution environment hosted in a cloud-based server that provides three services: (a) storage for home energy use data, (b) an application runtime for executing applications that analyse this data, and (c) trusted web-based services for interaction with the gateway, other cloud-based services, and user devices (described in more detail below). A VHome is owned by the consumer and hosted by a VHome SaaS provider in an IaaS cloud [1]. We describe the participation incentives for the consumer and these providers in Section 8.

A VHome is built from a virtual execution environment (VEE) provided by an IaaS provider. This could be a virtual machine [21] or a virtual container (private server) [53]. In Figure 2, the virtual execution environment is shown as a dotted home. Within the VEE, AR denotes the application runtime (such as a Java Virtual Machine) and DB denotes data stored in a database.

We envision that data stored in the database can be accessed by two types of applications. *Native* applications run on the VHome AR, and are certified to be "safe" using an approach described in more detail in Section 5.3. In contrast, *cloud-based applications* (denoted as CBA in Figure 2) pull energy data out of the VHome, which may violate consumer privacy. Therefore, access by a CBA to private data is mediated by privacy protection mechanisms (PPMs) that pre-process data before it is transferred out of the VHome. PPMs can implement privacy models such as differential privacy and k-anonymity [55] by employing mechanism such as obfuscation, noise addition, and homomorphic encryption [29]. An example of a PPM is to add random noise values to a meter reading, with the amplitude of the noise decreas-

ing with reading granularity, so that monthly readings may have little or no added noise, but per-second readings would have large amounts of added noise. Access Control Mechanisms (ACMs) additionally allow consumers to restrict and revoke CBA access to the APIs by scope and duration. For example, an ACM may allow a CBA to access only hourly meter readings and only from a specified day of the year. Moreover, this access may expire after 15 minutes.

In addition to the native and cloud-based applications, our system contains special-purpose trusted applications we call *Web Services (WS)*. As a trusted component of the VHome they have free access to the APIs and hence to the energy data. They perform three tasks. First, they periodically accept data (typically, but not necessarily, bulk data) uploaded from the gateway and store it in the database. Second, they fetch real-time data from the gateway when requested by the consumer. This allows bulk data to be transferred from the home to the VHome once a day, yet provide real-time data access when necessary. Third, they provide a control interface to the consumer for various administrative tasks, such as downloading and running native VHome applications, configuring ACMs, configuring Privacy Protection Mechanisms (PPMs), requesting VHome software updates, the migration or discarding of data, and configuring gateway actions.

5.3 Applications

We now discuss native and cloud-based applications in more detail. Note that the main difference between native and cloud-based applications is that native applications execute in a tightly-controlled runtime environment. Moreover, their bytecode is available for analysis. This allows the system to eliminate the privacy leakage that is possible due to these applications. In contrast, cloud-based applications cannot be tightly controlled. Therefore, the only way to preserve privacy when giving data to these applications is to modify the data itself, which we accomplish using the PPMs.

We envision that both classes of applications would be developed by third-party developers, much like those who participate in Apple's App Store. Developers would use standardized APIs, such as those described next, to access con-

be a dedicated, networked hardware device, or an integrated part of other home services' hardware such as a cable modem or set-top box, or it may be software deployed on a household computer.

---

[1]This separation could be used in many other domains such as data management in healthcare, or banking.

sumer data. Consumers would either download native applications to the VHome to execute within the VHome runtime or can use ACMs to give cloud-based applications access to their data (after processing by PPMs). Applications can have user interfaces (UIs) to enable their invocation from PCs, smart phones, or other Internet-enabled devices.

Native Applications (NAs): The leakage of private data from native applications can be restricted using one of the following approaches. In the first approach, a native application's executable is scanned to assure consumers that the application is incapable of network communication. Thus, the application cannot leak data out of the container, which guarantees privacy. In our preliminary implementation, we restrict native applications to be written in Java and not invoke native APIs. Our current thinking is that an application can be certified as safe if its bytecode does not use the Java.net API. This can be easily checked when a native application is submitted for inclusion into the application store.

The second approach is used for native applications that need to use the network API to access remote hosts, such as to scrape consumer energy data from a utility website. To deal with such applications, network communications from a native application are restricted to a specific IP address (or host name). For instance, a native application could be restricted to communicate only with the host name corresponding to a utility's web server. Moreover, read or write access from a native application to a database table can also be restricted. In the example data scraper application, it could be restricted to only write to the database, not read from it. As we show in Section 6, these restrictions on database access are easy to accomplish in our system. We can also restrict the set of web services APIs that Native Applications can access. This further limits their ability to compromise privacy.

Certified native applications are suitable for data mining, analytics, visualization, appliance control and home automation. Native applications can also obtain consumers' energy data from utilities and store it in the VHome DB, making them ideal for data consolidation, (e.g., maintaining copies of consumers' smart meter data recorded by their utility companies).

Cloud-Based Applications (CBAs): Unlike native applications, cloud-based applications are hosted using third parties' hosting services. The main purpose of a cloud-based application is to allow sharing and comparison of energy data between different VHomes. ACMs provide fine-grained access control over data (e.g., time series) which means VHome owners chose which part(s) of her data are shared with a CBA, and when is it shared (e.g., periodically). The challenge here lies in preserving privacy while allowing meaningful computations and comparisons. While certified native applications can be given access to raw data, data given to a CBA must be pre-processed using techniques that ensure that privacy is preserved. Examples of such pre-processing include obfuscation, noise addition and homomorphic encryption [29]. These actions are implemented by and configured using the PPMs. Similar to NAs, CBAs can be published on the application store, and VHome owners provide CBAs with their VHome URL and explicit authorization to read all or parts of their data.

5.4 User interfaces

The gateway, a VHome's web services (WS), and cloud-based applications all allow user interaction. These interactions are mediated using user interfaces implemented on a user device, such as a web browser, a mobile application, or other mediums like e-mail or SMS. User interfaces simplify the management and use of VHomes and applications using graphical interfaces. Examples of such user interfaces are those used to download native applications to a VHome, to configure the permissions granted to a CBA by a consumer, and to control appliances in the home from a mobile device.

6. IMPLEMENTATION DETAILS

This section presents the details of a prototype implementation of our system.

6.1 Gateway

We implement a software-based gateway using the Microsoft HomeOS [26] platform. HomeOS is a .NET based platform designed to provide centralized control of devices in the home (such as light switches, thermostats, cameras, and televisions). It provides developers with homogeneous abstractions to orchestrate such devices. We use these features for monitoring and controlling appliances and to enable the uploading of data to the VHome.

Figure 3 provides an overview of HomeOS [27]. It is comprised of software modules called *drivers* that communicate with devices and allows higher level modules (such as applications) to actuate the devices. Additionally, a *platform* module manages and coordinates all other modules. In our gateway we extend HomeOS by implementing some additional modules, described next.

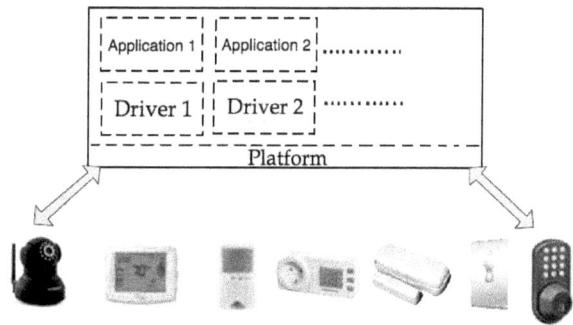

Figure 3: Overview of the HomeOS Platform.

Driver Modules
Each driver module monitors and controls an individual appliance using a sensor. We implement driver modules for the Aeon appliance sensor [1] and the CC Envi [4] power and temperature sensor. The Aeon sensor is installed in-series with an appliance and communicates to the gateway over Z-Wave [17]. The module is invoked by the coordinator module (described later) for polling data or controlling the sensor, and transmits the respective Z-Wave frames to the desired sensor. The Envi sensor measures the active power from a home every 6 seconds using a Hall-effect [7] transducer that is clipped around the split-phase wires at the home's main electricity supply. Measurements are transmitted wirelessly to the Envi console which is connected via a USB cable to

the gateway machine (an inexpensive netbook in our proto-type). The netbook stores the data on disk and transmits it periodically (e.g., once per day) to a VHome. Figure 4 shows the netbook running the gateway software, along with the Envi console.

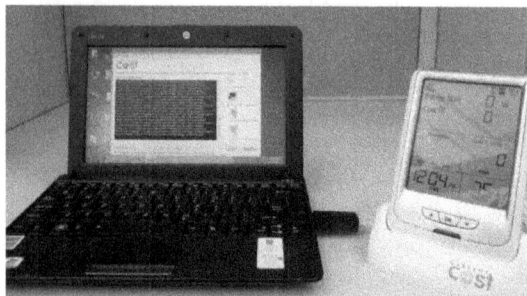

Figure 4: Gateway and Envi console.

Communication Module

This module provides communication between the VHome and the gateway. We considered several alternatives for communication including TCP, HTTP, and SSH. In the end, we decided to use XMPP [54], the protocol underlying the Jabber chat client, as our transport protocol because it uses a simple RPC mechanism that is secure, extensible, and provides real-time communication. Most importantly, XMPP ensures seamless communication from the VHome to the gateway despite the presence of NAT devices and firewalls at the home gateway.

Coordinator Module

This module records and processes energy data generated by the sensors' driver modules and caches it temporarily on the gateway. Periodic data uploads are received by the VHome's web services and are not sensitive to intermittent losses of connectivity. As a result, transient losses in network connectivity, lasting less than a few days, are easily tolerated. To facilitate coordination of sensor data and control between the gateway and the VHome, each sensor is assigned a *class ID* and an *object ID*. Sensors of the same type are assigned with the same class ID, but distinct object IDs. This allows the VHome to identify each sensor using the {class ID, object ID} tuple. The coordinator module uses the communication module to listen for commands from the VHome to control sensors. For example, if Aeon sensors have the class ID 1, and the one interfaced with the electric heater has an object ID 2, then the VHome issues the following command using XML to order the gateway to turn it off.

```
<setStatus classID=1 objectID=2>
  <power>0.0</power>
</setStatus>
```

The gateway performs the action and responds with the sensor's new status as an acknowledgement. Similarly other actions (e.g., dimming lights, managing AC temperature setpoints) can be performed using the VHome.

6.2 VHome

Each component of our prototype VHome, (e.g., the APIs, Web Services (WS), Access Control Mechanisms (ACMs), Privacy Protection Mechanisms (PPMs), and Native Applications (NAs)), is implemented as a Java Web Application

(or *webapp*) using the Java API for RESTful Web Services (i.e., JAX-RS framework) [22], and can be deployed in a Java Web Container. We use Apache Tomcat as the web container, and MySQL as the relational datastore which we instantiate in a virtual machine using the Amazon EC2 [2] cloud. Our choice of Java was calculated to ensure VHome portability across cloud providers (e.g., Windows Azure [15] and RootBSD [12]) and our implementation is configurable to use any relational cloud datastore (e.g., SQL Azure [16] and Amazon RDS [3]).

Similar to the organization used for sensors (described in Section 6.1), data is organized into classes where each class describes a unique type of data stream and has a unique class ID, a class name (e.g., heating), a descriptor (e.g., space heaters in the home), and a rating (e.g., 500 W). Data streams either emanate from the sensors at home or could be external (e.g., smart meter readings from utility's website). Particular streams of a class are identified as objects using a unique object ID within their class, and have an object name (e.g., master bedroom heater), a descriptor (e.g., installed on 01/01/2011, warranted until 01/01/2020) and a granularity (e.g., 60, indicating data is produced every 60 seconds).

Privacy Protection Mechanisms (PPMs) are implemented as webapps and access a data stream or streams via APIs. They can create new privacy-preserving streams, which can then be shared with cloud-based applications (CBAs). For instance, we implement *aggregation* as an example PPM where energy consumption time series data (e.g., produced every second) is aggregated to compute daily or weekly consumption values which are less revealing in nature.

The webapp instantiating the VHome APIs implements them as a set of TLS-Secure Representational State Transfer (REST) [45] URIs, which are then used by native and cloud-based applications to access data or control sensors and thus appliances. Table 1 provides a brief overview of the APIs' URIs, which native and cloud-based applications can invoke using HTTP GET or POST requests. Results are returned using JavaScript Object Notation (JSON). Applications can add or modify data streams subject to the Access Control Mechanisms (ACMs). Applications can potentially compute a hash of a data stream (e.g., MD5) and sign it (e.g., using user's private key) to ensure data integrity to some extent.

The ACM webapp regulates applications' access to all or a subset of APIs, configured using the WS. By default all APIs are regulated and therefore, require a valid access token to return results. A VHome owner could decide to not regulate the *ListAllClasses* API, which in turn could result in a privacy breach. The ACM webapp implements OAuth 2.0 [32], a token based authentication and authorization standard for securing API access. It uses the VHome DB to store data concerning access controls (e.g., tokens, access lists, and more) which is only accessible to the ACM webapp.

Figure 5 illustrates this access process for a cloud-based application (CBA) hosted as a web portal. To access any API, the CBA is first required to obtain a *one time authorization grant* from the ACM webapp by providing its identity (identifier, name, or host-URL) and a list of APIs that it requires access to and the parameters to the APIs. For instance, a CBA requiring access to the bedroom space heater consumption data (e.g., with class ID 1 and object ID 2) from January–March 2012 would request access to the data stream API as:

Function (regulated by default)	Description
ListAllClasses	Returns all attributes of all classes of data in the VHome DB.
ListClass/param/value *param*: class ID, class name or rating.	Returns all attributes of class with given parameter values.
ListObject/param1/value1/param2/value2 *param1*: class ID, class name or rating. *param2*: object ID, object name or granularity.	Returns all attributes of object with given parameter values.
AddClass/className/x/descriptor/y/rating/z	Adds class with given class name, descriptor, and rating.
AddObject/classID/x/objectName/y/descriptor/z/granularity/w	Adds object with given class ID, object name, descriptor and granularity.
AddStream/classID/x/objectID/y/	Adds time series data to the given data stream.
GetStream/classID/x/objectID/y/	Returns the complete time series data stream.
GetStream/classID/x/objectID/y/TS/t_1/t_2	Returns the complete time series (or TS) between timestamps t_1 and t_2.
GetStream/classID/x/objectID/y/Val/v_1/v_2	Returns the complete time series between data values (or Val) v_1 and v_2.
GetStatus/classID/x/objectID/y	Returns the current power consumption of device with given class ID and object ID.
SetStatus/classID/x/objectID/y/status/p	Sets the power consumption of device with given class ID and object ID to p.

Table 1: VHome API used by NAs and CBAs to access data.

```
https://<VHome URL>/GetStream/classID/1/
        objectID/2/TS/1325394000/1333252799
```

where TS indicates time series data, and 1325394000 and 1333252799 are the epoch timestamps at 01-01-2012 00:00:00 and 31-03-2012 23:59:59, respectively.

This allows restricting the scope of data access to certain data streams and/or certain segments of a stream's time series defined using timestamp and/or data values. The ACM webapp then redirects the user to the Web Services (WS) webapp so as to authenticate the user as the VHome owner. After authentication, the scope and nature of the requested access is described to the user, and her authorization for the access is requested. The WS implements this as a simple notification in a web-browser, which can be relayed to other remote UIs such as email, SMS, or mobile application notification. An example of such a notification from an application named "EXAMPLE" is:

> The application named EXAMPLE is requesting
> access to bedroom space heater data
> for Jan 1 to Mar 31, 2012.
> Allow or Deny ?

The CBA then has to exchange the one time authorization grant before it expires and obtain an *access token* and an (optional) *refresh token*. By using the access token, a CBA can use the required APIs until the token expires after which a new access token may be obtained using the refresh token. All tokens are valid for periods configured by the owner of the data. By matching the CBA's credentials (e.g., URL) to those registered while issuing the authorization grant the ACM validates each API access and prevents use of stolen access tokens. Further, if at any point the user decides to revoke (or pause) a CBA's access to data, she can simply revoke the access token and possibly the refresh token for that CBA. Our prototype implements the authorization grant, access token and refresh tokens in the form

of randomized 128-bit MD5 [46] codes, where the webapp maintains a lookup table for storing their scope and expiry times. Avoiding the encapsulation of scope and duration within the token circumvents token processing overhead for each API access. The authorization grant and access token issuing endpoints are published as GET/POST URIs by the VHome, and use JSON for token and error-message exchanges with CBAs. Our prototype implementation assumes one user per VHome, thus managing multiple users is not currently addressed.

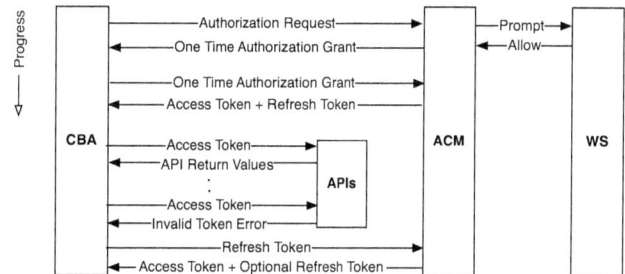

Figure 5: API access for a CBA.

The Web Services (WS) webapp is critical to a VHome. It enables a number of components and allows users to configure them. Firstly it coordinates periodic data uploads from the gateway over XMPP and transmits control commands to the gateway's coordinator module. Secondly it provides the consumer with a control portal to install native applications on the VHome. Native Applications, being JAVA web applications, are then profiled by the WS webapp for use of the JAVA.net interface and for the VHome APIs they require. The owner of the data (the consumer) can restrict the applications' ability read from and write to the database by disallowing or restricting the scope of the APIs. Likewise, consumers can configure ACM settings such as token formats and expiry periods, and chose which APIs it restricts. Similar to native applications, certified PPMs can be added to the VHome through this portal which can be run to create additional data streams. Lastly, the WS webapp allows users to purge native applications or data, and revoke access tokens of any CBA they want.

Our prototype VHome implementation is open source and can be found at `https://vhome.codeplex.com` while the gateway implementation using HomeOS is hosted at `http://homeos.codeplex.com/`.

6.3 Applications

We have created a sample application store as a web portal where users browse for native applications. It transfers the desired native applications' executables to the VHome' WS which installs them on the VHome, and can then be accessed using remote UIs. We now describe a few applications that we have built using our system which, without our architecture, cannot be implemented in a data privacy preserving form.

Data Scraper
This application obtains consumers' smart meter data from the utility provider and stores it in the VHome DB. It is implemented using the JAVA DOM interface as a VHome native application. Our prototype application is downloaded from the sample application store, to run on the VHome, where it scrapes data from the utility company's web portal and stores it in the VHome DB. When first using this application it obtains, from the consumer, their identification and password used to access the utility company's portal. The utility company in our prototype is Waterloo North Hydro [14]. The application also allows consumers to set automated periodic data scraping actions to ensure that data is obtained before it is discarded by the utility company's portal (i.e., after three months) and the consumer is relieved of manually retrieving the data. The application allows the data to be retained by the consumer even after it is no longer available on the utility company's portal. The data is stored as a data stream in the *Smart Meter* class which can then be accessed by other applications through APIs. Access to more than a year's worth of data can provide excellent opportunities for data analytics because in many climates seasonal changes must be accounted for when examining consumption histories. This application demonstrates how our architecture allows consumers to take ownership of their data, thus meeting the goal of *data ownership*.

Interactive Monitoring and Control
We have implemented a VHome native application that interfaces with the VHome web services to allow the consumers to use a web-browser to monitor and control home appliances in real-time. Native applications have no network access and can only be viewed by invoking the trusted VHome webapp container. We implement an Android smartphone application that invokes the VHome native application via the VHome webapp container and provides a smartphone application interface. This means consumers can use VHome native applications via web-browsers or with applications installed on their mobile devices (e.g., smartphones, or tablets). Figure 6 shows snapshots of different panels in the Android smartphone application. Screenshot-1 (on the left) shows the home monitor, which allows consumers to view current conditions of the home as reported by the CC

Envi console. In addition, the consumption data stored at the VHome is used to compute and display the day's and week's consumption. Screenshot-2 (in the center) shows the control panel which allows users to turn on or turn off different appliances connected to Aeon ZWave sensors and displays their current consumption. Further, it allows users to share the amount of energy they conserve by turning off appliances, on social networks such as Facebook and Twitter and to compete with their friends. This applications permits consumers to define events for which they wish to receive notifications. For instance, consumers can set a threshold for energy consumption and "abnormal consumption" notifications are issued if it is exceeded. Screenshot-3 (on the right) shows a past trend of aggregate energy consumption measured using the CC Envi sensor. This trend data can be used by consumers to better understand abnormal consumption notifications. Our prototype implements SMS, E-Mail and mobile application notifications. Since the VHome is cloud resident, energy data can be processed in the cloud and viewed using any Internet-enabled device, with relatively low latency.

Energy Data Analytics
In many parts of the world the price of electricity depends on the time of the consumption. In Ontario, a day is divided into peak, mid-peak, and off-peak hours, each with different rates [9]. We implement a VHome native application that processes a home's electricity consumption measured using the CC Envi sensor to determine how much energy is consumed during different hours of the day, its respective costs under the pricing scheme, and the total cost. It uses smart meter data obtained and stored by the data scraper application to verify a consumer's utility bill. Such simple analytics also provides consumers with meaningful insight into their hourly and daily consumption patterns, warns them of potential errors in their utility bills and can help them to time shift non time-critical consumption. This shows how our architecture meets the goal of *data analytics*.

Abnormal Energy Consumption Detection
We implement a VHome NA which informs consumers about abnormalities in their energy consumption. For instance, consider a scenario where residents forget to turn off their oven while they are away. Using the VHome APIs the application periodically obtains the energy consumption values from the gateway, measured by the CC Envi sensor. It then compares the values to a predicted value computed using an Auto-Regressive Moving Average model. If the measured value is higher than the predicted value by a threshold (e.g., 1 kW) then the application sends the consumer a notification message via e-mail, SMS, or the Android smartphone application. The consumer can then either use the Android application (described above) or reply to the email, SMS to take appropriate action. We defer the use of more complex abnormality detection techniques to future work.

7. EVALUATION

In this section, we compare our architecture to the different existing and proposed systems that can be used to store and analyse energy data. Table 2 compares these solutions by denoting which of the requirements of a consumer-centric solution they meet (i.e., the goals from Section 3). Commercial software solutions–Google Powermeter [5] and Microsoft Hohm [8] being centralized web services are scal-

Figure 6: Three screenshots of the Android smartphone application.

able but provide a fixed set of analytics with no data consolidation or privacy. Both services are now defunct leaving consumers with no data access. Utilities' web portals act similarly and share/discard data at their discretion but can ensure integrity of only smart meter data. Opower [10] provides consumers with some data analytics on their monthly utility bills, but they provide no real-time access to data or choice of analytics. The Greenbutton [6] initiative has standardized energy data formats, so that consumers can access their data and analyse it themselves (denoted "Greenbutton (Self)"). This allows consumers to choose analytics tools, run integrity checks on data, and provides data privacy, but it burdens them with data maintenance. Alternatively, consumers can delegate the data retrieval and maintenance to third parties (denoted "Greenbutton (Third Party)"). Third parties can manage, analyse and host consumers' data, but such unconditional access to raw data provides little data privacy.

Our architecture meets the requirements for a consumer-centric, privacy preserving, architecture for energy data analysis as explained below.

Consolidation: Native applications allow data to be read from any data source and stored in the VHome database. This allows consumers to easily consolidate their data from multiple sources by using one native application per data source.

Durability: The cloud-based storage of data allows for data durability. Instead of relying on a single computer in the consumer's home to store data, and relying on the consumer to remember to back up that data and to ensure that they have off-site backups, data is stored in the cloud. The data on cloud-based storage is replicated, backed-up, and maintained by professionals, guaranteeing durability.

Portability: Our solution has been designed to be portable across a variety of cloud-based providers and databases. This ensures application and data portability, as discussed in detail in Section 6.

Privacy: Our system provides data privacy in several ways. First, data within a VHome is not accessible by entities outside the VHome, eliminating most types of

privacy violations. Privacy leakage from native applications is prevented by certifying native applications, by checking Java byte code submitted to the application store to ensure that they are either not using network APIs, or when they need to, are only communicating with the specified hosts. The details of this process are described in Section 6.2. Privacy leakage from cloud-based applications is mitigated, to some extent, by data encryption or obfuscation by privacy preserving mechanisms. Of course, this protection of data privacy assumes that Iaas and VHome SaaS providers are trusted.

Flexibility: Our system allows consumers to download and install analytic algorithms of their choice. They can also send data to be processed by any cloud-based application by giving them time-limited, scope-limited access to their data.

Integrity: Native applications allow integrity for data stored directly in the VHome (e.g., via gateway) but cannot ensure integrity for data imported from external sources (e.g., smart meter data from utility company servers).

Scalability: Cloud-based servers allow massive scaling of both data set sizes and computation, unlike the use of home-based computers.

Extensibility: The native application store allows a consumer to extend their VHome with new analytic algorithms. Consumers can also send their data to any cloud-based application for analysis.

Good Performance: The use of a cloud to store data minimizes data access latencies by avoiding the use of the typically lower-bandwidth home access link. In addition it provides access to server systems with more memory and processing power than may be available on many consumer's home machines.

Universal Access: Cloud-resident data allows consumers to get real-time access to their data on their Internet-enabled mobile devices.

Thus, our system meets all of our design criteria, while our prototype implementation demonstrates the feasibility

Goals \ Solutions	Hohm [8], Powermeter [5]	Utility web portals [13,14]	Opower [10]	Greenbutton [6] (Self)	Greenbutton [6] (Third Party)	VHome
Consolidation				✓	✓	✓
Durability						✓
Portability				✓	✓	✓
Privacy				✓		✓
Flexibility				✓	✓	✓
Integrity		*		✓		*
Scalability	✓	✓	✓		✓	✓
Extensibility				✓	✓	✓
Performance	✓	✓			✓	✓
Universal access	✓	✓			✓	✓

Table 2: Comparative analysis with existing solutions, * denotes a partial solution.

of such a system using existing hardware, software and cloud infrastructure.

8. DISCUSSION

In some sense, ensuring that meter data remains private is moot, because utility companies already collect this data and share it with whomever they choose (e.g., Google PowerMeter and Microsoft Hohm) without seeking customers' permission. This situation, however, is likely to change in two ways in the future.

First, we anticipate that many jurisdictions, following the lead set by the province of Ontario (in Canada), will place severe restrictions on the sharing of meter data, thereby freezing innovation in data analytics and customized recommendations. Although this is being countered by proposals such as the Green Button initiative [6], which release usage data back to consumers, we believe that consumers are just not capable of doing their own data analytics, and are loath to share this data with third parties due to privacy concerns. Second, besides meter data, we anticipate that future consumers will generate many other equally private data streams including health-monitoring data. Our architecture balances privacy and innovation for applications that analyse these other data streams.

The entities that participate in our system are the utility companyies that collect smart meter data, consumers, IaaS providers, VHome SaaS providers, and application developers. We believe that each entity has an incentive to participate in the system.

Utilities: Utilities are under great pressure from legislatures to release usage data, as evidenced by the Green Button initiative. They also benefit from energy conservation measures, in that these reduce their need for costly generation capacity upgrades.

Consumers: Consumers are increasingly aware of the costs of the world's rampant energy consumption. In some cases consumers will be motivated to better understand and reduce their consumption by trying to improve the health of the planet and in other cases they will be motivated by trying to reduce their utility bills. They currently lack the infrastructure and tools to understand how to achieve reductions without giving up their privacy.

IaaS providers: IaaS providers are paid for their services, so they have a monetary incentive for participation.

VHome SaaS providers: We believe that VHome SaaS providers can be compensated for their efforts in two ways. First, some consumers may wish to pay for their own VHomes, so that they can maintain an archive of past usage and get recommendations for intelligent energy use. This is similar to those consumers who pay a monthly fee for services such as DropBox. Second, vendors of "green" energy-efficient products could subsidize the cost of VHomes, because recommendations for the use of their products, such as energy-efficient air conditioners, washing machines and LED lights, translate to increased sales.

Application developers: Application developers have the same incentives in our architecture as with the Apple App store: a mass audience for their work, so that a well-developed application can make its developer a lot of money. Certain applications may also be commissioned by equipment vendors, as discussed above.

In hindsight, our approach may appear to be obvious, merging the cloud with sensor data streams, an approach already implemented by systems such as Pachube [11]. However, there are three aspects of our work that are not obvious. First, we show how to use virtualized execution environments, in conjunction with an object-level framework, to provide practical solutions for the seemingly conflicting requirements of ensuring data privacy while fostering application development. Second, our approach enables the development of an ecosystem of energy management applications in much the same way as Apple's App Store provides an ecosystem for the development of iPhone and iPad applications. Third, our approach is diametrically opposed to the utility-centric view that is widely prevalent in the Smart Grid community. Instead of designing an architecture that caters to the needs of utilities, our approach places control firmly in the hands of consumers.

9. LIMITATIONS

Our prototype implementation demonstrates that it is possible to provide an infrastructure that enables the analysis of consumer energy consumption data while preserving their privacy but it suffers from certain limitations. Our current implementation is limited to dealing with time series and does not support other potential forms of energy data. Because it provides consumers with much greater control over their data, the consumer is faced with many decisions. Such cognitive overload may be eased by learning users' perceptions from user studies to help make decision-making simple and intuitive. Although fully realizable in our architecture, our current prototype implementation does not include mechanisms for ensuring integrity of data streams (e.g., by maintaing signed hashes). Our architecture also requires a trusted certification mechanism to certify applications and requires VHome SaaS and IaaS cloud providers to be non mallicious. Lastly, to gain data privacy, the consumer may have to bear the cost of hosting the VHome in the cloud and purchasing analytic applications.

10. CONCLUSIONS

We describe how cloud hosting services can be leveraged to ensure that consumers retain ownership and fine-grained control over their energy consumption data while enabling third party applications to analyse that data in a privacy preserving fashion. In addition, our cloud-based architecture provides applications with low-latency data access, long-term, durable storage, and access to the significant computational and storage resources needed to process growing volumes of energy data being collected. We believe our architecture is essential for the development, management and automation of applications that provide intelligent, privacy-preserving, energy data analytics. We defer the study of the scalability and performance of our prototype implementation for future work.

11. ACKNOWLEDGMENTS

The authors wish to thank the European Institute of Network Sciences (EINS), Natural Sciences and Engineering Research Council of Canada (NSERC), and MITACS for their financial support, and Ratul Mahajan from Microsoft Research for sharing HomeOS with us.

12. REFERENCES

[1] Aeon Labs Smart Energy Switch. www.aeon-labs.com.
[2] Amazon Elastic Compute Cloud (EC2). www.aws.amazon.com/ec2.
[3] Amazon Relational Database Service (RDS). www.aws.amazon.com/rds.
[4] Current Cost Envi CC-128. www.currentcost.com.
[5] Google Powermeter. www.google.com/powermeter.
[6] Green Button Initiative. www.greenbuttondata.org.
[7] Hall effect. http://en.wikipedia.org/wiki/Hall_effect.
[8] Microsoft Hohm. www.microsoft-hohm.com.
[9] Ontario time-of-use pricing. www.ontario-hydro.com.
[10] OPower Inc. www.opower.com.
[11] Pachube-Cosm Ltd. www.cosm.com.
[12] RootBSD Cloud Provider. www.rootbsd.net.
[13] San Diego Gas and Electric. www.sdge.com.
[14] Waterloo North Hydro Corp. www.wnhwebpresentment.com/app/.
[15] Windows Azure. www.windowsazure.com/.
[16] Windows SQL Azure. www.windowsazure.com/en-us/home/features/data-management/.
[17] Z-Wave Alliance. www.z-wavealliance.org.
[18] G. Ács and C. Castelluccia. I have a DREAM!: differentially private smart metering. In *Proc. of IH*, 2011.
[19] Y. Agarwal, R. Gupta, D. Komaki, and T. Weng. Buildingdepot: an extensible and distributed architecture for building data storage, access and sharing. In *Proc. of the Fourth ACM Workshop on Embedded Sensing Systems for Energy-Efficiency in Buildings*, BuildSys, 2012.
[20] P. Arjunan, N. Batra, H. Choi, A. Singh, P. Singh, and M. B. Srivastava. SensorAct: a privacy and security aware federated middleware for building management. In *Proceedings of the Fourth ACM Workshop on Embedded Sensing Systems for Energy-Efficiency in Buildings*, BuildSys '12, pages 80–87, New York, NY, USA, 2012. ACM.
[21] P. Barham, B. Dragovic, K. Fraser, S. Hand, T. Harris, A. Ho, R. Neugebauer, I. Pratt, and A. Warfield. Xen and the art of virtualization. *ACM SIGOPS Operating Systems Review*, 2003.
[22] B. Burke. *RESTful Java with Jax-RS*. O'Reilly Media, 2009.
[23] R. Cáceres, L. Cox, H. Lim, A. Shakimov, and A. Varshavsky. Virtual individual servers as privacy-preserving proxies for mobile devices. In *Proc. of ACM MobiHeld, 2009*.
[24] H. Choi, S. Chakraborty, Z. M. Charbiwala, and M. B. Srivastava. Sensorsafe: a framework for privacy-preserving management of personal sensory information. In *Proceedings of the 8th VLDB international conference on Secure data management*, SDM'11, pages 85–100, Berlin, Heidelberg, 2011. Springer-Verlag.
[25] S. Dawson-Haggerty, X. Jiang, G. Tolle, J. Ortiz, and D. Culler. sMAP: a simple measurement and actuation profile for physical information. In *Proc. of ACM SenSys*, 2010.
[26] C. Dixon, R. Mahajan, S. Agarwal, A. Brush, B. Lee, S. Saroiu, and V. Bahl. An operating system for the home. *Proc. NSDI*, 2012.
[27] C. Dixon, R. Mahajan, S. Agarwal, A. J. Brush, B. Lee, S. Saroiu, and V. Bahl. The home needs an operating system (and an app store). In *Proc. of HOTNETS*, 2010.
[28] C. Elsmore, A. Madhavapeddy, I. Leslie, and A. Chaudhry. Confidential carbon commuting: exploring a privacy-sensitive architecture for incentivising 'greener' commuting. In *Proceedings of the First Workshop on Measurement, Privacy, and Mobility*, 2012.
[29] F. D. Garcia and B. Jacobs. Privacy-friendly energy-metering via homomorphic encryption. In *Proc. of the 6th International Conference on Security and Trust Management*, 2010.
[30] U. Greveler, B. Justus, and D. Loehr. Multimedia Content Identification Through Smart Meter Power Usage Profiles. In *Proc. of CPDP*, 2012.
[31] H. Haddadi, R. Mortier, S. Hand, I. Brown, E. Yoneki, J. Crowcroft, and D. McAuley. Privacy analytics. *SIGCOMM Comput. Commun. Rev.*, Apr. 2012.
[32] E. Hammer-Lahav, D. Recordon, and D. Hardt. The OAuth 2.0 authorization protocol. *draft-ietf-oauth-v2-18*, 8, 2011.
[33] J. Kannan, P. Maniatis, and B.-G. Chun. A Data Capsule Framework For Web Services: Providing Flexible Data Access Control To Users. *CoRR*, 2010.

[34] J. Kannan, P. Maniatis, and B.-G. Chun. Secure data preservers forweb services. In *Proceedings of the 2nd USENIX conference on Web application development*, WebApps'11, pages 3–3, Berkeley, CA, USA, 2011. USENIX Association.

[35] Y. Kim, E.-H. Ngai, and M. Srivastava. Cooperative state estimation for preserving privacy of user behaviors in smart grid. In *Proc. of SmartGridComm*, 2011.

[36] F. Li, B. Luo, and P. Liu. Secure Information Aggregation for Smart Grids Using Homomorphic Encryption. In *Proc. of SmartGridComm*, 2010.

[37] W. Liu, K. Liu, and D. Pearson. Consumer-centric smart grid. In *Innovative Smart Grid Technologies (ISGT), 2011 IEEE PES*, pages 1–6. IEEE, 2011.

[38] D. McAuley, R. Mortier, and J. Goulding. The Dataware manifesto. In *Proc. of COMSNETS*, 2011.

[39] A. Molina-Markham, P. Shenoy, K. Fu, E. Cecchet, and D. Irwin. Private memoirs of a smart meter. In *Proc. of ACM BuildSys*, 2010.

[40] R. Mortier, C. Greenhalgh, D. McAuley, A. Spence, A. Madhavapeddy, J. Crowcroft, and S. Hand. The personal container, or your life in bits. *Digital Futures*, 2010.

[41] R. Mortier, C. Greenhalgh, D. McAuley, A. Spence, A. Madhavapeddy, J. Crowcroft, and S. Hand. The Personal Container, or Your Life in Bits. *Digital Futures*, 2010.

[42] M. Mun, S. Hao, N. Mishra, K. Shilton, J. Burke, D. Estrin, M. Hansen, and R. Govindan. Personal data vaults: a locus of control for personal data streams. In *Proc. of ACM CoNext, 2010*.

[43] S. R. Rajagopalan, L. Sankar, S. Mohajer, and H. V. Poor. Smart Meter Privacy: A Utility-Privacy Framework. *CoRR*, 2011.

[44] A. Rial and G. Danezis. Privacy-preserving smart metering. In *Proc. of ACM WPES*, 2011.

[45] L. Richardson and S. Ruby. *RESTful web services*. O'Reilly Media, Incorporated, 2007.

[46] R. Rivest. The md5 message-digest algorithm. 1992.

[47] C. Rottondi, G. Vertical, and A. Capone. A security framework for smart metering with multiple data consumers. In *Proc. of IEEE INFOCOM 2012 Workshop: Green Networking and Smart Grid*, 2012.

[48] A. Rowe, M. E. Bergeés, G. Bhatia, E. Goldman, R. Rajkumar, J. H. Garrett, J. M. F. Moura, and L. Soibelman. Sensor andrew: large-scale campus-wide sensing and actuation. *IBM J. Res. Dev.*, Jan. 2011.

[49] S. Ruj, A. Nayak, and I. Stojmenovic. A Security Architecture for Data Aggregation and Access Control in Smart Grids. *CoRR*, 2011.

[50] A. Shakimov, H. Lim, R. Caceres, L. Cox, K. Li, D. Liu, and A. Varshavsky. Vis-a-Vis: Privacy-preserving online social networking via Virtual Individual Servers. In *Proc. of COMSNETS, 2011*.

[51] E. Shi, T.-H. H. Chan, E. G. Rieffel, R. Chow, and D. Song. Privacy-Preserving Aggregation of Time-Series Data. In *NDSS*, 2011.

[52] K. Shilton, J. Burke, D. Estrin, R. Govindan, M. Hansen, J. Kang, and M. Mun. Designing the personal data stream: Enabling participatory privacy in mobile personal sensing. TPRC, 2009.

[53] S. Soltesz, H. Pötzl, M. Fiuczynski, A. Bavier, and L. Peterson. Container-based operating system virtualization: a scalable, high-performance alternative to hypervisors. In *Proc. of EuroSys*, 2007.

[54] P. St-Andre. Extensible Messaging and Presence Protocol (XMPP). *IETF Network Working Group, RFC3920*, 2004.

[55] L. Sweeney. k-anonymity: A model for protecting privacy. *International Journal of Uncertainty, Fuzziness and Knowledge-Based Systems*, 2002.

[56] T. Winter and P. Thubert et al. RPL: Ipv6 Routing Protocol for Low power and Lossy Networks. Internet Draft, IETF, 2010.

[57] R. Wishart, D. Corapi, A. Madhavapeddy, and M. Sloman. Privacy butler: A personal privacy rights manager for online presence. In *Proc. of Workshop of Smart Environments, 8th IEEE PerCom*, 2010.

Automatic Socio-Economic Classification of Households Using Electricity Consumption Data

Christian Beckel[*], Leyna Sadamori[*†], Silvia Santini[†]

[*]Institute for Pervasive Computing, ETH Zurich, Switzerland
[†]Wireless Sensor Networks Lab, TU Darmstadt, Germany

{beckel,sadamori}@inf.ethz.ch, santinis@wsn.tu-darmstadt.de

ABSTRACT

Interest in analyzing electricity consumption data of private households has grown steadily in the last years. Several authors have for instance focused on identifying groups of households with similar consumption patterns or on providing feedback to consumers in order to motivate them to reduce their energy consumption. In this paper, we propose to use electricity consumption data to classify households according to pre-defined "properties" of interest. Examples of these properties include the floor area of a household or the number of its occupants. Energy providers can leverage knowledge of such household properties to shape premium services (e.g., energy consulting) for their customers. We present a classification system – called CLASS – that takes as input electricity consumption data of a private household and provides as output the estimated values of its properties. We describe the design and implementation of CLASS and evaluate its performance. To this end, we rely on electricity consumption traces from 3,488 private households, collected at a 30-minute granularity and for a period of more than 1.5 years. Our evaluation shows that CLASS – relying on electricity consumption data only – can estimate the majority of the considered household properties with more than 70% accuracy. For some of the properties, CLASS's accuracy exceeds 80%. Furthermore, we show that for selected properties the use of a priori information can increase classification accuracy by up to 11%.

Categories and Subject Descriptors

H.4 [**Information Systems Applications**]: Miscellaneous

Keywords

Energy consumption analysis; smart electricity meters; household classification; machine learning;

1. INTRODUCTION

Electricity providers typically need to collect, process, and store consumption data of their customers for billing purposes. In the past, such consumption data has been collected at a very coarse granularity (e.g., monthly or even yearly). More recently, technological developments as well as changing market rules and regulatory frameworks have made the collection of more fine-grained data (e.g., hourly readings) both an opportunity and a necessity [1, 2, 3]. Electricity providers in Europe and elsewhere are thus putting in place new – or adapting existing – infrastructures to collect, process, and store the expected large amounts of electricity consumption data [1]. Beyond enabling billing according to dynamic pricing policies [3], the availability of such fine-grained data also opens up other opportunities. Several authors have for instance focused on using this data to provide feedback to customers about their electricity consumption [4, 5, 6]. Some studies have also looked at the possibility of deriving the consumption of single household appliances by analyzing the aggregated consumption curve of a private household [7, 8, 9]. Others have focused on discovering consumption curves with similar temporal patterns in order to better tune load prediction algorithms [10, 11].

In this paper, we argue that through the analysis of electricity consumption data it is possible to infer with high probability specific characteristics of private households – like their floor area or the number of persons living in them. We refer to these characteristics as the *properties* of a household. In a previous study, we have defined a set of properties that can be considered both useful to know as well as likely to be inferable with reasonable accuracy from electricity consumption data [12]. In this paper, we build upon our previous work and present the design, implementation, and evaluation of CLASS – a system that estimates the value of the properties of a household for which electricity consumption data is available.

The possibility to derive information about household properties in an automated fashion enables the development of novel services for electricity consumers. Indeed, the ongoing liberalization of electricity markets worldwide makes it easier for consumers to change their providers. These, in turn, need to increasingly offer premium services in order to attract new customers – as well as to retain existing ones. Through in-depth interviews with employees of four different Swiss energy providers, we have identified *energy consulting* as a representative example of such services [12]. The goal of energy consulting is to provide practical recommendations to customers in order to allow them to reduce their overall energy consumption, thus saving money. By providing such services, providers can not only retain existing customers or attract new ones, but also comply with the ever increasing societal and political pressure to reduce energy consumption in general.

Currently, the development of energy consulting is hampered by the difficulty of identifying customers who might both be interested in and profit from a consulting session. Furthermore, energy consulting sessions are still expensive as they typically involve a personal visit of a consultant to the customers' home. Our interviews have outlined that both the choice of target customers as well as the preparation – and thus effectiveness – of energy consulting sessions would be significantly improved if information about the properties of the customer's household was available in advance. For instance, households occupied by elderly people are more likely to be interested in – and be receptive to – consultants' recommendations [12]. Information about household properties is however typically not available, as providers know surprisingly little about their customers. This is particularly true in European countries in which open information repositories like public tax registers do not exist. Given this restriction, such information has to be acquired through expensive and time-consuming customer surveys. CLASS addresses this problem by providing a way to automatically estimate the properties of private households using electricity consumption data only – which energy providers already have for the purpose of billing. In this way, CLASS enables the development of energy consulting as well as of other premium services. In our work, we implicitly assume that providers will handle electricity consumption data according to existing regulations and industry practices with respect to the processing and propagation of personal data (e.g., [13, 14]).

We formulate the problem of determining the value of a property for a specific household as a classification problem. For a property that can take K different values, the household is accordingly assigned to one of K classes. CLASS takes as input a set of features computed over the consumption data of a household and returns as output – for each property – the class to which the household belongs. For instance, in our design the property `floor_area` can take one of three values: small ($\leq 100\ m^2$), medium ($> 100\ m^2$ and $\leq 200\ m^2$), or large ($> 200\ m^2$). CLASS can thus classify a household as a member of one of the three classes *small*, *medium* or *large*, thereby providing an estimation of the value of the corresponding property.

We have analyzed the performance of our classification system for twelve different properties and using four different classifiers. We have trained and evaluated the system using electricity consumption data of 3,488 households. The data traces have been collected over a period of about 1.5 years and at a granularity of 30 minutes (i.e., a data sample represents the average electricity consumption in kilowatts during a 30-minute interval). For all used data traces, we know the actual values of all properties for which the classification system is evaluated. This information – along with the consumption data itself – has been gathered in the context of a smart metering study conducted by the Irish Commission for Energy Regulation (CER). During this study, the electricity consumption data of 4,225 private households and 485 small and medium enterprises has been collected.

All participants of the study have also provided information about themselves by filling in detailed questionnaires. The twelve household properties used for the evaluation of our system have been extracted from these questionnaires as described in [12]. Both the electricity consumption data as well as the questionnaires – to which we refer to as the *CER data set* – have been recently made available to the public.[1]

Our results show that CLASS makes automatic classification of household properties feasible and reliable. In particular, CLASS's classification accuracy is higher than 70% for eight of the twelve properties considered in this study. For two properties, the accuracy exceeds 80%. Furthermore, the use of a priori information allows to further improve CLASS's performance by up to 11%. To the best of our knowledge, this is the first study that provides a quantitative analysis of the possibility to extract households' properties from electricity consumption data. This is also due to the fact that large, labeled data sets of electricity consumption data have only very recently been made available to the public.

The remainder of this paper is organized as follows. We review related work in section 2 and then present the design of CLASS in section 3. In sections 4 and 5 we present the setup and results of our evaluation. We discuss the limitations of our approach and interesting directions for future work in section 6.

2. RELATED WORK

Measuring and analyzing fine-grained electricity consumption data has gained attention of many researchers. One of the applications often referred to is non-intrusive load monitoring (NILM) [7, 8, 9, 15, 16]. NILM researchers aim at deriving detailed consumption information – such as the time period when certain appliances are running or how much electricity they consume – from the aggregated electricity consumption of the household. In [7], Zeifman et al. provide a review of established techniques, highlighting different types of features that can be extracted from the consumption data, as well as requirements to the data itself. With such detailed information available, behavioral studies increasingly investigate its use to motivate a more thrifty use of energy [5, 6]. Reinhardt et al. perform a different approach as they detect which appliances are currently running based on consumption data measured at individual power outlets [16]. In practice, NILM approaches suffer from the requirements to high frequency consumption measurements [8] – typically in the order of 1 Hz to multiple kilohertz – as well as from intensive training procedures [7]. Our work is substantially different to NILM as we do not aim at determining the consumption of individual appliances but at identifying high-level characteristics of a household – like the floor area or the number of people living in it. For this purpose, data with a much coarser granularity (e.g., one measurement every 30 minutes) is sufficient, although – in contrast to NILM – such a study requires a comparison of hundreds or thousands of households.

Focusing on data measured with a granularity of 15, 30, or 60 minutes, other related approaches detect consumption patterns of households over a long time frame [11, 17, 18, 19]. To energy providers, applications based on such data are of particular interest as this is the type of data that was col-

---

[1]www.ucd.ie/issda/data/commissionforenergyregulation/

lected during most of the smart meter trials so far. To identify patterns in electricity consumption data, for instance, De Silva et al. propose a data mining framework and introduce an incremental learning algorithm that predicts future electricity usage of private households [11]. Abreu et al. similarly focus on supporting the energy provider's supply management as they employ pattern recognition techniques to recognize specific consumption patterns such as daily routines or consumption baselines [17]. By clustering households based on their electricity consumption, Verdu et al. can recognize consumers with an "atypical" consumption pattern [18]. The authors employ a clustering technique – based on self-organizing maps – to classify (commercial) customers by their type. The results presented in these papers are based on relatively small data sets containing traces from 5–30 households. In [19], Birt et al. disaggregate the total electricity consumption into load categories. Evaluating hourly consumption data from 327 households, the authors observe correlations between the consumption of a household and the air temperature both in winter and summer. These correlations allow to determine the electricity consumed by heating and cooling systems, respectively. Our work differs from all approaches above in that we detect properties of households rather than consumption patterns or usage categories. We further rely on a significantly larger data set as we analyze consumption traces from 3,488 households.

Many authors have also investigated the problem of clustering consumers into groups by their consumption pattern. Knowledge about the characteristics of such clusters can be used to develop novel tariff schemes, improve network management, or to perform load forecasting. An early example of an approach that clusters consumption data from a large number of households is provided by Chicco et al. [20]. The authors cluster electricity consumption data of 471 non-residential customers. Evaluating the clusters along with the current tariffs of each of the customers, they detect examples of inefficient billing practices (e.g., in case there is a poor correlation between discriminatory factors and actual load patterns) [20]. Many approaches related to this use socalled self-organizing maps (SOMs) to cluster a large number of different households based on their electricity consumption [21, 22, 23]. For instance, Figueiredo et al. use this type of unsupervised learning to identify clusters of households with similar consumption behavior [21]. The authors further create rules that form a decision tree to automatically assign new households to one of the clusters by following that tree. Their results are based on electricity consumption traces from 165 households in Portugal collected at a 15-minute granularity. Other researchers also include different type of information to the analysis of plain electricity consumption data [22, 23]. Sanchez et al. [22], for instance, compute specific features out of the data and feed these features to a self-organizing map (SOM) along with additional information obtained through questionnaires. Instead of using electricity consumption to train the SOM, Räsänen et al. rely on properties of the dwelling [23]. This way, the authors make households in the same cluster comparable with respect to their dwelling characteristics. Ultimately, identifying the households with a high (annual) electricity consumption in each cluster, Räsänen et al.'s method allows to create targeted energy saving recommendations.

In a previous study, we also relied on SOMs to cluster groups of households based on their electricity consump-

tion [12]. In contrast to the SOM-based approaches presented above, we also evaluated the correlation of household properties with electricity consumption data. We performed this study to identify properties that have a high correlation and are considered particularly useful by domain experts – which we learned by conducting interviews with energy consultants from four different energy providers in Switzerland. In this paper we now investigate these properties in detail using a supervised learning method.

McLoughlin et al. analyze the correlation between the electricity consumption of a household and its demographic, socio-economic, and dwelling characteristics [24]. The data set used in this study is significantly larger than others previously considered and – although not explicitly stated – is most likely to be the same data set that we use for our investigation. Applying a multiple linear regression model on the data the authors show that certain properties (e.g., number of bedrooms, age of the head of household) have a large influence on the electricity consumption. The authors further show a relationship between maximum demand and the penetration of certain household appliances. Kolter et al. utilize such correlations and propose a regression method that leverages knowledge of dwelling properties such as size, insulation, or location, to estimate the most likely energy consumption level of a household based on its characteristics [25]. This estimation can then be compared against a household's actual consumption – as well as against that of similar households.

Our work – in contrast to all other approaches described in this section – estimates household properties by analyzing the household's electricity consumption data. To this end, we rely on a supervised learning method, using household properties along with the consumption data to train our system.

3. SYSTEM DESIGN

The upper part of figure 1 shows the two main components of our classification system: *feature extraction* and *classification*. The feature extraction component takes as input the electricity consumption trace of a household and computes over it a set of *features*. The classification component takes these features as input and uses them to classify the household according to previously specified properties. As depicted in the lower part of figure 1, the design of CLASS comprises three steps, which are explained in the following sections in detail. In section 3.1, we describe the set of features that is computed during the feature extraction. Sections 3.2 and 3.3 address the design of the classification component. First, we define *class labels* from which a classifier takes a possible estimate for the household properties. Second, we choose a set of classification algorithms that perform the classification and motivate our choice.

3.1 Feature Extraction

CLASS takes as input an electricity consumption trace of a private household. In its current implementation, CLASS assumes this trace refers to one week (including weekend) and that one sample every 30 minutes is available. Nonetheless, CLASS can be easily adapted to operate with data of different granularity. One week of data at a 30-minute granularity implies that an input trace contains 336 data samples. In [12], we have already defined a comprehensive set of features that can be computed on electricity consumption

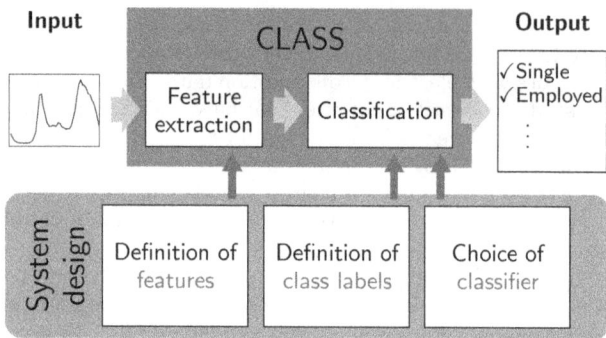

Figure 1: Overview of the design of our classification system. The upper part of the picture shows CLASS's main components: *feature extraction* and *classification*. The bottom part shows the design steps necessary to specify these components.

Household property	Classes and their labels
Number of appliances and entertainment devices (#devices)	Low (#devices $\leq$ 8) Medium (8 < #devices $\leq$ 11) High (11 < #devices)
Number of bedrooms (#bedrooms)	Very low (#bedrooms $\leq$ 2) Low (#bedrooms = 3) High (#bedrooms = 4) Very high (4 < #bedrooms)
Type of cooking facility (cooking)	Electrical Not electrical
Employment of chief income earner (employment)	Employed Not employed
Family (family)	Family (#adults = 1 and #children > 0) No family
Floor area (floor_area)	Small (floor_area $\leq$ 100 m^2) Medium (100 m^2 < floor_area and floor_area $\leq$ 200 m^2) Big (200 $m^2 \leq$ floor_area)
Children (#children)	Children (#children $\geq$ 1) No children (#children = 0)
Age of building (age_house)	Old (30 < age_house) New (age_house $\leq$ 30)
Number of residents (#residents)	Few (#residents $\leq$ 2) Many (3 $\leq$ #residents)
Single (single)	Single (#adults = 1 and #children = 0) No single
Retirement status of chief income earner (retirement)	Retired Not retired
Social class of chief income earner according to NRS social grades (social_class)	A or B C1 or C2 D or E

data. For its definition, we have considered both existing work in electricity consumption data analysis [20, 21, 22] as well as own considerations relative to the problem at hand. In this study, we use the same set of features as input of CLASS's classification algorithms.

The 22 features can be divided in four groups: consumption figures, ratios, temporal properties, and statistical properties. Each group contains 9, 5, 4, and 4 features, respectively. Consumption figures are aggregates of the electricity consumption data at different periods of the day (e.g., in the morning from 6 a.m. to 10 a.m. or during weekends). These features allow to compare households with respect to their average electricity consumption during these periods. Ratios are features calculated as the ratio of two different consumption figures (e.g., average consumption in the evening and during lunch time). These ratios allow to capture relevant patterns like whether cooking typically takes place over lunch-time, in the evening, or both. Temporal features capture the time instant of the first occurrence of an event, such as the daily electricity consumption exceeding a specific threshold. Finally, statistical properties include features such as the variance or the total number of peaks of the consumption trace over a day. For more details about the feature set used in this study, we refer the reader to [12].

3.2 Class Labels

As we rely on supervised machine learning methods, the classification requires a set of class labels, from which the classifier takes a possible estimate for the label of an unclassified household. The left column of table 1 shows the list of twelve properties considered in this study. The right column of table 1 shows the class labels we have accordingly defined for each property. The class labels represent the possible values that an estimate of a household property can take. In other words, given the consumption trace of a specific household, CLASS will return as output an estimate of the value of each of the properties listed in table 1. In this way, we turn the problem of estimating values of household properties into a classification problem.

We have compiled the set of properties listed in table 1 using the results presented in our previous work [12]. For example, we have carried out in-depth interviews with energy consultants of four different Swiss energy providers. The goal of the interviews was to derive a list of properties consultants need to know in order to: (1) identify customers potentially interested in energy consulting; (2) prepare for energy consulting sessions. This application-driven approach for identifying interesting properties has for instance outlined that knowing whether a household is occupied by a single person or a family is particularly relevant to consultants. Accordingly, the list in table 1 includes the properties single and family. For the evaluation of our system, it is also necessary to verify whether the ground truth of the desired household property is available in the CER data set. For instance, the property family requires information about the number of adults (#adults) as well as the number of children (#children) living in a specific household. Similarly, #residents describes the number of persons living in the household. The availability of this ground truth data allows us to verify whether and with which accuracy CLASS can classify a household as belonging to the class *Single / No single* or *Family /No family*. Similar considerations apply to all other properties listed in table 1.

We should note at this point that while for some of the properties there exists a "natural" definition of class labels (e.g., *Single / No single*), for other properties – like property age_house – several different sets of class labels can

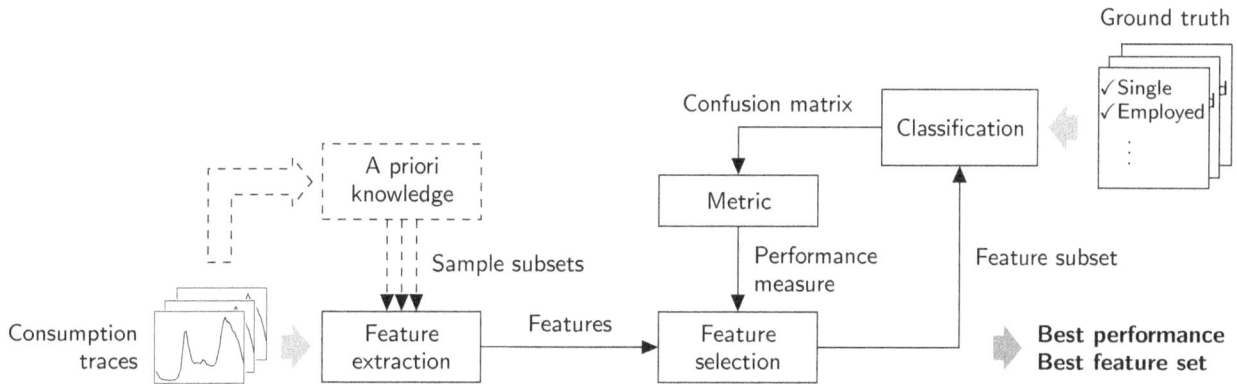

Figure 2: Evaluation system used to assess the performance of CLASS. The system allows to evaluate the performance of all considered classifiers in terms of their ability to estimate unknown household properties from electricity consumption data. The system also computes the feature set that allows to obtain the best classification results. Optionally, the system can also integrate a priori knowledge.

be defined. For these properties, we define two (or more) class labels according to the following criteria. Either we define labels such that the number of households per class is distributed (roughly) equally or we define labels according to qualitative considerations gathered during the aforementioned interviews. Class labels corresponding to the properties #devices, #residents, and social_class have been defined using the first criterion. Class labels for properties #bedrooms, age_house, floor_area using the second.

3.3 Classification

As mentioned before, the design of CLASS relies on supervised machine learning methods to classify the households. To this end, CLASS integrates a set of classifiers that "learn" the correlation between a given input feature vector and the corresponding (correct) class label. Hereby, a subset of input data for which the actual class labels are known is used to train the classifier and evaluate its performance. The performance is based on the number of correctly predicted class labels for each of the properties listed in table 1. For each property, CLASS then relies on the classifier that predicted most labels correctly in order to estimate the class label of an unclassified household.

There is a variety of classification algorithms available in literature (see e.g., [26] for an overview) that can be used within CLASS. The choice of classifiers for our system depends on several factors including ease of implementation, computational complexity, and achievable classification accuracy. In order to obtain a comprehensive picture of CLASS's performance, we have selected four well-known, representative classifiers: the k-Nearest Neighbor (kNN) classifier [26], the Linear Discriminant Analysis (LDA) classifier, the Mahalanobis classifier [27], and the Support Vector Machine (SVM) [28].

For a detailed description of the four classifiers, the reader is referred to [26, 27, 28]. Here, we outline the specific trade-offs exposed by these classifiers and provide few details about their implementation within CLASS. Besides its simplicity, a main advantage of the kNN classifier is that it does not make any assumption on the distribution of the input data, which also does not need to be linearly separable. On the other side, the kNN classifier has high computational and memory requirements. For the LDA classifier,

we assume a (multivariate) Gaussian distribution of the input data samples. This causes the parameters of the discriminant functions that partition the feature space to be dependent on the mean and covariance of the distributions for each class. The linear functions of the LDA classifier are obtained by assuming a common (pooled) covariance matrix for all classes, constructed by averaging the covariance matrices of each class. The need to assume Gaussianity of the input data is a major drawback of the LDA classifier. On the other side, it has very low requirements in terms of computation and memory usage. The Mahalanobis classifier is conceptually similar to the LDA classifier. One of the main differences is that the former relies on stratified covariance matrices, instead of a pooled covariance matrix. This results in quadratic discriminant functions, which typically allows the Mahalanobis classifier to have better classification performance than the LDA classifier. However, its performance is also more sensitive to the estimation accuracy of the stratified covariance matrices. SVMs are widely used in classification applications [26], which is due to their flexibility and thus, applicability to many classification problems of different natures. A major strength of SVMs is their ability to compute decision boundaries without assuming specific distributions of the input data (like the kNN classifier but unlike the LDA or Mahalanobis classifiers). Further, SVMs are able to cope with data that is not linearly separable, since they support non-linear decision boundaries. A major drawback is the computationally expensive training phase.

4. EVALUATION SETUP

After having discussed CLASS's design in the previous section, we now discuss how we evaluate its classification performance. In particular, we introduce the *metrics* used to quantify this performance and discuss the use of *feature selection algorithms*. Furthermore, we outline how the use of *a priori information* can help improving CLASS's overall performance. Figure 2 depicts our evaluation setup.

4.1 Metrics

The most common metric used to evaluate the performance of a classifier is the *accuracy* [29, 30]. For a two-class classification problem, this metric is computed as a function of four other quantities: the *true positives* (TP), *true negatives* (TN), *false positives* (FP), and *false negatives* (FN). Given a "target" class A and another class B, TP indicates the number of samples of A that are correctly classified as A. TN denotes the number of samples of B that are correctly classified as B. FP counts the number of samples of B that are incorrectly classified as A and, finally, FN indicates the number of samples of A that are incorrectly classified as B. These four values form the so-called *confusion matrix*, which is schematically depicted in figure 3.

		Predicted		Total
		Class A	Class B	
True	Class A	TP	FN	$S_A = TP + FN$
	Class B	FP	TN	$S_B = FP + TN$

Figure 3: Generic confusion matrix.

The accuracy of a classifier is defined as the ratio between the number of correct classifications and the total number of test samples [30]. Using the notation introduced above it can be computed as:

$$ACC = \frac{TP + TN}{TP + TN + FP + FN}. \quad (1)$$

In a general classification problem with K classes (K-class problem), the confusion matrix consists of K^2 elements. The accuracy is then computed as the ratio between the sum of the K diagonal elements (i.e., the number of correct classifications) and the sum of all test samples.

We compare the accuracy of CLASS to that of two simple classifiers to which we refer to as the *Random Guess* (RG) and *Biased Random Guess* (BRG) classifiers. In a K-class problem, the RG classifier assigns the input data sample uniformly at random to one of the K possible classes. The accuracy of the RG classifier is thus $ACC_{rg} = \frac{1}{K}$. The BRG classifier biases its random decision so that the proportion of households assigned to each of the available classes is equal to the actual distribution of the input data. If S_k denotes the number of test samples of the k-th class of a K-class problem and S is the total number of test samples, the accuracy of the BRG classifier is computed as: $ACC_{brg} = \sum_{k=1}^{K} (\frac{S_k}{S})^2$.

We further evaluate the performance of CLASS in terms of *precision* and *recall* [29]. In a two-class problem, precision is defined as

$$PRC = \frac{TP}{TP + FP} \quad (2)$$

and denotes the probability that a sample classified as class A truly belongs to class A. For instance, a precision of 80% in the classification of the property `single` indicates that if a household is classified as a single-person household, it is truly a single-person household with 80% probability. Recall is defined as

$$RCL = \frac{TP}{TP + FN} \quad (3)$$

and denotes the probability of a sample being classified as class A given that the sample belongs to class A. For instance, a recall of 80% in the classification of the property `single` indicates that 80% of the households that are actually single-person households are also classified as such.

Considering the values of precision and recall separately might be misleading. Indeed, a classifier that assigns all test samples to a class A – irrespectively of the true membership to class A or B – has a recall of 100%. Most likely, however, the same classifier will also have very low precision. Hence, metrics that combine both, precision and recall, are often used when evaluating the performance of a classifier. A commonly used metric is the F_1 *score*, which is defined as [29]:

$$F_1 = 2 * \frac{PRC * RCL}{PRC + RCL}. \quad (4)$$

We should note that since precision, recall and F_1 score assume that there exists a "target" class, they can only be computed for two-class problems. Accuracy can instead be computed for generic K-class classifiers.

4.2 Feature Selection

As shown in figure 2, the feature set described in section 3.1 and computed by CLASS's feature extraction component, undergoes a *feature selection* step before being used for the actual classification task. We include this step in our evaluation since using the complete feature set as input data might result in suboptimal classification performance [31]. Thus, while training the classifiers we search for the *optimal feature set*, i.e., the subset of features that maximizes classifiers's performance. This optimal set can be determined by performing an exhaustive search that examines all possible subsets of the feature set [31]. The subset that provides the best overall performance is then selected as the optimal feature set. However, the computational complexity of such an exhaustive search grows exponentially with the number of features [32]. So-called *feature selection algorithms* can thus be used to approximate an exhaustive search, although the optimal set provided by these algorithms might differ from the "true" optimum. CLASS implements two feature selection algorithms: *Sequential Forward Selection* (SFS) [31] and *Parallel Sequential Forward Selection* (P-SFS) [33, Sec. 4.2]. P-SFS typically performs better than – or at least as good as – SFS but requires slightly more computational and memory resources. The rationale behind both these algorithms is to avoid an exhaustive search by iteratively increasing the size of the feature set. Thereby, only the feature that – at each iteration – allows to maximize a given performance metric, is retained. In this study, we use accuracy as the performance metric of choice, although any other of the metrics mentioned above could also be used. For a more detailed description of SFS and P-SFS we refer the reader to [31] and [33, Sec. 4.2], respectively.

4.3 A Priori Knowledge

In some settings, partial information about household properties – e.g., their floor area – might be known a priori. Such information can for instance be obtained from public tax record repositories (as done, e.g., by Kolter et al. [25]) or through dedicated customer surveys. We thus evaluate the performance of CLASS also assuming that the values of selected properties are known in advance. Integrating this a priori information in the evaluation of CLASS allows to measure whether knowledge of selected properties might be valuable or not to improve CLASS's performance. Providers could then accordingly take action to retrieve the information that can be collected at lower cost and allows for the highest performance gains.

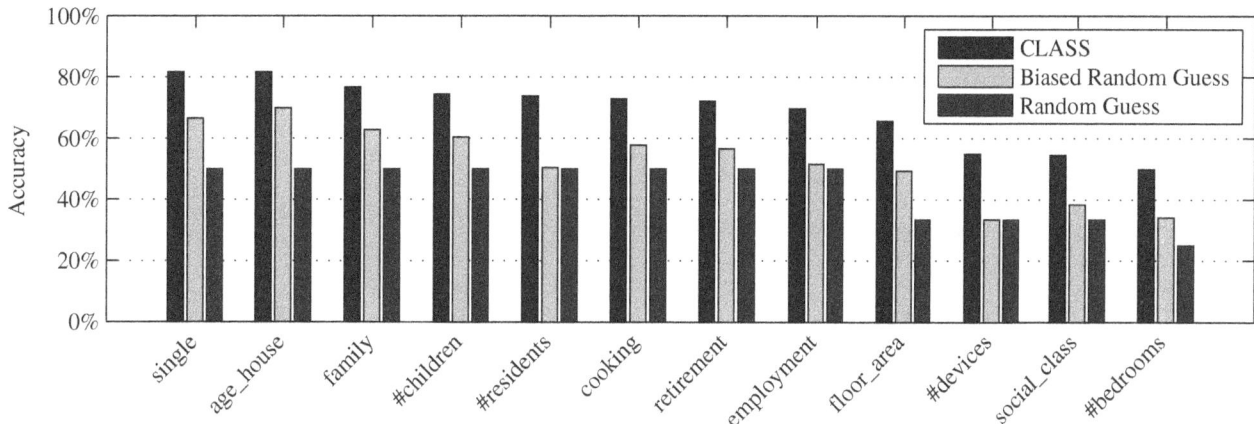

Figure 4: Accuracy of CLASS compared to the random guess and biased random guess classifiers.

4.4 Implementation Details

The evaluation of CLASS relies on 3,488 out of the 4,225 electricity consumption traces of private households included in the CER data set. The 737 neglected data traces are those for which questionnaire answers were not available or participants withdrew from the trial beforehand. From the 1.5 years described by the data we selected consumption traces from week 26, which is the second week of January (Jan 11, 2010 – Jan 17, 2010). We also run our experiments on all the remaining 72 weeks, showing that results obtained for week 26 are representative for the whole data set. We used the data of this week both for training and testing. To avoid data being part of both the training and test set at the same time, we use four-fold cross-validation. On average, the traces of 2,616 households are used for training and the remaining 872 households for testing. Each household is thus used three times for training and once for testing.

The results presented in section 5 are obtained using our own Matlab implementation of CLASS. We evaluate the classification performance for each of the twelve properties listed in table 1. All experiments are repeated independently for each of the four classifiers (kNN, LDA, Mahalanobis, SVM) described in section 3.3. For the implementation of the kNN, LDA, and Mahalanobis classifiers we use Matlab's Statistics toolbox[2]. For kNN, we choose $k = 5$ and use the Euclidian distance as the distance metric. We further employ a k-d tree for space partitioning to perform the neighbor search to avoid a linear search. As for the SVM classifier, we rely on the publicly available C-SVM implementation *LIBSVM*[3]. Setting $C = 1$, we follow a common choice to penalize outliers in the training data. For the P-SFS algorithm, we run three branches in parallel.

5. CLASSIFICATION RESULTS

In this section, we present the classification performance achieved by CLASS using the CER data set. In particular, we present results in terms of accuracy, F_1 score, precision and recall. Furthermore, we analyze the influence that the availability of a priori information has on the classification

performance. All experiments have been carried out using the setup described in the previous section.

5.1 Accuracy

Figure 4 shows the accuracy of CLASS for all the twelve household properties listed in table 1. For each property, the accuracy obtained by CLASS is compared to that of the Biased Random Guess (BRG) and Random Guess (RG) classifiers described in section 4.1. The accuracy of CLASS is selected as the highest among those obtained by the four classifiers described in section 3.3 (kNN, LDA, Mahalanobis, and SVM). We postpone the discussion of the accuracy obtained by the single classifiers to section 5.3. Please note that – as indicated in table 1 – the property `#bedrooms` can assume four different values. Thus, the estimation of this property corresponds to a four-class classification problem. Accordingly, the estimation of properties `floor_area`, `#devices`, and `social_class` is a three-class problem and that of the eight remaining properties a two-class one. For these "two-class properties" (the first eight from left in figure 4), the accuracy of CLASS ranges from 70% (`employment`) to 82% (`single` and `age_house`). For the three-class properties, the accuracy varies between 55% (`social_class`) and 66% (`floor_area`). Finally, the accuracy obtained for the only four-class property (`#bedrooms`) is 50%.

The accuracy of the Random Guess classifier ranges between 25% and 50% and that of the Biased Random Guess classifier between 34% and 70%. Overall, the performance increase – in terms of accuracy – obtained by CLASS with respect to the RG and BRG classifiers is 20%–32% and 12%–23%, respectively. The better performance of the BRG classifier is due to the fact that it assumes prior class probabilities to be known. CLASS can thus estimate with high accuracy ($> 80\%$) the properties `single` and `age_house` and achieves good performance (accuracy $\geq 70\%$) for the other six two-class properties (`family`, `#children`, `#residents`, `cooking`, `retirement`, `employment`). This implies that electricity providers can estimate with good accuracy the values of selected properties of their customers' households by analyzing their electricity consumption data. Providers can then in turn use this information to develop and deploy premium services for their customers. For instance, knowledge of the values of the properties `age_house` and `#devices`

---

[2]www.mathworks.de/products/statistics
[3]www.csie.ntu.edu.tw/~cjlin/libsvm

can be useful to prepare consulting sessions and elaborate household-specific energy saving tips.

We recall that the results presented in this section have been computed using one week of electricity consumption data as input (week 26 of the CER data set). To verify that the obtained results are representative for the whole CER data set, we assessed CLASS's performance using the consumption traces of all of the 72 weeks of the CER trial. For the property single, for instance, the resulting standard deviation of the accuracy is 0.7%, 1.2%, 1.3%, and 0.6% for the kNN, LDA, Mahalanobis, and SVM classifier, respectively. Furthermore, the accuracy obtained for this property with input data from week 26 is – with respect to that obtained for all other 72 weeks – the 38th highest for kNN, 45th highest for LDA, 71st for Mahalanobis, and 24th for SVM. Similar results apply to all other household properties considered in this study. Given these considerations, we conclude that there is no loss of generality in presenting results relative to one week of data only (presenting all results is impractical due to space constraints).

5.2   Precision and Recall

Figure 5 shows the values of precision and recall of all two-class properties listed in table 1. For the property #residents both its class labels *Few* (≤ 2) and *Many* (≥ 3) are considered, as this information is particularly interesting to energy consultants. Similarly, the precision and recall of both class labels of the property cooking (*Electrical / Not electrical*) are shown. The further right on the plot a property is located, the higher its precision; the further on the top, the higher its recall. Accordingly, the closer a data point is to the top right corner, the higher is the F_1 score (see equation 4) of the corresponding property. To improve the readability of the plot, we have also plotted 10 different level curves (from $F_1 = 10\%$ to $F_1 = 100\%$) of the F_1 score. For each property, the figure shows the precision and recall values corresponding to the highest F_1 score among those achieved with the four classifiers used within CLASS.

The graph shows that, for instance, CLASS achieves the best performance ($\sim 90\%$) in terms of F_1 score when classifying households according to property age_house and in particular for the class label *Old*. The recall in this case is 99.89% and the precision 81.61%. This implies that if the age of the building to which the household belongs to is actually *Old* (i.e., older than 30 years, as indicated in table 1), then CLASS will classify the household correctly in 99.89% of the cases. On the other side, if CLASS classifies a household as *Old*, then in 81.61% of the cases the household is actually *Old*. Similar considerations apply to other properties with a high F_1 score, like #children and cooking for the class labels *No Children*, and *Electrical*, respectively. Figure 5 also shows that households with less than three residents (property #residents, label *Few*) and those having an employed chief income earner (property employment, label *Employed*) are classified with a precision of 71% and 73%, respectively, and with a recall of approximately 80%. The recall value implies that if an electricity provider is interested in selecting, for instance, the specific group of households that have only few (i.e., ≤ 2, as indicated in table 1) residents, then CLASS would correctly identify 80% of all households of this type. In other words, the provider would be able to address 80% of the customers that live

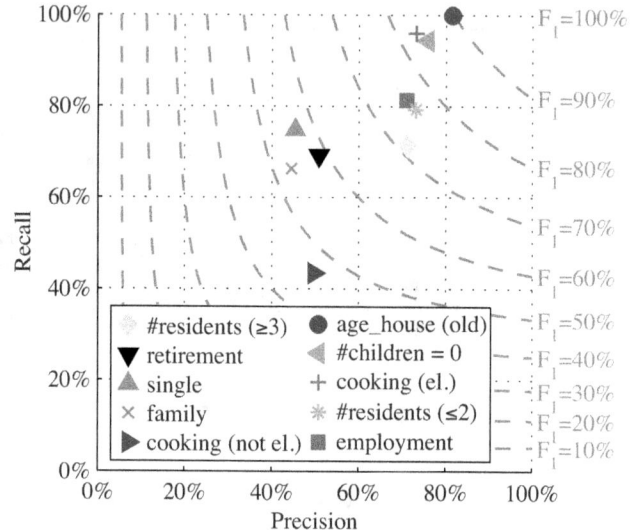

Figure 5: F_1 score obtained for all two-class properties. Each marker shows the highest value among those obtained by the considered classifiers.

in households with few residents. This can be particularly useful to elaborate, e.g., personalized marketing campaigns.

Relatively low F_1 scores are obtained for properties retirement, single, and family. In particular, the precision for the property single is just roughly 45%. This means that if CLASS classifies a household as *Single*, then in only 45% of the cases this classification is correct. On the other side, the recall value for the property single (label *Single*) is 75%, i.e., 75% of the households occupied by a single person are also classified as *Single*. One of the reasons for this low precision is that only 27% of the households in the CER data set are actually single-person households. As in any classification problem, such "rare samples" are hard to detect. The fact that the accuracy for the property single is 80% (as discussed in section 5.1) shows the importance of considering the performance of a classifier also in terms of precision and recall. The classifiers can indeed be tuned so that the one or the other metric is maximized. The results discussed in this section have been obtained by using accuracy as the figure of merit to maximize during feature selection. However, if the goal of the classification is to identify households belonging to a target group (e.g., single-person households), then the classifiers must be tuned to maximize the F_1 score. Evaluating the change in performance obtained when configuring CLASS to maximize the F_1 score is a part of our future work.

5.3   Classifier Comparison

Figure 6 shows the performance in terms of accuracy for each of the four classifiers implemented in CLASS and for each of the considered household properties. Figure 7 reports the corresponding F_1 score. The SVM classifier provides the highest accuracy for eight of the twelve household properties and the second highest for the remaining four properties. This strength of the SVM classifier is due to its ability to classify properties even in presence of non-linearly separable input data. It becomes particularly vis-

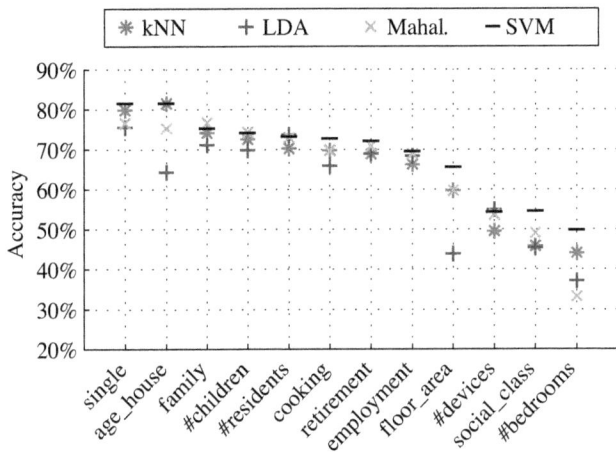

Figure 6: Accuracy of different classifiers for all household properties.

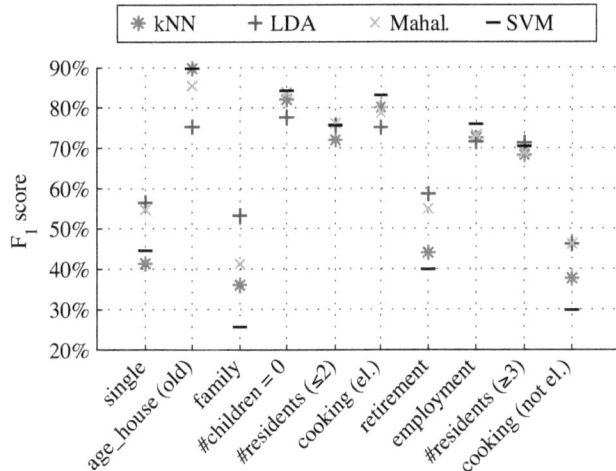

Figure 7: F₁ score of different classifiers for all household properties.

ible for multi-class properties (`floor_area`, `#devices`, `social_class`, and `#bedrooms`) as SVM clearly outperforms the other classifiers. LDA shows the worst performance for seven out of twelve household properties. For the properties `age_house` and `floor_area`, the accuracy provided by the LDA classifier is significantly lower than the accuracy provided by the three other classifiers. LDA's poor performance is due to the fact that the input data is not normally distributed (but it is assumed to be so by LDA). The kNN and Mahalanobis classifiers provide performances close to those of SVM for most properties but the multi-class ones.

The performances of the classifiers in terms of F₁ score are less homogeneous. Figure 7 shows that the LDA classifier provides the highest F₁ score for five out of ten properties. However, it also performs worst for four properties. Similar considerations apply to the performance of the other classifiers. When classifying the property `single`, for example, the F₁ score of the kNN and SVM classifiers is significantly lower than that of the LDA and Mahalanobis classifiers. We believe that the differences in classification performance with respect to the F₁ score can – at least partially – be attributed to the different number of households available for each class and property. In particular, the LDA classifier outperforms the other classifiers in terms of F₁ score when the classification requires identifying households that are outnumbered by the households of other classes. Hence, LDA is particularly good at detecting rare samples in the CER data set.

The results show that the choice of a particular classifier depends on the specific target application. In general, SVM stands out as the classifier that, for the given data set, allows to achieve the best performance in terms of accuracy. However, other classifiers might allow for slightly better performance for selected properties (e.g., Mahalanobis for `family`). If the application involves identifying households of a specific group, the classifier should be selected based on application-specific requirements (e.g., low false positive rates) as well as on the estimated number of households per class.

5.4 A Priori Knowledge

As already mentioned in section 4.3, the use of a priori knowledge can help increasing CLASS's classification per-

formance. Due to space constraints, we present here results obtained for two representative scenarios only. First, we investigate how the accuracy of the property `single` changes when information about other properties is available. In other words, we analyze which information should preferably be gathered (e.g., through customer surveys) in order to improve the accuracy of the property `single`. Second, we analyze how the accuracy obtained by CLASS for different properties changes when knowledge of one specific *a priori property* is known. In particular, the a priori property used below indicates the age of one of the residents and is indicated as `age_person`.[4]

The black segments in figure 8 show the baseline accuracies achieved when classifying property `single` when no a priori information is available. In contrast to section 5.1, we restrict the set of households to the ones where a priori information is provided by the questionnaires. This is noteworthy as property `floor_area` is only provided by a subset of households. Classifying property `single` on this subset results in a significantly higher accuracy (86%) compared to the subsets denoted by the other a priori properties (82%), which contain almost all households in the data set.

We then run CLASS to determine how the accuracy changes when each of the a priori properties listed on the x-axes is assumed to be known. To this end, we first select households belonging to a known class and then run the classification on this reduced data set. For instance, one of the a priori properties we consider for this experiment is `floor_area`. We select households with a floor area bigger than 180 m² and run CLASS over the data corresponding to these households. The cross labeled as $> 180\ m^2$ in figure 8 shows the accuracy obtained in this experiment. Accordingly, the cross labeled as $< 180\ m^2$ shows the accuracy obtained when households with a floor area smaller than $180\ m^2$ are considered as input data set. The stern marker in between the two crosses indi-

---

[4]The CER data set contains information about the age of the person who filled in the questionnaires for each household. From this information, we derive whether a person younger than 35, older than 65, or in between 35 and 65 years old is a resident of the household.

cates the mean accuracy weighted by the number of samples of the two classes ($> 180\ m^2$ and $< 180\ m^2$). Hence, this value represents the overall accuracy that is achieved when the property `floor_area` is known a priori. The same procedure is applied to obtain the accuracy when other properties and specific class labels are given. Other properties include `age_person` (class labels as defined above), `#bedrooms` (class labels $1-3$ and > 3), `ownership` (which describes whether or not the residents of a household own the house – class labels *Own* and *Rent*), `social_class` (as defined in table 1), and `house_type`, which indicates whether the residents's house is *Terraced*, *Detached*, *Semi-Detached*, or a *Bungalow*.

Figure 8 shows that when the property `floor_area` is known a priori the accuracy in the classification of a household as *Single* or *Not single* increases by 5.1% to 91% for households with floor area larger than $180\ m^2$. However, it decreases by 1.9% to 84% for households with floor area smaller than $180\ m^2$. The average accuracy increase with respect to the case in which no a priori knowledge is given is 0.7%. A similar effect can be observed when knowing property `#bedrooms` a priori. Restricting the set of households to the ones with more than three bedrooms increases accuracy when classifying property `single` by 7.7% to 89%. On the other side, the accuracy when classifying households with one to three bedrooms is 75.9% and thus 5.4% lower than the baseline. Overall, knowing the number of bedrooms a priori increases accuracy by 0.7%. When `age_person`, `ownership`, `social_class`, or `house_type` are used as a priori properties, the average accuracy changes also vary between -0.4% and 1.6%. Interestingly, knowing whether a house is owned or rented (indicated by the property `ownership`) decreases accuracy in both cases. This effect is possible since the distribution of the data changes when the set of households is separated into multiple subsets.

The results presented in figure 8 allow to draw the conclusion that – considering the accuracy for single-person households – an accuracy increase can also be obtained if it is known whether: at least one person younger than 65 lives in the household (property `age_person`); the household has more than three bedrooms (property `#bedrooms`); the house has more than $180\ m^2$; the social class of the chief income earner of the household is *A*, *B*, *C1*, or *C2* (property `social_class`); the household is a (semi-)detached house (property `house_type`). A more detailed analysis considering other properties than `single` is left to future work.

Figure 9 shows the accuracy obtained when classifying the properties `#bedrooms`, `#devices`, `employment`, `floor_area`, `#residents`, `single`, and `social_class` in five different experiments labeled as *None*, *Avg*, < 35, $35 - 65$, and > 65. Experiment > 65 refers to the case in which classification is run for the set of households that have at least one resident older than 65. In other words, the experiment is run assuming the `age_person` property (class > 65) is known a priori. Accordingly, experiments < 35 and $35 - 65$ are run using as input data only electricity consumption traces of households having at least one resident younger than 35 or in between 35 and 65, respectively. The results labeled as *Avg* represent the weighted accuracy of the accuracy of experiments < 35, $35 - 65$, and > 65. The weight for each experiment is computed as the ratio of number of households included in each experiment and the total number of households in the CER data set for which the `age_person` property is given. Thus, *Avg* indicates the accuracy obtained by using `age_person`

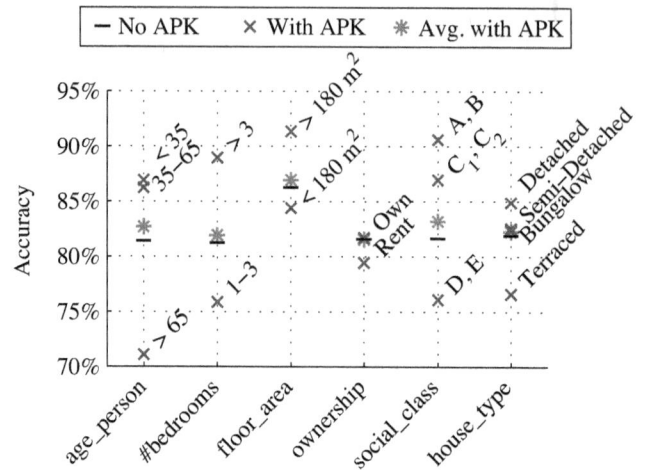

Figure 8: Accuracy obtained by CLASS when classifying the property `single` using different types of a priori knowledge (APK).

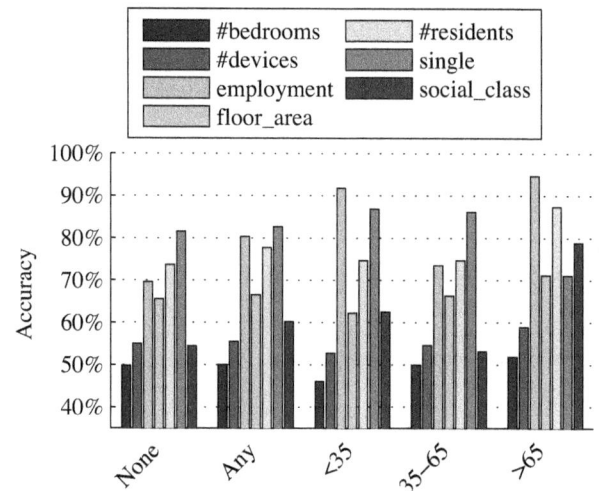

Figure 9: Accuracy obtained by CLASS when the value of the property `age_person` is known a priori.

as a priori property. For the sake of comparison, *All* shows the accuracy obtained by CLASS when all households are included into classification (i.e., no a priori information is available).

The consequences of using `age_person` as a priori property property depend both on the type of a priori knowledge available as well as on the property to be classified. For instance, knowing whether a person in between 35 and 65 lives in the households (experiment $35 - 65$), has almost no effect of the accuracy of the properties shown in figure 9. Instead, significant accuracy gains can be achieved when the a priori information tells whether at least one resident is younger than 35 or older than 65. In the second case, the accuracy of properties `employment`, `floor_area`, `#residents`, and `social_class` increases by 25.7%, 5.5%, 14.0%, and 24.6%, respectively. At the same time, the accuracy of the property `single` decreases by 10.3%. For the case in which it is known that at least one person younger than 35 lives in the household, the accuracy of properties `employment`, `single`, `#residents`, and `social_class` increases by 22.8%, 5.5%, 1.4%, and 8.3%, respectively. However, the accuracy of the property `floor_area` decreases by 3.3%. As for the accuracy of the remaining properties (`#bedrooms`, `#devices`), only marginal changes are obtained. The weighted accuracy (illustrated in figure 9 as Avg) can be improved by 11.3%, 4.4%, and 6.0% for properties `employment`, `#residents`, and `social_class` to a total of 80.3%, 77.8%, and 60.2%, respectively.

Considering these results, acquiring information about the age of at least one of the residents of a household is definitely valuable to learn other properties of the same household. Performing classification using all households traces (Avg) allows to improve the accuracy of properties `employment`, `#residents`, and `social_class`. For these properties the performance gain can be further strengthened by selecting only households with at least one resident older than 65. However, this additional gain is only achievable for the selected subset of households.

6. LIMITATIONS AND OUTLOOK

The paper shows that classification of households based on their electricity consumption is feasible for the household properties we investigated. Though, we expect that there is still room to optimize CLASS and improve classification performance. First, extending the feature set to provide information from more sources or more detailed information to the classifiers can further improve classification performance. For instance, we plan to integrate historical values of the outdoor temperature in Ireland as well as vacation information. We also plan to exploit more statistical properties on the load curves and to compute features for individual days rather than using a weekly average. With respect to feature selection, we will implement the sequential floating feature selection method (SFFS), which is considered as one of the best performing feature selection methods [34]. To increase the performance of CLASS for use cases in which high precision and recall scores are more relevant, we will change the figure of merit during feature selection to the F_1 score. Finally, we will investigate the boosting effect by combining classifiers, where – in the ideal case – some classifiers can compensate for the weaknesses of others.

Besides optimization of precision and accuracy, it is a part of future work to provide more detailed information with the classification results. Until now, for each household property, CLASS estimates the class a household (most likely) belongs to. First, adding a confidence value to each estimation allows energy providers to select households that most likely match a given class. Second, we plan to integrate regression models to estimate continuous variables (e.g., for properties `#residents`, `#devices`, `#bedrooms`) in addition to our class-based analysis.

Future work also contains an evaluation on the requirements for energy providers with respect to the measurement infrastructure. The results of this study are obtained using a single week of 30-minute electricity consumption data. Using data from multiple weeks, both in summer and winter, potentially increases classification performance. However, it also requires more data being measured by the energy providers. Similarly, we will study to what extent classification performance suffers from a decrease in granularity, for instance using hourly, daily, or monthly measurements or

even diluted data. This, in turn, simplifies data collection for energy providers as it reduces data storage and communication overhead. Finally, we investigate the contribution of individual features to classification performance, in order to provide – for each household property – a list of key features that is required (or sufficient) to classify the property.

7. CONCLUSIONS

This paper presents the design, development and evaluation of CLASS, a system that classifies private households according to pre-specified properties. Examples of such properties include the floor area of a household or the number of its occupants. CLASS analyzes the electricity consumption data of the household and estimates the values of its properties. We evaluate the performance of CLASS on data from 3,488 households and for twelve different properties. Our results outline that CLASS can classify with good accuracy ($> 70\%$) eight out of these twelve properties. The performance in terms of precision and recall as well as for different types of classifiers is also analyzed. Furthermore, we evaluate the impact that the availability of a priori information has on CLASS's classification performance. The quantitative results presented in this paper refer to the performance obtained by CLASS on the CER data set and are not necessarily representative also for other data sets. However, our results show that automatic classification of households based on electricity consumption data is feasible.

8. ACKNOWLEDGEMENTS

This work has been partially supported by the Hans L. Merkle Foundation and by the LOEWE research initiative of the State of Hesse, Germany, within the Priority Program Cocoon.

References

[1] J. Vasconcelos, "Survey of regulatory and technological developments concerning smart metering in the European Union electricity market," *EUI RSCAS Policy Papers*, no. 1, 2008.

[2] Directive 2009/72/EC of the European Parliament and of the Council of 13 July 2009 concerning common rules for the internal market in electricity and repealing Directive 2003/54/EC.

[3] Directive 2012/27/EU of the European Parliament and of the Council of 25 October 2012 on energy efficiency, amending Directives 2009/125/EC and 2010/30/EU and repealing Directives 2004/8/EC and 2006/32/EC.

[4] F. Mattern, T. Staake, and M. Weiss, "ICT for green – How computers can help us to conserve energy," in *1st International Conference on Energy-Efficient Computing and Networking (e-Energy)*, pp. 1–10, ACM, 2010.

[5] C. Fischer, "Feedback on household electricity consumption: a tool for saving energy?," *Energy Efficiency*, vol. 1, no. 1, pp. 79–104, 2008.

[6] S. Darby, "The effectiveness of feedback on energy consumption. A review for DEFRA of the literature on metering, billing and direct displays," 2006.

[7] M. Zeifman and K. Roth, "Nonintrusive appliance load monitoring: Review and outlook," in *IEEE International Conference on Consumer Electronics (ICCE)*, pp. 239–240, 2011.

[8] H. Kim, M. Marwah, M. Arlitt, G. Lyon, and J. Han, "Unsupervised disaggregation of low frequency power measurements," in *SIAM International Conference on Data Mining (SDM)*, pp. 747–758, 2010.

[9] G. Hart, "Nonintrusive appliance load monitoring," *Proceedings of the IEEE*, vol. 80, no. 12, pp. 1870–1891, 1992.

[10] F. Nogales, J. Contreras, A. Conejo, and R. Espínola, "Forecasting next-day electricity prices by time series models," *IEEE Transactions on Power Systems*, vol. 17, no. 2, pp. 342–348, 2002.

[11] D. De Silva, X. Yu, D. Alahakoon, and G. Holmes, "A data mining framework for electricity consumption analysis from meter data," *IEEE Transactions on Industrial Informatics*, vol. 7, pp. 399–407, 2011.

[12] C. Beckel, L. Sadamori, and S. Santini, "Towards automatic classification of private households using electricity consumption data," in *4th ACM Workshop on Embedded Sensing Systems for Energy-Efficiency in Buildings (BuildSys)*, pp. 169–176, 2012.

[13] Directive 95/46/EC of the European Parliament and of the Council of 24 October 1995 on the protection of individuals with regard to the processing of personal data and on the free movement of such data.

[14] Council of the Organization for Economic Co-operation and Development (OECD), "OECD guidelines on the protection of privacy and transborder flows of personal data," 1980.

[15] J. Liang, S. K. K. Ng, G. Kendall, and J. W. M. Cheng, "Load signature study—Part I: Basic concept, structure, and methodology," *IEEE Transactions on Power Delivery*, vol. 25, no. 2, pp. 551–560, 2010.

[16] A. Reinhardt, P. Baumann, D. Burgstahler, M. Hollick, H. Chonov, M. Werner, and R. Steinmetz, "On the accuracy of appliance identification based on distributed load metering data," in *2nd IFIP Conference on Sustainable Internet and ICT for Sustainability (SustainIT)*, 2012.

[17] J. M. Abreu, F. P. Câmara, and P. Ferrão, "Using pattern recognition to identify habitual behavior in residential electricity consumption," *Energy and Buildings*, vol. 49, pp. 479–487, 2012.

[18] S. V. Verdú, M. O. García, C. Senabre, A. G. Marín, and F. J. G. Franco, "Classification, filtering, and identification of electrical customer load patterns through the use of self-organizing maps," *IEEE Transactions on Power Systems*, vol. 21, no. 4, pp. 1672–1682, 2006.

[19] B. Birt, G. Newsham, I. Beausoleil-Morrison, M. Armstrong, N. Saldanha, and I. Rowlands, "Disaggregating categories of electrical energy end-use from whole-house hourly data," *Energy and Buildings*, vol. 50, pp. 93–102, 2012.

[20] G. Chicco, R. Napoli, P. Postolache, M. Scutariu, and C. Toader, "Customer characterization options for improving the tariff offer," *IEEE Transactions on Power Systems*, vol. 18, no. 1, pp. 381–387, 2003.

[21] V. Figueiredo, F. Rodrigues, Z. Vale, and J. Gouveia, "An electric energy consumer characterization framework based on data mining techniques," *IEEE Transactions on Power Systems*, vol. 20, no. 2, pp. 596–602, 2005.

[22] I. Sánchez, I. Espinós, L. Moreno Sarrion, A. López, and I. Burgos, "Clients segmentation according to their domestic energy consumption by the use of self-organizing maps," in *6th International Conference on the European Energy Market (EEM)*, pp. 1–6, IEEE, 2009.

[23] T. Räsänen, J. Ruuskanen, and M. Kolehmainen, "Reducing energy consumption by using self-organizing maps to create more personalized electricity use information," *Applied Energy*, vol. 85, no. 9, pp. 830–840, 2008.

[24] F. McLoughlin, A. Duffy, and M. Conlon, "Characterising domestic electricity consumption patterns by dwelling and occupant socio-economic variables: An Irish case study," *Energy and Buildings*, vol. 48, pp. 240–248, 2012.

[25] J. Kolter and J. Ferreira, "A large-scale study on predicting and contextualizing building energy usage," in *25th Conference on Artificial Intelligence (AAAI)*, 2011.

[26] C. M. Bishop, *Pattern Recognition and Machine Learning*. Springer, 2006.

[27] T. W. Anderson, *An Introduction to Multivariate Statistical Analysis*. John Wiley & Sons, 1984.

[28] C. Cortes and V. Vapnik, "Support-vector networks," *Machine Learning*, vol. 20, no. 3, pp. 273–297, 1995.

[29] I. Witten, E. Frank, and M. Hall, *Data Mining: Practical Machine Learning Tools and Techniques*. Morgan Kaufmann, 2011.

[30] E. Alpaydin, *Introduction to Machine Learning*. MIT press, 2004.

[31] A. Whitney, "A direct method of nonparametric measurement selection," *IEEE Transactions on Computers*, vol. 100, no. 9, pp. 1100–1103, 1971.

[32] A. Jain and D. Zongker, "Feature selection: Evaluation, application, and small sample performance," *IEEE Transactions on Pattern Analysis and Machine Intelligence*, vol. 19, no. 2, pp. 153–158, 1997.

[33] R. Fandos, *ADAC System Design and its Application to Mine Hunting Using SAS Imagery*. PhD thesis, TU Darmstadt, 2012.

[34] P. Pudil, J. Novovičová, and J. Kittler, "Floating search methods in feature selection," *Pattern Recognition Letters*, vol. 15, no. 11, pp. 1119–1125, 1994.

Computational Environmental Ethnography: Combining Collective Sensing and Ethnographic Inquiries to Advance Means for Reducing Environmental Footprints

Johanne Mose Entwistle
Alexandra Institute A/S
Denmark

Henrik Blunck, Niels Olof Bouvin, Kaj Grønbæk,
Mikkel B. Kjærgaard, Matthias Nielsen, Marianne G. Petersen,
Majken K. Rasmussen, Markus Wüstenberg
Department of Computer Science, Aarhus University, Denmark

ABSTRACT
We lack an understanding of human values, motivations and behavior in regards to new means for changing people's behavior towards more sustainable choices in their everyday life. Previous anthropological and sociological studies have identified these objects of study to be quite complex and to require new methods to be unfolded further. Especially behavior within the privacy of people's homes has proven challenging to uncover through the use of traditional qualitative and quantitative social scientific methods (e.g. interviews, participatory observations and questionnaires). Furthermore, many research experiments are attempting to motivate environmental improvements through feedback via, e.g., room displays, web pages or smart phones, based on (smart) metering of energy usage, or for saving energy by automatic control of, e.g., heating, lighting or appliances. However, existing evaluation methods are primarily unilateral by opting for either a quantitative or a qualitative method or for a simple combination—and therefore do not provide detailed insight into the potentials and impacts of such solutions. This paper therefore proposes a combined quantitative and qualitative collective sensing and anthropologic investigation methodology we term Computational Environmental Ethnography, which provides quantitative sensing data that document behavior while facilitating qualitative investigations to link the data to explanations and ideas for further sensing. We propose this methodology to include the establishment of base lines, comparative experimental feedback, traceable sensor data with respect for different privacy levels, visualization of sensor data, qualitative explanations of recurrent and exceptional patterns in sensor data, taking place as part of an innovative process and in an iterative interplay among complementing disciplines, potentially including also partners from industry. Experiences from using the methodology in a zero-emission home setting, as well as an ongoing case investigating transportation habits are discussed.

Categories and Subject Descriptors
H5.m. Information interfaces and presentation

General Terms: Measurement, Design, Human Factors.

Keywords: Evaluation methods, anthropology, environmental behavior, collective sensing, eco-feedback, smart control

1. INTRODUCTION
Climate change and other environmental concerns have led the European Union to decide that by 2020 developed countries should collectively reduce their greenhouse gas emissions by 20% [18]. This requires that companies, cities, and society at large are mobilized and engaged to take collective action towards a sustainable future.

To help reach such goals, research is exploring methods to help companies, cities and the society at large to decrease their environmental impact. Two of the methods considered in connection with computing technologies are: 1) increasing awareness: changing people's behavior with computing technologies encouraging more environmentally friendly behavior, e.g. using eco-feedback technologies for water or electricity [27]; 2) infrastructure improvements: optimize or redesign the infrastructure in buildings and cities using computing technologies to decrease the environmental footprint of human behavior, e.g., through automation [19]. Both methods have advantages and drawbacks, and active research is assessing their respective effectiveness. However, in either case, whether developing a new incentive for changing behavior, or optimizing infrastructure, the respective initiative should be grounded in an understanding of everyday life of humans [26, 41] and afterwards be evaluated to assess its impact and to understand problems and prospects. This complex of design and evaluation calls for a combination of qualitative and quantitative investigations.

Much existing work has approached such evaluations solely from the perspective of computer science: E.g., Barker et al. [11] present a quantitative evaluation of a system for flattening peak energy consumption, and Erickson et al. [7] present a quantitative and qualitative evaluation of an eco-feedback technology for water consumption. Furthermore, often when quantitative and qualitative methods are used together they are applied in a disconnected fashion in independent phases, and qualitative methods are limited to pre- and post-interviews. Furthermore, current approaches that combine quantitative and qualitative methods are primarily retrospective and developed and applied to generate knowledge for academic use alone—instead of being part of an innovation process, as proposed here, where the main purpose is to generate shared knowledge across academic disciplines and with industry and public sector parties in order to develop future innovative and effective solutions. An amplification of the single scientific perspective on the domain is given by Froehlich et al. [20] who compare work in different disciplines and argue for more interaction among the disciplines.

The emerging domain of environmental ethnography concerning energy consumption [28] within anthropology intends to study human behaviors and values relating to environmental issues with an aim of understanding these in their social and cultural context [49]. As such, environmental ethnography offers to provide new contextual insights and a better understanding of human motivations, barriers and values in the context of evaluating means for eco-footprint reduction.

The objects of study and the theories applied for understanding these objects are overlapping with environmental sociology, but there is a difference in methodology: Where sociology in comparison leans towards quantitative methods such as questionnaires and/or structured interviews [42, 44], environmental ethnography will take a more qualitative approach applying methods such as participatory observation and in-depth semi-structured interviews in its evaluation [25][46][47][48]. The objects of study also overlap with research on sustainability within environmental psychology, which focuses on the internal psychological mechanisms involved in human behavior, assuming the systematic use of information and consistent behavior based on specific intentions [53]; over the years several models [26][45] have been proposed that explain the factors impacting human pro-environmental behavior. In contrast, environmental ethnography views behavior in a social and cultural context, and explores how this context affects behavior, values, and motivation.

Another recent trend is the use of increasingly widespread sensing technologies and computational techniques and models to quantify and model environmental related human behavior. This we capture in the concept of collective sensing, which encompasses the use of both mobile and stationary sensors available in the users' environment, and which can be utilized to quantify user behavior that is relevant to environmental impact. Collective sensing creates new opportunities for mapping human behavior at different scales from individuals to societies, see, e.g., [29][2][3]. We argue for including such methods in the evaluation of new means for reducing environmental footprints. To do so, we propose an inter-disciplinary methodology of *Computational Environmental Ethnography (CEE)* that combines methods from the disciplines of computer science / engineering and anthropology. The methodology enables evaluations that not only quantify the environmental impact but also provide an understanding of the mechanisms behind the impact, such as human motivations, on the basis of both qualitative and quantitative technical investigations. The methodology furthermore allows us to evaluate the experienced human value, such as comfort, health, etc. With its systemized methodology and incorporation of anthropological data, CEE thus goes beyond the combination of consumption data and survey study, as conducted by researchers within environmental sociology [42]. The combination and interplay of different disciplines and types of data give evaluators access to additional and novel insights, which are not obtainable through the use of either qualitative or quantitative/technical studies alone. But it is important to stress that this new knowledge does not only arise because of additional or more diverse data, but that the mentioned interplay of disciplines and types of data provides a valuable synergy effect: The inter-disciplinary methodology facilitates that new types of questions may be asked, and therefore new types of knowledge be generated, and it ensures a more challenged and thus valid interpretations and understanding of the different types of data.

In this paper, we introduce CEE as an inter-disciplinary and combined quantitative and qualitative methodology, and we illustrate its usage in several case studies, focusing on a case for smart home controls for reducing footprints at home. Additionally, we present and discuss how CEE can enable more insightful evaluations in an ongoing case exploring transportation habits.

2. RELATED WORK

In this section we review existing method usage for studying means to reduce environmental footprints in regards to human behavior, barriers, motivations and values; e.g., for studying smart control systems, eco-feedback technologies, or for studying existing technologies or practices. As listed in Table 1 we classify work according to: the used research method; the involved research domains: computer science (CS), engineering (Eng), environmental ethnography (Env. Ethno.), environmental psychology (Env. Psyc.), environmental sociology (Env. Socio.), and architecture (Arch); the application area; the main research contributions; and how the research methods were applied.

The reviewed articles include work targeting consumption of water, gas, electricity, heating, ventilation and air conditioning in the home and public buildings as well as transportation. Most of the reviewed articles use quantitative methods (10), many a combination (8), and three articles use solely qualitative methods. In the articles that combine quantitative and qualitative methods, the methods are used in independent steps, and the qualitative method used is pre- or post-interviews. Furthermore, only one article involves several research domains. With a single disciplinary perspective the work closest to ours are the ones of Gram et al. [42] and Bates et al. [1]. In both cases input data is sensed (here, regarding electricity usage) as well as gathered in interviews with participants and then combined in analysis. In contrast to our work, though, their processes are i) not evolving, but only assess a status quo by selective data gatherings, ii) use a short evaluation period, and iii) do not utilize the method of participatory observations. In comparison, the methodology described in this paper proposes an inter-disciplinary methodology involving anthropology, engineering, and computer science, and utilizes both several qualitative and quantitative methods to gather more holistic findings. This, we argue, is valuable in order to make an actual impact on the environmental challenges ahead.

3. COMPUTATIONAL ENVIRONMENTAL ETHNOGRAPHY

CEE represents an iterative methodology combining primarily quantitative collective sensing with qualitative ethnographic inquires. The methodology is applicable in foremost two use scenarios: 1) in an explorative manner in a study of existing practices or in a development process with relatively few participants to ground a design process; 2) as an evaluation methodology when an environmental control/feedback system has been implemented or an environmental campaign has been initiated.

Figure 1 shows the interplay between the two inquiry disciplines. On the one hand, collective sensing analysis may inspire the ethnographic inquires to ask certain questions. Furthermore, if privacy restrictions allow, the collective sensing data concerning individual participants may be analyzed and used as basis for qualitative inquiries about their behavioral patterns.

On the other hand, the ethnographic inquiries may generate hypotheses for collective sensing data collection and also suggest indicators for verifying user claims about behavior in the qualitative interviews. The above items are just examples of the potential interplays between the inquiring disciplines. We propose to arrange the process of Computational Environmental Ethnographic data gathering and analysis in a lifecycle, structured by a set of stages, where in all but the final stage, the process is governed by either collective sensing or ethnographic inquires. We will provide more thorough explanations of such a lifecycle in Section 3.3 and the subsequent sections.

Table 1 – research methods, research domains and application areas for existing work.

Authors	Research Methods	Research Domains	Area	Research Contributions	Research Method usage
Srikantha et al. [15]	Quantitative	CS / Eng.	City	Methods for lowering peak power consumption using appliance elasticity	Quantitative method evaluations on data sets of sensor data
Gram-Hanssen et al. [42]	Quantitative/Qualitative	Env. Socio.	Home	Methods for identifying patterns of domestic electricity use and explanations for variations	Quantitative method evaluations on data sets of sensor data compared with variables from questionnaires and interviews.
Gram-Hanssen [43]	Qualitative	Env. Socio.	Home	The inter-relatedness of water consumption and cleanliness practices	Qualitative analysis
Bartiaux [41]	Quantitative/Qualitative	Env. Socio.	Home	Research on residential energy consumption challenging conception of consumer rationality	Quantitative survey and qualitative in-depth interviews
Rahayu et al. [14]	Quantitative	CS	Home	Methods for disaggregating individual usage events from electricity data	Quantitative method evaluations on data sets of sensor data
Froehlich et al. [2]	Quantitative	CS / Eng.	Home	Single point methods for disaggregating usage from water, gas and electricity data	Quantitative method evaluations on collected data sets of sensor data
Bates et al. [1]	Quantitative / Qualitative	CS	Home	Service-oriented analysis and sensing of resource usage	Quantitative analysis of collected sensor data and qualitative follow-up interviews.
Brush et al. [5]	Qualitative	CS	Home	Study results in regards to living with existing home automation systems	Qualitative methods including home tour and semi-structured interviews
Erickson et al. [7]	Qualitative / Quantitative	CS	Home	Eco-feedback technology for water consumption using smart water meters	Quantitative methods for analysis of log data and surveys and qualitative interviews
Kjeldskov et al. [8]	Qualitative / Quantitative	CS	Home	Eco-feedback tool for understanding and reducing electricity usage	Quantitative methods for analyzing tool logs and usage data and qualitative interviews
Costanza et al. [6]	Qualitative / Quantitative	CS	Home	Interactive energy annotation systems focusing on activities	Quantitative methods for analyzing usage logs of usage and qualitative semi structured interviews
Barker et al. [11]	Quantitative	CS	Home	Method for flattening electricity demand by scheduling background loads.	Quantitative method evaluations on collected data sets of sensor data
Scott et al. [12]	Quantitative	CS	Home	System for smarter scheduling of heating using occupancy prediction	Quantitative system evaluation in deployments and with collected data.
Lu et al. [13]	Quantitative	Arch. / CS	Home	Methods for smarter scheduling of HVAC using occupancy sensing	Quantitative method evaluation with collected data.
Winett et al. [50]	Quantitative	Env. Psyc.	Home	Assessment of the effect of monetary rebates and eco-feedback on household electricity consumption.	Controlled experiment of rebates and eco-feedback. Quantitative questionnaires combined with electricity meter readings.
Kuznetsov et al. [10]	Qualitative / Quantitative	CS	Home / Public	Design implications for persuasive water-conservations displays	Qualitative interviews and quantitative performance evaluations
Kim et al. [16]	Quantitative	CS / Eng.	Home / Public	Vibration-based water flow rate monitoring system and findings	Quantitative method evaluations on data sets of sensor data
Aswani et al. [17]	Quantitative	CS / Eng.	Public	Learning occupancy and temperature model-predictive HVAC control	Quantitative method evaluations on data sets of sensor data
Mun et al. [3][9]	Quantitative	CS / Eng.	Trans.	Methods for estimating transportation emissions using mobile sensing	Quantitative evaluations on collected data sets of sensor data
Froehlich et al. [4]	Quantitative / Qualitative	CS	Trans.	Methods and design of eco-feedback technologies for transportation events	Quantitative surveys, experience sampling and sensing evaluations and qualitative interviews
Hargreaves [46]	Qualitative	Env. Ethno.	Work	Practice, process and power in the workplace	Ethnographic case study

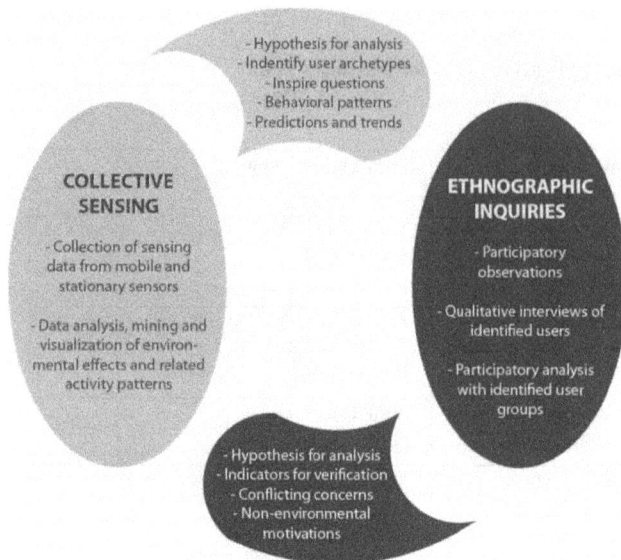

Figure 1: Interplay between collective sensing and ethnographic inquiries.

With an inter-disciplinary and dialectic methodology new insights, which could not have been gained through separate quantitative technical analysis or qualitative approaches alone, are brought into play. The systematic ongoing interplay between methods from the different disciplines ensures that the findings and conclusions of the evaluation are cross-examined (triangulated) and thus validated on a wide range of criteria.

We have developed this methodology to raise opportunities for these different disciplines to challenge, unfold, and validate each other's analyses and results. E.g., qualitative results are validated both through quantitative sensor data and qualitative anthropological data. The sensor data provides a breadth by being collected 24/7/365, while the anthropological data are collected periodically, and at different times of day to ensure a certain dispersion in activities, but also so that it allows for comparing e.g. several mornings, through which patterns can emerge.

3.1 Anthropological Part of the Methodology

A previous anthropological study of energy consumption in the home involving 24 Danish families concludes that changing people's energy consumption habits is a complex process necessitating a reassessment of our current perception of energy consumption as a simple and practical practice [25]. Furthermore, it calls for new methods to research into the complexity of *how* and *why* people consume energy that takes into account what people *say* they do and what they actually *do*.

Activities with environmental impact, e.g. energy consumption in the home, are a '*mediated consumption*'. It is never a direct intentional action. In other words: energy is something we all consume more or less subconsciously in the pursuit of other and greater ends: ends, to which we all subconsciously or consciously ascribe emotional, social and/or cultural meaning as well as practical meaning. E.g., we use light for reading, we take care of our family by cooking for them, and we are relaxed and entertained by TVs and computers—all of which consume energy. But we do not do these things for the sake of consuming energy, and the consumption involved in our practice is usually hidden, and for these two reasons we seldom understand our daily life and work practices in energy related terms. This invisibility also poses a challenge for researchers to research into energy consumption, because people

cannot necessarily tell us what, how and why they consume energy. So to research into the mechanisms of energy consumption we need to apply methods such as qualitative interviews and participatory observations; more details on uch ethnographic inquiries and methods will be given in Section 4.1. But as we have discovered in our previous studies, using these methods as stand-alone has its limitations. Especially actual behavior within the privacy of people's homes or cars has proven challenging to uncover through the use of traditional qualitative and quantitative social scientific methods (e.g. interviews, participatory observations and questionnaires).

3.2 Collective Sensing Part of the Methodology

Our methodology utilizes the many new opportunities provided by collective sensing for measuring both human behavior and environmental impacts in terms of electricity, heating, and water consumption, amount of waste, or amount of emitted pollution particles. Sensors able to measure such quantities are starting to become widely available and the growing political and commercial interest in greener living and green economy facilitates large-scale deployment of respective sensing installations as well as support for investigations including collective sensing initiatives using the widespread smartphones, see, e.g. [2][3][4][8][9].

Furthermore, the increasing abundance of sensors has enabled novel data processing techniques to extract from sensor data long-term and more accurate information about human behavior and environmental effects than obtainable by more traditional investigative means (such as interviews and human observation). These techniques include statistical aggregation, data mining, pattern extraction, spatio-temporal data structures and query processing, and sensor fusion techniques. Note that such techniques allow for, among others, human presence and activity recognition and thus allow to analyze not only environmental impacts, but also to infer human behavior and relate it to the measured environmental impacts accurately. Additionally, the participants' smartphones we utilize for sensor data gathering also offer a convenient channel for gathering situated participant feedback via experience sampling [30], where the user can be queried for, e.g., his current activity or subjective assessment of his current situation –as used also for environmental behavior profiling [4].

In the following we will briefly overview the uses and implications of the above techniques for investigations into both environmental behavior as well as impacts. Firstly, large amounts of continuously sensed data facilitate the application of statistical methods and data mining techniques for, among others, evidence-based testing of hypotheses as formulated by researchers. Herer, the high volume and frequency of data and also its complementing heterogeneous sources allow for a higher amount, accuracy and reliability, as is feasible to achieve by more traditional means such as sporadic human observations by researchers. Additionally, sensor-data-based testing is applicable also for verifying statements provided by interviewees regarding their actions and general behavior: Beyond verification, this also allows to correlate interviewee statements to their actual behavior, and thus also to uncover respective biases, and, in turn, allows for a more founded interpretation of interview and questionnaire data.

Secondly and beyond testing purposes, data mining and visual analysis techniques facilitate also the discovering and explanation of previously undetected or unexpected phenomena, and the shaping of new hypotheses. For instance, the distinguishing (and establishing of archetypes of) people with regards to their environmental behavior can be facilitated by clustering techniques when applied to the data inferred from sensing the environment and behavior of participants (and, optionally, to supplementing questionnaire data).

For other use-cases, data mining techniques can be paired with intuitive visual analysis tools in order to provide to researchers and domain experts in environmental investigations the following: i) global pictures of sensor-derived data such as footprints of energy consumption and greenhouse gas emission –be it per day or by year, per user or also by building population or company workforce, ii) the identification of temporary, spatially or thematically local phenomena such as irregularities, but also opportunities for lowering footprints—by visual highlighting and by allowing the investigators to 'zoom' in an explorative manner into data, selecting arbitrary combinations of data dimensions and levels of detail, iii) uncovering and assessing trends over time in regards to local as well as global aspects, iv) the possibility to predict the impact of changes in the underlying data via the ability to operate on (combinations of) sensed and modeled data, when aiming to assess the potential environmental effects of, e.g., infrastructural alterations, e.g., regarding transport options, or of eco-awareness campaigns, or of changes in human behavior in general. Note that the historical sensed data alone also provides a solid basis for such predictive data modeling.

Note, that among the non-trivial challenges to be addressed when using collective sensing is the protection of the privacy of the participants. This issue also exists, when interviewing participants, but during the longer collective sensing stages the participant is often less aware of the data gathering and of which kinds of private information he is revealing as it may be deduced from the sensed data, see [31][32] for related discussions.

3.3 Combining the Parts

We propose to conduct the inter-disciplinary evaluation by using methods of collective sensing and anthropology alongside each other in a set of stages in which the different disciplines play different roles. In the following, we motivate, using prototypical example cases, the usage of an iterative inter-disciplinary process in which the disciplines interact in a set of stages in order to a) achieve for the subject of investigation a picture as complete as possible and to b) allow for recording developments and changes in behavior and experiences throughout the exploration or evaluation. The iterative process may also be repeating creating a cycling process that revisits stages. In the subsequent sections we will then present in more detail concrete examples of how we apply the method to different actual cases.

A prototypical set of stages when using CEE for an evaluation include an initial *collective sensing and visualization* stage, where the technical disciplines prepare quantitative results, analysis, and visualizations to the anthropologists. This data can then provide the anthropologists with focus points for *participatory observation and analysis*, which is conducted in parallel to continued collective sensing. In this stage, the anthropologists make observations and uncover existing practices relating to the focus points. Afterwards the technical disciplines typically prepare updated quantitative results, analyses and visualizations with sensor data from the observation period. A comparative assessment of the quantitative and qualitative data can then be employed to identify focus points for interviews in a stage of *semi structured interviews and analysis*. The results of the interviews typically include a report about findings and questions for sensor data. The technical disciplines can then analyze the sensor data from the whole evaluation period in regards to the questions in a *technical analysis and assessment* stage. Finally, a stage of *inter-disciplinary analysis* can bring together both computer scientists / engineers and an-

thropologists in a common analysis of qualitative questions from the reports and the new questions that are derived from the technical analysis.

When the type of the intended investigation is an exploratory study, CEE could include an initial *survey and analysis stage* to quantify the importance of different environmental practices and aspects. This allows to already focus on studying the most important aspects and practices, when the technical disciplines set up the *collection sensing and visualization stage*. The analysis of the sensing results can then create focus points for unstructured interviews. The topics discovered through evaluating the unstructured interviews can then be more thoroughly explored, e.g., by refocusing the collective sensing, and/or by augmenting it with experience sampling [30] to gather qualitative input on the respective focus topics during continued sensing. Finally, a stage of *inter-disciplinary analysis* could bring together both computer scientists / engineers and anthropologists in a common analysis of qualitative questions from the reports and the new questions that are derived from the technical analysis.

These examples of stages sketch proposals of how to combine collective sensing and anthropological methods to optimally foster interdisciplinary analysis in a given scenario. The goal is to combine methods, firstly, to initiate and facilitate the exchange of questions, hypotheses, analysis results as illustrated in Figure 1 and, secondly, to ground these in the two disciplines of computer science and anthropology, such that findings are questioning current practices and are interdisciplinary by nature.

4. COMPLETED CASE: TEST HOUSE

In the following, we present a case description of how Computational Environmental Ethnography has been successfully applied—namely to the "Test House" case, as part of the project *Minimum Configuration Home Automation*.

The driving idea behind the Test House case is to build a house that consumes less energy than it produces in its lifecycle. Furthermore, the intention is to build a house that gives the inhabitants the experience of a healthy and comfortable indoor climate and which supports the inhabitants in living their daily lives.

One of the associated project aims was to validate these different intentions using different types of evaluations: These included, first of all, a technical evaluation of the energy performance of the house; in other words, an assessment of whether the house and its inhabitants live up to predefined criteria and standards that had been estimated through theoretical calculations using the software tool Be06 [24]. This evaluation was based on vast amounts of energy consumption and production data collected during the test year. Besides this purely technical assessment, an additional aim was to obtain qualitative models of explanations for the actual energy consumption in the house and a qualitative evaluation of the quality of the house as experienced by the family living in the house.

For these purposes, the Test House was set up as a living lab in which a family, consisting of a couple, their son and two daughters, lived for just over a year to test the house and its extensive home automation technology. The test family had agreed to being monitored through sensors, meters, interviews and observations. 24 hours a day the following data was logged: CO_2-levels, temperature, humidity, light intensity, presence, energy consumption and production, as well as interaction with the home automation system. Additionally, every quarter two days of participant observation were carried out and followed up by interviews.

4.1 The Proposed CEE Methodology Applied

The inter-disciplinary evaluation that is envisioned in CEE is part of a forward development process and not just a retrospective evaluation. Therefore, the evaluation itself is an iterative process. A complete iteration of CEE evaluation is split into six stages as illustrated in Figure 2 and is conducted over a period of three months[1] The stages conducted in each of the iterations follow the prototypical scheme as described in Section 3.3. In the figure, the stages in their order are illustrated; for each stage it is signaled, by color, which party—technicians, anthropologists or both—carry the main responsibility for the outcome of the respective stage. Finally, the intertwined arrangement of the stages is intended to illustrate that the process's division into stages is not completely rigid, but allows instead for dynamic re-allocating of time frames for individual stages, for overlapping of stages, and specifically for extended interactive transitions between stages, when considered necessary or useful.

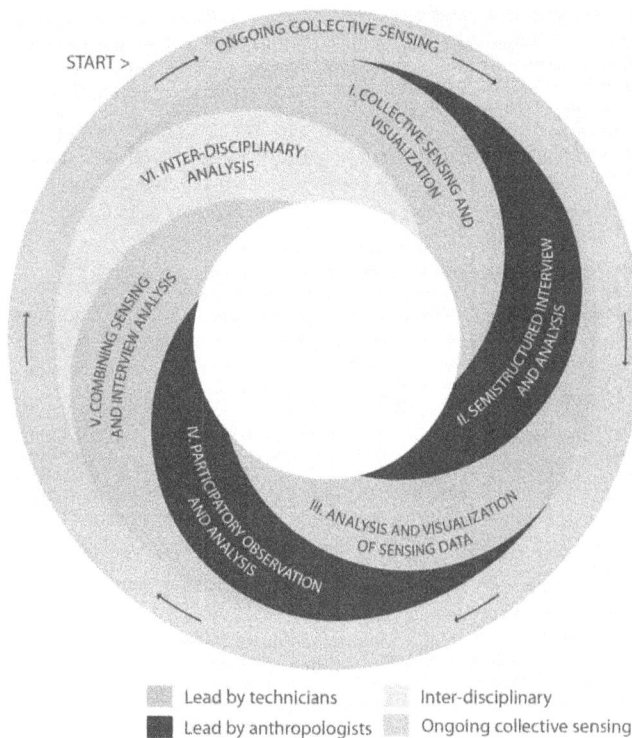

Figure 2 – Iterative Cycle of Computational Environmental Ethnographic analysis, as applied in the Test House case in four iterations.

Individual stages are summarized in Table 2, together with respective prototypical activities undertaken within them. We will in the following focus not on individual project results from the Test House case, but on how and to which end the methodology was applied in the case study in each stage. Furthermore, we will present the anthropological activities in greater detail, whereas we present the technical activities, for which general descriptions of the techniques to apply have been given in Section 3.2, in lesser detail.

Table 2

Stage	Action Details
S1: Collective Sensing and Visualization.	Sensing setup; initial assessment and visualization, focused on identifying interesting phenomena and benchmarking energy-efficiency
S2: Participant Observation and Analysis	Observations recording behavior, routines, interaction with the house and analysis of data
S3: Analysis and Visualization of Sensing Data	Analysis and visualization of behavior and energy-efficiency data; e.g. on floor plans, in focusable visualizations; merging with data from S2; extracting impacts for S4
S4: Semi-structured Interviews and Analysis	Interviews based on findings from S2 and S3, focusing on behavior and related meaning; compiling findings, with feedback from the family
S5: Combining Sensing and Interview Analysis	Combined analysis challenging the findings and questions compiled in S4 through analysis of sensor data
S6: Inter-disciplinary analysis	Analysis covering reasons for behaviors and energy impacts; Equitably involving all disciplines; summarizing findings

4.1.1 Stage 1: Collective Sensing and Visualization

In the first stage, computer scientist / engineers set up the collection of sensor data, and resulting data from a yet limited period is analyzed with the purpose of identifying interesting behavior or fluctuation, e.g., a high number of manual overrides of (a functional feature of) the home automation system, high/low indoor climate values, frequent opening and closing of windows). It is also assessed whether the inhabitants' use of the house is appropriate energy-wise. Results of the analysis are visualized and presented to the anthropologists in the form of statistical numbers, graphs, and floor plan overlays. Figure 3 shows an example of a visualization of: manual overrides, the nature of the overrides, their time stamp, presences, CO_2-levels and temperatures.[2]

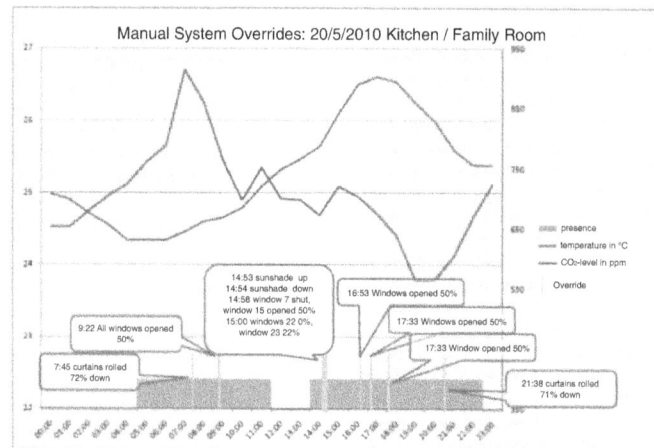

Figure 3 - a visualization of manual overrides, their nature, and respective times, presences, CO2-levels and temperatures.

---

[1] Live in homes in general—as well as specifically with respect to energy consumption, and related behaviour—changes quite substantially with the year's seasons. This is one of the reasons why we choose to have an iteration for each quarter of the year.

[2] Note, that data output includes also inferred data; for instance, while the open vs. closed state of curtains and windows was sensed directly, such state was for outside doors inferred from CO_2 sensor data.

These results allowed the subsequent participant observations to be more focused—in this case, observations on overriding practices and social, cultural and practical contexts for these practices. Identifying this focus crucial would not have been possible without the collected sensor data and the respective preliminary technical analysis of this data.

4.1.2 Stage 2: Participant Observation and Analysis

In order to conduct participatory observations, on a routine observation day the anthropologist arrives at the Test House in the morning, before the family gets out of bed, and observes their morning. The anthropologist leaves the home together with the family and returns with them in the afternoon. During this period, family life is observed and the anthropologist participates in cooking, laundering, dishwashing, homework, playtime, and meals. Furthermore, the anthropologist has followed the family outside the home for shopping and picking up the kids from day care. The participant observation records also the 'behavior' of the house and system, and also how the family reacts to and interacts with the house and the system. The data is recorded in field notes, on floor plans, and via photos [34][52]. In this case smartphone sensing was not utilized but could potentially have provided an additional information channel for the behavior of the participants, e.g., outside of the house.

Participant observation uncovers the practices of daily life of the family and thereby the practical and socio-cultural context in which the family interacts with the house and control system. This context is important to include in order to understand not just what the family does and how often, but also under which circumstances. This contributes to an understanding of why they act as they do, and what meaning they attribute to their own behavior and to the house's and the system's behavior [48]. This understanding is crucial, if we want to affect the way people behave, and this cannot be achieved through sensor data alone.

A central challenge inherent in participant observation within the home is that the presence of an anthropologist/guest in the home often will affect behavior in the home in ways that not necessarily apply for the method of participant observations when used in other empirical fields. It turned out, though, that a family with children has very tight routines in the morning and around mealtimes, and this gave the impression that not much was out of the ordinary even though the anthropologist was present. Furthermore, reflection on and working consciously with different positions and roles (such as guest, au-pair, kid's play-mate, etc.) in the home will help overcome parts of this challenge for the anthropologist [35][37][38].

Another challenge, even harder to overcome, is that the anthropologist is usually limited in her access to home life in more ways than one. In this case, the agreement with the family foresaw two days of observations per quarter. Additional observations could probably have been agreed on and carried out without bothering the family too much, but only to a limited extend, since most families are not willing to have an anthropologist/guest present in their home continuously.

Nonetheless, the main advantage in using participant observation, in the type of evaluation described here for the Test House case, is that this open and holistic approach enables and facilitates new questions that are not known to be relevant in the outset of the evaluation. Participant observation furthermore gives an embodied experience [36] to the anthropologist, which is helpful in the following interviews as it provides shared points of reference with the family and at the same time this experience allows anthropol-

ogists to ask the 'right questions' and interpret the answers within the context they arise [33].

4.1.3 Stage 3: Analysis and Visualization of Sensing Data

In this stage, computer scientists / engineers collect, analyze and energy-assess data from the periods of participant observation, and visualize the results for the anthropologist through graphs and floor plan overlays. These are presented and discussed in a meeting and hereby the qualitative data from the participant observations is contextualized with sensor data of, e.g., temperature or CO_2-levels. Sensor data (and data inferred from it) thus serve as a type of extended observation data in the sense that it registers actions and conditions that are not necessarily all registered by the anthropologist.

The energy and energy behavior related assessments are interesting here not as judgments of the family's behavior but as a means to understand why the family has a need to override, e.g., standards set by engineers for indoor climate (as to be provided by the home automation system). When the energy assessment is compared with the qualitative study this produces knowledge that enables us to question and challenge the dominating assumptions of indoor climate, intelligent control, and comfort values. This knowledge is crucial if we are to design more energy efficient buildings, because inhabitants will override and domesticate technology if it does not meet their needs, regardless of the good intentions behind the system.

4.1.4 Stage 4: Semi-structured Interviews and Analysis

Questions derived from participant observations and technical analysis provide focus for the interviews, but the interviews are still open, so that the family gets the opportunity of contributing with relevant information, not necessarily included in the interview-guide [39]—thus ensuring, that the main focus is given to what *is* important and not on what anthropologists *believe* to be important [34].

The interviews are carried out in the home with each parent individually. An interview takes approximately 1.5 hours, and the technical data visualizations are used, if relevant, as a tool for dialogue. Such visualizations proofed to be a good tool for getting the family to reflect on their practices of daily life—as concluded for the Test House case, but also observed by other researchers [51].

The family is encouraged in the interviews to be as concrete as possible and to give examples, e.g., of how they interact with the system. Therefore the interviews often involve home tours in which the family elaborates on their practices of daily life and on other issues. The interviews are taped, optionally with video, and transcribed, and the anthropologist subsequently analyses the compiled data.

Generally, the interview provides further insights in what meaning the family ascribe to their behavior in their home and their experiences with the house and its technology. In interplay with the sensor data and observational data it allows to reflect on and discuss conscious and subconscious behavior. As an example, the interviews provide insights in which significance it has for the family that they can override the automatic system. This knowledge could not have been generated through participant observation and sensor data alone – just as the actual identification of the significance and complex practice of overriding could not have emerged through interviews alone.

The different types of data employed provide different insights in practices of daily life and meaning, because they have been gathered through different methodologies, and therefore from different positions or views, each of which allows to uncover specific knowledge items, which cannot be identified by the complementing view alone [23]. This triangulation of data, i.e., the incorporation of different view points, also strengthens the validity of the obtained results.

As an outcome of this stage, the anthropologist compiles a report with the findings categorized into the emerged evaluation themes. The report is read and commented on by the family. Their changes and comments are included in the report when considered relevant. The report is sent to the engineers with questions to be answered by means of sensing and subsequent sensor data analysis.

4.1.5 Stage 5: Combining Sensing and Interview Analysis

Following the end of the evaluation period (every three months) the computer scientists / engineers analyze and assess the sensor data from the entire period (as well as from earlier periods to assess trends). In this analysis they are guided by the questions posed in the qualitative report from the anthropologists. For instance, the report contains an assumption that the family's overriding of the system is closely connected to sunshine (or, on occasion, to desires such as to listen to birds singing), and that thus the system's screening does not always meet the needs of the family. This assumption is then in this stage unfolded and challenged through the gathered sensor data and its analysis, which also provides insights into the circumstances (CO_2-levels, humidity, temperature, light intensity, desires) under which the family actually overrides the system. This provides useful knowledge to develop for the home automation a control of screening that better meets the needs of inhabitants.

4.1.6 Stage 6: Inter-disciplinary analysis

The inter-disciplinary analysis in this stage brings together both computer scientists / engineers, and anthropologists in a common analysis of qualitative questions from the report and new questions that are derived from the analysis. The different competences and organizational affiliations ensure that expertise on user behaviors, the house, the technology in the house, sensor data, energy assessment etc. is represented. This means that different participants can question each other's knowledge and hence develop models of explanations and conclusions that are validated by different disciplines and types of data.

An example of this is the theme of overriding (shown in sensor data and observed) which leads to a focus on the family's experience of high local indoor temperatures and their dissatisfaction with the automation system for not eliminating this problem (uncovered in interviews and home tours). Through the interdisciplinary analysis it is uncovered that the family's earlier changes in desired default temperature in combination with the current control logic of the system causes the windows to remain shut even though the family feels there is a need or desire for the windows to be open. This causes dissatisfaction with both the indoor climate and the control system. It also causes the family to override the system and to open the windows in the house—which comes with the risk of cooling the house locally and thus causing it to heat up even more. Furthermore, there is a risk of overcooling the house in general or locally, thereby leading to a higher energy consumption for re-heating the house. Without interdisciplinary analysis this conclusion and model of explanation could not have unfolded.

The model of explanation is valuable for the energy assessment of the house, because it helps explain why energy consumption is higher than calculated. But furthermore it also tells us about the relation between human and technology: It shows us that the automation system is too complex and not transparent. The users cannot see through the consequences of their actions, e.g. overrides, which leaves them frustrated and dissatisfied. From this follow further hypotheses and questions for research in the project regarding the matter of to which extent and by which means users should be enabled to configure the home themselves—which is a very crucial knowledge area when developing home automation systems.

Following the inter-disciplinary analysis the final and compiled evaluation report is collaboratively produced, and visualizations of sensor data support the qualitative findings and conclusions that have been unfolded through inter-disciplinary data and analysis.

The end of the report lists a number of questions related to the needs and challenges experienced by the users and the project. These questions are formulated in a constructive and positive manner, phrasing them in the form of 'How might we...?' questions, e.g.: "How might we ensure a indoor climate that meets the users needs for lower temperatures, while maintaining extensive inflow of natural light and view?"

Insights from the report and questions like these feed into the further development of home automation systems and sustainable architecture that better meet user's needs. We further discuss and summarize the performance of the methodology as applied in this case in Section 6, together with its performance in the case described in the following section.

5. ONGOING CASE: EXPLORING TRANSPORTATION NEEDS

In this section we present how to use CEE as an exploratory methodology in the area of transportation in the context of the research project EcoSense (ecosense.au.dk). We start by providing the background for the case, and subsequently describe how and to which end the stages of CEE are applied in this case.

5.1.1 Better Understanding of Transportation Needs

The transportation case aims to increase the sustainability of personal transportation by identifying and changing transportation behaviors of individuals. By 2008, transportation by road accounted for 22.7% of the total CO_2 equivalent emissions stemming from fuel consumption in total of all OECD countries [54]. Furthermore, for transportation by road its share as well as its absolute emission figures have not been reduced significantly in the last 20 years: Transportation by road is by far the largest contributor to CO_2 equivalent emissions caused by fuel combustion in transportation, outnumbering any other type of transportation by a factor 10. While these figures identify transportation by road as a major contributor to CO_2 equivalent emissions from fuel consumption, it does not address the behaviors in transportation that these figures are a result of. And without gaining knowledge of the actual behavior, efforts into decreasing these figures on an individual level will have uncertain outcomes. As transportation behaviors are a matter, which is not easily understood solely by means of quantitative data collection, this case is an obvious subject for CEE.

Within the EcoSense project these challenges were addressed with a focus on personal transportation in a community of neighboring companies. In an exploratory manner we aim to identify for this community, what means are prudent for lowering CO_2 equivalent emissions. Due to the process being exploratory, there is not a prescribed outcome apart from that it should support increased sustainability in transportation behaviors in the community. Possible means for this include, but are not limited to, advanced ride-sharing services, company funded shuttle-busses, shared electric vehicles, and better local traffic planning.

In late 2011 the community of neighboring companies, employing ca. 10,000 people, commissioned a structured questionnaire survey to investigate the transportation behavior of the employees in the community across numerous metrics, such as travel distance between home and work, common errands during transportation, and transportation mode. While this survey provided an overview in hitherto unseen detail (ca. 5.000 respondents) of commute between the community area and its hinterland, it was severely limited by its questionnaire format. It also was constrained to a single temporal point of impact, and therefore must be either updated periodically or be considered invalidated over time.

This reasoning inspired the development and deployment of a smartphone application, which could replace a periodical survey with continuous pure quantitative data collection in a collective sensing manner, see Figure 4. In the first version the user is asked to input corrections to automatic estimations of transportation modes, but in the long term the application's transportation mode detection is planned to be improved and that thereby transportation behavior will be recorded more autonomously, while being framed in an application with functionality that would serve as a reason for users to download, install and use it.

Figure 4 – Screenshots of the smartphone application for sensing transportation events.

5.1.2 Applying CEE for exploration

An initial test of the smartphone application was conducted within a small company. However, the pickup among users was lower than expected. In reaction to that, rather than opting out of utilizing a smartphone application, it was decided to combine the smartphone based sensing with ethnographic expertise, and in effect doing computational environmental ethnography. The reason for this was two-fold: first, the decision for CEE enables the development of a smartphone application that is grounded in actual needs of users instead of desired collected data, and secondly, the study will gain the potential to link collected data about transportation behavior with the reasons underlying it.

Furthermore, while both the questionnaire survey and the smartphone application did provide meaningful results, we deemed their accuracy improvable and their scope extendible.

Therefore, we saw the need for applying a methodology, such as CEE, that integrates quantitative and qualitative studies in a unified approach. However, because this case had been launched before the formalization of CEE as a methodology, the application deviated from the prototypical order given in section 3.3—by way of swapping of stages 2 (Semi-structured Interviews and Analysis) and 4 (Participatory Observation and Analysis).

The status of the project and the application of CEE is that the two first stages have been completed, namely the aforementioned collective sensing via a smartphone application, followed by a semi-structured interview and analysis stage. This section will describe the two completed stages, as well as outline the four planned subsequent stages.

1. Collective Sensing: The initial collective sensing application has been deployed as described earlier, and the respective sensing initialized in this stage allows continuous monitoring and evaluation of transportation behaviors, adaption and evolution over time and throughout the CEE methodology's lifecycle. This stage also includes an initial analysis of the data gathered so far, which serves as a baseline.

2. Semi-structured Interviews and Analysis: As a follow up to the collective sensing deployment, an interview-survey was conducted amongst participants from the community of neighboring companies, focusing on concerns of the participating individuals' privacy and on sharing of sensing data in face of technology that enables to track people with fine-grained detail. One of the major insights gained was that even though privacy and willingness to share location-related data was a concern to the participants, they would be willing to trade in such data, if it was done in a transaction that presented them with sufficient gain. Such gains could be constituted by the availability of better services, monetary benefits, etc. Furthermore, to better address the ease of use for participants and their concerns, it was decided to revise the smartphone application design used for collective sensing based on the interview-survey.

3. Analysis and Visualization of Sensing Data: The status of the process is the pending deployment of the re-designed smartphone application. The re-design aims to provide potential users in the community of neighboring companies with a service of value substantial enough, so that it will incite widespread pick-up and regular use of the smartphone application and result in a more representative data collection. Following the app deployment and while collection is ongoing, the core procedure in this stage will be to perform a more thorough analysis of the data. To this end, focusable and focused visualizations will be pivotal in understanding the collected data in order to analyze and identify potentially relevant patterns in transportation behaviors as well as subsequently, and more importantly, rendering the data understandable for computer scientists, anthropologists, and possibly experts from other fields, in order to decide upon the focus for the following participatory observation activities.

4. Participatory Observation and Analysis: As designers can only speculate about actual usage behavior, and as collectively sensed data can provide only a limited reflection on behaviors, it becomes essential to study transportation behaviors by means such as participatory observation. This activity addresses the risks that a) the collectively sensed data becomes detached from the transportation behavior that it intends to reflect, and b) that the CEE process diverges inexpediently from what remains to be identified as prudent ways for lowering CO_2 equivalent emissions. This activity is therefore a crucial re-grounding of technically sensed reality in individual users' experienced reality.

5. Combining Sensing, Interview, and Observation Analysis: This stage will combine the insights gained in the previous stages, in order to distill likely diverging insights into a framework for the understanding of transportation behaviors in the community of neighboring companies. In this case-scenario, we deemed it crucial to communicate findings to external parties, such as decision makers or experts from disciplines that have hitherto not taken part in the process. Thus, the reason for keeping this stage separate from the following stage of interdisciplinary analysis is that before a combined analysis effort can take place, it is vital that the set of insights and evidences gathered or to be gathered are assessed and important items are carefully selected. Thus, the assessment, selection and preparation of data, analysis and visualizations for presentation should be considered in this stage with thorough consideration of details to in- or exclude.

6. Interdisciplinary Analysis: In the last stage the data selected and prepared in the previous stage is to be subjected to an interdisciplinary analysis. This stage marks the conclusion of the CEE methodology's lifecycle but it can serve also, and here more importantly, as the stage at which future directions and hypotheses are formulated for continued development, as in a second cycle of the CEE methodology's process.

In this case study, we have explored how we can use the CEE methodology to comprehensively understand the transportation habits of a community of neighboring companies. By not only sensing transportation habits alone, or relying on questionnaires and surveys alone, but actively using the gathered data (and visualizations thereof) to inform the participatory observations and analyses, and vice versa, a deeper understanding of people's transportation habits can be gained. Overall, the case exemplifies how early evaluation and development of means for lowering transportation eco-footprints in a community can be steered and guided into promising future directions, including a grounding in the actual motivations of the targeted individuals.

The outcome of the first lifecycle of applying the CEE methodology will be a body of knowledge including how the transportation in the given community is a composition of the behaviors and habits of individuals. This knowledge will be pivotal in a subsequent lifecycle of the CEE methodology, which will shift from being exploratory to focusing on development, deployment and evaluation of concrete means for increasing sustainability in transportation in the community. The outcome should not, however, be viewed as a recipe for success, but rather a set of guidelines that will substantially inform the design and development and that addresses actual needs of targeted individuals and therefore will have a high pickup rate.

6. Discussion and Conclusions

We presented Computational Environmental Ethnography (CEE) as a methodology to advance—in evaluation and iterative development—means to reduce environmental footprints. We showed how the methodology combines collective sensing methods from the disciplines of computer science/engineering with ethnographic inquiries from the discipline of anthropology.

Within CEE, collective sensing allows us to continuously gather data, and thereby provides the means to assess both human behavior as well as its impacts on the environment—in an accuracy, scope and density that would not be achievable by human observation or questionnaire data alone. Anthropological inquiries, including participatory observations and qualitative interviewing, allow for a detailed understanding of humans' situation and behavior, including the motivations and concerns that govern environmental behavior.

We argue, that for environmental science interdisciplinary methods and the types of data they collect can deliver—in their complementing combination and interplay—a holistic picture, descriptions and interpretations of the chosen subject of study. The interdisciplinary methodology we proposed explicitly challenges all involved disciplines, and thereby enables knowledge gains, that would not have been achievable through the use of solely qualitative or solely quantitative/technical studies.

We demonstrated and discussed the methodology's applicability and its benefits in practice by evaluating it and its application in several cases. Concrete benefits of the methodology beyond the ones mentioned above were distilled and validated from these evaluations: Benefits for correlating data inferred from sensors and detailed human observations included: 1) relating sensor data to human concerns, and 2) the interpretation of human experiences and statements from interviews, uncovering potential discrepancies between such statements and objective data. Further benefits arise from the methodology's division into several stages, which are governed alternatingly by one of the involved disciplines. This delivers the intended understanding between the involved parties and the correlation between their respective types of data, hypotheses, developments and results. Additionally, as shown for the addressed cases and related footprint-reducing means, using the methodology ensured that neither evaluation of cases, nor the development or the advancing of means for footprint reduction led into dead-ends: Instead, using the methodology, dead-ends and fruitless deviations where identified and avoided at early stages. This was ensured in particular for those proposed means that were a) conflicting with humans' motivations and needs rather than making use of them – which became apparent quickly through ethnographic enquires, especially through participatory observations; or that b) proved inferior in (combinations of) energy saving and adoption rate – which became apparent via analysis of collective sensing input.

Overall, in the covered cases both evaluations and development processes remained focused throughout and did not deviate towards unrealistic or false assumptions and thus did not lose real-world applicability. We attribute this to the methodology's inherent and continuous challenging of one discipline's results by another one, and by the resulting purposeful and rapid shaping of holistic pictures and interpretations that in turn lead to improved designs or approaches.

For future work we will focus on further integration and evaluation of tools supporting the methodology, i.e., by developing questionnaire support within the smartphone sensing apps, and data visualizations that cover also the results of ethnographic inquiries. Such new innovations in tools supporting the comparison and matching of quantitative and qualitative data will contribute to the scalability of the methodology to large-scale investigations and ongoing continuous probing of environmental behavior and the advancing of means to improve it.

7. ACKNOWLEDGMENTS

We would like to thank the Danish Strategic Research Council for the support of the EcoSense project, in which this research took place. Furthermore, we would like to thank the Danish EBST, User Driven Innovation programme for supporting the MCHA-project, which is used as a main case in the paper.

8. REFERENCES

[1] O. Bates, A. K. Clear, A. Friday, M. Hazas, J. Morley: Accounting for Energy-Reliant Services within Everyday Life at Home. In Pervasive 2012: 107-124

[2] J. Froehlich, E.C. Larson, S. Gupta, G. Cohn, M. S. Reynolds, S. Patel: Disaggregated End-Use Energy Sensing for the Smart Grid. IEEE Pervasive Computing 10(1): 28-39 (2011)

[3] M. Mun, S. Reddy, K. Shilton, N. Yau, J. Burke, D. Estrin, M. H. Hansen, E.Howard, R. West, P. Boda: PEIR, the personal environmental impact report, as a platform for participatory sensing systems research. MobiSys 2009: 55-68

[4] J. Froehlich, T. Dillahunt, P. V. Klasnja, J. Mankoff, S. Consolvo, B. L. Harrison, J. A. Landay: UbiGreen: investigating a mobile tool for tracking and supporting green transportation habits. In CHI 2009: 1043-1052

[5] A. J. Bernheim Brush, B. Lee, R. Mahajan, S. Agarwal, S. Saroiu, C. Dixon: Home automation in the wild: challenges and opportunities. In CHI 2011: 2115-2124

[6] Costanza, E., Ramchurn, S. D. and Jennings, N. R. (2012) Understanding domestic energy consumption through interactive visualisation: a field study. In Ubicomp 2012: 216-225.

[7] T. Erickson, M. Podlaseck, S. Sahu, J. Dai, T. Chao, M. R. Naphade: The dubuque water portal: evaluation of the uptake, use and impact of residential water consumption feedback. CHI 2012: 675-684

[8] J. Kjeldskov, M. B. Skov, J. Paay, R. Pathmanathan: Using mobile phones to support sustainability: a field study of residential electricity consumption. In CHI 2012: 2347-2356

[9] S. Reddy, M. Mun, J.Burke, D. Estrin, M. H. Hansen, M. B. Srivastava: Using mobile phones to determine transportation modes. TOSN 6(2) (2010)

[10] S. Kuznetsov, E. Paulos, UpStream: Motivating Water Conservation with Low-Cost Water Flow Sensing and Persuasive Displays. In CHI 2010

[11] S. K. Barker, A. K. Mishra, D. E. Irwin, P. J. Shenoy, J. R. Albrecht: SmartCap: Flattening peak electricity demand in smart homes. In PerCom 2012: 67-75

[12] J. Scott, A. J. B. Brush, J. Krumm, B. Meyers, M. Hazas, P. Hodges, N. Villar: PreHeat: controlling home heating using occupancy prediction. In Ubicomp 2011: 281-290

[13] J. Lu, T. I. Sookoor, V. Srinivasan, G. Gao, B. Holben, J. A. Stankovic, E. Field, K. Whitehouse: The smart thermostat: using occupancy sensors to save energy in homes. In SenSys 2010: 211-224

[14] D. A. P. Rahayu, B. Narayanaswamy, S.Krishnaswamy, C. Labbé, D. P. Seetharam: Learning to be energy-wise: discriminative methods for load disaggregation. In e-Energy 2012: 10

[15] P. Srikantha, C. Rosenberg, S. Keshav: An analysis of peak demand reductions due to elasticity of domestic appliances. In e-Energy 2012: 28

[16] Y. Kim, H. Park, M. B. Srivastava: A longitudinal study of vibration-based water flow sensing. TOSN 9(1): 8 (2012)

[17] A. Aswani, N. Master, J. Taneja, D. E. Culler, C. Tomlin: Reducing Transient and Steady State Electricity Consumption in HVAC Using Learning-Based Model-Predictive Control. In IEEE 100(1): 240-253 (2012)

[18] European Commission, Leading global action to 2020 and beyond, 2009, Visited January 2013, http://ec.europa.eu/clima/publications/docs/post_2012_en.pdf

[19] M. Hazas, A. J. Bernheim Brush, and J. Scott, Sustainability does not begin with the individual. Interactions, 19 (5): 14-17 (2012)

[20] J. Froehlich, L. Findlater, J. A. Landay: The design of eco-feedback technology. In CHI 2010: 1999-2008

[21] Stern, E.. What shapes European Evaluation? Evaluation 10(1) (2004)

[22] C. Geertz. The interpretation of cultures: selected essays. New York: Basic Books, pp. 3-30 (1973)

[23] Entwistle, J. Mose and Søndergaard, A. The Montage Workshop - the Recreation of Realization. In EPIC 2009

[24] Building Energy Assessment Tool – Be06, http://www.sbi.dk/be06, visited 2012

[25] Entwistle, J. M. (2009) Energy Consumption in the Home Alexandra Institute 2009, www.alexandra.dk

[26] Kollmuss, A. and Agyeman, J.. Mind the Gap: Why do people act environmentally and what are the barriers to pro-environmental behavior?, Environmental Education Research, 8(3): 239-260 (2002)

[27] Darby, S.. The Effectiveness Of Feedback On Energy Consumption. Tech Report from the Environmental Change Institute, University of Oxford, UK (2006)

[28] T. L. Tudor, S. W. Barr, and A. W. Gilg. A Novel Conceptual Framework for Examining Environmental Behavior in Large Organizations: A Case Study of the Cornwall National Health Service (NHS) in the United Kingdom. Environment and Behavior. 40: 426-450 (2008)

[29] M. B. Kjærgaard, M. Wirz, D. Roggen, and G. Tröster. Detecting pedestrian flocks by fusion of multi-modal sensors in mobile phones. In UbiComp 2012. ACM, 240-249.

[30] Froehlich, J., M. Y. Chen, S. Consolvo, B. Harrison, and J. A. Landay. "MyExperience: a system for in situ tracing and capturing of user feedback on mobile phones." In MobiSys 2007: 57-70

[31] Andersen, M. S., and M. B. Kjærgaard. Towards a New Classification of Location Privacy Methods in Pervasive Computing. In Mobiquitous 2012: 150-161.

[32] A. Oulasvirta, A. Pihlajamaa, J. Perkiö, D. Ray, T. Vähäkangas, T. Hasu, N. Vainio, P. Myllymäki. Long-term Effects of Ubiquitous Surveillance in the Home. Ubicomp 2012

[33] Barth, F, "Socialantropologien som Grunnvitenskap" København. Folkeuniversitet: 4-5 (1980) [In Danish]

[34] Emerson, R. M, Fretz, R. I, Shaw, L. L. Writing Ethnographic Fieldnotes. University of Chicago Press (1995)

[35] O., Ton. Informed participation and participating informants. Canberra Anthropology 20(1-2): 96-108 (1997)

[36] Baarts, C. Håndværket. Opbygning af viden. In Hastrup, K (Red) En grundbog I antropologisk metode. Hans Reitzels Folag. 35-50 (2003)

[37] Rosaldo, R. "Grief and the Headhunters Rage (1984)

[38] Wadel, C. Feltarbeid i egen kultur-en innføring i kvalitativt orienter samfunnsforskning, Flekkefjord (1991) [In Danish]

[39] Kvale, S. En Introduktion til det Kvalitative Forskningsinterview. Munksgaard (1996) [In Danish]

[40] Shove E. Comfort, cleanliness and convenience: the social organisation of normality. Berg (2003)

[41] Bartiaux, F. Does environmental information overcome practice compartmentalisation and change consumers' behaviours? Journal of Cleaner Production 16 (2008)

[42] Gram-Hanssen, K.. Different Everyday Lives – Different Patterns of Electricity Use. American Council for an Energy Efficient Economy 2004: 1-13

[43] Gram-Hanssen, K. Teenage Consumption of cleanliness: how to make it sustainable. Sustainability: Science, Practice & Policy 3(2): 15-23 (2007)

[44] Stern, P. C. Toward a Coherent Theory of Environmentally Significant Behavior. Journal of Social Issues 56(3): 407-424 (2000)

[45] Lutzenhiser, L. A Cultural Model of Household Energy Consumption. Energy 17(1):47-60 (1992)

[46] Hargreaves, T.. Making Pro-Environmental Behaviour Work: An Ethnographic Case Study of Practice, Process and Power in the Workplace. PhD Thesis, (2008)

[47] Galtung, L. and Dinesen, M...Innovativ Evaluering og Observation. CEPRA striben (2013) [In Danish]

[48] Gulløv, E. & Højlund, S.. Feltarbejde blandt Børn. Metodologi og etik I etnografisk børneforskning" København. Gyldendal (2010) [In Danish]

[49] McDermott, R. P. The acquisition of a child by learning disability. In Understanding Practice. Perspectives on activity and context (1996)

[50] Winett, R. A.; Kagel, J. H.; Battalio, R. C.; Winkler, R. C. Effects of monetary rebates, feedback, and information on residential electricity conservation. Journal of Applied Psychology 63(1): 73-80 (1978)

[51] Eun Kyoung Choe, Sunny Consolvo, Jaeyeon Jung, Beverly Harrison, Shwetak N. Patel, and Julie A. Kientz. Investigating receptiveness to sensing and inference in the home using sensor proxies. In UbiComp 2012. ACM, 61-70.

[52] Spradley, J. Participant Observation. Wadsworth Thomson Learning (1997)

[53] Gram-Hanssen, Kirsten. Standby Consumption in Households Analyzed With a Practice Theory Approach. Journal of industrial Ecology. Yale (2009)

[54] Transportation emission statistics for OECD http://www.internationaltransportforum.org/jtrc/environment/CO2/OECD.pdf, Visited Jan 2013.

Keynote Talk

The Nimble Distribution Grid: Challenges and Vision

Alexandra von Meier
California Institute for Energy and
Environment
University of California, Berkeley

Distributed Control of Electric Vehicle Charging

Omid Ardakanian
University of Waterloo
oardakan@uwaterloo.ca

Catherine Rosenberg
University of Waterloo
cath@uwaterloo.ca

S. Keshav
University of Waterloo
keshav@uwaterloo.ca

ABSTRACT

Electric vehicles (EVs) are expected to soon become widespread in the distribution network. The large magnitude of EV charging load and unpredictable mobility of EVs make them a challenge for the distribution network. Leveraging fast-timescale measurements and low-latency broadband communications enabled by the smart grid, we propose a distributed control algorithm that adapts the charging rate of EVs to the available capacity of the network ensuring that network resources are used efficiently and each EV charger receives a fair share of these resources. We obtain sufficient conditions for stability of this control algorithm in a static network, and demonstrate through simulation in a test distribution network that our algorithm quickly converges to the optimal rate allocation.

Categories and Subject Descriptors

I.2.8 [**Artificial Intelligence**]: Problem Solving, Control Methods, and Search—*control theory*

Keywords

Electric vehicle charging, congestion control, distributed control

1. INTRODUCTION

Unlike gasoline-powered cars, battery electric vehicles and plug-in hybrid electric vehicles, both referred to as 'EVs', are powered by electricity stored in their on-board battery which is charged when EVs are plugged in to chargers located at homes or public charging stations. EV chargers can impose a significant load on the distribution network: with AC Level 2 charging, EVs can be charged at up to 80A at 240V [2], a load of 19.2kW, whereas a typical North American home has an average load of only 1kW. Therefore, a single EV being charged at the peak Level 2 rate could impose an instantaneous load as large as that imposed by nearly twenty average homes. Consequently, the large-scale introduction of EVs is likely to greatly affect the electrical grid's distribution system [5,7,13]. It can cause overloading of distribution branches and transformers, and voltage drop at distant buses. Persistent overloading can cause damage to conductors, overheat transformers, and degrade their insulation. Excessive voltage drop can cause damage to electrical appliances. Both may eventually lead to a loss of reliability due to the invocation of the protection system.

The branch and transformer congestion problem is known as a major barrier to large-scale adoption of EVs [13]. Using lower-level charging does reduce the impact on the grid but only at the expense of greatly increasing the duration of the charging process and the inefficiency of the network due to under-utilization. A compromise solution is therefore to exploit elasticity of EV charging load to control charging rates such that bottleneck lines and transformers are fully utilized, and the line voltage level remains within a predetermined range.

Existing approaches to control EV charging load either use a central controller to coordinate charging [10,13,20] or cast the control problem in the form of a distributed optimization [3,8,15]; in both approaches the charging schedule is computed well ahead of time and decisions are made on the basis of system-level considerations, such as mitigating distribution system losses or maximizing the load factor. In the centralized approach, the central controller uses power flow analysis to compute a charging schedule that does not congest any part of the distribution network. This analysis requires an accurate model of the distribution network and the expected locations of all EVs. In many cases, such a model is either not available or not up-to-date. Critically, both approaches also need to predict the future demand from non-EV loads, the number of charging EVs at each time slot, their locations, and their initial state of charge. The safety margin built in to hedge against prediction errors makes both approaches quite conservative.

Inspired by the design of the Internet, which offers best-effort services to elastic applications that back off in case of congestion [19], our approach is to quickly adapt EV charging rates to the condition of the network [4]. Specifically, we propose a distributed control algorithm so that every charger can independently set its charging rate based on *congestion signals* it receives from measurement nodes installed on its path to the subtransmission substation. This algorithm ensures that EV chargers receive a *proportionally fair* [11] share of the available capacity of the distribution network, and lines and transformers are not overloaded. Therefore, an

EV can be charged at the maximum rate when its feeding branches and transformers are lightly loaded (*e.g.*, during off-peak hours), and it is charged at a relatively low rate when they are heavily loaded (*e.g.*, during on-peak hours). This is a best-effort service; hence, in the rare event that the grid is overly congested some EVs might not be fully charged by their deadlines. Note that this is in line with today's practice where operators shed the load in favor of protecting their transmission and distribution assets.

We make the following three specific contributions:

- We formulate the EV charging control problem as an optimization problem where the objective function is chosen such that proportional fairness is achieved at its solution.

- We decompose the optimization problem into a set of distributively solvable subproblems and derive stable control laws by iteratively solving these subproblems.

- We illustrate using numerical simulations that this control converges in a small number of iterations to the solution of the utility maximization problem.

In this paper, we focus on a *static* network scenario in which the residential load is constant and a fixed number of EVs are connected to chargers. We explain in Section 3.2 that the dynamic case can be decomposed into a series of static snapshots. Thus, the control algorithm developed for a static scenario can also be used in a network with variable home loads and number of plugged-in EVs.

We make the critical assumption that the congestion level of every branch is measured in real time and is communicated to downstream EV chargers with a reasonably low delay. This requires an infrastructure which allows fast measurement and communication. The future smart grid is likely to have a considerable number of measurement and control devices that are interconnected by a ubiquitous low-latency broadband communication network [6]. This allows us to use fast time scale measurements and communication to rapidly adapt the charging rates of EVs to the available capacity of the network such that sustained overloading of branches is avoided [4].

2. RELATED WORK

Potential impacts of introducing a large number of EVs to the distribution network have been explored extensively in the literature and many scheduling algorithms have been proposed to shift the EV charging load to off-peak hours, thereby avoiding branch congestion and voltage drop in the distribution network. Most existing work suggests a centralized control for EV charging load. However, as discussed in a recent white paper [22], coordinating control at different levels becomes infeasible with such centralized control. This highlights the need for distributed control of EV charging and other responsive loads.

The closest lines of work to ours are by Gan *et al.* [8] and Ma *et al.* [15] which use distributed control to obtain a day-ahead charging schedule for EVs. In [8], it is assumed that the distribution transformers and EV chargers are instrumented with computation and communication devices, EVs are charged at a fixed rate, and the charging process cannot be interrupted. Based on these assumptions, the EV charging control problem is formulated as a discrete optimization

problem with the objective of flattening the aggregate demand served by a transformer. A stochastic distributed control algorithm is proposed to find an approximate solution to this optimization problem; it is shown that this algorithm almost surely converges to one of the equilibrium charging profiles. In [15], a decentralized algorithm is proposed to find the EV charging strategy that minimizes individual charging costs. It is shown that the optimal strategy obtained using this algorithm converges to the unique Nash equilibrium strategy when there is an infinite population of EVs. In the case of homogeneous EV populations, this Nash equilibrium strategy coincides with the valley-filling maximizing strategy (*i.e.*, the globally optimal strategy).

Our approach differs from the approach of these two papers in three ways. First, we control charging of EVs in real time, whereas they compute a day-ahead charging schedule based on predictions. Second, their goal is to simply flatten the load at a single point in the network, whereas we deal with line and transformer overloading in the entire distribution network. Third, these algorithms do not guarantee fair allocation of available network capacity to EVs while this is an important property of our control mechanism.

The idea of real time distributed control of the EV charging was first introduced in a vision paper [4]; a measurement and signalling architecture was proposed, and three possible distributed congestion control schemes for EV charging were outlined. This paper builds upon the architecture proposed in [4]. We make more precise assumptions about the underlying system, and propose a distributed algorithm that would enable fair, timely, and efficient charging of EVs starting from a static centralized optimization problem.

Our EV charging control problem has the same mathematical formulation as the rate control problem which was extensively studied in the context of the Internet [12, 14, 17, 21, 23, 24] and was used to analyze stability and other properties of the TCP congestion control protocol (see [4] for a comprehensive comparison of congestion control in a packet-switched network and the power distribution network). The common goal is to determine the available resources and allocate these resources among users to maximize a global objective function of the users' utilities that takes fairness into account. Control rules are found by solving the optimization problem in a distributed fashion exploiting the hidden decomposition structure of the optimization problem (see [18] for an introduction to network utility optimization and decomposition theory). Depending on the choice of the objective function, the solution to this maximization problem provides different notions of fairness, namely proportional fairness, max-min fairness, minimum potential delay fairness, and the more general notion of utility proportional fairness.

3. BACKGROUND AND ASSUMPTIONS

The electrical grid consists of generation, transmission, and distribution systems. The electricity generated by power plants is transmitted over long distances by the transmission network, *i.e.*, a mesh network of high voltage lines and step up transformers. Near demand centers, the voltage is stepped down to the primary distribution voltage at sub-transmission substations. The distribution system, which is the focus of this paper, is responsible for delivering electricity from these substations to consumers. A radial distribution system has a tree structure and is comprised of nodes

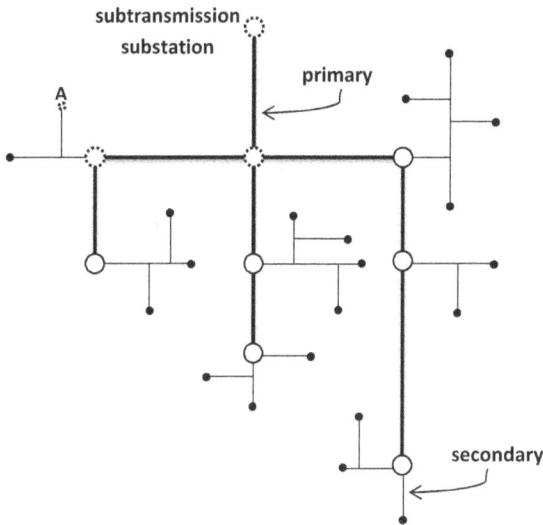

Figure 1: A one-line diagram displays the tree structure of a radial distribution network. Dotted circles represent upstream buses that supply a load connected to the pole-top transformer denoted by A.

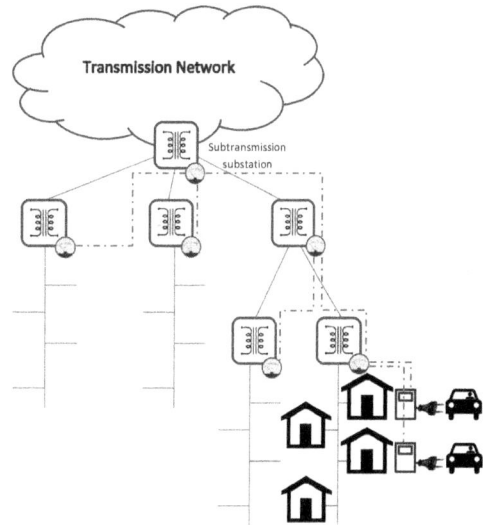

Figure 2: An illustration of a smart distribution network consisting of MCC nodes, and communication links (dashed lines).

(or buses) interconnected by lines (or feeders)[1]. A *one-line diagram* of a typical North American radial distribution system [16] is depicted in Figure 1. This diagram indicates each circuit irrespective of the number of phases with a single line. A load bus, represented by a circle in Figure 1, is a single point of connection on a circuit, and represents the location where power is consumed. In a distribution network a load bus is either a transformer connection to the subtransmission system (the *root* of the distribution tree), or a distribution substation from which distribution feeders originate and supply downstream distribution transformers, which step down the voltage to the secondary distribution voltage, and various loads, including homes and EV chargers. Loads are usually connected to leaf nodes of the distribution tree (*e.g.*, pole-top transformer A in Figure 1) and are not drawn in the one-line diagram.

3.1 Measurement and Communication

Currently, only a few measurement nodes are installed at buses. In the smart grid, we assume that many more such measurement nodes will be installed at load buses and pole-top transformers, as shown in Figure 2. A measurement node could be a current transformer which measures the current flowing through a line, a voltage transformer which measures the line voltage, or a sensor which measures the winding temperature of a transformer. These nodes are capable of continuously measuring the parameter of interest and computing an average every few milliseconds. This permits us to compute the *congestion state* of a line or a transformer. We define the congestion state as the difference between the *nominal setpoint* of the line or the transformer and its loading level. The nominal setpoint should be chosen such that continuous loading of a line or a transformer at this level

does not cause any damage; thus, the nominal setpoint is always less than or equal to the nameplate rating. The congestion state is used to compute the *congestion price* of a line or a transformer (see Section 5). Congestion prices are used by EV chargers to set their rates as described in Section 6. They are communicated to EV chargers in a signalling cycle as explained in Section 3.3.

We distinguish between two types of congestion: *line* congestion and *transformer* congestion. A line is said to be congested when the current passing through it exceeds its rating. Likewise, a transformer is said to be congested when it is loaded higher than its nameplate rating.

In a distribution network, the protection system consisting of fuses, relays, and circuit breakers, disconnects the load shortly after a persistent line or transformer congestion is detected. We define *persistent* congestion as a congestion event which lasts for a certain number of cycles of AC power; this is a characteristic of the distribution network's protection system. When persistent congestion is detected, a protective relay initiates the tripping action after a certain delay, which is an operating characteristic of the relay, to avoid further damage to equipment. This results in a power outage in the area supplied by the congested line/transformer. Therefore, we need to allocate charging rates to EV chargers without triggering the protection system (*i.e.*, without causing persistent congestion).

We assume that each measurement device is supplemented with a communication and control module. We refer to the entire device as a 'Measurement, Communication, and Control (MCC)' node. These MCC nodes form a logical tree (Figure 2). In a smart grid, a communication network will connect these MCC nodes to EV chargers (or other controllable loads) to enable the transmission of control signals.

3.2 Assumptions

We now lay out the assumptions that we make in this paper.

- Line or transformer overloading can be attributed en-

---

[1]Most distribution systems are radial; in cases where the network topology is a mesh, normally-open switches ensure that power flows only on a radial sub-graph. Thus, in this paper we assume that the distribution network forms a tree.

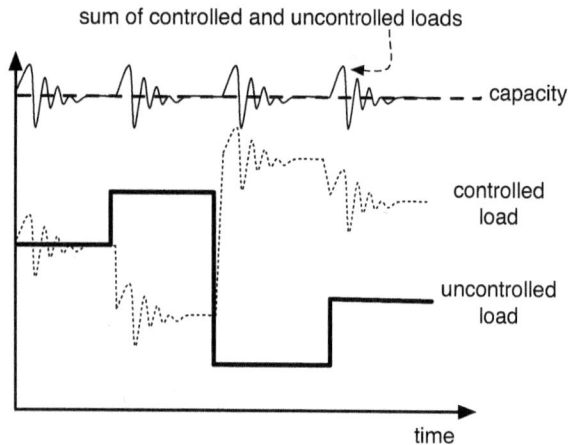

Figure 3: The proposed control algorithm adapts charging rates of EV chargers to the available capacity of the network.

tirely to its downstream loads. Therefore, if a measurement at a node indicates congestion, it can be ameliorated by sending a congestion signal to the node's descendants.

- An overload can be detected within a few cycles of its occurrence, before the invocation of protection mechanisms.

- It is not possible to infer congestion implicitly at the end nodes[2]. Therefore, congestion must be explicitly signalled.

- The distribution system can be subject to a transient congestion for a short durations of up to τ^c seconds before protection mechanisms, such as protective relays are brought to bear[3]. This gives a congestion signalling mechanism breathing room to reduce transient EV overloads. Moreover, we can choose system parameters conservatively such that an overshoot above a nominal setpoint, which is always lower than the nameplate rating, triggers congestion signalling, but does not trigger the protection system.

- The communication network is broadband, reliable, and has a low latency. We also assume that all EV chargers experience roughly the same delay when they receive signalling packets from MCC nodes (this is never exactly true but simplifies the analysis). We denote the feedback delay by d.

- The speed-of-light propagation delay between any charger and its connected substation is small, and typically less than 1ms. Thus, it is reasonable to ignore it in our model.

---

[2]In recent work, it has been shown that local sensing of the line voltage or frequency at end nodes can be used to implicitly infer the aggregate demand or the power imbalance at higher levels in the distribution network [9]. Developments arising from this pioneering work may allow implicit congestion sensing even in the electrical grid.

[3]We note that τ^c is typically inversely proportional to the magnitude of the overload current.

- EV chargers are owned by electric utilities. Furthermore, they are tamper-resistant and always set their rate based on signals they receive from MCC nodes.

- An EV battery can be charged at any rate less than the maximum Amperage rating of its charger, independent of its state of charge[4].

- The load from an EV charger can be increased or decreased at a fast time scale (on the order of milliseconds) with negligible effect on the EV battery lifetime.

- Rate updates at EV chargers are synchronized. Measurements of the congestion state at MCC nodes are also synchronized. This can be achieved by a broadcast time signal.

- The timescale of changes of uncontrolled loads and arrival and departure of EVs is slower than the timescale of rate updates in our distributed algorithm (Figure 3), and hence we can decouple them. This permits us to study our control problem using a model that describes a snapshot of the system in which uncontrolled loads are constant and the number of plugged-in EVs is fixed.

These assumptions imply that it is feasible to design and implement a control algorithm that changes the EV charging rate rapidly in response to the congestion state of the distribution system. With our proposed approach, if an EV is charging at a rate that overloads the distribution system, its rate can be decreased before τ^c, averting damage and invocation of grid self-protection mechanisms.

3.3 System Operation

We now describe the operation of our system. Every T_c milliseconds, the root MCC node initiates a signalling cycle by sending its congestion price to its direct children. Upon receiving the congestion price(s), an intermediate MCC node sends its own congestion price, computed using its latest recorded congestion state (as discussed in Section 5.3.1), along with the received price(s) to its children. A signalling cycle ends when EV chargers receive the congestion prices from all their parents. In Section 5.3.2, we explain how EV chargers use congestion prices of MCC nodes located on their path to set their charging rate.

4. OPTIMIZATION PROBLEM

In this section we formulate the control problem as a centralized static optimization problem. The global objective function is chosen such that the solution to this optimization problem satisfies the definition of proportional fairness.

4.1 Utility Function

We attribute a utility, i.e., a measure of satisfaction, to an EV owner whose EV is connected to a charger denoted by s. This charger is capable of charging EVs at rates that are in $[0, m_s]$, where m_s is the peak charging rate that it supports. Since the departure of EVs from homes and charging stations is non-deterministic, it is reasonable to assert that EV owners are greedy and prefer to finish charging their EVs as

---

[4]With some battery technologies, the charging rate decreases as the state of charge increases. We do not consider this in our present analysis.

soon as possible to avoid *range anxiety*. Hence, we define the utility of a greedy EV owner by its charging rate x_s, *i.e.*, the EV owner's satisfaction is proportional to the rate at which their EV is charged.

4.2 Primal Problem

Our objective is to allocate the available capacity of the network fairly among EV chargers without straining the distribution network; this is generally done by solving a constrained optimization problem. There exist different fairness criteria as discussed by [11], and they are differentiated by the global objective function of the optimization problem. We adopt the notion of proportional fairness which satisfies axioms of fairness specified in game theory[5]. A proportionally fair allocation is in fact a Nash bargaining (arbitrated) solution [23]. It can be shown that proportional fairness is achieved if we maximize the value of a global objective function which is the sum of the logarithm of the utility functions. For notational simplicity, we denote $\log(x_s)$ by U_s. Observe that U_s is infinitely differentiable, increasing, and strictly concave on its interior domain.

Our optimization problem would therefore become a maximization of the sum of the U_s of those chargers that are charging an EV, subject to physical constraints imposed by chargers, lines, and transformers. Since lines and transformers supply the aggregate load imposed by both homes and EV chargers, and the home loads are supposedly uncontrolled, it is necessary to subtract the load of every home from the rated capacity of its feeding transformers and lines to obtain the available capacity of every line and transformer. In effect, based on the last assumption of Section 3.2, we consider a snapshot of the system in which home loads are constant and a fixed number of EVs are plugged in to chargers.

The control problem can then be formulated as follows:

$$\max_{x} \sum_{s \in S} U_s(x_s) \tag{1}$$

subject to
$$0 \leq x_s \leq m_s \quad \forall s \in \mathcal{S}$$
$$\sum_{s:R_{sl}=1} x_s \leq c_l \quad \forall l \in \mathcal{L},$$

where $\mathcal{S}$ and $\mathcal{L}$ are the sets of active EV chargers, and distribution lines and transformer respectively, c_l is the available capacity of line or transformer l, and R is a $|\mathcal{S}|$ by $|\mathcal{L}|$ matrix encoding the topology of the network

$$R_{sl} = \begin{cases} 1 & \text{if line/transformer } l \text{ supplies charger } s \\ 0 & \text{otherwise} \end{cases}$$

We say that a line or a transformer supplies an EV charger when it is located on the path from the subtransmission substation to that charger.

This problem is a convex optimization problem as it maximizes an objective function which is the sum of concave functions (and is therefore concave), and each constraint defines a convex set. We denote a rate allocation, *i.e.*, a vector of charging rates of all chargers, by $x = <x_1, \cdots, x_{|\mathcal{S}|} >$. A rate allocation is *feasible* if it satisfies all constraints of the optimization problem.

---

[5]We have chosen proportionally fair resource allocation since it is the only one that provides a scale invariant Pareto optimal solution.

The two constraints of the above optimization problem can be written in matrix form:

$$0 \preceq x \preceq m$$
$$xR \preceq c,$$

where $x, c,$ and m are vectors with $|\mathcal{S}|$, $|\mathcal{L}|$, and $|\mathcal{S}|$ components respectively, and $\preceq$ is the generalized inequality for vectors. We refer to the second constraint as the *coupling constraint*; it couples charging rates of different EV chargers supplied by the same line or transformer.

In the next section we write the dual problem and apply the dual decomposition method to obtain a set of distributively solvable subproblems. We then design a distributed algorithm which solves the dual problem by solving these subproblems locally and independently.

5. CONTROLLER DESIGN

The centralized optimization problem formulated in the previous section can be solved to find a rate allocation that is proportionally fair. This requires full knowledge of the topology of the distribution network, the available capacity of lines and transformers, and the number and the location of plugged-in EVs. The distributed approach to solve the optimal control problem has three key advantages over the centralized approach. First, it gives autonomy to local controllers thereby increasing robustness of the control system. Second, it significantly reduces the communication overhead and is therefore more scalable. Third, it decreases the overall latency of control as control decisions are made locally.

Our plan is therefore to design a distributed control algorithm by solving the Lagrangian dual of the centralized optimization problem. We apply the dual decomposition method to obtain distributively solvable subproblems that are controlled at the higher level by a master problem through congestion prices. The proposed algorithm requires solving the master problem and these subproblems in an iterative fashion. From a control theory standpoint, solutions to these problems constitute our controls and congestion prices are feedback.

5.1 Dual Problem

The Lagrangian function of our optimization problem is

$$g(\lambda) = \max_{0 \preceq x \preceq m} \{\sum_{s \in S} \log x_s + \sum_{l \in \mathcal{L}} \lambda_l(c_l - y_l)\}, \tag{2}$$

where $\lambda = (\lambda_1, \ldots, \lambda_{|\mathcal{L}|})$ is a vector of Lagrangian multipliers associated with the coupling constraints, and

$$y_l = \sum_{s:R_{sl}=1} x_s \quad \forall l \in \mathcal{L}$$

Thus, the Lagrangian dual problem would be

$$\min_{\lambda} \max_{0 \preceq x \preceq m} \{\sum_{s \in S} \log x_s + \sum_{l \in \mathcal{L}} \lambda_l(c_l - y_l)\} \tag{3}$$

subject to $\quad \lambda_l \geq 0 \quad \forall l \in \mathcal{L},$

which is equivalent to

$$\min_{\lambda} \left\{ \sum_{l \in \mathcal{L}} \lambda_l c_l + \max_{0 \preceq x \preceq m} \{\sum_{s \in S} f_s(x_s; \lambda)\} \right\} \tag{4}$$

subject to $\quad \lambda_l \geq 0 \quad \forall l \in \mathcal{L},$

where

$$f_s(x_s; \lambda) = \log x_s - x_s \sum_{l:R_{sl}=1} \lambda_l, \qquad (5)$$

In the above equation, $f(x; \lambda)$ represents f as a function of x parameterized by λ. Since $f_s(x_s; \lambda)$ is the sum of two concave functions of x_s, it is also concave and has a unique maximum.

We note that (4) is derived from (3) by using the following equation.

$$\sum_{l \in \mathcal{L}} \left(\lambda_l \sum_{s:R_{sl}=1} x_s \right) = \sum_{s \in \mathcal{S}} \left(x_s \sum_{l:R_{sl}=1} \lambda_l \right) = xR\lambda^T$$

An important remark is that strong duality holds as all inequality constraints are affine. Therefore, we can write the following KKT optimality conditions

$$\frac{1}{\hat{x}_s} = \sum_{l:R_{sl}=1} \hat{\lambda}_l \qquad \forall s \in \mathcal{S} \qquad (6)$$

$$0 \le \hat{x}_s \le m_s \qquad \forall s \in \mathcal{S} \qquad (7)$$

$$\hat{\lambda}_l \ge 0 \qquad \forall l \in \mathcal{L} \qquad (8)$$

$$\hat{\lambda}_l(\hat{y}_l - c_l) = 0 \qquad \forall l \in \mathcal{L} \qquad (9)$$

where $\hat{x}$ and $\hat{\lambda}$ are the unique optimizers of the Lagrangian dual problem. The first condition says that the gradient of Lagrangian vanishes at the optimal point, and the last condition, i.e., the complementary slackness condition, implies that either the optimal Lagrangian multiplier is zero, or the corresponding line or transformer is *fully utilized*, i.e., the line or transformer loading reached its nominal setpoint. Combining the first three conditions gives us the following relation between $\hat{x}$ and $\hat{\lambda}$.

$$\hat{x}_s = min \left\{ \frac{1}{\sum_{l:R_{sl}=1} \hat{\lambda}_l}, m_s \right\} \qquad (10)$$

5.2 Dual Decomposition

Writing the Lagrangian dual problem in the form of (4) reveals its hidden decomposition structure [18]. Particularly, each EV charger can locally solve a subproblem given by

$$\max_{0 \le x_s \le m_s} f_s(x_s; \lambda), \qquad (11)$$

provided that it knows the sum of the Lagrangian multipliers corresponding to the lines and transformers that are supplying its load. It turns out that Lagrangian multipliers play the role of congestion prices (or *shadow prices* [12]) in our problem.

These subproblems are controlled by a master problem by means of congestion prices. The master problem is responsible for updating the congestion prices and can be written in the following form

$$\min_{\lambda \succeq 0} \left\{ \sum_{l \in \mathcal{L}} \lambda_l c_l + \sum_{s \in \mathcal{S}} f_s(\hat{x}_s; \lambda) \right\}. \qquad (12)$$

where $f_s(\hat{x}_s; \lambda)$ is the optimal value of (11). Observe that the objective function of the master problem is linear in λ and its derivative with respect to a Lagrangian multiplier is given by

$$\frac{\partial g}{\partial \lambda_l}(\lambda) = c_l - y_l$$

5.3 Control Laws

Our approach is to solve the dual optimization problem using a distributed algorithm which has two separate parts. The first part adjusts congestion prices of lines and transformers by periodically measuring the available capacity and solving the master problem at each MCC using the gradient projection method. The second part updates charging rates of EVs by solving the subproblems.

In the following we derive control laws for updating congestion prices and adjusting charging rates by solving the master problem and the subproblems respectively. These control laws constitute the distributed algorithm outlined in Section 6. In Section 7, we specify sufficient conditions for convergence of this algorithm to primal and dual optimal values.

5.3.1 A Control Law for Updating the Congestion Price

Since the dual function is differentiable, we can adopt the gradient method with a projection onto the positive orthant to solve the master problem (12). The following algorithm updates congestion prices in each iteration in opposite direction to the gradient of the dual function.

$$\lambda_l(t+1) = \max\{\lambda_l(t) - \kappa(c_l - y_l(t)), 0\} \quad \forall l \in \mathcal{L} \qquad (13)$$

Here κ is a sufficiently small positive constant which determines the responsiveness and stability of control, and has to be selected carefully. Note that it is not necessary to estimate c_l and y_l at an MCC node to compute $c_l - y_l$. This is because $c_l - y_l$ is equal to the line or transformer loading subtracted from its nominal setpoint denoted ξ_l. The nominal setpoint of a line or transformer is known a priori and its loading can be measured by the corresponding MCC node.

5.3.2 A Control Law for Adjusting the Charging Rate

We denote the latest congestion price vector that an EV charger received by $\lambda(\bar{t})$. The subproblem (11) can be easily solved by finding the stationary point of $f_s(x_s; \lambda)$.

$$f'_s(x_s(t); \lambda(\bar{t})) = \frac{1}{x_s(t)} - \sum_{l:R_{sl}=1} \lambda_l(\bar{t}) \stackrel{\text{set to}}{=} 0 \rightarrow$$

$$x_s(t) = \min \left\{ \frac{1}{\sum_{l:R_{sl}=1} \lambda_l(\bar{t})}, m_s \right\} \qquad (14)$$

Note that $x_s(t)$ would be the rate of EV charger s for the interval $[t, t+1)$, and adjusting the charging rates impacts the loading of upstream feeders and transformers immediately[6]. More specifically, $y_l(t)$ is given by

$$y_l(t) = \sum_{s:R_{sl}=1} x_s(t) \qquad \forall l \in \mathcal{L} \qquad (15)$$

We end this section by a remark that the unit of time in (13) and (14) is T_c milliseconds, and therefore $\bar{t}$ equals $t - \frac{d}{T_c}$ as congestion prices are received by EV chargers after d milliseconds.

---

[6]There is a fundamental difference between congestion control protocols in the Internet and our EV charge control protocol. In computer networks, when traffic sources change their rates it is only reflected on link utilization after a delay, known as the forward delay. However, there is no forward delay in our problem as power flows in the grid at the speed-of-light.

6. DISTRIBUTED CHARGING CONTROL ALGORITHM

We now describe the algorithms that operate at the MCC nodes and at EV chargers and implement the control laws derived in Section 5.

6.1 Normal Operation

During normal operation, our distributed charging control algorithm measures the congestion state of a line or a transformer and computes the corresponding congestion price based on (13). This price is sent to descendant EV chargers in a signalling cycle initiated by the root MCC node every T_c milliseconds. Recall we assume that MCC nodes are synchronized and therefore update their congestion prices at the same time (see Algorithm 1).

Algorithm 1: Congestion price update at MCC node l with nominal setpoint ξ_l

---

input: $\xi_l, \kappa(> 0)$

while *true* **do**

  Measure load

  congestion state $\leftarrow \xi_l -$ load

  price $\leftarrow \max\{$price $- \kappa \times$ congestion state, $0\}$

  Send price along with all received prices to children

  Wait until the next **clock tick**

end

After receiving congestion prices from upstream MCC nodes, every charger computes its charging rate using (14) and starts charging at this rate. We assume that EV chargers are also synchronized, i.e., they adjust their rates at the same time.

Algorithm 2: Rate adjustment at EV charger s

---

input: m_s, new congestion prices

while *true* **do**

  $\lambda \leftarrow$ *vector of* new congestion prices

  aggregate price $\leftarrow \sum_{l\in ascendants} \lambda_l$

  rate $\leftarrow \min\{\frac{1}{\text{aggregate price}}, m_s\}$

  Start charging the battery at rate

  Wait until the next **clock tick**

end

Note that the clock ticks every T_c in both algorithms.

6.2 Emergency Response Mode

To assure grid reliability at all times, we have designed a 'fail-safe' load reduction mechanism to respond to sudden load spikes. This mechanism operates both at the MCC nodes and at EV chargers. If any MCC node detects that its overall load has exceeded the nominal setpoint of ξ_l by a factor $\eta, \eta > 1$ for a duration that exceeds $\tau^c - d$, it sends an emergency shutdown signal to all descendant EV chargers. In response, all EV chargers reduce their charging rate to 0. Since this response is guaranteed to move the system from an overloaded state to an underloaded state within time τ^c, it avoids triggering the protection system, and at the same time does not affect our proofs of system stability and convergence, discussed next.

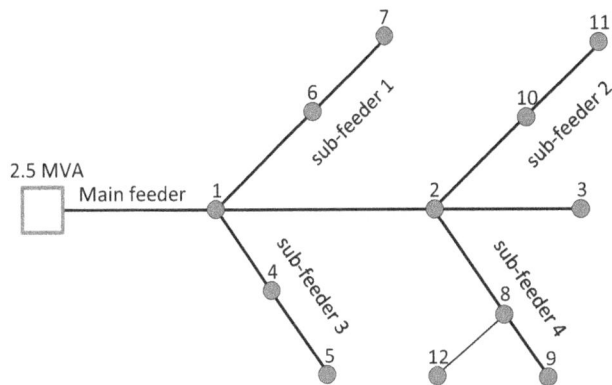

Figure 4: The one-line diagram of our test distribution network.

7. CONVERGENCE ANALYSIS

This section studies the conditions under which the proposed distributed control algorithm converges to the solution of the centralized optimization problem (1) in a static context, i.e., no EVs arrive or depart and the change in the magnitude of uncontrollable loads is negligible. Recall that the primal optimum is equal to the dual optimum as strong duality holds. Therefore, we instead show that the distributed control algorithm converges to the solution of (3).

Our proof technique is to extend Theorem 1 in [14] to derive the following result.

Theorem 1. Starting from any initial rates $0 \preceq x \preceq m_s$ and congestion prices $\lambda \succeq 0$, the distributed algorithm introduced in Section 6 converges to the primal-dual optimal values if

(1) $T_c \geq d$

(2) $0 < \kappa < \frac{2}{\overline{m}^2 \overline{L}\overline{S}}$

where $\overline{m} := \max_s m_s$, $\overline{L} := \max_s \sum_l R_{sl}$, and $\overline{S} := \max_l \sum_s R_{sl}$.

Recall that $\overline{L}$ is the maximum number of lines and transformers which are instrumented with MCC nodes and are supplying the load of an EV charger, $\overline{S}$ is the total number of EVs being charged in the distribution network, and $\overline{m}$ is the maximum charging rate supported by an EV charger.

To prove this theorem, we extend the proof in [14] by showing that the first condition is satisfied when the communication delay is bounded by d. Recall that our algorithm requires MCC nodes to update congestion prices periodically; thus, congestion prices are unchanged between two consecutive updates which are T_c milliseconds apart. Assuming that $t - \frac{d}{T_c}$ is the last time that congestion prices are updated, we have $\lambda(t - \frac{d}{T_c}) = \lambda(t)$ if $T_c \geq d$. Substituting $\lambda(t - \frac{d}{T_c})$ with $\lambda(t)$ in (14), our algorithm becomes identical to the one presented in [14].

8. RESULTS

The behavior of our control algorithm is determined by the value of two critical parameters κ and T_c. The first parameter determines the step size in changing congestion prices: the larger this value, the more responsive the system. The

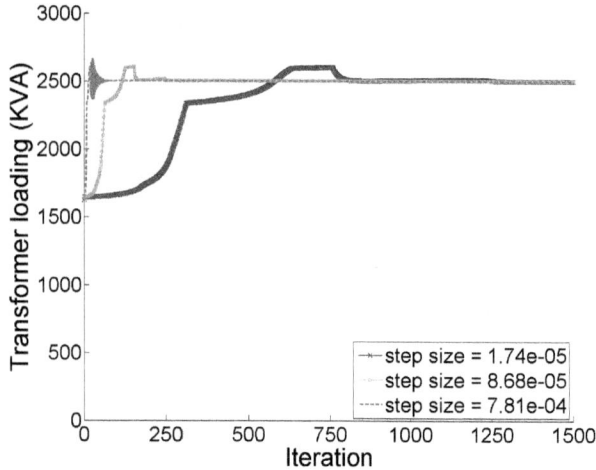

Figure 5: The value of the step size κ determines how the loading of the substation transformer changes over time (Scenario A).

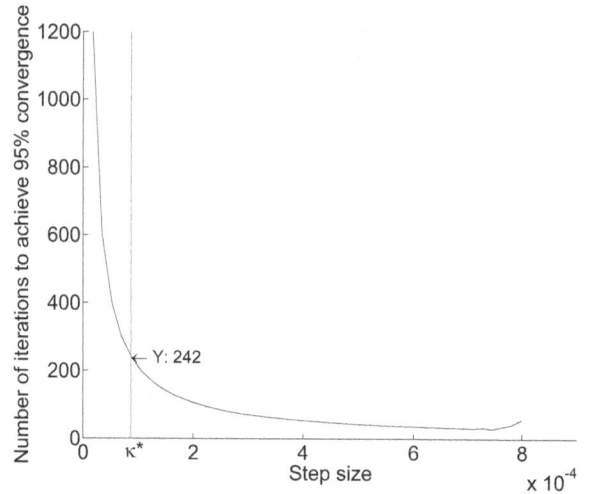

Figure 6: The number of iterations it takes to achieve 95% convergence (Scenario A).

second parameter determines the frequency of control: the larger this value, the less the control overhead, but the longer the time that the system may be experiencing transient congestion (to prevent triggering the protection system, we require $T_c < \tau^c - d$). Theorem 1 states sufficient conditions under which the proposed distributed algorithm converges. However, the space of possible parameters defined by Theorem 1 is large and depending on their values the control system exhibits different characteristics. In this section, we simulate our algorithm in a test distribution network to investigate the sensitivity of the control to the choice of these two parameters, and decide on their optimal values considering overall system constraints.

8.1 Test Distribution System

Our analysis is based on a 4.16kV radial distribution system comprised of 13 buses. This system is obtained by simplifying the IEEE 13-bus test feeder [1]. In particular, we do not model in-line transformers, voltage regulators, shunt capacitor banks, and protection switches. Moreover, we assume that the three phase system is balanced, *i.e.*, loads are distributed as evenly as possible between the phases. Thus, we perform a *per-phase analysis* of the network, assuming that all lines and transformers are single-phase.

Figure 4 depicts our test distribution network consisting of a main feeder connecting buses 1, 2, and 3 (the circles labelled with these numbers in the figure) to the substation and four sub-feeders branching out from the main feeder. The main feeder is rated at 730A and sub-feeders 1 to 4 are rated at 340A, 340A, 140A, and 230A respectively. The network is supplied by a single phase 2.5MVA transformer which is installed at the substation and steps down the voltage from 115kV to 4.16kV. The voltage is further reduced to the secondary distribution voltage by pole-top transformers that are connected to buses. For sake of simplicity, in this paper we do not model these transformers and feeders radiating from them, although the analysis would be exactly similar to what we do in this section.

We assume that the nominal setpoint of each line or transformer is equal to its nameplate rating, and every bus in the test distribution network is instrumented with an MCC node. Spot loads are connected to selected load buses. We assume that they draw constant current from the network and the power factor at the substation bus is one. This permits us to ignore the reactive power flow in our simulations. Furthermore, it is assumed that EV chargers that support charging at up to 16A are also connected to selected load buses.

Table 1 summarizes our simulation scenarios. A simulation scenario describes a snapshot of the system in which constant current spot loads are connected to buses, and a number of EVs are charged by chargers which are downstream of these buses. We choose Scenario A so that the substation transformer is the only congested resource when allocating proportionally fair rates to EV chargers. In contrast, in Scenario B, both the substation transformer and the line connecting buses 2 and 8 become congested simultaneously when allocating proportionally fair rates to EV chargers.

8.2 Sensitivity Analysis

In this section, we compare the number of iterations that it takes to achieve 95% convergence and the maximum deviation from the nominal setpoint value of a line or a transformer for different values of κ.

Consider the first simulation scenario. In this case, the maximum step size for which the convergence of the algorithm is guaranteed is $\kappa^* = \frac{2}{16^2 \times 18 \times 5} = 8.68 \times 10^{-5}$. As we increase the value of κ, the control system transitions from an over-damped system to an under-damped system and eventually to an unstable system for $\kappa > 8.05 \times 10^{-4}$ which is much larger than κ^*. Figure 5 shows how the loading of the substation transformer varies over different iterations for three different values of κ.

Changing the value of κ changes the system responsiveness, specifically, the number of iterations it takes to achieve the 95% convergence, where the 95% convergence is said to

The bus number	3		4		5		6		7		8		9		10		11		12	
Scenario A	30	2	40	2	40	2	10	2	20	2	70	0	20	2	100	2	20	2	40	2
Scenario B	30	2	40	2	40	2	10	2	20	2	160	0	20	2	10	2	20	2	40	2

Table 1: For each bus the first column shows the spot load connected to it (in Ampere) and the second column shows the number of EV chargers supplied by it.

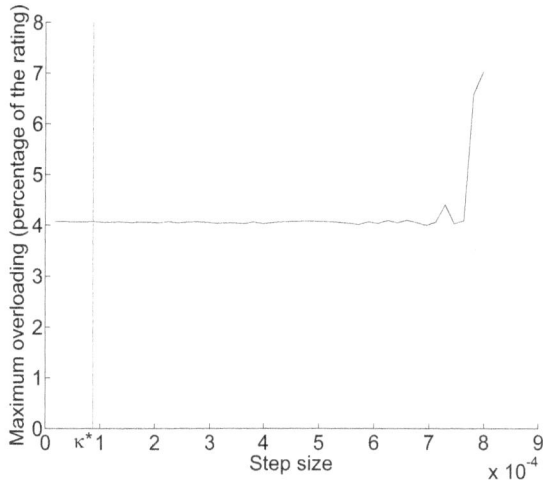

Figure 7: The maximum deviation from the rating of the substation transformer (Scenario A).

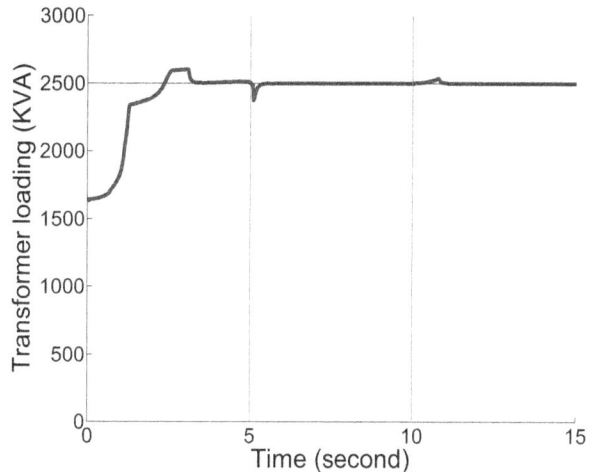

Figure 8: Loading of the substation transformer as the system transitions from Scenario A to Scenario B after 250 iterations and back to Scenario A again after another 250 iterations.

be achieved when the loading of a line or transformer remains in the boundary given by $\pm 5\%$ of its nominal set-point. Figure 6 shows that the number of iterations required for 95% convergence decays exponentially as we increase the value of κ. When the step size is equal to κ^*, it takes 242 iterations to achieve 95% convergence; assuming that T_c is 20 milliseconds the algorithm achieves 95% convergence in 4.8 seconds. Should we set the step size larger, for example $\kappa = 4 \times 10^{-4}$, it takes only 54 iterations to achieve 95% convergence which is equivalent to approximately one second (though, at this value of κ, convergence is not guaranteed!).

We find that the maximum deviation from the set point value is almost 4% for a wide range of step size values including κ^* (see Figure 7). However, once κ is large, the system starts oscillating around the set point value and the maximum deviation could be relatively high.

Now consider the second simulation scenario. We find that similar to Scenario A, the number of iterations it takes to achieve 95% convergence decays exponentially as we increase the value of κ. If we set the step size to κ^*, the algorithm converges after 354 iterations and the maximum deviation is about 1%.

Lastly, we validate the assumption that if the timescale of changes of uncontrolled loads and arrival and departure of EVs is slower than the timescale of rate updates in our algorithm, the stability result obtained for a static network scenario can be extended to the dynamic case. To see this, suppose that uncontrolled loads vary every 5 seconds in our test distribution network, and the system can be described in an interval of 15 seconds by the following sequence of snapshots: Scenario A, Scenario B, and again Scenario A. Figure 8 and

Figure 9 show the substation transformer loading and the loading of the line connecting buses 2 and 8 respectively, assuming that the step size is κ^*, and T_c is 20 milliseconds. It can be readily seen that since κ and T_c are chosen according to Theorem 1, transitions from one scenario to another one does not destabilize the system.

As an extreme case, we consider the situation where the home load goes from zero (so that all EVs are charged at their maximum rate) to Scenario A, where the transformer is congested, and back to zero in the three time periods (Figure 10). Here, at time 5s, without the emergency response mechanism, the transient congestion at the transformer lasts for 30 iterations or 0.6s. However, by adding the emergency response mode which is invoked after $5T_c = 100\text{ms}$, the transient load does not exceed the design specification of $\tau^c = 200\text{ms}$, avoiding invocation of the grid protection system.

8.3 Proportional Fairness

The solution to the centralized optimization problem (3) indicates that the fair share of every EV charger in Scenario A must be 11.72A. Our results also show that when the algorithm converges, 11.72A is allocated to each charger. Figure 11 shows the charging rates of a few EV chargers over time.

In Scenario B, the proportionally fair rate of chargers connected to buses 9 and 12 is 2.5A, whereas the proportionally fair rate of all other chargers is 14.35A (see Figure 12). This is because chargers connected to buses 9 and 12 have both bottleneck resources on their path to the substation, and therefore their fair rate would be smaller than other charg-

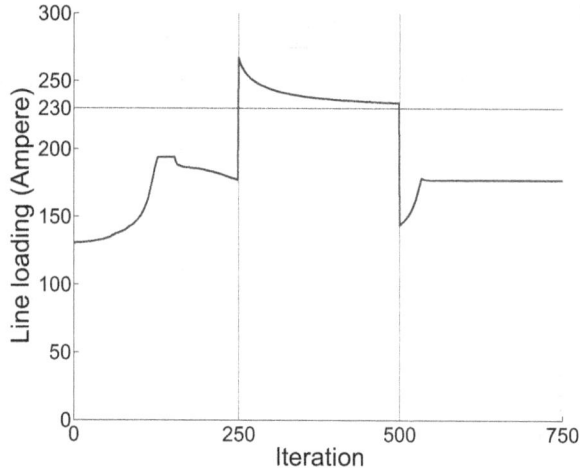

Figure 9: Loading of the line connecting buses 2 and 8 as the system transitions from Scenario A to Scenario B after 250 iterations and back to Scenario A again after another 250 iterations.

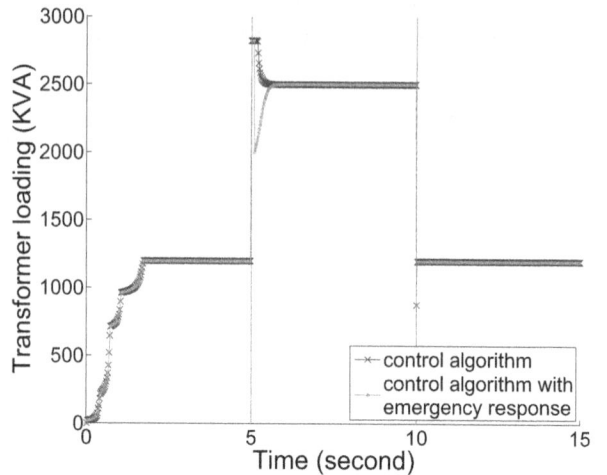

Figure 10: Loading of the substation transformer as the system transitions from zero home load to Scenario A after 250 iterations and back to zero home load again after another 250 iterations.

ers. Note that the optimal congestion price of non bottleneck lines and transformers is zero (see Equation 9).

8.4 Design Space

As discussed earlier, our design space is defined by the conditions of Theorem 1. Clearly, the step size should be equal to κ^* to increase responsiveness of the control system, thereby reducing the number of iterations it takes to achieve 95% convergence. Once we fix the value of κ, we decide on the value of T_c. Since the 95% convergence time is proportional to T_c, we prefer to set it as small as possible, i.e. equal to d.

However, in practice T_c is also bounded from below by another parameter. It turns out that sending congestion prices periodically from MCC nodes to EV chargers causes a significant communication overhead. In effect, the communication medium and the protocol chosen for transmission of these signals determine the fastest rate at which we can propagate updates in our algorithms. Hence, although the communication delay is of the order of a few milliseconds, it is not feasible to update congestion prices faster than every 20 milliseconds.

9. DISCUSSION AND CONCLUSION

Our work represents a novel approach to controlling the charging of electric vehicles. Instead of forecasting the number of EVs and the non-EV load several hours ahead and solving an optimization problem, we rely on fast measurements and communication to avert persistent congestion and the invocation of protection mechanisms. Moreover, by using a mathematical framework originally developed for rate control in the Internet, each EV charger can independently update its charging rate, yet the global allocation converges to the nominal operating setpoint, the allocated rates are proportionally fair, and the allocation is optimal. Additionally, the use of an emergency response mode averts protection events without compromising these benefits.

Our control algorithm can be used to implement demand response (via EV charging) in a distribution network. This requires defining a new type of congestion which corresponds to generation shortfall. Concretely, when supply falls short of demand, the system operator sends a signal to root MCC nodes. Consequently, each root MCC node indicates congestion by increasing its congestion price or sending an emergency shutdown signal to controlled loads in its distribution tree.

A possible extension to our work would be to investigate the voltage drop problem at distant buses. This requires adding other constraints to the centralized optimization problem. A voltage drop can be modeled as line congestion. Corresponding MCC nodes should update their congestion state based on the difference between the measured line voltage and their rated voltage.

Another direction for future work is to design an asynchronous distributed algorithm to control EV charging, similar to the algorithm proposed in [14]. This eliminates the need for time synchronization between EV chargers and MCC nodes and would give us a new bound on step size values for which convergence of the algorithm is guaranteed.

It is interesting to compare our distributed approach with a centralized one. They can be compared in terms of robustness, efficiency, communication overhead, and control and communication delay. The distributed approach is more robust as there is no single point of failure. The centralized approach is potentially more efficient if the primal optimization problem can be solved quickly. Both approaches are expected to have roughly the same overhead and delay. A further comparison of these two approaches is left for future work.

The main limitation of our approach is that it requires a heavy communication overhead; our design point is one message to each EV every 20ms. Reducing this overhead by decreasing the communication frequency, or by using an un-

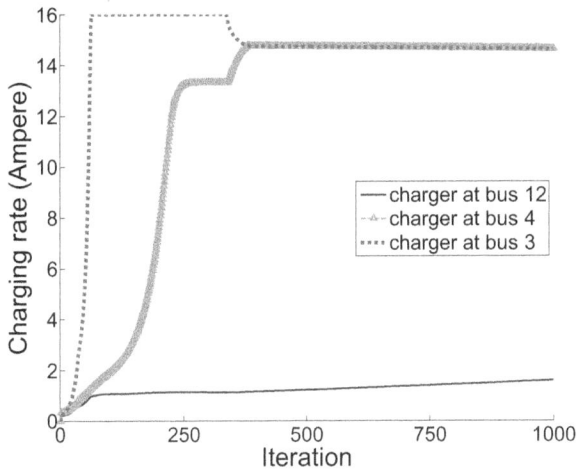

Figure 12: Changes to charging rates of three chargers over time obtained for Scenario B.

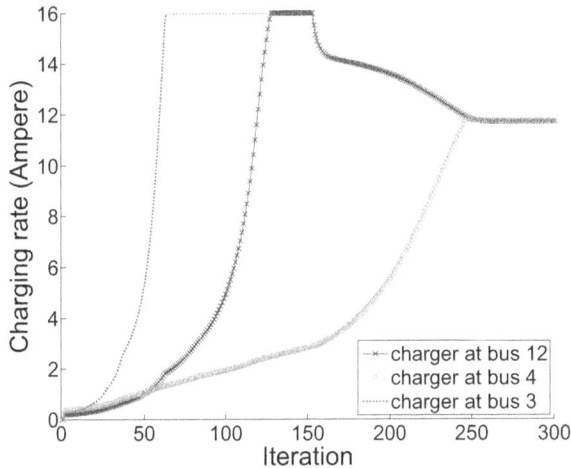

Figure 11: Changes to charging rates of three chargers over time for Scenario A.

derlying broadcast medium, such as 700 MHz cellular radio, is a clear direction for future work.

A second problem with our approach is that it does not scale well: the choice of κ, which controls system responsiveness, is constrained by the number of EVs and the maximum EV charging rate. As these increase, κ and system responsiveness decrease. What is needed is a less conservative bound for κ that has better scaling properties. Alternatively, we could use a larger value of κ than κ^* and then rely on the emergency response mode to address any resultant overloads. We plan to investigate these alternatives in future work.

Finally, our operating parameters, such as the protection threshold τ^c, safety margin η, and communication interval T_c would be better chosen from a realistic EV charging testbed. We hope to validate our choice of these parameters through a deeper engagement with electric utilities.

10. REFERENCES

[1] IEEE Distribution Test Feeders.
   http://ewh.ieee.org/soc/pes/dsacom/testfeeders.
[2] SAE J1772 Standard.
   http://standards.sae.org/j1772_201210.
[3] C. Ahn, C.-T. Li, and H. Peng. Optimal decentralized charging control algorithm for electrified vehicles connected to smart grid. *J. Power Sources*, 196(2):10369 – 10379, 2011.
[4] O. Ardakanian, C. Rosenberg, and S. Keshav. Real-time distributed congestion control for electrical vehicle charging. *Performance Evaluation Review*, 2012.
[5] A. G. Boulanger, A. C. Chu, S. Maxx, and D. L. Waltz. Vehicle Electrification: Status and Issues. *Proceedings of the IEEE*, 99(6):1–23, 2011.
[6] R. Brown. Impact of smart grid on distribution system design. In *IEEE PES General Meeting*, pages 1–4. IEEE, 2008.
[7] EPRI. Impact of Plug-in Hybrid Electric Vehicles on Utility Distribution . Technical Report, 2009.
[8] L. Gan, U. Topcu, and S. Low. Stochastic distributed protocol for electric vehicle charging with discrete charging rate. Technical report, California Institute of Technology, 2011.
[9] T. Ganu, J. Hazra, D. P. Seetharam, S. A. Husain, V. Arya, L. C. De Silva, R. Kunnath, and S. Kalyanaraman. nPlug: a smart plug for alleviating peak loads. In *ACM e-Energy*, 2012.
[10] X. Gong, T. Lin, and B. Su. Survey on the impact of electric vehicles on power distribution grid. In *Power Engineering and Automation Conference*, volume 2, pages 553–557. IEEE, 2011.
[11] F. Kelly. Charging and rate control for elastic traffic. *European transactions on Telecommunications*, 8(1):33–37, 1997.
[12] F. P. Kelly, A. K. Maulloo, and D. K. H. Tan. Rate control for communication networks: Shadow prices, proportional fairness and stability. *Journal of the Operational Research Society*, 49(3):237–252, 1998.
[13] J. Lopes, F. Soares, and P. Almeida. Integration of electric vehicles in the electric power system. *Proceedings of the IEEE*, 99(1):168–183, 2011.
[14] S. H. Low and D. E. Lapsley. Optimization flow control. I. Basic algorithm and convergence. *Networking, IEEE/ACM Transactions on*, 7(6):861–874, 1999.
[15] Z. Ma, D. S. Callaway, and I. A. Hiskens. Decentralized charging control of large populations of plug-in electric vehicles. *IEEE Trans. Control Systems Technology*, (99):1–12, 2011.
[16] A. Meier. *Electric Power Systems: A Conceptual Introduction*. Wiley-IEEE Press, 2006.
[17] F. Paganini, Z. Wang, J. Doyle, and S. Low. Congestion control for high performance, stability, and fairness in general networks. *Networking, IEEE/ACM Transactions on*, 13(1):43–56, 2005.
[18] D. Palomar and M. Chiang. A tutorial on decomposition methods for network utility

maximization. *Selected Areas in Communications, IEEE Journal on*, 24(8):1439–1451, 2006.

[19] S. Shenker. Fundamental design issues for the future internet. *IEEE Journal on Selected Areas in Communications*, 13:1176–1188, 1995.

[20] E. Sortomme, M. Hindi, S. MacPherson, and S. Venkata. Coordinated charging of plug-in hybrid electric vehicles to minimize distribution system losses. *IEEE Trans. Smart Grid*, 2(1):198–205, 2011.

[21] R. Srikant. *The Mathematics of Internet Congestion Control (Systems and Control: Foundations and Applications)*. Birkhauser, 2004.

[22] J. Taft and P. De Martini. Cisco Systems – Ultra Large-Scale Power System Control Architecture. `http://www.cisco.com/web/strategy/docs/energy/control_architecture.pdf`.

[23] H. Yaïche, R. R. Mazumdar, and C. Rosenberg. A game theoretic framework for bandwidth allocation and pricing in broadband networks. *IEEE/ACM Trans. Networking*, 8(5):667–678, 2000.

[24] L. Ying, G. Dullerud, and R. Srikant. Global stability of internet congestion controllers with heterogeneous delays. *Networking, IEEE/ACM Transactions on*, 14(3):579–590, 2006.

Real-Time Deferrable Load Control: Handling the Uncertainties of Renewable Generation

Lingwen Gan
California Inst. of Tech.
lgan@caltech.edu

Adam Wierman
California Inst. of Tech.
adamw@caltech.edu

Ufuk Topcu
University of Pennsylvania
utopcu@seas.upenn.edu

Niangjun Chen
California Inst. of Tech.
ncchen@caltech.edu

Steven H. Low
California Inst. of Tech.
slow@caltech.edu

ABSTRACT

Real-time demand response is essential for handling the uncertainties of renewable generation. Traditionally, demand response has been focused on large industrial and commercial loads, however it is expected that a large number of small residential loads such as air conditioners, dish washers, and electric vehicles will also participate in the coming years. The electricity consumption of these smaller loads, which we call deferrable loads, can be shifted over time, and thus be used (in aggregate) to compensate for the random fluctuations in renewable generation. In this paper, we propose a real-time distributed deferrable load control algorithm to reduce the variance of aggregate load (load minus renewable generation) by shifting the power consumption of deferrable loads to periods with high renewable generation. At every time step, the algorithm minimizes the expected variance to go with updated predictions. We prove that suboptimality of the algorithm vanishes as time horizon expands. Further, we evaluate the algorithm via trace-based simulations.

Categories and Subject Descriptors

J.2 [**Physical Science and Engineering**]: Engineering

Keywords

smart grid, deferrable load control, demand response, model predictive control

1. INTRODUCTION

The electricity grid is expected to change dramatically over the coming decades. Conventional coal and nuclear generation is being rapidly substituted by renewable generation such as wind and solar [5]. However, these renewables are difficult to predict. For example, wind generation prediction has a root-mean-square error of around 18% of the nameplate capacity looking 24 hours ahead [17]. Such high uncertainty in generation calls the traditional control strategy of "generation follows demand" into question.

Real-time demand response programs seek to induce dynamic demand management of customers' electricity load in response to power supply conditions, e.g., by reducing or deferring power consumption in response to requests from the utility. Such programs have the potential to compensate for the uncertainties in renewables in real-time so as to ease the incorporation of renewable energy into the grid, and so are recognized as priority areas for the future smart grid by both the National Institute of Standards and Technology [27] and the Department of Energy [11].

The success of demand response depends on the willingness and ability of consumers' electrical loads to be deferred over time. Such *deferrable loads* are expected to take many forms, e.g., plug-in electric vehicles, dryers, air conditioners, etc. The penetration of deferrable loads is expected to grow significantly in the coming years as a result of increasing penetration of electric vehicles and smart appliances [12]. This expected increase highlights the potential for scheduling deferrable loads in order to compensate for the random fluctuations of renewable energy.

However, realizing the potential of deferrable loads requires the coordination of a large number of distributed loads. Current approaches for achieving such coordination include 1) direct load control (DLC) by load serving entities (LSE) [14, 15, 21, 25], and 2) time-of-use pricing and other complex pricing structures [2, 7, 23]. DLC is the focus of this paper since the LSE has full control over the loads. Specifically, this paper focuses on decentralized DLC algorithms. The motivation for this approach is that, as the penetration of deferrable loads grows, the scale of the task of controlling deferrable loads will prevent centralized control and so distributed, decentralized coordination will become necessary.

Related work.

There is a growing body of work on decentralized direct load control algorithms. This literature focuses on both evaluating algorithms in simulation-based evaluations [1, 22, 26] and on deriving theoretical performance guarantees [14, 25]. For example, [25] proposes a decentralized charging strategy for electric vehicles (EV) that is optimal if all EVs are identical, and [14] provides an algorithm for the setting when EVs are not necessarily identical.

Typically, the algorithms proposed in the literature, e.g., [1, 14, 22, 25, 26], have not considered uncertainties in renewable generation and deferrable load arrivals. However, of course, only predictions of these quantities are known ahead of time in practice, and the impact of prediction errors can be dramatic, e.g., see Figure 3.

Only very recently have algorithms that consider the uncertainties in renewable generation and deferrable load arrivals been proposed. Most of this work focuses on simulation-based studies, e.g., [6,9,10]; however some work does derive analytic performance guarantees [4,8,24,29]. For example, reference [8] proposes an algorithm that achieves the optimal competitive ratio in the case where renewable generation is precisely known (and constant) and [24] proposes an algorithm with some worst-case performance guarantees. Note that, while the algorithms proposed in [8,24] are analyzed with a "worst-case" perspective, this paper focuses on the "average-case" perspective.

Summary and contributions.

The goal of this paper is to provide a real-time algorithm for decentralized deferrable load control in the context of uncertain predictions about both future loads and future renewable generation. More specifically, in this paper we propose a novel extension of the "optimal deferrable load control problem" studied in [14]. This extension incorporates uncertainty about both deferrable and non-deferrable loads, in addition to inexact predictions of renewable generation; and then uses this problem to derive a new algorithm for deferrable load control. Further, we perform both analytic and trace-based performance analysis of the algorithm in order to quantify the impact of prediction uncertainties on deferrable load control. In particular, the contributions of the work are threefold.

First, we model renewable generation prediction as a Wiener filtering process [31] (Section 2), that is able to model any zero mean, stationary prediction evolutions. Additionally, the formulation includes a very general model for deferable loads that allows for heterogeneous deadlines and maximum charging rates, as well as stochastic arrivals.

Second, in the context of this model, we introduce a real-time algorithm for deferrable load control with uncertainty (Section 3.2). The real-time algorithm essentially solves a series of optimal control problems whose horizon lengths shrink with time. At any time, the algorithm uses only the information that is available, i.e., specifications of deferrable loads that have already arrived and predictions on future loads and renewable generation. In this sense, the algorithm we propose is a (non-trivial) extension of the algorithm proposed in [14], which applies only in the case of exact knowledge of loads and renewables. A key technique introduced by the algorithm is the concept of a "pseudo deferrable load," which is simulated at the utility to represent future deferrable load arrivals.

Third, we perform a detailed performance analysis of our proposed algorithm. The performance analysis uses both analytic results and trace-based experiments to study (i) the reduction in expected load variance achieved via deferrable load control, and (ii) the value of using real-time control via our algorithm when compared with static (open-loop) control. *The theorems in Section 4 characterize the impact of prediction inaccuracy on deferrable load control.* These analytic results highlight that as the time horizon expands, the expected load variance obtained by our proposed algorithm approaches the optimal value (Corollary 3). Also, as the time horizon expands, the algorithm obtains an increasing variance reduction over the optimal static algorithm (Corollary 5, 6). Furthermore, in Section 5 we provide trace-based experiments using data from Southern California Edison and Alberta Electric System Operator to validate the analytic results. These experiments highlight that our proposed algorithm obtains a small suboptimality under high uncertainties of renewable generation, and has significant performance improvement over the optimal static control.

2. MODEL OVERVIEW AND NOTATION

This paper studies the design and analysis of real-time control algorithms for scheduling deferrable loads to compensate for the random fluctuations in renewable generation. In the following we present a model of this scenario that serves as the basis for our algorithm design and performance evaluation. The model includes renewable generation, non-deferrable loads, and deferrable loads, which are described in turn. The key differentiation of this model from that of [14] is the inclusion of uncertainties (prediction errors) on future renewable generation and loads.

Throughout, we consider a discrete-time model over a finite time horizon. The time horizon is divided into T time slots of equal length and labeled $1, \ldots, T$. In practice, the time horizon could be one day and the length of a time slot could be 10 minutes.

2.1 Renewable generation and non-deferrable load

Renewable generation like wind is stochastic and difficult to predict. Similarly, non-deferrable load including lights are hard to predict at a low aggregation level.

Since the focus is on scheduling deferrable loads, we aggregate renewable generation and non-deferrable load into one process termed the *base load*, $b = \{b(\tau)\}_{\tau=1}^{T}$, which is defined as the difference between non-deferrable load and renewable generation, and is a stochastic process.

To model the uncertainty of base load, we use a causal filter based model described as follows, and illustrated in Figure 1. In particular, the base load at time τ is modeled as a random deviation $\delta b = \{\delta b(\tau)\}_{\tau=1}^{T}$ around its expectation $\bar{b} = \{\bar{b}(\tau)\}_{\tau=1}^{T}$. The process $\bar{b}$ is specified externally to the model, e.g., from historical data and weather report, and the process $\delta b(\tau)$ is further modeled as an uncorrelated sequence of identically distributed random variables $e = \{e(\tau)\}_{\tau=1}^{T}$ with mean 0 and variance σ^2, passing through a causal filter. Specifically, let $f = \{f(\tau)\}_{\tau=-\infty}^{\infty}$ denote the impulse response of this causal filter and assume that $f(0) = 1$, then $f(\tau) = 0$ for $\tau < 0$ and

$$\delta b(\tau) = \sum_{s=1}^{T} e(s) f(\tau - s), \quad \tau = 1, \ldots, T.$$

At time $t = 1, \ldots, T$, a prediction algorithm can observe the sequence $e(s)$ for $s = 1, \ldots, t$, and predicts b as[1]

$$b_t(\tau) = \bar{b}(\tau) + \sum_{s=1}^{t} e(s) f(\tau - s), \quad \tau = 1, \ldots, T. \quad (1)$$

Note that $b_t(\tau) = b(\tau)$ for $\tau = 1, \ldots, t$ since f is causal.

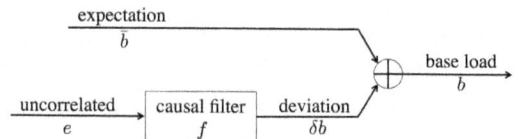

Figure 1: Diagram of the notation and structure of the model for base load, i.e., non-deferrable load minus renewable generation.

This model allows for non-stationary base load through the specification of $\bar{b}$ and a broad class of models for uncertainty via f and e. In particular, two specific filters f that we consider in detail later in the paper are:

---

[1]This prediction algorithm is a Wiener filter [31].

114

(i) A filter with finite but flat impulse response, i.e., there exists $\Delta > 0$ such that

$$f(t) = \begin{cases} 1 & \text{if } 0 \leq t < \Delta \\ 0 & \text{otherwise;} \end{cases}$$

(ii) A filter with an infinite and exponentially decaying impulse response, i.e., there exists $a \in (0, 1)$ such that

$$f(t) = \begin{cases} a^t & \text{if } t \geq 0 \\ 0 & \text{otherwise.} \end{cases}$$

These two filters provide simple but informative examples for our discussion in Section 4.

2.2 Deferrable load

To model deferrable loads we consider a setting where N deferrable loads arrive over the time horizon, each requiring a certain amount of electricity by a given deadline. Further, a real-time algorithm has imperfect information about the arrival times and sizes of these deferrable loads.

More specifically, we assume a total of N deferrable loads and label them in increasing order of their arrival times by $1, \ldots, N$, i.e., load n arrives no later than load $n + 1$ for $n = 1, \ldots, N - 1$. Further, we define $N(t)$ as the number of loads that arrive before (or at) time t for $t = 1, \ldots, T$ and fix $N(0) := 0$. Thus, load $1, \ldots, N(t)$ arrive before or at time t for $t = 1, \ldots, T$ and $N(T) = N$.

For each deferrable load, its arrival time and deadline, as well as other constraints on its power consumption, are captured via upper and lower bounds on its possible power consumption during each time. Specifically, the power consumption of deferrable load n at time t, $p_n(t)$, must be between given lower and upper bounds $\underline{p}_n(t)$ and $\overline{p}_n(t)$, i.e.,

$$\underline{p}_n(t) \leq p_n(t) \leq \overline{p}_n(t), \quad n = 1, \ldots, N, \ t = 1, \ldots, T. \quad (2)$$

These are specified externally to the model. For example, if an electric vehicle plugs in with Level II charging, then its power consumption must be within $[0, 3.3]$kW. However, if it is not plugged in (has either not arrived yet or has already departed) then its power consumption is 0kW, i.e., within $[0, 0]$kW. Further, we assume that a deferrable load n must withdraw a fixed amount of energy P_n by its deadline, i.e.,

$$\sum_{t=1}^{T} p_n(t) = P_n, \quad n = 1, \ldots, N. \quad (3)$$

Finally, the N deferrable loads arrive randomly throughout the time horizon. Define

$$a(t) := \sum_{n=N(t-1)+1}^{N(t)} P_n \quad (4)$$

as the total energy request of all deferrable loads that arrive at time t for $t = 1, \ldots, T$. We assume that $\{a(t)\}_{t=1}^{T}$ is a sequence of independent identically distributed random variables with mean λ and variance s^2. Further, define

$$A(t) := \sum_{\tau=t+1}^{T} a(\tau) \quad (5)$$

as the total energy requested after time t for $t = 1, \ldots, T$.

In summary, at time $t = 1, \ldots, T$, a real-time algorithm has full information about the deferrable loads that have arrived, i.e., $\underline{p}_n$, $\overline{p}_n$, and P_n for $n = 1, \ldots, N(t)$, and knows the expectation of future deferrable load total energy request $\mathbf{E}(A(t))$. However, a real-time algorithm has no other knowledge about deferrable loads that arrive after time t.

2.3 The deferrable load control problem

We can now formally state the deferrable load control problem that is the focus of this paper. Recall that the objective of real-time deferrable load control is to compensate for the random fluctuations in renewable generation and non-deferrable load in order to "flatten" the *aggregate load* $d = \{d(t)\}_{t=1}^{T}$, which is defined as

$$d(t) = b(t) + \sum_{n=1}^{N} p_n(t), \quad t = 1, \ldots, T. \quad (6)$$

In this paper, we focus on minimizing the *sample path variance* of the aggregate load d, $V(d)$, as a measure of "flatness", that is defined as

$$V(d) = \frac{1}{T} \sum_{t=1}^{T} \left(d(t) - \frac{1}{T} \sum_{\tau=1}^{T} d(\tau) \right)^2. \quad (7)$$

We can now formally specify the optimal deferrable load control (ODLC) problem that we seek to solve:

$$\textbf{ODLC:} \quad \min \quad \frac{1}{T} \sum_{t=1}^{T} \left(d(t) - \frac{1}{T} \sum_{\tau=1}^{T} d(\tau) \right)^2 \quad (8)$$

$$\text{over} \quad p_n(t), d(t), \quad \forall n, t$$

$$\text{s.t.} \quad d(t) = b(t) + \sum_{n=1}^{N} p_n(t), \quad \forall t;$$

$$\underline{p}_n(t) \leq p_n(t) \leq \overline{p}_n(t), \quad \forall n, t;$$

$$\sum_{t=1}^{T} p_n(t) = P_n, \quad \forall n.$$

In the above ODLC, the objective is simply the sample path variance of the aggregate load, $V(d)$, and the constraints correspond to equations (6), (2), and (3), respectively. We chose $V(d)$ as the objective for ODLC because of its significance for microgrid operators [20]. However, additionally, [14] has proven that the optimal solution does not change if the objective function $V(d)$ is replaced by $f(d) = \sum_{t=1}^{T} U(d(t))$ where $U : \mathbb{R} \to \mathbb{R}$ is strictly convex. Hence, we can use $V(d)$ without loss of generality.

3. ALGORITHM DESIGN

Given the optimal deferrable load control (ODLC) problem defined in (8), the first contribution of this paper is to design an algorithm that solves ODLC in real-time, given uncertain predictions of base and deferrable loads.

There are two key challenges for the algorithm design. First, the algorithm has access only to uncertain predictions at any given time, i.e., at time t the algorithm only knows deferrable loads 1 to $N(t)$ rather than 1 to N, and only knows the prediction b_t instead of b itself. Second, even if there was no uncertainty in predictions, solving the ODLC problem requires significant computational effort when there are a large number of deferrable loads.

Motivated by these challenges, in this section we design a decentralized algorithm with strong performance guarantees even when there is uncertainty in the predictions. The algorithm builds on the work of [14], which provides a decentralized algorithm for the case without uncertainty in predictions. We present the details of the algorithm from [14] in Section 3.1 and then present a modification of the algorithm to handle uncertain predictions in Section 3.2.

Algorithm 1 Deferrable load control without uncertainty

Input: The utility knows the base load b and the number N of deferrable loads. Each load $n \in \{1, \ldots, N\}$ knows its energy request P_n and power consumption bounds $\overline{p}_n$ and $\underline{p}_n$. The utility sets K, the number of iterations.

Output: Deferrable load schedule $p = (p_1, \ldots, p_N)$.

(i) Set $k \leftarrow 0$ and intitialize the schedule $p^{(k)}$ as

$$p_n^{(k)}(t) \leftarrow 0, \quad t = 1, \ldots, T, \; n = 1, \ldots, N.$$

(ii) The utility calculates the average load $g^{(k)} = d^{(k)}/N$,

$$g^{(k)}(t) \leftarrow \frac{1}{N}\left(b(t) + \sum_{n=1}^{N} p_n^{(k)}(t)\right), \quad t = 1, \ldots, T,$$

and broadcasts $g^{(k)}$ to all deferrable loads.

(iii) Each load n updates a new schedule $p_n^{(k+1)}$ by solving

$$\begin{aligned}
\min \quad & \sum_{\tau=1}^{T} g^{(k)}(\tau)p_n(\tau) + \frac{1}{2}\left(p_n(\tau) - p_n^{(k)}(\tau)\right)^2 \\
\text{over} \quad & p_n(1), \ldots, p_n(T) \\
\text{s.t.} \quad & \underline{p}_n(\tau) \le p_n(\tau) \le \overline{p}_n(\tau), \quad \forall \tau; \\
& \sum_{\tau=1}^{T} p_n(\tau) = P_n,
\end{aligned}$$

and reports $p_n^{(k+1)}$ to the utility.

(iv) Set $k \leftarrow k+1$. If $k < K$, go to Step (ii).

3.1 Load control without uncertainty

We start with the case where the algorithm has complete knowledge (no uncertainty) about base load and deferrable loads. In this context, the key algorithmic challenge is to solve the ODLC problem in (8) via a decentralized algorithm. Such a decentralized algorithm was proposed in [14], and we summarize the algorithm and its analysis here.

Algorithm definition: The algorithm from [14] is given in Algorithm 1. It is iterative and the superscripts in brackets denote the round of iteration. In each iteration $k \ge 0$, there are two key steps: Step (ii) and (iii). In Step (ii), the utility calculates the average load $g^{(k)}$ and broadcasts it to all deferrable loads. Note that the utility only needs to know the reported schedule $p_n^{(k)}$, the base load b, and the number of deferrable loads N. It does not need to know the constraints of the deferable loads. In Step (iii), each deferrable load n updates $p_n^{(k+1)}$ by solving a convex optimization. The objective function has two terms. The first term can be interpreted as the electricity bill if the electricity price was set to $g^{(k)}$. The second term vanishes as iterations continue.

Algorithm convergence results: Importantly, though Algorithm 1 is iterative, it converges very fast. In fact, the simulations in [14] stop the iterations after 15 rounds (i.e., K=15) in all cases because convergence is already achieved. Further, Algorithm 1 provably solves the ODLC problem given in (8) when there is no uncertainty, i.e., when $N(t) = N$ and $b_t = b$ for $t = 1, \ldots, T$ [14]. More precisely, let $\mathcal{O}$ denote the set of optimal solutions to (8), and define $d(p, \mathcal{O}) := \min_{\hat{p} \in \mathcal{O}} \|p - \hat{p}\|$ as the distance from a deferrable load schedule p to optimal deferrable load schedules $\mathcal{O}$.

PROPOSITION 1 ([14]). *When there is no uncertainty, i.e., $N(t) = N$ and $b_t = b$ for $t = 1, \ldots, T$, the deferrable load schedules $p^{(k)}$ obtained by Algorithm 1 converge to optimal schedules to ODLC, i.e., $d(p^{(k)}, \mathcal{O}) \to 0$ as $k \to \infty$.*

A particular class of optimal solutions will be of interest

to us later in the paper, so we define them here. Specifically, we call a feasible deferrable load schedule $p = (p_1, \ldots, p_N)$ *valley-filling*, if there exists some constant $C \in \mathbb{R}$ such that $\sum_{n=1}^{N} p_n(t) = (C - b(t))^+$ for $t = 1, \ldots, T$.

PROPOSITION 2 ([14]). *If a valley-filling deferrable load schedule exists, then it solves ODLC. Further, in such cases, all optimal schedules to ODLC have the same aggregate load.*

Note that valley-filling schedules tend to exist if there is a large number of deferrable loads, in such settings optimal solutions to ODLC are valley-filling.

3.2 Load control with uncertainty

Algorithm 1 provides a decentralized approach for solving the ODLC problem; however it assumes exact knowledge (certainty) about base load and deferrable loads. In this section, we adapt Algorithm 1 to the setting where there is uncertainty in base load and deferrable load predictions, while maintaining strong performance guarantees. In particular, in this section we assume that at time t, only the prediction b_t is known, not b itself, and only information about deferrable loads 1 to $N(t)$ and the expectation of future energy requests $\mathbf{E}(A(t))$ are known.

Algorithm definition: To adapt Algorithm 1 to deal with uncertainty, the first step is straightforward. In particular, it is natural to replace the base load b by its prediction b_t in Algorithm 1 to deal with the unavailability of b.

However, dealing with unavailable future deferrable load information is trickier. To do this we use a pseudo deferrable load, which is simulated at the utility, to represent future deferrable loads. More specifically, let $q = \{q(\tau)\}_{\tau=t}^{T}$ with $q(t) = 0$ denote the power consumption of the pseudo load, and assume that it requests $\mathbf{E}(A(t))$ amount of energy, i.e.,

$$\sum_{\tau=t}^{T} q(\tau) = \mathbf{E}(A(t)). \tag{9}$$

We also assume that q is point-wise upper and lower bounded by some upper and lower bounds $\overline{q}$ and $\underline{q}$, i.e.,

$$\underline{q}(\tau) \le q(\tau) \le \overline{q}(\tau), \quad \tau = t, \ldots, T. \tag{10}$$

Note that $\underline{q}(t) = \overline{q}(t) = 0$. The bounds $\underline{q}$ and $\overline{q}$ should be set according to historical data. Here, for simplicity, we consider them to be $\underline{q}(\tau) = 0$ and $\overline{q}(\tau) = \infty$ for $\tau = t+1, \ldots, T$.

Given the above setup, the utility solves the following problem at every time slot $t = 1, \ldots, T$, to accommodate the availability of only partial information.

ODLC-t:
$$\begin{aligned}
\min \quad & \sum_{\tau=t}^{T}\left(d(\tau) - \frac{1}{T-t+1}\sum_{s=t}^{T} d(s)\right)^2 \tag{11} \\
\text{over} \quad & p_n(\tau), q(\tau), d(\tau), \quad n \le N(t), \tau \ge t \\
\text{s.t.} \quad & d(\tau) = b_t(\tau) + \sum_{n=1}^{N(t)} p_n(\tau) + q(\tau), \quad \tau \ge t; \\
& \underline{p}_n(\tau) \le p_n(\tau) \le \overline{p}_n(\tau), \quad n \le N(t), \; \tau \ge t; \\
& \sum_{\tau=t}^{T} p_n(\tau) = P_n(t), \quad n \le N(t); \\
& \underline{q}(\tau) \le q(\tau) \le \overline{q}(\tau), \quad \tau \ge t; \\
& \sum_{\tau=t}^{T} q(\tau) = \mathbf{E}(A(t))
\end{aligned}$$

where $P_n(t) = P_n - \sum_{\tau=1}^{t-1} p_n(\tau)$ is the energy to be consumed at or after time t, for $n = 1, \ldots, N(t)$ and $t = 1, \ldots, T$.

Algorithm 2 Deferrable load control with uncertainty

Input: At time t, the utility knows the prediction b_t of base load and the number $N(t)$ of deferrable loads. Each deferrable load $n \in \{1, \ldots, N(t)\}$ knows its future energy request $P_n(t)$ and power consumption bounds $\overline{p}_n$ and $\underline{p}_n$. The utility sets K, the number iterations.

Output: At time t, output the power consumption $p_n(t)$ for deferrable loads $1, \ldots, N(t)$.

At time slot $t = 1, \ldots, T$:

(i) Set $k \leftarrow 0$. Each deferrable load $n \in \{1, \ldots, N(t)\}$ initializes its schedule $\{p_n^{(0)}(\tau)\}_{\tau=t}^T$ as

$$p_n^{(0)}(\tau) \leftarrow \begin{cases} p_n^{(K)}(\tau) & \text{if } n \leq N(t-1) \\ 0 & \text{if } n > N(t-1) \end{cases}, \quad \tau = t, \ldots, T$$

where $p_n^{(K)}$ is the schedule of load n in iteration K of the previous time slot $t-1$.

(ii) The utility solves

$$\min \quad \sum_{\tau=t}^T \left(b_t(\tau) + \sum_{n=1}^{N(t)} p_n^{(k)}(\tau) + q(\tau) \right)^2$$

$$\text{over} \quad q(t), \ldots, q(T)$$

$$\text{s.t.} \quad \underline{q}(\tau) \leq q(\tau) \leq \overline{q}(\tau), \quad \tau \geq t;$$

$$\sum_{\tau=t}^T q(\tau) = \mathbf{E}(A(t))$$

to obtain a pseudo schedule $\{q^{(k)}(\tau)\}_{\tau=t}^T$. The utility then calculates the average aggregate load per deferrable load $g^{(k)}$ as

$$g^{(k)}(\tau) \leftarrow \frac{1}{N(t)} \left(b_t(\tau) + \sum_{n=1}^{N(t)} p_n^{(k)}(\tau) + q^{(k)}(\tau) \right)$$

for $\tau = t, \ldots, T$, and broadcasts $\{g^{(k)}(\tau)\}_{\tau=t}^T$ to deferrable loads $n = 1, \ldots, N(t)$.

(iii) Each deferrable load $n = 1, \ldots, N(t)$ solves

$$\min \quad \sum_{\tau=t}^T g^{(k)}(\tau) p_n(\tau) + \frac{1}{2} \left(p_n(\tau) - p_n^{(k)}(\tau) \right)^2$$

$$\text{over} \quad p_n(t), \ldots, p_n(T)$$

$$\text{s.t.} \quad \underline{p}_n(\tau) \leq p_n(\tau) \leq \overline{p}_n(\tau), \quad \tau \geq t;$$

$$\sum_{\tau=t}^T p_n(\tau) = P_n(t),$$

to obtain a new schedule $\{p_n^{(k+1)}(\tau)\}_{\tau=t}^T$, and reports $\{p_n^{k+1}(\tau)\}_{\tau=t}^T$ to the utility.

(iv) Set $k \leftarrow k + 1$. If $k < K$, go to Step (ii).

(v) Deferrable load $n \in \{1, \ldots, N(t)\}$ sets $p_n(t) \leftarrow p_n^K(t)$ and $P_n(t+1) \leftarrow P_n(t) - p_n(t)$.

Now, adjusting Algorithm 1 to solve ODLC-t gives Algorithm 2, which is real-time and shrinking-horizon. Note that if base load prediction is exact (i.e., $b_t = b$ for $t = 1, \ldots, T$) and all deferrable loads arrive at the beginning of the time horizon (i.e., $N(t) = N$ for $t = 1, \ldots, T$), then ODLC-1 reduces to ODLC and Algorithm 2 reduces to Algorithm 1.

Algorithm convergence results: We provide analytic guarantees on the convergence and optimality of Algorithm 2. In particular, we prove that Algorithm 2 solves ODLC-

t at every time slot t. Specifically, let $\mathcal{O}(t)$ denote the set of optimal schedules to ODLC-t, and define $d(p, \mathcal{O}(t)) := \min_{(\hat{p}, \hat{q}) \in \mathcal{O}(t)} \|p - \hat{p}\|$ as the distance from a schedule p to optimal schedules $\mathcal{O}(t)$ at time t, for $t = 1, \ldots, T$.

THEOREM 1. *At time $t = 1, \ldots, T$, the deferrable load schedules $p^{(k)}$ obtained by Algorithm 2 converge to optimal schedules to ODLC-t, i.e., $d(p^{(k)}, \mathcal{O}(t)) \to 0$ as $k \to \infty$.*

The theorem is proved in Appendix A.1. Though iterative, Algorithm 2 converges fast, similar to Algorithm 1. In the simulations, setting $K = 15$ is enough for all test cases.

Similar to Proposition 2, "t-valley-filling" provides a simple characterization of the solutions to ODLC-t. Specifically, at time $t = 1, \ldots, T$, a feasible schedule (p, q) is called *t-valley-filling*, if there exists some constant $C(t) \in \mathbb{R}$ such that

$$q(\tau) + \sum_{n=1}^{N(t)} p_n(\tau) = (C(t) - b_t(\tau))^+, \quad \tau = t, \ldots, T. \quad (12)$$

Given this definition of t-valley-filling, the following corollary follows immediately from Proposition 2.

COROLLARY 1. *At time $t = 1, \ldots, T$, a t-valley-filling deferrable load schedule, if exists, solves ODLC-t. Furthermore, in such cases, all optimal schedules to ODLC-t have the same aggregate load.*

This corollary serves as the basis for the performance analysis we perform in Section 4. Remember that t-valley-filling schedules tend to exist in cases where there are a large numbers of deferrable loads.

4. PERFORMANCE EVALUATION

To this point, we have shown that Algorithm 2 makes "optimal" decisions with the information available at every time slot, i.e., it solves ODLC-t at time t for $t = 1, \ldots, T$. However, these decisions are still suboptimal compared to what could be achieved if exact information was available. In this section, our goal is to understand the impact of uncertainty on the performance. In particular, we study two questions:

(i) How do the uncertainties about base load and deferrable loads impact the expected sample path load variance obtained by Algorithm 2?

(ii) What is the improvement of using the real-time control provided by Algorithm 2 over using the optimal static control?

Our answers to these questions are below. Throughout, we focus on the special, but practically relevant, case when a t-valley-filling schedule exists at every time $t = 1, \ldots, T$. As we have mentioned previously, when the number of deferrable loads is large this is a natural assumption that holds for practical load profiles. The reason for making this assumption is that it allows us to use the characterization of optimal schedules given in (12). In fact, without loss of generality, we further assume $C(t) \geq b_t(\tau)$ for $\tau = t, \ldots, T$, under which (12) implies

$$d(t) = C(t) = \frac{1}{T - t + 1} \left(\sum_{\tau=t}^T b_t(\tau) + \mathbf{E}(A(t)) + \sum_{n=1}^{N(t)} P_n(t) \right)$$
$$(13)$$

for $t = 1, \ldots, T$. Thus, equation (13) defines the model we use for the performance analysis of Algorithm 2.

The expected load variance of Algorithm 2.

We start by calculating the expected load variance, $\mathbf{E}(V)$, of Algorithm 2. The goal is to understand how uncertainty about base load and deferrable loads impacts the load variance. Note that, if there are no base load prediction errors and deferrable loads arrive at the beginning of the time horizon, then Algorithm 2 obtains optimal schedules that have zero load variance. In contrast, when there are base load prediction errors and stochastic deferrable load arrivals, the expected load variance is given by the following theorem.

To state the result, recall that $\{f(t)\}_{t=-\infty}^{\infty}$ is the causal filter modeling the correlation of base load and define $F(t) := \sum_{s=0}^{t} f(s)$ for $t = 0, \ldots, T$.

THEOREM 2. *The expected load variance $\mathbf{E}(V)$ obtained by Algorithm 2 is*

$$\mathbf{E}(V) = \frac{s^2}{T} \sum_{t=2}^{T} \frac{1}{t} + \frac{\sigma^2}{T^2} \sum_{t=0}^{T-1} F^2(t) \frac{T-t-1}{t+1}. \quad (14)$$

The theorem is proved in Appendix A.4.

Theorem 2 explicitly states the interaction of the variability of base load prediction (σ) and deferrable load prediction (s) with the horizon length T. Besides, it highlights the correlation of base load prediction error through F. More specifically, the expected load variance $\mathbf{E}(V)$ tends to 0 as the uncertainties in base load and deferrable loads vanish, i.e., $\sigma \to 0$ and $s \to 0$.

COROLLARY 2. *The expected load variance $\mathbf{E}(V) \to 0$ as $\sigma \to 0$ and $s \to 0$.*

Another remark about Theorem 2 is that the two terms in (14) correspond to the impact of the uncertainties in deferrable loads and base load respectively. In particular, Theorem 2 is proved in Section A.4 by analyzing these two cases separately and then combining the results. Specifically, the following two lemmas are the key pieces in the proof of Theorem 2, but are also of interest in their own right.

LEMMA 1. *If there is no base load prediction error, i.e., $b_t = b$ for $t = 1, \ldots, T$, then the expected load variance obtained by Algorithm 2 is*

$$\mathbf{E}(V) = s^2 \frac{\sum_{t=2}^{T} \frac{1}{t}}{T} \approx s^2 \frac{\ln T}{T}.$$

The lemma is proved in Appendix A.2.

LEMMA 2. *If there are no deferrable load arrivals after time 1, i.e., $N(t) = N$ for $t = 1, \ldots, T$, then the expected load variance obtained by Algorithm 2 is*

$$\mathbf{E}(V) = \frac{\sigma^2}{T^2} \sum_{t=0}^{T-1} F^2(t) \frac{T-t-1}{t+1}.$$

The lemma is proved in Appendix A.3.

Lemma 1 highlights that the more uncertainty in deferrable load arrival, i.e., the larger s, the larger the expected load variance $\mathbf{E}(V)$. On the other hand, the longer the time horizon T, the smaller the expected load variance $\mathbf{E}(V)$.

Similarly, Lemma 2 highlights that a larger base load prediction error, i.e., a larger σ, results in a larger expected load variance $\mathbf{E}(V)$. However, if the impulse response $\{f(t)\}_{t=-\infty}^{\infty}$ of the modeling filter of the base load decays fast enough with t, then the following corollary highlights that the expected load variance actually tends to 0 as time horizon T increases despite the uncertainty of base load predictions.

COROLLARY 3. *If there are no deferrable load arrivals after time 1, i.e., $N(t) = N$ for $t = 1, \ldots, T$, and $|f(t)| \sim O(t^{-1/2-\alpha})$ for some $\alpha > 0$, then the expected load variance obtained by Algorithm 2 satisfies $\mathbf{E}(V) \to 0$ as $T \to \infty$.*

The corollary is proved in Appendix A.5.

The improvement of Algorithm 2 over static control.

The goal of this section is to quantify the improvement of real-time control via Algorithm 2 over the optimal static (open-loop) control. To be more specific, we compare the expected load variance $\mathbf{E}(V)$ obtained by the real-time control Algorithm 2, with the expected load variance $\mathbf{E}(V')$ obtained by the optimal static control, which only uses base load prediction at the beginning of the time horizon (i.e., $\bar{b}$) to compute deferrable load schedules. We assume $N(t) = N$ for $t = 1, \ldots, T$ in this section since otherwise any static control cannot obtain a schedule for all deferrable loads. Thus, the interpretation of the results that follow is as a quantification of the value of incorporating updated base load predictions into the deferrable load controller.

To begin the analysis, note that $\mathbf{E}(V)$ for this setting is given in Lemma 2. Further, it can be verified that the optimal static control is to solve ODLC with b replaced by $\bar{b}$, and the corresponding expected load variance $\mathbf{E}(V')$ is given by the following lemma.

LEMMA 3. *If there is no stochastic load arrival, i.e., $N(t) = N$ for $t = 1, \ldots, T$, then the expected load variance $\mathbf{E}(V')$ obtained by the optimal static control is*

$$\mathbf{E}(V') = \frac{\sigma^2}{T^2} \sum_{t=0}^{T-1} \left(T(T-t) f^2(t) - F^2(t) \right).$$

The lemma is proved in Appendix A.6.

Next, comparing $\mathbf{E}(V)$ and $\mathbf{E}(V')$ given in Lemma 2 and 3 shows that Algorithm 2 always obtains a smaller expected load variance than the optimal static control. Specifically,

COROLLARY 4. *If there is no deferrable load arrival after time 1, i.e., $N(t) = N$ for $t = 1, \ldots, T$, then*

$$\mathbf{E}(V') - \mathbf{E}(V) = \frac{\sigma^2}{T} \sum_{t=1}^{T} \frac{1}{2t} \sum_{m=0}^{t-1} \sum_{n=0}^{t-1} (f(m) - f(n))^2 \geq 0.$$

The corollary is proved in the extended version [16].

Corollary 4 highlights that Algorithm 2 is guaranteed to obtain a smaller expected load variance than the optimal static control. The next step is to quantify how much smaller $\mathbf{E}(V)$ is in comparison with $\mathbf{E}(V')$.

To do this we compute the ratio $\mathbf{E}(V')/\mathbf{E}(V)$. Unfortunately, the general expression for the ratio is too complex to provide insight, so we consider two representative cases for the impulse response $f(t)$ of the causal filter in order to obtain insights. Specifically, we consider examples (i) and (ii) from Section 2.1. Briefly, in (i) $f(t)$ is finite and in (ii) $f(t)$ is infinite but decays exponentially in t. For these two cases, the ratio $\mathbf{E}(V')/\mathbf{E}(V)$ is summarized in the following corollaries.

COROLLARY 5. *If there is no deferrable load arrival after time 1, i.e., $N(t) = N$ for $t = 1, \ldots, T$, and there exists $\Delta > 0$ such that*

$$f(t) = \begin{cases} 1 & \text{if } 0 \leq t < \Delta \\ 0 & \text{otherwise,} \end{cases}$$

then

$$\frac{\mathbf{E}(V')}{\mathbf{E}(V)} = \frac{T/\Delta}{\ln(T/\Delta)} \left(1 + O\left(\frac{1}{\ln(T/\Delta)} \right) \right).$$

The corollary is proved in the extended version [16].

COROLLARY 6. *If there is no deferrable load arrival after time 1, i.e., $N(t) = N$ for $t = 1, \ldots, T$, and there exists $a \in (0,1)$ such that*

$$f(t) = \begin{cases} a^t & \text{if } t \geq 0 \\ 0 & \text{otherwise,} \end{cases}$$

118

then

$$\frac{\mathbf{E}(V')}{\mathbf{E}(V)} = \frac{1-a}{1+a} \frac{T}{\ln T} \left(1 + O\left(\frac{\ln \ln T}{\ln T}\right)\right).$$

The corollary is proved in the extended version [16].

Corollary 5 highlights that, in the case where f is finite, if we define $\lambda = T/\Delta$ as the ratio of time horizon to filter length, then the load reduction roughly scales as $\lambda/\ln(\lambda)$. Thus, the longer the time horizon is in comparison to the filter length, the larger expected load variance reduction we obtain from using Algorithm 2 as compared with the optimal static control.

Similarly, Corollary 6 highlights that, in the case where f is infinite and exponentially decaying, the expected load variance reduction scales with T as $T/\ln T$ with coefficient $(1-a)/(1+a)$. Thus, the smaller a is, which means the faster f dies out, the more load variance reduction we obtain by using real-time control. This is similar to having a smaller Δ in the previous case.

5. EXPERIMENTAL RESULTS

In this section we use trace-based experiments to explore the generality of the analytic results in the previous section. In particular, the results in the previous section characterize the expected load variance obtained by Algorithm 2 as a function of prediction uncertainties, and quantify the improvement of Algorithm 2 over the optimal static (open-loop) controller. However, the analytic results make simplifying assumptions on the form of uncertainties and solution schedules (equation (13)). Therefore, it is important to assess the performance of the algorithm using real-world data.

5.1 Experimental setup

The numerical experiments we perform use a time horizon of 24 hours, from 20:00 to 20:00 on the following day. The time slot length is 10 minutes, which is the granularity of the data we have obtained about renewable generation.

Base load.

Recall that base load is a combination of non-deferrable load and renewable generation. The non-deferrable load traces used in the experiments come from the average residential load in the service area of Southern California Edison in 2012 [28]. In the simulations, we assume that non-deferrable load is precisely known so that uncertainties in the base load only come from renewable generation. In particular, non-deferrable load over the time horizon of a day is taken to be the average over the 366 days in 2012 as in Figure 2(a), and assumed to be known to the utility at the beginning of the time horizon. In practice, non-deferrable load at the substation feeder level can be predicted within 1–3% root-mean-square error looking 24 hours ahead [13].

The renewable generation traces we use come from the 10-minute historical data for total wind power generation of the Alberta Electric System Operator from 2004 to 2009 [3]. In the simulations, we scale the wind power generation so that its average over the 6 years corresponds to a number of penetration levels in the range between 5% and 30%, and pick the wind power generation of a randomly chosen day as the renewable generation during each run. Figure 2(b) shows the wind power generation for four representative days, one for each season, after scaling to 20% penetration.

We assume that the renewable generation is not precisely known until it is realized, but that a prediction of the generation, which improves over time, is available to the utility. The modeling of prediction evolution over time is according

to a martingale forecasting process [18,19], which is a standard model for an unbiased prediction process that improves over time.

Specifically, the prediction model is as follows: For wind generation $w(\tau)$ at time τ, the prediction error $w_t(\tau) - w(\tau)$ at time $t < \tau$ is the sum of a sequence of independent random variables $n_s(\tau)$ as

$$w_t(\tau) = w(\tau) + \sum_{s=t+1}^{\tau} n_s(\tau), \quad 0 \le t < \tau \le T.$$

Here $w_0(\tau)$ is the wind prediction without any observation, i.e., the expected wind generation $\bar{w}(\tau)$ at the beginning of the time horizon (used by static control).

The random variables $n_s(\tau)$ are assumed to be Gaussian with mean 0. Their variances are chosen as

$$\mathbf{E}(n_s^2(\tau)) = \frac{\sigma^2}{\tau - s + 1}, \quad 1 \le s \le \tau \le T$$

where $\sigma > 0$ is such that the root-mean-square prediction error $\sqrt{\mathbf{E}(w_0(T) - w(T))^2}$ looking T time slots (i.e., 24 hours) ahead is 0%–22.5% of the nameplate wind generation capacity.[2] According to this choice of the variances of $n_s(\tau)$, root-mean-square prediction error only depends on how far ahead the prediction is, in particular as in Figure 2(c). This choice is motivated by [17].

Deferrable loads.

For simplicity, we consider the hypothetical case where all deferrable loads are electric vehicles. Since historical data for electric vehicle usage is not available, we are forced to use synthetic traces for this component of the experiments. Specifically, in the simulations the electric vehicles are considered to be identical, each requests 10kWh electricity by a deadline 8 hours after it arrives, and each must consume power at a rate within $[0, 3.3]$kW after it arrives and before its deadline.

In the simulations, the arrival process starts at 20:00 and ends at 12:00 the next day so that the deadlines of all electric vehicles lie within the time horizon of 24 hours. In each time slot during the arrival process, we assume that the number of arriving electric vehicles is uniformly distributed in $[0.8\lambda, 1.2\lambda]$, where λ is chosen so that electric vehicles (on average) account for 5%–30% of the non-deferrable loads. While this synthetic workload is simplistic, the results we report are representative of more complex setups as well.

Uncertainty about deferrable load arrivals is captured as follows. The prediction $\mathbf{E}(A(t))$ of future deferrable load total energy request is simply the arrival rate λ times the length of the rest of the arrival process $T' - t$ where T' is the end of the arrival process (12:00), i.e.,

$$\mathbf{E}(A(t)) = \lambda(T' - t), \quad t = 1, \dots, T'.$$

If $t > T'$, i.e., the deferrable load arrival process has ended, then $\mathbf{E}(A(t)) = 0$.

Baselines for comparison.

Our goal in the simulations is to contrast the performance of Algorithm 2 with a number of common benchmarks to tease apart the impact of real-time control and the impact of different forms of uncertainty. To this end, we consider four controllers in our experiments:

---

[2]Average wind generation is 15% of the nameplate capacity, so the root-mean-square prediction error looking T time slots ahead is 0%–150% the average wind generation.

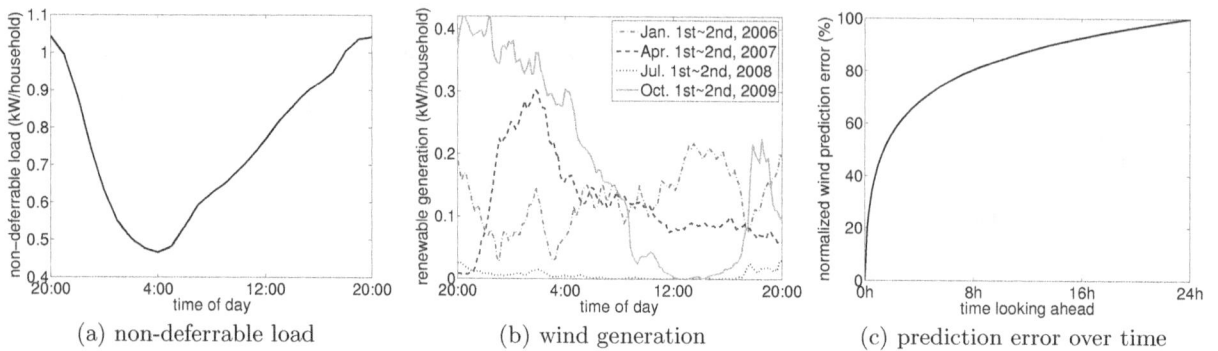

(a) non-deferrable load   (b) wind generation   (c) prediction error over time

Figure 2: Illustration of the traces used in the experiments. (a) shows the average residential load in the service area of Southern California Edison in 2012. (b) shows the total wind power generation of the Alberta Electric System Operator scaled to represent 20% penetration. (c) shows the normalized root-mean-square wind prediction error as a function of the time looking ahead for the model used in the experiments.

(i) *Offline optimal control:* The controller has full knowledge about the base load and deferrable loads, and solves the ODLC problem offline. It is not realistic in practice, but serves as a benchmark for the other controllers since offline optimal control obtains the smallest possible load variance.

(ii) *Static control with exact deferrable load arrival information:* The controller has full knowledge about deferrable loads (including those that have not arrived), but uses only the prediction of base load that is available at the beginning of the time horizon to compute a deferrable load schedule that minimizes the expected load variance. This static control is still unrealistic since a deferrable load is known only after it arrives. But, this controller corresponds to what is considered in prior works, e.g., [14, 15, 25].

(iii) *Real-time control with exact deferrable load arrival information.* The controller has full knowledge about deferrable loads (including those that have not arrived), and uses the prediction of base load that is available at the current time slot to update the deferrable load schedule by minimizing the expected load variance to go, i.e., Algorithm 2 with $N(t) = N$ for $t = 1, \ldots, T$. The control is unrealistic since a deferrable load is known only after it arrives; however it provides the natural comparison for case (ii) above.

(iv) *Real-time control without exact deferrable load arrival information, i.e., Algorithm 2.* This corresponds to the realistic scenario where only predictions are available about future deferrable loads and base load. The comparison with case (iii) highlights the impact of deferrable load arrival uncertainties.

The performance measure that we show in all plots is the "suboptimality" of the controllers, which we define as

$$\eta := \frac{V - V^{\text{opt}}}{V^{\text{opt}}},$$

where V is the load variance obtained by the controller and V^{opt} is the load variance obtained by the offline optimal, i.e., case (i) above. Thus, the lines in the figures correspond to cases (ii)-(iv).

5.2 Experimental results

Our experimental results focus on two main goals: (i) understanding the impact of prediction accuracy on the expected load variance obtained by deferrable load control

algorithms, and (ii) contrasting the real-time (closed-loop) control of Algorithm 2 with the optimal static (open-loop) controller. We focus on the impact of three key factors: wind prediction error, the penetration of deferrable load, and the penetration of renewable energy.

The impact of prediction error.

To study the impact of prediction error, we fix the penetration of both renewable generation (wind) and deferrable loads at 10% of non-deferrable load, and simulate the load variance obtained under different levels of root-mean-square wind prediction errors (0%–22.5% of the nameplate capacity looking 24 hours ahead). The results are summarized in Figure 3(a). It is not surprising that suboptimality of both the static and the real-time controllers that have exact information about deferrable load arrivals is zero when the wind prediction error is 0, since there is no uncertainty for these controllers in this case.

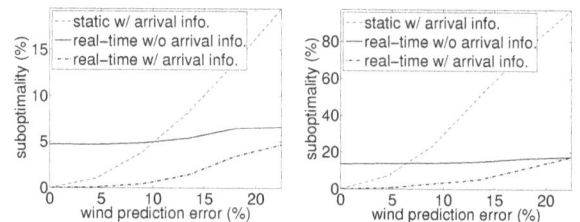

(a) Wind and deferrable load penetration are both 10%.   (b) Wind and deferrable load penetration are both 20%.

Figure 3: Illustration of the impact of wind prediction error on suboptimality of load variance.

As prediction error increases, the suboptimality of both the static and the real-time control increases. However, notably, the suboptimality of real-time control grows much more slowly than that of static control, and remains small (<4.7%) if deferrable load arrivals are known, over the whole range 0%–22.5% of wind prediction error. At 22.5% prediction error, the suboptimality of static control is 4.2 times that of real-time control. This highlights that real-time control mitigates the influence of imprecise base load prediction over time.

Moving to the scenario where deferrable load arrivals are not known precisely, we see that the impact of this inexact information is less than 6.6% of the optimal variance. However, real-time control yields a load variance that is surprisingly resilient to the growth of wind prediction error, and eventually beats the optimal static control at around 10% wind prediction error, even though the optimal static control has exact knowledge of deferrable loads and the adaptive control does not.

As prediction error increases, the suboptimality of the real-time control with or without deferrable load arrival information gets close, i.e., the benefit of knowing additional information on future deferrable load arrivals vanishes as base load uncertainty increases. This is because the additional information is used to overfit the base load prediction error.

The same comparison is shown in Figure 3(b) for the case where renewable and deferrable load penetration are both 20%. Qualitatively the conclusions are the same, however at this higher penetration the contrast between the resilience of adaptive control and static control is magnified, while the benefit of knowing deferrable load arrival information is minified. In particular, real-time control without arrival information beats static control with arrival information, at a lower (around 7%) wind prediction error, and knowing deferrable load arrival information does not reduce suboptimality of real-time control with 22.5% wind prediction error.

The impact of deferrable load penetration.

Next, we look at the impact of deferrable load penetration on the performance of the various controllers. To do this, we fix the wind penetration level to be 20% and wind prediction error looking 24 hours ahead to be 18%, and simulate the load variance obtained under different deferrable load penetration levels (5%–30%). The results are summarized in Figure 4(a).

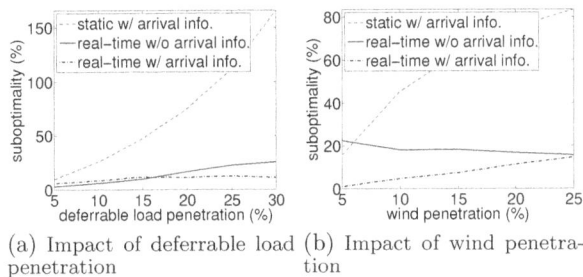

(a) Impact of deferrable load penetration (b) Impact of wind penetration

Figure 4: Suboptimality of load variance as a function of (a) deferrable load penetration and (b) wind penetration. In (a) the wind penetration is 20% and in (b) the deferrable load penetration is 20%. In both, the wind prediction error looking 24 hours ahead is 18%.

Not surprisingly, if future deferrable loads are known and uncertainty only comes from base load prediction error, then the suboptimality of real-time control is very small ($<11.2\%$) over the whole range 5%–30% of deferrable load penetration, while the suboptimality of static control increases with deferrable load penetration, up to as high as 166% (14.9 times that of real-time control) at 30% deferrable load penetration.

However, without knowing future deferrable loads, the suboptimality of real-time control increases with the deferrable load penetration. This is because larger amount of deferrable loads introduces larger uncertainties in deferrable load arrivals. But the suboptimality remains smaller than that of static control over the whole range 5%–30% of deferrable load penetration. The highest suboptimality 25.7%

occurs at 30% deferrable load penetration, and is less than 1/6 of the suboptimality of static control, which assumes exact deferrable load arrival information.

The impact of renewable penetration.

Finally, we study the impact of renewable penetration. To do this we fix the deferrable load penetration level to be 20% and the wind prediction error looking 24 hours ahead to be 18%, and simulate the load variance obtained by the 4 test cases under different wind penetration levels (5%–25%). The results are summarized in Figure 4(b).

A key observation is that if future deferrable loads are known and uncertainty only comes from base load prediction error, then the suboptimality of real-time control grows much slower than that of static control, as wind penetration level increases. As explained before, this highlights that real-time control mitigates the impact of base load prediction error over time. In fact, the suboptimality of real-time control is small ($<15\%$) over the whole range 5%–25% of wind penetration levels. Of course, without knowledge of future deferrable loads, the suboptimality of real-time control becomes bigger. However, it still eventually outperforms the optimal static controller at around 6% wind penetration, despite the fact that the optimal static controller is using exact information about deferrable loads.

6. CONCLUDING REMARKS

We have proposed a real-time algorithm for decentralized deferrable load control that can schedule a large number of deferrable loads to compensate for the random fluctuations in renewable generation. At any time, the algorithm incorporates updated predictions about deferrable loads and renewable generation to minimize the expected load variance to go. Further, we have derived an explicit expression for the expected aggregate load variance obtained by the algorithm by modeling the base load prediction updates as a Wiener filtering process. Additionally, we have highlighted the importance of the expression by using it to evaluate the improvement of real-time control over static control. Interestingly, the sub-optimality of static control is $O(T/\ln T)$ times that of real-time control in two representative cases of base load prediction updates. The qualitative insights from the analytic results were validated using trace-based simulations, which confirm that the algorithm has significantly smaller sub-optimality than the optimal static control.

There remain many open questions on deferrable load control. For example, is it possible to reduce the communication and computation requirements of the proposed algorithm by assuming achievability of t-valley-filling? How to extend the algorithm to a receding horizon implementation? Additionally, how to apply the technique used here to incorporate prediction evolution for other demand response settings.

7. ACKNOWLEDGEMENT

This work was supported by NSF NetSE grant CNS 0911041, ARPA-E grant DE-AR0000226, Southern California Edison, National Science Council of Taiwan, R.O.C, grant NSC 101-3113-P-008-001, Resnick Institute, Okawa Foundation, NSF CNS 1312390, NSF grant CNS 0846025, and DoE grant DE-EE000289.

8. REFERENCES

[1] S. Acha, T. C. Green, and N. Shah. Effects of optimised plug-in hybrid vehicle charging strategies on electric distribution network losses. In *IEEE PES Transmission and Distribution Conference and Exposition*, pages 1–6, 2010.

[2] D. J. Aigner and J. G. Hirschberg. Commercial/industrial customer response to time-of-use electricity prices: Some experimental results. *The RAND Journal of Economics*, 16(3):341–355, 1985.

[3] Alberta Electric System Operator. Wind power / ail data, 2009. http://www.aeso.ca/gridoperations/20544.html.

[4] J.-Y. L. Boudec and D.-C. Tomozei. Satisfiability of elastic demand in the smart grid. *arXiv preprint arXiv:1011.5606*, 2010.

[5] California Public Utilities Commission. Zero net energy action plan, 2008. http://www.cpuc.ca.gov/NR/rdonlyres/6C2310FE-AFE0-48E4-AF03-530A99D28FCE/0/ZNEActionPlanFINAL83110.pdf.

[6] M. Caramanis and J. Foster. Management of electric vehicle charging to mitigate renewable generation intermittency and distribution network congestion. In *IEEE CDC*, pages 4717–4722, 2009.

[7] L. Chen, N. Li, S. H. Low, and J. C. Doyle. Two market models for demand response in power networks. In *IEEE SmartGridComm*, pages 397–402, 2010.

[8] S. Chen and L. Tong. iems for large scale charging of electric vehicles: architecture and optimal online scheduling. In *IEEE SmartGridComm*, pages 629–634, 2012.

[9] A. Conejo, J. Morales, and L. Baringo. Real-time demand response model. *IEEE Transactions on Smart Grid*, 1(3):236–242, 2010.

[10] S. Deilami, A. Masoum, P. Moses, and M. Masoum. Real-time coordination of plug-in electric vehicle charging in smart grids to minimize power losses and improve voltage profile. *IEEE Transactions on Smart Grid*, 2(3):456–467, 2011.

[11] Department of Energy. The smart grid: an introduction, 2008. http://energy.gov/sites/prod/files/oeprod/DocumentsandMedia/DOE_SG_Book_Single_Pages%281%29.pdf.

[12] Department of Energy. One million electric vehicles by 2015, 2011. http://www1.eere.energy.gov/vehiclesandfuels/pdfs/1_million_electric_vehicles_rpt.pdf.

[13] E. A. Feinberg and D. Genethliou. Load forecasting. In *Applied Mathematics for Restructured Electric Power Systems*, Power Electronics and Power Systems, pages 269–285. Springer US, 2005.

[14] L. Gan, U. Topcu, and S. H. Low. Optimal decentralized protocol for electric vehicle charging. In *IEEE CDC*, pages 5798–5804, 2011.

[15] L. Gan, U. Topcu, and S. H. Low. Stochastic distributed protocol for electric vehicle charging with discrete charging rate. In *IEEE PES General Meeting*, pages 1–8, 2012.

[16] L. Gan, A. Wierman, U. Topcu, N. Chen, and S. H. Low. Real-time deferrable load control: handling the uncertainties of renewable generation, 2013. Technical report, available at http://www.its.caltech.edu/~lgan/index.html.

[17] G. Giebel, R. Brownsword, G. Kariniotakis, M. Denhard, and C. Draxl. *The State-Of-The-Art in Short-Term Prediction of Wind Power*. ANEMOS.plus, 2011.

[18] S. C. Graves, D. B. Kletter, and W. B. Hetzel. A dynamic model for requirements planning with application to supply chain optimization. *Manufacturing & Service Operation Management*, 1(1):50–61, 1998.

[19] S. C. Graves, H. C. Meal, S. Dasu, and Y. Qiu. Two-stage production planning in a dynamic environment, 1986. http://web.mit.edu/sgraves/www/papers/GravesMealDasuQiu.pdf.

[20] N. Hatziargyriou, H. Asano, R. Iravani, and C. Marnay. Microgrids. *IEEE Power and Energy Magazine*, 5(4):78–94, 2007.

[21] Y.-Y. Hsu and C.-C. Su. Dispatch of direct load control using dynamic programming. *IEEE Transactions on Power Systems*, 6(3):1056–1061, 1991.

[22] M. Ilic, J. Black, and J. Watz. Potential benefits of implementing load control. In *IEEE PES Winter Meeting*, volume 1, pages 177–182, 2002.

[23] N. Li, L. Chen, and S. H. Low. Optimal demand response based on utility maximization in power networks. In *IEEE PES General Meeting*, pages 1–8, 2011.

[24] Q. Li, T. Cui, R. Negi, F. Franchetti, and M. D. Ilic. On-line decentralized charging of plug-in electric vehicles in power systems. *arXiv:1106.5063*, 2011.

[25] Z. Ma, D. Callaway, and I. Hiskens. Decentralized charging control for large populations of plug-in electric vehicles. In *IEEE CDC*, pages 206–212, 2010.

[26] K. Mets, T. Verschueren, W. Haerick, C. Develder, and F. De Turck. Optimizing smart energy control strategies for plug-in hybrid electric vehicle charging. In *IEEE/IFIP NOMS Wksps*, pages 293–299, 2010.

[27] National Institute of Standards and Technology. Nist framework and roadmap for smart grid interoperability standards, 2010. http://www.nist.gov/public_affairs/releases/upload/smartgrid_interoperability_final.pdf.

[28] Southern California Edison. 2012 static load profiles, 2012. http://www.sce.com/005_regul_info/eca/DOMSM12.DLP.

[29] A. Subramanian, M. Garcia, A. Dominguez-Garcia, D. Callaway, K. Poolla, and P. Varaiya. Real-time scheduling of deferrable electric loads. In *ACC*, pages 3643–3650, 2012.

[30] Wikipedia. Krasovskii-lasalle principle. http://en.wikipedia.org/wiki/Krasovskii-LaSalle_principle.

[31] Wikipedia. Wiener filter. http://en.wikipedia.org/wiki/Wiener_filter.

APPENDIX

A. PROOFS

In this section, we only include proofs of the main results due to space restrictions. The remainder of the proofs can be found in the extended version [16].

A.1 Proof of Theorem 1

For brevity and without loss of generality, we prove Theorem 1 for $t = 1$ only. Thus, we can abbreviate b_t and $N(t)$ by b and N respectively without introducing confusion.

For feasible p, q to ODLC-t and $p = (p_1, \ldots, p_N)$, define

$$L(p, q) = \sum_{\tau=1}^{T} \left(b(\tau) + \sum_{n=1}^{N} p_n(\tau) + q(\tau) \right)^2.$$

Since the sum of the aggregate load $\sum_{\tau=1}^{T} d(\tau)$ is a constant, minimizing the ℓ_2 norm of the aggregate load is equivalent to minimizing its variance. Hence, if subject to the same constraints, the minimizer of L is also the solution to ODLC-t. According to the proof of Proposition 1 in [14], we have

$$L(p^{(k+1)}, q^{(k)}) \leq L(p^{(k)}, q^{(k)})$$

for $k \geq 0$, and the equality is attained if and only if $p^{(k+1)} = p^{(k)}$ and $p^{(k)}$ minimizes $L(p, q^{(k)})$ over all feasible p, i.e., (the first order optimality condition)

$$\left\langle b + \sum_{n=1}^{N} p_n^{(k)} + q^{(k)}, \ p_n' - p_n^{(k)} \right\rangle \geq 0$$

for $n = 1, \ldots, N$ and all feasible p_n'. According to Step (ii) of Algorithm 2, it is straightforward that

$$L(p^{(k+1)}, q^{(k+1)}) \leq L(p^{(k+1)}, q^{(k)})$$

for $k \geq 0$, and the equality is attained if and only if $q^{(k+1)} = q^{(k)}$ and $q^{(k)}$ minimizes $L(p^{(k+1)}, q)$ over all feasible q, i.e., (the first order optimality condition)

$$\left\langle b + \sum_{n=1}^{N} p_n^{(k+1)} + q^{(k)}, \ q' - q^{(k)} \right\rangle \geq 0$$

for all feasible q'. It then follows that

$$L(p^{(k+1)}, q^{(k+1)}) \leq L(p^{(k)}, q^{(k)})$$

and the equality if attained if and only if $(p^{(k+1)}, q^{(k+1)}) = (p^{(k)}, q^{(k)})$, and

$$\left\langle b + \sum_{n=1}^{N} p_n^{(k)} + q^{(k)}, \; p_n' - p_n^{(k)} \right\rangle \geq 0,$$

$$\left\langle b + \sum_{n=1}^{N} p_n^{(k)} + q^{(k)}, \; q' - q^{(k)} \right\rangle \geq 0$$

for all feasible p and q, i.e., $(p^{(k)}, q^{(k)})$ minimizes $L(p,q)$. Then by Lasalle's Theorem [30], we have $d(p^{(k)}, \mathcal{O}(t)) \to 0$ as $k \to \infty$. ∎

A.2 Proof of Lemma 1

When $b_t = b$ and $\mathbf{E}(a(t)) = \lambda$ for $t = 1, \ldots, T$, the model (13) for Algorithm 2 reduces to

$$d(t) = \frac{1}{T - t + 1} \left(\sum_{\tau=t}^{T} b(\tau) + \lambda(T - t) + \sum_{n=1}^{N(t)} P_n(t) \right) \quad (15)$$

for $t = 1, \ldots, T$. Then

$$(T - t + 1)d(t) = \sum_{\tau=t}^{T} b(\tau) + \lambda(T - t) + \sum_{n=1}^{N(t)} P_n(t)$$

$$(T-t+2)d(t-1) = \sum_{\tau=t-1}^{T} b(\tau) + \lambda(T-t+1) + \sum_{n=1}^{N(t-1)} P_n(t-1)$$

for $t = 2, \ldots, T$. Subtract the two equations and simplify using the fact that $b(t-1) + \sum_{n=1}^{N(t-1)} \left(P_n(t-1) - P_n(t) \right) = b(t-1) + \sum_{n=1}^{N(t-1)} p_n(t-1) = d(t-1)$ and the definition of $a(t)$ to obtain

$$d(t) - d(t-1) = \frac{1}{T - t + 1}(a(t) - \lambda)$$

for $t = 2, \ldots, T$. Substituting $t = 1$ into (15), it can be verified that $d(1) = \lambda + \sum_{\tau=1}^{T} b(\tau)/T + (a(1) - \lambda)/T$, therefore

$$d(t) = \lambda + \frac{1}{T} \sum_{\tau=1}^{T} b(\tau) + \sum_{\tau=1}^{t} \frac{1}{T - \tau + 1}(a(\tau) - \lambda)$$

for $t = 1, \ldots, T$. The average aggregate load is

$$u = \frac{1}{T} \sum_{t=1}^{T} d(t) = \lambda + \frac{1}{T} \left(\sum_{\tau=1}^{T} b(\tau) + \sum_{\tau=1}^{T} (a(\tau) - \lambda) \right).$$

Hence,

$$\mathbf{E}(d(t) - u)^2$$
$$= \mathbf{E} \left(\sum_{\tau=1}^{t} \frac{1}{T - \tau + 1}(a(\tau) - \lambda) - \frac{1}{T} \sum_{\tau=1}^{T} (a(\tau) - \lambda) \right)^2$$
$$= \mathbf{E} \left(\sum_{\tau=1}^{t} \frac{\tau - 1}{T(T - \tau + 1)}(a(\tau) - \lambda) - \frac{1}{T} \sum_{\tau=t+1}^{T} (a(\tau) - \lambda) \right)^2$$
$$= \frac{s^2}{T^2} \left(\sum_{\tau=1}^{t} \frac{(\tau - 1)^2}{(T - \tau + 1)^2} + T - t \right)$$

for $t = 1, \ldots, T$. The last equality holds because $(a(\tau) - \lambda)$ are independent for all τ and each of them have mean zero and variance s^2. It follows that

$$\mathbf{E}(V) = \frac{1}{T} \sum_{t=1}^{T} \mathbf{E}(d(t) - u)^2$$
$$= \frac{s^2}{T^3} \left(\sum_{t=1}^{T} \sum_{\tau=1}^{t} \frac{(\tau - 1)^2}{(T - \tau + 1)^2} + \sum_{t=1}^{T} (T - t) \right)$$
$$= \frac{s^2}{T^3} \left(\sum_{\tau=1}^{T} \frac{(\tau - 1)^2}{T - \tau + 1} + \sum_{t=1}^{T} (T - t) \right)$$
$$= \frac{s^2}{T^3} \left(\sum_{t=1}^{T} \frac{(T - t)^2}{t} + \sum_{t=1}^{T} \frac{(T - t)t}{t} \right)$$
$$= s^2 \frac{\sum_{t=2}^{T} \frac{1}{t}}{T} \sim s^2 \frac{\ln T}{T}. \quad ∎$$

A.3 Proof of Lemma 2

In the case where no deferrable arrival after $t = 1$, i.e., $N(t) = N$ for $t = 1, \ldots, T$, the model (13) for Algorithm 2 reduces to

$$(T - t + 1)d(t) = \sum_{\tau=t}^{T} b_t(\tau) + \sum_{n=1}^{N} P_n(t) \quad (16)$$

for $t = 1, \ldots, T$. Substitute t by $t - 1$ to obtain

$$(T - t + 2)d(t - 1) = \sum_{\tau=t-1}^{T} b_{t-1}(\tau) + \sum_{n=1}^{N} P_n(t - 1)$$

for $t = 2, \ldots, T$. Subtract the two equations to obtain

$$(T - t + 1)d(t) - (T - t + 2)d(t - 1)$$
$$= \sum_{\tau=t}^{T} e(t)f(\tau - t) - b(t - 1) - \sum_{n=1}^{N} p_n(t - 1)$$
$$= e(t)F(T - t) - d(t - 1),$$

which implies

$$d(t) - d(t - 1) = \frac{1}{T - t + 1} e(t)F(T - t)$$

for $t = 2, \ldots, T$. Substituting $t = 1$ into (16) and recalling the definition of b_t in (1), it can be verified that

$$d(1) = \frac{1}{T} \left(\sum_{n=1}^{N} P_n + \sum_{\tau=1}^{T} \bar{b}(\tau) \right) + \frac{1}{T} e(1)F(T - 1).$$

Therefore,

$$d(t) = \frac{1}{T} \left(\sum_{n=1}^{N} P_n + \sum_{\tau=1}^{T} \bar{b}(\tau) \right) + \sum_{\tau=1}^{t} \frac{1}{T - \tau + 1} e(\tau)F(T - \tau)$$

for $t = 1, \ldots, T$. The average aggregate load is

$$u = \frac{1}{T} \left(\sum_{n=1}^{N} P_n + \sum_{t=1}^{T} \bar{b}(t) \right) + \frac{1}{T} \sum_{\tau=1}^{T} e(\tau)F(T - \tau).$$

Hence,

$$\mathbf{E}(d(t) - u)^2$$

$$= \mathbf{E}\left(\sum_{\tau=1}^{t} \frac{1}{T-\tau+1} e(\tau) F(T-\tau) - \sum_{\tau=1}^{T} \frac{1}{T} e(\tau) F(T-\tau)\right)^2$$

$$= \mathbf{E}\left(\sum_{\tau=1}^{t} \frac{\tau-1}{T(T-\tau+1)} e(\tau) F(T-\tau)\right.$$

$$\left. - \sum_{\tau=t+1}^{T} \frac{1}{T} e(\tau) F(T-\tau)\right)^2$$

$$= \frac{\sigma^2}{T^2}\left(\sum_{\tau=1}^{t} \frac{(\tau-1)^2}{(T-\tau+1)^2} F^2(T-\tau) + \sum_{\tau=t+1}^{T} F^2(T-\tau)\right)$$

for $t = 1, \ldots, T$. The last equality holds because $e(\tau)$ are uncorrelated random variables with mean zero and variance σ^2. It follows that

$$\mathbf{E}(V) = \frac{1}{T}\sum_{t=1}^{T} \mathbf{E}(d(t) - u)^2$$

$$= \frac{\sigma^2}{T^3}\sum_{t=1}^{T}\left(\sum_{\tau=1}^{t} \frac{(\tau-1)^2}{(T-\tau+1)^2} F^2(T-\tau) + \sum_{\tau=t+1}^{T} F^2(T-\tau)\right)$$

$$= \frac{\sigma^2}{T^3}\sum_{\tau=1}^{T} F^2(T-\tau)\frac{(\tau-1)^2}{T-\tau+1} + \frac{\sigma^2}{T^3}\sum_{\tau=2}^{T}(\tau-1)F^2(T-\tau)$$

$$= \frac{\sigma^2}{T^2}\sum_{\tau=1}^{T} F^2(T-\tau)\frac{\tau-1}{T-\tau+1} = \frac{\sigma^2}{T^2}\sum_{t=0}^{T-1} F^2(t)\frac{T-t-1}{t+1}. \blacksquare$$

A.4 Proof of Theorem 2

Similar to the proof of Lemma 1 and 2, use the model (13) to obtain

$$d(t) = \lambda + \frac{1}{T}\sum_{\tau=1}^{T}\bar{b}(\tau) + \sum_{\tau=1}^{t} \frac{1}{T-\tau+1}\left(e(\tau)F(T-\tau) + a(\tau) - \lambda\right)$$

for $t = 1, \ldots, T$ and

$$u = \lambda + \frac{1}{T}\sum_{\tau=1}^{T}\bar{b}(\tau) + \sum_{\tau=1}^{T} \frac{1}{T}\left(e(\tau)F(T-\tau) + a(\tau) - \lambda\right).$$

Hence,

$$\mathbf{E}(d(t) - u)^2$$

$$= \mathbf{E}\left(\sum_{\tau=1}^{t} \frac{1}{T-\tau+1} e(\tau) F(T-\tau) - \sum_{\tau=1}^{T} \frac{1}{T} e(\tau) F(T-\tau)\right)^2$$

$$+ \mathbf{E}\left(\sum_{\tau=1}^{t} \frac{1}{T-\tau+1}\left(a(\tau) - \lambda\right) - \sum_{\tau=1}^{T} \frac{1}{T}\left(a(\tau) - \lambda\right)\right)^2.$$

The first term is exactly that in Lemma 2, and the second term is exactly that in Lemma 1. Hence, the expected load variance is

$$\mathbf{E}(V) = \frac{\sigma^2}{T^2}\sum_{t=0}^{T-1} F^2(t)\frac{T-t-1}{t+1} + \frac{s^2}{T}\sum_{t=2}^{T} \frac{1}{t}. \qquad \blacksquare$$

A.5 Proof of Corollary 3

If $|f(t)| \sim O(t^{-1/2-\alpha})$ for some $\alpha > 0$, then $|f(t)| \leq Ct^{-1/2-\alpha}$ for some $C > 0$ and all $t \geq 1$. Without loss of

generality, assume that $0 < \alpha < 1/2$ and $C \geq (1-2\alpha)/(1+2\alpha)$. Then $F(0) = 1$ and

$$|F(t)| = \left|\sum_{\tau=0}^{t} f(\tau)\right| \leq 1 + \sum_{\tau=1}^{t} C\tau^{-1/2-\alpha} \leq \frac{2C}{1-2\alpha} t^{1/2-\alpha}$$

for $t = 1, \ldots, T$. The last inequality holds because $C \geq (1-2\alpha)/(1+2\alpha)$. Therefore it follows from Lemma 2 that

$$\mathbf{E}(V) \leq \frac{\sigma^2}{T}\sum_{s=0}^{T-1} F^2(s)\frac{1}{s+1}$$

$$\leq \frac{\sigma^2}{T} + \frac{\sigma^2}{T}\sum_{s=1}^{T-1} \frac{4C^2}{(1-2\alpha)^2} s^{1-2\alpha}\frac{1}{s+1}$$

$$\leq \frac{\sigma^2}{T} + \frac{\sigma^2}{T}\frac{4C^2}{(1-2\alpha)^2}\sum_{s=1}^{T-1} \frac{1}{s^{2\alpha}}$$

$$\leq \frac{\sigma^2}{T} + \frac{4\sigma^2 C^2}{(1-2\alpha)^2 T} + \frac{4\sigma^2 C^2}{(1-2\alpha)^3 T^{2\alpha}}.$$

Hence, $\mathbf{E}(V) \to 0$ as $T \to \infty$. $\blacksquare$

A.6 Proof of Lemma 3

The aggregate load d obtained by the optimal static algorithm is

$$d(t) = \frac{1}{T}\left(\sum_{n=1}^{N} P_n + \sum_{\tau=1}^{T}\bar{b}(\tau)\right) - \bar{b}(t) + b(t)$$

$$= \frac{1}{T}\left(\sum_{n=1}^{N} P_n + \sum_{\tau=1}^{T}\bar{b}(\tau)\right) + \sum_{\tau=1}^{T} e(\tau)f(t-\tau)$$

for $t = 1, \ldots, T$. Hence,

$$\mathbf{E}(d(t) - u)^2$$

$$= \mathbf{E}\left(\sum_{\tau=1}^{T} e(\tau)\left(f(t-\tau) - \frac{1}{T}F(T-\tau)\right)\right)^2$$

$$= \frac{\sigma^2}{T^2}\sum_{\tau=1}^{T} T^2 f^2(t-\tau) - 2Tf(t-\tau)F(T-\tau) + F^2(T-\tau)$$

for $t = 1, \ldots, T$. It follows that

$$\mathbf{E}(V') = \frac{1}{T}\sum_{t=1}^{T} \mathbf{E}(d(t) - u)^2$$

$$= \frac{\sigma^2}{T}\sum_{t=1}^{T}\sum_{\tau=1}^{T} f^2(t-\tau) - \frac{2\sigma^2}{T^2}\sum_{\tau=1}^{T} F(T-\tau)\sum_{t=1}^{T} f(t-\tau)$$

$$+ \frac{\sigma^2}{T^2}\sum_{\tau=1}^{T} F^2(T-\tau)$$

$$= \frac{\sigma^2}{T}\sum_{t=1}^{T}\sum_{\tau=0}^{t-1} f^2(\tau) - \frac{\sigma^2}{T^2}\sum_{\tau=1}^{T} F^2(T-\tau)$$

$$= \frac{\sigma^2}{T}\sum_{\tau=0}^{T-1}(T-\tau)f^2(\tau) - \frac{\sigma^2}{T^2}\sum_{\tau=0}^{T-1} F^2(\tau)$$

$$= \frac{\sigma^2}{T^2}\sum_{t=0}^{T-1}\left(T(T-t)f^2(t) - F^2(t)\right). \qquad \blacksquare$$

124

Constrained Tâtonnement for Fast and Incentive Compatible Distributed Demand Management in Smart Grids

Shweta Jain
CSA, IISc
Bangalore, India
jainshweta@csa.iisc.ernet.in

Balakrishnan Narayanaswamy
IBM Research
Bangalore, India
murali.balakrishnan@in.ibm.com

Y. Narahari
CSA, IISc
Bangalore, India
hari@csa.iisc.ernet.in

Saiful A Hussain
Universiti Brunei Darussalam
saiful.husain@ubd.edu.bn

Voo Nyuk Yoong
Universiti Brunei Darussalam
nyukyoong.voo@ubd.edu.bn

ABSTRACT

Growing fuel costs, environmental awareness, government directives, an aggressive push to deploy Electric Vehicles (EVs) (a single EV consumes the equivalent of 3 to 10 homes) have led to a severe strain on a grid already on the brink. Maintaining the stability of the grid requires automatic agent based control of these loads and rapid coordination between them. In the literature, a number of iterative pricing, signaling and tâtonnement (or bargaining) approaches have been proposed to allow smart homes, storage devices and the autonomous agents that control them to be responsive to the state of the grid in a distributed manner. These existing approaches are not scalable due to slow convergence and moreover the approaches are not incentive compatible. In this paper, we present a tâtonnement framework for resource allocation among intelligent agents in the smart grid, that non-trivially generalizes past work in this area. Our approach based on the work in server load balancing involves communicating carefully chosen, centrally verifiable constraints on the set of actions available to agents and cost functions, leading to distributed, incentive compatible protocols that converge in a constant number of iterations, *independent of the number of users*. These protocols can work on the top of prior approaches and result in a substantial speed-up, while ensuring that it is in the best interests of the agents to be truthful. We demonstrate this theoretically and through extensive simulations for three important scenarios that have been discussed in the literature. We extend the techniques to account for capacity limits in each time slot, the EV charging problem and the distributed storage control problem. We establish the generality and usefulness of this technique and making the case that it should be incorporated into future smart grid protocols.

Categories and Subject Descriptors

I.2.11 [**Distributed Artificial Intelligence**]: Multiagent systems

General Terms

Algorithms

Keywords

Distributed Algorithms; Demand Management; Electric Vehicle; Smart Distributed System

1. INTRODUCTION

Power utilities worldwide face two major challenges - peak demand and power (supply - demand) imbalance. Peak demand is a period in which the demand for electrical power is at a significantly higher than average supply level. In order to satisfy a large peak demand, generation and distribution companies have to make large capital expenditures in new generation stations and larger capacity lines and transformers. In addition, in free market situations, this forces companies to purchase electricity on the expensive 'spot market' [15]. Satisfying peak demand requires generation companies to install and use expensive peaking power plants (that are seldom run), which in turn increases the spot prices substantially. For example, it is estimated that a 5% lowering of demand would have resulted in a 50% price reduction during the peak hours of the California electricity crisis in 2000/2001 [18]. Peaks also lead to substantial energy losses. The second major problem faced by utilities is that of supply-demand imbalance. In current electricity markets "demand exhibits virtually no price responsiveness and supply faces strict production constraints and very costly storage .... Extreme volatility in prices and profits will be the outcome."[7].

In the midst of these difficulties faced by utilities, growing fuel costs, environmental awareness and government directives have increased the push to deploy Electric Vehicles (EVs). These EVs also can provide substantial amounts of local storage - a single EV can store and consume the equivalent of 3 to 5 homes straining the grid [10]. While EVs have environmental benefits [11], they will have substantial impacts on the economics [14], design, operation, and stability of the grid [20]. At the same time distributed storage, either in the form of EVs or batteries have been suggested for utility and consumer cost reduction [33, 36, 32].

Fortunately, the introduction of new communication [30] and control [12] infrastructure into the electricity grid is expected to allow increased prosumer participation in the smart grid. This participation will in fact be necessary to reduce costs. However, the infeasibility of continuous human intervention and consumption control has led to the model of autonomous software agents, representing the consumers, that intelligently optimize and schedule energy usage [33, 24]. Our focus in this paper is on the *design of negotiation protocols* between electricity distribution utilities and smart consumption agents that exploit this advanced communication and control infrastructure to allow coordination of consumption and help maintain the stability of the grid. Our work lies at the intersection of game theory [25], mechanism design [26] and the design of distributed algorithms [23].

1.1 Electricity Congestion Control Protocols

While distributed algorithms can serve a number of purposes in the smart grid, ranging from frequency control and maintaining voltage stability to providing various ancillary services [21], we focus on the problem of congestion control as an illustrative example. However, the protocols we suggest should be applicable in many other scenarios.

As mentioned in the introduction, shifts in demand and supply continue to strain a grid already on the brink. Thus, in order to limit peaks, minimize losses and reduce costs, utilities need to exploit the patience and flexibility of consumers and any available local storage. For example, when EVs are not in use, they can play the role of a battery allowing users to time shift their electricity purchases. When consumers have access to storage, they can potentially store energy during off-peak hours and use this energy to satisfy demand during the peak. Peak limitation through storage adoption may be more acceptable to consumers than consumption limitation since it does not require behaviour modification. However, increased storage adoption can lead to 'herding' effects where co-incident charging of batteries creates new and larger peaks, increasing costs [8] and losses. Thus, utilities need to encourage users to not only shift but also coordinate consumption, through what are often called congestion control mechanisms.

Congestion control schemes in the power systems literature (often based on successful protocols developed for internet congestion control [19]) that discourage consumption when the grid is loaded, fall into four basic categories: day ahead or time of use (TOU) pricing, dynamic pricing, signaling and back-off strategies and tâtonnement (or negotiation).

While many of these methods have been proposed in the literature they all come with their own limitations. TOU pricing (particularly in situations where users have significant storage, patience or flexibility) leads to the herding problem described above, where large amounts of consumption are shifted to low price regions creating new peaks [8]. Dynamic pricing can lead to customer confusion, and "consumers generally shy away from markets when products are complicated, supply is uncertain, prices are volatile, and information is lacking" [9]. Signaling based approaches, where consumers respond directly to signals from the utility, require that consumers are willing to give up control of their consumption which may be unlikely, or require that users reveal their preferences, which in turn leads to privacy and security concerns [2]. Randomized back-off type approaches where users back-off on consumption when the grid is loaded (motivated by CSMA protocols developed for wireless communication) require trust that the agent will respond as expected to signals. This is not incentive compatible as the response strategies are unverifiable and it is in the agents' interest to choose small back-off windows which allow them to consume electricity with minimum inconvenience.

Tâtonnement [34], where consumption agents and the utility exchange prices and consumption profiles until convergence, is then perhaps the most promising protocol for user consumption scheduling and pre-commitment. While they were not explicitly categorized as such, tâtonnement forms the basis of much work in this area including [24, 27, 33, 22, 35, 17]. Unfortunately, the strategies proposed in the literature either do not converge with parallel or require synchronous serial updates (where no two agents can update consumption at the same time) to ensure convergence. This implies substantial communication overhead and a number of iterations linear in the number of agents, as explicitly noted in [24, 27], which rapidly becomes infeasible as the number of users in the grid increases.

Since typical distribution grids could have millions of users and tens of millions of devices, *developing protocols for multi-agent consumption coordination that overcome these synchronization, convergence, and incentive compatibility issues is the primary focus of our work*. In particular we develop constrained tâtonnement protocols for fast and incentive compatible distributed demand management in smart grids

1.2 Our Contributions

In this paper, we consider a game theoretic resource allocation setting with a central distribution company that wishes to allocate a limited quantity of an expensive electricity resource to a set of autonomous consumption agents, at minimum cost. Typically, because the cost of generation is quadratic in the quantity generated and because of peak and congestion effects, the distribution company would prefer balanced consumption across the slots, subject to physical capacity constraints and the requirements of the users. The allocation to each agent is decided in an iterative manner. In each iteration, the center signals information to the agents (such as prices for consumption in each time slot) and the agents respond with new consumption levels. The center then updates its signals and the process is iterated until convergence. These kinds of protocols have been widely studied in the game theoretic settings, where they are called tâtonnement protocols [34] and more recently in the power systems literature [19, 24, 16, 17, 27, 33].

In this paper, we show how a protocol that adds verifiable constraints on the user consumption changes during the tâtonnement process, can be used to non-trivially improve the iterative strategies proposed earlier for demand response [24, 27], EV charging [19] and distributed storage management [17, 33]. We show that when the proposed protocols are used, user selfish best response strategies converge within a poly-log (in an approximation parameter that depends on the desired accuracy) number of rounds *independent of the number of agents*. In this direction, our contributions are to develop the following.

- **Tâtonnement game model:** We describe the problem of peak minimizing or minimum cost electricity resource allocation in smart grids as a tâtonnement game between a single seller and multiple buyers. We explain how solutions presented in the literature, such as [24, 27] require a linear number of sequential updates to converge and others, such as [3, 19], may not be incentive compatible.

- **Fast, incentive compatible distributed protocols:** We map the energy demand management problem to a server allocation problem studied in [4] and develop a new protocol where, in addition to signals and prices, *centrally verifiable constraints* are also communicated to the users. We demonstrate with theory and through extensive simulations, that this leads to incentive compatible, distributed protocols that converge in a *constant* number of iterations, *independent* of the number of agents which represents a substantial speed-up for the situations studied in [24, 27].

- **Distributed demand management protocols incorporating capacity constraints and micro storage:** We show how *constrained* tâtonnement is useful in different resource allocation problems that are practically relevant in electricity grids, including load scheduling when users have local storage (the problem studied in [33]) or the distribution utility has capacity constraints in different time slots. Here, we demonstrate that these extensions require a *rethinking of how prices are to be set when new constraints need to be satisfied*.

A highlight of the protocol is its generality as it can be used on top of the distributed optimization techniques suggested in the literature and at the same time achieves faster convergence and ensures incentive compatibility.

The rest of this paper is organized as follows. In Section 2, we describe the problems facing energy utility companies in more detail, review some of the recent literature on agent based models for electricity consumption scheduling and their limitations with respect to convergence and incentive compatibility. In Section 3, we explain how a parallel to theoretical work on a server load balancing can be applied to the problem of electricity resource allocation to design new distributed protocols for energy consumption scheduling. We also show how extending these techniques to the problem of load balancing with capacity constraints in each time slot (a problem of importance in electricity grids with day ahead markets) requires a novel structure of the prices announced to the users. Finally, we show how these protocols also can be used for demand management when users have access to local storage - either from batteries or EVs - allowing fast convergence.

2. DISTRIBUTED STORAGE AND DEMAND SIDE MANAGEMENT

The modern electricity grid is expected to provide a reliable supply to consumers (the US grid operates at an aver-

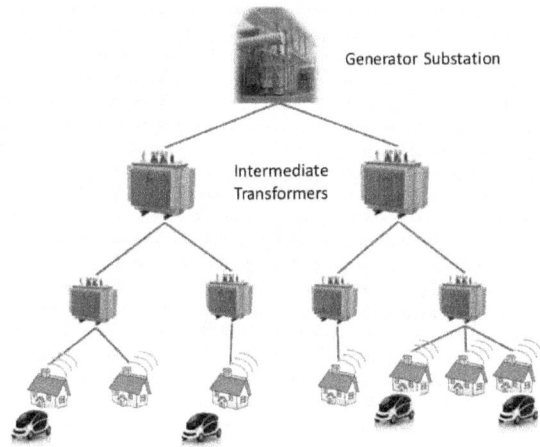

Figure 1: Smart Grid Structure

age of three nines 99.9% reliability) at any cost. "In order to supply demand that varies daily and seasonally, and given that demand is largely uncontrollable and interruptions very costly, installed generation capacity must be able to meet maximum (peak) demand" [31]. As a result, "the average utilisation of the generation capacity is below 55% and the lowest marginal cost plant would operate at about 85% load factor, while a plant with the highest fuel cost would operate only a few hours per year". Reducing this peak to average ratio is one of the primary goals of energy utilities, and can be achieved through customer demand response programs.

Demand response in electricity markets can be defined as the changes in electric consumption by users from their normal consumption patterns in response to signals (e.g. pricing) from the distribution authority [1]. In response to these signals, users can defer or reduce loads in coordination with other consumption agents to reduce the overall load on the grid. However, this response could lead to temporary user inconvenience and this inconvenience should be limited as far as possible to encourage participation in demand response programs.

One avenue to shift consumption or shave peaks with minimal user inconvenience is through the use of energy storage. Local storage would allow users to coordinate charging and consumption and then use energy from the storage when the grid is over loaded. While storage is expensive, distributed storage is already prevalent. For example, due to a chronic shortage of electricity of more than 10% [13], outages are common throughout India, causing customers to purchase storage and local generation. At the same time EVs are expected to become more common throughout the developed world [14]. When these EVs are parked, they can be plugged in to absorb energy and store it for later consumption, acting as large local storage.

2.1 Electricity distribution system model

Traditionally, most electricity distribution grids are radial in nature [5] and hence can be represented in the form of a tree with the nodes representing the buses and the undirected edges representing the electrical line connections, as in Figure 1. This tree can further be modeled as a rooted tree with the root node representing the generator sub-station and the leaf nodes representing the loads (the customers buying electricity). This rooted tree structure is used as a model of the distribution grid in this paper.

The loads are parametrized by a *power requirement* which is the total amount of power they want to consume across some time duration called a *consumption period*, such as a day. The consumption period is further divided into smaller duration called *slots*, corresponding to hours, minutes or even seconds. With each load, a *strategy vector* is associated which is a vector of powers consumed by the load across all slots in a consumption period. Each load is free to consume any amount of power in the individual slots as long as the total power consumed across all slots in a consumption period equals the specified requirement.

User preference can also captured by further allowing the loads to be active (consume power) only in some subset of slots during any consumption period. That is, for every load, we can further associate a binary 'indicator' vector indicating which subset of slots in any consumption period the load can consume power. An example where such a model can be applied is in the case of EV (Electric Vehicle) charging by individual customers at their homes. Some people may have to use their cars for work and can charge only during off-work hours. However, such users will be in-different as to 'how' the car gets charged during the off-work hours as long as it is charged by beginning of the work hour next day. In this example, the consumption period could be one day and the slots' duration can refer to one hour. The power requirement of the EV is specified by its storage capacity. Hence any scheduler has complete freedom in choosing an appropriate strategy vector so long as the the total power requirement is met and the user's preference (indicator) is respected. Note that in general, as we will discuss later, this model is also general enough to model arbitrary energy consumption when users have access to local storage and is not limited to the EV charging described here.

2.2 Existing Demand Response Protocols and their Limitations

As described in Section 1.1, demand response protocols fall into four basic categories: day ahead or time of use (TOU) pricing, dynamic pricing, signaling, and tâtonnement. In TOU and dynamic pricing [1], consumers are charged based on the state of the grid, either predicted day ahead (TOU pricing) or real time (dynamic pricing). This requires a good estimate of user requirements and preferences at the central utility, which may be unrealistic, and have been seen to lead to large swings in prices and consumption, which discourage user participation and lead to grid instability.

The third approach to this problem is motivated by parallels to the development of the internet [19] and in particular internet protocols such as TCP [3], which assume agents respond to signals from the central authority (essentially by fiat) and will curtail load as required. Deciding on which users should or could reduce consumption requires them to reveal their preferences and consumption patterns which raises privacy concerns [2]. Back-off strategies borrowed from the wireless domain, where users delay consumption when the grid is overloaded, have also similarly been suggested for peak power reduction. However, these are not incentive compatible since it is often not in the users (selfish) interest to delay consumption and since the back-off strategies are not verifiable by the utility. That is, selfish users may be incentivized to choose small back-off windows result-

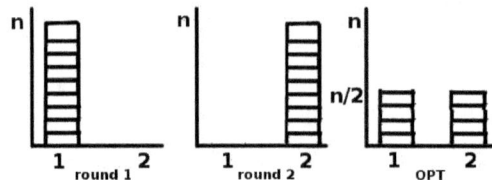

Figure 2: Example demonstrating that parallel updates may not converge and that sequential updates could require a linear number of iterations

ing in a breakdown of the protocol. In this paper we develop incentive compatible protocols by introducing centrally verifiable constraints that actually speed up convergence.

In tâtonnement [34], the price and consumption information is updated iteratively until the aggregate user demand matches the supply. Tâtonnement in smart grids has been proposed both from an optimization viewpoint and from a game theoretic perspective. Primal dual algorithms allow (primal) consumption to be set iteratively in response to (dual) prices [16]. Similar approaches have also been suggested [17] for the grid level supply demand matching problem, (known as Economic Dispatch in the power systems community). The game theoretic approach to these resource allocation problems assumes that consumers are selfish utility maximizing agents who maximize welfare (utility-price) in response to pricing signals [22, 1, 24].

Even with electricity pricing and distributed storage, herding effects [8] can actually cause an increase in the size of electricity peaks. That is, users with storage will purchase large amounts of electricity when prices are low, creating new peaks. In addition, storage and demand response can increase the unpredictability of demand. This particularly problematic in electricity markets, where, because of physical power system constraints distribution companies have to commit one day ahead to purchase levels [15].

Thus, an important question is how to incentivize users to coordinate their use of storage to flatten power consumption peaks while allowing distribution companies to gain estimates of the overall demand profile. Many variants of tâtonnement, where the central authority and consumption agents exchange consumption information and pricing signals until convergence, have been proposed in the multi-agent and power systems literature. This game theoretic perspective has been advanced both with storage in [33, 27] and without in [24] (an excellent survey covering these and other game theoretic approaches can be found in [29]). A similar approach, from an primal-dual optimization perspective (without storage) was suggested in [16].

However, all these protocols have some convergence issues that we seek to address here. For example, the protocols in [24, 16, 33, 27] do not converge if users update their consumption profiles in parallel. To see why this is the case, consider a simple example where we have two slots and large number n of users as shown in Figure 1. Suppose initially all users consume in the first slot. With parallel updates the users will alternate between State 1 and State 2. The optimal solution is the one where load on each slot is $n/2$ which will take $n/2 = O(n)$ iterations with serial updates. Thus, the protocol requires a linear number (in the number of consumers) of synchronous, serial updates, which is infeasible in large grids with millions of users and adds substantial signaling overhead. In this paper, we develop a parallel to the problem of server load balancing and show that using an algorithmic technique called bounded best response we can develop protocols that converge to near-optimum solution in a number of rounds independent of the number of users.

2.3 The Grid Model

User Agents and Preferences: Let $\mathcal{N} = \{1, 2, \ldots, n\}$ represent the set of autonomous agents or users, each of which controls its own load. We consider the model with discrete time intervals (i.e. time is divided into hourly slots or in slots of 15 minutes in a day). We divide a day into H time slots $\mathcal{H} = \{1, 2, 3, \ldots, H\}$. An agent i is available in time slots $\mathcal{S}_i$. This setting can represent, for example, the case where an agent controls an EV and has a preferable time slot to charge its vehicle. If the agent wishes to go to the office at $8:00$ A.M. in the morning and return at $7:00$ P.M. then the agent would charge the vehicle between slots $7:00$ P.M. to $8:00$ A.M. For simplicity, we assume that an agent has the flexibility to consume any amount of electricity in a time slot, but has a certain number of total units required for a particular day. The model is also general enough to be extended to the situation where multiple appliances are controlled by a single agent, or when per slot limits on appliance consumption are placed.

Consumption profiles: Based on the signals received from the central utility, each user i decides on an amount of electricity to be consumed in slot h denoted by l_i^h, which we call the user's consumption profile. This results in a total consumption of $L^h = \sum_{i \in \mathcal{N}} l_i^h$ in each time slot which is the overall consumption profile.

Distribution company optimization objectives: There are a number of different objectives that a central electricity distribution utility may be interested in. One important goal is to minimize the Peak to Average Ratio (PAR),

$$\text{PAR} = \frac{H \max_h L_h}{\sum_{i=1}^{H} L_h} \qquad (1)$$

subject to the user consumption constraints, namely that (i) For any time $h \notin \mathcal{S}_i$, we have $l_i^h = 0$ and (ii) For each user i, there is a fixed requirement for charging, denoted by ξ_i such that $\sum_{h \in \mathcal{S}_i} l_i^h = \xi_i$ where ξ_i is the unit of electricity a user wants to consume which is fixed. Note that the denominator in calculating PAR is a constant in our model, as the energy requirement of each load agent is considered to be fixed and the problem reduces to that of minimizing $\max_{h \in H} L_h$, which is a peak load minimization problem.

There are number of possible (convex) cost functions that the central distribution company may wish to minimize. For example, the utility may be interested in minimizing a quadratic function of the total load $\sum_h a L_h^2 + b L_h + c$, since the generation cost of many generators can be modeled as a quadratic function [15]. Alternatively, peak minimization and I^2R loss minimization are often important goals [5]. The common feature of all these cost functions is that moving consumption from a heavily loaded time slot to a less loaded slot decreases the total cost. Constrained tâtonnement protocols can be shown to work for any such cost function, though we focus on the peak load minimization problem here.

Note that since cost function that utility company consider is convex monotonic increasing function in total load L_h,

minimizing the maximum peak is identical to minimizing the total cost which is given by $\sum_h C(L_h)$ where C is the cost function that the utility company is paying.

2.4 Smart Grid Settings Investigated

The primary problem we are interested in in this work is of user, appliance or storage load sifting to meet system constraints and minimize costs and losses. Our aim is to show that by choosing constraints and prices appropriately we can improve the convergence of protocols suggested for a number of different smart grid scenarios [24, 16, 22, 33, 27] demonstrating the wide applicability of this idea. In this paper, we consider three settings that illustrate different aspects of the constrained tâtonnement protocol. Three of these were previously studied in the literature for which linear time algorithms were suggested and we demonstrate the improvements possible here.

Setting 1: Basic tâtonnement game: Here, we model autonomous loads and no local storage or per slot capacity constraints, motivated by [24, 27]. This allows us to highlight the parallel to the problem of load balancing and also demonstrate the proofs of fast incentive compatible, convergence for the basic peak minimization or convex cost minimization problem.

Setting 2: Demand management with capacity constraints: This scenario is unique to our work (from a distributed multi-agent perspective) and we demonstrate that it is possible to ensure distributed constraint satisfaction while maintaining fast incentive compatibility. In particular, we consider a utility that participates in the day ahead energy market [15]. In electricity markets, due to large generator start up times and slow ramping rates of the most efficient generators, electricity distributors and generators commit day ahead to levels of consumption and supply in each time slot over the next 24 hours. These are essentially contracts between generation and distribution companies and come with (large) penalties for any deviation from the committed levels. After considering these day ahead commitments and the predicted consumption of users who do not participate in demand management programs, a distribution company is then left with the problem of incentivizing users to schedule consumption subject to capacity constraints. We show, using the intuition from Setting 1, that we can develop a protocol that guarantees fast convergence for this important practical problem. An interesting point we make here is that for fast convergence with capacity limits, the form of the prices announced to each user have to be defined carefully : a direct extension of the previous methods does not work correctly. In particular we show that constrained tâtonnement guarantees convergence to a solution which satisfies capacity constraints, provided one exists, in a number of iterations independent of the number of users *if prices are structured appropriately*.

Setting 3: Distributed micro-storage management: This setting is motivated by [33], where each agent has access to limited local storage, possibly in the form of a parked EV or battery backup. The agent then wishes to use this storage to minimize costs, while the central utility would like to exploit user storage to minimize peaks. The fundamental question then is how the central utility should update prices in the presence of distributed storage. For this class of problems again a distributed linear time algorithm was presented in [33]. We show that constrained tâtonnement allows fast convergence, even in the presence of selfish users who use storage to minimize their local costs.

Settings 2 and 3 serve to highlight the generality of the protocol and how it can be used for a wide variety of user and distribution company settings. We show through simulations on real consumption traces how to extend the protocol to the situation of a large number of distributed demand management agents with local storage and evaluate the practicality of our protocols.

3. CONSTRAINED TÂTONNEMENT FOR DISTRIBUTED ENERGY RESOURCE ALLOCATION

We now formalize our game theoretic model of the energy resource allocation problem between a central distribution utility and autonomous consumption agents in a setting we refer to as a tâtonnement game. This setting allows us to analyze the situation where a set of autonomous consumption agents selfishly try to maximize their utilities independently with co-ordination enforced by the utility through pricing and constraints.

3.1 Game Theoretic Model

Game theory is the study of strategic decision making. More formally, it is "the study of mathematical models of conflict and cooperation between intelligent rational decision-makers." [25]. The players in our game are the autonomous agents who are trying to optimize their consumption profile. Strategy for a player i is the load vector $l_i = \{l_i^1, l_i^2, \ldots, l_i^T\}$. We also denote l_{-i} as the load vectors of all other players except i. The payoff to each player i can then be defined as

$$u_i(l_i, l_{-i}) = -\frac{\xi_i}{\sum_j \xi_j} \sum_t c_t \quad (2)$$

The first term $(\frac{\xi_i}{\sum_j \xi_j})$ in the payoff denotes the fraction of load agent i is using and c_t is the price the electricity company charges at time slot t. Thus $\sum_t c_t$ denotes the total amount that the electricity company receives. In our model, we simply set $c_t = C(L_t)$ where $L_t = \sum_i l_i^t$ (total amount of load consumed at time t), $C(L_t)$ is the cost, electricity company is paying for L_t unit of load at time t. We assume that $C(L_t)$ is a strictly convex increasing function. Thus the objective of each player i, would be to consume the load l_i so as to maximize their utility defined in (2)

We will be interested in two properties of a final consumption solution. A load profile $(l_1^*, l_2^*, \ldots, l_n^*)$ is a *Nash equilibrium* if:

$$u_i(l_i^*, l_{-i}^*) \geq u_i(l_i, l_{-i}^*) \ \forall \ l_i \ \forall \ i \quad (3)$$

Nash equilibrium in some sense denotes the stability of the result of our protocol where no user has an incentive to deviate unilaterally. The other important property of a solution is Pareto optimality. A load profile $(l_1^*, l_2^*, \ldots, l_n^*)$ is Pareto optimal solution if $\nexists (l_1, l_2, .., l_n)$ such that:

$$u_i(l_i^*, l_{-i}^*) \ \leq \ u_i(l_i, l_{-i}) \ \forall \ i \text{ and} \quad (4)$$
$$u_j(l_j^*, l_{-j}^*) \ < \ u_j(l_j, l_{-j}) \text{ for some } j \quad (5)$$

We then have the following theorems from [24]:

THEOREM 3.1. *A load profile* $(l_1^*, l_2^*, \ldots, l_n^*)$ *is a Nash equilibrium iff it is the solution of optimization problem*

$$min \sum_h C(L_h).$$

THEOREM 3.2. *If load profile* $(l_1^*, l_2^*, \ldots, l_n^*)$ *is a solution of optimization problem* $min \sum_h C(L_h)$, *then it is a Pareto optimal profile.*

From the two theorems it can be seen that any algorithm which minimizes the cost function $\sum_h C(L_h)$ will also result in Nash equilibrium and Pareto optimal solution. In the following sections we develop a distributed algorithm in which player play selfishly subject to centrally verifiable constraints to reach to the optimal solution.

3.2 Constrained Tâtonnement

In the basic Setting 1, autonomous loads are co-ordinated by price updates. We first discuss how the parallel can be established to the problem of server load balancing studied [4] and then prove fast convergence. We then extend the results to Setting 2 and 3 as how to impose capacity constraints and how to include storage in the model. Interestingly, we show that imposing capacity constraints directly does not work and one needs to design the cost functions with care.

Recall that we consider a setting where agents have appliance that have certain consumption requirements and are only available to be run in a certain sub-set of the time slots. This models a number of appliance like EVs and even appliances like heaters and air conditioners that have explicit or implicit storage capabilities. The tâtonnement process we study is primarily a model for investigating stability of equilibrium [34]. In each iteration, prices for each slot are announced (by the central utility), and agents state how much electricity they would like to purchase (demand) in each time slot at that price. No transactions take place at non-equilibrium prices. Instead, prices are lowered for slots with positive prices and excess supply and raised for slots with excess demand. Even though they were not explicitly named as such, the protocols suggested in [24, 27, 33] and many other are essentially tâtonnement processes, which require a linear number of sequential updates to ensure convergence. We show that, with suitable modifications to this basic protocol, it is possible to get bounds on the number of updates required for convergence that depend only on the approximation factor, independent of the number of agents. This also highlights the fact that even settling for an approximate solution can often lead to a substantial speed-up in tâtonnement protocols.

There is a strong parallel between our problem of energy resource allocation and the work on server scheduling [4]. In [4], they consider the problem of allocating jobs to servers to minimize peak loads on servers. In our model, jobs correspond to the loads or appliances controlled by the agents and each machine corresponds to one of the time slots. The total load on a machine corresponds to the amount of energy consumed in each time slot. To emphasize the parallel, we will also use load to refer to the energy consumed.

3.3 Setting 1 : Basic Setting

Let S_i denotes the set of time slots in which player i prefers to consume some load and p_{ih} be the fraction of load which the ith user consumes at h^{th} hour and d denotes the iteration of the algorithm. We have $\sum_{h \in S_i} p_{ih} = 1, l_i^h = \xi_i p_{ih} \; \forall h \in S_i$. Each user in this setting will try to balance the load subject to $\sum_{t \in S_i} l_i^t = \xi_i \; \forall i$. In the basic setting, the two constraints for the tâtonnement protocol that the distribution company broadcasts and a user must satisfy in each update are the

(i) **Inertia constraints** whereby an agent i may move a fraction of load from hour j to hour r where $j, r \in S_i$ only if $L_j \geq (1+\epsilon)^3 L_r$ and the

(ii) **Bounded Step constraints**, $\eta \leq p_{ih} \leq 1$ for all i where $\eta = \frac{\epsilon}{H^2}$ and $p_{ih}^d/(1+\epsilon) \leq p_{ih}^{d+1} \leq (1+\epsilon)p_{ih}^d$.

Here ϵ is a parameter that we select later depending on the accuracy required. Essentially, the inertia rule allows users to make a move only if there is significant difference in cost (ensuring convergence) and the bounded step rule prevents large swings and cycles in the dynamics of the protocol. The formal outline of the protocol is in 1. The protocol does not requires communication between the agents and thus required less overhead.

Protocol 1 The constrained tâtonnement protocol for distributed demand management

1: **while** Any change in user consumption profiles **do**
2:     Distribution utility broadcasts total load in each time slot and corresponding cost functions for each slot
3:     Users update consumption profile based on the cost function, internal preferences, Bounded Step constraints and Inertia constraints.
4:     Users transmit new consumption profiles to the distribution company
5:     The distribution utility checks that inertia constraints and bounded step constraints are not violated
6: **end while**

Distributed Incentive Compatibility: In Protocol 1, every user is in fact unaware of other users' loads, leading to a distributed protocol. The user only sees the total load present on the slots in which she is consuming energy and tries to balance consumption according to her preferences, the cost functions announced by the central utility for those slots and the constraints making the protocol incentive compatible. Privacy problems do not arise since the load profiles of users are not required to be public knowledge.

The central authority announces prices of $(L_h)^2$, for each $h \in S_i$ to each agent i. Thus, in iteration $d+1$, user i is faced with the following optimization problem,

$$\underset{\gamma_1,\ldots,\gamma_H}{\text{minimize}} \quad \sum_{t \in S_i} (L_t^d - \gamma_t)^2 \qquad (6)$$

$$\text{s.t.} \quad \gamma_t = \sum_{t' \in R_t} x_{t,t'} \; \forall \, t, \; \sum_{t \in S_i} \gamma_t = 0$$

$$\gamma_t \leq p_{it}^d \, \xi_i \, \frac{\epsilon}{1+\epsilon}, \; -\gamma_t \leq p_{it}^d \, \xi_i \, \epsilon \qquad (7)$$

where γ_t represents the amount of load agent i moves from (or to) time t. $R_t = \{t' : L_t \geq (1+\epsilon)^3 L_{t'} \text{ or } L_{t'} \geq (1+\epsilon)^3 L_t\}$ denotes the set of time slots t' for time t such that user can either shift his load from t to t' or from t' to t and finally $x_{t,t'}$ is a design variable which is positive when the load is shifted from t to t' and negative when load is shifted from t' to t. R_t and $x_{t,t'}$ arise from the inertia constraints and the bound on γ_t comes from the bounded step constraints.

Let OPT be the the maximum load on any slot in the

optimal assignment that minimizes the peak load over all $h \in H$. To allow the protocol to work we assume that each user starts with at least an η fraction of its total load in each time slot $t \in S_i$. Since, $OPT \geq \frac{1}{H} \sum_i \xi_i$, the effect of this assumption on the final solution is small. Two lemmas follow immediately,

LEMMA 3.1. *For all hours j and iterations d, we have* $\frac{L_j^d}{1+\epsilon} \leq L_j^{d+1} \leq (1+\epsilon)L_j^d$

LEMMA 3.2. *The dynamics are acyclic.*

PROOF. By the inertia constraint, load can move from hour j to hour r only if $\frac{L_j}{L_r} \geq (1+\epsilon)^3$. Since agents are maximally greedy, the load on slot j will decrease at least by $\frac{L_j}{1+\epsilon}$ and due to Lemma 3.1, the load of slot r will increase to at most $L_r(1+\epsilon)$, resulting in a total decrease of load by at least a factor of $(1+\epsilon)$. The sum of squares of loads decreases over time and thus the dynamics are acyclic. $\square$

Thus, if both inertia and bounded constraints are satisfied at each iteration then situations like in Figure 2 can be avoided. Using these lemmas, it is possible to derive the following result,

THEOREM 3.3. *The constrained tâtonnement protocol converges to a $(1+\delta)$ approximation in $O(\frac{1}{\delta^6} \log^6 \frac{H}{\delta})$ iterations with $\epsilon = O(\frac{\delta^2}{\log H})$.*

Proof follows from the results in [4]. Thus, the protocol converges in a number of iterations which is independent of the number of agents unlike the prior approaches discussed in Section 2.2.

3.3.1 Simulations for Setting 1

To obtain the load profiles of users in all our experiments we used the model and data collected as part of an extensive sensing and survey based study of U.K. homes, where electricity demand was recorded over the period of a year within 22 homes in the UK [28]. For each agent i, the time slots are chosen i.i.d. following the average demand distribution in Figure 3. The number of slots for each agent i, $|S_i|$, is drawn from a multinomial distribution with mean 4 (except in Figures 7 and 8 with varying mean).

We first investigate the effect of various parameters on the convergence and optimality of the constrained tâtonnement protocol. For studying the effect of δ and expected number of slots, we fixed the number of users to be 50. We typically ran 500 experiments to obtain 95% confidence bounds. Let COST be the cost of the final solution obtained by the protocol and OPT be the optimal solution achieved by running the optimal problem centrally. From Theorem 3.3, we see that the required approximation factor δ trades off accuracy and rate of convergence. Figures 4 and 5 show the change in the number of rounds and the COST/OPT ratio of the protocol with different δ's. As δ increases, the protocol converges faster though the final COST increases, however the increase is not substantial. In particular, we see that to achieve a cost within 1% of the optimal, required only about 60 iterations. *Note that from Figure 6 this number does not depend on the number of agents.* In addition, Figure 4 shows the speed-ups possible with just 2000 agents, hinting at the promise of when large smart grids with millions of users.

Figure 6 shows that the number of iterations required for convergence with the proposed protocol is independent of

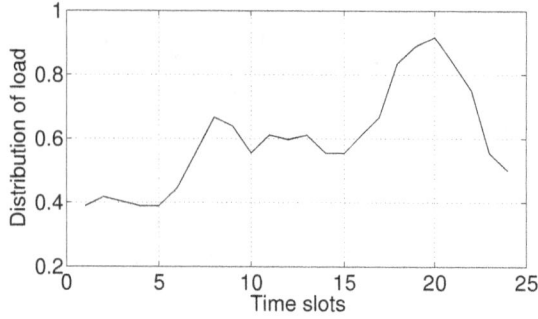

Figure 3: Distribution of the load on a hourly basis

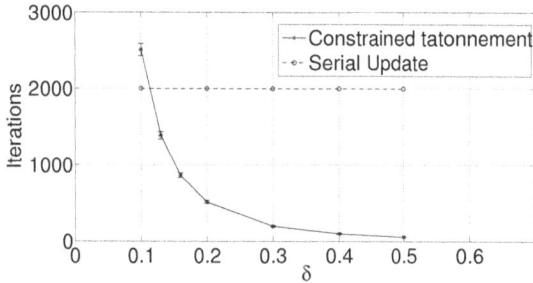

Figure 4: Decrease in the number of iterations with increase in the approximation factor δ

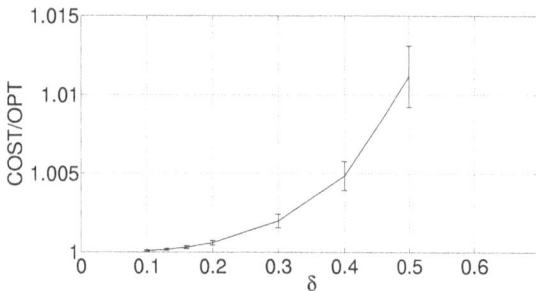

Figure 5: Sub optimality of final solution with desired approximation factor δ

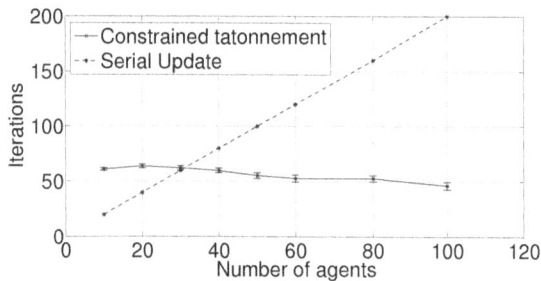

Figure 6: Comparison of sequential updates and the constrained tâtonnement protocol

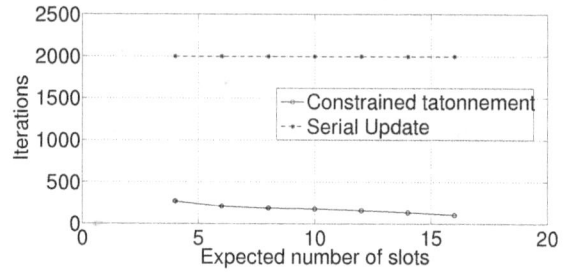

Figure 7: Faster convergence of constrained tâtonnement with increased user flexibility $E[|\mathcal{S}_i|]$

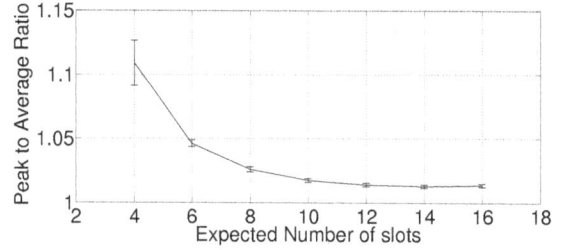

Figure 8: Decrease in peak to average ratio with increased user flexibility $E[|\mathcal{S}_i|]$

the number of users, in agreement with Theorem 3.3. This is in contrast to the linear increase in number of iterations for the sequential protocols in [24, 33]. For comparison, we kept the value of $\delta = 0.5$, thus $\epsilon = 0.0787$. Figures 7 and 8 show how the number of rounds and peak average ratio vary with the increase in expected number of slots, $E[|\mathcal{S}_i|]$, a representation of the patience of the users. We see that when users are more flexible, the protocol converges much faster as they have more choices to shift, also resulting in decrease in peak load. However, we note that the standard protocols do not benefit from increased user flexibility.

3.4 Setting 2 : Incorporating Capacity Constraints for Each Hour

As described in Sections 1 and 2, distribution companies make day ahead commitments and so want users to balance load subject to per slot capacity constraints of the form $L_t = \sum_{i=1}^{n} l_i^t \leq C_t, \forall t$. Now, depending on the requirements of the distribution utility, various types of cost functions could be communicated to the users. However, incorporating the capacity constraint does not follow as a direct extension to the protocol described in the previous section.

Non-Triviality: Suppose our goal is to minimize $\max_t L_t^2$ subject to the capacity limits and we directly apply Protocol 1 developed in the previous section, adding an additional constraint that the users are allowed to move a fraction of load from slot j to slot r if r is not at its capacity. Consider a scenario where $L_r < C_r$, $L_j > C_j$, $L_j < L_r$ then the user would shift load from a time j to time r because it has unfilled capacity, but that results in violating the inertia constraint.

Our solution to this problem is to use *scaling*, whereby we use constrained tâtonnement to balance L_t/C_t, $\forall\ t$. Thus, the cost function announced to the users is a scaled quadratic cost, This scaling can be done by defining a 'scaled world' where, $l'_{ih} = l_{ih}/C_h, l'_{ih} = \xi'_{ih}p_{ih}, \xi'_{ih} = \xi_i/C_h \forall h, i, C'_h = 1, L'_h = L_h/C_h \ \forall h$ and $,\sum_h p_{ih} = 1 \ \forall i$. Now, shifting γ unit of load by user i from hour j to hour r, in round $d+1$ implies:

$$l'^{d+1}_{ij} = l'^d_{ih} - \gamma \text{ and } l'^{d+1}_{ir} = l'^d_{ir} + \gamma \frac{C_j}{C_r} \qquad (8)$$

Note that since the end user is unaware of the scaling, utility company should be providing the simple rules to the end customer for simplicity. So that users won't have to be worried about the capacity constraints. (8) ensures that if a user shifts his load from low capacity to high capacity then his utility increases more compared to shifting load from high capacity to low capacity. Note that we can enforce (8) by designing the cost function in a proper way so that user the does not have to take into account complication shifting. The two constraints can then be modified as (i) **Inertia constraints:** A fraction of load of user i may move from hour j to hour r where $j, r \in S_i$ only if $L_j \geq (1+\epsilon)^3 L_r \frac{C_j}{C_r}$ and (ii) **Bounded step constraints:** $\eta \leq p_{ih} \leq 1$ for all i where $\eta = \frac{\epsilon}{H^2}$ and $p^d_{ih}/(1+\epsilon) \leq p^{d+1}_{ih} \leq (1+\epsilon)p^d_{ih}$. The bounded step rule remains the same as it imposes constraints only on the fraction of load and not the absolute load, but the inertia constraints change because of the scaling. We will now see that after imposing the modified constraints and following the rules given in (8) the dynamics will remain acyclic.

LEMMA 3.3. *By applying modified inertia rule and bounded step constraint, dynamics are acyclic. Moreover if there exist a feasible solution(i.e. exist a load profile such that all the slots are within capacity limit satisfying the constraints) then at the end of the algorithm we have $L_t < (1+\epsilon)^3 C_t \ \forall t$ i.e. the load on each slots are within some approximate fraction of its capacities.*

PROOF. First we will prove that the dynamics are acyclic. If a user i shifts ρ fraction of load from j^{th} hour to r^{th} hour at iteration $d+1$, then he will be charged according to the following updation rules: $p^{d+1}_{ij} = p^d_{ij} - \rho p^d_{ij}, \ p^{d+1}_{ir} = p^d_{ir} + \rho p^d_{ij} \frac{C_j}{C_r}$
But by bounded step rule we have

$$p^d_{ij} - \rho p^d_{ij} \geq \frac{p^d_{ij}}{1+\epsilon}, \ p^d_{ir} + \rho p^d_{ij} \frac{C_j}{C_r} \leq p^d_{ir}(1+\epsilon) \qquad (9)$$

By equation (9) we have:

$$\rho p^d_{ij} \leq p^d_{ij} \frac{\epsilon}{1+\epsilon}, \rho p^d_{ij} \leq \frac{C_r}{C_j} p^d_{ir} \epsilon \qquad (10)$$

Note that in order to shift ρp^d_{ij} amount of load from time j to r, user i has to choose

$$\rho p^d_{ij} = min\{p^d_{ij} \frac{\epsilon}{1+\epsilon}, \frac{C_r}{C_j} p^d_{ir} \epsilon\}$$

Consider 2 cases:
- Case 1: if $p^d_{ij} \frac{\epsilon}{1+\epsilon} > \frac{C_r}{C_j} p^d_{ir} \epsilon$ then
$p^{d+1}_{ir} = p^d_{ir} + p^d_{ir}\epsilon$
Thus load on machine r increases by at most $(1+\epsilon)$ fraction.
- Case 2: $p^d_{ij} \frac{\epsilon}{1+\epsilon} < \frac{C_r}{C_j} p^d_{ir} \epsilon$ then
$p^{d+1}_{ij} = \frac{p^d_{ij}}{1+\epsilon}$
Thus load on machine j decreases at least by $\frac{1}{1+\epsilon}$ factor.

Next we will show that once the algorithm converges all the slots are within approximate capacity limits. Suppose when the algorithm converges $\exists \ t$ such that $L_t > C_t(1+\epsilon)^3$. Since there exists a feasible solution
$\Rightarrow \exists i, t_j$ s.t. $L_{t_j} < C_{t_j}$ and $t, t_j \in S_i$. Since $\frac{L_t}{C_t} > \frac{L_{t_j}}{C_{t_j}}$, it would be beneficial for agent i to shift the load from t to t_j. Thus we can assume that the inertia rule fails
$\Rightarrow \frac{L_t}{C_t} < \frac{L_{t_j}}{C_{t_j}}(1+\epsilon)^3 < (1+\epsilon)^3$ which leads to contradiction. $\square$

Thus, the central utility announces quadratic scaled prices such that agent i is faced with the following optimization problem in iteration $d+1$,

$$\underset{\gamma_1,...,\gamma_H}{\text{minimize}} \quad \sum_{t \in S_i}(L_t - \gamma_t - \frac{(\sum_{t' \in S_i}L_{t'})C_t}{\sum_{t' \in s_i}C_{t'}})^2 \qquad (11)$$

$$\text{s.t.} \quad \gamma_t = \sum_{t' \in R'_t} x_{t,t'} \forall t, \sum_{t \in S_i} \gamma_t = 0$$

$$\gamma_t \leq p^d_{it} \ \xi_i \ \frac{\epsilon}{1+\epsilon}, -\gamma_t \leq p^d_{it} \ \xi_i \ \epsilon$$

where $R'_t = \{t' : L_t \geq (1+\epsilon)^3 L_{t'} \ C_t/C_{t'} \text{ or } L_{t'} \geq (1+\epsilon)^3 L_t \ C_{t'}/C_t\}$

This quadratic cost function ensures that it is in best interest of every player i to balance the load $L_t/C_t \ \forall t \in S_i$. If such prices and constraints are used then we have the following theorem,

THEOREM 3.4. *The constrained tâtonnement dynamics in the 'scaled world' converges to $(1+\delta)$ approximation in $O(\frac{1}{\delta^6}log^6 \frac{H}{\delta})$ iterations for $\epsilon = O(\frac{\delta^2}{logH})$.*

We then have $L'_{max} \leq (1+\delta)OPT'$ where OPT' is the optimal load in the scaled world. In the original world, let C'_* and C''_* be the capacity of slot having the OPT' and L'_{max} load respectively. Then we have

$$\frac{L_{max}}{C_{max}} \leq \frac{L_{max}}{C'_*} \leq (1+\delta)\frac{OPT}{C''_*} \leq (1+\delta)\frac{OPT}{C_{min}} \qquad (12)$$

where C_{max} and C_{min} are the largest and smallest capacities, respectively. Hence in the original (unscaled) world we have the following corollary:

COROLLARY 3.1. *The scaled constrained tâtonnement protocol results in a $(1+\delta)*(\frac{C_{max}}{C_{min}})$ approximation in $O(\frac{1}{\delta^5}log^6 \frac{H}{\delta})$ iterations for $\epsilon = O(\frac{\delta^2}{logH})$.*

Here again we see that an approximately optimal solution *that satisfies the capacity constraints* is reached in a number of iterations independent of the number of agents.

3.4.1 Simulations for Setting 2

With updated rules we now illustrate the dynamics of the scaled constrained tâtonnement protocol where users solve (11). An important point to emphasize is that whenever there exists a solution that satisfies the capacity constraints, $L_t \leq C_t \forall t$, the protocol will output a feasible solution. For simulations in this section we fixed number of agents to be 50, $\delta = 0.2$. Expected number of slots for every users are selected by uniform distribution between 2 to 24 and capacities of each slot are also chosen uniformly between C_{min} and C_{max}. We define cost of the solution as $\sum_t L^2_t$ whereas OPT is defined as minimum value of $\sum_t L^2_t$ s.t. $L_t \leq C_t$.

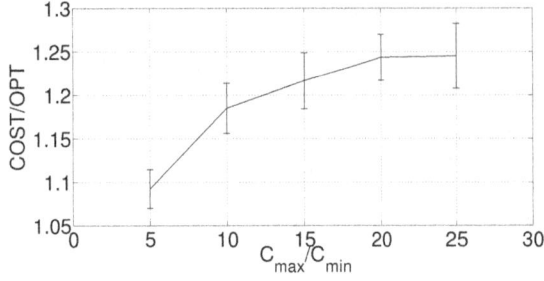

Figure 9: Increased sub-optimality (COST/OPT) with increasing bandwidth (C_{max}/C_{min})

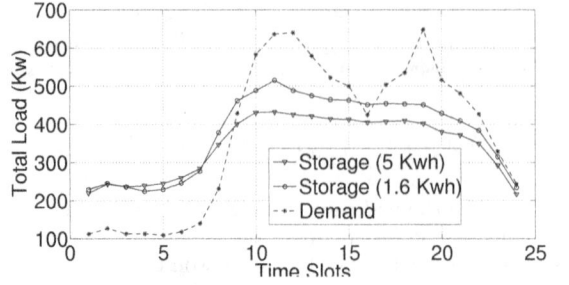

Figure 11: Flattening of overall load profile though constrained tâtonnement with micro-storage

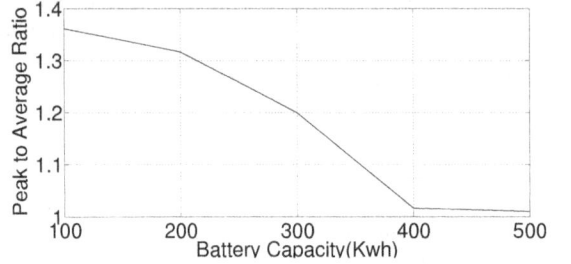

Figure 10: Decreasing of sub-optimality (COST/OPT) with increased load factor ($\frac{\sum_h C_h}{\sum_{i,h} l_i^h}$)

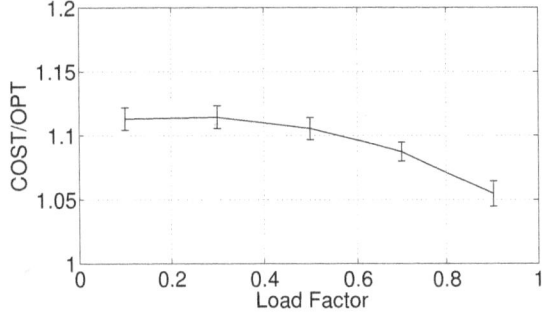

Figure 12: Decrease in peak to average ratio with increased per user storage capacity

Figure 9 shows that as the ratio of the maximum capacity and minimum capacity, also called the bandwidth ($\frac{C_{max}}{C_{min}}$) increases, the ratio between protocol cost and optimal cost also increases, in agreement with Corollary 3.1. The results also show that even though we are balancing L_t/C_t we are not that far from the true optimal solution which balances L_t over all t.

We next try to understand the performance of the protocol under different loads. We define the 'load factor' of the system to be the fraction $\frac{\sum_h C_h}{\sum_{i,h} l_i^h}$. Thus, a load factor of 1 means that the system is completely loaded to capacity. For the purpose of simulation in this setting we try to find the values of ξ_i by fixing the load factor on every slot. After finding the $\xi_i's$ according to the load factor we then run our algorithm after placing random load from each user on each time slots. Figure 10 shows that when the system is operating at or near capacity, the protocol performs particularly well. The reason is that when the load factor is smaller it results in larger scaling of the total load available for the slots which has higher capacity.

3.5 Setting 3 : Distributed Micro-Storage Management

In the storage problem, each household has storage of capacity B_{max} (We considered here that each user has same storage capacity but the algorithm can be easily extended for different storage capacity as users are running optimization problem at their end independent of other users) and a consumption profile dm such that dm_t amount of load is required at each time slot t. Each user i then has t additional constraints, corresponding to the requirement that the

electricity consumed in each time slot must be drawn from the grid prior to t. For this problem we directly apply constrained tâtonnement with the center announcing quadratic costs, $(L_h)^2$ where in addition to the inertia and bounded step constraints each user has to satisfy their own internal storage constraints. The optimization problem faced by user i in iteration $d+1$ is,

$$\underset{\gamma_1,\dots,\gamma_H}{\text{minimize}} \quad \sum_{t\in\mathcal{H}} (L_t^d - \gamma_t)^2 \tag{13}$$

$$\text{s.t.} \quad 0 \le \sum_{t=1}^{\tau} (l_i^t - \gamma_t - dm_t) \le B_{max} \;\forall \tau, \gamma_t = \sum_{t'\in R_t} x_{t,t'} \;\forall t$$

$$\sum_{t\in\mathcal{H}} \gamma_t = 0, \; \gamma_t \le p_{it}^d \xi_i \frac{\epsilon}{1+\epsilon}, \; -\gamma_t \le p_{it}^d \xi_i \epsilon$$

Note that the goal here is to find the storage profile given the load profile. One interesting direction can be to combine setting 1 and setting 3 so that a user can update the load and storage profiles in parallel.

3.5.1 Simulations for Setting 3

We have simulated daily energy consumption profiles based on the models in [28] to obtain dm_t. This model results in realistic load curves as seen in Figures 3 and 11. The broken line in Figure 11 represents the initial total load profile without storage and the solid lines represent the load profiles with storage at the end of the tâtonnement protocol(with battery capacity 1.6 KWh and 5 KWh). We can see from Figure 11 that the peak load decreases substantially with constrained tâtonnement. Figure 12 shows that as the storage capacity each user has increases, the peak load decreases. Thus, constrained tâtonnement prevents the herding that was observed in the agent models of [8, 33]. As shown in

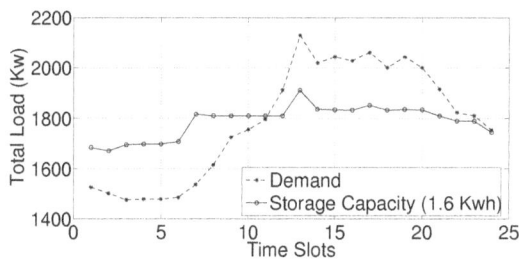

Figure 13: Flattening of overall load profile with storage

figure 11, using storage capacity of 1.6 Kwh results in 27.6% peak reduction whereas increasing storage capacity to 5 Kwh results in 33.6% reduction in peak. To demonstrate the generality of our results, we also conduct experiments on the microgrid data set from [6]. This data set consists of electrical data over a single 24-hour period from 443 unique homes. In Figure 13 we see that substantial reductions in peaks are possible with co-ordinated charging of local storage.

4. RELATIONSHIP TO PRIOR WORK

This work builds on a few different research directions that we review here for clarity. The first is the line of research pioneered in [19] where they "argue that the concepts and technologies pioneered by the Internet - the fruits of the past four decades of Internet research - can make fundamental contributions to the architecture and operations of the future grid." They identify the similarities between the cyber-physical systems that constitute the internet and the next generation smart grid and propose that many of the techniques studied, developed and proven effective in the internet could be used to design protocols for the smart grid. In particular they outline the possible benefits of both proactive and reactive congestion control, made possible by the large scale deployment of smart meters throughout the grid. [3] suggest the use of these techniques for real time congestion management. Extending this line of work we introduce game theoretic principles, modelling users as self interested agents who may lie to maximize their own welfare. In addition, we conduct a rigorous analysis of the number of iterations and sub-optimality of these algorithms.

The second line of work is based on agent based models of storage [33] and demand [27, 24] management. In particular, [33, 24] consider game theoretic concepts such as individual rationality and solve for the Nash equilibrium. However, the protocols suggested require a linear number of iterations to converge making them infeasible in large grids or even for large simulations as mentioned in [33]. Work such as [22, 16] show how how to capture the utilities of a large number of different appliances using simple constraints and create a distributed optimization problems which can then be solved by tâtonnement. These approaches form the motivation for our Setting 3. *Our primary contribution to this line of research is to show that it is possible to add simple, centrally verifiable constraints to all of these distributed demand management problems to obtain convergence in a number of iterations independent of the number of users, a substantial speed-up over the linear number of iterations required in [24, 33].* This improvement is particularly important, as mentioned earlier, in modern distribution grids with millions of users and possibly tens of millions of smart appliances.

5. CONCLUSIONS AND FUTURE WORK

In this paper, we presented a tâtonnement framework for resource allocation among intelligent agents in the smart grid, that non-trivially generalizes past work in this area. Using a parallel to work in server load balancing we developed fast, incentive compatible, distributed protocols by communicating *constraints* along with pricing signals. We adapt the theoretical work to a variety of practical scenarios and show that incorporating novel constraints in practical settings requires a rethinking of how prices and constraints are set. Extensive simulations validate the improvements possible with constrained tâtonnement. A highlight of our simulations is that in some scenarios performance was substantially better than expected from the (worst case) theoretical bounds. In particular, fewer iterations were required to achieve a smaller optimality gap than expected in Figures 4 and 7. Improving the bounds with better models of agents and their consumption is an interesting direction of future work. In addition, special tâtonnement strategies can be (and in some cases have been) applied to other important resource allocation problems in next generation cyber-physical systems where our work may find application. We would also like to extend our work to some more complicated setting where agents have the preferences over available slots and to include maximum charging rates of the battery in setting 3. In setting 2, we assumed that there exist a load profile consumption which satisfies the capacity constraints. In future, we would like to present the solution concept where such consumption profile does not exist. For example, extending these protocols to handle deadlines for the completion of certain loads which an agent may have. Finally, a better understanding of the practical overheads and comparison of different protocols is an important direction of future work.

6. REFERENCES

[1] M. Albadi and E. El-Saadany. Demand response in electricity markets: An overview. In *IEEE Power Engineering Society General Meeting*, 2007.

[2] R. Anderson and S. Fuloria. On the security economics of electricity metering. *Proceedings of the WEIS*, 2010.

[3] O. Ardakanian, C. Rosenberg, and S. Keshav. Real-time distributed congestion control for electrical vehicle charging. *Greenmetrics*, 2012.

[4] B. Awerbuch, Y. Azar, and R. Khandekar. Fast load balancing via bounded best response. In *SODA*, pages 314–322, 2008.

[5] P. Barker and R. De Mello. Determining the impact of distributed generation on power systems. i. radial distribution systems. In *IEEE Power Engineering Society Summer Meeting*, volume 3, pages 1645 –1656 vol. 3, 2000.

[6] S. Barker, A. Mishra, D. Irwin, E. Cecchet, P. Shenoy, and J. Albrecht. Smart*: An open data set and tools for enabling research in sustainable homes. 2012.

[7] S. Borenstein. The trouble with electricity markets: Understanding California's restructuring disaster. *The Journal of Economic Perspectives*, 16(1):191–211, 2002.

[8] T. Carpenter, S. Singla, P. Azimzadeh, and S. Keshav. The impact of electricity pricing schemes on storage adoption in ontario. In *e-Eenergy*, 2012.

[9] H. Chao. Competitive electricity markets with consumer subscription service in a smart grid. *Journal of Regulatory Economics*, pages 1–26, 2012.

[10] K. Clement-Nyns, E. Haesen, and J. Driesen. The impact of charging plug-in hybrid electric vehicles on a residential distribution grid. *Power Systems, IEEE Transactions on*, 25(1):371–380, 2010.

[11] EPRI and NRDC. Environmental assessment of plug-in hybrid electric vehicles. volume 1: Nationwide greenhouse gas emissions. Technical Report 1015 325, July 2007.

[12] H. Farhangi. The path of the smart grid. *IEEE Power and Energy Magazine*, 8(1):18–28, 2010.

[13] S. Ghosh. Electricity consumption and economic growth in india. *Energy Policy*, 30(2):125–129, 2002.

[14] S. Hadley and A. Tsvetkova. Potential impacts of plug-in hybrid electric vehicles on regional power generation. *The Electricity Journal*, 22(10), 2009.

[15] C. Harris. *Electricity Markets: Pricing, Structures and Economics*. Wiley, Sussex, England, 2006.

[16] L. Huang, J. Walrand, and K. Ramchandran. Optimal smart grid tariffs. In *Information Theory and Applications Workshop (ITA)*, pages 212 –220, 2012.

[17] M. Ilic, L. Xie, and J. Joo. Efficient coordination of wind power and price-responsive demand Part I: Theoretical foundations. *IEEE Transactions on Power Systems*, 26(4):1875–1884, 2011.

[18] International Energy Agency. The power to choose - enhancing demand response in liberalised electricity markets findings of IEA demand response project. 2003.

[19] S. Keshav and C. Rosenberg. How internet concepts and technologies can help green and smarten the electrical grid. *ACM SIGCOMM Computer Communication Review*, 41(1):109–114, 2011.

[20] M. Kintner-Meyer, K. Schneider, and R. Pratt. Impacts assessment of plug-in hybrid vehicles on electric utilities and regional us power grids, Part 1: Technical analysis. *Pacific Northwest National Laboratory*, 2007.

[21] P. Kundur, N. Balu, and M. Lauby. *Power system stability and control*, volume 4. McGraw-hill New York, 1994.

[22] N. Li, L. Chen, and S. Low. Optimal demand response based on utility maximization in power networks. In *IEEE Power and Energy Society General Meeting*, pages 1 –8, july 2011.

[23] N. A. Lynch. *Distributed algorithms*. Morgan Kaufmann, 1996.

[24] A. Mohsenian-Rad, V. Wong, J. Jatskevich, R. Schober, and A. Leon-Garcia. Autonomous demand-side management based on game-theoretic energy consumption scheduling for the future smart grid. *IEEE Transactions on Smart Grid*, 1(3), 2010.

[25] R. Myerson. *Game theory: analysis of conflict*. Harvard University Press, 1997.

[26] Y. Narahari, D. Garg, R. Narayanam, and H. Prakash. *Game theoretic problems in network economics and mechanism design solutions*. Springer, 2009.

[27] S. Ramchurn, P. Vytelingum, A. Rogers, and N. Jennings. Agent-based control for decentralised demand side management in the smart grid. In *AAMAS*, pages 5–12, February 2011.

[28] I. Richardson, M. Thomson, D. Infield, and C. Clifford. Domestic electricity use: A high-resolution energy demand model. *Energy and Buildings*, 42(10):1878–1887, 2010.

[29] W. Saad, Z. Han, H. V. Poor, and T. Basar. Game theoretic methods for the smart grid. *CoRR*, abs/1202.0452, 2012.

[30] V. Sood, D. Fischer, J. Eklund, and T. Brown. Developing a communication infrastructure for the smart grid. In *Electrical Power & Energy Conference (EPEC), 2009 IEEE*, pages 1–7. Ieee, 2009.

[31] G. Strbac. Demand side management: Benefits and challenges. *Energy Policy*, 36(12):4419 – 4426, 2008.

[32] R. Urgaonkar, B. Urgaonkar, M. Neely, and A. Sivasubramaniam. Optimal power cost management using stored energy in data centers. In *Proceedings of the ACM SIGMETRICS*, pages 221–232. ACM, 2011.

[33] P. Vytelingum, T. Voice, S. Ramchurn, A. Rogers, and N. Jennings. Agent-based micro-storage management for the smart grid. In *AAMAS*, pages 39–46, 2010.

[34] D. Walker. Walras's theories of tatonnement. *The Journal of Political Economy*, 95(4):758–774, 1987.

[35] L. Xie and M. Ilic. Model predictive dispatch in electric energy systems with intermittent resources. In *IEEE International Conference on Systems, Man and Cybernetics*, pages 42 –47, oct. 2008.

[36] T. Zhu, A. Mishra, D. Irwin, N. Sharma, P. Shenoy, and D. Towsley. The case for efficient renewable energy management for smart homes. *Proceedings of the Third Workshop on Embedded Sensing Systems for Energy-efficiency in Buildings (BuildSys)*, November 2011.

Dynamic Provisioning in Next-Generation Data Centers with On-site Power Production

Jinlong Tu, Lian Lu and Minghua Chen
Department of Information Engineering
The Chinese University of Hong Kong

Ramesh K. Sitaraman
Department of Computer Science
University of Massachusetts at Amherst
& Akamai Technologies

ABSTRACT

The critical need for clean and economical sources of energy is transforming data centers that are primarily energy consumers to also energy producers. We focus on minimizing the operating costs of next-generation data centers that can jointly optimize the energy supply from on-site generators and the power grid, and the energy demand from servers as well as power conditioning and cooling systems. We formulate the cost minimization problem and present an offline optimal algorithm. For "on-grid" data centers that use only the grid, we devise a deterministic online algorithm that achieves the best possible competitive ratio of $2 - \alpha_s$, where α_s is a normalized look-ahead window size. The competitive ratio of an online algorithm is defined as the maximum ratio (over all possible inputs) between the algorithm's cost (with no or limited look-ahead) and the offline optimal assuming complete future information. We remark that the results hold as long as the overall energy demand (including server, cooling, and power conditioning) is a convex and increasing function in the total number of active servers and also in the total server load. For "hybrid" data centers that have on-site power generation in addition to the grid, we develop an online algorithm that achieves a competitive ratio of at most $\frac{P_{\max}(2-\alpha_s)}{c_o+c_m/L}\left[1+2\frac{P_{\max}-c_o}{P_{\max}(1+\alpha_g)}\right]$, where α_s and α_g are normalized look-ahead window sizes, $P_{\max}$ is the maximum grid power price, and L, c_o, and c_m are parameters of an on-site generator.

Using extensive workload traces from Akamai with the corresponding grid power prices, we simulate our offline and online algorithms in a realistic setting. Our offline (resp., online) algorithm achieves a cost reduction of 25.8% (resp., 20.7%) for a hybrid data center and 12.3% (resp., 7.3%) for an on-grid data center. The cost reductions are quite significant and make a strong case for a joint optimization of energy supply and energy demand in a data center. A hybrid data center provides about 13% additional cost reduction over an on-grid data center representing the additional cost

benefits that on-site power generation provides over using the grid alone.

Categories and Subject Descriptors

F.1.2 [**Modes of Computation**]: Online computation; G.1.6 [**Optimization**]: Nonlinear programming; I.1.2 [**Algorithms**]: Analysis of algorithms; I.2.8 [**Problem Solving, Control Methods, and Search**]: Scheduling

General Terms

Algorithms, Performance

Keywords

data centers; dynamic provisioning; on-site power production; online algorithm

1. INTRODUCTION

Internet-scale cloud services that deploy large distributed systems of servers around the world are revolutionizing all aspects of human activity. The rapid growth of such services has lead to a significant increase in server deployments in data centers around the world. Energy consumption of data centers account for roughly 1.5% of the global energy consumption and is increasing at an alarming rate of about 15% on an annual basis [21]. The surging global energy demand relative to its supply has caused the price of electricity to rise, even while other operating expenses of a data center such as network bandwidth have decreased precipitously. Consequently, the energy costs now represent a large fraction of the operating expenses of a data center today [9], and decreasing the energy expenses has become a central concern for data center operators.

The emergence of energy as a central consideration for enterprises that operate large server farms is drastically altering the traditional boundary between a data center and a power utility (c.f. Figure 1). Traditionally, a data center hosts servers but buys electricity from an utility company through the power grid. However, the criticality of the energy supply is leading data centers to broaden their role to also generate much of the required power on-site, decreasing their dependence on a third-party utility. While data centers have always had generators as a short-term back-up for when the grid fails, on-site generators for sustained power supply is a newer trend. For instance, Apple recently announced that it will build a massive data center for its iCloud services with 60% of its energy coming from its on-site generators that use "clean energy" sources such as fuel cells with biogas and solar panels [25]. As another example, eBay recently announced that it will add a 6 MW facility to its existing data center in Utah that will be largely powered by on-site fuel cell generators [17]. The trend for *hybrid* data centers that generate electricity on-site (c.f. Figure 1) with reduced reliance on the grid is driven by the confluence of several factors. This trend is also mirrored in the broader power industry where the centralized model for power generation with few large power plants is giving way to a more distributed generation model [11] where many smaller on-site generators produce power that is consumed locally over a "micro-grid".

A key factor favoring on-site generation is the potential for cheaper power than the grid, especially during peak hours. On-site generation also reduces transmission losses that in turn reduce the effective cost, because the power is generated close to where it is consumed. In addition, another factor favoring on-site generation is a requirement for many enterprises to use cleaner renewable energy sources, such as Apple's mandate to use 100% clean energy in its data centers [6]. Such a mandate is more easily achievable with the enterprise generating all or most of its power on-site, especially since recent advances such as the fuel cell technology of Bloom Energy [7] make on-site generation economical and feasible. Finally, the risk of service outages caused by the failure of the grid, as happened recently when thunderstorms brought down the grid causing a denial-of-service for Amazon's AWS service for several hours [18], has provided greater impetus for on-site power generation that can sustain the data center for extended periods without the grid.

Our work focuses on the key challenges that arise in the emerging hybrid model for a data center that is able to simultaneously optimize *both* the generation and consumption of energy (c.f. Figure 1). In the traditional scenario, the utility is responsible for energy provisioning (**EP**) that has the goal of supplying energy as economically as possible to meet the energy demand, albeit the utility has no detailed knowledge and no control over the server workloads within a data center that drive the consumption of power. Optimal energy provisioning by the utility in isolation is characterized by the unit commitment problem [31, 36] that has been studied over the past decades. The energy provisioning problem takes as input the demand for electricity from the consumers and determines which power generators should be used at what time to satisfy the demand in the most economical fashion. Further, in a traditional scenario, a data center is responsible for capacity provisioning (**CP**) that has the goal of managing its server capacity to serve the incoming workload from end users while reducing the total energy demand of servers, as well as power conditioning and various cooling systems, but without detailed knowledge or control over the power generation. For instance, dynamic provisioning of server capacity by turning off some servers during periods of low workload to reduce the energy demand has been studied in recent years [23, 28, 10, 27].

The convergence of power generation and consumption within a single data center entity and the increasing impact of energy costs requires a new integrated approach to both energy provisioning (**EP**) and capacity provisioning (**CP**). A key contribution of our work is formulating and developing algorithms that simultaneously manage on-site power generation, grid power consumption, and server capacity with the goal of minimizing the operating cost of the data center.

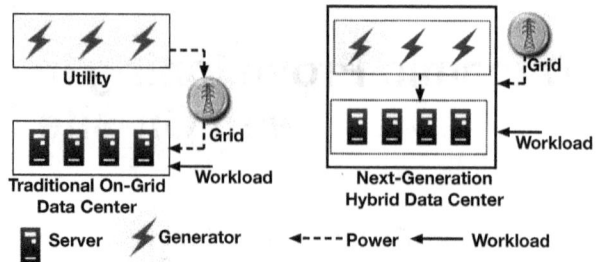

Figure 1: While an "on-grid" data center derives all its power from the grid, next-generation "hybrid" data centers have additional on-site power generation.

Online vs. Offline Algorithms. In designing algorithms for optimizing the operating cost of a hybrid data center, there are three time-varying inputs: the server workload $a(t)$ generated by service requests from users and the price of a unit energy from the grid $p(t)$, and the total power consumption function g_t for each time t where $1 \leq t \leq T$. We begin by investigating *offline* algorithms that minimize the operating cost with perfect knowledge of the entire input sequence $a(t)$, $p(t)$ and g_t, for $1 \leq t \leq T$. However, in real-life, the time-varying input sequences are not knowable in advance. In particular, the optimization must be performed in an *online* fashion where decisions at time t are made with the knowledge of inputs $a(\tau), p(\tau)$ and g_τ, for $1 \leq \tau \leq t+w$, where $w \geq 0$ is a small (possibly zero) look-ahead window. Specifically, an online algorithm has no knowledge of inputs beyond the look-ahead window, *i.e.*, for time $t + w < \tau \leq T$. We assume the inputs within the look-ahead are perfectly known when analyzing the algorithm performance. In practice, short-term demand or grid price can be estimated rather accurately by various techniques including pattern analysis and time series analysis and prediction [19, 14]. As is typical in the study of online algorithms [12], we seek theoretical guarantees for our online algorithms by computing the *competitive ratio* that is ratio of the cost achieved by the online algorithm for an input to the optimal cost achieved for the same input by an offline algorithm. The competitive ratio is computed under a worst case scenario where an adversary picks the worst possible inputs for the online algorithm. Thus, a small competitive ratio provides a strong guarantee that the online algorithm will achieve a cost close to the offline optimal even for the worst case input.

Our Contributions. A key contribution of our work is to formulate and study data center cost minimization (**DCM**) that integrates energy procurement from the grid, energy production using on-site generators, and dynamic server capacity management. Our work jointly optimizes the two components of **DCM**: energy provisioning (**EP**) from the grid and generators and capacity provisioning (**CP**) of the servers.

- We theoretically evaluate the benefit of joint optimization by showing that optimizing energy provisioning (**EP**) and capacity provisioning (**CP**) separately results in a factor loss of optimality $\rho = LP_{\max} / (Lc_o + c_m)$ compared to optimizing them jointly, where $P_{\max}$ is the maximum grid power price, and L, c_o, and c_m are the capacity, incremental cost, and base cost of an on-site generator respectively. Further, we derive an efficient offline optimal algorithm for hybrid data centers that

Competitive Ratio	On-grid	Hybrid	
No Look-ahead	2	$\frac{2P_{\max}}{c_o+c_m/L}$	$1+2\frac{P_{\max}-c_o}{P_{\max}}$
With Look-ahead	$2-\alpha_s$	$\frac{P_{\max}(2-\alpha_s)}{c_o+c_m/L}$	$1+2\frac{P_{\max}-c_o}{P_{\max}(1+\alpha_g)}$

Table 1: Summary of algorithmic results. The on-grid results are the best possible for any deterministic online algorithm.

jointly optimize **EP** and **CP** to minimize the data center's operating cost.

- For on-grid data centers, we devise an online deterministic algorithm that achieves a competitive ratio of $2-\alpha_s$, where $\alpha_s \in [0,1]$ is the normalized look-ahead window size. Further, we show that our algorithm has the best competitive ratio of any deterministic online algorithm for the problem (c.f. Table 1). For the more complex hybrid data centers, we devise an online deterministic algorithm that achieves a competitive ratio of $\frac{P_{\max}(2-\alpha_s)}{c_o+c_m/L}\left[1+2\frac{P_{\max}-c_o}{P_{\max}(1+\alpha_g)}\right]$, where α_s and α_g are normalized look-ahead window sizes. Both online algorithms perform better as the look-ahead window increases, as they are better able to plan their current actions based on knowledge of future inputs. Interestingly, in the on-grid case, we show that there exists *fixed* threshold value for the look-ahead window for which the online algorithm matches the offline optimal in performance achieving a competitive ratio of 1, *i.e.*, there is no additional benefit gained by the online algorithm if its look-ahead is increased beyond the threshold.

- Using extensive workload traces from Akamai and the corresponding grid prices, we simulate our offline and online algorithms in a realistic setting with the goal of empirically evaluating their performance. Our offline optimal (resp., online) algorithm achieves a cost reduction of 25.8% (resp., 20.7%) for a hybrid data center and 12.3% (resp., 7.3%) for an on-grid data center. The cost reduction is computed in comparison with the baseline cost achieved by the current practice of statically provisioning the servers and using only the power grid. The cost reductions are quite significant and make a strong case for utilizing our joint cost optimization framework. Furthermore, our online algorithms obtain almost the same cost reduction as the offline optimal solution even with a small look-ahead of 6 hours, indicating the value of short-term prediction of inputs.

- A hybrid data center provides about 13% additional cost reduction over an on-grid data center representing the additional cost benefits that on-site power generation provides over using the grid alone. Interestingly, it is sufficient to deploy a partial on-site generation capacity that provides 60% of the peak power requirements of the data center to obtain over 95% of the additional cost reduction. This provides strong motivation for a traditional on-grid data center to deploy at least a partial on-site generation capability to save costs.

Due to space limitations, all proofs are in our technical report [39].

2. THE DATA CENTER COST MINIMIZATION PROBLEM

We consider the scenario where a data center can jointly optimize energy production, procurement, and consumption so as to minimize its operating expenses. We refer to this data center cost minimization problem as **DCM**. To study **DCM**, we model how energy is produced using on-site power generators, how it can be procured from the power grid, and how data center capacity can be provisioned dynamically in response to workload. While some of these aspects have been studied independently, our work is unique in optimizing these dimensions simultaneously as next-generation data centers can. Our algorithms minimize cost by use of techniques such as: (i) dynamic capacity provisioning of servers – turning off unnecessary servers when workload is low to reduce the energy consumption (ii) opportunistic energy procurement – opting between the on-site and grid energy sources to exploit price fluctuation, and (iii) dynamic provisioning of generators - orchestrating which generators produce what portion of the energy demand. While prior literature has considered these techniques in isolation, we show how they can be used in coordination to manage both the supply and demand of power to achieve substantial cost reduction.

Notation	Definition
T	Number of time slots
N	Number of on-site generators
β_s	Switching cost of a server ($)
β_g	Startup cost of an on-site generator ($)
c_m	Sunk cost of maintaining a generator in its active state per slot ($)
c_o	Incremental cost for an active generator to output an additional unit of energy ($/Wh)
L	The maximum output of a generator (Watt)
$a(t)$	Workload at time t
$p(t)$	Price per unit energy drawn from the grid at t ($P_{\min} \le p(t) \le P_{\max}$) ($/Wh)
$x(t)$	Number of active servers at t
$s(t)$	Total server service capability at t
$v(t)$	Grid power used at t (Watt)
$y(t)$	Number of active on-site generators at t
$u(t)$	Total power output from active generators at t (Watt)
$g_t(x(t),a(t))$	Total power consumption as a function of $x(t)$ and $a(t)$ at t (Watt)

Note: we use bold symbols to denote vectors, *e.g.*, $\boldsymbol{x} = \langle x(t)\rangle$. Brackets indicate the unit.

Table 2: Key notation.

2.1 Model Assumptions

We adopt a discrete-time model whose time slot matches the timescale at which the scheduling decisions can be updated. Without loss of generality, we assume there are totally T slots, and each has a unit length.

Workload model. Similar to existing work [13, 34, 16], we consider a "mice" type of workload for the data center where each job has a small transaction size and short duration. Jobs arriving in a slot get served in the same slot. Workload can be split among active servers at arbitrary granularity like a fluid. These assumptions model a "request-response" type of workload that characterizes serving web

139

content or hosted application services that entail short but real-time interactions between the user and the server. The workload to be served at time t is represented by $a(t)$. Note that we do not rely on any specific stochastic model of $a(t)$.

Server model. We assume that the data center consists of a sufficient number of homogeneous servers, and each has unit service capacity, *i.e.*, it can serve at most one unit workload per slot, and the same power consumption model. Let $x(t)$ be the number of active servers and $s(t) \in [0, x(t)]$ be the total server service capability at time t. It is clear that $s(t)$ should be larger than $a(t)$ to get the workload served in the same slot. We model the aggregate server power consumption as $b(t) \triangleq f_s(x(t), s(t))$, an increasing and convex function of $x(t)$ and $s(t)$. That is, the first and second order partial derivatives in $x(t)$ and $s(t)$ are all non-negative. Since $f_s(x(t), s(t))$ is increasing in $s(t)$, it is optimal to always set $s(t) = a(t)$. Thus, we have $b(t) = f_s(x(t), a(t))$ and $x(t) \geq a(t)$.

This power consumption model is quite general and captures many common server models. One example is the commonly adopted standard linear model [9]:

$$f_s(x(t), a(t)) = c_{idle}x(t) + (c_{peak} - c_{idle})a(t),$$

where c_{idle} and c_{peak} are the power consumed by an server at idle and fully utilized state, respectively. Most servers today consume significant amounts of power even when idle. A holy grail for server design is to make them "power proportional" by making c_{idle} zero [32].

Besides, turning a server on entails switching cost [28], denoted as β_s, including the amortized service interruption cost, wear-and-tear cost, *e.g.*, component procurement, replacement cost (hard-disks in particular) and risk associated with server switching. It is comparable to the energy cost of running a server for several hours [23].

In addition to servers, power conditioning and cooling systems also consume a significant portion of power. The three[1] contribute about 94% of overall power consumption and their power draw vary drastically with server utilization [33]. Thus, it is important to model the power consumed by power conditioning and cooling systems.

Power conditioning system model. Power conditioning system usually includes power distribution units (PDUs) and uninterruptible power supplies (UPSs). PDUs transform the high voltage power distributed throughout the data center to voltage levels appropriate for servers. UPSs provides temporary power during outage. We model the power consumption of this system as $f_p(b(t))$, an increasing and convex function of the aggregate server power consumption $b(t)$.

This model is general and one example is a quadratic function adopted in a comprehensive study on the data center power consumption [33]: $f_p(b(t)) = C_1 + \pi_1 b^2(t)$, where $C_1 > 0$ and $\pi_1 > 0$ are constants depending on specific PDUs and UPSs.

Cooling system model. We model the power consumed by the cooling system as $f_c^t(b(t))$, a time-dependent (*e.g.*, depends on ambient weather conditions) increasing and convex function of $b(t)$.

This cooling model captures many common cooling systems. According to [24], the power consumption of an out-

side air cooling system can be modelled as a time-dependent cubic function of $b(t)$: $f_c^t(b(t)) = K_t b^3(t)$, where $K_t > 0$ depends on ambient weather conditions, such as air temperature, at time t. According to [33], the power draw of a water chiller cooling system can be modelled as a time-dependent quadratic function of $b(t)$: $f_c^t(b(t)) = Q_t b^2(t) + L_t b(t) + C_t$, where $Q_t, L_t, C_t \geq 0$ depend on outside air and chilled water temperature at time t. Note that all we need is $f_c^t(b(t))$ is increasing and convex in $b(t)$.

On-site generator model. We assume that the data center has N units of homogeneous on-site generators, each having an power output capacity L. Similar to generator models studied in the unit commitment problem [20], we define a generator startup cost β_g, which typically involves heating up cost, additional maintenance cost due to each startup (*e.g.*, fatigue and possible permanent damage resulted by stresses during startups), c_m as the sunk cost of maintaining a generator in its active state for a slot, and c_o as the incremental cost for an active generator to output an additional unit of energy. Thus, the total cost for $y(t)$ active generators that output $u(t)$ units of energy at time t is $c_m y(t) + c_o u(t)$.

Grid model. The grid supplies energy to the data center in an "on-demand" fashion, with time-varying price $p(t)$ per unit energy at time t. Thus, the cost of drawing $v(t)$ units of energy from the grid at time t is $p(t)v(t)$. Without loss of generality, we assume $0 \leq P_{\min} \leq p(t) \leq P_{\max}$.

To keep the study interesting and practically relevant, we make the following assumptions: (i) the server and generator turning-on cost are strictly positive, *i.e.*, $\beta_s > 0$ and $\beta_g > 0$. (ii) $c_o + c_m/L < P_{\max}$. This ensures that the minimum on-site energy price is cheaper than the maximum grid energy price. Otherwise, it should be clear that it is optimal to always buy energy from the grid, because in that case the grid energy is cheaper and incurs no startup costs.

2.2 Problem Formulation

Based on the above models, the data center total power consumption is the sum of the server, power conditioning system and the cooling system power draw, which can be expressed as a time-dependent function of $b(t)$ ($b(t) = f_s(x(t), a(t))$):

$$b(t) + f_p(b(t)) + f_c^t(b(t)) \triangleq g_t(x(t), a(t)).$$

We remark that $g_t(x(t), a(t))$ is increasing and convex in $x(t)$ and $a(t)$. This is because it is the sum of three increasing and convex functions. *Note that all results we derive in this paper apply to any $g_t(x, a)$ as long as it is increasing and convex in x and a.*

Our objective is to minimize the data center total cost in entire horizon $[1, T]$, which is given by

$$\text{Cost}(x, y, u, v) \triangleq \sum_{t=1}^{T} \{v(t)p(t) + c_o u(t) + c_m y(t) \quad (1)$$
$$+ \beta_s[x(t) - x(t-1)]^+ + \beta_g[y(t) - y(t-1)]^+\},$$

which includes the cost of grid electricity, the running cost of on-site generators, and the switching cost of servers and on-site generators in the entire horizon $[1, T]$. Throughout this paper, we set initial condition $x(0) = y(0) = 0$.

We formally define the data center cost minimization problem as a non-linear mixed-integer program, given the workload $a(t)$, the grid price $p(t)$ and the time-dependent func-

---

[1]The other two, networking and lighting, consume little power and have less to do with server utilization. Thus, we do not model the two in this paper.

tion $g_t(x,a)$, for $1 \le t \le T$, as time-varying inputs.

$$\min_{x,y,u,v} \quad \text{Cost}(x,y,u,v) \tag{2}$$
$$\text{s.t.} \quad u(t) + v(t) \ge g_t(x(t), a(t)), \tag{3}$$
$$u(t) \le Ly(t), \tag{4}$$
$$x(t) \ge a(t), \tag{5}$$
$$y(t) \le N, \tag{6}$$
$$x(0) = y(0) = 0, \tag{7}$$
$$\text{var} \quad x(t), y(t) \in \mathbb{N}^0, u(t), v(t) \in \mathbb{R}_0^+, \ t \in [1, T],$$

where $[\cdot]^+ = \max(0, \cdot)$, $\mathbb{N}^0$ and $\mathbb{R}_0^+$ represent the set of non-negative integers and real numbers, respectively.

Constraint (3) ensures the total power consumed by the data center is jointly supplied by the generators and the grid. Constraint (4) captures the maximal output of the on-site generator. Constraint (5) specifies that there are enough active servers to serve the workload. Constraint (6) is generator number constraint. Constraint (7) is the boundary condition.

Note that this problem is challenging to solve. First, it is a non-linear mixed-integer optimization problem. Further, the objective function values across different slots are correlated via the switching costs $\beta_s[x(t) - x(t-1)]^+$ and $\beta_g[y(t) - y(t-1)]^+$, and thus cannot be decomposed. Finally, to obtain an online solution we do not even know the inputs beyond current slot.

Next, we introduce a proposition to simplify the structure of the problem. Note that if $(x(t))_{t=1}^T$ and $(y(t))_{t=1}^T$ are given, the problem in (2)-(7) reduces to a linear program and can be solved independently for each slot. We then obtain the following.

PROPOSITION 1. *Given any $x(t)$ and $y(t)$, the $u(t)$ and $v(t)$ that minimize the cost in (2) with any $g_t(x, a)$ that is increasing in x and a, are given by: $\forall t \in [1, T]$,*

$$u(t) = \begin{cases} 0, & \text{if } p(t) \le c_o, \\ \min\left(Ly(t), g_t(x(t), a(t))\right), & \text{otherwise}, \end{cases}$$

and

$$v(t) = g_t(x(t), a(t)) - u(t).$$

Note that $u(t), v(t)$ can be computed using *only* $x(t), y(t)$ at current time t, thus can be determined in an online fashion.

Intuitively, the above proposition says if the on-site energy price c_o is higher than the grid price $p(t)$, we should buy energy from the grid; otherwise, it is the best to buy the cheap on-site energy up to its maximum supply $L \cdot y(t)$ and the rest (if any) from the more expensive grid. With the above proposition, we can reduce the non-linear mixed-integer program in (2)-(7) with variables $\boldsymbol{x}$, $\boldsymbol{y}$, $\boldsymbol{u}$, and $\boldsymbol{v}$ to the following integer program with only variables $\boldsymbol{x}$ and $\boldsymbol{y}$:

DCM :

$$\min \quad \sum_{t=1}^T \left\{ \psi\left(y(t), p(t), d_t(x(t))\right) + \beta_s[x(t) - x(t-1)]^+ \right.$$
$$\left. + \beta_g[y(t) - y(t-1)]^+ \right\} \tag{8}$$
$$\text{s.t.} \quad x(t) \ge a(t),$$
$$(6), (7),$$
$$\text{var} \quad x(t), y(t) \in \mathbb{N}^0, \ t \in [1, T],$$

where $d_t(x(t)) \triangleq g_t(x(t), a(t))$, for the ease of presentation in later sections, is increasing and convex in $x(t)$ and $\psi(y(t), p(t), d_t(x(t)))$ replaces the term $v(t)p(t) + c_o u(t) + c_m y(t)$ in the original cost function in (2) and is defined as

$$\psi(y(t), p(t), d_t(x(t))) \tag{9}$$
$$\triangleq \begin{cases} c_m y(t) + p(t) d_t(x(t)), & \text{if } p(t) \le c_o, \\ c_m y(t) + c_o Ly(t) + & \text{if } p(t) > c_o \text{ and} \\ \quad p(t)\left(d_t(x(t)) - Ly(t)\right), & d_t(x(t)) > Ly(t), \\ c_m y(t) + c_o d_t(x(t)), & \text{else.} \end{cases}$$

As a result of the analysis above, it suffices to solve the above formulation of **DCM** with only variables $\boldsymbol{x}$ and $\boldsymbol{y}$, in order to minimize the data center operating cost.

2.3 An Offline Optimal Algorithm

We present an offline optimal algorithm for solving problem **DCM** using Dijkstra's shortest path algorithm [15]. We construct a graph $G = (V, E)$, where each vertex denoted by the tuple $\langle x, y, t \rangle$ represents a state of the data center where there are x active servers, and y active generators at time t. We draw a directed edge from each vertex $\langle x(t-1), y(t-1), t-1 \rangle$ to each possible vertex $\langle x(t), y(t), t \rangle$ to represent the fact that the data center can transit from the first state to the second state. Further, we associate the cost of that transition shown below as the weight of the edge:

$$\psi(y(t), p(t), d_t(x(t))) + \beta_s[x(t) - x(t-1)]^+$$
$$+ \beta_g[y(t) - y(t-1)]^+.$$

Next, we find the minimum weighted path from the initial state represented by vertex $\langle 0, 0, 0 \rangle$ to the final state represented by vertex $\langle 0, 0, T+1 \rangle$ by running Dijkstra's algorithm on graph G. Since the weights represent the transition costs, it is clear that finding the minimum weighted path in G is equivalent to minimizing the total transitional costs. Thus, our offline algorithm provides an optimal solution for problem **DCM**.

THEOREM 1. *The algorithm described above finds an optimal solution to problem **DCM** in time $O\left(M^2 N^2 T \log(MNT)\right)$, where T is the number of slots, N the number of generators and $M = \max_{1 \le t \le T} \lceil a(t) \rceil$.*

PROOF. Since the numbers of active servers and generators are at most M and N, respectively, and there are $T + 2$ time slots, graph G has $O(MNT)$ vertices and $O(M^2 N^2 T)$ edges. Thus, the run time of Dijkstra's algorithm on graph G is $O\left(M^2 N^2 T \log(MNT)\right)$. $\square$

Remark: In practice, the time-varying input sequences ($p(t)$, $a(t)$ and g_t) may not be available in advance and hence it may be difficult to apply the above offline algorithm. However, an offline optimal algorithm can serve as a benchmark, using which we can evaluate the performance of online algorithms.

3. THE BENEFIT OF JOINT OPTIMIZATION

Data center cost minimization (**DCM**) entails the joint optimization of both server capacity that determines the energy demand and on-site power generation that determines the energy supply. Now consider the situation where the data center optimizes the energy demand and supply separately.

First, the data center dynamically provisions the server capacity according to the grid power price $p(t)$. More formally, it solves the *capacity provisioning* problem which we refer to as **CP** below.

$$\textbf{CP}: \quad \min \quad \sum_{t=1}^{T} \left\{ p(t) \cdot d_t(x(t)) + \beta_s [x(t) - x(t-1)]^+ \right\}$$
$$\text{s.t.} \quad x(t) \geq a(t),$$
$$x(0) = 0,$$
$$\text{var} \quad x(t) \in \mathbb{N}^0, \ t \in [1, T].$$

Solving problem **CP** yields $\bar{x}$. Thus, the total power demand at time t given $\bar{x}(t)$ is $d_t(\bar{x}(t))$. Note that $d_t(\bar{x}(t))$ is not just server power consumption, but also includes consumption of power conditioning and cooling systems, as described in Sec. 2.2.

Second, the data center minimizes the cost of satisfying the power demand due to $d_t(\bar{x}(t))$, using both the grid and the on-site generators. Specifically, it solves the *energy provisioning* problem which we refer to as **EP** below.

$$\textbf{EP}:$$
$$\min \quad \sum_{t=1}^{T} \left\{ \psi\left(y(t), p(t), d_t(\bar{x}(t))\right) + \beta_g [y(t) - y(t-1)]^+ \right\}$$
$$y(0) = 0,$$
$$\text{var} \quad y(t) \in \mathbb{N}^0, \ t \in [1, T].$$

Let $(\bar{x}, \bar{y})$ be the solution obtained by solving **CP** and **EP** separately in sequence and (x^*, y^*) be the solution obtained by solving the joint-optimization **DCM**. Further, let $C_{\text{DCM}}(x, y)$ be the value of the data center's total cost for solution (x, y), including both generator and server costs as represented by the objective function (8) of problem **DCM**. The additional benefit of joint optimization over optimizing independently is simply the relationship between $C_{\text{DCM}}(\bar{x}, \bar{y})$ and $C_{\text{DCM}}(x^*, y^*)$. It is clear that $(\bar{x}, \bar{y})$ obeys all the constraints of **DCM** and hence is a feasible solution of **DCM**. Thus, $C_{\text{DCM}}(x^*, y^*) \leq C_{\text{DCM}}(\bar{x}, \bar{y})$. We can measure the factor loss in optimality ρ due to optimizing separately as opposed to optimizing jointly on the worst-case input as follows:

$$\rho \triangleq \max_{\text{all inputs}} \frac{C_{\text{DCM}}(\bar{x}, \bar{y})}{C_{\text{DCM}}(x^*, y^*)}.$$

The following theorem characterizes the benefit of joint optimization over optimizing independently.

THEOREM 2. *The factor loss in optimality ρ by solving the problem **CP** and **EP** in sequence as opposed to optimizing jointly is given by $\rho = LP_{\max}/(Lc_o + c_m)$ and it is tight.*

The above theorem guarantees that for *any* time duration T, *any* workload a, *any* grid price p and *any* function $g_t(x, a)$ as long as it is increasing and convex in x and a, solving problem **DCM** by first solving **CP** then solving **EP** in sequence yields a solution that is within a factor $LP_{\max}/(Lc_o + c_m)$ of solving **DCM** directly. Further, the ratio is tight in that there exists an input to **DCM** where the ratio $C_{\text{DCM}}(\bar{x}, \bar{y})/C_{\text{DCM}}(x^*, y^*)$ equals $LP_{\max}/(Lc_o + c_m)$.

The theorem shows in a quantitative way that a larger price discrepancy between the maximum grid price and the on-site power yields a larger gain by optimizing the energy provisioning and capacity provisioning jointly. Over the

	Cooling & Power Conditioning	Optimization Type	Competitive Ratio
LCP [23]	No	obj: convex var: continuous	3
CSR [27]	No	obj: linear var: integer	$2 - \alpha_s$
GCSR this work	Yes	obj: convex and increasing var: integer	$2 - \alpha_s$

Note that α_s is the normalized look-ahead window size, whose representations are different under the different settings of [27] and our work.

Table 3: Comparison of the algorithm GCSR proposed in this paper, CSR in [27], and LCP in [23].

past decade, utilities have been exposing a greater level of grid price variation to their customers with mechanisms such as time-of-use pricing where grid prices are much more expensive during peak hours than during the off-peak periods. This likely leads to larger price discrepancy between the grid and the on-site power. In that case, our result implies that a joint optimization of power and server resources is likely to yield more benefits to a hybrid data center.

Besides characterizing the benefit of jointly optimizing power and server resources, the decomposition of problem **DCM** into problems **CP** and **EP** provides a key approach for our online algorithm design. Problem **DCM** has an objective function with mutually-dependent coupled variables x and y indicating the server and generator states, respectively. This coupling (specifically through the function $\psi(y(t), p(t), d_t(x(t)))$) makes it difficult to design provably good online algorithms. However, instead of solving problem **DCM** directly, we devise online algorithms to solve problems **CP** that involves only server variable x and **EP** that involves only the generator variables y. Combining the online algorithms for **CP** and **EP** respectively yields the desired online algorithm for **DCM**.

4. ONLINE ALGORITHMS FOR ON-GRID DATA CENTERS

We first develop an online algorithm for **DCM** for an *on-grid* data center, where there is no on-site power generation, a scenario that captures most data centers today. Since *on-grid* data center has no on-site power generation, solving **DCM** for it reduces to solving problem **CP** described in Sec. 3.

Problems of this kind have been studied in the literature (see *e.g.*, [23, 27]). The difference of our work from [23, 27] is as follows (also summarized in Table 3). From the modelling aspect, we explicitly take into account power consumption of both cooling and power conditioning systems, in addition to servers. From the formulation aspect, we are solving a different optimization problem, *i.e.*, an integer program with convex and increasing objective function. From the theoretical result aspects, we achieve a small competitive ratio of $2 - \alpha_s$, which quickly decreases to 1 as look-ahead window w increase.

Recall that **CP** takes as input the workload a, the grid price p and the time-dependent function g_t, $\forall t$ and outputs the number of active servers x. We construct solutions to **CP** in a divide-and-conquer fashion. We will first de-

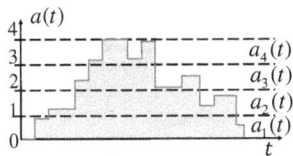

Figure 2: An example of how workload a is decomposed into 4 sub-demands.

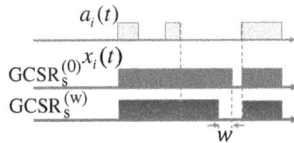

Figure 3: An example of $a_i(t)$ and corresponding solution obtained by $\mathbf{GCSR_s^{(w)}}$.

compose the demand a into sub-demands and define corresponding sub-problem for each server, and then solve capacity provisioning *separately* for each sub-problem. Note that the key is to correctly decompose the demand and define the subproblems so that the combined solution is still optimal. More specifically, we slice the demand as follows: for $1 \leq i \leq M = \max_{1 \leq t \leq T} \lceil a(t) \rceil$, $1 \leq t \leq T$,

$$a_i(t) \triangleq \min \left\{ 1, \max \left\{ 0, a(t) - (i-1) \right\} \right\}.$$

And the corresponding sub-problem $\mathbf{CP_i}$ is defined as follows.

$$\mathbf{CP_i}: \quad \min \quad \sum_{t=1}^{T} \left\{ p(t) \cdot d_t^i \cdot x_i(t) + \beta_s [x_i(t) - x_i(t-1)]^+ \right\}$$
$$\text{s.t.} \quad x_i(t) \geq a_i(t),$$
$$x_i(0) = 0,$$
$$\text{var} \quad x_i(t) \in \{0, 1\}, \ t \in [1, T],$$

where $x_i(t)$ indicates whether the i-th server is on at time t and $d_t^i \triangleq d_t(i) - d_t(i-1)$. d_t^i can be interpreted as the power consumption due to the i-th server at t.

Problem $\mathbf{CP_i}$ solves the capacity provisioning problem with inputs workload a_i, grid price p and d_t^i. The key reason for our decomposition is that $\mathbf{CP_i}$ is easier to solve, since a_i take values in $[0, 1]$ and exactly one server is required to serve each a_i. Generally speaking, a divide-and-conquer manner may suffer from optimality loss. Surprisingly, as the following theorem states, the individual optimal solutions for problems $\mathbf{CP_i}$ can be put together to form an optimal solution to the original problem $\mathbf{CP}$. Denote $C_{\mathrm{CP_i}}(\boldsymbol{x_i})$ as the cost of solution $\boldsymbol{x_i}$ for problem $\mathbf{CP_i}$ and $C_{\mathrm{CP}}(\boldsymbol{x})$ the cost of solution $\boldsymbol{x}$ for problem $\mathbf{CP}$.

THEOREM 3. *Consider problem* $\mathbf{CP}$ *with any* $d_t(x(t)) = g_t(x(t), a(t))$ *that is convex in* $x(t)$. *Let* $\bar{\boldsymbol{x}}_i$ *be an optimal solution and* $\boldsymbol{x}_i^{on}$ *an online solution for problem* $\mathbf{CP_i}$ *with workload* $\boldsymbol{a}_i$, *then* $\sum_{i=1}^{M} \bar{\boldsymbol{x}}_i$ *is an optimal solution for* $\mathbf{CP}$ *with workload* $\boldsymbol{a}$. *Furthermore, if* $\forall \boldsymbol{a}_i, i$, *we have* $C_{\mathrm{CP_i}}(\boldsymbol{x}_i^{on}) \leq \gamma \cdot C_{\mathrm{CP_i}}(\bar{\boldsymbol{x}}_i)$ *for a constant* $\gamma \geq 1$, *then* $C_{\mathrm{CP}}(\sum_{i=1}^{M} \boldsymbol{x}_i^{on}) \leq \gamma \cdot C_{\mathrm{CP}}(\sum_{i=1}^{M} \bar{\boldsymbol{x}}_i)$, $\forall \boldsymbol{a}$.

Thus, it remains to design algorithms for each $\mathbf{CP_i}$. To solve $\mathbf{CP_i}$ in an online fashion one need only orchestrate one server to satisfy the workload a_i and minimize the total cost. When $a_i(t) > 0$, we must keep the server active to satisfy the workload. The challenging part is what we should do if the server is already active but $a_i(t) = 0$. Should we turn off the server immediately or keep it idling for some time? Should we distinguish the scenarios when the grid price is high versus low?

Inspired by "ski-rental" [12] and [27], we solve $\mathbf{CP_i}$ by the following "break-even" idea. During the idle period, *i.e.*,

---

Algorithm 1 $\mathbf{GCSR_s^{(w)}}$ for problem $\mathbf{CP_i}$

1: $C_i = 0, x_i(0) = 0$
2: **at** current time t, **do**
3: Set $\tau' \leftarrow \min\{t' \in [t, t+w] \mid C_i + \sum_{\tau=t}^{t'} p(\tau) d_\tau^i \geq \beta_s\}$
4: **if** $a_i(t) > 0$ **then**
5: $\quad x_i(t) = 1$ and $C_i = 0$
6: **else if** $\tau' = $ NULL or $\exists \tau \in [t, \tau'], a_i(\tau) > 0$ **then**
7: $\quad x_i(t) = x_i(t-1)$ and $C_i = C_i + p(t) d_t^i x_i(t)$
8: **else**
9: $\quad x_i(t) = 0$ and $C_i = 0$
10: **end if**

$a_i(t) = 0$, we accumulate an "idling cost" and when it reaches β_s, we turn off the server; otherwise, we keep the server idling. Specifically, our online algorithm $\mathbf{GCSR_s^{(w)}}$ (Generalized Collective Server Rental) for $\mathbf{CP_i}$ has a look-ahead window w. At time t, if there exist $\tau' \in [t, t+w]$ such that the idling cost till τ' is at least β_s, we turn off the server; otherwise, we keep it idling. More formally, we have Algorithm 1 and its competitive analysis in Theorem 4. A simple example of $\mathbf{GCSR_s^{(w)}}$ is shown in Fig. 3.

Our online algorithm for $\mathbf{CP}$, denoted as $\mathbf{GCSR^{(w)}}$, first employs $\mathbf{GCSR_s^{(w)}}$ to solve each $\mathbf{CP_i}$ on workload $\boldsymbol{a}_i$, $1 \leq i \leq M$, in an online fashion to produce output $\boldsymbol{x}_i^{on}$ and then simply outputs $\sum_{i=1}^{M} \boldsymbol{x}_i^{on} = \boldsymbol{x}^{on}$ as the output for the original problem $\mathbf{CP}$.

THEOREM 4. $\mathbf{GCSR_s^{(w)}}$ *achieves a competitive ratio of* $2 - \alpha_s$ *for* $\mathbf{CP_i}$, *where* $\alpha_s \triangleq \min(1, w d_{\min} P_{\min} / \beta_s) \in [0, 1]$ *is a "normalized" look-ahead window size and* $d_{\min} \triangleq \min_t \{d_t(1) - d_t(0)\}$. *Hence, according to Theorem 3,* $\mathbf{GCSR^{(w)}}$ *achieves the same competitive ratio for* $\mathbf{CP}$. *Further, no deterministic online algorithm with a look-ahead window w can achieve a smaller competitive ratio.*

A consequence of Theorem 4 is that when the look-ahead window size w reaches a break-even interval $\Delta_s \triangleq \beta_s / (d_{\min} P_{\min})$, our online algorithm has a competitive ratio of 1. That is, having a look-ahead window larger than Δ_s will not decrease the cost any further.

5. ONLINE ALGORITHMS FOR HYBRID DATA CENTERS

Unlike on-grid data centers, hybrid data centers have on-site power generation and therefore have to solve both capacity provisioning ($\mathbf{CP}$) and energy provisioning ($\mathbf{EP}$) to solve the data center cost minimization ($\mathbf{DCM}$) problem. We design an online algorithm that we call $\mathbf{DCMON}$ solving $\mathbf{DCM}$ as follows.

1. Run algorithm $\mathbf{GCSR}$ from Sec. 4 to solve $\mathbf{CP}$ that takes workload $\boldsymbol{a}$, grid price $\boldsymbol{p}$ and time-dependent function g_t, $\forall t$ as input and produces the number of active servers $\boldsymbol{x}^{on}$.

2. Run algorithm $\mathbf{CHASE}$ described in Section 5.2 below to solve $\mathbf{EP}$ that takes the energy demand $d_t(x^{on}(t)) = g_t(x^{on}(t), a(t))$ and grid price $p(t)$, $\forall t$ as input and decides when to turn on/off on-site generators and how much power to draw from the generators and the grid. Note that a similar problem has been studied in the

microgrid scenarios for energy generation scheduling in our previous work [26]. In this paper, we adapt algorithm **CHASE** developed in [26] to our data center scenarios to solve **EP** in an online fashion.

For the sake of completeness, we first briefly present the design behind **CHASE** in Sec. 5.1 and the algorithm and its intuitions in Sec. 5.2. Then we present the combined algorithm **DCMON** in Sec. 5.3.

5.1 A useful structure of an offline optimal solution of EP

We first reveal an elegant structure of an offline optimal solution and then exploit this structure in the design of our online algorithm **CHASE**.

5.1.1 Decompose EP into sub-problems EP_is

For the ease of presentation, we denote $e(t) = d_t(x^{on}(t))$. Similar as the decomposition of workload when solving **CP**, we decompose the energy demand e into N sub-demands and define sub-problem for each generator, then solve energy provisioning *separately* for each sub-problem, where N is the number of on-site generators. Specifically, for $1 \leq i \leq N$, $1 \leq t \leq T$,

$$e_i(t) \triangleq \min\{L, \max\{0, e(t) - (i-1)L\}\}.$$

The corresponding sub-problem **EP$_i$** is in the same form as **EP** except that $d_t(\bar{x}(t))$ is replaced by $e_i(t)$ and $y(t)$ is replaced by $y_i(t) \in \{0, 1\}$. Using this decomposition, we can solve **EP** on input e by simultaneously solving simpler problems **EP$_i$** on input e_i that only involve a single generator. Theorem 5 shows that the decomposition incurs no optimality loss. Denote $C_{EP_i}(y_i)$ as the cost of solution y_i for problem **EP$_i$** and $C_{EP}(y)$ the cost of solution y for problem **EP**.

THEOREM 5. *Let $\bar{y}_i$ be an optimal solution and y_i^{on} an online solution for* **EP$_i$** *with energy demand e_i, then $\sum_{i=1}^{N} \bar{y}_i$ is an optimal solution for* **EP** *with energy demand e. Furthermore, if $\forall e_i, i$, we have $C_{EP_i}(y_i^{on}) \leq \gamma \cdot C_{EP_i}(\bar{y}_i)$ for a constant $\gamma \geq 1$, then $C_{EP}(\sum_{i=1}^{N} y_i^{on}) \leq \gamma \cdot C_{EP}(\sum_{i=1}^{N} \bar{y}_i), \forall e$.*

5.1.2 Solve each sub-problem EP_i

Based on Theorem 5, it remains to design algorithms for each **EP$_i$**. Define

$$r_i(t) = \psi(0, p(t), e_i(t)) - \psi(1, p(t), e_i(t)). \quad (10)$$

$r_i(t)$ can be interpreted as the one-slot cost difference between not using and using on-site generation. Intuitively, if $r_i(t) > 0$ (resp. $r_i(t) < 0$), it will be desirable to turn on (resp. off) the generator. However, due to the startup cost, we should not turn on and off the generator too frequently. Instead, we should evaluate whether the *cumulative* gain or loss in the future can offset the startup cost. This intuition motivates us to define the following cumulative cost difference $R_i(t)$. We set initial values as $R_i(0) = -\beta_g$ and define $R_i(t)$ inductively:

$$R_i(t) \triangleq \min\{0, \max\{-\beta_g, R_i(t-1) + r_i(t)\}\}, \quad (11)$$

Note that $R_i(t)$ is only within the range $[-\beta_g, 0]$. An important feature of $R_i(t)$ useful later in online algorithm design is that it can be computed given the past and current inputs. An illustrating example of $R_i(t)$ is shown in Fig. 4.

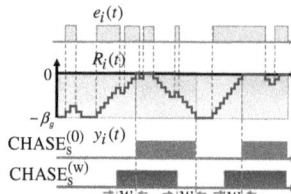

Figure 4: An example of $e_i(t)$, $R_i(t)$ and the corresponding solution obtained by **CHASE$_s^{(w)}$** for **EP$_i$**.

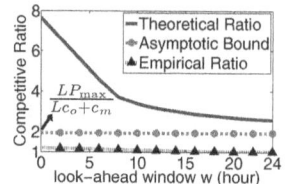

Figure 5: Theoretical and empirical ratios of algorithm **DCMON$^{(w)}$** vs. look-ahead window size w.

Intuitively, when $R_i(t)$ hits its boundary 0, the cost difference between not using and using on-site generation within a certain period is at least β_g, which can offset the startup cost. Thus, it makes sense to turn on the generator. Similarly, when $R_i(t)$ hits $-\beta_g$, it may be better to turn off the generator and use the grid. The following theorem formalizes this intuition, and shows an optimal solution $\bar{y}_i(t)$ for problem **EP$_i$** at the time epoch when $R_i(t)$ hits its boundary values $-\beta_g$ or 0.

THEOREM 6. *There exists an offline optimal solution for problem* **EP$_i$** *, denoted by $\bar{y}_i(t)$, $1 \leq t \leq T$, so that:*

- *if $R_i(t) = -\beta_g$, then $\bar{y}_i(t) = 0$;*

- *if $R_i(t) = 0$, then $\bar{y}_i(t) = 1$.*

5.2 Online algorithm CHASE

Our online algorithm **CHASE$_s^{(w)}$** with look-ahead window w exploits the insights revealed in Theorem 6 to solve **EP$_i$**. The idea behind **CHASE$_s^{(w)}$** is to track the offline optimal in an online fashion. In particular, at time 0, $R_i(0) = -\beta_g$ and we set $y_i(t) = 0$. We keep tracking the value of $R_i(t)$ at every time slot within the look-ahead window. Once we observe that $R_i(t)$ hits values $-\beta_g$ or 0, we set the $y_i(t)$ to the optimal solution as Theorem 6 reveals; otherwise, keep $y_i(t) = y_i(t-1)$ unchanged. More formally, we have Algorithm 2 and its competitive analysis in Theorem 7. An example of **CHASE$_s^{(w)}$** is shown in Fig. 4.

The online algorithm for **EP**, denoted as **CHASE$^{(w)}$**, first employs **CHASE$_s^{(w)}$** to solve each **EP$_i$** on energy demand e_i, $1 \leq i \leq N$, in an online fashion to produce output y_i^{on} and then simply outputs $\sum_{i=1}^{N} y_i^{on}$ as the output for the original problem **EP**.

Algorithm 2 CHASE$_s^{(w)}$ for problem EP$_i$

1: **at** current time t, **do**
2: Obtain $(R_i(\tau))_{\tau=t}^{t+w}$
3: Set $\tau' \leftarrow \min\{\tau \in [t, t+w] \mid R_i(\tau) = 0 \text{ or } -\beta_g\}$
4: **if** $\tau' = $ NULL **then**
5:     $y_i(t) = y_i(t-1)$
6: **else if** $R_i(\tau') = 0$ **then**
7:     $y_i(t) = 1$
8: **else**
9:     $y_i(t) = 0$
10: **end if**

THEOREM 7. **CHASE$_s^{(w)}$** *for problem* **EP$_i$** *with a look-*

ahead window w has a competitive ratio of

$$1 + \frac{2\beta_g \left(LP_{\max} - Lc_o - c_m\right)}{\beta_g LP_{\max} + wc_m P_{\max} \left(L - \frac{c_m}{P_{\max} - c_o}\right)}.$$

Hence, according to Theorem 5, $\mathbf{CHASE}^{(\mathbf{w})}$ achieves the same competitive ratio for problem $\mathbf{EP}$.

5.3 Combining GCSR and CHASE

Our algorithm $\mathbf{DCMON}^{(\mathbf{w})}$ for solving problem $\mathbf{DCM}$ with a look-ahead window of $w \geq 0$, i.e., knowing grid prices $p(\tau)$, workload $a(\tau)$ and the function $g_\tau, 1 \leq \tau \leq t + w$, at time t, first uses $\mathbf{GCSR}$ from Sec. 4 to solve problem $\mathbf{CP}$ and then uses $\mathbf{CHASE}$ in Sec. 5.2 to solve problem $\mathbf{EP}$. An important observation is that the available look-ahead window size for $\mathbf{GCSR}$ to solve $\mathbf{CP}$ is w, i.e., knows $p(\tau)$, $a(\tau)$ and g_τ, $1 \leq \tau \leq t + w$, at time t; however, the available look-ahead window size for $\mathbf{CHASE}$ to solve $\mathbf{EP}$ is only $[w - \Delta_s]^+$, i.e., knows $p(\tau)$ and $e(\tau) = d_\tau(x^{on}(\tau))$, $1 \leq \tau \leq t + [w - \Delta_s]^+$, at time t (Δ_s is the break-even interval defined in Sec. 4). Detailed explanation on this is relegated to our technical report [39].

Thus, a bound on the competitive ratio of $\mathbf{DCMON}^{(\mathbf{w})}$ is the product of competitive ratios for $\mathbf{GCSR}^{(\mathbf{w})}$ and $\mathbf{CHASE}^{([\mathbf{w}-\mathbf{\Delta_s}]^+)}$ from Theorems 4 and 7, respectively, and the optimality loss ratio $LP_{\max} / (Lc_o + c_m)$ due to the offline-decomposition stated in Sec. 3, which is given in the following Theorem.

THEOREM 8. $\mathbf{DCMON}^{(\mathbf{w})}$ for problem $\mathbf{DCM}$ has a competitive ratio of

$$\frac{P_{\max} (2 - \alpha_s)}{c_o + c_m/L} \left[1 + \frac{2 \left(LP_{\max} - Lc_o - c_m\right)}{LP_{\max} + \alpha_g P_{\max} \left(L - \frac{c_m}{P_{\max} - c_o}\right)} \right]. \tag{12}$$

The ratio is also upper-bounded by

$$\frac{P_{\max} (2 - \alpha_s)}{c_o + c_m/L} \left[1 + 2\frac{P_{\max} - c_o}{P_{\max}} \cdot \frac{1}{1 + \alpha_g} \right],$$

where $\alpha_s = \min(1, w/\Delta_s) \in [0, 1]$ and $\alpha_g \triangleq \frac{c_m}{\beta_g} [w - \Delta_s]^+$ $\in [0, +\infty)$ are "normalized" look-ahead window sizes.

As the look-ahead window size w increases, the competitive ratio in Theorem 8 decreases to $LP_{\max} / (Lc_o + c_m)$ (c.f. Fig. 5), the inherent approximation ratio introduced by our offline decomposition approach discussed in Section 3. However, the real trace based empirical performance of $\mathbf{DCMON}^{(\mathbf{w})}$ without look-ahead is already close to the offline optimal, i.e., ratio close to 1 (c.f. Fig. 5).

6. EMPIRICAL EVALUATION

We evaluate the performance of our algorithms by simulations based on real-world traces with the aim of (i) corroborating the empirical performance of our online algorithms under various realistic settings and the impact of having look-ahead information, (ii) understanding the benefit of opportunistically procuring energy from both on-site generators and the grid, as compared to the current practice of purchasing from the grid alone, (iii) studying how much on-site energy is needed for substantial cost benefits.

6.1 Parameters and Settings

Workload trace: We use the workload traces from the Akamai network [1, 30] that is the currently the world's largest content delivery network. The traces measure the workload of Akamai servers serving web content to actual end-users. Note that our workload is of the "request-and-response" type that we model in our paper. We use traces from the Akamai servers deployed in the New York and San Jose data centers that record the hourly average load served by each deployed server over 22 days from Dec. 21, 2008 to Jan. 11, 2009. The New York trace represents 2.5K servers that served about 1.4×10^{10} requests and 1.7×10^{13} bytes of content to end-users during our measurement period. The San Jose trace represents 1.5K servers that served about 5.5×10^9 requests and 8×10^{12} bytes of content. We show the workload in Fig. 6, in which we normalize the load by the server's service capacity. The workload is quite characteristic in that it shows daily variations (peak versus off-peak) and weekly variations (weekday versus weekend).

Grid price: We use traces of hourly grid power prices in New York [2] and San Jose [3] for the same time period, so that it can be matched up with the workload traces (c.f. Fig. 6). Both workload and grid price traces show strong diurnal properties: in the daytime, the workload and the grid price are relatively high; at night, on the contrary, both are low. This indicates the feasibility of reducing the data center cost by using the energy from the on-site generators during the daytime and use the grid at night.

Server model: As mentioned in Sec. 2, we assume the data center has a sufficient number of homogeneous servers to serve the incoming workload at any given time. Similar to a typical setting in [32], we use the standard linear server power consumption model. We assume that each server consumes 0.25KWh power per hour at full capacity and has a power proportional factor (PPF=$(c_{peak} - c_{idle})/c_{peak}$) of 0.6, which gives us $c_{idle} = 0.1KW$, $c_{peak} = 0.25KW$. In addition, we assume the server switching cost equals the energy cost of running a server for 3 hours. If we assume an average grid price as the price of energy, we get about $\beta_s = \$0.08$.

Cooling and power conditioning system model: We consider a water chiller cooling system. According to [5], during this 22-day winter period the average high and low temperatures of New York are $41°F$ and $29°F$, respectively. Those of San Jose are $58°F$ and $41°F$, respectively. Without loss of generality, we take the high temperature as the daytime temperature and the low temperature as the nighttime temperature. Thus, according to [33], the power consumed by water chiller cooling systems of the New York and San Jose data centers are about

$$f_{c,NY}^t(b) = \begin{cases} (0.041b^2 + 0.144b + 0.047)b_{\max}, & \text{at daytime,} \\ (0.03b^2 + 0.136b + 0.042)b_{\max}, & \text{at nighttime,} \end{cases}$$

and

$$f_{c,SJ}^t(b) = \begin{cases} (0.06b^2 + 0.16b + 0.054)b_{\max}, & \text{at daytime,} \\ (0.041b^2 + 0.144b + 0.047)b_{\max}, & \text{at nighttime,} \end{cases}$$

where $b_{\max}$ is the maximum server power consumption and b is the server power consumption normalized by $b_{\max}$. The maximum server power consumption of the New York and San Jose data centers are $b_{\max}^{NY} = 2500 \times 0.25 = 625KW$ and $b_{\max}^{SJ} = 1500 \times 0.25 = 375KW$. Besides, the power con-

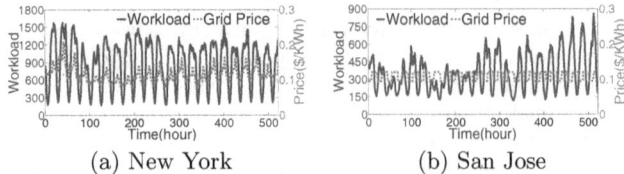

(a) New York   (b) San Jose

Figure 6: Real-world workload from Akamai and the grid power price.

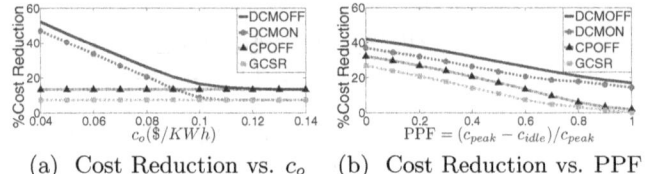

(a) Cost Reduction vs. c_o (b) Cost Reduction vs. PPF

Figure 7: Variation of cost reduction with model parameters.

sumed by the power conditioning system, including PDUs and UPSs, is $f_p(b) = (0.012b^2 + 0.046b + 0.056)b_{\max}$ [33].

Generator model: We adopt generators with specifications the same as the one in [4]. The maximum output of the generator is 60KW, *i.e.*, $L = 60KW$. The incremental cost to generate an additional unit of energy c_o is set to be \$0.08/K-Wh, which is calculated according to the gas price [2] and the generator efficiency [4]. Similar to [37], we set the sunk cost of running the generator for unit time $c_m = \$1.2$ and the startup cost β_g equivalent to the amortized capital cost, which gives $\beta_g = \$24$. Besides, we assume the number of generators $N = 10$, which is enough to satisfy all the energy demand for this trace and model we use.

Cost benchmark: Current data centers usually do not use dynamic capacity provisioning and on-site generators. Thus, we use the cost incurred by static capacity provisioning with grid power as the benchmark using which we evaluate the cost reduction due to our algorithms. Static capacity provisioning runs a fixed number of servers at all times to serve the workload, without dynamically turning on/off the servers. For our benchmark, we assume that the data center has complete workload information ahead of time and provisions exactly to satisfy the peak workload and uses only grid power. Using such a benchmark gives us a conservative evaluation of the cost saving from our algorithms.

Comparisons of Algorithms: We compare four algorithms: our online and offline optimal algorithms in on-grid scenarios, *i.e.*, **GCSR** and **CPOFF**, and hybrid scenarios, *i.e.*, **DCMON** and **DCMOFF**.

6.2 Impact of Model Parameters on Cost Reduction

We study the cost reduction provided by our offline and online algorithms for both on-grid and hybrid data centers using the New York trace unless specified otherwise. We assume no look-ahead information is available when running the online algorithms. We compute the cost reduction (in percentage) as compared to the cost benchmark which we described earlier. When all parameters take their default values, our offline (resp. online) algorithms provide up to 12.3% (resp., 7.3%) cost reduction for on-grid and 25.8% (resp., 20.7%) cost reduction for hybrid data centers (c.f. Fig. 7. The default value of c_o is \$0.08/KWh.). Note that the online algorithms provide cost reduction that are 5% smaller than offline algorithms on account of their lack of knowledge of future inputs. Further, note that cost reduction of a hybrid data center is larger than that of a on-grid data center, since hybrid data center has the ability to generate energy on-site to avoid higher grid prices. Nevertheless, the extent of cost reduction in all cases is high providing strong evidence for the need to perform energy and server capacity optimizations.

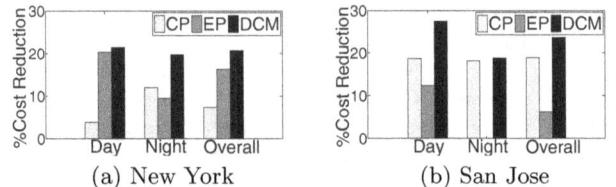

(a) New York   (b) San Jose

Figure 8: Relative values of CP, EP, and DCM.

Data centers may deploy different types of servers and generators with different model parameters. It is then important to understand the impact on cost reduction due to these parameters. We first study the impact of varying c_o (c.f. Fig. 7). For a hybrid data center, as c_o increases the cost of on-site generation increases making it less effective for cost reduction (c.f Fig. 7a). For the same reason, the cost reduction of a hybrid data center tends to that of the on-grid data center with increasing c_o as on-site generation becomes less economical.

We then study the impact of power proportional factor (PPF). More specifically, we fix $c_{peak} = 0.25KW$, and vary PPF from 0 to 1 (c.f. Fig. 7b). As PPF increases, the server idle power decreases, thus dynamic provisioning has lesser impact on the cost reduction. This explains why **CP** achieves no cost reduction when PPF=1. Since **DCM** also solves **CP** problem, its performance degrades with increasing PPF as well.

6.3 The Relative Value of Energy versus Capacity Provisioning

In this subsection, we use both New York and San Jose traces. For a hybrid data center, we ask which optimization provides a larger cost reduction: energy provisioning (**EP**) or server capacity provisioning (**CP**) in comparison with the joint optimization of doing both (**DCM**). The cost reductions of different optimization are shown in Fig. 8.

For the New York scenario in Fig. 8a, overall, we see that **EP**, **CP**, and **DCM** provide cost reductions of 16.3%, 7.3%, and 20.7%, respectively. However, note that during the day doing **EP** alone provides almost as much cost reduction as the joint optimization **DCM**. The reason is that during the high traffic hours in the day, solving **EP** to avoid higher grid prices provides a larger benefit than optimizing the energy consumption by server shutdown. The opposite is true during the night where **CP** is more critical than **EP**, since minimizing the energy consumption by shutting down idle servers yields more benefit.

For the San Jose scenario in Fig. 8b, overall, **EP**, **CP**, and **DCM** provide cost reductions of 6.1%, 19%, and 23.7%, respectively. Compared to the New York scenario, the reason why **EP** achieves so little cost reduction is that the grid power is cheaper and thus on-site generation is not that eco-

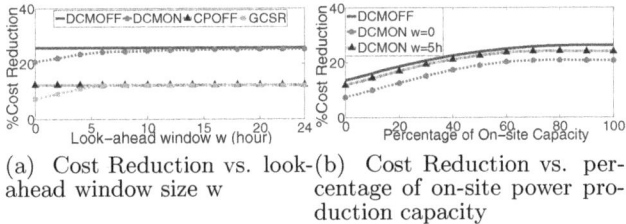

(a) Cost Reduction vs. look-ahead window size w (b) Cost Reduction vs. percentage of on-site power production capacity

Figure 9: Variation of cost reduction with look-ahead and on-site capacity.

nomical. Meanwhile, **CP** performs closer to **DCM**, which is because the workload curve is highly skew (shown in Fig. 6b) and dynamic provisioning for the server capacity saves a lot of server idling cost as well as cooling and power conditioning cost.

In a nutshell, **EP** favors high grid power price while workload with less regular pattern makes **CP** more competitive.

6.4 Benefit of Looking Ahead

We evaluate the cost reduction benefit of increasing the look-ahead window. From Fig. 9a, we observe that while the performance of our online algorithms are already good when there is no look-ahead information, they quickly improve to the offline optimal when a small amount of look-ahead, *e.g.*, 6 hours, is available, indicating the value of short-term prediction of inputs. Note that while the competitive ratio analysis in Theorem 8 is for the worst case inputs, our online algorithms perform much closer to the offline optimal for realistic inputs.

6.5 How Much On-site Power Production is Enough

Thus far, in our experiments, we assumed that a hybrid data center had the ability to supply all its energy from on-site power generation ($N = 10$). However, an important question is how much investment should a data center operator make in installing on-site generator capacity to obtain largest cost reduction.

More specifically, we vary the number of on-site generators N from 0 to 10 and show the corresponding performances of our algorithms. Interestingly, in Fig. 9b, our results show that provisioning on-site generators to produce 80% of the peak power demand of the data center is sufficient to obtain all of the cost reduction benefits. Further, with just 60% on-site power generation capacity we can achieve 95% of the maximum cost reduction. The intuitive reason is that most of time the demands of the data center are significantly lower than their peaks.

7. RELATED WORK

Our study is among a series of work on dynamic provisioning in data centers and power systems [38, 22, 35].

In particular, for the capacity provisioning problem, [23] and [27] propose online algorithms with performance guarantee to reduce servers operating cost under convex and linear mixed integer optimization scenarios, respectively. Different from these two, our work designs online algorithm under non-linear mixed integer optimization scenario and we take into account the operating cost of servers as well as power conditioning and cooling systems. [24, 40] also mod-

el cooling systems, but focus on offline optimization of the operating cost.

Energy provisioning for power systems is characterized by unit-commitment problem (UC) [8, 31], including a mixed-integer programming approach [29] approach and a stochastic control approach [36]. All these approaches assume the demand (or its distribution) in the entire horizon is known *a priori*, thus they are applicable only when future input information can be predicted with certain level of accuracy. In contrast, in this paper we consider an online setting where the algorithms may utilize only information in the current time slot.

In addition to the difference of our work and existing works in the two problems (*i.e.*, capacity provisioning and energy provisioning), our work is also unique in that we jointly optimize both problems while existing works focus on only one of them.

8. CONCLUSIONS

Our work focuses on the cost minimization of data centers achieved by jointly optimizing *both* the supply of energy from on-site power generators and the grid, and the demand for energy from its deployed servers as well as power conditioning and cooling systems. We show that such an integrated approach is not only possible in next-generation data centers but also desirable for achieving significant cost reductions. Our offline optimal algorithm and our online algorithms with provably good competitive ratios provide key ideas on how to coordinate energy procurement and production with the energy consumption. Our empirical work answers several of the important questions relevant to data center operators focusing on minimizing their operating costs. We show that a hybrid (resp., on-grid) data center can achieve a cost reduction between 20.7% to 25.8% (resp., 7.3% to 12.3%) by employing our joint optimization framework. We also show that on-site power generation can provide an additional cost reduction of about 13%, and that most of the additional benefit is obtained by a partial on-site generation capacity of 60% of the peak power requirement of the data center.

This work can be extended in several directions. First, it is interesting to study how energy storage devices can be used to further reduce the data center operating cost. Second, another interesting direction is to generalize our analysis to take into account deferable workloads. Third, extension from homogeneous servers and generators to heterogeneous setting is also of great interest.

9. ACKNOWLEDGMENTS

The work described in this paper was partially supported by China National 973 projects (No. 2012CB315904 and 2013CB336700), several grants from the University Grants Committee of the Hong Kong Special Administrative Region, China (Area of Excellence Project No. AoE/E-02/08 and General Research Fund Project No. 411010 and 411011), and two gift grants from Microsoft and Cisco.

10. REFERENCES

[1] Akamai tech. http://www.akamai.com.
[2] Nationalgrid. https://www.nationalgridus.com/.
[3] Pacific gas and electric company. http://www.pge.com/nots/rates/tariffs/rateinfo.shtml.
[4] Tecogen. http://www.tecogen.com.

[5] The weather channal. http://www.weather.com/.

[6] Apple's onsite renewable energy, 2012. http://www.apple.com/environment/renewable-energy/.

[7] Distributed generation, 2012. http://www.bloomenergy.com/fuel-cell/distributed-generation/.

[8] C. Baldwin, K. Dale, and R. Dittrich. A study of the economic shutdown of generating units in daily dispatch. *IEEE Trans. Power Apparatus and Systems*, 1959.

[9] L. Barroso and U. Holzle. The case for energy-proportional computing. *IEEE Computer*, 2007.

[10] A. Beloglazov, R. Buyya, Y. Lee, and A. Zomaya. A taxonomy and survey of energy-efficient data centers and cloud computing systems. *Advances in Computers*, 2011.

[11] A. Borbely and J. Kreider. *Distributed generation: the power paradigm for the new millennium*. CRC Press, 2001.

[12] A. Borodin and R. El-Yaniv. *Online computation and competitive analysis*. Cambridge University Press, 1998.

[13] J. Chase, D. Anderson, P. Thakar, A. Vahdat, and R. Doyle. Managing energy and server resources in hosting centers. In *Proc. ACM SIGOPS*, 2001.

[14] A.J. Conejo, M.A. Plazas, R. Espinola, and A.B. Molina. Day-ahead electricity price forecasting using the wavelet transform and arima models. *Power Systems, IEEE Transactions on*, 2005.

[15] E. Dijkstra. A note on two problems in connexion with graphs. *Numerische mathematik*, 1959.

[16] R. Doyle, J. Chase, O. Asad, W. Jin, and A. Vahdat. Model-based resource provisioning in a web service utility. In *Proc. USITS*, 2003.

[17] K. Fehrenbacher. ebay to build huge bloom energy fuel cell farm at data center. 2012. http://gigaom.com/cleantech/ebay-to-build-huge-bloom-energy-fuel-cell-farm-at-data-center/.

[18] K. Fehrenbacher. Is it time for more off-grid options for data centers?. 2012. http://gigaom.com/cleantech/is-it-time-for-more-off-grid-options-for-data-centers/.

[19] Daniel Gmach, Jerry Rolia, Ludmila Cherkasova, and Alfons Kemper. Workload analysis and demand prediction of enterprise data center applications. In *Workload Characterization, 2007. IISWC 2007. IEEE 10th International Symposium on*, 2007.

[20] S. Kazarlis, A. Bakirtzis, and V. Petridis. A genetic algorithm solution to the unit commitment problem. *IEEE Trans. Power Systems*, 1996.

[21] J. Koomey. Growth in data center electricity use 2005 to 2010. *Analytics Press*, 2010.

[22] M. Lin, Z. Liu, A. Wierman, and L. Andrew. Online algorithms for geographical load balancing. In *Proc. IEEE IGCC*, 2012.

[23] M. Lin, A. Wierman, L. Andrew, and E. Thereska. Dynamic right-sizing for power-proportional data centers. In *Proc. IEEE INFOCOM*, 2011.

[24] Z. Liu, Y. Chen, C. Bash, A. Wierman, D. Gmach,

Z. Wang, M. Marwah, and C. Hyser. Renewable and cooling aware workload management for sustainable data centers. In *Proc. ACM SIGMETRICS*, 2012.

[25] J. Lowesohn. Apple's main data center to go fully renewable this year. 2012. http://news.cnet.com/8301-13579_3-57436553-37/apples-main-data-center-to-go-fully-renewable-this-year/.

[26] L. Lu, J. Tu, C. Chau, M. Chen, and X. Lin. Online energy generation scheduling for microgrids with intermittent energy sources and co-generation. In *Proc. ACM SIGMETRICS*, 2013.

[27] T. Lu and Chen M. Simple and effective dynamic provisioning for power-proportional data centers. In *Proc. IEEE CISS*, 2012.

[28] V. Mathew, R. Sitaraman, and P. Shenoy. Energy-aware load balancing in content delivery networks. In *Proc. IEEE INFOCOM*, 2012.

[29] J. Muckstadt and R. Wilson. An application of mixed-integer programming duality to scheduling thermal generating systems. *IEEE Trans. Power Apparatus and Systems*, 1968.

[30] E. Nygren, R. Sitaraman, and J. Sun. The Akamai Network: A platform for high-performance Internet applications. 2010.

[31] N. Padhy. Unit commitment-a bibliographical survey. *IEEE Trans. Power Systems*, 2004.

[32] D. Palasamudram, R. Sitaraman, B. Urgaonkar, and R. Urgaonkar. Using batteries to reduce the power costs of internet-scale distributed networks. In *Proc. ACM Symposium on Cloud Computing*, 2012.

[33] S. Pelley, D. Meisner, T. Wenisch, and J. VanGilder. Understanding and abstracting total data center power. In *Workshop on Energy-Efficient Design*, 2009.

[34] E. Pinheiro, R. Bianchini, E. Carrera, and T. Heath. Load balancing and unbalancing for power and performance in cluster-based systems. In *Workshop on compilers and operating systems for low power*, 2001.

[35] A. Qureshi, R. Weber, H. Balakrishnan, J. Guttag, and B. Maggs. Cutting the electric bill for internet-scale systems. In *Proc. ACM SIGCOMM*, 2009.

[36] T. Shiina and J. Birge. Stochastic unit commitment problem. *International Trans. Operational Research*, 2004.

[37] M. Stadler, H. Aki, R. Lai, C. Marnay, and A. Siddiqui. Distributed energy resources on-site optimization for commercial buildings with electric and thermal storage technologies. *Lawrence Berkeley National Laboratory*, 2008.

[38] R. Stanojevic and R. Shorten. Distributed dynamic speed scaling. In *Proc. IEEE INFOCOM*, 2010.

[39] J. Tu, L. Lu, M. Chen, and R. Sitaraman. Dynamic provisioning in next-generation data centers with on-site power production. Technical report, Department of Information Engineering, CUHK, 2013. http://arxiv.org/abs/1303.6775.

[40] H. Xu, C. Feng, and B. Li. Temperature aware workload management in geo-distributed datacenters. In *Proc. ACM SIGMETRICS, extended abstract*, 2013.

MultiGreen: Cost-Minimizing Multi-source Datacenter Power Supply with Online Control

Wei Deng, Fangming Liu, and Hai Jin
Services Computing Technology and System Lab
School of Computer Science and Technology
Huazhong University of Science and Technology, Wuhan, China
fmliu@hust.edu.cn

Chuan Wu
Department of Computer Science
The University of Hong Kong
Hong Kong
cwu@cs.hku.hk

Xue Liu
School of Computer Science
McGill University
Montreal, Canada
xueliu@cs.mcgill.ca

ABSTRACT

Faced by soaring power cost, large footprint of carbon emission and unpredictable power outage, more and more modern Cloud Service Providers (CSPs) begin to mitigate these challenges by equipping their Datacenter Power Supply System (DPSS) with multiple sources: (1) smart grid with time-varying electricity prices, (2) uninterrupted power supply (UPS) of finite capacity, and (3) intermittent green or renewable energy. It remains a significant challenge how to operate among multiple power supply sources in a complementary manner, to deliver reliable energy to datacenter users over time, while minimizing a CSP's operational cost over the long run. This paper proposes an efficient, *online* control algorithm for DPSS, called MultiGreen. MultiGreen is based on an innovative two-timescale Lyapunov optimization technique. Without requiring *a priori* knowledge of system statistics, MultiGreen allows CSPs to make online decisions on purchasing grid energy at two time scales (in the long-term market and in the real-time market), leveraging renewable energy, and opportunistically charging and discharging UPS, in order to fully leverage the available green energy and low electricity prices at times for minimum operational cost. Our detailed analysis and trace-driven simulations based on one-month real-world data have demonstrated the optimality (in terms of the tradeoff between minimization of DPSS operational cost and satisfaction of datacenter availability) and stability (performance guarantee in cases of fluctuating energy demand and supply) of MultiGreen.

Categories and Subject Descriptors

D.4.7 [**OPERATING SYSTEMS**]: Organization and Design—*Distributed systems*

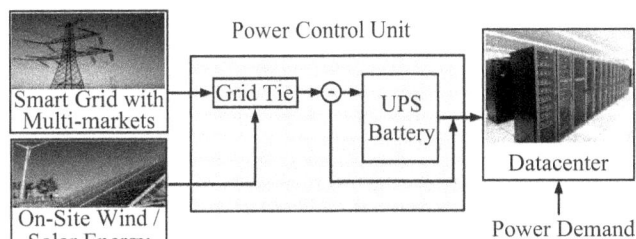

Figure 1: An illustration of the datacenter power supply system (DPSS).

Keywords

Cloud Computing; Datacenter; Power Supply; Renewable Energy; Energy Efficiency; Online Control; Lyapunov Optimization

1. INTRODUCTION

The proliferation of cloud computing services has promoted massive, geographically distributed datacenters. Cloud service providers (CSPs) are typically facing three major power-related challenges: (1) Skyrocketing power consumption and electricity bills, *e.g.*, Google with over $1,120GWh$ and $\$67M$, and Microsoft with over $600GWh$ and $\$36M$ [31]. (2) Serious environmental impact, as IT carbon footprints can occupy 2% of the global CO_2 emissions reportedly [18]. (3) Unexpected power outages, *e.g.*, Amazon experienced an outage in October 2012 in its US-East-1 region, which was triggered by a series of failures in the power infrastructure [17].

To address these challenges, modern CSPs begin to equip their datacenter power supply systems (DPSS) with multiple power sources in a complementary manner, as illustrated in Fig. 1. *First*, modern datacenters obtain their primary power from a smart grid. Smart grids typically provide different pricing schemes at different timescales, such as a long-term-ahead pricing market and the real-time market [1, 15, 19, 21, 37]. *Next*, datacenters are equipped with uninterrupted power supplies (UPSs) to guard against possible power failures. The supply of UPSs may mostly keep a datacenter running for $5 \sim 30$ minutes upon a power outage [36]. *Finally*, CSPs are also starting to green their datacenter operations by integrating on-site renewable energy, such as solar and wind energies [8, 10, 22–24, 35, 41]. The renewable energy is connected to the grid via a grid-tie device, which combines electricity produced from the renewable sources and the grid on the same circuit for power supply [4, 10]. The amount of renewable energy produced could vary significantly over time [18]. UPS can be used to store energy during periods of high levels of renewable energy generation and/or low electricity prices in the grid markets. When the renewable energy is insufficient or prices from the grid are high, the UPS battery can be discharged to provide power [11–13, 20, 36, 38].

An important problem faced by CSPs is how to minimize the long-term cost of running their datacenters. Several key decisions need to be made in an online fashion when operating such a DPSS. (1) How much power to be purchased from the grid's long-term market and the real-time market, respectively? (2) How to efficiently utilize the available renewable energy? (3) How to opportunistically use the UPS battery to store excess power generated/purchased and supply power when needed? It is challenging to optimally utilize the multiple sources to reliably power a datacenter, while minimizing its operational cost. On the demand side, power demand in a datacenter is time-varying, due to variant resource usage of different applications running in the datacenter; on energy supply side, the grid may offer long-term prices and real-time prices, which change over time; further, the unpredictable nature of renewable energy adds onto the supply uncertainties.

There have been a number of works investigating datacenter power supply in cases of varying power demand, renewable energy supply and electricity prices from smart grids. These work may either assume *a priori* knowledge of the power demand [1, 19, 20, 37], or require a substantial amount of statistics of the system dynamics, in order to predict the future demand based on different forecast techniques [10, 15, 21, 41]. Some only optimize single-day or single-household power supply [14, 21], while others do not consider the interactions among renewable energy usage, multi-timescale grid power purchasing and energy storage from the prospective of a datacenter operator [14, 19, 20, 27, 29, 30, 33]. In contrast, we seek to design an efficient online control strategy for long-term cost minimizing operation of the DPSS under dynamic power demand and uncertain renewable energy supply in a synergetic manner, without requiring *a priori* knowledge or stationary distribution of system statistics.

Specifically, based on a stochastic optimization model that characterizes time-varying power demand and renewable energy supply, finite UPS capacity and two-timescale grid markets, we derive an *online* DPSS control algorithm –MultiGreen – by applying a two-stage Lyapunov optimization technique [9, 26, 39]. MultiGreen decides the amount of energy to purchase from the grid's long-term market in intervals of longer periods of time, as the basic energy supply to address demand dynamics and real-time price fluctuations in the future interval; MultiGreen also decides the amount of energy to purchase from the real-time market, as well as the amount of energy to store into or discharge from the UPS batteries, in shorter time scales. The online decisions are set to best utilize the available renewable energy produced over time and

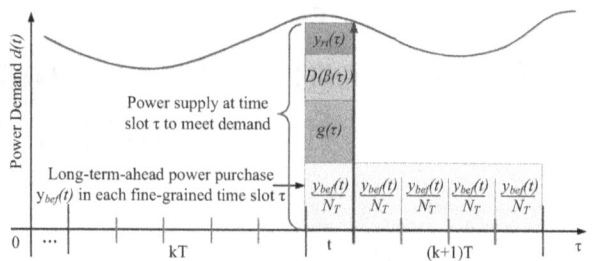

Figure 2: An illustration of the system model. Each coarse-grained time slot is divided into $N_T = 5$ fine-grained time slots.

the periods with low electricity prices, in order to minimize the operational cost in the long run of the datacenter.

A salient feature of MultiGreen is that, even without requiring any *a priori* knowledge of the system dynamics, it can arbitrarily approach the optimal offline cost which is computed with full knowledge of the system over its long run within a provable $O(1/V)$ gap. The algorithmic parameter V serves as a control knob, by adjusting which CSPs can control the tradeoff between the minimization of the DPSS operational cost and satisfaction of the constraints of datacenter availability and UPS lifetime. We analyze the performance of our online control algorithm with rigorous theoretical analysis. Further, we demonstrate the optimality (in terms of the tradeoff between minimization of DPSS operational cost and satisfaction of datacenter availability constraint) and stability (performance guarantee in cases of fluctuating power demand and supply) achieved by MultiGreen, using extensive simulations based on one-month worth of traces from live power systems.

2. SYSTEM MODEL AND OBJECTIVE

As illustrated in Fig. 2, we consider a DPSS system operating in a discrete-time model. Time is divided into $K(K \in \mathbb{N}_+)$ coarse-grained slots of length T in accordance with the interval length of grid's long-term market, *e.g.*, days or hours [21]. Each coarse-grained time slot is further divided into $N_T(N_T \in \mathbb{N}_+)$ fine-grained time slots, *e.g.*, $N_T = 5$ in Fig. 2. Empirically, each fine-grained time slot is 15 or 60 minutes long per which the datacenter can adjust power control strategies in a more prompt fashion [19, 39].

2.1 Online Control Decisions

We assume that datacenter energy demand $d(t)$ and renewable energy generation $g(t)$ are random variables. We don't assume they follow any specific probability distribution functions. We assume a datacenter provider can buy electricity from the grid's long-term contracts or buy electricity in spot markets. While it is not common in today's markets for datacenters to directly buy energy from energy markets, this may change in the future. In fact, Google formed a subsidiary Google Energy LLC, and get approval by FERC (US Federal Energy Regulatory Commission) 3 years ago [16]. This approves that Google LLC can buy and even sell electricity. As we can imagine, in the future, more datacenter providers will participate in electricity markets, as cloud-scale datacenters grow rapidly and can draw tens to hundreds of megawatts. The operation of DPSS includes four key control decisions in two timescales:

2.1.1 Long-term-ahead Energy Purchase

At the beginning (the first fine-grained time slot) of each coarse-grained time slot $t = kT(k = 1, 2, ..., K)$, the DPSS observes the energy demand $d(t)$ and renewable energy $g(t)$ generated during time slot t. Then, the DPSS makes a decision on how much energy $y_{bef}(t)$ to be purchased at a price $p_{lt}(t)$ (with an upper bound price P_{max}) in the long-term market. The DPSS averagely schedules energy $y_{lt}(t) = y_{bef}(t)/N_T$ to be used in each fine-grained time slot in this coarse-grained time slot. For example, if the DPSS decides to purchase $100KW$ in the day-ahead grid market according to the observation of the current demand and renewable energy production, it will schedule $20KW$ for each fine-grained time slot in the next day when $N_T = 5$.

2.1.2 Real-time Energy Balancing

Since the primary costs for renewable energy generation are construction costs such as deploying solar panels and wind turbines, their operational cost is negligible [35], and we focus on operational cost minimization in this paper. Thus, the renewable energy is assumed to be harvested free after deployment, and we preferentially use it. When the renewable energy is generated during time slot t, we use it to meet energy demand as much as possible. If there is excess renewable energy, we use the battery to store it.

Specifically, at each fined-grained time slot $\tau \in [t, t+T-1]$, the actual energy demand $d(\tau)$ and available renewable energy $g(\tau)$ can be readily observed by the DPSS. If the long-term-ahead purchasing and the renewable energy are enough to meet the current energy demand, i.e., $y_{lt}(t)+g(\tau) \geq d(\tau)$, then no real-time energy discharging/purchasing is needed. Otherwise, the DPSS has to make a decision on whether to discharge energy $D(\beta(\tau))$ from the battery. If the discharged power is still not enough, the DPSS decides how much additional energy $y_{rt}(\tau)$ to be purchased from the grid's real-time market at the real-time price $p_{rt}(\tau)(\leq P_{max})$ to fulfill the current demand. Any superfluous energy is used to charge the battery $R(\tau)$. Thus, we have:

$$y_{lt}(t) + y_{rt}(\tau) + D(\beta(\tau)) + g(\tau) - R(\tau) = d(\tau), \quad (1)$$

where $D(\beta(\tau))$ denotes the amount of UPS energy discharged at the depth-of-discharge (DoD) level of $\beta(\tau)$ ($\beta(\tau) \in [0,1]$). DoD is a measure of how much energy has been withdrawn from the battery, expressed as a percentage of the full discharging capacity. For example, let D_{max} denote the maximum energy that we can discharge per time, then $D(\beta(\tau)) = \beta(\tau)D_{max}$. The battery is either charged or discharged or not in use at each time slot, i.e., $R(\tau) \cdot D(\beta(\tau)) = 0$.

2.2 Online Control Constraints

There are a series of constraints that the above decision-making should satisfy:

2.2.1 Balancing procurement accuracy and cost

In practice, the price of electricity in the grid's real-time market tends to be higher on average than that in the grid's long-term market, i.e., $\mathbb{E}p_{rt}(\tau) > \mathbb{E}p_{lt}(t)$ [1, 21, 37], as upfront payment is associated with cheaper contract prices in the long-term market. Hence, when procuring energy in the two-timescale markets, the DPSS should make the best tradeoff between procurement accuracy and power cost. Additionally, we assume that the maximal amount of power that the datacenter can draw from the grid at each time is

limited by P_{grid}:

$$0 \leq y_{lt}(t) + y_{rt}(\tau) \leq P_{grid}. \quad (2)$$

2.2.2 Guaranteeing datacenter availability

Let $m(\tau)$ denote the UPS energy level at time τ. We assume that the efficiencies of UPS charging and discharging are the same, denoted by $\eta \in [0,1]$, e.g., $\eta = 0.8$ means that only 80% of the charged or discharged energy is useful when charging or discharging. The dynamics of UPS level $m(\tau)$ can be expressed as:

$$m(\tau + 1) = m(\tau) + \eta R(\tau) - D(\beta(\tau))/\eta. \quad (3)$$

We assume that under any feasible control algorithm, the UPS battery always reserves a minimum energy level M_{min}, to guarantee reliable datacenter operation in case of power outage. For instance, the energy buffer eBuff [11] always retains five-minute-worth of reserved energy in UPS to ensure datacenter availability. We assume that UPS has a capacity of M_{UPS}, thus we have:

$$M_{min} \leq m(\tau) \leq M_{UPS}. \quad (4)$$

Typically, M_{min} can power the peak demand of a datacenter for about a minute, while M_{UPS} can for $5 \sim 30$ minutes [36].

2.2.3 UPS lifetime and operational cost

In practice, UPS is constrained by the maximum amounts of energy for recharge and discharge per time:

$$0 \leq D(\beta(\tau)) \leq D_{max}, 0 \leq R(\tau) \leq R_{max}, \quad (5)$$

where D_{max} and R_{max} are the maximum energy that we can recharge and discharge UPS per time, respectively.

It has been practically shown that the UPS lifetime is a decreasing function of DoD and charge/discharge cycles [11]. The cost of operating the battery is a function of how often/much it is charged and discharged. We assume that the costs of each recharging and discharging operation are the same, denoted as C_r. If a new UPS costs C_{ups} to purchase and it can sustain L_{ups} total cycles of charging and discharging at the maximum DoD, then $C_r = C_{ups}/L_{ups}$. If the lifetime of the UPS is $Life_{ups}$, then the maximum allowable discharging and charging number over the long run $[0, t-1]$ where $t \in KT$ is $N_{max} = L_{ups} \cdot \frac{KT}{Life_{ups}}$. N_{max} satisfies:

$$0 \leq \sum_{\tau=0}^{t-1} a(\tau) \leq N_{max}, \quad (6)$$

where $a(\tau)$ denotes whether the UPS is used in time slot τ or not, that is $a(\tau) = 1$ if $D(\tau) > 0$ or $R(\tau) > 0$, $a(\tau) = 0$ otherwise. Hence, at time slot t, the operational cost of UPS operation is $a(t)C_r$.

2.3 Stochastic Cost Minimization Formulation

At each fine-grained time slot τ, the DPSS operational cost is the sum of costs for grid energy purchasing and UPS charging/discharging. Therefore, $Cost(\tau) \triangleq y_{lt}(t)p_{lt}(t) + y_{rt}(\tau)p_{rt}(\tau) + a(\tau)C_r$. We seek to design an online DPSS control algorithm that can systematically make online decisions by solving the following stochastic cost minimization problem **P1**:

$$\min_{y_{bef}, y_{rt}, D(\beta), R} \quad Cost_{av} \triangleq \liminf_{t \longrightarrow \infty} \frac{1}{t} \sum_{\tau=0}^{t-1} \mathbb{E}[Cost(\tau)] \quad (7)$$

$$\text{s.t.} \quad \forall t : \text{constraints } (1)(2)(4)(5)(6).$$

Since the battery can be charged to store energy or discharged to serve demand, the current control decisions are coupled with the future decisions. For example, current decisions may leave insufficient battery capacity for storing renewable energy produced in the near future, or overuse the battery and threaten datacenter availability. To solve this optimization problem, the commonly used dynamic programming technique and Markov decision process suffer from a curse of dimensionality and require significant knowledge of the demand and supply over the long term [2,15]. In contrast, the recently developed Lyapunov optimization framework [9,26] is shown to enable the design of online control algorithms for such constrained optimization of time-varying systems without requiring *a priori* knowledge of the future workload and costs. In particular, our above two-timescale power delivery model fits well the two-stage Lyapunov optimization framework [39], that enables us to perform two levels of control strategies at two levels of time granularity. Therefore, we design our online control algorithm based on the two-timescale Lyapunov optimization technique.

2.4 An Optimal Offline Algorithm

Now we present a polynomial-time optimal offline solution for problem **P1** as a benchmark for comparison. In the theoretically optimal scenario, DPSS knows all future system statistics including energy demand $d(t)$, renewable energy production $g(t)$ and grid energy prices $p_{lt}(t), p_{rt}(t)$, $\forall t \in [0, 1, ..., KT]$. First, we present the following straightforward Lemma 1 about the optimal real-time energy purchasing without proof for brevity.

Lemma 1. *In every optimal solution of the optimization problem **P1**, it holds $\forall \tau$, $y_{rt}(\tau) \equiv 0$ or $p_{rt}(\tau) \equiv 0$.*

The above lemma implies that real-time energy purchasing is unnecessary in the optimal solution, where all the future statistics are known in advance. Thus, solving the optimization problem **P1** is equivalent to solving K single time-slot problems **P2** as follows, $\forall t \in [0, 1, ..., KT]$, at the first fine-grained time slot of each coarse-grained time slot over the long horizon KT.

$$\min_{y_{bef}, D(\beta), R} \quad y_{lt}(t) p_{lt}(t) + a(t) C_r \qquad (8)$$

$$\text{s.t.} \quad y_{lt}(t) + g(t) + D(\beta(t)) - R(t) = d(t),$$

$$0 \le y_{lt}(t) \le P_{grid},$$

$$\forall t : \text{constraints } (4)(5)(6).$$

P2 only includes linear terms, and hence can be solved in polynomial time using standard linear programming techniques, *e.g.*, interior point methods [3].

3. ONLINE ALGORITHM DESIGN

Now we develop an online algorithm to achieve near-optimal solution without *a priori* knowledge of power demand and renewable energy generation.

3.1 A Lyapunov Optimization Solution

To guarantee datacenter availability and deliver reliable energy to datacenter power demand, we should guarantee constraint (4) of battery level. We use an auxiliary variable $X(t)$ to track the battery level, defined as follows:

$$X(t) = m(t) - VP_{max}/T - M_{min} - D_{max}/\eta, \qquad (9)$$

where $V \ge 0$ is a control parameter to be specified later, which affects the distance to the optimal value and is related to the battery capacity. The intuition behind $X(t)$ is that by carefully tuning the weights V for decision-making, we can ensure that whenever charging/discharge the battery, the energy level in the battery always lies in the feasible region $[M_{min}, M_{UPS}]$. Recall that $m(t)$ is the actual battery level in time slot t and evolves according to Eq. (3). The dynamics of $X(t)$ is given as:

$$X(t+1) = X(t) + \eta R(\tau) - D(\beta(\tau))/\eta. \qquad (10)$$

In Theorem 2, we will prove that for any time slot t, the battery level $m(t)$ is always in the safe range, since $X(t)$ is deterministically lower and upper bounded.

With the queue, we transform the inequality constraint (4) into a queue stability problem [26]. As a scalar measure of the queue length, we define a quadratic *Lyapunov function* as:

$$L(t) \triangleq \frac{1}{2} X^2(t). \qquad (11)$$

This represents a scalar metric of queue congestion in the system. To keep the system stable by persistently pushing the Lyapunov function towards a lower congestion state, we introduce the T-slot conditional *Lyapunov drift* as:

$$\triangle_T (t) \triangleq \mathbb{E}[L(t+T) - L(t)|X(t)]. \qquad (12)$$

Then following the Lyapunov *drift-plus-penalty* framework [9], we add a function of the expected operational cost over T slots (*i.e.*, the penalty function) to (12) to obtain the drift-plus-penalty term. Our control algorithm is designed to make decisions on $y_{bef}(t), y_{rt}(t), D(\beta(t))$ and $R(t)$ to minimize an upper bound on the following drift-plus-penalty term in every time frame of length T:

$$\triangle_T (t) + V\mathbb{E}\{\sum_{\tau=t}^{t+T-1} Cost(\tau)|X(t)\}, \qquad (13)$$

where the control parameter V is chosen by the CSP to tune the tradeoff between DPSS cost minimization and datacenter availability (battery level). For instance, if V is set to be larger and more emphasis is put to cost minimization, then UPS will be overly used for certain times and thus datacenter availability only achieves a weak satisfaction. A key derivation step is to obtain an upper bound on this term. The following Theorem 1 gives the analytical bound on the drift-plus-penalty term.

Theorem 1. *(Drift-plus-Penalty Bound) Let $V > 0$ and $t = kT(k \in \mathbb{Z}_+)$. Considering the quadratic Lyapunov function Eq. (11), we assume that $\mathbb{E}[L(0)] < \infty$. Under all possible energy management actions to ensure the constraints in problem **P1**, the drift-plus-penalty of the cloud datacenter system satisfies:*

$$\triangle_T (t) + V\mathbb{E}\{\sum_{\tau=t}^{t+T-1} Cost(\tau)|X(t)\} \qquad (14)$$

$$\le \quad B_1 T + V\mathbb{E}\{\sum_{\tau=t}^{t+T-1} Cost(\tau)|X(t)\}$$

where $B_1 = \frac{1}{2}\max\{R_{max}^2\eta^2, D_{max}^2/\eta^2\}$.

PROOF. Let $t = kT(k \in \mathbb{Z}_+)$ and $\tau \in [t, t+T-1]$. Squaring the queue update Eq. (10) yields: $X^2(\tau+1) = X^2(\tau) + 2X(\tau)[R(\tau)\eta - D(\beta(\tau))/\eta] + [(R(\tau)\eta - D(\beta(\tau))/\eta]^2$.

As $D(\beta(\tau)) \in [0, D_{max}], R(\tau) \in [0, R_{max}]$, we obtain:

$$[X^2(\tau+1) - X^2(\tau)]/2$$
$$= [R(\tau)\eta - \frac{D(\beta(\tau))}{\eta}]X(\tau) + [(R(\tau)\eta - \frac{D(\beta(\tau))}{\eta}]^2/2$$
$$\leq X(\tau)[R(\tau)\eta - \frac{D(\beta(\tau))}{\eta}] + \max\{R^2_{max}\eta^2, \frac{D^2_{max}}{\eta^2}\}/2.$$

Taking expectations over $d(t), g(t), p_{lt}(t)$ and $p_{rt}(t)$, conditioning on $X(t)$, we get the 1-slot conditional Lyapunov drift $\triangle_1(\mathbf{Q}(t))$:

$$\triangle_1(t) \leq B_1 + \mathbb{E}\{X(\tau)[R(\tau)\eta - D(\beta(\tau))/\eta]|X(t)\},$$

where $B_1 = \frac{1}{2}\max\{R^2_{max}\eta^2, D^2_{max}/\eta^2\}$. Summing the above inequality over $\tau \in [t, t+1, ..., t+T-1]$, we obtain the following inequality:

$$\triangle_T(\mathbf{Q}(t))$$
$$\leq B_1 T + \mathbb{E}\{\sum_{\tau=t}^{t+T-1} X(\tau)[R(\tau)\eta - D(\beta(\tau))/\eta]|X(t)\}$$

Adding the operational cost $V\mathbb{E}\{\sum_{\tau=t}^{t+T-1} Cost(\tau)|X(t)\}$ to both sides, we prove the theorem. $\square$

Remarks: Our control algorithm is then constructed following the "minimizing drift-plus-penalty" principle of the Lyapunov optimization technique: at every time slot, choose a set of feasible energy purchasing and UPS battery charging/discharging actions to minimize the right-hand-side (RHS) of (14). The parameter V is chosen to enforce different weights to time-averaged operational cost $Cost_{av}$ and queue drift $\triangle_T(t)$ for the CSP to tune the tradeoff between DPSS cost minimization and datacenter availability. The operational cost achieved can be smaller if datacenter availability is just weakly satisfied, e.g., slightly overcharge the UPS battery.

3.2 Relaxed Optimization Problem

The key principle of Lyapunov optimization framework is to choose online control policies to minimize the right-hand-side (RHS) of (14) in Theorem 1, i.e., an upper bound of the drift-plus-penalty framework in (13). However, to minimize the RHS of (14), the DPSS needs to know the concatenated queue backlog $X(t)$ over future time frame $\tau \in [t, t+T-1]$. The queue $X(t)$ depends on UPS battery level $m(t)$, the energy demand $d(t)$ and available renewable energy $g(t)$. The highly variable nature of energy demand, renewable energy and electricity prices has been a major obstacle to make accurate decisions. In practice, system operators can use forecast techniques to predict the future statistics. However, the 90th percentile forecast error for 1-hour-ahead prediction of the renewable energy can be as high as 22.2% [10]. For a 25% penetration of wind energy in a smart grid, the day-ahead forecast error of wind energy generation can result in an additional operational cost of \$4.41 per MWh for the operator [40].

Therefore, we instead approximate near-future queue backlog statistics using the current values, i.e., $X(\tau) = X(t)$ for $t < \tau \leq t+T-1$. This significantly reduces the computational complexity and eliminates the need for any forecast technique in our algorithm, while only bringing a slight "loosening" of the upper bound on the drift-plus-penalty term, as proved in Corollary 1. For this approximation, we will show that our algorithm can still approach the optimal

performance with a proven bound in Theorem 3 in Sec. 4 and simulations in Sec. 5.3.

Corollary 1. *(Loosening Drift-plus-Penalty Bound) Let $V > 0$ and $t = kT$ for some nonnegative integer k. Replacing the concatenated queue $X(\tau)$ with $X(t)$, the drift-plus-penalty satisfies:*

$$\triangle_T(t) + V\mathbb{E}\{\sum_{\tau=t}^{t+T-1} Cost(\tau)|X(t)\} \qquad (15)$$
$$\leq B_2 T + \mathbb{E}\{\sum_{\tau=t}^{t+T-1} X(t)[R(\tau)\eta - D(\beta(\tau))/\eta]|X(t)\}$$

where $B_2 = B_1 + T(T-1)[R^2_{max}\eta^2 - D^2_{max}/\eta^2]/2$, and B_1 is given in Theorem 1.

PROOF. According to Eq. (10), for any $\tau \in [t, t+T-1]$, we can get that:

$$X(t) - (\tau-t)D_{max}/\eta \leq X(\tau) \leq X(t) + (\tau-t)\eta R_{max},$$

Therefore, recalling each term in Eq. (14), we have:

$$\sum_{\tau=t}^{t+T-1} X(\tau)[R(\tau)\eta - D(\beta(\tau))/\eta]$$
$$\leq \sum_{\tau=t}^{t+T-1} [X(t) + (\tau-t)\eta R_{max}]R(\tau)\eta$$
$$- \sum_{\tau=t}^{t+T-1} [X(t) - (\tau-t)D_{max}/\eta]D(\beta(\tau))/\eta$$
$$= \sum_{\tau=t}^{t+T-1} X(t)[R(\tau)\eta - D(\beta(\tau))/\eta]$$
$$+ \sum_{\tau=t}^{t+T-1} (\tau-t)[R_{max}R(\tau)\eta^2 - D_{max}D(\beta(\tau))/\eta^2]$$
$$\leq \sum_{\tau=t}^{t+T-1} X(t)[R(\tau)\eta - D(\beta(\tau))/\eta]$$
$$+ T(T-1)[R^2_{max}\eta^2 - D^2_{max}/\eta^2]/2.$$

Therefore, by defining $B_2 = B_1 + T(T-1)[R^2_{max}\eta^2 - D^2_{max}/\eta^2]/2$, substituting the above inequality into (14), we prove the theorem. $\square$

3.3 Two-timescale Online Control Algorithm

Comparing RHS of (14) with RHS of (15), we can see that the RHS of (15) gives a larger upper bound than the RHS of (14). We seek to minimize the RHS of (15), to derive the online decisions. The control decision $y_{bef}(t)$ should be made at the beginning of each coarse-grained time slot while $y_{rt}(\tau), D(\beta(\tau))$, and $R(\tau)$ are made at each fine-grained time slot. Thus, we can separate the problem into two independent sub-problems **P3** and **P4** as given in our MultiGreen Algorithm 1, to make decisions in the two timescales, respectively. At each coarse-grained time slot $t = kT$, MultiGreen decides how much energy $y_{bef}(T) = N_T y_{lt}(t)$ to be purchased from the grid's long-term energy market. The decision should make sure that the current energy demand is met and the battery stores enough energy for the future need. At each real-time slot $\tau \in [t, t+T-1]$, MultiGreen decides real-time market procurement $y_{rt}(\tau)$, and UPS battery discharging $D(\beta(\tau))$ and charging $R(\tau)$ to supply energy when needed or store additional energy, so as to match the power demand and supply. At the end of each time slot, MultiGreen updates its queue statistics.

Remarks: MultiGreen is computationally efficient. Each time it only needs to solve two linear programs with four variables $(y_{bef}(t), y_{rt}(\tau), D(\beta(\tau)), R(\tau))$ and four linear constraints in (1)(2)(3)(5)(6). We can easily solve the two sub-problems **P4** and **P5** using classical linear programming

Algorithm 1: The Online Algorithm MultiGreen.

1) *Long-term-ahead Energy Planing*: At each coarse-grained time slot $t = kT(k \in \mathbb{Z}_+)$, the DPSS decides the optimal power procurement $y_{bef}(t)$ in the grid's long-term market to minimize the following problem **P3**:

$$\min \quad \mathbb{E}\Big\{\sum_{\tau=t}^{t+T-1} V\Big[y_{lt}(t)p_{lt}(t) + y_{rt}(\tau)p_{rt}(\tau)\Big]|X(t)\Big\}$$
$$+\mathbb{E}\Big\{\sum_{\tau=t}^{t+T-1} X(t)[R(\tau)\eta - D(\beta(\tau))/\eta]|X(t)\Big\}$$
$$\text{s.t.} \quad (1)(2).$$

2) *Real-time Energy Balancing*: Then the DPSS averagely schedules energy $y_{lt}(t) = y_{bef}(t)/N_T$ to be used for each fine-grained time slot $\tau \in [t, t+T-1]$. The DPSS decides real-time energy procurement $y_{rt}(\tau)$, and UPS battery discharging $D(\beta(\tau))$ and charging $R(\tau)$ to minimize the following optimization problem **P4**:

$$\min \quad V y_{rt}(\tau)p_{rt}(\tau) + X(t)\Big[R(\tau)\eta - \frac{D(\beta(\tau))}{\eta}\Big]$$
$$\text{s.t.} \quad (1)(2)(5)(6)$$

3) *Queue Update*: Update the actual and virtual queues using Eq. (3) and Eq. (10).

---

approaches, *e.g.*, interior point methods [3]. MultiGreen makes online control decisions $y_{bef}(t), y_{rt}(\tau), D(\beta(\tau))$ and $R(\tau)$ solely based on the current available statistics at each time slot, including queue statistics, energy demand, volume of the available renewable energy, energy prices and UPS energy level. These statistics typically only require a few bits to store, and take very little time to calculate and transmit. Besides, interior point methods have a low computational complexity (usually polynomial time) in practice [3]. Though advanced prediction techniques can complement MultiGreen to make more accuracy decisions, a tradeoff exists between the benefits of decision accuracy and complexity of implementing the forecast techniques. Note that, MultiGreen is more suitable for delay-sensitive energy demand than delay-tolerant demand. That is, MultiGreen seeks to address energy demand when it is generated immediately. No energy demand should be delayed to a future time to address. We leave it as the future work to design a smart power supply system for mixed workloads.

4. PERFORMANCE ANALYSIS

In this section, we analyze our MultiGreen algorithm in terms of performance bound and robustness.

4.1 Performance Bound

We first analyze the gap between the result achieved by MultiGreen, if accurate knowledge of $X(\tau)$ in the future coarse-grained interval is employed rather than our approximation. We assume that the theoretical offline optimal objective function value is ϕ^{opt} of the cost minimization problem **P1**.

Theorem 2. *(Performance Bound): The time-averaged cost $Cost_{av}^{Green}$ achieved by the MultiGreen algorithm based*

---

on accurate knowledge of $X(\tau)$ in the future coarse-grained interval satisfies the following bound with any given control parameter $V(V > 0)$:

(1) The time-average cost $Cost_{av}^{Green}$ achieved by MultiGreen satisfies the following bound:

$$Cost_{av}^{Green} \triangleq \liminf_{t \to \infty} \frac{1}{t}\sum_{\tau=0}^{t-1} \mathbb{E}[Cost(\tau)] \le \phi^{opt} + \frac{B_2}{V}, \quad (16)$$

where B_2 is given in Corollary 1.

(2) The UPS battery level $m(t)$ is always in the range $[M_{min}, M_{UPS}]$. Datacenter availability is satisfied.

(3) All control decisions are feasible.

PROOF. (1) Let $t = kT(k \in \mathbb{Z}_+)$ and $\tau \in [t, t+T-1]$. From the optimal offline policy in Sec. 2.4, we know that there is an optimal solution ϕ^{opt}. Since MultiGreen minimizes the RHS of Eq. (15), plugging the policy π into the RHS of Eq. (15), we have:

$$\triangle(t) + V\mathbb{E}\{\sum_{\tau=t}^{t+T-1} Cost_{av}^{Green}(\tau)|X(t)\}$$
$$\le B_2 T + V\phi^{opt}.$$

Taking the expectation of both sides and rearranging the terms, we get:

$$E\{L(t+T) - L(t)\} + VTE\{Cost_{av}^{Green}(t)\}$$
$$\le B_2 T + VT\phi^{opt}.$$

Summing the above over $t = kT, k = 0, 1, 2, ...K-1$, using the fact that $L(t) > 0$, and dividing both sides by VKT, we obtain:

$$\frac{1}{KT}\mathbb{E}\{\sum_{\tau=0}^{KT-1} Cost_{av}^{Green}(\tau)\} \le \phi^{opt} + \frac{B_2}{V}.$$

Taking the limit as $K \to \infty$, we complete the proof.

(2) We first observe that subproblem **P4** has the following properties related to battery operation:

Lemma 2. *If $X(t) > 0$, then $R(t) = 0$; if $X(t) < -VP_{max}/T$, then $D(\beta(t)) = 0$.*

We first prove that $-VP_{max}/T - D_{max}/\eta \le X(t) \le M_{UPS} - VP_{max}/T - M_{min} - D_{max}/\eta$. We prove the result using induction. Since $X(0) = m(0) - VP_{max}/T - M_{min} - D_{max}/\eta$, $M_{min} \le m(0) \le M_{UPS}$, we know that $-VP_{max}/T - D_{max}/\eta \ge X(0) \le M_{UPS} - VP_{max}/T - M_{min} - D_{max}/\eta$.

Now we first consider $0 < X(t) \le M_{UPS} - VP_{max}/T - M_{min} - D_{max}/\eta$, then $R(t) = 0$. Since there is no battery recharging and the maximum discharged energy is D_{max}/η each time, we have: $-VP_{max}/T - D_{max}/\eta < -D_{max}/\eta < X(t+1) \le X(t) \le M_{UPS} - VP_{max}/T - M_{min} - D_{max}/\eta$.

Next, suppose $-VP_{max}/T < X(t) \le 0$, then $D(\beta(t)) = 0$. The maximum charging and recharging energy each time are $R_{max}\eta$ and D_{max}/η, respectively. Thus we obtain: $-VP_{max}/T - D_{max}/\eta < X(t+1) \le X(t) + R_{max}\eta \le M_{UPS} - VP_{max}/T - M_{min} - D_{max}/\eta$.

Finally, we consider the case of $-VP_{max}/T - D_{max}/\eta \le X(t) \le -VP_{max}/T$. Since $X(t) < -VP_{max}/T$, $D(\beta(t)) = 0$. Then $-VP_{max}/T - D_{max}/\eta \le X(t) \le X(t+1) \le -VP_{max}/T \le -VP_{max}/T + M_{UPS} - M_{min} - D_{max}/\eta$.

Then, from Eq. 9, we have: $-VP_{max}/T - D_{max}/\eta \le X(t) = m(t) - VP_{max}/T - M_{min} - D_{max}/\eta \le M_{UPS} - VP_{max}/T - M_{min} - D_{max}/\eta$. It is easy to see that $M_{min} \le m(t) \le M_{UPS}$.

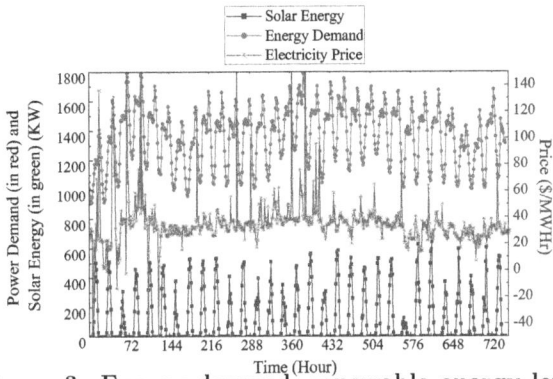

Figure 3: Energy demand, renewable energy levels and energy prices in January 2012.

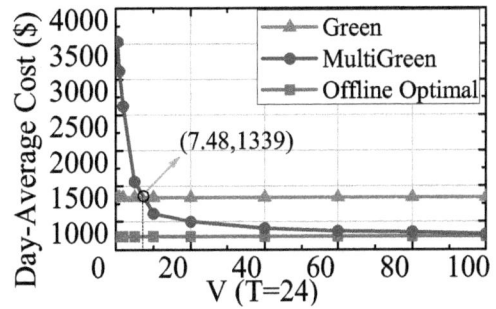

Figure 4: Impact of parameter V.

(3) Since MultiGreen makes decisions to satisfy all the constraints in problem **P3** and **P4**, combining the constraints together, all the constraints of problem **P1** are satisfied. Therefore, MultiGreen control decisions are feasible to problem **P1**. $\square$

Remarks: MultiGreen can approach the optimal solution of problem **P1** within a deviation of B_2/V. As CSPs increase the value of V, they can push the average cost to be arbitrarily close to the minimum value, according to a desired tradeoff between DPSS cost minimization and datacenter availability. The length of time slot T decides how frequently MultiGreen performs energy procurement and battery charging and discharging. We will carry out detailed evaluations in Sec. 5.2.2 to show that even infrequent actions can still achieve significant cost reduction.

4.2 Robustness Analysis

Since MultiGreen approximates future queue backlog $X(\tau)$ using its current level, an important question remained to be answered is: *is the performance still bounded if MultiGreen makes its decisions based on an approximated queue backlog $\widehat{X}(\tau)$ that is different from the actual value $X(\tau)$?* The dynamic UPS energy levels reflect the variation of energy demand and renewable energy supply. The following Theorem 3 demonstrates the robustness of MultiGreen in its performance to uncertainties of energy demand and supply.

Theorem 3. *(Robustness): We assume that the estimated virtual battery level $\widehat{X}(\tau)$ and its actual level $X(\tau)$ satisfy $|\widehat{X}(\tau) - X(\tau)| \leq \theta$. Then, if we use this approximated UPS battery level in the MultiGreen algorithm, we can obtain:*

$$Cost_{av}^{Green} \triangleq \liminf_{t \to \infty} \frac{1}{t} \sum_{\tau=0}^{t-1} \mathbb{E}[Cost(\tau)] \leq \phi^{opt} + \frac{B_\varepsilon}{V}, \quad (17)$$

where $B_\varepsilon = B_2 + T\theta(D_{max} + R_{max} + M_{UPS} + M_{min})$ and B_2 is given in Corollary 1. Here, D_{max} and R_{max} are the maximum amounts of UPS energy that we can recharge and discharge, respectively; M_{ups} and M_{min} are the maximum and minimum UPS energy levels, respectively.

PROOF. Let $\varepsilon_X(t) = X(t) - X(\tau)$, that is a function of the variation of demand $\varepsilon_d(t)$ and renewable supply $\varepsilon_g(t)$. Given $\widehat{X}(t)$ and $\varepsilon_X(t)$, when minimizing the RHS of Eq. (15), we

try to minimize $f(\widehat{X}(t))$ defined as below:

$$f(\widehat{X}(t)) \triangleq \sum_{\tau=t}^{t+T-1} X(t)[R(\tau)\eta - D(\beta(\tau))/\eta]$$

$$= f(X(\tau)) + \sum_{\tau=t}^{t+T-1} \varepsilon_X(t)[R(\tau)\eta - D(\beta(\tau))/\eta]$$

$$\leq f(X(\tau)) + T\theta(R_{max}\eta + D_{max}/\eta)$$

Substituting the above result into the inequality (15), we know that (15) holds with $X(\tau)$ replaced by $\widehat{X}(t)$, and B_2 replaced by $B_\varepsilon = B_2 + T\theta(R_{max}\eta + D_{max}/\eta)$. The rest of proof is similar to the proof of Theorem 2. $\square$

Remarks: Comparing (16) in Theorem 2 and (17) in Theorem 3, we can see that the upper bound in (17) is looser, *i.e.*, V needs to be set to a larger value when the future demand and supply are estimated, in order to obtain the same time-averaged operational cost as when the accurate future information is known. The larger the uncertainty θ is, the larger V should be. This implies that the robustness of MultiGreen in cases of inaccurate future information, can be achieved at the cost of weaker datacenter availability.

5. PERFORMANCE EVALUATION

We evaluate MultiGreen through trace-driven simulations with realistic parameters and one-month data on datacenter energy demand, renewable energy production and electricity prices.

5.1 Real-World Trace and Experimental Setup

Real-World Traces: To simulate the intermittent availability of renewable energy, we use solar energy data from the Measurement and Instrumentation Data Center (MIDC) [25]. Specifically, we use the meteorological data from Jan. 1st, 2012 to Jan. 31th, 2012 from central U.S.. To simulate the varying electricity prices, we use the electricity prices in central U.S. between Jan. 1st, 2012 and Jan. 31th, 2012, from the New York Independent System Operator (NYISO) [28]. Similar to [38], we use the energy demand from a Google Cluster including Web search and Webmail services. We regulate the data to our assumed datacenter by removing demand peaks above P_{grid}. The traces are shown in Fig. 3.

System Parameters: According to results in recent empirical studies, we assume that the limits of UPS charging and discharging rates are $D_{max} = R_{max} = 0.5MW$, and charging/discharging costs are $C_r = C_d = 0.1$ dollars [36]. The minimum battery level M_{min} is 5-minute worth of energy of the UPS [11]. The maximum number of UPS charge/discharge cycles is $L_{UPS} = 5,000$ with a 4-year lifetime constraint [11]. The efficiency of UPS charg-

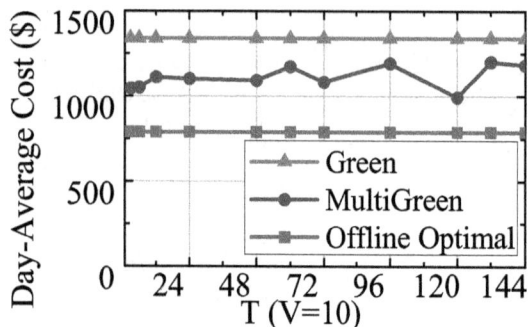

Figure 5: Impact of parameter T.

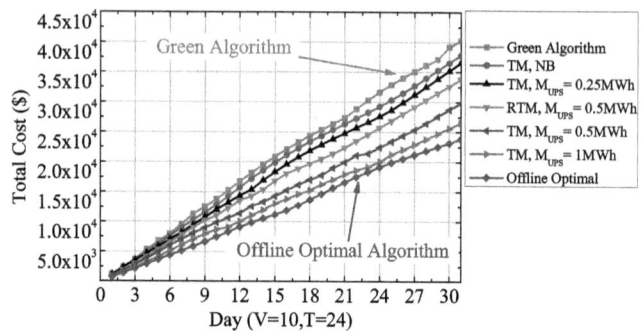

Figure 6: Impact of battery capacity and grid market structure. Two Markets—TM, Real-Time Market—RTM, No Battery—NB.

ing/discharging is $\eta = 0.8$ [20]. We set the grid energy limit as $P_{grid} = 2MW$ [36].

Algorithms for Comparison: We compare MultiGreen with the offline optimal algorithm (*Optimal*) and an online algorithm *Green* that solely leverages renewable energy production, without exploiting time-varying electricity prices. The Green algorithm also tries to maximize the usage of renewable energy, *i.e.*, leveraging UPS battery to store excess renewable energy production for future need. However, the Green algorithm ignores the two-timescale grid markets, and does not store grid energy in UPS when the electricity prices are low and supply energy when the electricity prices are high.

5.2 Analysis of Sensitivity on Critical Factors

From Theorem 2, we note that the performance of MultiGreen depends on parameters V and T, battery capacity and the energy prices in the two-timescale grid markets. We conduct sensitivity analysis on these critical factors to characterize their impact on the DPSS operational cost.

5.2.1 Impact of Control Parameter V

As shown in Fig. 4, to simulate a 1-day-ahead power market, we fix T to be 24 time slot and each fine-grained time slot is 1 hour, *i.e.*, $N_T = 24$. We conduct experiments with different values of V, which show that as V increases from 0.5 to 100, MultiGreen achieves a time-averaged cost that becomes closer to the optimal solution. This quantitatively confirms Theorem 2 that *MultiGreen can approach the optimal solution within a diminishing gap of $O(1/V)$*. In contrast, the Green algorithm has a constant cost that is irrelevant with V. Interestingly, the crossover between cost curves of MultiGreen and Green clearly captures the trade-off between the average operational cost and constraint satisfaction. When $V < 7.48$, MultiGreen has a higher cost and a higher level of constraint satisfaction than Green. On the other hand, when $V > 7.48$, due to more frequent battery charging/discharging, MultiGreen has a lower cost and a lower level of constraint satisfaction than Green. By choosing an appropriate value of V, *e.g.*, $V = 10$, MultiGreen can achieve a significantly lower cost compared with Green while guaranteeing acceptable satisfaction of constraints on datacenter availability and UPS lifetime.

5.2.2 Impact of Coarse-grained Time Frame T

In Fig. 5, we fix V to be 10 and vary T from 3 time slots (3 hours) to 144 time slots (6 days), which is a sufficient-long range for exploring the impact of different timescales of the grid's long-term market. We observe that T has rel-

atively less impact on the cost of operating the DPSS. The fluctuation of the time-averaged cost is more notable when T becomes longer. The rationale is that the term B_ε in Theorem 3 is proportional to T, which means that the uncertainties of energy demand and renewable energy increase with the increase of T. Nevertheless, the time-averaged cost only fluctuates within $[-9.7\%, +8.5\%]$. This corroborates Theorem 3 that, even with *infrequent decisions* of the DPSS operations, MultiGreen can still achieve significant cost reduction.

5.2.3 Impact of Battery Capacity and Grid Markets

In Fig. 6, we compare the time-averaged total cost under different battery sizes ($M_{UPS} \in \{0, 0.25, 0.5, 1\}MWh$) over the 31-day period with $V = 10$ and $T = 24$. It shows that the time-averaged total cost decreases with the increase of the UPS battery capacity. The rationale is that an UPS with larger capacity can store more superfluous renewable energy generated, or more energy purchased from the grid when the price is low, to serve the demand, resulting in lower overall costs.

In Fig. 6, we also compare the case with energy purchase in two-timescale markets with the case where only the real-time market exists, both with $V = 10, T = 24, M_{UPS} = 0.5MWh$. We can observe that the existence of the grid's long-term market can bring in additional cost reduction. The reason is that DPSS can purchase certain amount of energy beforehand in the grid's long-term market with relative lower prices

In addition, we can observe that even without the UPS battery, the MultiGreen algorithm with the two-timescale markets can reduce the cost by 10.06%, compared to the Green algorithm. With two markets, when we increase the battery size from 0 to $1MWh$, the average operational cost reduction ranges from 10.06% to 34.21%. The benefit brought by energy storage is higher than that of exploiting the two markets. When the battery size is large enough, MultiGreen can approach the optimal offline algorithm.

5.3 Characterizing Algorithm Robustness

As mentioned in Sec. 3, our MultiGreen algorithm approximates the future queue statistics as the current values. Now we explore the influence of approximation errors on the performance of MultiGreen. We add a random approximation error to the datacenter energy demand, solar energy generation and energy prices, *e.g.*, uniformly distributed $\pm 50\%$

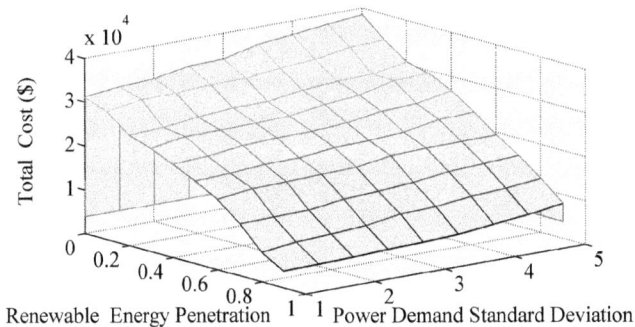

Figure 7: DPSS operational cost at various levels of renewable energy penetration and energy demand variation.

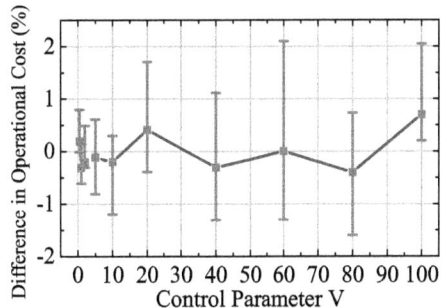

Figure 8: The impact of approximation errors in operational cost reduction.

errors [39]. We let MultiGreen make all the control decisions based on the data set with such random errors under different values of V. In Fig. 8, we show the differences in percentage between the DPSS operational costs achieved with approximated values and the results we obtained using the original traces. We observe that the difference fluctuates within $[-1.3\%, 2.1\%]$ for all values of V. Thus, MultiGreen is robust to inaccurate future information.

Further, we study the impacts of renewable energy penetration (the percentage of renewable energy in the total datacenter energy supply) and the variation of datacenter energy demand on the total cost. In Fig. 7, x-axis represents *renewable energy penetration* in the range of $[0,100\%]$. Y axis represents the standard deviation of the demand, *i.e.*, $\sqrt{\sum_{t=0}^{KT-1}[d(t) - \mathbb{E}(\vec{d})]^2 \times p_{d(t)}}$, where $\mathbb{E}(\vec{d})$ is the expectation of the series of demand $d(t)$ over time length $t \in [0, KT]$, and $p_{d(t)}$ is the distribution probability of $d(t)$. We assume that the random variable of the datacenter energy demand is uniformly distributed ($p_{d(t)} = 1/KT$). As expected, Fig. 7 shows that with the increase of penetration of renewable energy, the DPSS operational cost decreases significantly. The rationale is that renewable energy is harvested cost-free (we do not consider the construction cost). In contrast, as the variation of demand increases, the operational cost increases slightly. The rationale is that intensive variation incurs large approximation errors.

6. RELATED WORK

In this secton, we discuss the research most pertinent to this work as follows. The first category of works is exploiting renewable energy in datacenters. Many large IT companies recently consider greening their datacenters with renewable energy [7,8,10,22–24,35,41]. However, the intermittent nature of renewable energy poses significant challenges to make use of them. Some works make the traffic "follow the renewables" to execute workload when/where renewable energy is available [10,22–24,41] or carbon footprint is low [7]. However, these approaches require prediction of renewable energy production when scheduling workload, or sacrifice performance to avoid wasting renewable energy. Other works supply renewable energy to deferrable loads to align demand with intermittent available renewable energy [27,29,30]. But they are from the prospective of renewable energy providers and do not consider energy storage and multiple markets in the smart grid.

The second category of works is leveraging energy storage in datacenters. Recently, UPS shows its benefits to reduce electricity costs in datacenters [11–13,36,38]. Datacenters can store energy in the UPS when energy prices are low and discharge UPS when prices are high, to reduce the power drawn from the grid [13,36]. Moreover, UPS can shave peaks [11,12]. During periods of low demand, UPS batteries store energy, while stored energy can be used to temporarily augment the grid supply during hours of peak load. However, these works focus on studying the benefits of UPS battery for power cost reduction, and no renewable energy and grid markets are considered. On the contrary, we leverage UPS to study how to manage multiple power supplies of a datacenter in an integrated way.

Third stream of works is on multiple timescale dispatch, pricing and scheduling in smart grid. Nair *et al.* [1] studied the optimal energy procurement from long-term, intermediate, and real-time markets under intermittent renewable energy supplies. Jiang *et al.* [19] proposed an optimal multi-period power procurement and demand response algorithm without energy storage. "Risk-limiting-dispatching" is proposed in [37] to manage integrated renewable energy. However, the above three approaches assume that the demand can be known ahead. Jiang *et al.* [21] solved the optimal day-ahead procurement and real-time demand response problem using dynamic programming, while He *et al.* [15] formulated the multi-timescale power dispatch and scheduling problem as a Markov decision problem. Both these approaches need substantial system statistics and are computationally expensive. We mitigate these disadvantages by applying two-stage Lyapunov optimization that makes online control decisions without *a priori* knowledge or any stationary distribution of energy prices, demand and supply. Recently the authors in [13,32,33,39,41] distributed requests across multiple data centers to reduce electricity costs by leveraging both time diversity and location diversity of electricity prices in the smart grid. In contrast, we study how to reduce the operational cost in a datacenter powered by multiple power sources rather than how to distribute requests across datacenters.

In addition, interest has been growing in power management in smart grids and datacenters using Lyapunov optimization [5,6,9,26,42]. On smart grids, several works have proposed optimal power management based on single-stage Lyapunov optimization. However, they either focused on managing individual household demand [14] or did not consider the interaction between renewable energy and energy storage [14,19,20,27]. In contrast, we manage the uncertain datacenter demand and multi-source energy supply in a systematic fashion using two-stage Lyapunov optimization. Although [34, 39] have used two-stage Lyapunov to design a two-timescale algorithm and a T-Step Lookahead algorithm, both of them study how to schedule jobs or distribute requests in solely grid-powered geographical datacenters rather than how to supply multi-source energy in a datacenter with uncertain demand.

7. CONCLUSION

In this paper, we study an important problem of how to minimize the operational cost of datacenters by using mutiple energy resouces. We propose MultiGreen, an online control algorithm applying the two-stage Lyapunov optimization technique, which optimally schedules multiple energy supply sources to power a datacenter, in a cost minimizing fashion. Without requiring *a priori* knowledge of system statistics, MultiGreen can deliver reliable energy to datacenters while minimizing the operational cost in the datacenter's long-run operation. Both mathematical analyses and trace-driven evaluations demonstrate the optimality and robustness of MultiGreen. Especially, it can approach the offline optimal cost within a diminishing gap of $O(1/V)$, which is mainly decided by the UPS battery capacity, grid market structure and DPSS operation frequency for energy purchasing and UPS charging/discharging.

8. ACKNOWLEDGEMENTS

The Corresponding Author is Fangming Liu. The research was supported by a grant from The National Natural Science Foundation of China (NSFC) under grant No.61133006.

9. REFERENCES

[1] S. Adlakha and A. Wierman. Energy Procurement Strategies in the Presence of Intermittent Sources. *Technical report*, 2012.

[2] D. Bertsekas. *Dynamic Programming and Optimal Control*, volume 1. Athena Scientific Belmont, MA, 1995.

[3] G. Dantzig. *Linear Programming and Extensions*. Princeton university press, 1998.

[4] N. Deng, C. Stewart, and J. Li. Concentrating Renewable Energy in Grid-tied Datacenters. In *Proc. of IEEE International Symposium on Sustainable Systems and Technology (ISSST)*, May 2011.

[5] W. Deng, F. Liu, H. Jin, and X. Liao. Online Control of Datacenter Power Supply Under Uncertain Demand and Renewable Energy. In *Proc. of IEEE ICC*, June 2013.

[6] W. Deng, F. Liu, H. Jin, and C. Wu. SmartDPSS: Cost-Minimizing Multi-source Power Supply for Datacenters with Arbitrary Demand. In *Proc. of ICDCS*, July 2013.

[7] P. Gao, A. Curtis, B. Wong, and S. Keshav. It's Not Easy Being Green. In *Proc. of ACM SIGCOMM*, Aug. 2012.

[8] Y. Gao, Z. Zeng, X. Liu, and P. R. Kumar. The Answer Is Blowing in the Wind: Analysis of Powering Internet Data Centers with Wind Energy. In *Proc. of INFOCOM (Mini Conference)*, Apr. 2013.

[9] L. Georgiadis, M. Neely, M. Neely, and L. Tassiulas. *Resource Allocation and Cross-layer Control in Wireless Networks*. Now Pub, 2006.

[10] Í. Goiri, K. Le, T. Nguyen, J. Guitart, J. Torres, and R. Bianchini. GreenHadoop: Leveraging Green Energy in Data-Processing Frameworks. In *Proc. of EuroSys*, Apr. 2012.

[11] S. Govindan, A. Sivasubramaniam, and B. Urgaonkar. Benefits and Limitations of Tapping into Stored Energy for Datacenters. In *Proc. of ACM ISCA*, June 2011.

[12] S. Govindan and B. Urgaonkar. Leveraging Stored Energy for Handling Power Emergencies in Aggressively Provisioned Datacenters. In *Proc. of ACM ASPLOS*, Mar. 2012.

[13] Y. Guo, Z. Ding, Y. Fang, and D. Wu. Cutting Down Electricity Cost in Internet Data Centers by Using Energy Storage. In *Proc. of GLOBECOM*, Dec. 2011.

[14] Y. Guo, M. Pan, and Y. Fang. Optimal Power Management of Residential Customers in the Smart Grid. *IEEE Transactions on Parallel and Distributed Systems*, 23(9):1593–1606, 2012.

[15] M. He, S. Murugesan, and J. Zhang. Multiple Timescale Dispatch and Scheduling for Stochastic Reliability in Smart Grids with Wind Generation Integration. In *Proc. of INFOCOM*, Apr. 2011.

[16] http://en.wikipedia.org/wiki/Google_Energy.

[17] http://www.datacenterknowledge.com/archives /2012/10/22/amazon-cloud-outage-affecting-many sites/.

[18] http://www.greenpeace.org. How Clean is Your Cloud. 2012.

[19] L. Huang, J. Walrand, and K. Ramchandran. Optimal Power Procurement and Demand Response with Quality-of-usage Guarantees. *Arxiv preprint arXiv:1112.0623*, 2011.

[20] L. Huang, J. Walrand, and K. Ramchandran. Optimal Demand Response with Energy Storage Management. *Arxiv preprint arXiv:1205.4297*, 2012.

[21] L. Jiang and S. Low. Multi-period Optimal Procurement and Demand Responses in the Presence of Uncrtain Supply. In *Proc. of IEEE Conference on Decision and Control (CDC)*, 2011.

[22] C. Li, A. Qouneh, , and T. Li. iSwitch: Coordinating and Optimizing Renewable Energy Powered Server Clusters. In *Proc. of ISCA*. ACM, June 2012.

[23] Z. Liu, Y. Chen, C. Bash, A. Wierman, D. Gmach, Z. Wang, M. Marwah, and C. Hyser. Renewable and Cooling Aware Workload Management for Sustainable Data Centers. In *Proc. of SIGMETRICS*, June 2012.

[24] Z. Liu, M. Lin, A. Wierman, S. Low, and L. Andrew. Geographical Load Balancing with Renewables. In *Proc. of ACM GreenMetrics*, June 2011.

[25] MIDC. http://www.nrel.gov/midc/.

[26] M. Neely. Stochastic Network Optimization with Application to Communication and Queueing Systems. *Synthesis Lectures on Communication Networks*, 3(1):1–211, 2010.

[27] M. Neely, A. Tehrani, and A. Dimakis. Efficient Algorithms for Renewable Energy Allocation to Delay Tolerant Consumers. In *Proc. of SmartGridComm*, Oct. 2010.

[28] NYISO. http://www.nyiso.com.

[29] A. Papavasiliou. *Coupling Renewable Energy Supply*

with Deferrable Demand. PhD thesis, University of California, Berkeley, 2012.

[30] A. Papavasiliou and S. Oren. Supplying Renewable Energy to Deferrable Loads: Algorithms and Economic Analysis. In *Proc. of Energy Society General Meeting*, July 2010.

[31] A. Qureshi. *Power-Demand Routing in Massive Geo-Distributed Systems*. PhD thesis, Massachusetts Institute of Technology, 2010.

[32] L. Rao, X. Liu, L. Xie, and W. Liu. Minimizing Electricity Cost: Optimization of Distributed Internet Data Centers in a Multi-electricity-market Environment. In *Proc. of INFOCOM*, Mar. 2010.

[33] L. Rao, X. Liu, L. Xie, and Z. Pang. Hedging Against Uncertainty: A Tale of Internet Data Center Operations Under Smart Grid Environment. *IEEE Transactions on Smart Grid*, 2(3):555–563, 2011.

[34] S. Ren, Y. He, and F. Xu. Provably-Efficient Job Scheduling for Energy and Fairness in Geographically Distributed Data Centers. In *Proc. of ICDCS*, June 2012.

[35] C. Stewart and K. Shen. Some Joules Are More Precious Than Others: Managing Renewable Energy in the Datacenter. In *Proc. of the Workshop on Power Aware Computing and Systems*, 2009.

[36] R. Urgaonkar, B. Urgaonkar, M. Neely, and A. Sivasubramanian. Optimal Power Cost Management Using Stored Energy in Data Centers. In *Proc. of ACM SIGMETRICS*, June 2011.

[37] P. Varaiya, F. Wu, and J. Bialek. Smart Operation of Smart Grid: Risk-limiting Dispatch. *in Proc. of the IEEE*, 99:(1):40–57, 2011.

[38] D. Wang, C. Ren, A. Sivasubramaniam, B. Urgaonkar, and H. Fathy. Energy Storage in Datacenters: What, Where, and How much? In *Proc. of SIGMETRICS*, June 2012.

[39] Y. Yao, L. Huang, A. Sharma, L. Golubchik, M. Neely, et al. Data Centers Power Reduction: A Two Time Scale Approach for Delay Tolerant Workloads. In *Proc. of IEEE INFOCOM*, Mar. 2012.

[40] R. Zavadil, J. King, N. Samaan, J. Lamoree, M. Ahlstrom, B. Lee, D. Moon, C. Finley, D. Savage, R. Koehnen, et al. Final Report–2006 Minnesota Wind Integration Study. *The Minnesota Public Utilities Commission*, 2006.

[41] Y. Zhang, Y. Wang, and X. Wang. GreenWare: Greening Cloud-Scale Data Centers to Maximize the Use of Renewable Energy. In *Proc. of ACM Middleware*, Dec. 2011.

[42] Z. Zhou, F. Liu, H. Jin, B. Li, B. Li, and H. Jiang. On Arbitrating the Power-Performance Tradeoff in SaaS Clouds. In *Proc. of IEEE INFOCOM*, Apr. 2013.

Modelling and Real-Trace-Based Evaluation of Static and Dynamic Coalescing for Energy Efficient Ethernet

Angelos Chatzipapas
angelos.chatzipapas@imdea.org

Vincenzo Mancuso
vincenzo.mancuso@imdea.org

Institute IMDEA Networks, and University Carlos III of Madrid
Madrid, Spain

ABSTRACT

The IEEE Standard 802.3az, namely Energy Efficient Ethernet (EEE), has been recently introduced to reduce the power consumed in LANs. Since then, researchers have proposed various traffic shaping techniques to leverage EEE in order to boost power saving. In particular, packet coalescing is a promising mechanism which can be used on top of EEE to tradeoff power saving and packet delay. In this paper, we analyze the interesting and special case of 1000Base-T EEE links, in which power saving operations are triggered only when links are inactive in both transmission directions. We are the first to provide an analytical model for EEE 1000Base-T which accounts for the bidirectional nature of LAN traffic. Our model allows to compute the power saving achieved by EEE, with and without packet coalescing, by using a few significant traffic descriptors. Furthermore, we use real traffic traces to investigate on the performance of static as well as dynamic coalescing schemes. Our results show that dynamic coalescing does not significantly outperform static coalescing in terms of power save and delay.

Categories and Subject Descriptors

C.2.5 [**Computer-Communication Networks**]: Local and Wide-Area Networks—*Ethernet*; G.3 [**Probability and Statistics**]: Queueing Theory

Keywords

EEE; Bidirectional Gigabit Links; Dynamic Coalescing

1. INTRODUCTION

During the past years a lot of effort has been invested to increase processing, communication, switching speed and data storage with little effort to optimize the power consumption. According to [1], about 14 TWh were consumed in 2005 by the telecom core network in EU-25[1] and the yearly consumption is expected to increase to about 30 TWh by 2020. Although this power consumption is useful for the human beings, it is also potentially harmful for our environment since it produces an augmented amount of CO_2 emissions and highly contributes to the greenhouse effect. The current threat to the environment could turn into a much more serious threat in the near future, since there is

[1]The first 25 countries that joined the European Union.

a growing demand of new generation devices that require connection to the Internet (such as televisions, white goods, etc.). In addition, existing network connected devices are now increasing their bandwidth demands (e.g., Web servers, databases, etc.). Indeed, the Internet traffic might grow with the number of data centers in the network and the number of users that demand higher amounts of traffic such as bigger files, videos, TV over IP etc. Hence, as the data traffic demand rises, especially in developing countries, more and more energy consumption is expected for networking.

In order to protect the environment and obtain lower service cost, Internet Service Providers and Network Operators are currently deploying new strategies to reduce energy consumptions. In this context, our work investigates on the recently approved Energy Efficient Ethernet (EEE) standard for power saving in Local Area Networks. Indeed, according to [2], the authors estimate significant reductions of about 4 TWh per year over one billion devices.

Legacy Ethernet is a power-unaware standard which consumes a constant amount of power independently from the actual traffic flowing through the wires. However, low speed Ethernet cards consume about 200 mW, which is not a significant consumption considering that a server or a home PC consumes tens to thousands of Watts. Therefore, so far Ethernet power saving strategies did not rise the interest of researchers and developers, due to the irrelevance of potential savings for low speed connections. In contrast, new high speed Gigabit interface cards may consume up to 20 W [3] which makes reasonable the introduction of a power saving mechanism. In fact, taking into account that the amount of Web Hosting Centers and server farms has been extremely increased due to the new trends and services (YouTube, Facebook, Twitter etc.), there are now billions of running interfaces that consume a constant amount of power. In addition, Ethernet links are basically inactive most of the time. Therefore, a new power aware Ethernet standard (standardized late 2010) was introduced to minimize the power consumption of the links when low traffic is present, namely IEEE 802.3az, or EEE [4].

While some effort has been put in understanding the behavior of EEE links where power saving can be activated independently in each traffic direction, e.g., [5, 6, 7], in this paper, we are the first to present an analytical model for bidirectional EEE links, e.g., EEE links in which power saving operations can only be activated when there is no traffic in both link directions. The latter (namely, the *bidirectional EEE case*) is a very relevant case, since the EEE standard adopts this bidirectional behavior for 1000Base-T cards, which are the most commonly adopted and diffused gigabit network cards, as of today. Notably, packet coalescing techniques have been proposed to boost EEE performance when traffic is not scarce and packet arrivals have short spacing. The basic idea behind coalescing is to aggregate packets in a buffer of limited size until either the buffer is full or a timeout expires.

We propose a model that uses simple statistical parameters (such as mean interarrival time and its variance) to estimate the power consumption of a bidirectional EEE link over time, and the packet delay incurred in crossing the link when coalescing techniques are adopted.

We collected real traces from a large web hosting center and extracted from them the traffic parameters needed by our model. We also used real traces to validate our model in terms of EEE power saving end packet delay by simulating the EEE and packet coalescing behaviors with real input traffic, using a modified ns-3 simulator [8]. In line with other studies focusing on EEE, e.g., [5, 9], we show that, without coalescing, EEE enables non-negligible power saving only when the offered traffic is rather low (few percents of the link capacity) and packet arrivals are bursty.

Another fundamental contribution of our work consists in the performance evaluation of packet coalescing strategies for EEE. Not only we evaluate the power saving enhancements achievable by means of static coalescing approaches, but we also propose dynamic strategies to adapt the coalescing parameters to the traffic characteristics. Our performance analysis shows that both the size of the coalescing queue and the duration of the coalescing timeout should adapted to the offered traffic. However, we show that a dynamic coalescing approach can be used, which seamlessly adapts to any traffic conditions, and achieves nearly optimal results in terms of power saving and packet delay. Nonetheless, our thorough investigation shows that also static coalescing can achieve nearly optimal results, thus questioning the importance of exploring more complex approaches based on run time adaptation of the coalescing parameters.

The rest of the paper is organized as follows. Section 2 discusses the related work. Section 3 presents the EEE standard and describes the behavior of 1 Gbps EEE links. Section 4 describes an analytical model for the estimation of power consumption and packet delay in bidirectional EEE links. Section 5 introduces dynamic coalescing algorithms to adapt the coalescing parameters to the time-varying traffic conditions. In Section 6 we use real-trace-based simulations to validate our model and present an extensive performance evaluation of EEE with and without coalescing schemes. Section 7 summarizes and concludes the paper.

2. RELATED WORK

Modeling of EEE. Various analytical models exist in the literature for *unidirectional* links. In [5] the authors propose an analytical model that allows to compute fast the potential EEE power saving and it performs very well for EEE links with no coalescing, using simple statistical parameters for unidirectional traffic. In [10], using parameters such as the packet arrival time and the service rate of the *coalescer*, the model is able to compute the mean queue length, the mean packet delay and the delay for the downstream queue for 10 Gbps links. A two state analytical model for 10GBase-T links is presented in [6] that estimates the power consumption of EEE links. This is a quick modeling tool, but it is not very accurate in case of small transition inter-

vals, since it divides the time into discrete intervals equal to multiples of frame transmission time. Herrería-Alonso *et al.* [7] propose and analyze a model for both legacy EEE and burst transmission with 10 Gbps cards. Their model estimates the power saving using the arrival rate for Poisson traffic and the average service rate. They also propose a model with GI/G/1 queues for both frame and burst transmissions. The model allows to compute the average delay of packets and the power saving of the link using Poisson and deterministic traffic, but it is specifically designed for the case of 10 Gbps links and thus it cannot be used to estimate the power saving of the widely used 1 Gbps links [9].

EEE performance evaluation. A few number of EEE evaluations and extension proposals have appeared during the last few years to study and improve EEE's performance. Reviriego *et al.* proved initially the inefficiency of EEE by simulating the standard on ns-2 for 100Base-T, 1000Base-T and 10GBase-T links [11]. In [12] the authors provide a first evaluation on newly released EEE NIC cards for 1 Gbps links. They measure the power consumption of the cards with real traffic and prove that: (*i*) the power consumption during transitions is similar to the power consumption in "Active" state and, (*ii*) great power saving can be achieved but for very low loads, reporting saving up to 30% for 100 Mbps links and up to 70% for 1 Gbps links. We have previously shortly reported on traffic measurements and potential power saving under EEE gigabit links in [13].

Coalescing in EEE links. One of the first evaluations of packet coalescing for EEE is presented in [2] for 10 Gbps Ethernet links. The results show that packet coalescing outperforms legacy EEE in terms of power consumption and it overcomes the major problem of EEE, namely the overhead due to protocol state transitions (which correspond to hardware operational states). However, EEE introduces additional delay for the packets to cross the Ethernet link. For the measurements in [2], only two pairs of timeout-buffer values are used (buffer of 10 packets with a 12 μs timeout and buffer of 100 packets with $120\mu s$ are used). The authors of [14] perform more extensive simulations on packet coalescing by using a timeout of 10 μs for testing 100 Mbps, 1000 Mbps and 10 Gbps Ethernet links. Coordinated Transmission with EEE in 10 Gbps links is analyzed in [15]. Reviriego *et al.* show that for links with loads less than 50% this method can reduce the power consumption by powering down some PHY layer components and can allow longer cable lengths than the standard default 100 m. For LAN switches, a synchronized coalescing method is proposed by Mostowfi and Christensen [16], achieving potential power saving of about 40% for realistic TCP parameters (results were achieved via simulation). A new dynamic scheme is proposed in [17] but it lacks of full evaluation and comparison of the various parameters for both static and dynamic schemes. In all the above performance evaluation works, it is assumed that traffic is unidirectional and power saving is operated independently over the two link directions.

Our contribution. With our work, we add to the literature a model that considers the bidirectional behavior of 1 Gbps links and we evaluate its accuracy by means of EEE simulations based on real traffic traces. Additionally, we propose a complete performance evaluation of coalescing algorithms and show that static tuning of coalescing parameters can achieve near-optimal results in terms of power saving and packet delay.

3. EEE WITH GIGABIT ETHERNET

Energy Efficient Ethernet 802.03az [4] was standardized in September 2010. It aims to provide significant power saving in LANs. Formerly, the evolution of LANs led towards higher link speeds for faster communication and higher bandwidth, in order to satisfy the increased demand for data (link speeds from 10 *Mbps* to 10 *Gbps*) without power consumption concerns. In fact, the electricity consumption of relatively "old" network interfaces remained in very low levels so the main concern of Ethernet component producers was not to save power. For example, in 100 *Mbps* Ethernet links, the Ethernet devices consume about 200 *mW* of power [11]. However, higher speed Ethernet links (1 *Gbps* or faster) require several Watts of power consumption [3] for each network interface. Considering a usual server that consumes around 200 *W*, a simple Ethernet device contributes to $\sim$ 10% of this amount. Indeed, data centers and web hosting centers have a huge number of network interface cards which eventually generate a high cost (in terms of electricity bills). Thus, the idea of reducing the power consumption of Ethernet devices appears in the foreground.

Legacy Ethernet consumes a constant amount of power either with or without traffic, which makes it totally inefficient with typical Ethernet traffic profiles. This behavior results in a huge waste of power since it is well known that Ethernet links are inactive most of the time with utilization factors from 5% for a home PC to 30% for heavy loaded data servers [13, 16, 18]. EEE aims to reduce this waste of power and approach *power proportionality*, i.e., a power consumption proportional to the served traffic. The EEE standard introduces four new states for the Ethernet link, namely state "Active" (A) which corresponds to the busy period, state "Low Power Idle" (LPI) in which there is no traffic and the link consumes substantially less power than in state A ($\sim$ 90% less power according to [12]), and states "Sleep" (S) and "WakeUp" (W) which correspond to the time spent during switching from state A to LPI and vice versa, respectively [5]. EEE specifications and state transition schemes are different for 100Base-T, 1000Base-T and 10GBase-T links. In particular, since we focus on commonly deployed 1 *Gbps* links, in the following subsection we describe in detail how EEE 1000Base-T links behave, since they represent the only case in which the EEE standard accounts for the bidirectional behavior of traffic.

3.1 Gigabit EEE Link Operation

Behavior of 1Gbps EEE links. EEE 1000Base-T links can be in one of the following four states: Active (A), Sleep (S), WakeUp (W) and Low Power Idle (LPI). The state transition diagram is illustrated in Figure 1. Frame transmissions in either of the traffic directions only occur in state A. When the two network cards connected to the link complete transmitting all the buffered frames, the link enters state S as a transition to state LPI. If no frame arrives for T_s seconds while in state S, the link enters state LPI, during which power consumption is minimized. A frame arrival in state LPI results in the link transitioning to state W which lasts T_w seconds. After this wake interval, the link transitions to state A and any of the network interfaces connected to the link can transmit. Standard values for T_s and T_w are 182 μs and 16 μs, respectively. Thus, the fact that a frame arrival in the sleep interval (state S) causes an immediate transition to state A, avoids incurring in delays of up to 182

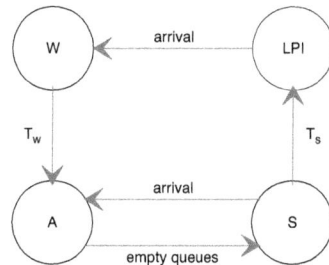

Figure 1: State transition diagram for EEE 1000Base-T.

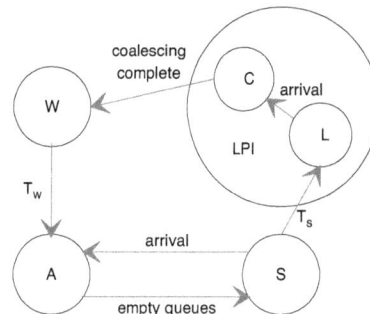

Figure 2: Modified state transition diagram for EEE 1000Base-T with coalescing.

μs, which can be quite large if compared to the transmission time of a single packet (e.g., 12 μs for a 1500-byte packet).

Behavior of 1Gbps EEE links with coalescing: When coalescing techniques are used, the transition from state LPI to state W is delayed. Therefore, we fictitiously split state LPI into two states, as shown in Figure 2: state L, which represents state LPI when there is no packet queued in the coalescing buffers—and which is equivalent to LPI in systems with no coalescing—and state C, in which coalescing buffers are not empty but neither the coalescing timer expired nor the coalescing buffers were completely full. Indeed, the system transitions from state C to state W as soon as one of the coalescing buffers gets full or the coalescing timeout expires. We remark that the newly introduced state C is fictitious, since the transition $L \to C$ does not represents a change of state for the EEE Ethernet hardware. State C simply represents the extension of an LPI interval due to coalescing operations, counting from the arrival of the first packet in one of the two coalescing queues. The time spent in state C, namely τ_c, is a random variable which depends on the size N_c of the coalescing buffers, and it is limited by the coalescing timeout T_c.

3.2 Efficiency Problems of EEE

As shown in previous works, such as [11, 14], the problem of EEE links (and especially in 1000Base-T links) is the transitioning time. When the traffic is scarce and packets are spaced rather that bursty, the EEE mechanism rarely allows the link to complete the transition to LPI. Thereby, the link spends more time transitioning than transmitting. Specifically, 1 *Gbps* links require about 12 μs for transmitting a big packet, while for transitioning they spend at least $T_s + T_w = 182 + 16 = 198$ μs. For smaller packets the analogy is even worst. For this reason, packet coalescing is proposed to avoid frequent state transitions and prolong the duration of state LPI. Packet coalescing tends to approach *energy proportionality* at the cost of additional delay to the packets. However, we will show in the next Section that the delay is negligible if compared to the energy benefits.

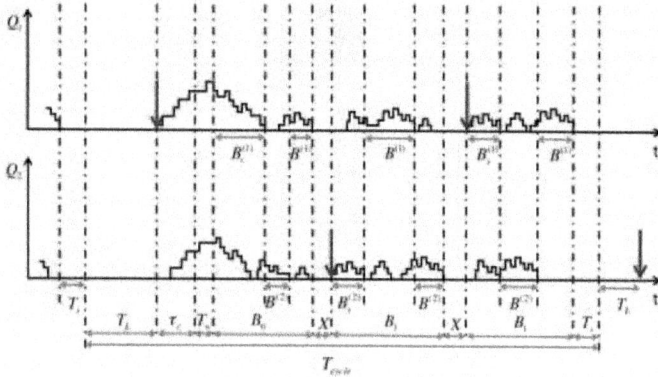

Figure 3: System cycle with coalescing.

4. A MODEL FOR BIDIRECTIONAL EEE GIGABIT LINKS

The goal of using EEE is to reduce the power consumption of Ethernet links; therefore, it is of great interest to evaluate the performance of EEE under different working conditions, and so models and simulators are needed to produce such a performance evaluation. We have developed an analytical model for EEE with bidirectional traffic with and without coalescing. We further modified the ns-3 simulator [8] to validate the model and evaluate the performance of EEE.

4.1 System Cycle and Power Saving

We model the behavior of the EEE link with coalescing as two $M/G/1$ queues, namely Q_1 and Q_2, in which the packet service rate is non-zero only when the EEE is in state A, where it equals a constant R corresponding to the link speed. Frames arrive to the transmission queues, representing the two network cards connected to the link, according to two independent Poisson processes with rates λ_1 and λ_2, respectively. We denote by $S_p^{(i)}$ the size of a single frame and by $E[S_p^{(i)}]$ the average frame size in link direction $i \in \{1, 2\}$.

Since events that cause state transitions can occur in either of the two link directions, for sake of tractability, we assume that these events are uncorrelated.

4.1.1 Cycle Analysis

The sample path of the queue can be viewed as a sequence of cycles as illustrated in Figure 3. A cycle starts with the packet arrival that induces the transition from state L to state C. Then, when the coalescing timer T_c expires or the number of coalesced packets reaches N_c, the link transitions to state W. Note that both transitions $L \to C$ and $C \to W$ can be caused by arrivals in either link directions. Note also that setting $T_c = 0$ and/or $N_c = 1$ yields the legacy EEE operation with no coalescing (i.e., the duration of state C is 0). In particular, transition $L \to C$ happens because of an arrival to queue Q_1 (namely, an arrival in direction 1) with probability $P_1 = \lambda_1/(\lambda_1 + \lambda_2)$, or because of an arrival to Q_2 (i.e., in direction 2) with probability $P_2 = 1 - P_1$.

The coalescing interval is followed by a busy period with the link in state A, whose duration is denoted by B_0. This initial busy period is followed by a random number ψ of sleep/active interval pairs, with each pair corresponding to an arrival in state S in either direction 1 or 2, before a time T_s has elapsed.

Note that the sleep time is then reduced to the random

time interval between the start of state S and the next frame arrival, i.e., it is upper bounded by T_s. Finally, a sleep interval of duration T_s precedes the idle period T_L, whose random duration corresponds to the time interval before the beginning of a new cycle, i.e., before the next arrival in the system. We denote the length of a cycle by T_{cycle} and its average by $E[T_{cycle}]$.

Using results from renewal theory [19], we can focus on the system cycle, and compute the fraction of time spent in each link state as the ratio between the average time in each state in a cycle and the average cycle duration. We denote the average fraction of time spent in state α as η_α, for $\alpha \in \{A, C, L, S, W\}$. The following Theorem shows how to compute the average duration of a system cycle.

THEOREM 1. *For bidirectional EEE links in which arrivals in S are served immediately, the average cycle duration is given by:*

$$E[T_{cycle}] = (T_w + E[\tau_c]) \left[1 + \frac{1}{\lambda_1 + \lambda_2} \left(\frac{\lambda_1 \rho_1}{1 - \rho_1} + \frac{\lambda_2 \rho_2}{1 - \rho_2} \right) \right]$$
$$+ \frac{e^{(\lambda_1 + \lambda_2)T_s}}{\lambda_1 + \lambda_2} \left[1 + \frac{\rho_1}{1 - \rho_1} + \frac{\rho_1^2(2 - \rho_1)(\lambda_1 \rho_2 + \lambda_2)}{2\lambda_1(1 - \rho_1 \rho_2)(1 - \rho_1)^2} \right.$$
$$\left. + \frac{\rho_2}{1 - \rho_2} + \frac{\rho_2^2(2 - \rho_2)(\lambda_2 \rho_1 + \lambda_1)}{2\lambda_2(1 - \rho_1 \rho_2)(1 - \rho_2)^2} \right] \quad (1)$$

with $\rho_i = \lambda_i/\mu_i$, and $\mu_i = R/E[S_p^{(i)}]$, $i \in \{1, 2\}$.

PROOF. Consider the different intervals included in T_{cycle}, starting with the beginning of an L interval. The cycle is composed by the following elements: (i) an interval in state L, with random duration T_L until the first arrival to Q_1 or Q_2; (ii) a coalescing interval C of duration $\tau_c \leq T_c$; (iii) a wake-up interval of fixed duration T_w; (iv) an interval B_0 lasting till the first epoch at which both queues Q_1 and Q_2 are empty; (v) an interval $X < T_s$ with exactly one arrival at time X, followed by an interval B_1 lasting until both queues are empty again; (vi) and finally a sleep interval of fixed duration T_s which triggers a new state L. Element (v) is optional, since it occurs only if there is one arrival within T_s seconds after B_0. Moreover, element (v) can repeat $\psi \geq 0$ times, until the idle interval following B_1 is longer than T_s. Each repetition of X and B_1 exhibits the same distribution because of the memoryless property of Poisson arrivals. However, busy intervals B_0 and B_1 can start either because of Q_1 or Q_2 activities, the two cases leading to different *conditional* average durations, as we will discuss in the following. Overall, the total cycle duration is:

$$T_{cycle} = T_L + \tau_c + T_w + B_0 + \psi(X + B_1) + T_s. \quad (2)$$

Note that $T_L + \tau_c$ is the time spent in state LPI during a cycle, while $B_0 + \psi B_1$ is the total time during which the link is active, and $T_s + \psi X$ is the time spent in state S. Let us now compute the average value for each of the elements composing the system cycle.

Interval T_L. State L lasts until the first packet arrival to Q_1 or Q_2. Since both arrival processes are Poisson, the first arrival behaves as the first of a Poisson flow with rate $\lambda_1 + \lambda_2$, i.e., with the following probability distribution:

$$f_{T_L}(t) = (\lambda_1 + \lambda_2)e^{-(\lambda_1 + \lambda_2)t}, \quad t \geq 0; \quad (3)$$

and its average is then:

$$E[T_L] = \frac{1}{\lambda_1 + \lambda_2}. \quad (4)$$

Interval τ_c. For large values of N_c, we can approximate the duration of state C as the minimum time before $N_c - 1$ packet arrivals occur in either direction 1 or 2, and T_c. Let us denote with τ_{c_1} and τ_{c_2} the time before $N_c - 1$ arrivals appear in direction 1 or 2, respectively. Thereby, the time spent in coalescing is $\tau_c = \min\{\tau_{c_1}, \tau_{c_2}, T_c\}$. Recalling that the cumulative distribution function (CDF) of the minimum of n independent random variables is given by the following formula:

$$F_{\min_{i=1...n}\{X_i\}}(x) = 1 - \prod_{i=1}^{n}(1 - F_{X_i}(x)), \quad (5)$$

and considering that the distributions of τ_{c_1}, τ_{c_2}, and T_c are as follows:

$$F_{\tau_{c_1}}(t) = u(t)\left[1 - \sum_{k=0}^{N_c-2}\frac{(\lambda_1 t)^k}{k!}e^{-\lambda_1 t}\right]; \quad (6)$$

$$F_{\tau_{c_2}}(t) = u(t)\left[1 - \sum_{k=0}^{N_c-2}\frac{(\lambda_2 t)^k}{k!}e^{-\lambda_2 t}\right]; \quad (7)$$

$$F_{T_c}(t) = u(t - T_c); \quad (8)$$

where $u(t)$ is the unit step function, then the CDF of τ_c is given by:

$$F_{\tau_c}(t) = 1 - u(T_c - t)\left[\sum_{k=0}^{N_c-2}\frac{(\lambda_1 t)^k}{k!}e^{-\lambda_1 t}\right]\left[\sum_{k=0}^{N_c-2}\frac{(\lambda_2 t)^k}{k!}e^{-\lambda_2 t}\right], \quad (9)$$

where we used $1 - u(t) = u(-t)$. The average coalescing time is then as follows:

$$E[\tau_c] = \int_0^{T_c} t \cdot dF_{\tau_c}(t). \quad (10)$$

However, assuming that we can tune N_c and T_c in a way that the coalescing queues do not fill completely with probability almost one, then $E[\tau_c] \simeq T_c$. Note that the latter assumption is realistic for actual implementations since the power saving increases with N_c, although delay increases too and we want to bound it to T_c.

Interval T_w. The wake-up interval which precedes the first busy interval has fixed duration T_w.

Interval B_0. This interval is composed of various sub-parts, as shown in Figure 3. After the link transitions to state A from state C, at least one queue is not empty. The system remains busy until *both* queues are empty again. Let us observe the system from the viewpoint of the queue that received the packet that caused the beginning of the the coalescing state (transition $L \to C$). We denote with $B_c^{(i)}$, $i \in \{1, 2\}$, the first busy period seen at queue i only, after the transition $C \to A$. This is the busy period of an $M/G/1$ queue, for which the average depends on the arrival rate λ_i, the mean service time $E[S_p^{(i)}]/R$, and the queue size $Z_c^{(i)}$ at the beginning of the busy period [19]:

$$E[B_c^{(i)}] = \frac{E\left[Z_c^{(i)}\right]E\left[S_p^{(i)}\right]/R}{1 - \rho_i} = \frac{E\left[Z_c^{(i)}\right]\rho_i}{\lambda_i(1 - \rho_i)}. \quad (11)$$

If queue i is the one who received the packet that triggered the transition $L \to C$, then $E\left[Z_c^{(i)}\right] = 1 + \lambda_i(T_w + E[\tau_c])$, i.e., the initial queue size equals the arrival that triggers the coalescing timer, plus the average number of Poisson arrivals during the average coalescing time $E[\tau_c]$ and the

wake-up interval T_w. The probability that queue Q_1 is the one who initiates the coalescing procedure is simply given by the probability of having a Poisson arrival with rate λ_1 before a Poisson arrival with rate λ_2, counting from the beginning of state L. I.e.:

$$Pr(Q_1 \text{ triggers coalescing}) = \frac{\lambda_1}{\lambda_1 + \lambda_2}; \quad (12)$$

$$Pr(Q_2 \text{ triggers coalescing}) = \frac{\lambda_2}{\lambda_1 + \lambda_2}. \quad (13)$$

With no loss of generality, assume now that queue Q_1 triggers the coalescing. Therefore, as observed from Q_1, the system goes through a busy period $B_c^{(1)}$, at the end of which Q_1 is empty with probability 1, while Q_2 is empty with probability $1 - \rho_2$. If Q_2 is not empty, let us observe the followup in the evolution of the system from the viewpoint of Q_2: there is a busy period $B^{(2)}$ for Q_2, at the end of which Q_2 will be empty, while Q_1 can be empty with probability $1 - \rho_1$. The process can replicate by alternating busy intervals $B^{(1)}$ and $B^{(2)}$, i.e., we alternate the observation of the system from the viewpoint of a queue or the other, until the observation of the queue status at the end of a busy period reveals that both queues are empty. At that point we have a transition $A \to S$. Busy periods $B^{(i)}$ have different average duration with respect to $B_c^{(i)}$. In fact, the initial backlog of the queue in $B^{(i)}$ is not $Z_c^{(i)}$. Using the Pollaczek-Khinchin mean formula to estimate the average backlog of an $M/G/1$ queue at a random observation point, we would have $\lambda_i E[S_p]/R + \frac{\lambda_i^2 E[S_p^2]/R^2}{2(1-\rho_i)}$ as initial backlog. However, we are interested in the conditional initial backlog $Z^{(i)}$, given that the observed queue is not empty (otherwise there is no busy period), which happens with probability ρ_i. Therefore we have $E\left[Z^{(i)}\right] = 1 + \frac{\rho_i}{2(1-\rho_i)}$, and the observed busy periods are given by:

$$E\left[B^{(i)}\right] = \rho_i\frac{2 - \rho_i}{2\lambda_i(1 - \rho_i)^2}. \quad (14)$$

Following the above procedure, one can compute the average duration of the first period after state C during which either queue 1 or 2 are busy:

$$E[B_0] = \frac{1}{\lambda_1 + \lambda_2}\left[\lambda_1 E\left[B_c^{(1)}\right] + \lambda_2 E\left[B_c^{(2)}\right]\right.$$
$$\left. + \frac{\rho_1(\lambda_1\rho_2 + \lambda_2)E\left[B^{(1)}\right]}{1 - \rho_1\rho_2} + \frac{\rho_2(\lambda_2\rho_1 + \lambda_1)E\left[B^{(2)}\right]}{1 - \rho_1\rho_2}\right]. \quad (15)$$

Interval X. The interval X between the end of a busy period and the beginning of the next busy period is exponentially distributed with rate $\lambda_1 + \lambda_2$, given that the next arrival occurs within T_s seconds:

$$f_X(t) = \frac{(\lambda_1 + \lambda_2)e^{-(\lambda_1 + \lambda_2)t}}{1 - e^{-(\lambda_1 + \lambda_2)T_s}}, \quad t \in [0, T_s].$$

Accordingly, the average of X is as follows:

$$E[X] = \frac{1}{\lambda_1 + \lambda_2} - \frac{T_s}{e^{(\lambda_1 + \lambda_2)T_s} - 1}.$$

Interval B_1. Similarly to the case of B_0, this interval is composed by various subparts. The first part is $B_s^{(i)}$, which is the first busy interval on either Q_1 or Q_2 after the period X. Since packets are served immediately when they arrive in state S, the initial backlog is exactly 1. For the rest, the computation of $B_s^{(i)}$ is analogue to the one of $B_c^{(i)}$:

$$E[B_s^{(i)}] = \frac{\rho_i}{\lambda_i(1-\rho_i)}.$$

The following alternating busy intervals are exactly like for the case of B_0, i.e., we have intervals $B^{(1)}$ and $B^{(2)}$.

Considering that each interval B_1 starts because of an arrival in Q_1 or Q_2 before T_s expires, and since arrivals for Q_1 and Q_2 in state S are independent and both follow a Poisson process, the probability of starting an interval B_1 due to arrivals for Q_i is $\frac{\lambda_i}{\lambda_1+\lambda_2}$.

Putting together the pieces, the average of B_1 is as follows:

$$E[B_1] = \frac{1}{\lambda_1+\lambda_2}\left[\lambda_1 E\left[B_s^{(1)}\right] + \lambda_2 E\left[B_s^{(2)}\right]\right.$$
$$\left. + \frac{\rho_1(\lambda_1\rho_2+\lambda_2)E\left[B^{(1)}\right]}{1-\rho_1\rho_2} + \frac{\rho_2(\lambda_2\rho_1+\lambda_1)E\left[B^{(2)}\right]}{1-\rho_1\rho_2}\right]. \quad (16)$$

Number of repetitions ψ. Busy intervals B_1 occur if the residual interarrival time at the end of the previous busy interval is shorter than T_s. Since arrivals are Poisson, the probability of having no arrivals in any link direction in T_s is $P_0 = e^{-(\lambda_1+\lambda_2)T_s}$. Thereby, the number $\psi \geq 0$ of busy periods of type B_1 in a cycle, i.e., not counting B_0, can be seen as the number of consecutive successes of a geometric random variable ψ with success probability $1 - P_0$. Hence, its average value is:

$$E[\psi] = \frac{1 - P_0}{P_0} = e^{(\lambda_1+\lambda_2)T_s} - 1.$$

Interval T_s. The sleep interval which follows the last busy interval before entering state L has fixed duration T_s.

Average cycle duration. Putting together the results obtained for the cycle components, after some algebraic elaboration, result (1) follows.

□

COROLLARY 1. *The fraction of time spent in LPI is:*

$$\eta_{LPI} = \frac{\frac{1}{\lambda_1+\lambda_2} + E[\tau_c]}{E[T_{cycle}]}. \quad (17)$$

PROOF. The time spent in LPI corresponds to states L and C, i.e., intervals T_L and τ_c, as described in the proof of Theorem 1; therefore the proof follows. □

COROLLARY 2. *The fraction of time spent in WakeUp state is given by:*

$$\eta_W = \frac{T_w}{E[T_{cycle}]}. \quad (18)$$

PROOF. The time spent in $WakeUp$ state is constant in each cycle and corresponds to the interval T_w. Therefore the proof follows. □

COROLLARY 3. *The fraction of time spent in Sleep state is given by:*

$$\eta_S = \frac{E[X]E[\psi] + T_s}{E[T_{cycle}]}. \quad (19)$$

PROOF. The time spent in *Sleep* state corresponds to intervals X, which are repeated ψ times, plus a complete sleep

interval of T_s seconds, occurring once in a cycle, just before entering state L. Considering the proof of Theorem 1, the proof follows. □

COROLLARY 4. *The fraction of time spent in Active state is given by:*

$$\eta_A = \frac{E[B_0] + E[B_1]E[\psi]}{E[T_{cycle}]}. \quad (20)$$

PROOF. The time spent in *Active* state is given by the sum of busy intervals during which at least one network interface transmits. Therefore, considering the proof of Theorem 1, the proof follows. □

4.1.2 Power Saving

Using the results of the analysis carried out for the cycle duration, we can now compute the power saving factor Φ achieved by EEE with or without coalescing.

THEOREM 2. *The average power consumption achieved by EEE is proportional to the fraction of time spent in LPI:*

$$\Phi \propto \eta_{LPI}. \quad (21)$$

PROOF. The average power consumption over a system cycle, is computed by considering that the power consumption in states W (namely $P^{(W)}$) and S (namely $P^{(S)}$) is practically the same as in state A (namely $P^{(A)}$), while in LPI (i.e., $P^{(LPI)}$ in states L and C), the power consumption decreases by a factor $k \simeq 10$, as experimentally shown by Reviriego *et al.* [12]. In legacy gigabit cards, the power consumption $P^{(legacy)}$ is practically constant and equals the one consumed in state A in an EEE card. Therefore we have the following expression for the power saving factor Φ:

$$\Phi = 1 - \frac{\sum_{\alpha\in\{A,S,LPI,W\}}\eta_\alpha P^{(\alpha)}}{P^{(legacy)}} \simeq \frac{k-1}{k}\cdot\eta_{LPI}. \quad (22)$$

Φ is thus shown to be proportional to η_{LPI} with a proportionality factor $(k-1)/k \simeq 0.9$. □

The power saving achieved with EEE is proportional to η_{LPI} and the values of Φ and η_{LPI} are similar. As a consequence, in the rest of the paper we will refer to *power saving performance* either in case of actual power saving figures or when discussing η_{LPI} values.

4.2 Packet Delay

As concerns the average frame delay we differentiate between packet arrivals in the different states A, S, L, C, W.

Delay D_A of packets arriving in A. If a packet arrives in state A, we can use results for M/G/1 queues. Therefore, the average waiting time can be simply computed by means of the P-K formula [19]:

$$D_A^{(i)} = \frac{\lambda_i E[\sigma_i^2]}{2(1-\rho_i)}, \quad i\in\{1,2\}, \quad (23)$$

where $\sigma_i = S_p^{(i)}/R$ is the random service time, with $E[\sigma_i] = 1/\mu_i$. The statistics of σ_i depends on the packet size distribution. For the sake of simplicity, here we assume that packet size is constant and equal to $1/\mu_i$, so that we use $E[\sigma_i^2] = 1/\mu_i^2$, which yields $D_A^{(i)} = \frac{\rho_i}{2\mu_i(1-\rho_i)}$.

Note that the resulting delay is different in the two link directions.

Delay D_S of packets arriving in S. If a packet arrives while the device is in state S, the device will directly transition to state A and the packet is immediately served. Thus, the delay is $D_S = 0$.

Delay D_L of packets arriving in L. Only one packet can arrive while the device is in state L, which triggers an immediate transmission to state C. The delay of this packet is at most $T_c + T_w$, in case of scarce traffic which yields the expiration of the coalescing timeout. More in general, the average queuing delay experienced by this packet is the sum of the average coalescing time, given in Eq. (10), plus the constant wake-up time T_w:

$$D_L = E[\tau_c] + T_w. \tag{24}$$

Delay D_C of packets arriving in C. When a packet arrives in state C, it suffers from (i) the residual coalescing interval, (ii) the constant wake-up interval T_w, and (iii) the time to process and transmit packets already present in the queue at the arrival epoch of the new packet.

Considering that Poisson arrivals are uniformly distributed over time, we estimate the average residual coalescing time as $E[\tau_c]/2$. Correspondingly, the average queue size at the arrival epoch is $\lambda_i E[\tau_c] + \frac{\lambda_i}{\lambda_1+\lambda_2}$ for an arrival in direction $i \in \{1, 2\}$, where the second term represents the probability that the packet triggering the $L \to C$ transition belongs to direction i. The average cumulative serving time for those packets is then $\rho_i\left(\frac{E[\tau_c]}{2} + \frac{1}{\lambda_1+\lambda_2}\right)$. Therefore, the delay suffered by a packet arriving in state C is, on average:

$$D_C^{(i)} = T_w + \frac{E[\tau_c]}{2} + \rho_i\left(\frac{E[\tau_c]}{2} + \frac{1}{\lambda_1+\lambda_2}\right), \quad i \in \{1,2\}. \tag{25}$$

Considering that the delay D_C is affected by the arrival rate, this delay assumes different average values for packets sent in the two different link directions.

Delay D_W of packets arriving in W. If a packet arrives while the device is in state W, the delay is composed of (i) the average residual wake-up time (for uniformly distributed Poisson arrivals, this equals $T_w/2$), and (ii) the required time to serve all packets arrived earlier, since the beginning of state C: $\frac{\lambda_i}{\lambda_1+\lambda_2} + \lambda_i(E[\tau_c] + T_w/2)$ packets, on average (again the first term is due to the packet which triggers the $L \to C$ transition). Therefore, the delay is:

$$D_W^{(i)} = \frac{T_w}{2} + \rho_i\left(\frac{1}{\lambda_1+\lambda_2} + E[\tau_c] + T_w/2\right), \quad i \in \{1,2\}. \tag{26}$$

This delay is different for different traffic directions, as it was the case for D_A and D_C.

Average delay of a packet. To find the average delay that a packet can suffer, we need to compute the probability that a packet arrives in any of the possible EEE states. For each link direction, these probabilities can be seen as the average number of packets received in each of the different states divided by the average number of arrivals in a system cycle.

There is only one packet per cycle arriving in state L: it belongs to link direction i with probability $\lambda_i/(\lambda_1 + \lambda_2)$, which is then the average number of packets received in state L in direction i, namely $n_L^{(i)}$. Similarly, since there are ψ arrivals per cycle in state S, the average number of packet arrivals in direction i in state S is given by $n_S^{(i)} = \frac{\lambda_i}{\lambda_1+\lambda_2}E[\psi]$.

Since arrivals follow a Poisson process, packets received in link direction i during the fixed-length wake-up interval,

which is present only once in a cycle, are $n_W = \lambda_i T_w$, on average. Similarly, the number of arrivals in direction i during the coalescing interval (state C) is $n_C = \lambda_i E[\tau_c]$, on average. Eventually, arrivals in direction i in state A are the total number of (Poisson) arrivals in a cycle less the arrivals in the other states, i.e., $n_A^{(i)} = \lambda_i E[T_{cycle}] - n_S^{(i)} - n_W^{(i)} - n_L^{(i)} - n_C^{(i)}$. As a result, the corresponding average delay that a packet suffers in direction i is as follows:

$$D_i = \frac{\sum_{\alpha \in \{A,S,L,C,W\}} n_\alpha^{(i)} D_\alpha^{(i)}}{\lambda_i E[T_{cycle}]}, \quad i \in \{1,2\}. \tag{27}$$

5. DYNAMIC COALESCING STRATEGIES

In this section we present two classes of algorithms that can be used to dynamically match the coalescing parameters to the traffic conditions. These algorithms will allow us to explore the performance of dynamic packet coalescing, which has not been addressed so far in the literature.

The rationale behind proposing dynamic coalescing is that static configurations might incur in high delays when the traffic is low. For instance, to increase power saving, coalescing techniques try to avoid short and frequent transmission bursts by means of large T_c and N_c values. However, as soon as the traffic intensity causes frequent coalescing timeouts, the latency often exceeds T_c. Therefore, in a static configuration, one cannot use very large values of T_c when the traffic intensity can be low with non-negligible probability. Similarly, large values of N_c increase the achievable power saving, but cause high latency due to queue backlog to be served after the coalescing period. Using a static configuration for all traffic conditions might incur in the following problem: when the traffic is low the coalescing timer expires frequently, while when the traffic is high one or both coalescing buffers get full too quickly. In the former case, the experienced delay might be too high, while in the latter case, the achieved power saving might be far from optimal.

In the following subsections we present two classes of dynamic coalescing algorithms for EEE and evaluate their behavior in terms of power saving and packet delay. The first algorithm, Dynamic Timeout Algorithm, makes use of tunable coalescing timeout of duration T_c, while the second one, Dynamic Queue Size Algorithm, uses tunable coalescing buffers of variable size N_c.

5.1 Dynamic Timeout

In this class of algorithms, as indicated by its name, the adjustable parameter is the coalescing timeout T_c, while the coalescing buffers have fixed size N_c. We recall that T_c is defined as the maximum interval that EEE network cards can remain in state C after the arrival of the first packet in state L, thus extending the normal EEE power saving interval. As stated before, when the coalescing timeout expires, both Ethernet interfaces start transmitting the queued packets to the other link edge. The goal of this first class of algorithms is to keep the system in state C for as long as it is needed to fill up at least one coalescing buffer, i.e., to adjust T_c so that the coalescing operation (i.e., the duration of state C) lasts $\sim T_c$ seconds and the maximum coalescing gain is achieved by filling up the coalescing buffer.

The algorithm's behavior is described by means of pseudocode in Algorithm 1. In this algorithm, the value of T_c keeps changing when the transition $C \to A$ occurs. If the transition occurs because of a timeout expiration, then T_c is incremented, unless it reaches a maximum value $T_c^{\max}$. Its maximum value may depend on various factors but the most important is the maximum delay tolerance that is allowed either by applications, or by quality of service constraints. Similarly, if the transition $C \to A$ occurs because one of the coalescing buffers gets full, then T_c is decremented, unless it reaches its minimum allowable value $T_c^{\min}$. Therefore, the algorithm tries to adjust T_c in a way that state C lasts approximately T_c seconds while trying both to avoid unnecessary long timeouts and to fill up coalescing buffers.

In Algorithm 1, increments and decrements of T_c can follow different strategies. In particular, we consider that both increments and decrements can be either additive or multiplicative. Therefore, to fully specify the behavior of Algorithm 1, we need to specify (i) whether additive or multiplicative increments and decrements are used, (ii) the steps used in case of additive operation (δ_{up} for increments and/or δ_{down} for decrements), or the multiplicative factors used in case of multiplicative operation (γ_{up} for increments and/or γ_{down} for decrements), and (iii) the range $[T_c^{\min}, T_c^{\max}]$ in which T_c can be adjusted.

The performance of Algorithm 1 depends on the target N_c value fixed in the system. Throughout our experiments, we use $T_c^{\min} = 0.1 \, ms$ and $T_c^{\max} = 100 \, ms$, while we tested all combinations of additive and multiplicative increments and decrements, with various values for δ_{up}, δ_{down}, γ_{up}, and γ_{down}.

5.2 Dynamic Queue Size

The second algorithm is similar to the first, but it adjusts the coalescing buffer size N_c instead of the coalescing timeout T_c. In Algorithm 2, N_c is dynamically and automatically tuned in order to adapt the coalescing operation to achieve

---

Data: δ_{up} or γ_{up}, δ_{down} or γ_{down}, $T_c^{\min}$, $T_c^{\max}$, N_c
if *Transition $C \to A$* **then**
$\quad$ **if** *Coalescing timeout expired* **then**
$\quad\quad$ **if** $T_c < T_c^{\max}$ **then**
$\quad\quad\quad$ increase T_c;
$\quad\quad$ **end**
$\quad$ **else**
$\quad\quad$ **if** $T_c > T_c^{\min}$ **then**
$\quad\quad\quad$ decrease T_c;
$\quad\quad$ **end**
$\quad$ **end**
$\quad$ restart the coalescing timeout with new T_c;
end

Algorithm 1: Dynamic Timeout Algorithm.

Data: δ_{up} or γ_{up}, δ_{down} or γ_{down}, $N_c^{\min}$, $N_c^{\max}$, T_c
if *Transition $C \to A$* **then**
$\quad$ **if** *Coalescing timeout expired* **then**
$\quad\quad$ **if** $N_c > N_c^{\min}$ **then**
$\quad\quad\quad$ decrease N_c;
$\quad\quad$ **end**
$\quad$ **else**
$\quad\quad$ **if** $N_c < N_c^{\max}$ **then**
$\quad\quad\quad$ increase N_c;
$\quad\quad$ **end**
$\quad$ **end**
$\quad$ restart the T_c;
end

Algorithm 2: Dynamic Queue Algorithm.

---

a target coalescing delay T_c. I.e., the target of Algorithm 2 is to keep the system in state C for about T_c seconds and during this interval, accumulate as many packets as possible.

In Algorithm 2, when the traffic intensity is such low that the coalescing timeout expires before N_c packets are queued in any of the two coalescing buffers, the algorithm decreases N_c. The minimum value for N_c is $N_c^{\min} = 2$, since smaller values would not result in any coalescing operation. Conversely, when the traffic increases and at least one coalescing buffer fills up before the coalescing timeout expires, the algorithm increases N_c up to its maximum value $N_c^{\max}$. Similarly to what stated for Algorithm 1, increments and decrements can be either additive or multiplicative, with parameters that we keep calling δ_{up}, δ_{down}, γ_{up}, and γ_{down} like in the previous algorithm.

The performance of Algorithm 2 depends on the target value of T_c fixed in the system. Throughout our experiments, we use $N_c^{\min} = 2 \, pkt$ and $N_c^{\max} = 10000 \, pkt$, while we tested all combinations of additive and multiplicative increments and decrements, with various values for δ_{up}, δ_{down}, γ_{up}, and γ_{down}.

In the next Section we compare the two classes of dynamic coalescing algorithms with static algorithms.

6. PERFORMANCE EVALUATION

In this section we use ns-3 simulations and the model described in Section 4 to assess the performance of coalescing techniques over 1 *Gbps* EEE links. Note that the model cannot be used for dynamic coalescing, so that we will use simulations for the performance assessment of the *Dynamic Timeout* and the *Dynamic Queue Size* algorithms. However, we will show that static coalescing performs almost as well as dynamic coalescing, which reveals that our model can be used to quickly predict potential EEE performance with and without dynamic coalescing.

We investigate on the power saving and on the delay of the packets by using different δ and γ parameters for the dynamic algorithms presented in Section 5, and for various values of N_c and T_c for both static and dynamic approaches. Notably, for our evaluation we use real traffic traces that we captured at two of the firewall interfaces of InterHost, a large web hosting center in Madrid. Therefore, our results correspond to realistic traffic patterns, as well as realistic power saving and packet delay figures achieved by means of EEE and coalescing techniques.

To run simulations we modified the ns-3 simulator to implement EEE and packet coalescing. First, we designed and coded a novel Ethernet channel object in ns-3. Such an Ethernet channel can simulate the bidirectional behavior of Ethernet links. Second, we added on the net devices the EEE functionality, i.e., we defined the EEE states. Third, we implemented packet coalescing and coordinated packet transmission so that a simulated EEE link enters state L only when both traffic directions are inactive for T_s seconds, and exits state C only when coalescing operations are complete either because the coalescing timer expires or one of the coalescing buffers fills up.

6.1 Model vs. Simulation

Here we show that model and simulation return similar results, thus validating the model proposed in Section 4. High precision timestamped real bidirectional traces have been collected and used as input to the simulator. High precision

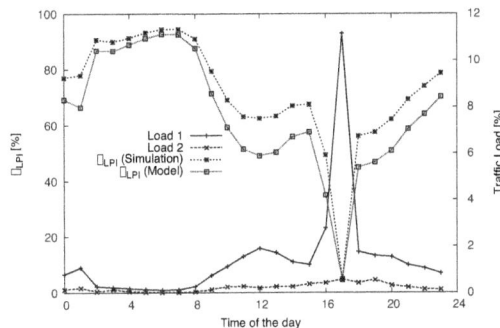

(a) Load and power saving opportunities (η_{LPI}).

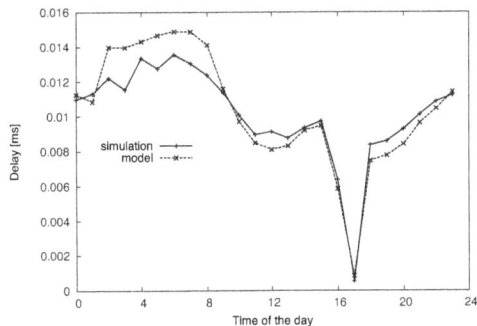

(b) Delay in the most delayed link direction.

Figure 4: EEE model and simulation results with real traces, sampling a one-day traffic pattern (without coalescing).

timestamping is very important for such kind of simulations as we have shown in [13]. As concerns the model, we used as input a few simple but significant statistical parameters, i.e., the average load, arrival rate and packet size in the two link directions.

6.1.1 Without Coalescing

We first consider a plain EEE scenario, with no coalescing. In Figure 4, we depict η_{LPI} and the packet delay computed via our modified ns-3 simulator and via the analytical model that we presented in Section 4.

Figure 4a illustrates the EEE power saving (in terms of η_{LPI}) for a typical day, and the measured traffic sampled once every 60 minutes. The load is very low most of the time, i.e., less than 1%, but we observed an exceptional peak value of medium traffic ($\sim 10\%$) at 5pm. This traffic behavior is in line with typical data center loads as described in [20].

We can observe that the model we previously presented estimates the power saving with a good accuracy in either low (0–1%) or medium ($\sim 10\%$) traffic conditions. Moreover, in Figure 4a, EEE can save most of the time more than 50% of power. In extreme cases (but not infrequently, since they constitute about 1/3 of the day's samples) the time spent in power saving (LPI) exceeds 90%. Remarkably, EEE enables substantial power saving ($\eta_{LPI} > 90\%$) when the traffic is limited to few percents of the link capacity. However, EEE does not appear to be able to save much power when the traffic surges to relatively low peaks, e.g., to 10% of the link capacity. With higher loads, we can safely assume that a plain EEE approach would bring negligible power saving. This justifies the research for EEE enhancements that would allow significant savings even when traffic reaches typical utilization levels such as 10% of the link capacity.

The accuracy of our model, in terms of delay performance, is evaluated through Figure 4b over the same traffic traces studied in Figure 4a. The Figure reports the delay suffered in the link direction over which the delay due to EEE state transitions is higher. Results achieved with model and simulation differs by at most $3\mu s$, which is a negligible quantity. The Figure also shows that packet delay is higher when the traffic offered to the link is lower. Moreover, delays due to EEE are never relevant, since they are comparable to or shorter than the duration a single packet transmission time.

In general, we have shown that our model can be safely used to estimate EEE performances in terms of both power saving and delay when no coalescing is adopted. In the next subsection we will show the correctness of our model also for the case of EEE with static coalescing. Before that, in light of our traffic measurements and power saving estimates, we enlighten the potential economical benefit of EEE.

Economical relevance of EEE power saving. From a purely economical point of view, our results for EEE are considerable. For instance, let us consider a large data center, e.g., the one of Google [21]. The data center includes over 40,000 servers, and each server has about 3 connected network ports, on average [22]. Each port may consume about 2 Watts using legacy NICs [3]. A typical load distribution consist in having about 40% of the link at almost zero load, 48% at about 10% load, and the rest of the links operated at higher load [20]. Thus, considering our results and that the average cost of electricity in Europe is about 0.15 $Euros$ per KWh, we can roughly estimate that EEE would reduce the annual electricity bill by 135,000 Euros. This is a rough estimate, and it does not include additional savings due to the reduced need for air conditioning.

Our preliminary results are particularly interesting since: (i) we can expect (and we will actually show it in the following) that packet coalescing can further boost the power saving achievable by means of EEE; and (ii) faster Ethernet cards, i.e., 10 and 100 $Gbps$, consume even more power (at least two and five times more, respectively), so that the potential for power saving is greater for higher data rates.

6.1.2 With Static Coalescing

To validate our model for EEE with static packet coalescing, in Table 1 we list some representative results achieved with model and simulation. We use the traffic traces collected at two different links in InterHost, Madrid (Spain). One of the links had low traffic ($< 10\%$) and the second one had traffic peaks of more than 30%. For each trace we report loads, arrival rates and average packet sizes for each link direction. We simulate the static coalescing algorithm for multiple combinations of N_c and T_c, and report the achieved results for the fraction of time spent in LPI (η_{LPI}), and for the average packet delay in the two directions (D_1 and D_2). We also use the average traffic descriptors reported in the Table to evaluate η_{LPI}, D_1 and D_2 through our model.

We observe that we have high traffic in one direction and low traffic in the other direction in all traces but the last, where the traffic is almost balanced. We note that high loads correspond to large packet sizes, which, in turn, correspond to the typical behavior of TCP traffic, i.e., large packets in one direction and small acknowledgements in the other direction. From Table 1, we can also observe that the model approximates very well both the time spent in LPI, η_{LPI}, and the average delay in the two link directions. η_{LPI} is

Table 1: η_{LPI} and average delays computed with the model and via simulation

ρ_1 [%]	ρ_2 [%]	$S_p^{(1)}$ [bytes]	$S_p^{(2)}$ [bytes]	λ_1 [pkts/s]	λ_2 [pkts/s]	T_c [ms]	N_c [pkts]	η_{LPI} [%] simul.	η_{LPI} [%] model	D_1 [ms] simul.	D_1 [ms] model	D_2 [ms] simul.	D_2 [ms] model
0.20	0.06	802	281	310	268	5	50	97.45	96.84	3.157	3.067	3.383	3.063
						5	100	97.45	96.84	3.157	3.067	3.383	3.062
						10	50	98.29	98.11	5.925	5.657	6.140	5.652
						10	100	98.30	98.11	5.942	5.660	6.147	5.652
						20	100	98.92	98.91	11.130	10.725	11.306	10.710
0.71	39.69	67	1512	13187	32815	5	50	0.001	0.001	0.002	0.006	0.002	0.011
						5	100	1.69	1.74	0.001	0.026	0.001	0.039
						10	50	1.72	0.88	0.003	0.006	0.004	0.011
						10	100	1.79	1.74	0.005	0.027	0.005	0.039
						20	100	1.85	1.75	0.007	0.027	0.009	0.039
0.87	29.64	83	1477	13048	25084	5	50	7.87	4.56	0.024	0.046	0.019	0.060
						5	100	7.97	8.64	0.028	0.174	0.023	0.225
						10	50	7.93	4.57	0.041	0.046	0.029	0.060
						10	100	8.71	8.65	0.068	0.174	0.058	0.225
						20	100	9.20	8.66	0.101	0.175	0.082	0.225
0.98	53.66	69	1500	17769	44719	5	50	0.10	0.03	0.000	0.000	0.001	0.004
						5	100	0.14	0.06	0.001	0.000	0.003	0.005
						10	50	0.10	0.03	0.000	0.000	0.001	0.004
						10	100	0.14	0.06	0.001	0.000	0.003	0.005
						20	100	0.14	0.06	0.001	0.000	0.003	0.005
3.72	0.37	1180	165	3938	2778	5	50	85.03	90.90	1.896	2.442	2.123	2.363
						5	100	85.29	90.90	1.954	2.442	2.153	2.364
						10	50	88.26	94.08	3.971	4.946	3.819	4.786
						10	100	90.10	94.10	4.331	4.972	4.453	4.811
						20	100	91.90	95.82	8.926	9.031	8.256	9.707
5.06	0.50	1170	165	5408	3809	5	50	78.76	88.10	1.727	2.380	1.869	2.277
						5	100	79.21	88.10	1.815	2.380	1.922	2.277
						10	50	82.05	91.63	3.854	4.334	3.160	4.146
						10	100	85.52	92.18	4.079	4.914	4.109	4.701
						20	100	87.54	94.17	7.700	8.029	6.910	8.642
9.74	6.28	1165	855	10452	9182	5	50	65.61	64.14	1.390	1.601	1.252	1.550
						5	100	70.55	66.28	1.947	1.850	1.685	1.792
						10	50	69.70	64.43	1.593	1.632	1.634	1.581
						10	100	79.90	75.90	3.610	3.879	3.663	3.757
						20	100	79.84	76.05	3.710	3.928	3.670	3.805

estimated with ±5% deviation in most of the cases (90%), while average delay estimations are subject to an error which is of the order of 10% for high delay cases (several ms), and a few μs for the case of low delays (tens of μs or less).

Overall, under any of the tested traffic conditions and coalescing configurations, the results achieved through the model are accurate enough with respect to simulations. However, running simulations requires much more time and computational resources than the model. Hence, our model results in a suitable tool for the quick evaluation of (i) EEE potentials and (ii) coalescing configuration effectiveness, under any traffic condition.

6.2 Static vs. Dynamic Coalescing

We now compare the performance of static and dynamic coalescing. We use simulation only, since our model was not designed to predict the behavior of dynamic coalescing schemes. However, we will show that static and dynamic coalescing achieve very similar results, so that developing a model for EEE with dynamic coalescing is unnecessary.

6.2.1 Static Coalescing

Table 1 reveals that coalescing enables high power saving opportunities in a variety of load conditions. In particular, even in presence of loads of the order of 10%, power can be saved 60 to 80% of the time. High value of N_c and T_c would even allow for relevant power saving ($\sim$ 10%) under high traffic ($\sim$ 30%). The Table also reveals that delays grow fast with N_c and T_c. However, under high load, the achieved average delay assumes values not greater than a few hundreds of μs. High delays can be suffered only when the traffic is low and the coalescing parameters are high.

Notably, using $N_c \in [50, 100]$ packets and $T_c = 10\,ms$ yields

high power saving under all reported cases, with quite limited and acceptable delays. Hence, static coalescing appears to be near-optimal under a wide range of traffic conditions.

In addition to what reported in Table 1, we tested many possible combinations for many traffic traces, with N_c ranging from 2 to 10000 packets, and T_c ranging from 0.1 to 100 ms. Due to space limitations, we omit here a complete description of our results with static coalescing and—since configurations resulting in high delay are undesirable—we limit our discussion only to cases with average delay below 1 ms. Specifically, we selected four different traffic traces, corresponding to the most typical traffic loads. Figure 5 reports the results for both static and dynamic algorithms under the following traffic conditions; Figure 5a is for $\rho_1 = 0.2\%$ and $\rho_2 = 0.06\%$, Figure 5b for $\rho_1 = 5.1\%$ and $\rho_2 = 0.1\%$, Figure 5c for $\rho_1 = 10.9\%$ and $\rho_2 = 0.2\%$ and Figure 5d for $\rho_1 = 39.7\%$ and $\rho_2 = 0.7\%$. Since we are interested in the potential performance of coalescing in terms of power saving and delay, Figure 5 plots the values for the delay in the most loaded link direction as a function of the achieved η_{LPI}. Each blue "*" marker in the Figure is obtained under a different coalescing configuration for N_c and T_c.

Figures 5a to 5d show that higher power saving corresponds to higher delay. However, delay can be minimized in exchange of small power saving reduction. Under low loads, as in Figure 5a, the value of η_{LPI} stays well above 90% in all cases, which means that coalescing of a very few packets (e.g., $N_c = 2$), combined with a short coalescing timeout (e.g., $T_c = 1\,ms$), is more than enough to achieve high power saving with very limited packet delay. Conversely, under high load (e.g., Figure 5d), there is no static configuration that can bring high power saving with bounded delay.

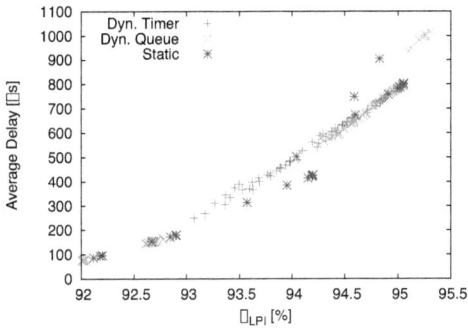

(a) $\rho_1 = 0.2\%$, $\rho_2 = 0.06\%$.

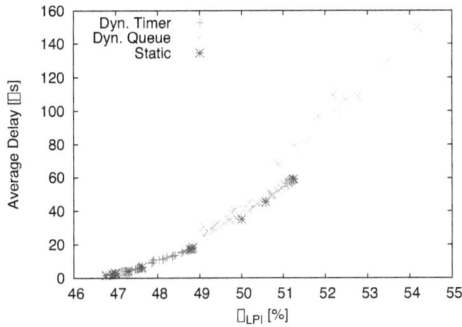

(b) $\rho_1 = 5.1\%$, $\rho_2 = 0.1\%$.

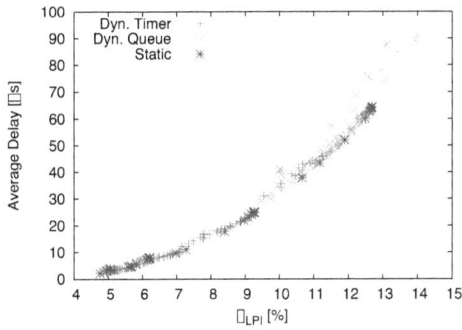

(c) $\rho_1 = 10.9\%$, $\rho_2 = 0.2\%$.

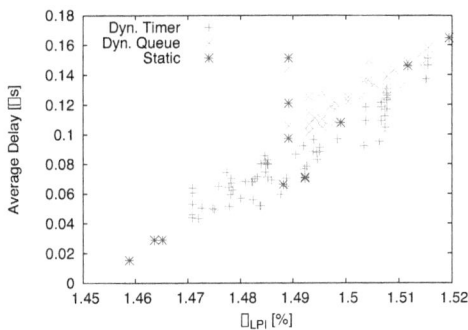

(d) $\rho_1 = 39.7\%$, $\rho_2 = 0.7\%$.

Figure 5: η_{LPI} vs. delay performance using different coalescing algorithms under various traffic conditions.

6.2.2 Dynamic Timeout

The Dynamic Timeout Algorithm "helps" the EEE coalescer to adapt its T_c value to the traffic conditions. In Figure 5, in addition to results for static coalescing, we plot the average delay (the maximum in the two link directions) versus η_{LPI} for $N_c \in [2, 10000]$, δ from 0.1 to 10 ms, and γ from 1 to 100%. Adopted traffic traces are the same as for the evaluation of static coalescing. Similarly to the static coalescing case, we report results for the cases in which the average added delay per packet is below 1 ms only, which is a relatively small and acceptable value. Tested values for additive parameters δ larger than 1 ms yielded delays higher than 1 ms, so we are not reporting those tests in the Figures. The red "+" markers plotted in Figure 5 correspond to the different tested configurations.

For all four traces used in Figure 5, we can see that Dynamic Timeout achieves almost the same results as for the static coalescing. Under low load conditions η_{LPI} is slightly higher than 90% while for loads equal to $\sim 5\%$, $\sim 10\%$ and $\sim 40\%$ η_{LPI} is $\sim 50\%$, $\sim 10\%$ and $\sim 1.5\%$, respectively. Within each Figure, it is possible to observe that there are no huge differences in the power saving and delay performance achieved under different configurations for the same traffic trace analysis (less than 10% for η_{LPI} and up to a few hundreds of μs for the delay).

We conclude that dynamically adjusting the coalescing timeout does not outperform static coalescing when the coalescing delay has to be kept low under all traffic conditions.

6.2.3 Dynamic Queue Size

The Dynamic Queue Size algorithm adapts N_c to the traffic conditions and slightly outperforms the other tested algorithms in terms of power saving. This behavior is shown in Figure 5, which also includes results for the Dynamic Queue Size algorithm, tested for $T_c \in [0.1, 100]$ ms, additive parameters δ from 1 to 10 packets, and multiplicative parameters γ from 1 to 100%. Similarly to what we did for the other algorithms, we simulate EEE with coalescing using the Dynamic Queue Size algorithm over the same traffic traces as before. Each green "×" marker in Figures 5a to 5d corresponds to a different configuration for the Dynamic Queue Size algorithm.

In terms of achievable power saving, results for the Dynamic Queue Size case are, in general, slightly better than for static coalescing and for the Dynamic Timeout algorithm. Specifically, in Figure 5a, under very low load, η_{LPI} increases due to Dynamic Queue Size are relatively low. Instead, under medium/high loads (see Figures 5b and 5c) we observe an increase of a few percents in η_{LPI}, with respect to the other algorithms. Eventually, under very high loads (e.g., see Figure 5d), η_{LPI} is very small under any algorithm and configuration. In all cases, the (slightly) higher power saving is achieved at the expenses of slightly higher delay.

We can safely infer that dynamically adjusting the coalescing buffer size does not allow to achieve significantly better performance with respect to static coalescing. Indeed, having shown that static coalescing and dynamic coalescing algorithms achieve similar power saving and delay tradeoffs under a variety of traffic configurations, we conclude that static coalescers are preferable for real implementation due to their low complexity.

171

7. CONCLUSIONS

In this paper, we have presented a model for bidirectional EEE links with static or no coalescing. We have shown that the model can be used to accurately estimate power saving and packet delay over EEE gigabit links. This model is unique in the existing literature, in which only unidirectional EEE links are accounted for.

We have modified the ns-3 simulator to implement (i) the EEE standard and (ii) static as well as dynamic coalescing algorithms, and thus validate our model. Furthermore, we have proposed an exhaustive performance evaluation on the impact of packet coalescing techniques over EEE power saving and delay performances. Specifically, we have tested EEE and coalescing algorithms by means of real packet traces we collected at the firewall interfaces of a large web hosting center in Madrid, Spain.

Notably, we have shown that static coalescing algorithms, in which the coalescing queue size N_c and the coalescing timeout T_c are fixed, can achieve results as good as dynamic coalescing algorithms, in which either N_c and T_c can be dynamically adapted to the traffic characteristics. Hence, our model can be used to estimate the potential power saving-delay tradeoff of EEE with static or dynamic coalescing.

Our study has shown that the sole EEE standard (without coalescing) works fine under scarce traffic ($< 1\%$). In contrast, as soon as the traffic exceeds a few percents of the link capacity, EEE needs to be endowed with packet coalescing to achieve significant power saving. Thanks to coalescing, significant economy can be achieved with link loads as high as $40-50\%$, while plain EEE would not allow to achieve detectable power saving with loads higher than a few percents of the link capacity. According to our results, static packet coalescing techniques appear to be nearly-optimal, and thus preferable to dynamic techniques due to their complexity.

8. ACKNOWLEDGMENTS

This research was supported in part by the Spanish MICINN grant TEC2011-29688-C02-01.

The authors would like to thank InterHost S.A. for allowing them to collect anonymized traffic traces in their web hosting center in Madrid, Spain.

9. REFERENCES

[1] European Commission - (DG INSFO), "Study of the impacts of ICT on energy efficiency," Sept. 2008.

[2] K. Christensen, P. Reviriego, B. Nordman, M. Bennett, M. Mostowfi, and J. A. Maestro, "IEEE 802.3az: The Road to Energy Efficient Ethernet," *IEEE Communications Magazine*, vol. 48, no. 11, pp. 50–56, Nov. 2010.

[3] R. Sohan, A. Rice, A. Moore, and K. Mansley, "Characterizing 10 Gbps Network Interface Energy Consumption," in *Proceedings of IEEE LCN 2010*, Oct. 2010.

[4] IEEE Std. 802.3az, "Energy Efficient Ethernet," 2010.

[5] M. Ajmone Marsan, A. Fernandez Anta, V. Mancuso, B. Rengarajan, P. Reviriego Vasallo, and G. Rizzo, "A Simple Analytical Model for Energy Efficient Ethernet," *IEEE Communications Letters*, no. 99, pp. 1–3, June 2011.

[6] D. Larrabeiti, P. Reviriego, J. A. Hernandez, J. A. Maestro, and M. Uruena, "Towards an Energy Efficient 10 Gb/s optical Ethernet: Performance analysis and viability," *Optical Switching and Networking*, vol. 8, no. 3, pp. 131–138, Mar. 2011.

[7] S. Herrería-Alonso, M. Rodríguez-Pérez, M. Fernández-Veiga, and C. López-García, "How efficient is energy-efficient ethernet?" in *Proceedings of ICUMT*, Oct. 2011.

[8] NS-3 website: http://www.nsnam.org/.

[9] S. Herrería-Alonso, M. Rodríguez-Pérez, M. Fernández-Veiga, and C. López-García, "A GI/G/1 Model for 10Gb/s Energy Efficient Ethernet Links," *IEEE Transactions on Communications*, vol. 60, no. 11, pp. 3386–3395, Nov. 2012.

[10] M. Mostowfi and K. Christensen, "An Energy-Delay Model for a packet coalescer," in *Proceedings of IEEE Southeastcon*, Mar. 2012.

[11] P. Reviriego, J. A. Hernandez, D. Larrabeiti, and J. A. Maestro, "Performance Evaluation of Energy Efficient Ethernet," *IEEE Communications Letters*, vol. 13, no. 9, pp. 697–699, Sept. 2009.

[12] P. Reviriego, K. Christensen, J. Rabanillo, and J. A. Maestro, "Initial Evaluation of Energy Efficient Ethernet," *IEEE Communications Letters*, vol. 15, no. 5, pp. 578–580, May 2011.

[13] V. Mancuso and A. Chatzipapas, "On IEEE 802.3az Energy Efficiency in Web Hosting Centers," *IEEE Communications Letters*, vol. 16, no. 11, pp. 1880–1883, Nov. 2012.

[14] P. Reviriego, J. A. Maestro, J. A. Hernandez, and D. Larrabeiti, "Burst Transmission for Energy Efficient Ethernet," *IEEE Computer Society*, vol. 14, no. 4, pp. 50–57, July 2010.

[15] P. Reviriego, K. Christensen, and A. Sanchez-Macian, "Using Coordinated Transmission with Energy Efficient Ethernet," in *Proceedings of IFIP Networking*, May 2011.

[16] M. Mostowfi and K. Christensen, "Saving Energy in LAN Switches: New Methods of Packet Coalescing for Energy Efficient Ethernet," in *Proceedings of IGCC*, July 2011.

[17] S. Herrería-Alonso, M. Rodríguez-Pérez, M. Fernández-Veiga, and C. López-García, "Bounded energy consumption with dynamic packet coalescing," in *Proceedings of IEEE NOC*, June 2012, pp. 1–5.

[18] B. Nordman, "EEE Savings Estimates," IEEE 802.3 Energy Efficient Ethernet Study Group, May 2007. [Online]. Available: http://www.ieee802.org/3/eee_study/public/may07/nordman_2_0507.pdf

[19] L. Kleinrock, *Queueing Systems: Theory.* John Wiley and Sons, 1975, vol. 1.

[20] T. Benson, A. Anand, A. Akella, and M. Zhang, "Understanding data center traffic characteristics," *SIGCOMM Comput. Commun. Rev.*, vol. 40, no. 1, pp. 92–99, Jan. 2010.

[21] Google, "(Google Data Center Video Tour)," http://www.google.com/about/datacenters/events/2009-summit.html#tab0=4, Apr. 2009.

[22] S. Bapat, "The Future of Data Centers (... and the Stuff That Goes In Them)," in *Proceedings of 1st Berkeley Symposium on Energy Efficient Electronic Systems*, June 2009.

Inferring Connectivity Model from Meter Measurements in Distribution Networks

Vijay Arya
IBM Research – India
vijay.arya@in.ibm.com

T. S. Jayram
IBM Research – Almaden
jayram@us.ibm.com

Soumitra Pal
CSE, IIT - Bombay, India
mitra@cse.iitb.ac.in

Shivkumar Kalyanaraman
IBM Research – India
shivkumar-k@in.ibm.com

ABSTRACT

We present a novel analytics approach to infer the underlying interconnection between various metered entities in a radial distribution network. Our approach uses a time series of power measurements collected from different meters in the distribution grid and infers the underlying network between these meters. The collected measurements are used to set up a system of linear equations based upon the principle of conservation of energy. The equations are analyzed to estimate a tree network that optimally fits the time series of meter measurements. We study experimentally the number of measurements needed to infer the true underlying connectivity with the help of both synthetic and real smart meter measurements in the noiseless setting.

Categories and Subject Descriptors

G.1.6 [**Mathematics of Computing**]: Optimization;
G.3 [**Mathematics of Computing**]: Time series analysis;
H.4 [**Information Systems Applications**]: Miscellaneous;
I.6.5 [**Model Development**]: Modeling methodologies

General Terms

Measurement, Algorithms, Design, Experimentation

Keywords

Power distribution grids, radial networks, connectivity model, phase identification, compressed sensing, sparsity

1. INTRODUCTION

The *connectivity model* of a distribution grid gives the underlying interconnection between various assets and customers in the grid downstream of a substation *i.e.* which distribution transformer is connected to which feeder, which customer is connected to which distribution transformer,

and which customer is connected to which phase, and so on. The accuracy of this information generally deteriorates over time due to repairs, maintenance, and balancing efforts. As a consequence, most energy distributors have partial or inaccurate knowledge of the connectivity information from the substations down to the consumers.

The connectivity model is important as it is needed in the operations and maintenance of distribution networks. Many solutions that automate the management of distribution networks require the connectivity model as input. For instance, it is required by the outage management system (OMS) to accurately record and respond to outages. During a storm or other disruptive events, when an asset (e.g. a transformer) fails, the connectivity model is needed to identify the customers and interconnected assets that may have been affected by the failure. Additionally many energy distributors have obligations to accurately report customer outages to regulatory bodies which is difficult without an accurate connectivity model. The distribution management system (DMS) needs the connectivity model to conduct accurate power-flow calculations. Inaccurate models in these calculations lead to faulty voltage profiles of the distribution grid, which affects the efficiency and reliability of energy delivery.

A large fraction of all losses in a power system occur in the distribution network and the growing imbalance between supply and demand is driving the deployment of solutions that can improve the overall efficiency of energy delivery. An accurate connectivity model can enable many of these solutions. For instance, solutions for energy auditing and loss localization use the connectivity model to localize energy losses from theft and inefficiencies in the distribution network. The phase balancing solution requires the connectivity model in order to balance the load on the three phases of a feeder so that losses incurred while delivering energy to the customer are minimized. Additionally when customers have behind the meter resources such as distributed generation and storage, the connectivity model is required to ensure a balanced and reliable infusion of energy back into the grid via the distribution network [1–5].

Energy distributors worldwide are undertaking smart grid transformations and instrumenting their distribution networks with meters that communicate measurements with greater frequency and report in real-time on the electrical behavior of the network [6]. Different meters are being deployed as part of the AMI initiative – feeder meters, transformer meters, and consumer smart meters. One of the

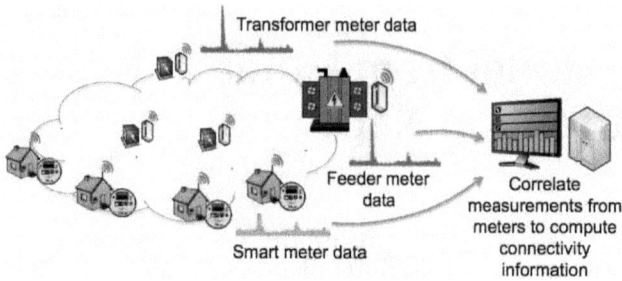

Figure 1: System Overview: A time series of power measurements from various meters is used to estimate a connectivity model between the metered assets in the distribution grid.

expectations from this deployment is the automated inference of the connectivity model. Distributors expect that in future, expensive field inspections would no longer be necessary to determine the connectivity model and that it would be inferred automatically. One possible solution is to use power-line communications wherein asset meters talk to each other by reading and writing signals onto the power-line. However this is expensive as it incurs significant capital and maintenance costs. Besides the installation affects grid operations, which is inconvenient as well.

In this work, we present a novel analytics approach [7] (Fig. 1) to estimate the connectivity model of a radial[1] distribution network. Our techniques are novel as they are purely based upon a time series of power measurements collected by various meters in the distribution grid. The measurements are used to set up a system of linear equations based upon the principle of conservation of energy *i.e.* during any time interval, the load measured by a feeder meter must be equal to the sum of loads measured by all customer meters connected to that feeder plus any errors. The errors arise due to imperfect synchronization of measurements at different meters, different sampling rates, unmetered loads such as street lights, and unknown and time-varying transmission line losses (errors are discussed in section 4.1). The equations are analyzed to estimate a tree connectivity model between the meters which is consistent with the observed time series measurements. We view the main contributions of this work as follows:

1. A novel analytics approach is proposed to estimate the connectivity model of a distribution network from a time series of meter measurements based upon mathematical optimization.

2. We develop novel integer programming formulations for noiseless and noisy variants of the inference problem. We propose continuous relaxations of these integer programs (with and without sparsity) that may be used with increasing number of measurements to retrieve the connectivity model efficiently. We study the conditions for uniqueness of solution to these mathematical programs as a function of the number of meter measurements.

3. The methods, tools, and implementation using mathematical programming are outlined together with ex-

perimental results using synthetic and real smart meter measurements.

The rest of the paper is organized as follows. Section 2 introduces the distribution network and its connectivity model. Section 3 presents related work followed by section 4 which describes the analytics approach to infer the connectivity model. Section 5 presents the mathematical inference problem resulting from the analytics technique as well as solutions for noiseless and noisy variants of the problem. Section 6 presents preliminary experimental results. Section 7 presents conclusions and directions for future work.

2. THE DISTRIBUTION SYSTEM AND ITS CONNECTIVITY MODEL

This section describes the major components of a distribution system which are relevant to this work, how these are interconnected, and the connectivity model that we seek to infer from meter measurements.

2.1 Distribution systems 101

The majority of electrical power is generated at large power plants as a 3-phase AC voltage and reaches the distribution network via a transmission system.

A distribution network starts from a *distribution substation* that serves customers in a geographical area. A distribution substation feeds one or more 3-phase *feeders* that carry power either directly to a few industrial customers or to several low-voltage *distribution transformers* (DTs) that step down the voltage further to serve residential *customers*. Depending on several factors including the type and density of customers, common feeder voltages include 33kV, 27kV, 13.2kV, 8.3kV, and 4.8kV. A substation may serve a couple of thousand customers while a feeder may serve a few hundred customers.

A 3-phase feeder consists of three transmission lines that carry AC power with their voltage waveforms shifted by 120°. These phases are usually labelled as A, B, and C. In north america, each DT receives power by tapping onto one of the three phases of a feeder and is therefore single-phase. A DT serves about 1-10 single-phase residential customers. Some DTs that serve larger 3-phase loads such as super-markets and office buildings are 3-phase themselves. In other countries such as Australia, the DTs are generally 3-phase. These DTs have a higher capacity and serve about 50-200 customers. However the residential customer is single-phase as usual and each customer is connected to one of the three phases of the DT[2].

Barring a few exceptions, most distribution networks are radial *i.e.* they have a tree structure. Even though the feeders from a substation may be connected together for contingencies, the configuration of circuit breakers is such that the operational network is always a tree and there exists only one path for power to flow from the feeder to the customer [1,2].

2.2 The connectivity model

Figure 2 shows an example of a distribution network under a substation (left) and the corresponding connectivity model that we seek to recover (right).

---

[1] A radial network has a tree structure

---

[2] Rest of the paper assumes a north american setting for ease of exposition though our solutions are applicable to other grids as well.

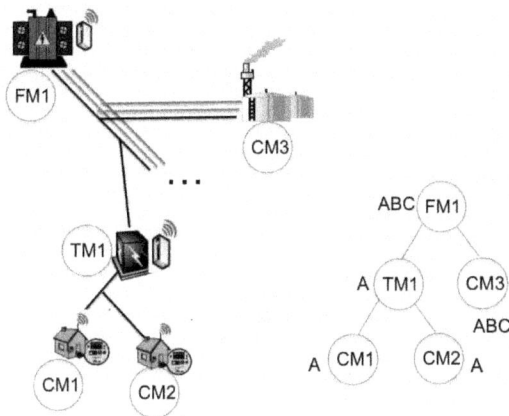

Figure 2: An example of a distribution network and its connectivity model

The figure shows a setting where a distributor has metered a feeder, the downstream DTs, and its customers (the meters are shown in circles). The connectivity model shows that meters TM1 and CM3 are downstream of FM1. Similarly CM1 and CM2 are downstream of TM1 and so on. The phases of meters are shown, for example CM1 and CM2 are phase A. The tree network between the meters is essentially a representation of the connectivity model between the metered assets *i.e.* the feeders, DTs, and customers. We essentially seek to infer the above tree network between the meters, along with the phase of single-phase meters (A, B, or C), purely based upon their power measurements.

While the set of customers under each substation is known and remains unchanged[3], accurate knowledge about the interconnections between the feeders, DTs, and customers is variable. These interconnections generally change over the years due to repairs, maintenance, and rebalancing efforts. For e.g. a DT may be assigned to another phase during a loss of phase[4] fault or in order to balance the load on the three phases. A customer could be assigned to a different DT if the connected DT fails. Some DTs or larger loads may also be moved from one feeder to the other to ensure power quality to the customer. Some of these interconnections may also be altered while restoring service to customers after storm related events. As a consequence several distributors have inaccurate or out of date connectivity models of their distribution network.

After undertaking AMI deployment which is generally capital intensive, most energy distributors seek to maximize their ROI using meter measurements. Automated inference of the connectivity model is a natural choice as it also enables a number of other solutions to manage distribution grids. Some distributors may have feeder and customer measurements and seek to improve the accuracy of customer phase connectivity information. Others who have measurements from DTs as well may seek to improve the accuracy of the connectivity information between the DTs and customers. Some distributors may have undergone a full AMI deployment and seek to automate the inference of the connectivity

---

[3]Many utilities in India print the substation number on the customer's electricity bill.

[4]A fault that results in loss of power on one of the 3 phases

model in order to avoid expensive manual inference after maintenance and repairs.

3. RELATED WORK

The literature on automated inference of the connectivity model is limited and restricted to the identification of customer phase.

Caird [8] discloses a system and method for phase identification with suitably enhanced automated meters that can detect phases based upon a unique signal injected into the phase line. The disadvantage of signal injection methods or in general those that rely on power-line communication (PLC) is that they require enhanced hardware to transmit and receive signals at different points of the grid, increasing capital and maintenance costs. If a PLC compliant meter were to cost \$5 more than a regular smart meter, then a utility with 2M customers would need to spend an additional \$10M excluding grid meters. Moreover in north america, feeders from substations can run for tens of miles before reaching a DT. Therefore PLC-based solutions become impractical and expensive as the signal does not propagate without repeaters. Our approach on the other hand simply relies on meter measurements and therefore doesn't require any additional hardware other than conventional meters. Moreover there is no requirement for interventions through signal injection or physical access to record measurements.

Dilek's [9] work on phase prediction in power circuits is similar in spirit to our approach to infer customer phase. The author employs a heuristic Tabu search on power flow measurements to determine the phase of attached loads. However there are several differences as well. Unlike [9], our work formulates a clean mathematical optimization problem for both noiseless and noisy variants of the problem. We discuss different types of errors, uniqueness of optimal solution, and mathematical relaxations that can be used to obtain solutions efficiently with increasing number of measurements. Whereas the approach of [9] is tested only on a few loads, we present experimental results for larger number of customers.

Our prior [10, 11] work describes an analytics approach to infer the phase of a customer for a variant of problem encountered in Australia, where a DT is 3-phase and its customers are single-phase. In this work, we show that a similar approach is applicable even to the north american market, although the problem instances may become large. Here the customers and DTs are single-phase while the upstream feeder is 3-phase.

The phase connectivity problem is a small subset of the entire connectivity model puzzle. In this work, we show that the energy conservation principle can be generalized to infer the entire tree connectivity model. We also propose a new sparsity based optimization formulation that is useful for larger problem instances where the number of measurements may be less relative to the number of customers.

Our work can be regarded as an early example of the tomography technique [12–14] applied in the smart grid context where measurements collected from meters in the distribution network are used to estimate its connectivity model.

4. THE ANALYTICS APPROACH

In our approach, the connectivity model is determined using a time series of synchronized measurements collected from the feeder, DTs, and customers. The principle of conservation of energy implies that during any time interval:

(a) the total load on a feeder equals the sum of loads of all downstream customers on that feeder, (b) the load on a DT equals the sum of loads of all downstream customers connected to it, and (c) the load on each phase of a feeder equals the sum of loads of all customers (equivalently all DTs) on that phase. Since customer loads vary with time and across different customers, the connectivity model can in fact be recovered by analyzing a time series of load measurements from the feeders, DTs, and customers collected over multiple time steps.

Time	Meter Measurements (Wh)				
Interval	M1	M2	M3	M4	M5
(0, 10]	8	3.5	1	2.5	4.5
(10, 20]	20.5	13.5	5	8.5	7
(20, 30]	18.5	14.5	9	5.5	4
⋮	⋮	⋮	⋮	⋮	⋮

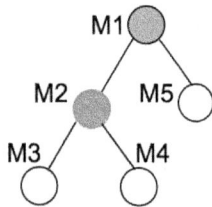

Figure 3: An example of the mathematical problem that arises in the analytics approach. (top): a time series of power-flow measurements from different meters. (bottom): the true connectivity model to be recovered from measurements.

The table in Fig 3 shows an example of a time series of power-flow measurements taken from all the meters under a feeder over multiple time intervals. Each row of the table corresponds to one time series measurement from all meters, which we simply refer to as one measurement. For instance, the measurement on the first row shows that during the first 10 minutes M1 recorded 8 watt-hours (Wh) of load while M5 records 4.5Wh of load, and so on. We wish to determine the connectivity model using the above time series measurements. Now observe that, for each row in the table, the loads measured by M3 and M4 add up to the load measured by M2. Similarly M1 = M3 + M4 + M5 = M2 + M5. From this, one can conclude that the tree network between the meters is most likely the one shown in Fig 3.

In this example, M1 meters a feeder which serves a DT and a large customer, metered respectively by M3 and M5. The DT serves two residential customers, metered by M3 and M4. The table shows the total Wh measured by M1. Since M1 is a feeder meter, it records the Wh on each phase separately. Similarly M5 a 3-phase customer and meters each phase separately. M2, M3, and M4 are single-phase.

In the above example, at each time step, the load measured by a parent meter is *exactly* equal to the sum of loads measured by its child meters. In practice however, due to energy losses on the transmission lines and other errors explained in the next section, the load measured at a parent meter is only *approximately* equal to the sum of loads measured at its child meters.

Fig. 4 shows the general problem that results from the analytics approach. Given a list of meters under a feeder

and their power-flow measurements, the tree connectivity between the meters needs to be recovered. In general, the list of meters may also be a forest instead of a single tree.

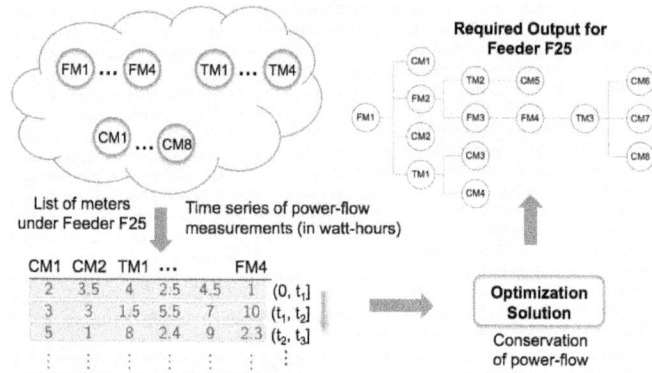

Figure 4: Analytics Approach: Given a time series of power-flow measurements from different meters, the optimization solution exploits the principle of conservation of energy to infer a tree connectivity between the meters which is consistent with the measurements. FM_, TM_, and CM_ correspond to feeder, DT, and consumer meters respectively.

4.1 Measurement setup and errors

Consumer smart meters can record and report periodic measurements of energy consumed in watt-hours (Wh) over small time intervals such as $\Delta t = 30$ min or 1h, as setup by the utility. For e.g. Itron smart meters used by CenterPoint Energy in US record Wh over 15 min intervals. Government regulations require that the Wh reported by consumer meters be accurate, typically of the order of 99.5% accuracy.

The distribution grid is generally monitored using SCADA (supervisory control and data acquisition) systems [15] that record feeder measurements close to the substation. While feeder metering is quite common, the metering of DTs varies across distributors.

The energy measured by customer and grid meters over common time intervals follows the principle of conservation only approximately instead of exactly due to a number of different *errors*.

Synchronization. A customer meter records readings based on its internal clock which may be out of synch with respect to the true clock. For e.g. if a meter reports that 75Wh were consumed from 10:00 to 10:30AM and its clock lags the true clock by 1 sec, in reality the 75Wh were consumed from 10:00:01 to 10:30:01 AM. Therefore even if all consumer meters are setup to report over the same time intervals, each may suffer from a different clock drift and report Wh consumed over a slightly different time interval. Although these errors are expected to be minimal, they prevent the conservation equations to hold with equality.

Sampling rates of grid meters. The DT and feeder meters are much more complex devices. Unlike consumer meters, these are used for monitoring instead of billing. They measure power parameters such as voltage, current, and power factor. However based on the manufacturer and settings, they measure these parameters at different sampling rates and may report either an average or instantaneous value over small time intervals (e.g. 5, 10 min). Therefore the watt-hours computed from these parameters are estimates of the actual watt-hours supplied during a time interval and may contain errors (the *real* or *active* power (Wh) supplied can be computed as a product of rms voltage, rms current, and power factor measurements). Additionally, clock synchronization problems may occur at the DT and feeder meters.

Line losses and Unmetered loads. Two other important sources of errors include line losses and unmetered loads. Since the transmission lines from the substation to DTs and DTs to customers have a certain amount of electrical resistance, some of the transferred energy is lost as heat. These losses vary with ambient temperature, load, and age of the feeder. Lastly there may be a few unmetered loads in the system such as street and traffic lights which also introduce errors in the conservation equations.

5. THE MATHEMATICAL INFERENCE PROBLEM AND SOLUTION

Our goal is to infer the connectivity tree between the feeder, DT, and customer meters. The set of meters under each feeder forms a tree and the set of trees corresponding to all feeders under one substation effectively result in a forest. For ease of exposition, we begin by assuming that the set of all meters under one feeder is given and we wish to determine a tree connectivity between them (this assumption is not required and one could possibly start with the set of all meters under one substation as well).

In the meter connectivity tree corresponding to a feeder, the customer meters are the leaf nodes, the DT meters are the intermediate nodes, and the feeder meter is the source node. We determine this connectivity tree iteratively by working out the set of leaf meters in the subtree rooted at each non-leaf meter. Observe that by knowing the leaf nodes under each non-leaf node of a tree, one can uniquely recover the entire tree in a straightforward manner.

5.1 The leaf connectivity (LC) problem

We define a *leaf connectivity* (LC) problem that takes as input a time series of load measurements from a set of leaf meters ℓ and a non-leaf meter s and determines the *subset* $p \subseteq \ell$ of leaf meters present in the subtree rooted at s. We represent this as $p = \text{LC}(s, \ell)$.

We model the problem as follows. Let $n = |\ell|$ denote the number of leaf meters. Let m be the total number of load (Wh) measurements taken over time. Let s_k and L_{kj} denote the loads measured by the non-leaf meter s and leaf meter j in the interval k respectively, $1 \leq k \leq m$, $1 \leq j \leq n$. Now let $A = [L_{kj}]_{m \times n}$ denote the matrix of all leaf meter measurements. Let $b = [s_j]_{m \times 1}$ denote the vector of source meter measurements. Thus each row of A corresponds to one time series measurement and each column of A corresponds to load measurements from one meter over multiple time intervals.

Our goal is to identify the subset of leaf meters $p \subseteq \ell$ present in the subtree rooted at s. Therefore let x_j be an indicator variable such that

$$x_j = \begin{cases} 1 & \text{if leaf } j \text{ is in the subtree rooted at } s \\ 0 & \text{otherwise} \end{cases} \quad (1)$$

Let $X = [x_j]_{n \times 1}$. Now the principle of conservation of energy implies that:

$$b = AX + e \quad (2)$$

where $e = [\epsilon_k]_{m \times 1}$. $\epsilon_k \in \mathbb{R}$ is the error in the kth measurement that compensates for the difference between the load measured by s and sum of loads measured by all meters in p (errors are described in section 4.1). The model allows errors to vary across measurements.

The leaf connectivity problem is to determine the unknown binary vector $X \in \{0, 1\}^n$ given A, b, and unknown e from (2). Before describing our solutions to noiseless and noisy variants of this problem, we show how different connectivity problems can be mapped onto this basic problem and solved.

5.2 Instances of leaf connectivity (LC) problem

As explained in section 2.2, the accuracy of the connectivity model as well as the level of instrumentation in the grid is variable and depending on these, a distributor may ask different questions about the connectivity model. We consider a few examples to explain how these questions could be answered by solving different instances of the leaf connectivity (LC) problem of varying sizes.

Customer-feeder. The set of all customers under a feeder may be determined by calling LC with s as the feeder meter and ℓ as the set of all customer meters under the substation.

Customer-DT. If a distributor has instrumented the DTs and customers, then the customer to DT connectivity can be identified by calling LC with s as DT meter and ℓ as the set of all customer meters under a feeder.

Customer-phase. A distributor may have metered customers and feeders but not DTs. In this case, the set of customers connected to say, phase A of a feeder can be determined by calling LC with ℓ as the set of all customer meters under the feeder and s as the feeder meter that measures phase A. With 3-phase customers, each phase meter is added as a separate meter in ℓ.

DT-phase. The set of all DTs connected to say phase A of a feeder may be determined by regarding the feeder meter of phase A as s and ℓ as the set of all DTs and 3-phase customer meters under the feeder.

Tree connectivity model. A distributor may have metered the customers, feeders, and DTs. Given the set of meters under each substation, the customer-feeder connectivity is determined as above. Then for each feeder, ℓ is the set of all customer meters under the feeder. We make a call to LC for each DT, with s as the DT meter. Thus the entire tree connectivity between the meters may be recovered via a sequence of calls to LC and building a tree based on those solutions.

5.2.1 Problem sizes

The size of LC problem instances vary based on the type of connectivity information that needs to be retrieved. For example, the customer-feeder connectivity problem instances may be large since ℓ represents all meters under one substation. However, if the customers under each feeder are either known or recovered via prior calls to LC, then ℓ is limited by the number of customers or DTs under a feeder. For e.g., for the customer-phase connectivity problem, ℓ is bounded by the number of customers under a feeder. Thus for real problems, we expect $n = |\ell|$ to vary from a few hundred (per feeder) to a few thousand meters (per substation), although this may vary with each distributor's network.

5.3 The noiseless LC problem and solution

With no errors, the load at any non-leaf meter s exactly matches the sum of the loads of its leaf meters $p \subseteq \ell$. *i.e.* the vector $e = 0$ in (2).

5.3.1 Single Measurement (m = 1)

For one measurement from a single time step, if all meter readings can be converted to integers without loss of accuracy, the problem reduces to the *Subset-sum* problem [16]. We wish to find a subset $p \subseteq \ell$ of leaf meters whose Wh measurements sum up to the Wh measured by s. The subset-sum problem is NP-hard and can be solved in pseudo-polynomial time using a dynamic program. However, one measurement may not guarantee a unique solution.

5.3.2 Multiple Measurements (m > 1)

In the general case when we have a time series of measurements, (2) can be posed as a 0-1 integer linear program (ILP) with a zero objective function.

$$\text{(ILP1)} \qquad \min \ 0^T X$$
$$AX = b$$
$$x_j \in \{0, 1\}, \ 1 \leq j \leq n \qquad (3)$$

(3) can have multiple solutions when the number of time series measurements (*i.e.* constraints) is low. ILP solvers [17] can be used to obtain a solution to (3). However, ILPs are NP-hard and therefore some instances may require exponential time. We now propose three relaxations (Eq.(4), Eq.(6), and Eq.(7)) that can be used to retrieve the true underlying leaf connectivity solution in polynomial time given sufficient number of time series measurements m.

5.3.3 Linear systems relaxation (m = n)

$$AX = b$$
$$x_j \in \{0,1\} \quad x_j \in \mathbb{R}, \ 1 \leq j \leq n \qquad (4)$$

(4) is an unconstrained relaxation of (3) wherein we drop the binary constraints on X. It can be solved using linear algebra provided A has full rank i.e. the number of measurements m equals the number of leaf meters n. When this holds, the set of measurement equations (4) has a unique solution. Uniqueness implies that the true underlying binary solution to the LC problem can be retrieved simply as $X = A^{-1}B$.

5.3.4 Linear programming relaxation 1 (m ≤ n)

First we transform the binary variables in (3) from $\{0, 1\}$ to $\{-1, 1\}$ using the transformation $Y = 2X - 1_n$, where 1_n is a $n \times 1$ vector of all 1's. Observe that $X \in \{0, 1\}^n \Leftrightarrow Y \in \{-1, 1\}^n$. Therefore we have the following ILP that is equivalent to (3):

$$AY = 2b - A1_n, \quad Y \in \{-1, 1\}^n \qquad (5)$$

To solve the above, we use the following LP relaxation:

$$\text{(LP1)} \qquad \text{Min} \ \|Y\|_1$$
$$\text{s.t.} \ AY = 2b - A1_n$$
$$Y \in \{-1,1\} \quad Y \in [-1, 1]^n \qquad (6)$$

The objective function of LP1 is not strictly linear due to the L_1 norm, however this can be linearized using standard LP methods. The following lemma relates (6) to (5).

LEMMA 1. *If (6) returns an integer solution, that solution is the unique integer solution of both (6) and (5).*

We omit the proof for ease of exposition. The proof follows from the fact that the fractional solution $\in [-1, 1]^n$ has a lower L_1 norm than an integer solution $\in \{-1, 1\}^n$. Note that the converse of Lemma 1 is not true. (6) can return a fractional solution even if (5) has a unique integer solution. For instance this may happen when the number of measurements in A is less.

Mangasarian et al. [18, 19] have showed that when $m > n/2$, a system of the form (6) has a unique solution $\in [-1, 1]^n$ with high probability (w.h.p.). For our problem, this implies that as m crosses $n/2$, (6) will start to retrieve the true underlying solution to the LC problem w.h.p. We show the same using experiments as well.

5.3.5 Linear programming relaxation 2 (m ≤ n)

$$\text{(LP2)} \qquad \min \sum_j x_j$$
$$\text{s.t.} \ AX = b$$
$$X \in \{0,1\} \quad X \in [0, 1]^n \qquad (7)$$

LP2 minimizes the L_1 norm of X and attempts to retrieve the sparsest integer solution $\in [0, 1]^n$. However it may return a fractional or integer solution based on the number of measurements m in A.

Let $K = |p|$ denote the sparsity of the true underlying solution to the LC problem, $1 \leq K \leq n$. We show using experiments that (7) retrieves the true solution with high probability when $m > nH(K/n)/2$, where $H(z) = -z\log_2 z - (1 - z)\log_2(1 - z)$ is the binary entropy function. Note that $nH(K/n)/2 < n/2$ when $K \neq n/2$. Therefore when $K = |p| = n/2$, (7) requires about $n/2$ measurements to retrieve the true solution and in all other cases, less than $n/2$ measurements suffice. In [20], we derive the mathematical conditions for a K-sparse solution to be the unique sparsest solution to (7).

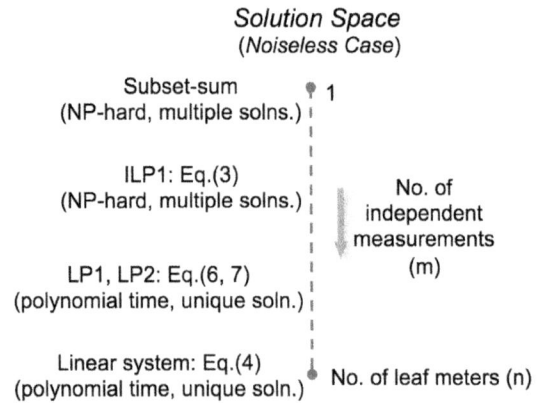

Solution Space
(Noiseless Case)

Subset-sum
(NP-hard, multiple solns.) ——————— 1

ILP1: Eq.(3)
(NP-hard, multiple solns.)

No. of independent measurements (m)

LP1, LP2: Eq.(6, 7)
(polynomial time, unique soln.)

Linear system: Eq.(4)
(polynomial time, unique soln.) — No. of leaf meters (n)

Figure 5: **Noiseless Problem: As the number of time series measurements increase, the leaf connectivity problem can be solved efficiently.**

5.3.6 Solution Space:

Fig. 5 shows the solution space for the noiseless LC problem. Given one measurement, the problem reduces to a variant of subset-sum. Given very few measurements, we may need to solve the ILP (3) that is NP-hard and may yield multiple solutions. When the number of measurements exceeds $nH(K/n)/2$, LP2 retrieves the true solution with high probability and when they exceed $n/2$, LP1 retrieves the true solution with high probability. LP1 and LP2 need only polynomial time. When $m = n$, one can retrieve the unique solution simply by solving the system of linear equations (4).

5.4 The noisy LC problem and solution

This is the general setting wherein errors can vary across measurements due to line losses and other factors discussed in section 4.1. We now propose two approaches to retrieve a solution to the noisy variant of the leaf connectivity problem, with and without sparsity-based regularisation, as in the noiseless setting.

We essentially determine a LC solution X that optimally fits the measurements by minimizing either the L_1 or L_2 norm of the error vector e. As long as errors do not grow significantly with measurements, these approaches are expected to retrieve the true underlying leaf connectivity solution with increasing number of measurements [11].

Without sparsity-based regularization:

$$(\text{ILP}_e/\text{LP}_e) \quad \min_X \quad \|b - AX\|_1 \quad \text{(Linear)}$$

$$X \in \{0, 1\}^n \ / \ X \in [0, 1]^n$$

$$(\text{IQP}_e/\text{QP}_e) \quad \min_X \quad \|b - AX\|_2^2 \quad \text{(Quadratic)}$$

$$X \in \{0, 1\}^n \ / \ X \in [0, 1]^n$$

With sparsity-based regularization:

$$(\text{ILP}_{es}/\text{LP}_{es}) \quad \min_X \quad \lambda \sum_j x_j + \|b - AX\|_1 \quad \text{(Linear)}$$

$$X \in \{0, 1\}^n \ / \ X \in [0, 1]^n$$

$$(\text{IQP}_{es}/\text{QP}_{es}) \quad \min_X \quad \lambda \sum_j x_j + \|b - AX\|_2^2 \quad \text{(Quadratic)}$$

$$X \in \{0, 1\}^n \ / \ X \in [0, 1]^n$$

Solutions to all the above integer linear and integer quadratic programs and their relaxations can be obtained using standard MIP solvers such as CPLEX [17]. The choice of quadratic vs. linear depends on the type of errors and computational performance. For gaussian errors, quadratic minimization generally yields the maximum likelihood estimates (MLE) of the true underlying solution. In terms of computation, all of the above relaxations can be solved in polynomial time while the integer versions are NP-hard. The quadratic programs above have n variables whereas the linear programs have $n + m$ variables. The additional m continuous variables are needed to linearize the L_1 norm in the objective functions.

The sparsity-based formulations are from compressed sensing [21–24]. These allow the retrieval of sparse underlying solutions with fewer number of measurements. These can be beneficial for large LC problem instances with $K = |p|$ small compared to $n = |\ell|$.

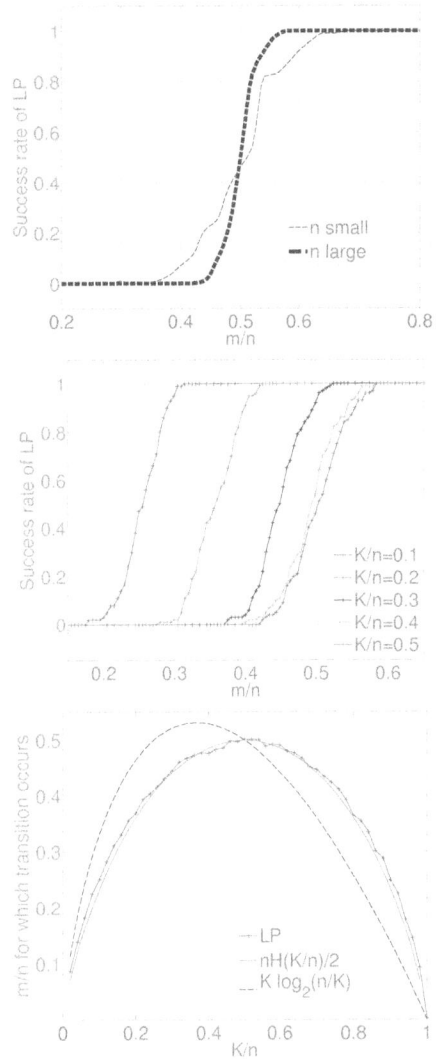

Figure 6: Success rates of LP1 and LP2 with random measurements for noiseless leaf connectivity problem. (top): LP1, (middle): LP2, and (bottom): number of measurements needed by LP2 as a function of sparsity. This is compared with the bounds based on binary entropy (accurate fit) and compressed sensing (inaccurate fit).

6. EXPERIMENTS

Our analytics approach exploits a key property of the measurements in the distribution grid – the variability of meter measurements over time and across meters. Each power-flow measurement in A constrains the solution further and as measurements grow, A maps each $X \in \{0, 1\}^n$ uniquely to a b. Therefore the success rate of the proposed mathematical formulations depends on the number and type of measurements in A. A related and important characteristic is the probability distribution of errors e. In this section, we present our results for the former and leave the later for investiagation in future work.

We conduct Monte-Carlo simulations using two types of data: (i) synthetic meter readings generated uniformly at random, and (ii) Anonymous half-hourly smart meter measurements under one substation. We randomly select leaf connectivity (LC) solutions $X \in \{0,1\}^n$ of varying sparsity to construct $b = AX$. We solve for X using the relaxations LP1 and LP2 of Sec. 5.3 by invoking CPLEX [17] from within MATLAB [25]. We compare the optimal solutions output by these programs with the true underlying LC solution as a function of number of measurements. We plot the success rate of these LPs over multiple runs of experiments.

Random measurements. In this case, the meter measurements (Wh) in $A_{m \times n}$ are generated from uniform distribution. Fig. 6(top) shows benchmark results for LP1. A transition behavior is observed at $m = n/2$. When the number of measurements m exceeds the number of leaf meters n, the system retrieves the true underlying solution w.h.p. The results are independent of the problem size, except that as the problem size increases, the transitions becomes step-like.

Fig. 6(middle) shows benchmark results for LP2. In this case, based on the sparsity K of the true X, the transitions occur before $m = n/2$. Recall that $K = |p| = \|X\|_1$. The worst case occurs for solutions with $K = n/2$ when about $m = n/2$ measurements are needed to retrieve the true solution. Fig. 6(bottom) shows the value of m/n when LP2 retrieves the true underlying solution as a function of sparsity. It shows that $m = nH(K/n)/2$ measurements are sufficient to retrieve the true underlying solution, where $H(z) = -z \log_2 z - (1 - z) \log_2(1 - z)$ is the binary entropy function. This is compared against the standard compressed sensing bound of $K\log(n/K)$ measurements.

Smart meter measurements. Uniqueness and recovery of the true solution is dependent on both the number of measurements and their variability. In the previous case, with random measurements, the variability of measurements across time and meters is assured. As a consequence, each new measurement is independent and adds a new constraint to the system. However real meter measurements may be correlated due to several reasons such as seasonal and diurnal patterns of consumption. Therefore a natural question to consider is if smart meter measurements jointly exhibit sufficient variability to constrain the true solution.

Fig 7 shows the results for two different sets of 100 smart meter measurements. Fig 7 (top) shows a transition behavior similar to the random case, except that the difference between LP1 and LP2 is now less. As before, the plot shows that $m = n/2$ meter measurements are sufficient to retrieve the true solution in the noiseless case. Fig 7 (bottom) shows another case where the meter measurements matrix A may not have the right properties to map solutions X uniquely. In this case, some of the meters have missing measurements and columns of A corresponding to these meters have zero entries. The plot shows that selection and filtering of measurements is an important criteria to consider for real deployments of our solution. Note that the plots above show the behavior of relaxations and not of the integer programs which may potentially perform better.

7. CONCLUSIONS AND FUTURE WORK

In this work, a novel analytics approach has been presented to infer the connectivity model of a distribution network in smart grids. The connectivity model is a key input required in the operations and maintenance of distribution networks. It enables the deployment of solutions that improve the overall efficiency and reliability of energy delivery to the consumer. Our analytics approach relies solely on

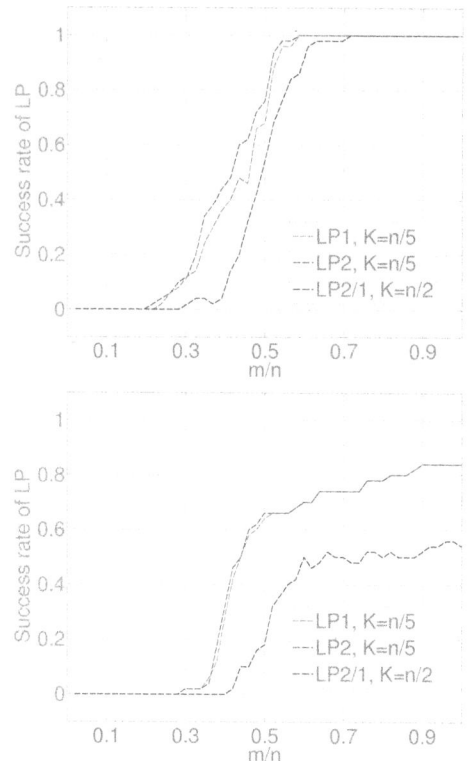

Figure 7: Success rates of LP1 and LP2 with two different sets of real smart meter measurements

meter measurements and is therefore readily usable and less expensive than solutions based on power-line communications.

We presented the inference problem underlying our approach and proposed a number of mathematical programming techniques that can be used to retrieve the connectivity tree between the meters by successively solving instances of the leaf connectivity problem. We presented preliminary experimental results that give an indication of the number of measurements needed to retrieve the true underlying solution in the noiseless setting.

There are a number of directions for future work in terms of addressing different variants of the connectivity problem, better optimization techniques, deriving conditions for uniqueness of solutions, and experiments with larger number of meter measurements for both noiseless and noisy settings.

In an other variant of the connectivity problem, there may be 2-phase and 3-phase customer meters that record aggregate consumption instead of per phase. In our optimization approach, the connectivity problem corresponding to each node is solved separately instead of jointly. The joint approach has benefits and may require fewer measurements as each meter can have only one parent in a tree. We plan to study these problems. In terms of experiments, we plan to conduct more experiments with smart meter measurements to gain insights about the measurement matrices that provide uniqueness properties. We also plan to conduct experiments with different types of errors and leverage from compressed sensing to investigate the number of measurements needed for uniqueness of solution.

180

8. REFERENCES

[1] J. D. D. Glover and M. S. Sarma, *Power System Analysis and Design*, 3rd ed. Pacific Grove, CA, USA: Brooks/Cole Publishing Co., 2001.

[2] W. H. Kersting, *Distribution system modeling and analysis*, 2nd ed. Boca Raton :: CRC Press,, 2007.

[3] J. Fan and S. Borlase, "The evolution of distribution," *Power and Energy Magazine, IEEE*, vol. 7, no. 2, pp. 63–68, 2009.

[4] J. Bouford and C. Warren, "Many states of distribution," *Power and Energy Magazine, IEEE*, vol. 5, no. 4, pp. 24–32, 2007.

[5] T. R. Blackburn, "Distribution transformers: Proposal to increase meps levels," http://www.energyrating.gov.au/wp-content/uploads/Energy_Rating_Documents/Library/Industrial_Equipment/Distribution_Transformers/200717-meps-transformers.pdf, Equipment Energy Efficiency Program (E3), Department of climate change and energy efficiency, Australia, Tech. Rep. 2007/17, 2007.

[6] T. Standish, "Digitization of the electric grid: The enabler of energy efficiency and economic gain," http://www.centerpointenergy.com/staticfiles/CNP/Common/SiteAssets/doc/, March 2009.

[7] V. Arya, T. S. Jayram, D. Seetharam, and S. Kalyanaraman, "Determining a connectivity model in smart grids," Patent Pending US 13/563373, Mar. 2012.

[8] K. Caird, "Meter Phase Identification," US Patent Application 20100164473, January 2010, Patent No. 12/345702.

[9] M. Dilek, "Integrated Design of Electrical Distribution Systems: Phase Balancing and Phase Prediction Case Studies," Ph.D. dissertation, Virginia Polytechnic Institute and State University, 2001.

[10] V. Arya, S. Kalyanaraman, D. Seetharam, V. Chakravarthy, K. Dontas, J. Kalagnanam, and C. Pavlovski, "Systems and Methods for Phase Identification," Patent Pending 13/036,628, Feb. 2011.

[11] V. Arya, D. Seetharam, S. Kalyanaraman, K. Dontas, C. Pavlovski, S. Hoy, and J. Kalagnanam, "Phase identification in smart grids," in *Smart Grid Communications (SmartGridComm), 2011 IEEE International Conference on*, oct. 2011, pp. 25–30.

[12] R. Castro, M. Coates, G. Liang, R. Nowak, and B. Yu, "Network Tomography: Recent Developments," *Statistical Science*, vol. 19, no. 3, pp. 499–517, 2004.

[13] A. Adams, T. Bu, T. Caceres, N. Duffield, T. Friedman, J.Horowitz, F. Lo Presti, S. Moon, V. Paxson, and D. Towsley, "The use of End-to-end Multicast Measurements for Characterising Internal Network Behavior," *IEEE Communications Magazine*, vol. 38, no. 5, pp. 152–158, May 2000.

[14] M. Coates, R. Castro, R. Nowak, M. Gadhiok, R. King, and Y. Tsang, "Maximum likelihood network topology identification from edge-based unicast measurements," *SIGMETRICS Perform. Eval. Rev.*, vol. 30, no. 1, pp. 11–20, Jun. 2002.

[15] S. A. Boyer, *Supervisory Control and Data Acquisition*, 2nd ed. ISA, 1999.

[16] T. H. Cormen, C. E. Leiserson, R. L. Rivest, and C. Stein, "35.5: The subset-sum problem," in *Introduction to Algorithms*. MIT Press, 2001.

[17] IBM ILOG, "CPLEX: High-performance software for Mathematical Programming and Optimization," http://www-01.ibm.com/software/integration/optimization/cplex-optimizer.

[18] O. L. Mangasarian and M. Ferris, "Uniqueness of Integer Solution of Linear Equations," *Optimization Letters*, pp. 1–7, 2010.

[19] O. Mangasarian and B. Recht, "Probability of unique integer solution to a system of linear equations," *European Journal of Operational Research*, vol. 214, no. 1, pp. 27 – 30, 2011.

[20] T. S. Jayram, S. Pal, and V. Arya, "Recovery of a sparse integer solution to an underdetermined system of linear equations," in *NIPS Workshop on Sparse Representation and Low-rank Approximation*, 2011.

[21] E. Candes and T. Tao, "Decoding by linear programming," *IEEE Transactions on Information Theory*, vol. 51, no. 12, pp. 4203 – 4215, 2005.

[22] D. L. Donoho, "Compressed sensing," *IEEE Transactions on Information Theory*, vol. 52, no. 4, pp. 1289–1306, 2006.

[23] E. J. Candés, J. Romberg, and T. Tao, "Robust uncertainty principles: exact signal reconstruction from highly incomplete frequency information," *IEEE Transactions on Information Theory*, vol. 52, no. 2, pp. 489–509, 2006.

[24] D. Donoho and J. Tanner, "Precise undersampling theorems," *Proceedings of the IEEE*, vol. 98, no. 6, pp. 913–924, 2010.

[25] MATLAB, *version 7.8.0.347 (R2009a)*. The MathWorks Inc., 2009.

Malicious False Data Injection in Hierarchical Electric Power Grid State Estimation Systems

Yangyue Feng
Information Security Group
Department of Mathematics
Royal Holloway, University of London
Egham, Surrey, UK
yangyue.feng@rhul.ac.uk

Chiara Foglietta
Engineering Department
University of "Roma TRE"
Via della Vasca Navale 79
00146 Roma Italy
fogliett@dia.uniroma3.it

Alessio Baiocco
Information Security Group
Department of Mathematics
Royal Holloway, University of London
Egham, Surrey, UK
alessio.baiocco@rhul.ac.uk

Stefano Panzieri
Engineering Department
University of "Roma TRE"
Via della Vasca Navale 79
00146 Roma Italy
panzieri@dia.uniroma3.it

Stephen D. Wolthusen
Information Security Group
Department of Mathematics
Royal Holloway, University of London
Egham, Surrey, UK

Norwegian Information
Security Laboratory
Gjøvik University College
N-2818 Gjøvik, Norway
stephen.wolthusen@rhul.ac.uk

ABSTRACT

The problem of *malicious* false data injection in power grid state estimators has recently gained considerable attention. Most of this attention, however, has been focused on the assumption of a centralised state estimator. In a next-generation smart grid environment incorporating distributed generation and highly variable demand induced by electric mobility, distributed state estimation is highly desirable to enhance overall grid robustness. We therefore consider the case of a bi-level *hierarchical state estimator*, which provides only partial observability to lower-tier state estimators.

Using a formal observability model, we consider the case of an active adversary able to modify a set of measurements and derive bounds on the maximum number of manipulated measurements that can be tolerated, the composition of *attack vectors*, and give a formulation for identifying minimal sets of additional measurements to tolerate k-measurement attacks in this hierarchical state estimator. This allows us a more rigorous formulation over existing models.

Categories and Subject Descriptors

C.3 [**Special-purpose and Application-based Systems**]: Real-time and embedded systems; G.2.3 [**Discrete Mathematics**]: Applications

General Terms

Theory, Security, Algorithm

Keywords

Hierarchical State Estimation, Partial Observability, Malicious Bad Data Injection

1. INTRODUCTION

State estimators seek to establish the state of a power network based on a set of measurements collected and aggregated periodically that is combined with the current estimate of the system's state. Robust and accurate state estimation has been an important element in the efficient and reliable operation of power networks over several decades, but has become more important with the migration to *smart grid* environments that must combine greater efficiency with a substantially more demanding combination of generators and loads. Larger numbers of loads such as electric vehicles requiring a *fast charge* cycle can have a significant impact, particularly when arising in a relatively small geographic area or concentrated over time, whilst similar increased demands also arise from local generation.

Although state estimation has conventionally been in the remit of transmission system operators (TSO), it is becoming increasingly desirable to conduct state estimation in multiple loci. This can be at the TSO level where multiple entities wish to retain a degree of autonomy over their state estimate, but also for distributed generation and micro-grid environments. Multi-area and distributed state estimation has hence attracted research since the 1970s [8], but this raises a number of questions on the reliability and trustworthiness of state measurements and estimates.

The ability to detect and compensate for *bad data* as may arise from faulty sensors or communication failure has been studied intensively as part of the design of state estimators, as has the analysis of *observability* in the presence of faults [1]. Common approaches include the detection of state changes and bad measurement values, allowing to fill in missing data where redundant information is available, and to report violations otherwise based on fault models. These approaches, however, rely on assumptions which are violated in the case of malicious manipulation of underlying sensor information or their communication. Given the severe consequences of bad state estimation particularly for the case of a smart grid operating relatively close to its safety margins and the need to perform particularly timely contingency analysis, this has recently gained attention with research focusing on the feasibility of several classes of attacks including forcing the state estimator to undesirable states as well as a formulation for criteria ensuring that malicious bad data injection can be detected [14, 12, 13].

As this existing body of work on malicious bad data has thus far been limited to centralised state estimation models, we argue that it is necessary to consider these problems also for the case of distributed state estimators. More specifically, we consider multi-area state estimators with limited redundancy in their measurements (i.e. overlap) to be of particular interest. In this paper we therefore describe a bi-level hierarchical state estimator relying only on tie-lines for overlap and study the problem of observability in this instance. This state estimator is a simplified instance of the k-level hierarchical state estimator detailed in our work but is sufficient for the problem detailed here. In particular, while we retain the hierarchical structure of the full model, we restrict ourselves to a linear time-invariant (LTI) formulation also commonly found in the literature as this is sufficient for the problem at hand.

Our contribution lies in the formulation of a partial observability model derived from LTI observability and study data injection attacks beginning with the case of data suppression for hierarchical bi-level state estimation at both the local and central state estimation levels and relate this to the network observability criterion. Based on this, we also derive criteria for redundancy required to tolerate such attacks resulting in denial of observability in the hierarchical model.

The remainder of this paper is organised as follows: We review related work and provide selected background in Section 2 followed by the bi-level hierarchical multi-area state estimator and a linearised implementation thereof in Section 3. A special form of malicious attack, namely false data injection and suppression, is discussed in Section 4, where we demonstrate the applicability of such attacks in our hierarchical model. In Section 5 we then give a formal definition of network observability, and study the denial of observability by this particular sort of attack. We also derive bounds for tolerable attacks and requirements for additional sensor placement for retaining observability. Finally, Section 6 concludes our work with a summary and a discussion of our on-going work in this area.

2. RELATED WORK

This work is related both to power network state estimation and the theory of observability as well as attack mechanisms; a review of basic theory underpinning state estimation as well as fundamental techniques and problems can be found in the book by Abur Gómez-Expósito [1], with a more concise summary of problems in centralised systems provided by the survey of Monticelli *et al.* [15].

A number of state estimation methods have been proposed both generically and for the case of power networks in particular with the main focus being the conventional centralised state estimation architecture and approaches based on power flow models. Nearly all such models rely on the simple but efficient weighted least squares (WLS) algorithm to perform the estimate as the computational complexity of the algorithm is a main consideration. A substantial body of work has emerged in recent decades also addressing problems of robustness and bad data identification as well as on observability of the underlying system state.

However, as already noted in section 1, the requirements inherent in smart grids and further demands arising from factors including the liberalisation of energy markets have resulted in increased interest in distributed state estimation. Whilst such models were first investigated in the mid-1970s (see the recent survey [8]), this has gained new interest as capabilities within the grid have increased at the same time as new requirements have emerged from the need to integrate variable and distributed generation as well as dynamic loads.

One instance of a multi-level hierarchical state estimation framework was proposed by Gómez-Expósito *et al.* [7], describing a model encompassing a substation, transmission level, and an in-between TSO (transmission system operator) level. A two-stage FWLS algorithm is given based on factorisation of measurement model which provides theoretical justification of this hierarchical state estimator. In further work by Gómez-Expósito *et al.* [8], multi-area concepts for state estimation are surveyed, while recent work by Gómez-Quiles *et al.* [9] reports a two-stage procedure for substation state estimation based on feeder-level decomposition and coordination in conjunction with a theoretical justification for this decomposition. Although a fully distributed model is desirable, current results require a significant reduction in model fidelity rendering such models to be of limited use in the study of malicious activity [19]. We therefore restrict our consideration to bi-level hierarchical multi-area estimation as our principal subject lies in the observability and controllability of the power network.

Secondly, network observability and bad data processing constitute two important functional prerequisites for the state estimation problem. Observability of a power system based on the Kalman criterion refers to the feature that there are sufficiently many available measurements and they are well distributed throughout the network in such a way that state estimation is possible, and the network is said to be observable. Usually, there are two ways to carry on the observability analysis: Topological analysis and numerical analysis, although it is common to see a hybrid of the two. In the work of Monticelli *et al.* [15, 16] such studies on observability are reported. Similarly, in the more recent work by Giani *et al.* [6], observability is formally modelled and countermeasures to deliberate malicious attacks are discussed based on known-secure phasor measurement units (PMUs) are suggested, although we note that PMU themselves may be the subject of spoofing attacks as described by Shepard et al. [17] based on manipulation of the time-base.

In addition to the unavoidable measurement errors, sensor, and telemetry failures, a number of attack mechanisms exist which may lead to the the state estimation obtaining inconsistent results, including ones chosen by the attacker. In the paper by Liu et al. [14], a class of attacks against power system state estimation, *false data injection attack* is presented. Liu et al. focus on the detectability for this class of attacks and suggest two attack scenarios, which are characterised by bounds on attacker capabilities. In one scenario, the attackers have limited access to some specific meters, whilst in another when the attackers are limited by the resources available. Detection and bounds on adversaries to launch such types of attack are given. In the work by Kosut et al. [12, 13], further studies on this type of attack are carried out; after introducing the bad data detection problem based on measurement residuals, the unobservable attack condition is connected with the classical network observability conditions. A further protection criterion against false data injection attack is proposed in [2] and optimal and suboptimal algorithms are developed.

In our work, we build on the aforementioned research of such undetectable false data injection attack for the case of multi-level, multi-area state estimators based and can describe conditions under which attacks may succeed. By considering a model with limited overlap, attacks can now be described also by their distribution, so we report on models of k-sparse attacks, i.e. where attackers can at most choose k meters to compromise, can be distributed over the grid and how attack vectors are composed on certain critical meters. We connect this attack directly with the classical network observability model and study how such attacks would cause the denial of network observability.

We note that recent work by Sou et al. [18] studies finding critical k-tuples of the measurements, using a *Min-Cut* optimisation procedure and also a mixed integer linear programming (MILP) calculation, which allows extension of this result to be applied in identifying critical measurements and considering meter placement. Combined with our study on the bounds of malicious false data injection attack, this can be utilised to identify which individual measurements are critical, and which may be substituted, namely, how to repair the loss of critical measurements by allowing redundancy to ensure continued satisfaction of the (distributed) observability criterion. Other work on decentralised power state estimation, FDIA on DC state estimation and hierarchical state estimation could be found in [11, 4, 20].

3. HIERARCHICAL STATE ESTIMATION

The traditional centralised state estimation is described with the following power flow model and solved with a standard weighted least-squares (WLS) based solution.

$$z = h(x) + e \quad (1)$$

where x is the state vector to be estimated (size n), which usually is composed of power voltage magnitudes and phase angle values; z is the known measurement vector (size $m > n$), which typically comprises power injections, branch power flows and voltage magnitudes; h is the vector of functions, usually non-linear, relating error free measurements to the state variables. Alternative formulations are based on an AC (alternating current) model, in which case the h func-

tion is non-linear and computationally expensive, and on a linearised DC (direct current) model, which is significantly simpler but at the cost of a loss of precision and accuracy.

In either case e is the vector of measurement errors, for which the following assumptions are made regarding the statistical properties of the measurement errors: $\mathbb{E}(e) = 0$; measurement errors are independent, i.e. $\mathbb{E}(e_i e_j) = 0$. Hence $cov(e) = \mathbb{E}[e \cdot e^T] = R$ defined as $diag\{\sigma_1^2, \sigma_2^2, \ldots, \sigma_m^2\}$. The standard deviation σ_i of each measurement i is calculated to reflect the expected accuracy of the corresponding measurement used.

As noted in section 2, the state estimation problem is usually solved as an unconstrained weighted least-squares (WLS) problem. The WLS estimator minimises the weighted sum of the squares of the residuals, expressed as

$$J(x) = \sum_{i=1}^{m} \frac{(z_i - h_i(x))^2}{R_i}$$
$$= [z - h(x)]^T R^{-1} [z - h(x)] \quad (2)$$

where $R = diag(R_i)$ is the weighting matrix.

This least square problem is solved with the Gauss-Newton algorithm [3]. At the minimum, the first-order optimality conditions must be satisfied. This can be expressed as:

$$g(x) = \frac{\partial J(x)}{\partial x} = -H^T(x)R^{-1}[z - h(x)] = 0 \quad (3)$$

where $H = \partial h/\partial x$ is the $m \times n$ measurement Jacobian matrix.

The first-order necessary condition for a minimum is that

$$\frac{\partial J(x)}{\partial x} = -H(x)^T R^{-1}[z - h(x)] = 0 \quad (4)$$

Expanding the non-linear function $g(x)$ into its Taylor series around the state vector x^k yields:

$$g(x) \cong g(x^k) + G(x^k)(x - x^k) = g(x^k) + G(x^k)\Delta x^{k+1} = 0$$
$$G(x^k)x = G(x^k)x^k - g(x^k)$$
$$x = x^k - \left[G(x^k)\right]^{-1} g(x^k) \quad (5)$$

where k is the iterative index; x^k is the solution vector at iteration k. The matrix $G(x^k)$ is called *gain matrix* and it is calculated as

$$G(x^k) = \frac{\partial g(x^k)}{\partial x} = H^T(x^k)R^{-1}H(x^k) \quad (6)$$

The gain matrix is sparse, positive definite and symmetric provided that the system is fully observable under the Kalman criterion. The matrix $G(x)$ is typically not inverted, but instead it is decomposed into its triangular factors and the following sparse linear set of equations are solved using forward/back substitutions at each iteration k:

$$G(x^k)\Delta x^{k+1} = H^T(x^k)R^{-1}[z - h(x_k)] \quad (7)$$

where $\Delta x^{k+1} = x^{k+1} - x^k$. This equation is also referred to as the *Normal Equations*. Iterations are terminated when an appropriate tolerance is reached on Δx_k.

Followed by the requirements inherent in smart grids and further demands arising from energy markets, distributed state estimation has raised more research interest. Ideally, a fully distributed state estimator over the power grid is desirable to enhance overall grid robustness. For the following, however, we consider a hierarchical concept of the network and state estimation as the basis of our further analysis, which is based on the general hierarchical model of WLS state estimation in [7]. In particular, we apply our security analysis on a two-level hierarchical model with multiple areas, which can be described as:

$$
\begin{aligned}
z_{1j} &= f_{1j}(y_{1j}) + e_{1j}, \qquad j = 1, 2 \\
z_{1b} &= f_{1b}(y_1) + e_{1b} \\
y_1 &= f_2(x) + e_2
\end{aligned}
\tag{8}
$$

where z_{1j} is the $m_{1j} \times 1$ measurement vector of area j, and z_{1b} is the $m_{1b} \times 1$ boundary measurement vector when overlapping areas occur, they denotes to the lower level one local measurements. y_{1j} stands for the $p_{1j} \times 1$ internal state vector of area j, of level one. y_1 is the $p_1 \times 1$ vector of system-wide state vector at level one, where p_1 is the sum of p_{11} and p_{12}; x is the $p_2 \times 1$ state vector of level two.

A Linearised Implementation of the Bi-Level Hierarchical Multi-Area State Estimator

We realise the bi-hierarchical state estimation model (8) using a DC power flow model, where equation (1) is represented by a linear regression model:

$$
z = Hx + e
\tag{9}
$$

The structure of the measurement Jacobian H is as follows:

$$
H(x) = \frac{\partial h(x)}{\partial x}
\tag{10}
$$

and in the WLS algorithm for finding the estimated value $\hat{x}$ of x, the first-order necessary optimality condition for a minimum WLS is

$$
\frac{\partial J(x)}{\partial x} = -H^T R^{-1} [z - Hx] = 0
\tag{11}
$$

with which the estimated value $\hat{x}$ is evaluated as

$$
H^T(x) R^{-1} H(x) \, \hat{x} = H^T(x) R^{-1} z
\tag{12}
$$

where the matrix $G(x) = H^T(x) R^{-1} H(x)$ is the gain matrix. In particular, the two-level hierarchical model with two areas and no border variables is described as the following, where F_{1j} and F_2 represent the Jacobian matrix of f_{1j} and f_2:

$$
\begin{aligned}
z_{1j} &= F_{1j} y_{1j} + e_{1j}, \qquad j = 1, 2 \\
y_1 &= F_2 x + e_2
\end{aligned}
\tag{13}
$$

For illustration purposes consider the commonly used IEEE 14-bus 20-line system as shown in Figure 1, which we have decomposed into a two-level, two-area state estimator relying only on tie-lines. The hierarchical structure of our decomposition is presented in Figure 2. The location of the top-level state estimator in our hierarchical model is arbitrary; more specifically, lower-tier state estimators are associated with their own regions, and the k-th level state estimators with the union of all regions forming the tree at whose root this region lies.

The network is decomposed into two non-overlapping areas, with tie-lines $(5, 8)$, $(4, 9)$, $(7, 9)$ to connect the $Area_1$ and $Area_2$. Each area is governed by its own local state estimator, which we call the lower level, and connected by communication links to the second level state estimator, which we call the top level. The lower level state estimators will run WLS algorithms at each sub-area, and feed the estimation into top level state estimator as measurements.

The connection between levels is as follows. The output of the low-level areas (the state vector) is the measurement

vector of the higher-level areas, and the gain matrix of the low-level areas is considered as the weighting matrix of the higher level.

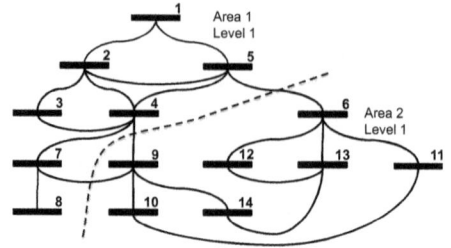

Figure 1: IEEE 14 bus bar system and a two-level, two-area hierarchical decomposition (I)

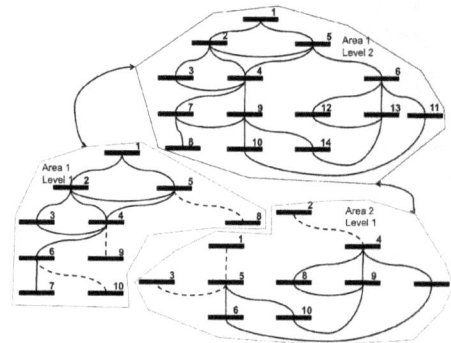

Figure 2: IEEE 14 bus bar system and a two-level, two-area hierarchical decomposition (II)

Usually, the measurement vector z comprises line power flows, bus power injections, and bus voltage magnitudes, denoted by P/Q_{inj}, P/Q_{flow} and V_{mag}, where P stands for real power and Q for reactive power. These measurements can be expressed in terms of the state variables either using rectangular or polar coordinates. When using the polar coordinates for a system containing N buses, the state vector will have $2N - 1$ elements: N bus voltage magnitudes and $N - 1$ phase angles, where the phase angle of one reference bus is set equal to an arbitrary value, such as 0. The state vector x will have the following form assuming bus 1 is chosen as reference, with θ for phase angles and V for voltage:

$$
x^T = [\theta_2, \theta_3, \ldots, \theta_N, V_1, V_2, \ldots, V_N]
\tag{14}
$$

The dimension of measurement vector z will have to guarantee the observability and the convergence of the state estimation process. When the system contains T lines, and on all the buses and lines there has been placed a meter, z could have $1 + 2N + 4T$ elements: a voltage taken from the reference bus, active and reactive power injections evaluated from the busbars, active and reactive power flows evaluated from the lines from two directions. It has the following form:

$$
\begin{aligned}
z^T = [\, &V_{mag1}, P_{inj1}, \ldots, P_{injN}, Q_{inj1}, \ldots, Q_{injN}, \\
&P_{flow1}, \ldots, P_{flowT}, P'_{flow1}, \ldots, P'_{flowT}, \\
&Q_{flow1}, \ldots, Q_{flowT}, Q'_{flow1}, \ldots, Q'_{flowT} \,]
\end{aligned}
\tag{15}
$$

The structure of the measurement Jacobian H is as follows:

$$H = \begin{bmatrix} \dfrac{\partial V_{mag}}{\partial \theta} & \dfrac{\partial V_{mag}}{\partial V} \\[2mm] \dfrac{\partial P_{inj}}{\partial \theta} & \dfrac{\partial P_{inj}}{\partial V} \\[2mm] \dfrac{\partial Q_{inj}}{\partial \theta} & \dfrac{\partial Q_{inj}}{\partial V} \\[2mm] \dfrac{\partial P_{flow}}{\partial \theta} & \dfrac{\partial P_{flow}}{\partial V} \\[2mm] \dfrac{\partial Q_{flow}}{\partial \theta} & \dfrac{\partial Q_{flow}}{\partial V} \end{bmatrix} \qquad (16)$$

The state estimator takes given busbar data and line data to calculate the measurements P/Q_{inj}, P/Q_{flow} and V_{mag}, constructs the Jacobian matrix H, and carries out the iterative solution algorithm for WLS state estimation problem which can be outlined as follows:

1. Start iteration, iteration index $k = 0$;

2. Initialize the state vector x^k, as a flat start;

3. Calculate the measurement function $h(x^k)$;

4. Build the measurement Jacobian $H(x^k)$;

5. Calculate the gain matrix, $G(x^k) = H(x^k)^T R^{-1} H(x^k)$;

6. Calculate the right hand side $t^k = H(x^k) R^{-1} (z - h(x^k))$;

7. Test for convergence, $\max \|\Delta x^k\|_2 \leq \epsilon$?

8. If not, update $x^{k+1} = x^k + \Delta x^k$, $k?k + 1$, and go to step 3. Else, stop.

The hierarchical state estimation solution involves a further iterative process among the two levels, with information exchange:

Multi-Area Level 1 The inputs are the vectors $z_{1j}, j = 1, 2$, where j is the index of the areas, and the weighting matrices R_{1j}. The output is the estimation of $\hat{y}_{1j}$ for each area, calculating the value:

$$\left[F_{1j}^T R_{1j}^{-1} F_{1j} \right] \hat{y}_{1j} = F_{1j}^T R_{1j}^{-1} z_{1j} \qquad (17)$$

Multi-Area Level 2 The inputs of this level are the vector $\hat{y}_1$ estimated from the Level 1 and the gain matrices as $G_{1j} = F_{1j}^T R_{1j}^{-1} F_{1j}$. The output is the estimated value of $\hat{x}$ using the equation

$$\left[F_2^T G_1^{-1} F_2 \right] \hat{x} = F_2^T G_1^{-1} \hat{y}_1 \qquad (18)$$

where y_1 and G_1 are defined, respectively, as juxtaposition of y_{1j} and G_{1j}. The output of this level is also the updated value of $\tilde{y}_1 = F_2 \hat{x}$, and it should update the values of the Level 1 outputs.

The information flow among the levels is essentially the transmission of results from low-level areas: the state vector of low-level areas ($\tilde{y_{1j}}$) and the gain matrices of low-level areas (G_{1j}). The state vector of low-level areas ($\tilde{y_{1j}}$) is a vector of dimension $2*n-1$, where n is the number of busbars in the low-level area. The gain matrix for each low level is

a square matrix of dimension $n \times n$, where n is the number of busbars in the low level areas. We don't consider, in the paper, the delays, the amount of information, the protocols and everything else related to the information exchange.

We base our analysis of a particular form of attacks against state estimation on the model given in this section.

4. MALICIOUS FALSE DATA INJECTION

In the power grid state estimation, robustness, accuracy and data integrity are essential in state estimator design. The accuracy of estimation relies on the integrity and precision of measurement data, and there exist a number of stochastically characterisable sources of bad data such as communication errors, device breakdown, or other random natural factors which we do not consider further in this paper. However, measurements could be maliciously manipulated to cause the state estimation to fail. In [14], a class of malicious attack against power grid state estimation, false data injection, is studied with the traditional centralised model of state estimator. The attacker could create arbitrary, non-random data into the collected measurements once he or she knows about the network topology and gets resources to launch the attack.

We carry out a similar study of false data injection for the hierarchical model of state estimation, and focus on the boundary of such attacks and the robustness of state estimator. We notice that, first, the classic centralised state estimation model is equivalent to the top level in the hierarchical SE model; secondly, the attacker could influence differently in the two-level model, such as influencing the convergence property of state estimator in a hierarchical case. We note that in a distributed or, as reported here, hierarchical state estimator, the attacker benefits from an additional degree of freedom in choosing the attack vector not found in centralised models, which we will consider in the following.

4.1 Undetectable False Data Injection

In order to study the maliciously injected data on power grid state estimation, Liu *et al.* [14] first described a type of attack, called false data injection attack (FDIA), where it is under the circumstances that the attackers grasps the configuration or target power system, the measurement Jacobian matrix H for a DC model. It is proven that this type of attack could bypass the bad data detection based on measurement residual method (Theorem 1, [14]). In another form, an attack vector a is unobservable or undetectable if it is chosen as a linear combination of the column vector of H, which forms the condition as

$$a = Hc, \qquad c \neq 0 \qquad (19)$$

It is also justified that if the attacker can compromise k specific meters where $k \leq m - n + 1$, the attack vector always exists (Theorem 2, [14]).

In [12] and [13], the above undetectable condition is related to classical network observability conditions (Theorem 1, [12]), which is stated as follows, and provides a quantitative way of analysing the connection of network topology and the existence of unobservable attack vector:

THEOREM 1. *(Kosut* et al.*) There exists an unobservable k-sparse attack vector a if and only if the network becomes unobservable when some k measurements are removed, i.e.,* there exists an $(m - k) \times n$ submatrix of H that does not have full column rank.

Under such circumstances, the DC power flow state estimation problem based on a linearised AC power flow model (1) is reformulated as the following,

$$z = Hx + a + e \qquad (20)$$

where $z \in R^m$ is the vector power flow measurements, $x \in R^n$ is the vector of system state, e the Gaussian measurement noise with zero mean and covariance matrix Σ_e, and the injected vector a is the malicious data injected by an adversary, with at most k non-zero entries ($\| a \|_0 \leq k$). A vector a is said to have sparsity k if $\| a \|_0 = k$.

4.1.1 Detectability and Observability

In the paper by Liu, Ning and Reiter [14], the bad measurement detection technique is based on calculating the measurement residual, the difference between the vector of observed measurements, and the vector of estimated measurements, i.e. $z - H\hat{x}$, and using its L_2-norm $\|z - H\hat{x}\|$ to detect the presence of bad measurement by comparing with a threshold τ. If there exists a non-zero k-sparse a for which $a = Hc$ for some $c \neq 0$ (as in the case of $c = 0$, a is trivial), then $z = Hx + a + e = H(x + c) + e$. In other words, x is observationally equivalent to $x + c$, and the injected vector a will lead the control centre to believe that the true network state x is $x + c$, for arbitrary c.

The difference between random bad data caused by natural factors and the maliciously injected data is that, it is unlikely that random bad data a will satisfy the unobservable condition $a = Hc$. However, an adversary could synthesise the attack vector to satisfy the unobservable condition. When an attack vector satisfies the condition, a equals to Hc for an arbitrary non-zero vector c, and this attack is called a *false data injection attack*. The attacker could bypass the bad data measurement detection and introduce arbitrary errors into the output of the state estimation [14].

4.1.2 False Data Injection in Multi-Area Hierarchical State Estimation

We extend the work introduced in the previous section in a straightforward manner to the hierarchical case, i.e. the model (20) under the multi-level, multi-area setting. We consider the two-level, r-area state estimation described in Section 3. Assume that at lowest level 1, the power network is partitioned into r non-overlapping areas S_j with n_j buses each and connected by tie-lines. Each area is governed by its own local state estimation and they are connected by communication links to a higher level state estimation, represented at level 2. We suppose in some area S_j, at level 1, there is a potential attack vector a_{0j} injected; or at level 1, an attack vector a_{0b} is injected to the border variables; and at level 2, an attack vector a_1 is injected. So the two-level multi-area model (8) from Section 3 is now changed into

$$\begin{aligned} y_{0j} &= f_{1j}(y_{1j}) + a_{0j} + e_{1j}, \qquad j = 1, \ldots, r \\ y_{0b} &= f_{1b}(y_1) + a_{0b} + e_{1b} \\ y_1 &= f_2(y_2) + a_1 + e_2 \end{aligned} \qquad (21)$$

If at area S_j, the attack vector is synthesized by the adversary to satisfy the unobservable condition. With the linearised regression counterpart of (21), this would be ex-

plained as follows, in the same way of condition (19):

$$a_{0j} = F_{1j}c \qquad \text{for some } c \neq 0 \qquad (22)$$

which furthermore derives to

$$y_{0j} = F_{1j}(y_{1j} + c) + e_{1j} \qquad (23)$$

At the area S_j, y_{1j} is observationally equivalent to $y_{1j} + c$, and the adversary's injection of data is undetectable.

Similarly, attack vectors could be injected to the border measurements, when the adversary synthesizes the vector a_{0b} to satisfy the unobservable condition:

$$a_{0b} = F_{1b}c' \qquad \text{for some } c' \neq 0 \qquad (24)$$

We observe that the estimated y_1 is observationally equivalent to $y_1 + c'$.

$$y_{0b} = F_{1b}(y_1 + c') + e_{1b} \qquad (25)$$

When the false data injection attack occurs at level 2, the injected vector is synthesised as the following condition:

$$a_1 = F_2c'' \qquad \text{for some } c'' \neq 0 \qquad (26)$$

which renders the second level estimated state variable y_2 observationally equivalent to $y_2 + c''$, with c'' the injected error.

$$y_1 = F_2(y_2 + c'') + e_2 \qquad (27)$$

Thus, we conclude that the false data injection attack could take place in the bi-level hierarchical state estimation with non-overlapping sub-areas, where the proof closely follows Theorem 1 in [14]; these attack vectors could bypass bad measurement detection if they are chosen to be linear composition of the column vectors of corresponding Jacobian matrices on the sub-areas, border areas, or level 2.

In Section 5, we prove that the FDIA causes the denial of network observability in the hierarchical SE model. We remark that first of all, the maliciously injected data attack at the top level in the hierarchical model is equivalent to the classic, centralised case, and secondly while our formulation is based on a two-level hierarchical model, this analysis is clearly extensible to arbitrary hierarchy levels, also with non-overlapping areas.

5. OBSERVABILITY OF ATTACKS

Suppose that our system is composed of the underlying state-based electric power grid, to which the partial observability is an *a priori* requisition. The topology of the power network considered here may be organised as a mesh or looped networks (in the case of local or micro-grid environments where distributed state estimation is of particular interest); the state estimator we are using is a hierarchical (or further distributed) state estimator that functions over non-overlapping sub-areas or sub-areas that share certain measurements, that could be later viewed as substations [8, 9]. We also consider the explicit use of an information flow network, such as the Internet, functioning together with the power grid, that provides information and communication which allows more user-end demand management. We first introduce a formal definition for network observability, considering a linear time-invariant (LTI) model that follows the classical algebraic approach by Kalman [10].

5.1 False Data Injection Against State Estimation

Our model stays with the confines of an LTI formulation, which limits the types of attacks we may consider given the limited accuracy and precision resulting from linearisation. For both the centralised case and the hierarchical case, one could consider the following types of attacks for an attacker able to choose k measurements to modify.

Denial of observability: This attack arises when an attacker can disable measurements in the power network and cause removal of measurements, which will need to be distinguished from conventional redundancy requirements. It will be proven in 5.3 that the removal of measurements will render the network unobservable. For an LTI model it is not easy to acquire precise data for detection since we cannot take dynamic effects into consideration, but one needs to clearly identify the result of the adversary being able to choose which (critical) measurements he or she can manipulate. We concentrate mainly on this aspect in this paper.

Denial of state estimator convergence: Due to the fact that between different levels of the hierarchical SE there will be communication and information exchange, the attacker could inject undetectable vectors so the WLS algorithm no longer could converge at a local estimator or a high-level estimator. We will need to bound the characteristics (number, distribution, possibly magnitude) of measurements an adversary may influence before we can no longer guarantee that the state estimators (top-level, all lower-tier, fraction of lower-tier, etc.) are no longer able to reach convergence. It is more crucial for the dynamic case, but is also relevant in an LTI formulation.

Forcing of state estimate: As in the above type of attack, the attacker could inject undetectable data and lead the state estimator to converge to an incorrect estimate; this is the type of attack most commonly considered in previous work. We will need to impose bounds on the characteristics of measurements (number, distribution, possibly magnitude) that an adversary may change at most before being able to force a state estimate (top-level, all lower-tier, fraction of lower-tier, etc.) to a specific value or range.

5.2 Partial Network Observability in an LTI Formulation

Formally, a system is said to be observable if, for any possible sequence of state and control vectors, the current state can be determined in finite time using only the outputs. An SE system is said to be observable if the available measurement set contains enough information to obtain a unique estimate of the system state variables. In the literature of control theory, observability is a measure for how well internal states of a system can be inferred by knowledge of its external outputs. As its mathematical dual, the concept of controllability denotes the ability to move a system around in its entire configuration space using only certain admissible manipulations. A theory of the network observability has been discussed in [16], for our model we will use a similar approach as presented in [5] for studying the observability of the power network.

If a state estimator is not observable, it means the current values of some of its states cannot be determined through output sensors: this implies that their value is unknown to the Energy Management System (EMS) controller and, consequently, that it will be unable to fulfil the control specifications referred to these estimates.

5.2.1 Observability for a Linear Time-Invariant Discrete-Time System

If we first consider our system as a linear, time-invariant, discrete-time system, which could be expressed through the following equations [5]:

$$x(k+1) = A_d x(k), \ x(0) = x_o \text{ unknown} \tag{28}$$

$$y(k) = C_d x(k) \tag{29}$$

where $x(k) \in R^n$, $y(k) \in R^p$, A_d and C_d are constant matrices, $x(k)$ the internal state variables, and $y(k)$ the output measurements. The natural question arising is whether we can learn the state space variables defined by (28) using only information from the output measurements (29), which leads to the following definition:

DEFINITION 1. *The LTI discrete-time system given by (28) and (29) is observable if for any state $x(k)$, there is a finite time k' such that $x(k)$ can be uniquely determined from $y(k)$ for $0 \leq k \leq k'$.*

This condition is equal to the following theorem. If we define the observability matrix as

$$\mathcal{O}(A_d, C_d) = \begin{bmatrix} C_d \\ C_d A_d \\ C_d A_d^2 \\ \vdots \\ C_d A_d^{n-1} \end{bmatrix}_{(np) \times n} \tag{30}$$

THEOREM 2. *(Gajic and Lelic) The linear discrete-time system (28) with measurements (29) is observable if and only if the observability matrix (30) has rank equal to n.*

5.2.2 Observability for a Linear Time-Invariant Continuous-Time System

The typical mathematical model normally applied for computing the observability and controllability in time-dependent linear control systems is given as follows:

$$\frac{\partial x(t)}{\partial t} = Ax(t) + Bu(t), x(t_0) = x_0 \tag{31}$$

$$y(t) = Cx(t) + Du(t) \tag{32}$$

where $x(t)$ is a vector $(x_1(t), x_2(t), ..., x_n(t))^T$ which represents the current state of a system with n nodes at time t. A is a $n \times n$ matrix showing the topology of the system signalling interactions between nodes. B is an *input* $n \times m$ matrix where $m \leq n$ and which represents that set of nodes intentionally controlled by an user or controller. For such a control, the controller needs to specify an *input* vector (i.e. $u(t) = (u_1(t), u_2(t), ..., u_m(t))^T$) to push the system to the desired state. And $y(t)$ is the measurement vector $(y_1(t), y_2(t), ..., y_p(t))^T$, C and D the *output* matrices.

DEFINITION 2. *The LTI continuous-time system given by (31) and (32) is observable if for any initial state $x(t_0)$, there is a finite time τ such that $x(t_0)$ can be uniquely determined from the input, output signals $u(t)$ and $y(t)$ for $0 \leq t \leq \tau$.*

This condition is equal to the following theorem. Similarly, the observability matrix is defined as

$$\mathcal{O}(A,C) = \begin{bmatrix} C \\ CA \\ CA^2 \\ \vdots \\ CA^{n-1} \end{bmatrix}_{(np) \times n} \tag{33}$$

When the system is observable if and only the matrix has rank n, which is represented by Theorem 3.

THEOREM 3. *(Gajic and Lelic) The linear continuous-time system with measurements is observable if and only if the observability matrix has full rank.*

For our system of the two-level two-area state estimator described in Section 3 with power flow equation (8), we fit its linearised counterpart (13) into the above LTI formulation, and have the following observability theorem to hold, which states that at the lower-tier area S_j, network observability is locally guaranteed if and only if the corresponding Jacobian matrix F_{1j} is full-rank. We refer this feature to the partial network observability provided at the first level.

THEOREM 4. *The two-level two-area system with measurements (8) is observable at the sub area S_j if and only if the matrix F_{1j} has full rank.*

PROOF. We take the matrix A_d as identity matrix I_d, and the matrix C_d as F_{1j} following Definition 1, it suffices to show that the observability matrix has full rank if and only if F_{1j} has full rank according to Theorem 2. □

The network observability condition at the first level could be extended to the 2nd-tier SE by the following corollary, while it is not guaranteed of the entire network observability if simply the 2nd-tier observability condition is satisfied.

COROLLARY 1. *The two-level two-area system with measurements (8) is observable at the 2nd level state estimation if and only if the matrix F_2 has full rank.*

5.3 Denial of Observability in Hierarchical State Estimator

As previously discussed in 5.1, the denial of observability in the hierarchical SE by false data injection is our main purpose of study in this paper. In this case, the attacker could manipulate measurements by disabling some, and cause the observability matrix to be rank-deficient, which leads the state estimation no longer able to carry on. This influence could occur in three situations, where the attacker could choose to switch off a meter at a non-overlapping area, at a shared area, or directly manipulate the top level estimator. Apart from giving the bounds of such adversary attack, we would like to identify which measurements are critical and required, and which can be substituted, i.e. the redundancy to ensure continued satisfaction of the observability criterion.

Suppose at lower level, i.e., level 1, the attacker has the capacity to switch off k meters[1]. In each individual area,

WLS algorithm is applied obtaining the estimation for each busbar of voltage and phase angle. Suppose in $Area_i$, the number of buses is N_i, there are n_i state variables to be estimated and m_i measurements. Then the local Jacobian matrix H_i is obtained as an $m_i \times n_i$ matrix. We arrive at the following constraints for an unobservable attack vector to be injected, with a similar approach in [12].

THEOREM 5. *At Level 1 in the i-th area, there exists an unobservable k-sparse attack vector a if and only if the sub-network becomes unobservable when some k measurements are removed, i.e., there exists an $(m_i - k) \times n_i$ submatrix of H_i that does not have full column rank.*

PROOF. ($\rightarrow$) Suppose $a = H_i c$ ($c \neq 0$) is a k-sparse unobservable attack vector of size m_i, without loss of generality, assume $^{(*)}$ the first $m_i - k$ entries of a are zero according to the sparsity definition. Let H' be the submatrix made of the first $m_i - k$ rows of H_i, then $H'c = 0$ from the previous assumption $^{(*)}$, which means that H' does not have full column rank.

($\leftarrow$) Suppose there exists an $(m_i - k) \times n_i$ submatrix of H_i that does not have full column rank, without loss of generality, let H' be this submatrix and consists of the first $m_i - k$ columns of H_i. Thus $H'c = 0$ for some $c \neq 0$, which means that $H_i c$ is the unobservable k-sparse vector by definition. □

At level 2, all the estimation values of voltage and phase angle from the level 1 WLS algorithm are collected, and put into calculation of measurements with which the WLS is applied again. The level 2 Jacobian matrix F_2 is obtained as a $p_1 \times p_2$ matrix, where p_1 is the sum of n_i's, i.e. the total number of level 1 estimated state variables. The adversary could directly manipulate F_2 to render the higher level state estimation unobservable, following Corollary 1.

THEOREM 6. *At Level 2, there exists an unobservable k-sparse attack vector a if and only if the higher level network becomes unobservable when some k measurements are removed, i.e., there exists a $(p_1 - k) \times p_2$ submatrix of F_2 that does not have full column rank.*

PROOF. Proof is analogous to that of Theorem 5. □

For the overlapping areas at the lower level, suppose $Area_i$ and $Area_j$ have l_{ij} shared measurements, which could represent the sub-areas that share l_{ij} meters. The attacker, once identify these measurements, could launch a false data injection attack by manipulating the shared meters, which will influence the lower level and top level state estimation. We have the following theorem describing the false data injection attacks over overlapping areas.

THEOREM 7. *At the overlapping area of $Area_i$ and $Area_j$ which contains l_{ij} measurements, there exists an unobservable k-sparse attack vector a if and only if the joint network becomes unobservable when some k measurements are removed from the joint area, i.e., there exists an $(l_{ij} - k) \times n$ submatrix of the joint area of H_i or H_j that does not have full column rank, where n is the smaller of n_i and n_j.*

PROOF. Let H_{ij} be the submatrix of size $l_{ij} \times n$ of both H_i and H_j where $n = min(n_i, n_j)$ that represents the joint area. With similar technique as the proof in Theorem 5 we can show that there exists an unobservable k-sparse attack

---

[1]In [14], the number of measurements is treated identical as number of meters. We follow this convention in the proofs in this section, since from a fixed topology of meter placement, it is easy to derive the coefficient for meter numbers from measurement numbers.

190

vector $a = H_{ij}c$ $(c \neq 0)$ if and only if the sub network H_{ij} becomes unobservable when some k measurements are removed from the joint area. We could thus construct attack vectors a_i and a_j to the $Area_i$ and $Area_j$ that contain a, and a_i and a_j are k-sparse, unobservable if and only if the joint area becomes unobservable under the attack vector a.

For simplicity we only show the construction of a_i. Suppose $a = H_{ij}c$ $(c \neq 0)$ and without loss of generality assume the first $l_{ij} - k$ entries are zero; we construct a_i as $[0_{m_i - l_{ij}} \mid a]$, and $a_i = H_i c_i$ $(c_i \neq 0)$ is the unobservable attack vector with sparsity k. $\square$

5.3.1 Bounds of Tolerated Meter Compromise and Additional Meters

The above theorems give an analysis on the existence condition of attack vectors. Alternatively, with a defence perspective, we would be interested to know in order to retain the network observability, what bounds could be imposed to the adversary of the false data injection attack. We derive the following theorem about the maximum number of manipulated measurements that can be tolerated, when the attacker could choose to compromise k meters and keep the data injected undetectable. First we give a lemma on the bounds of attacks at the local area of level 1.

LEMMA 1. *At an area of Level 1 of the state estimator where the Jacobian matrix H is of size $m \times n$, if the attacker can compromise k meters, the maximum number of meters to be compromised is $m - n$ in order to maintain the local network observable.*

PROOF. Suppose there is a k-sparse unobservable attack vector a where, by definition $a = Hc$ for some $c \neq 0$. The network is observable, from Theorem 4 H has full rank. Suppose a has switched off k measurements in H, without loss of generality, partitioning $H^T = [H'^T \mid H_a^T]$ where H' is the unaffected measurements with size $(m - k) \times n$ and $H_a = 0$ of size $k \times n$. To allow the estimation of state variables from remained measurements, H' has to have full rank n from Theorem 4. However, when $k \geq (m - n) + 1$, the row number of H' is strictly less than n, and the network is rendered unobservable. $\square$

Using Lemma 1, we may now derive the following result on the bounds of false data injection attack on the overall network, at Level 1, when there are no overlapping areas:

COROLLARY 2. *At the junction of $Area_i$ and $Area_j$ which does not contain joint measurements, where the Jacobian matrix H_i, H_j is of size $m_i \times n_i$, $m_j \times n_j$ respectively, if the attacker can compromise k meters, the maximum number of compromised meters to be tolerated is $min(m_i - n_i, m_j - n_j)$ in order to maintain the first level network observable.*

When the network has overlapping areas and joint measurements between the sub-networks, we come up with the following theorem, which indicates that the shared measurements at overlapping areas are more critical since manipulation on them takes effect in both areas, and thus should be provided with additional protective measures as the adversary might want to concentrate attacks on these areas. In an extreme case, an attack happening on just the joint measurements could render the whole network unobservable:

THEOREM 8. *At the junction of $Area_i$ and $Area_j$ which contains l_{ij} joint measurements, where the Jacobian matrix H_i, H_j is of size $m_i \times n_i$, $m_j \times n_j$ respectively, if the attacker can compromise k meters, the maximum number of compromised meters to be tolerated is $min(m_i - n_i, m_j - n_j)$ in order to maintain the first level network observable.*

Moreover, the attack satisfies that $H_i^s(x_i + c_i) = H_j^s(x_j + c_j)$ without considering measurement errors, where H_*^s is the submatrix of H_* for the shared area and c_* is the injected error in the $*$-th area.

PROOF. With Lemma 1 and Corollary 2, we have the bound of tolerable attacks for retaining the first level network observable as $min(m_i - n_i, m_j - n_j)$. And at the shared area the measurements z^s are the same while the states are different, which means $z^s = H_i^s(x_i + c_i) = H_j^s(x_j + c_j)$ without considering measurement errors according to (20). $\square$

In order to ensure the partial network observability, we also derive the following theorem about the minimum number of additional measurements to be redundantly placed.

THEOREM 9. *Under a k-sparse attack where the attacker could at most compromise k measurements in a local network, the minimum number d of additional measurements to ensure observability is $k - (m - n)$, suppose $k \leq m$.*

PROOF. Let the Jacobian matrix be H, from Theorem 4 H has full rank n (under the assumption $m \geq n$). Suppose the k-sparse attack a has switched off k measurements and turns the Jacobian matrix H into a partition $H^T = [H'^T \mid H_a^T]$ without loss of generality, and $H_a = 0$ of size $k \times n$. From the proof of Lemma 1, when $k \geq (m - n) + 1$, the unaffected measurements H' has row number strictly less than n, and the network is unobservable, i.e. the estimation of state variables from H' of size $(m - k) \times n$ is not available, one has to add measurements to compose a new Jacobian matrix $\hat{H}^T = [H'^T \mid H_d^T]$ with full rank n, with H_d of size $d \times n$. Thus the minimum number of additional measurements $d = n - (m - k) = k - (m - n)$. [2] $\square$

In other words, the number of additional measurements to keep the local network observable d is in the range that $k - (m - n) \leq d \leq max(n, k)$.

6. CONCLUSIONS

State estimation in a heterogeneous, intelligent, distributed power network such as the next-generation smart grid environment induces a number of new challenges also for security including the need to minimise the cost of such security requirements.

In this work we introduce a bi-level two-area state estimation model with the goal of leading to a deeper understanding of hierarchical and multi-area state estimation problems which are not suitably described by classical centralised state estimation models. In particular, we have extended a type of malicious false data injection attack presented in [14] against state estimation, and give some results on how it would cause denial of network observability in the hierarchical model, related to an LTI formulation of observability; bounds of tolerated attacks and redundant measurements are also studied. This work would be applicable to

---

[2]This is under the assumption that measurements on different meters are not linearly dependant.

the research of critical measurements identification, meter protection as well as meter placement.

In our on-going and future work, we would like to incorporate a number of extensions. On-going work concentrates on arbitrary (not only bi-level) hierarchical multi-area state estimation models with minimal constraints leading to a distributed formulation with an explicit AC model. This will allow us to study both the attacks described in the taxonomy in section 5.1 as well as novel attack mechanisms not currently discussed in the literature arising both from the different formulation but also from a more in-depth study of model properties also applicable to the centralised case. In particular, this work extends to the study of state estimator robustness under the circumstances of the two other types of false data injection attacks, namely the denial of state estimator convergence and forcing of state estimate. We also currently study approaches for detection of bad data injection attacks in the form of defender-attacker interactions and what can be characterised about measurements in the presence of such attacks as well as the inclusion of explicit information flows within the smart grid.

Acknowledgements *The research by Y.F., A.B. and S.W. is based in part upon work supported by the 7th Framework Programme of the European Union Joint Technology Initiatives Collaborative Project ARTEMIS under Grant Agreement 269374 (Internet of Energy for Electric Mobility). The research by C.F. and S.P. is partially supported by the 7th Framework Programme of the European Union STREP Project under Grant Agreement 285647 (COCKPITCI - Cybersecurity on SCADA: risk prediction, analysis and reaction tools for critical infrastructures, www.cockpitci.eu).*

7. REFERENCES

[1] ABUR, A., AND GÓMEZ-EXPÓSITO, A. *Power System State Estimation: Theory and Implementation.* CRC Press, Boca Raton, FL, USA, 2004.

[2] BI, S. AND ZHANG, Y. J. Defending mechanisms against false-data injection attacks in the power system state estimation. *GLOBECOM Workshops, 2011 IEEE* (5-9, Dec. 2011), 1162 – 1167.

[3] BJÖRCK, A. *Numerical Methods for Least Squares Problems.* SIAM, Philadelphia, 1996.

[4] BOBBA, R. B., ROGERS, K. M., WANG, Q., KHURANA, H., NAHRSTEDT, K., AND OVERBYE, T. J. Detecting False Data Injection Attacks on DC State Estimation. *Secure Control Systems Workshop, CPSWeek* (Apr. 2010).

[5] GAJIC, Z. AND LELIC, M. *Modern Control System Engineering.* Prentice Hall International, Intl. Series in Systems and Control Engineering, London, 1996.

[6] GIANI, A. AND BITAR, E. AND GARCIA, M. AND MCQUEEN, M. AND KHARGONEKAR, P. AND POOLLA, K. Smart grid data integrity attacks: characterizations and countermeasures. *Smart Grid Communications (SmartGridComm), 2011 IEEE International Conference on* (17-20, Oct. 2011), 232–237.

[7] GÓMEZ-EXPÓSITO, A., ABUR, A., DE LA VILLA JAÉN, A., AND GÓMEZ-QUILES, C. A Multilevel State Estimation Paradigm for Smart Grids. *Proceedings of the IEEE 99*, 6 (June 2011), 952–976. doi:10.1109/JPROC.2011.2107490.

[8] GÓMEZ-EXPÓSITO, A., DE LA VILLA JAÉN, A., GÓMEZ-QUILES, C., ROUSSEAUX, P., AND VAN CUTSEM, T. A Taxonomy of Multi-Area State Estimation Methods. *Electric Power Systems Research 81*, 4 (Apr. 2011), 1060–1069. doi:10.1016/j.epsr.2010.11.012.

[9] GÓMEZ-QUILES, C., GÓMEZ-EXPOSITO, A., AND DE LA VILLA JAÉN, A. State Estimation for Smart Distribution Substations. *IEEE Transactions on Smart Grid 3*, 2 (June 2012), 986–995. doi:10.1109/TSG.2012.2189140.

[10] KALMAN, R. E. On the General Theory of Control Systems. *Automatic Control, IRE Transactions on 4*, 3 (Dec. 1959), pp. 110.

[11] KEKATOS, V., AND GIANNAKIS, G. B. Decentralized Power System State Estimation. *Proc. IEEE GLOBECOM* (Dec. 2012).

[12] KOSUT, O. AND JIA, L. AND THOMAS, R. J. AND TONG, L. On Malicious Data Attacks on Power System State Estimation. *Universities Power Engineering Conference (UPEC), 2010 45th International* (Aug.-Sept. 2010), 1–6.

[13] KOSUT, O. AND JIA, L. AND THOMAS, R. J. AND TONG, L. Malicious Data Attacks on the Smart Grid. *IEEE Transactions on Smart Grid 2*, 4 (Dec. 2011), 645–658. doi:10.1109/TSG.2011.2163807.

[14] LIU, Y., NING, P., AND REITER, M. K. False Data Injection Attacks against State Estimation in Electric Power Grids. In *Proceedings of the 16th ACM Conference on Computer and Communications Security* (Chicago, IL, USA, Nov. 2009), S. Jha and A. D. Keromytis, Eds., ACM Press, pp. 21–32. doi:10.1145/1653662.1653666.

[15] MONTICELLI, A. Electric Power System State Estimation. *Proceedings of the IEEE 88*, 2 (Feb. 2000), 262–282. doi:10.1109/5.824004.

[16] MONTICELLI, A. AND WU, F. F. Network Observability: Theory. *IEEE Transactions on Power Apparatus and Systems PAS-104*, 5 (May 1985).

[17] SHEPARD, D. P., HUMPHREYS, T. E., AND FANSLER, A. A. Evaluation of the Vulnerability of Phasor Measurement Units to GPS Spoofing Attacks. In *Critical Infrastructure Protection VI: Sixth Annual IFIP WG 11.10 International Conference on Critical Infrastructure Protection* (Washington D.C., USA, 2012), J. Butts and S. Shenoi, Eds., IFIP Advances in Information and Communication Technology, Springer-Verlag. To appear.

[18] SOU, K. C., SANDBERG, H., AND JOHANSSON, K. H. Computing Critical k-Tuples in Power Networks. *IEEE Transactions on Power Systems 27*, 3 (Aug. 2012), 1511–1520. doi:10.1109/TPWRS.2012.2187685.

[19] XIE, L., CHOI, D.-H., KAR, S., AND POOR, H. V. Fully Distributed State Estimation for Wide-Area Monitoring Systems. *IEEE Transactions on Smart Grid 3*, 3 (Sept. 2012), 1154–1169. doi:10.1109/TSG.2012.2197764.

[20] ZONOUZ, S. A., AND SANDERS, W. H. A Kalman-Based Coordination for Hierarchical State Estimation: Agorithm and Analysis. *Proc. the 41st HICSS* (Jan. 2008), 187. doi:10.1109/HICSS.2008.23.

eBond: Energy Saving in Heterogeneous R.A.I.N

Marcus Hähnel, Björn Döbel, Marcus Völp and Hermann Härtig
{mhaehnel,doebel,voelp,haertig}@tudos.org
Technische Universität Dresden
Faculty of Computer Science
Institute of Systems Architecture
Operating Systems Group

ABSTRACT

Network energy is a significant, although not the largest, cost factor in medium to large scale server installations. On the other hand, most server installations work with redundant link and infrastructure layouts to reduce the risk of network outages. Introducing eBond, an energy-aware bonding network device, we exploit possible heterogeneities in these redundant layouts to adapt network device energy consumption to dynamic server bandwidth demands. Replaying the trace of a realistic scenario in a simulation of eBond with fine grain energy profiles measured at two network cards we achieve energy savings up to 75 % for the server-side network interconnect.

Categories and Subject Descriptors

Software and its engineering [**Operating systems**]: Power Management; Networks [**Network protocols**]: Network layer protocols

Keywords

energy; network; server; eBond; network card, bonding

1. INTRODUCTION

Energy demand is one of the larger cost drivers in large scale server installations. Modern data centers consume between 10 % and 15 % of their total operation power in network links and infrastructure [12]. This demand translates into a significant though not the highest cost factor on the power bill.

In this paper, we focus on optimizing network link energy in medium to large scale server settings by adjusting the power demand of server-side network cards to the actual bandwidth requirements of the servers. Our approach is based on the observation that network links are typically redundant to limit the risk of network outages. Rather than connecting servers with the same high-end network interface cards (NICs), for example two 10 gigabit Ethernet (10 GbE) NICs, we propose to introduce heterogeneity by also including more lightweight connections such as Gigabit Ethernet (GbE) NICs. In the rare case of failure of all high bandwidth connections, these inexpensive cards may still offer some limited bandwidth to an otherwise disconnected server. However for the more common case of a medium loaded data center, being able to scale energy consumption by switching between energy-demanding high bandwidth cards and low power connections gives room for significant energy and cost savings.

Our proposed setup is especially beneficial for installations with a high difference of demand during day and night times or other cyclic demand variations. These variations can often not be compensated by load balancing because latency requirements prohibit relocating the system load to other global regions. Prime examples include on-line gaming services such as OnLive [17] or Google's live search.

After providing the necessary background and relating our work to the works of others, Section 3 presents the setup and results of our study of two wired network cards. Although energy efficiency has been a hot topic for quite some time now, we found that recent network interface cards still offer only limited power scaling possibilities and that switching to a lower bandwidth card leaves room for power savings.

Motivated by these results and realizing that server resilience demands for alternate connections anyway, we developed eBond — an energy-aware bonding network device — which we introduce in greater detail in Section 4. eBond exploits the possibility to layout network infrastructure heterogeneously and builds on channel bonding, which sometimes is also called redundant array of inexpensive networks (or R.A.I.N) [6], to reroute traffic to low power infrastructure if the current server load tolerates the reduced bandwidth of this infrastructure. To evaluate the performance of eBond, we have implemented a network power simulator (see Section 5) to replay exactly the same network traces for different NIC characteristics. Section 6 presents the results of our evaluation using traces of two real-world scenarios. We show, that we can save up to 75 % of the energy used by the network cards when using our approach.

eBond integrates itself into our larger vision of energy-adaptive computing and networking. In Section 7 we conclude this paper highlighting our vision of energy-adaptive computing.

2. BACKGROUND & RELATED WORK

Channel bonding was first introduced in 2000 by the IEEE 802.3 group [8] and has since been used to improve outage resilience and network bandwidth. It does so by coupling redundant links into one virtual link [7]. In our setup, we will use transmit load balancing (mode 5) to redirect traffic between the 10 GbE and the GbE NIC. On the server side, eBond switches between these two cards depending on the amount of outgoing and incoming traffic. No special support is required by the cards or the switch with the exception that one of the two NICs has to be able to take over the MAC address of the respective other. In case eBond decides to power down one of the two NICs, the other card will take over and respond to all traffic sent to this MAC. While Imaizumi et al. [13] recognize the potential for energy savings in link aggregated setups, they do not extend this to heterogeneous device configurations. To the best of our knowledge, channel bonding has not been used before for server-level network link energy optimizations in heterogeneous setups.

Research on energy optimization typically focuses on the CPU [19, 2] or uses whole system measurements [18] to characterize a system's current demand. In these latter works, the power consumption of individual devices is often difficult to isolate, in particular as these devices are still lacking the power measurement equipment that was recently introduced in CPUs [14, 11].

In 2010 the IEEE ratified the IEEE 802.3az standard that promises energy-efficient Ethernet [3]. The approach taken there is to power down a port that is not used, providing a "sleep mode" for the Ethernet port. At the same time the device is never considered off-line, as a low power connection to the other side is kept alive and refreshes the sleep status, or wakes the port if required. While this can yield important improvements in energy consumption it is mainly beneficial for network connections that are idle for longer periods of time — a setting not always present in highly loaded web services.

An alternate approach was suggested by Gunaratne et al., trying to adapt the link rate depending on demand [9]. While this is an interesting approach it requires specialized hardware support and especially requires the processing units on the chip to adapt sufficiently depending on the link rate to reach relevant energy savings. We found that such a scaling was not possible with our cards.

Wireless network energy has been studied extensively in the setting of mobile devices [1]. The resulting models, which weigh throughput and power consumption with other factors prevalent in mobile settings such as battery lifetime, tend to become very complex [5].

Sohan et al.'s study on 10 GbE NIC energy consumption [20] confirms our findings that NIC power often does not scale well with bandwidth. They conclude that further hardware improvements are required to make the network energy scale. Our solution is entirely based on software assuming heterogeneity in redundant links, which server installations have to provide anyway. This provides a convenient intermediate solution while waiting for more efficient hardware designs to be able to scale power near-proportionally to bandwidth.

Of course, network link energy is only a part of the energy spent for data center networking. Heller et al.[12] focus on optimizing the infrastructure's energy costs by rerouting links to turn off unused switches and Gupta et al. have analyzed the feasibility of power management in those devices [10]. Our work is orthogonal to these results and may allow for additional savings.

3. A STUDY OF TWO WIRED NETWORK CARDS

Sohan et al.'s study [20] on network energy consumption provides power values for a wide range of cards. However, we needed higher resolution results with many measurement points to demonstrate the power savings of our setup with a reasonable high accuracy. To obtain these detailed energy models, we measured the power consumption of two exemplary network interface cards at varying bandwidths.

3.1 Methodology

To obtain the energy profiles, we created a direct private network link between two Intel Core-i5 PCs (i.e., no switches, etc.). The machines were connected using a single, 3 m long CAT6 network cable. To get precise, high-resolution power consumption measurements, we installed a riser card and cut the 3.3 V and the 12 V rails of the ribbon cable. Into this riser card we then plugged the to be measured network card. We employ a Yokogawa WT-210 [22] digital power meter, that is capable of measuring current and voltage at the same time with a sampling frequency of up to 10 Hz. Amperage was measured by routing all 3.3 V and 12 V rails through one dedicated Yokogawa power meter for each of the two voltage levels. The power meter provides integrated shunts, which are necessary for measuring currents. The voltage was taken between the riser card's corresponding voltage rail and one of the ground wires of the system's power supply using the voltage inputs of the Yokogawa power meters. This setup ensures the highest precision because also variations in voltage are recognized and factored into the total power consumption.

We measured the power consumption of an 1 Gbit Intel EXPI9301CTBLK network card with an E25869 (B) on card CPU as well as a 10 Gbit Intel Ethernet Server Adapter X520-T featuring an E76983 (A) CPU. The cards have a manufacturer claimed typical power rating of 1.9 W [16] and 18 W [15], respectively.

3.2 Challenges

Our first attempt to obtain power characteristics of our network cards was to run a microbenchmark, which gradually increased the bandwidth in steps of 1 Mbit/s after every measurement interval. At a first glance, this benchmark produced a seemingly nice profile and was reproducible with nearly identical results over several independent runs. However, it turned out that after we degraded the bandwidth again at the end of one run to obtain further results, the power consumption for this degraded bandwidth did not match the original power consumed when running our microbenchmark at this bandwidth. More important however, the power demand for this bandwidth did not adjust itself over time but continued to deviate while we increased our power sampling times. We are not absolutely certain what caused these deviations, but assume that this is due to some chip internal logic that adjusts parameters based on a history of previous usage. To compensate for this effect, we repeated our benchmark switching randomly (with a uniform distribution) between the to be measured bandwidth settings.

3.3 Results

Figure 1 shows the results of one measurement run. Each bandwidth level was held for 10 seconds. The figure shows, that the measured power did not deviate during a single interval but was also not constant at the same bandwidth in different intervals. There is no guarantee that a higher bandwidth leads to higher power levels. We experienced this same effect in our previous approach for generating network card profiles.

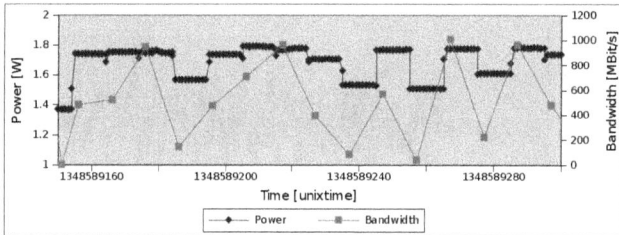

Figure 1: The individual power levels as seen during the microbenchmark using randomly selected bandwidth levels

The random distribution of the trace however, allowed us to gather detailed information, including minimum, maximum and average power consumption levels at the different bandwidths. The results of this extraction are the detailed power profiles of the network interface cards, one of which is shown in Figure 2a. Plotted is the power consumption for varying receive bandwidths. The Figure shows the 1 Gbit/s Ethernet card. Figure 2b shows a comparison between the average send and receive powers for the gigabit card, which shows nearly no difference between the two power levels. Even the dips are nearly the same. Figure 2c zooms into this effect by showing the difference between send and receive power ($P_{send} - P_{receive}$) on the various bandwidth levels. The variation is well below 0.02 W.

For the 10 Gbit/s card, we only show the receive power as the power needed for sending was virtually identical. We further performed our benchmark in steps of 10 Mbit/s intervals to reduce both profile creation and simulation time. A plot with the minimum, maximum and average power consumption calculated over multiple consecutive runs, is plotted in Figure 2d. For each of the displayed bandwidth values there were at least five measurements. In general, the trend of the power levels to increase can be seen with both cards, but the 10 Gbit/s device scales very poorly with load (please note the offset of the y axis). This adapter is also the only device using the 12 V rail, albeit there were no variations of the power levels on this rail when the bandwidth was changed.

One aspect not shown in the above Figures is when the cards are sending and receiving at the same time. For the 10 GbE device this case is nearly indistinguishable from the sending/receiving curves. There is nearly no increase in power when sending and receiving at a bandwidth, compared to only doing one of these operations. The Gigabit Ethernet adapter shows different characteristics between only sending, only receiving or performing both operations at the same time. In order to visualize this difference, Figure 3 illustrates the whole profile as a breakdown of send

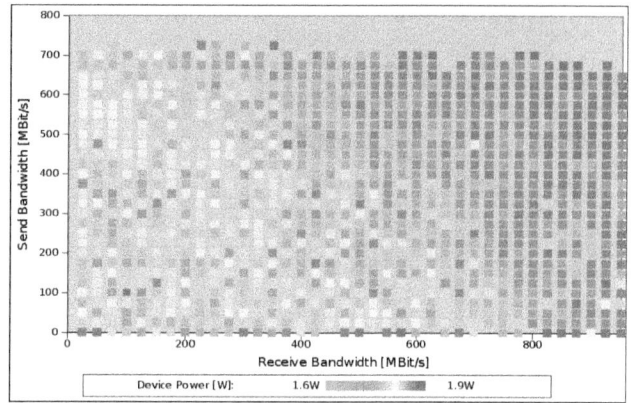

Figure 3: The individual power for different send/receive-bandwidth combinations, as determined by the microbenchmark

and receive bandwidth. We use the same breakdown in our simulator. The axes show the respective send and receive bandwidths. The colors indicate the power consumption at those bandwidth levels.

3.4 The Odd One Out

During our benchmarks we also tested one further network card: a gigabit Ethernet card manufactured by Intel (model number EXPI9300PTLPBLK). It belongs to the PT family of Intel gigabit network cards and is rated with a typical power consumption of 3.3 W. We found, that this card does not scale with bandwidth at all. If the interface is up it consumes a near constant 1.82 W increased only to 1.83 W when we draw the full bandwidth from the card. Curiously, it is also the only card we have seen to conserve energy when the interface is powered down without unplugging the cable. While the CT series card can achieve even lower power levels when the cable is unplugged, this procedure is infeasible in a data center environment. Power-saving methods must be controllable by software to be automated or at least remote controlled. Table 1 shows an overview of the relevant measurements of the three different cards in the most significant situations.

3.5 Summary

The results of our investigation goes in line with the measurements performed by Sohan et al. in their study on 10 GbE NIC energy consumption [20]. We found, that network cards still do not scale their power consumption with bandwidth requirements, at least not in a way that is comparable, in terms of saved power, with using dedicated lower power network cards.

Moreover, we found that the power characteristics of network cards vary widely even within similar cards from the same manufacturer. We have created a profile for a 10 GbE and a Gigabit Ethernet card, which characterizes the power requirements of these cards at different bandwidth levels. We also analyzed the capability of these network cards to switch themselves into lower power modes when the cable is unplugged or when the interface is disabled.

We were only able to evaluate add-on network cards, because the correct instrumentation of a mainboard's components is very hard. While we are currently investigating such

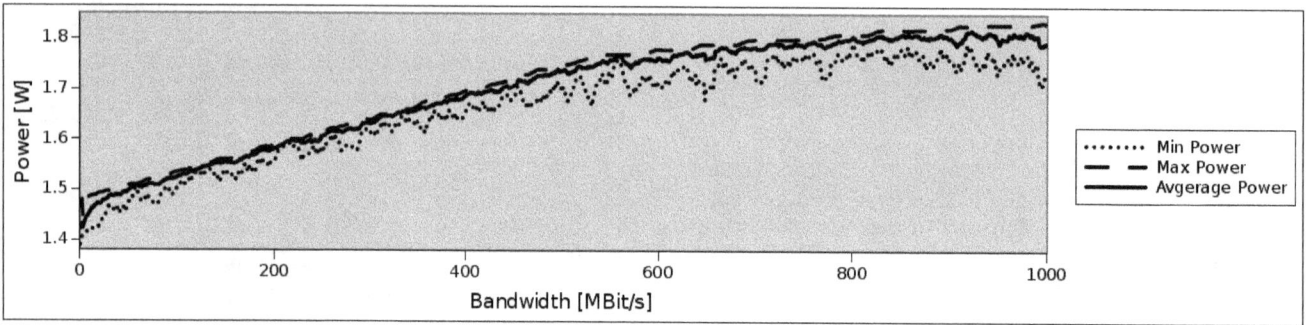

(a) Gigabit ethernet, receiving

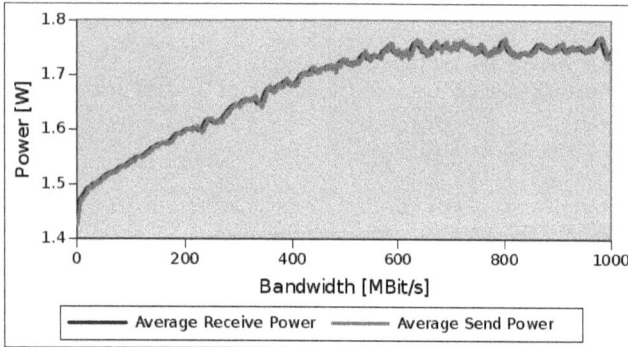

(b) Gigabit ethernet, average sending / receiving

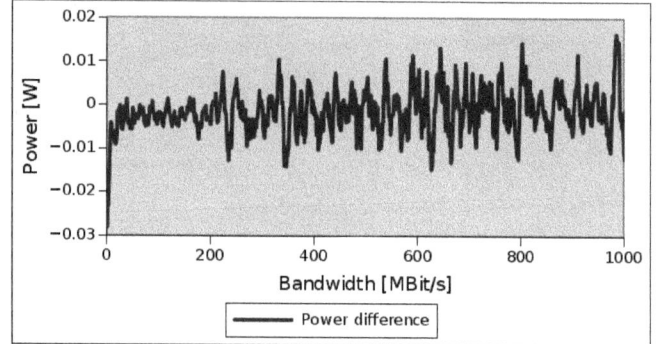

(c) Gigabit ethernet, average sending / receiving

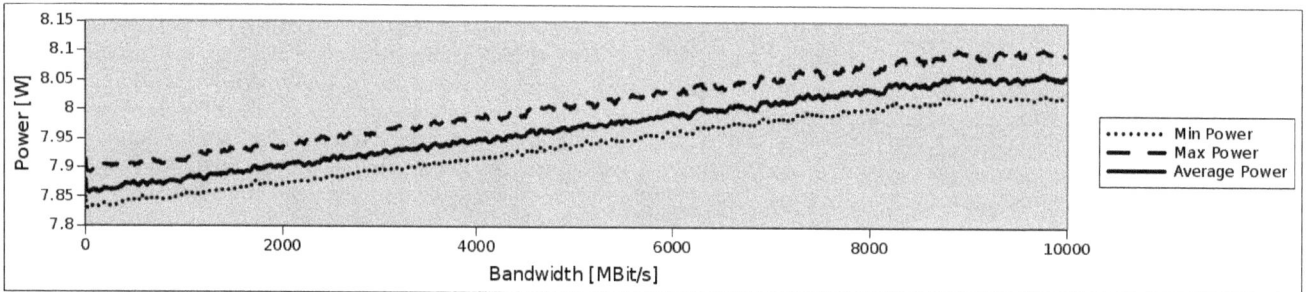

(d) 10 Gigabit ethernet, receiving

Figure 2: Minimum, maximum and average power consumption of the two measured network cards with randomly selected bandwidth settings.

a setup, it was not available in time for the publication. We expect, however, that such measurements would not change the general message of the measurements. The processing capabilities of server grade on-board network cards comparable to add-on chips, and, baring overhead introduced by PCIe, should not be significantly different. Especially, if we combine a possible on-board 10 GbE card with an PCIe Gigabit Ethernet card, we still expect there to be a difference in energy consumption that favours the Gigabit card, albeit at lower savings.

4. EBOND: ENERGY-EFFICIENT BONDING

Based on the observations made in the previous section, we propose eBond, an energy aware network scheduler for adjusting the servers' network energy consumption to their loads. To do so, eBond exploits features of the Linux bonding interface [6] operation of network cards.

After a short introduction to the concept, we present some scenarios where we think eBond will be beneficial to reduce energy costs while keeping the service at a high quality level. After that, we give a sketch of the algorithm for scheduling the network cards.

4.1 Concept

As many server installations are equipped with redundant network interfaces [4] we propose to employ a heterogeneous scheme, with high energy network cards to handle the expected peak network load and one or more low energy cards as additional connection and backup.

eBond always chooses the more energy efficient cards as long as these cards can sustain the requested bandwidth using channel bonding both for switching between cards and for sustaining bandwidth if the requirements exceed the bandwidth of a single card. The decision which cards to activate is based on the observations we made in the previous

196

	10 Gigabit X520-T2	Gigabit CT EXPI9301CTBLK	Gigabit PT EXPI9300PTLPBLK
Interface down, cable unplugged	7.35 W	0.08 W	0.7 W
Interface down, cable plugged in	7.88 W	1.35 W	0.7 W
Interface up, no transfer	7.88 W	1.35 W	1.82 W
Interface up, transfer at full duplex bandwidth	8.10 W	1.92 W	1.83 W

Table 1: The measured network cards at various modes

section. In the case of only two cards (one 10 GbE and one 1 GbE), this decision is to pick the GbE card whenever the bandwidth is below one gigabit per second.

Our system is prepared to work directly with the power profiles to also handle scenarios where the decision which card or combination of cards is more power efficient cannot be made as simple.

We now present the scenarios where the usage of the eBond interface is beneficial and then give the details of the eBond systems algorithm. To simplify the following discussion, we restrict ourselves to two card setups. In these setups, the bandwidth threshold sufficed as a card selection criterion. We show that eBond may save 75 % of NIC energy consumption in a realistic scenario and while only introducing a negligible amount of overload.

4.2 Scenarios

Clearly, channel bonding is most beneficial in scenarios where a significant portion of the requests can be handled by the low bandwidth card. These are scenarios with significantly higher peak bandwidth requirements or with large bandwidth variations.

One possible such scenario is that of a server that has regular variations in its load, like a weekly cycle or a day and night cycle. As most servers aim to serve users close to them, a day and night cycle should be observable for a lot of FTP servers or web services with local server infrastructure like Wikipedia or OnLive. Usually such web services are distributed across the globe to reduce access latency and to pay for cheaper, intra-continental traffic. But this also leads to the above mentioned day/night cycles in bandwidth, that can not simply be compensated by re-routing traffic from other continents, without loosing the low-latency property.

While we refer to the variations with the terms "day cycle" and "night cycle" for high and low bandwidth times respectively the concept is of course valid for other variation patterns as well. We will use these terms for the remainder of this paper without loss of generality.

In order to be useful, eBond also requires part of the bandwidth to be in the range of more than one NIC type. It is, for example, not beneficial to have traffic that varies between 2 Gbit/s and 10 Gbit/s, and only have 10 Gbit/s network cards available. In that case the main optimization would be to choose energy efficient 10 Gbit/s network cards. An optimal scenario for eBond would have a night cycle of well below 1 Gbit/s and a day traffic of well over 1 Gbit/s.

Further, we require the small bandwidth network card, which is to serve the night cycle, to have a lower power footprint than the high bandwidth card at at least one bandwidth range. Judging from our analysis in Section 3 and the results presented in [20] by Sohan et al, we claim that this is the case with most modern network equipment.

We do not necessarily require a strict ordering of the network cards' energy efficiency, there might be some overlap, where the high bandwidth card is more energy efficient than the low bandwidth card or that one card only covers the mid bandwidth ranges while the other covers high and low ranges at the same time. But we have not yet encountered such a setup in practice.

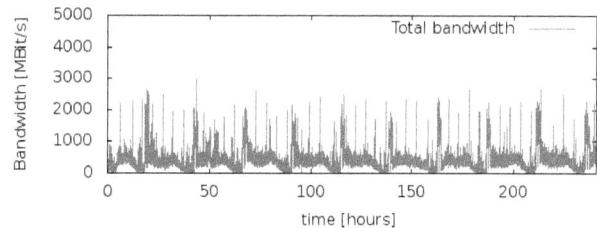

(a) Bandwidth requirement of a Debian/Ubuntu FTP server over 10 days

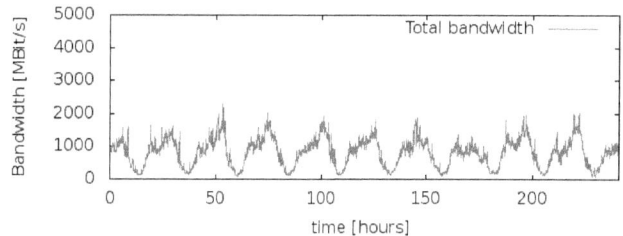

(b) Bandwidth of the uplink of a dormitory complex over 10 days

Figure 4: Bandwidth of example scenarios

For our later analysis we chose two scenarios that fit the criteria discussed above:

1. *Debian/Ubuntu FTP* represents a trace of our local Debian/Ubuntu mirror over 43 days. The trace recorded the bandwidth every 5 seconds, separating incoming and outgoing traffic. A plot of the total traffic is shown in Figure 4a.

2. *Uplink of a dormitory complex* captured a trace of up- and downstream bandwidth and was stored in rrd format. Because of this, the data is available in resolutions of 1, 5, 30 and 360 minutes for the most recent 2, 10, 60 and 720 days respectively. We used the data with 5 minute resolution as shown in Figure 4b for our experiments, because it represents the best trade-off between resolution and simulated time.

4.3 Implementation and Design

The initial version of eBond runs the algorithm shown in Algorithm 1.

Algorithm 1 Basic eBond algorithm

1: $cur_{card} \leftarrow default_{card}$

2: **while** True **do**
3: $\quad P_{min} \leftarrow \infty$
4: $\quad opt_{card} \leftarrow$ None

5: $\quad BW_{send} \leftarrow (sentBytes(t) - sendBytes(t-i))/i$
6: $\quad BW_{recv} \leftarrow (recvBytes(t) - recvBytes(t-i))/i$

7: $\quad \triangleright$ Find the optimal card for the bandwidth
8: $\quad$ **for all** $card \in Cards$ **do**
9: $\quad\quad P \leftarrow card.getPower(BW_{send}, BW_{recv})$
10: $\quad\quad$ **if** $P < P_{min}$ **then**
11: $\quad\quad\quad P_{min} \leftarrow P$
12: $\quad\quad\quad opt_{cart} \leftarrow card$
13: $\quad\quad$ **end if**
14: $\quad$ **end for**

15: $\quad opt_{card}.activate()$
16: $\quad$ **wait until** card **is ready**
17: $\quad cur_{card}.powerDown()$
18: $\quad cur_{card} \leftarrow opt_{card}$
19: $\quad$ **sleep** i
20: **end while**

In each iteration the up- and downstream bandwidth requirements are determined by querying the statistics of the bonding interface. The bandwidths are always calculated over a user defined sliding window. While the above algorithm assumes the window to be equal to the reconfiguration interval, it is also possible to use other window sizes.

Then for each available network card in the bonding interface the power levels at the required bandwidth levels are calculated and the minimum power network card is identified. The such selected card is activated and the previous card deactivated. After that, the eBond driver sleeps until the next reconfiguration interval.

The reconfiguration interval gives the minimum time between checks of the bandwidth. If sudden surges occur outside this timeframe, eBond may not be able to cope with these effects. This may lead to service level agreement (SLA) violations, as a low power network card may not be able to cope with the increased traffic.

Using this basic algorithm, we observed a flutter effect leading to either increased power consumption or an increase in SLA violations or both, depending on the workload. This happens when traffic is oscillating around the trip point between two cards. In this case, there may be a lot of changes between the active cards, leading to an actual increase in total power consumption, because both have to be driven at the same time for a short amount of time to ensure that no connections are lost. We introduced two methods to mitigate this effect.

First we introduce a hysteresis: We only switch to the lower power interface, if an increase in bandwidth by a factor we term *predictor* will still result in the new card being more energy efficient. An unsupported bandwidth has infi-

nite power consumption in the card, in order to avoid the card being chosen.

We set this predictor to 0.1 in some of our experiments to simulate a 10 % increased bandwidth requirement.

Still this does not fix repeated sudden high spikes in the network load. We address this issue by introducing a cool down time, in which the interface may not be switched down to a lower bandwidth card. This is based on the observation, that spikes often have a temporal locality. During this cool down time we only allow switching to a higher bandwidth interface.

This also limits the switching frequency and thus the time that two network cards are turned on at the same time.

Please be reminded, that this cool down is *not* equivalent to increasing the reconfiguration interval as this would prevent the interface from switching to a higher bandwidth when required. The cool down only limits switching down. As such it presents a trade-off between higher power consumption (longer cool down) and higher number of SLA violations (shorter cool down).

4.4 Advantages of the Bonding Interface

We chose the bonding interface for our work, because it presents an easy implementation of energy aware network card selection. Not only is bonding a technique that is available in most data center switch technology but it also allows us to not care about modifications of the routing tables or similar intricacies of the network stack, that appear when switching between network interfaces [1].

Also bonding is mostly already done in data centers, where availability is an important service selling point. Our proposal only improves on the existing bonding techniques by suggesting the usage of heterogeneous network cards as backup links. We exploit this heterogeneity as a potential for energy improvements. The price of our approach is a slightly increased mean time to recovery in the event of a link failure because the other card needs to be turned on first.

To disclose all risks we must also mention that our approach can lead to reduced available bandwidth should all the available high bandwidth links fail. This could be mitigated by providing more high bandwidth links, that could be powered down during normal operation.

This setup gives the operator also an easy choice to drive a single server in either traditional bonding modes or operate in energy efficiency mode using the eBond system.

The bonding interface further allows us to extend the system to any number of cards, and not be limited to two cards. It is even possible to use completely different cards and physical layers, as long as their drivers and switch technology support bonding.

5. SIMULATOR

We decided to use a simulator to evaluate the effectiveness of the eBond interface. There are several reasons behind this decision. The first is simulation time. We wanted to use real world profiles as were introduced in Section 4.2, which span several weeks of sever operation. This long time span is required, because we need to capture several day an night cycles, spikes and irregularities, to demonstrate the effectiveness of our eBond interface across a wide range of scenarios. To run these long term test cases with various parameters would take years. This is clearly unacceptable. Shorter periods either do not capture the effects we need to demonstrate the effectiveness of eBond, biasing the result in one or another direction, or are synthetic, making the evaluation less realistic compared to the load scenarios of real world servers.

The second reason is, that we did not have enough high precision measurement technology to capture two cards simultaneously on all power rails with sufficiently high accuracy.

We believe that our above shown method for capturing network card energy behavior presents us with sufficiently precise profiles for a simulation to capture the energy consumption of the network interfaces with high enough detail, and only limited error. This can also be seen in the amount of variation we saw in our profiles as detailed in Figures 2b-2d in Section 3.3.

The simulator is implemented as a python script, which evaluates the power consumption of the network cards based on the profiles taken as described in Section 3. The simulation interval is based on the data point interval of the scenarios' datasets, but never smaller than the eBond reconfiguration interval. The eBond algorithm, including cool down and predictor, is used to determine active network cards.

The configuration file for the simulator is the same that we also used for the eBond interface. No further settings are required. The same is valid for the energy profiles of the network card. This keeps the configuration overhead minimal and ensures consistency between the real eBond interface and the network simulator.

The simulator then evaluates the data and generates an energy profile that we show for our two demo scenarios in Section 6 as well as detailed statistics on SLA violations and network card usage.

The sources of the simulator, together with the NIC energy profiles, will be made available in time for the conference at our github repository [21].

6. EVALUATION

To determine the prospective energy savings of eBond we used the two scenarios introduced in Section 4.2.

We replayed 43 day traces of a Debian/Ubuntu FTP mirror and 10 day traces of a Dormitory network uplink in our network simulator using different network card scheduling policies. The simulator accumulates the network bandwidth used by the network cards during transfer and idle times. This power is also recorded in a trace, by matching each bandwidth adjustment against the power profiles presented in Section 3. We first present the detailed results of the FTP scenario and then provide a short summary of the results of the Dormitory uplink scenario.

6.1 Detailed FTP Scenario

The graphs in Figure 5 show energy characteristics of the FTP trace over a 10 day period in different scenarios. Figure 5a presents the average power consumed in the traditional setting where all load is served by a single 10 GbE card. When we compare this consumption to the bandwidth graph seen in Figure 4a we can see a clear optimization potential.

In Figure 5b, we present the power consumption graph of the two network links when combined into an energy-aware bonding device. We see that most of the time, the GbE card suffices to meet the bandwidth demand of the FTP

server, yielding a lower average power consumption. Whenever the FTP server's demand exceeds the capability of the GbE card, eBond switches to the 10 GbE card. Figure 5c is a zoomed in view of the first 12 hours of the graph.

Table 2 presents some statistics for the simulation, which confirm these results. We were able to save an average amount of 140 Wh or 74.7 % per day when compared to the single 10 GbE scenario's power demand. The scenario *eBond 3* presents the most aggressive power saving scheme, with no hysteresis or cool-down time, and no load prediction. While this has the most savings because it immediately switches to the most energy efficient card for the current load, it also induces a large number of service level agreement (SLA) violations. These happen, when a requested bandwidth could not be served by the current network card, which leads to lower bandwidth or increased latency from the view of the client.

The setups 1 and 2 present more reasonable configurations that balance energy savings against the number of SLA violations. The concrete parameters that deliver a balanced

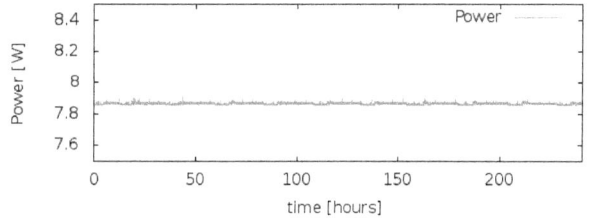

(a) 10 day power demand with only one 10 GbE card

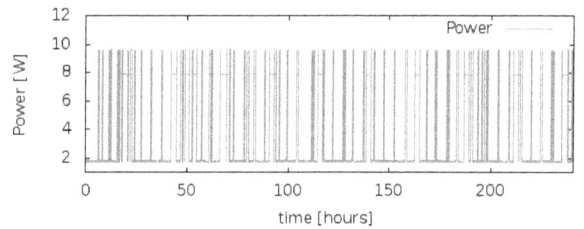

(b) 10 day power demand with eBond (10 GbE + GbE card)

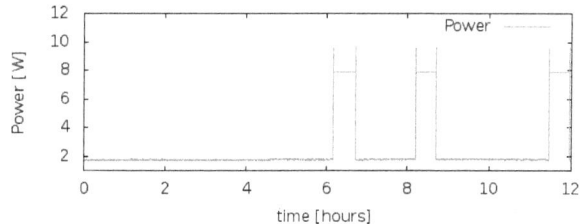

(c) 12 hour power demand with eBond (10 GbE + GbE card)

Figure 5: Figures showing the power demand of the network cards in the system for the FTP server scenario. Figure 5a is the setup without the eBond system, while Figures 5b and 5c show power usage with the eBond heterogeneous network channel bonding in the balanced configuration. Figure 5c shows a zoomed in section of Figure 5b.

	Single 10 GbE	eBond 1: high savings	eBond 2: balanced	eBond 3: aggressive
Simulated time	43 days	43 days	43 days	43 days
Prediction	-	10 %	10 %	0 %
Cool-down time	-	0 h	0.5 h	0 h
Total energy	8113 Wh	2055.8 Wh	2758.4 Wh	2033.8 Wh
time on 10 GbE	100 %	3.825 %	15.07 %	3.39 %
time on GbE	0 %	96.253 %	84.95 %	96.74 %
SLA violations	0 (0 s, 0 %)	195 (1035 s, 0.028 %)	103 (519 s, 0.014 %)	252 (1265 s, 0.034 %)
Saved energy	0 %	74.7 %	66 %	74.9 %

Table 2: Statistics of the Simulation for the Debian/Ubuntu FTP Server scenario. The SLA violations are given as the number of times the required network bandwidth could not be provided. In parentheses is the total time during which the bandwidth was lower than required together with the percentage of the total time this amounts to.

setup heavily depend on the load type and pattern and must be configured specific to the expected server workload.

These savings are already with two quite efficient cards. When considering the results of Sohan et al. [20] there may be even more potential for energy savings in existing server setups.

6.2 Uplink scenario

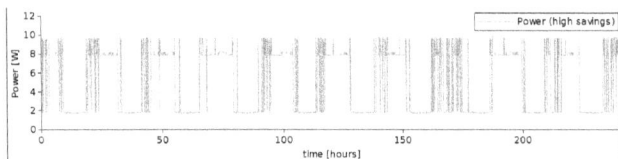

(a) 10 day power demand with high *savings* eBond profile

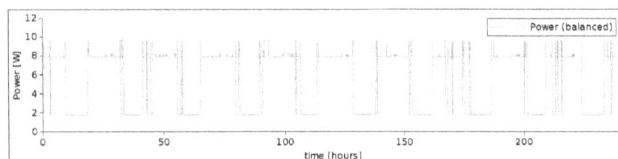

(b) 10 day power demand with *balanced* eBond profile

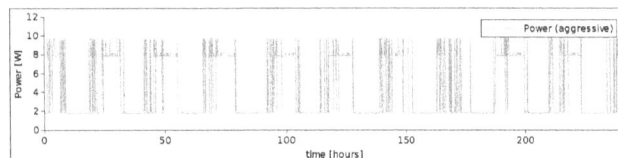

(c) 10 day power demand with *aggressive* eBond profile

Figure 6: Figures showing the power demand of the network cards in the system for the dormitory uplink scenario for the different eBond scheduling profiles as shown in Table 2

This second scenario is more stable with less spikes in the bandwidth as was shown in Figure 4b in Section 4.2. On the one hand, this makes the predicting the traffic easier and thereby causes a reduction of SLA violations compared to the FTP scenario. On the other hand, the scenario has also less potential for energy savings, as the required bandwidth does only drop to less than 1 Gbit/s during night time. We ran the same simulation as for the previous scenario using a 30 minute cooldown and 10 % prediction. The result was an energy graph as presented in Figure 6, with the subfig-

ures showing the different scheduling profiles of the network eBond network card scheduler.

The number of SLA violations has been greatly reduced due to the more predictable nature of the network usage compared to the FTP scenario. A comparison of SLA violation times expressed as percentages of the runtime are presented in Table 4.

	high savings	balanced	aggressive
FTP	74.7 %	66 %	74.9 %
Uplink	35.9 %	30 %	43.2 %

Table 3: Energy savings compared to the single NIC setup for the two scenarios under the 3 eBond policies as seen in Table 2

	high savings	balanced	aggressive
FTP	0.028 %	0.014 %	0.034 %
Uplink	0.07 %	0 %	0.14 %

Table 4: Percent of time, that the bandwidth requirement could NOT be satisfied (SLA Violations)

7. CONCLUSIONS AND FUTURE WORK

In this paper, we presented eBond — an energy-aware network bonding interface — to adjust network link energy to the current bandwidth demands of servers in medium to large scale data centers. Our approach exploits heterogeneity in the redundant layout of connections by switching between low power but also low bandwidth network interface cards and high bandwidth cards, which we found to be more demanding and less adaptive. No special infrastructure is required beside the redundant link layout that resilient server installations have to provide anyway. Our simulation of eBond with real-world network traces indicates power savings of up to 75 %. While the power savings depend on the concrete server load scenario our implementation allows for different, user configurable, profiles to select the network card scheduling behavior best fitted for the typical load situation of the network.

There are multiple directions we aim to investigate for future work. On the hardware side, more adaptive network cards, possibly integrating the low bandwidth circuitry next to the high bandwidth setup for better scalability are imaginable with an off-loaded eBond instance to select between. An integration of even more link types such as optical or wireless board-to-board interconnects would be highly interesting as well as other scenarios besides networking.

Further we plan to extend our research to whole heterogeneous network hierarchies, where we include switching technologies into our observations and use different bandwidth switches according to the demand of the attached subnetworks. This will allow us to venture even farther into the domain of whole-datacenter energy efficiency which we also extend in parallel by our work on QOS-based, energy-aware scheduling of resources on individual nodes of the network.

Acknoledgement

The authors would like to thank Waltenegus Dargie for providing his instrumentation setuo for the measurements. We further extend our thanks to Hannes Weissbach for help with the benchmarks and Adam Lackorzynski, Michael Kluge (ZIH), and Maximilian Marx for providing trace data. This work was partially funded by the German Research Council (DFG) through the Collaborative Research Center CRC 912 "Highly-Adaptive Energy- Efficient Systems" (HAEC), the Special Purpose Program "Dependable Embedded Systems" (SPP 1500) and the cluster of excellence "center for Advancing Electronics Dresden" and by the EU and the state Saxony through the ESF young researcher group "IMData".

8. REFERENCES

[1] AGARWAL, Y., PERING, T., WANT, R., AND GUPTA, R. Switchr: Reducing system power consumption in a multi-client, multi-radio environment. In *Proceedings of the 2008 12th IEEE International Symposium on Wearable Computers* (Washington, DC, USA, 2008), ISWC '08, IEEE Computer Society, pp. 99–102.

[2] BELLOSA, F. The benefits of event: driven energy accounting in power-sensitive systems. In *Proceedings of the 9th ACM SIGOPS European workshop: beyond the PC: new challenges for the operating system* (New York, NY, USA, 2000), ACM, pp. 37–42.

[3] CHRISTENSEN, K., REVIRIEGO, P., NORDMAN, B., BENNETT, M., MOSTOWFI, M., AND MAESTRO, J. Ieee 802.3az: the road to energy efficient ethernet. *Communications Magazine, IEEE 48*, 11 (2010), 50–56.

[4] CISCO. Design considerations for high availability and scalability in blade server environments. White Paper, 2009.

[5] FERRETTI, S., GHINI, V., MARZOLLA, M., AND PANZIERI, F. Modeling the energy consumption of multi-nic communication mechanisms. In *Proceedings of the 2012 IEEE Online Conference on Green Communications (GreenCom)* (2012), IEEE Computer Society.

[6] FOUDRIAT, E. C., MALY, K., MUKKAMALA, R., OVERSTREET, C. M., MATHEWS, L., AND BALAY, S. RAIN (redundant array of inexpensive networks): Expanding existing networks to support multitraffic performance. Tech. rep., Norfolk, VA, USA, 1994.

[7] GEOFF THOMPSON. Proposal for parallel path trunking in 802, 1998.

[8] GROUP, I. . Tutorial on Link Aggregation and Trunking. http://grouper.ieee.org/groups/802/3/trunk_study/tutorial/index.html, 1997.

[9] GUNARATNE, C., CHRISTENSEN, K., NORDMAN, B., AND SUEN, S. Reducing the energy consumption of ethernet with adaptive link rate (alr). *Computers, IEEE Transactions on 57*, 4 (2008), 448–461.

[10] GUPTA, M., GROVER, S., AND SINGH, S. A feasibility study for power management in lan switches. In *Network Protocols, 2004. ICNP 2004. Proceedings of the 12th IEEE International Conference on* (2004), pp. 361–371.

[11] HÄHNEL, M., DÖBEL, B., VÖLP, M., AND HÄRTIG, H. Measuring energy consumption for short code paths using rapl. In *Greenmetrics* (London, UK, June 2012), S. Low, J. L. Boudec, C. Rosenberg, and G. Zussman, Eds., ACM Sigmetrics.

[12] HELLER, B., SEETHARAMAN, S., MAHADEVAN, P., YIAKOUMIS, Y., SHARMA, P., BANERJEE, S., AND MCKEOWN, N. Elastictree: Saving energy in data center networks. In *7th USENIX Symposium on Networked Systems Design and Implementation* (San Jose, CA, USA, April 2010).

[13] IMAIZUMI, H., NAGATA, T., KUNITO, G., YAMAZAKI, K., AND MORIKAWA, H. Power saving mechanism based on simple moving average for 802.3ad link aggregation. In *GLOBECOM Workshops, 2009 IEEE* (30 2009-dec. 4 2009), pp. 1 –6.

[14] INTEL CORP. *Intel® 64 and IA-32 Architectures Software Developer Manual: RAPL MSR updates.* 2012, ch. 14, p. 32.

[15] INTEL CORP. Intel® Ethernet-Server-Adapter X520-T2. http://www.intel.de/content/www/de/de/network-adapters/gigabit-network-adapters/ethernet-x520-t2.html, 2012.

[16] INTEL CORP. Intel® Gigabit CT Desktop Adapter Product Brief. http://www.intel.com/content/www/us/en/network-adapters/gigabit-network-adapters/gigabit-ct-desktop-adapter-brief.html, 2012.

[17] ONLIVE. Onlive.com website. http://www.onlive.com/, 2012.

[18] PATHAK, A., HU, Y. C., AND ZHANG, M. Where is the energy spent inside my app?: fine grained energy accounting on smartphones with eprof. In *EuroSys* (2012), P. Felber, F. Bellosa, and H. Bos, Eds., ACM, pp. 29–42.

[19] SNOWDON, D. C., PETTERS, S. M., AND HEISER, G. Accurate on-line prediction of processor and memory energy usage under voltage scaling. In *Proceedings of the 7th International Conference on Embedded Software* (Salzburg, Austria, Oct 2007), pp. 84–93.

[20] SOHAN, R., RICE, A., MOORE, A. W., AND MANSLEY, K. Characterizing 10 Gbps network interface energy consumption. Tech. Rep. UCAM-CL-TR-784, University of Cambridge, Computer Laboratory, July 2010.

[21] TU DRESDEN OPERATING SYSTEMS GROUP. GitHub Page. https://github.com/TUD-OS.

[22] YOKOGAWA. WT-210 manua;. http://c418683.r83.cf2.rackcdn.com/uploaded/bu7604_00e_020_3.pdf, 2012.

The Energy Consumption of TCP

Raffaele Bolla
DITEN University of Genoa
Genoa - Italy
raffaelle.bolla@unige.it

Roberto Bruschi
CNIT
Genoa - Italy
roberto.bruschi@cnit.it

Olga Maria Jaramillo Ortiz
DITEN University of Genoa
Genoa - Italy
olga.jaramillo@unige.it

Paolo Lago
DITEN University of Genoa
Genoa - Italy
paolo@reti.dist.unige.it

ABSTRACT

In this paper, our objective is to experimentally evaluate how the incoming and outgoing traffic patterns provided by the TCP protocol impact on the energy efficiency of a networked device under a number of realistic scenarios. To this purpose, we set up a complex testbed that allowed us to perform a huge number of measurements on energy- and network-performance indexes on a state-of-the-art PC with power management capabilities, and the Linux operating system. The performed measurements and collected results gave us the chance not only (i) to provide a complete energy profiling of TCP connections, but also (ii) to analyze into detail the role of underlying network protocols and (iii) to clearly identify specific situations where current network protocols may trigger high inefficiencies in networked hosts.

Categories and Subject Descriptors

C.2.1 [**Computer-Communication Networks**]: Network architecture and Design – *Network communication;* C.2.5 [**Computer-Communication Networks**]: Local and Wide-Area Networks – *Internet (TCP);* C.4 [**Performance of systems**]: Performance attributes

General Terms

Measurement, Performance

Keywords

Green-networking, experimental testbed, Linux

1. INTRODUCTION

In the last few years, a number of studies have identified the eco-sustainability as one of the key aspects that may potentially constraint the Internet technology evolution, and its wide-adoption as public telecommunication infrastructure.

Today, telecom operators have enormous power requirements for supplying wire-line and wireless networking infrastructures, and appear to be among the major direct energy consumers in their nations. Estimates and projections from operators and third-parties [1] clearly suggest that such energy requirements will rapidly become no more sustainable, if no radical changes in Internet technology design will be undertaken. Moreover, this figure becomes even more impressive if we consider not only "networking" devices (e.g., routers, switches, etc.) inside telecom and home networks, but also the "networked" ones [2][3].

Consumer electronic is increasingly based on Internet technologies, and many appliances are becoming even smarter and "connected". Despite the proliferation of tablets and mobile phones, many other Customer Premises Equipment (CPE) are going to massively appear in the user homes (e.g., VoIP phones, set-top boxes, smart household appliances, etc.).

The number of Internet connected devices is forecasted to explode to over 15 billion by 2015 – twice the world's population [4]. Moreover, Cisco forecasts that by 2015 Internet traffic will get very close to the impressive threshold of 1 zettabytes a year, reaching approximately 966 exabytes a year.

Starting from this scenario, the sustainability of the Internet strongly and clearly relies on the efficiency of technologies and protocols working at the network edge, and more specifically inside networked devices [5].

In order to build a more sustainable Internet, on one hand, hardware platform for networked devices should include advanced power management schemes, just like today's PC technologies, which may allow certain proportionality between their energy consumption and actual workload. On the other hand, the network protocol stack and its implementation (often realized at software level of the networked devices) need to be optimized as much as possible, in order to best fit the dynamics and features of power management schemes. It is worth noting that optimization solutions, assuring just only very small power savings, may have a disruptive impact given the high density of networked hosts. However, the adoption of these power management capabilities in network devices affects the performance of the Internet traffic and therefore the device energy savings [6].

In this respect, we decided to focus on the Transmission Control Protocol (TCP), since we firmly believe that its optimization could have a key role in the future green Internet. This decision arises from a double-faced consideration:

- The TCP protocol is and will be present and widely used in networked devices. Recent studies showed that it still carries more than 80% of the Internet traffic [7].

- The TCP implements the flow and congestion control mechanisms, which directly manage the largest part of the packet-level traffic dynamics.

Packet-level dynamics are a very important aspect, especially because the processing of incoming and outgoing packets is a no negligible part of networked device workload. Moreover, as already shown in [8], packet-level time scales roughly coincide with (or at least are very close to) the ones of hardware power management primitives. Different traffic arrival patterns may trigger hardware power management mechanisms, and result into quite different energy efficiency levels [9].

In this paper, our objective is to deeply and experimentally evaluate how the incoming and outgoing TCP traffic patterns may impact on the energy efficiency of a networked device under a number of realistic scenarios. To this purpose, we set up a complex and complete testbed that allowed us to perform a huge number of white- and black-box measurements on energy- and network-performance indexes on a state-of-the-art PC with power management capabilities, and the Linux operating system. The performed measurements and collected results gave us the chance not only (i) to provide a complete energy profiling of TCP connections, but also (ii) to analyze into detail the role of underlying network protocols and (iii) to clearly identify specific situations where current network protocols may trigger high inefficiencies in networked hosts. The achieved results my constitute a solid basis for driving future researches towards more energy-efficient TCP optimizations and networked device architectures.

The paper is organized as follows. Section 2 discusses related work. The testbed and all its tools and components are described in section 3 Section 4 describes the power management primitives usually available in general-purpose PCs, and that are usually available in many consumer electronic devices. The collected results and their analysis are in section 5. Finally, the conclusions are drawn in section 6.

2. RELATED WORK

One of the earliest works on energy efficiency of TCP has been realized by Zorzi and Rao [10]. The authors analyzed the energy consumption performance of various TCP versions (Old Tahoe, Tahoe, Reno and New Reno); by selecting the right TCP version, the performance and the energy efficiency can be improved. The authors in [11], have conducted a similar work. They demonstrated that the performances of energy/throughput tradeoffs of varios TCP versions (Tahoe, Reno and New Reno) are fairly similar; with the Tahoe version the most energy conserving of the three.

Other contributions analyze the computational energy cost of TCP [9]. The authors present a detailed study of the energy cost of different TCP functions. Their experimental results showed that the TCP processing cost accounts for 15%, of which 20% to 30% is used to compute TCP checksums.

Considering the importance of a correct TCP performance on current greened networks, in [12] studied the behavior of various TCP versions (Tahoe, Reno, New Reno, Bic, Cubic, Vegas). Conducting several simulation tests, they demonstrated that is possible to achieve energy savings and maintaining good network performances by choosing the right TCP version.

The purpose of our work is in line with the concern of the research community to know the energy consumption of the TCP, but from another point of view. Our interest is to know how the different network conditions (different end-to-end bandwidth, end-to-end delays, and additional packet loss probabilities) affect the behavior of the TCP traffic triggering high inefficiencies in networked hosts.

3. THE TESTBED AND MEASUREMENT METHODOLOGY

Estimating the energy consumption due to the networking protocol stack, and especially of the TCP, is not a simple task, given the huge number of factors that may have a significant role in energy- and network-performance. As first, the energy consumption of a device strictly depends on its hardware architecture, and the available power management primitives. As second, network performance and working dynamics may heavily depend on the software level (both on applications and the operating system). Under this line of reasoning, it can be deduced that an experimental evaluation performed only with "black box" performance indexes may result in a noisy data collection, whose interpretation, generalization, and exploitation could be really hard, or even impossible.

A complete and general understanding of TCP impact on energy consumption can be clearly achieved only by a deep and careful characterization of the internal behaviors and dynamics of networked devices, too. Only in this way, it would be possible to map the measured aggregate/external performance to their real sources/causes (e.g., at hardware or software levels, due to features of network protocols, etc.). This approach may certainly allow generalizing the achieved results also to other hardware architecture, operating systems, etc. Thus, we set up a benchmarking environment (depicted in Fig. Figure 1) that allows collecting the largest possible set of internal performance indexes. All the used hardware devices, software and measurement tools were selected in order to ease the collection "white-box" measurements, and to provide enough flexibility for realizing different testing scenarios.

The testbed is substantially composed by five main HW platforms and measurement tools:

- *The System and Test (SUT)*: a commercial off-the-shelf PC, based on the Intel i5 processor, and running the Debian Linux operating system. As described in details in the following sub-section, the SUT has been provided of a number of Software/Hardware (SW/HW) probes to perform white-box measurements.
- *A high-end server*: used for TCP traffic generation, network performance indexes collection, and network emulation.
- *A Gigabit Ethernet switch*: used to separate the Ethernet links between the SUT and the server, and to allow Ethernet link speed changes only at the server, without involving the SUT (that is kept connected at 1 Gbps).
- *An AC watt-meter*: to measure the entire energy consumption of the SUT.
- *A multi-channel Data AcQuisition (DAQ)*: used to collect a high number of DC power consumption probes, placed inside

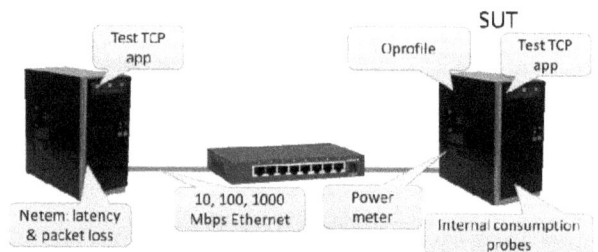

Figure 1. Experimental test-bed

the SUT as described in subsection 3.3.

The testing methodology is simple. A couple of applications for generating (server) and receiving (client) TCP traffic are placed on the SUT and on the server, respectively. Such applications generate continuously a number of TCP connections (upon completion of TCP connection, a new one with the same features is started), which are used to move traffic unidirectionally (from the "sender" to the "receiver"). Some network emulation tools are used to realize different network scenarios, by reducing the capacity between the sender and the receiver, to introduce a packet loss probability, and to increase the delay between the hosts. At the same time, all the SW and HW probes collect samples on energy consumption and on network performance. All the tests have been repeated 20 times, and their duration was fixed to 200 times the average connection life time.

The rest of this section is devoted to introduce the testbed elements above mentioned into details.

3.1 The System under Test

The SUT is a Linux workstation equipped with an Intel i5 processor running at 2.68GHz, with 4 physical cores and a maximum power consumption of 95W [12]. The workstation is equipped with 2 GB of DDR3 RAM, and an Intel PRO Gigabit Ethernet adapter. The Operating System (OS) is the Linux Debian 5.0.6, and the kernel version is the "vanilla" 3.4, which supports symmetric multi-processor (SMP). In order to collect "clean" internal measurements and avoiding background dynamics from other SW services, the OS was configured with the essential SW packages to run the tests here reported. So, every service/application useless for our scopes was not installed (e.g., the graphical interface environment – X-server, Gnome, etc. – was not installed).

All the configurations related to the networking stack and of the operative system have been kept as suggested in the Debian default settings. For instance, the TCP receive window has been kept fixed to 8 KB. Only, socket resource recycling option has been disabled, since, in the presence of cyclic TCP connection generations between the same pair of hosts (as in the next subsection), it can introduce not realistic behavior in the OS. The chosen TCP version is the CUBIC [14][15][1] (the default TCP version of the Linux kernel).

3.2 Traffic Generation and Network Emulation

In order to generate and receive TCP connections, we implemented a simple and very light-weight testing application, named "Tcptest" [16]. With respect to other well-known applications for generating TCP flows (e.g., TTCP, iPerf among the others), Tcptest allows more flexible flow generation (especially in the presence of multiple simultaneous connections), avoiding useless overhead operations (e.g., disk access, etc.).

Tcptest can be used both as client (receiver) and server (sender) of connections. Once a connection is established, Tcptest sends a desired number of segments at the maximum MTU. When the last segment is successfully received by the client, the TCP connection is immediately closed. The segments of all the generated connections contain replicas of a same pre-allocated pattern.

The amount of data sent, the number of simultaneous TCP connections, and other options are configurable through a simple command line interface. Tcptest is able to work in two modes:

• *One-shot*: the chosen number of TCP connections is generated only once.
• *Continuous*: the chosen number of TCP connections is generated in a persistent and cyclic way.

In both the modes above, Tcptest provides the average duration time of connections and other monitoring parameters made available by the Linux kernel. Continuous mode also provides the number of running and completed connections every second. For our purposes, except for some preliminary tests, we used Tcptest in continuous mode.

Regarding the network emulation, our objective was to act on the main network performance indexes that can heavily influence the TCP behavior: the available bandwidth capacity, the end-to-end delay, the packet loss probability. In order to emulate different bandwidth capacities, we disabled the Ethernet auto-negotiation at the server NIC, and made it working at 1 Gbps, 100 Mbps, and 10 Mbps. We can note how 10 Mbps is a very interesting bandwidth capacity because it roughly correspond to the speed of common broadband residential accesses (e.g., ADSL), today provided by many Telecom operators. Regarding packet losses and end-to-end delay, we used the NetEm module [17], available in latest releases of the Netfilter framework in the Linux kernel. The NetEm module was applied at the server for both incoming and outgoing traffic.

3.3 Power Consumption Probes

As previously sketched, in order to provide a complete characterization of the energy absorption dynamics, we decided to collect two main kinds of data:

• the energy absorption of the entire SUT on the AC plug;
• the energy absorption of the most significant HW

Figure 2. The SUT equipped with PCI-e, DDR3 and ATX risers connected to the DAQ system.

Figure 3. the ATX riser board prototype specifically realized for this work. It can sense the current and the voltage of every rails in 24 and 8 ATX power connectors, and also the energy supplies of up to 4 fans.

_____

[1] A number of tests were carried out also with all the other TCP versions available in the Linux kernel, but from the energy perspective we did not seen substantial difference.

subcomponents of the SUT (e.g., CPU, RAM, PCI network cards, etc.).

On one hand, the measures on the AC plug were simply collected by means of a Raritan Dominion PX-5297 wattmeter [18]; on the other hand, the measures on the internal subcomponents required a certain effort. In fact, PC HW does not usually include power probes, and the power supply to HW subcomponents is generally carried within multiple heterogeneous interfaces (e.g., different sockets for CPUs, DIMM and SO-DIMM slots for memories, PCI-e slots for network interface cards, etc.). In order to independently measure the absorption of HW sub-components, where possible, we used some special bus risers [19][20] equipped with current and voltage measurement probes. Unfortunately, this approach cannot be applied to all the physical interfaces in a PC, due to issues related to the much miniaturized physical connector dimensions, to the stability of electrical signals, etc. This last case applies for sure to CPU sockets.

However, in order to measure the energy consumption of CPU, we exploited the ATX standard [21] that defines the internal power supply for general purpose computing systems. ATX motherboards are supplied by means of a 24 pin connector, which includes among the others three main voltage rails at 3.3, 5 and 12 V, and a ground rail. The 12 V rail supplies many components of the motherboard and part of the CPU. An additional 6/8 pin ATX connector, carrying a second 12 V rail, is generally available on the motherboard to provide additional power to the CPU. In more detail, the 12 V rail on the 6/8 pin connector is commonly devoted to provide isolated power supply to the CPU cores only. Other subcomponents of the CPU (e.g., cache, bus and DDR controllers and other internal control units) are supplied by means of the 12 V rail on the 24 pin connector [12].

Owing the above mentioned standard characteristics of PC supplies, we decided to monitor the CPU energy consumption by sensing both the 12 V ATX rails. To this purpose, we designed and developed a riser board for ATX power connectors (see Fig. 3), which allows putting some current and voltage probes on the available supply rails (mainly the two 12 V rails, and the 5 and 3.3 V ones). Figure 4 shows the high-level architecture implemented by the ATX riser board. The current probes are realized by means of very small (10 mΩ) and precise Current Sensing Resistors (CSR). The voltage drop V_{sense}^{n} across its CSR (where n denote the n^{th}ATX voltage rail) is obviously proportional to the incoming current. To reduce the noise and eventual cross-talk on the measured V_{sense}^{n} values, we decided to closely connect the CSRs to current-sense amplifiers. The amplified CSR voltage drops and the ATX rail voltages are finally measured by an external DAQ device. In our tests, we used An Agilent U2356A multifunction DAQ [22], which can sample 64 channels at 500 Ksample/s with a 16 bit resolution

3.4 Internal SW Probes

The internal SW measurements have been carried out by using a specific tool (called profiler). This software tool can trace the percentage of CPU utilization for each source-code function of any SW application, service, kernel activity and module running on the PC. This powerful tool obviously allows us deeply analyzing the role and the computational efficiency of all the SW components, and it gives the chance of breaking down the energy consumption per SW functionality.

From a general point of view, many tools for SW profiling may perturb the system performance, since they generally require a relevant computational effort. We have experimentally verified

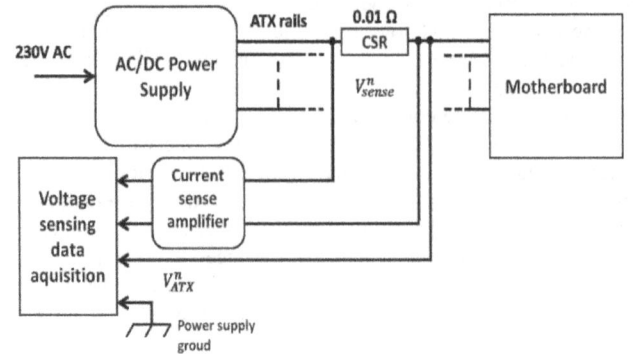

Figure 4. High-level scheme of the riser board for ATX voltage rails.

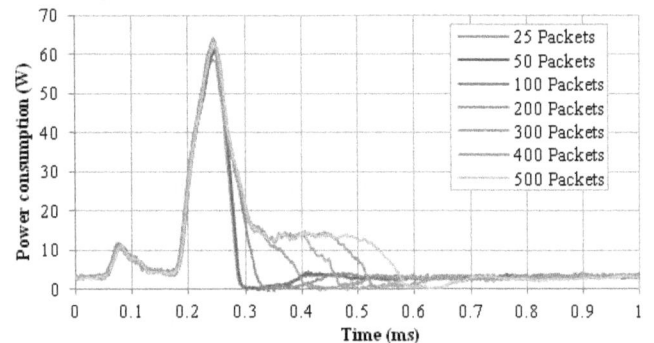

Figure 5. Energy consumption of the CPU when receiving a burst of a variable number of packets. These results have been obtained with the ATX riser board in section 3.3.

that one of the best is Oprofile [23], an open source tool that realizes a continuous monitoring of system dynamics with a frequent and quite regular sampling of CPU HW registers. Oprofile allows the effective evaluation of the CPU utilization of both Linux user- and kernel-space with a very low computational overhead. It is worth noting that Oprofile can collect HW register only during CPU activity periods, so that, when the CPU enters into low power idle modes (C-states – see section 4), no samples are collected. Thus, for evaluating the real CPU time utilization of SW functionalities, we measured also the CPU activity percentage.

3.5 Network Performance Probes

In order to evaluate TCP and network performance indexes, we enabled a packet sniffer (the well-known "tcpdump" application) on the server to collect entire traffic traces of performed testing sessions. Then, collected traffic traces have been analysed offline by means of the Linux "tcptrace" utility in order to find the number of packet losses, out-of-order packets, TCP retransmission due to timeout and fast retransmit, etc.

4. ANATOMY OF PC POWER MANAGEMENT

In general purpose computing systems, the Advanced Configuration and Power Interface (ACPI) [24] provides today a standardized interface between the hardware, where power management capabilities are realized, and the software. This standard interface completely hides the CPU internal techniques to reduce power consumption, which may differ depending on the processor, to the OSs and SW applications. The ACPI standard introduces two main different power saving mechanisms, namely performance (P-) and power (C-) states, which can be independently employed and tuned for the largest part of today's processors. However, due to the effectiveness and the simplicity of usage, the power management of modern OSs tends mainly to rely on C-states.

Regarding the C-states, the C_0 is the active power state where the CPU executes instructions, while C_1 through C_n states corresponds to sleeping or idle modes, where the processor consumes less power and dissipates less heat. As the sleeping state (C_1, ..., C_n) becomes deeper, the transition between the active and the sleeping state (and vice versa) requires longer time and more energy [25]. C-states are usually very effective primitives (more than P-states), since they literally allow to quickly shut off a number of internal CPU components, avoiding intrinsic energy wasting due to leakage current. However, when components are re-turned on, they generally need a no negligible additional start-up power. For instance, Fig. 5 shows the instantaneous power consumption of an Intel Core i5 processor when receiving IP packet bursts with different lengths. Here, the first packet is signaled to the CPU after approximately 50 μs. Then, at about 250 μs, we have a first HW start-up consumption due to the re-activation of CPU memory controllers, and then a huge spike of energy due to the wake-up of CPU cores. Finally, the period where the CPU really processes the incoming packets is clearly visible, and changing according to the number of packets in the burst.

C-state transitions are mainly driven by the OS and device I/O operations. When the CPU finishes serving its job backlog, the scheduler of the OS may decide to enter in a low power idle mode (C_1, ..., C_n). In such states, the CPU can wake only by some scheduled activities in the OS, or by an interrupt coming from a I/O component (like a NIC, a keyboard, etc.). In this respect, it is worth noting that network traffic dynamics (and then the TCP itself) can heavily influence interrupt generation from I/O hardware, and then the effectiveness of C-state transitions.

In the C_0 state, the ACPI allows the processor performance to be tuned by means of P-states. P-states modify the operating energy configuration by altering the working frequency and/or voltage, or throttling the clock of the processor. Thus, using P-states, a CPU can provide different power consumption and performance when active. Given issues in silicon electrical stability, the transition time between different P-states is generally very slow (more than 1 ms), and for this reason their frequent automatic tuning is not enabled by default in many OSs.

5. EXPERIMENTAL RESULTS

As previously sketched, TCP traffic is well-known to provide different performance and working behavior depending on network parameters, like end-to-end bandwidth, delay and packet loss probability. Thus, by exploiting the testbed and the tools introduced in section 3), we decided to perform a number of test sessions under different simple network scenarios, which may represent some common cases in the today's Internet. As shown in Table 1, the selected scenarios provide different end-to-end bandwidth (10, 100, and 1000 Mbps), end-to-end delays, and "additional" packet loss probabilities[2]. In more detail, the 10 Mbps bandwidth bottlenecks can represent typical situations in today's residential broadband network accesses (e.g., DSL); the 100 Mbps and the 1 Gbps cases in enterprise networks and data-

Table 1. Testing Scenarios

Testing Scenario	Bottleneck bandwidth [Mbps]	Additional packet loss probability [%]	Additional delay [ms]
a	1000	0%	0
b	1000	0.5%	0
c	1000	0%	10
d	10	0%	0
e	100	0%	0
f	10	0%	10
g	100	0%	10

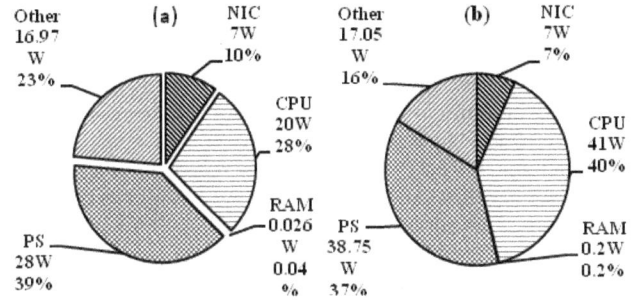

Figure 6. Energy consumption of the different internal HW components of the SUT during idle (a) and active (b) periods (RAM: memory, NIC: network interface card; PS: AC/DC power supply; Other: all the other components in the SUT, from motherboard chips to disks).

centers. In order to be able to deeply analyze the experimental results, and to precisely decompose the consumption sources in such a complex system, the testing scenarios have to remain as simple and deterministic as possible, avoiding the introduction of stochastic "disturbs" on network parameters (e.g., on bandwidth and delay, etc.).

In every testing scenario, we performed a number of tests by changing the length of the TCP connections. The chosen length values correspond to 10, 100, 1000, and 10000 segments of the maximum TCP transfer unit, equal to 1460 B. Without losing of generality in our analysis, we will report the results obtained with only a single TCP connection between the SUT and the server. Moreover, we used the TCP connection for unidirectional data transfers, not only for sake of simplicity, but also because it is a common approach in many applications.

The rest of this section is organized as follows. A first breakdown of the HW energy consumption sources of the SUT is in subsection 5.1. In subsection 5.2, the analysis of TCP performance and of energy consumption is discussed. In subsection 5.3, measures regarding the dynamics and the role of the various SW elements in the SUT are shown. Finally, Subsection 5.4 goes further into the details of TCP profiling trying to break down the CPU utilization on a per TCP functionality basis.

5.1 Identifying the energy-proportional HW components

The first measurements we collected were aimed at evaluating the power consumptions of different SUT HW components, and their proportionality to the actual workload. To this purpose, the power consumption of these internal elements were measured both when

---

[2] Additional to the ones induced by the TCP congestion control.

the system is idle (turned on, but performing no operations), and when one core is active and performing operations at the maximum speed. The obtained results are in Fig. 6, and, as expected, outline how the CPU is the most energy-hungry component, exhibiting also a clear dependency on the workload since it passes from 20W (28% of the overall consumption) in idle mode to 41 W (40% of the overall consumption) when active. Except for the AC/DC power supply (whose efficiency depends on the electric load of internal SUT components), all the other components exhibit negligible energy absorption variations (below 200 mW). For this reason, in following we will focus only on the energy consumption of the CPU, omitting the contributions of the other HW components.

5.2 The TCP Power Consumption and Performance

This subsection introduces the network and energy performance measures of the TCP behavior in the testing scenarios in Table 1. Regarding network performance, the average lifetimes of the TCP connections are reported in Fig. 7, and the frequency of segment losses due to fast-retransmits and time-out expirations are in Fig. 8. As expected, the lifetime of TCP connections heavily depends on the bottleneck bandwidth, on the Round-Trip-Time (RTT), and on packet losses. We can note how the addition of 0.5% packet loss probability increases the TCP connection lifetime in a significant way (case *b* with respect to *a*). This effect is more visible in shorter connections (10 and 100 pkts). The addition of the 10 ms delay in the RTT introduces even more evident performance decays than in the previous case (see cases *c*, *f*, *g* with respect to the *a*, *d*, *e* ones). Moreover, the additional delay results to have an impact proportional to the bottleneck bandwidth (i.e., it is proportionally more evident for connection crossing the 1 Gbps bottleneck, than in the 10 Mbps case). Finally, also the bottleneck speed has obviously a key role in the TCP performance, due to both the increase of transmission times of TCP segments and inefficiency of TCP congestion control. In fact, as shown in Fig. 8, the number of retransmissions, due to the reception of duplicated Acknowledgements (ACKs) or to timeout expirations, increases significantly in the 10 and 100 Mbps cases. This increase is a clear sign of the central role of TCP congestion control in the scenarios *d*, *e*, *f*, *g*. However, all the collected results are in line with many studies [14] on the TCP performance proposed in the past years.

Passing to the energy consumption, Fig. 9 reports the energy consumed by a TCP connection in all the considered testing scenarios. In more detail, we collected the average power consumption of the SUT during TCP tests (the Tcptest application has been made running in "continuous" mode – see section 3.2, – and each measured value comes from averaging the results of 20 test sessions, each one with approximately 200 TCP connections). The obtained power consumption values were then multiplied for the average TCP connection lifetimes in Fig. 7 in order to find the energy absorbed by the SUT during a single TCP connection. Then, Fig. 9 shows the difference between such energy absorptions and the ones the SUT would have experienced while in idle mode for the same time period. In other words, the values in Fig. 9 are the energy delta consumed by the SUT for sending or receiving data from a TCP connection.

Analyzing the data in Fig. 9, we can outline how, in presence of lower bottleneck speeds (cases *d*, *e*, *f* and *g*), the energy consumption increases in a significant way. This consumption

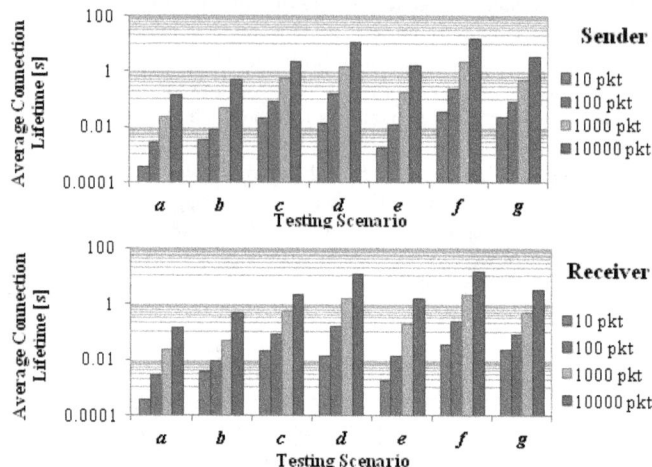

Figure 7. Average life time of TCP connections when the SUT acts as "sender" and "receiver".

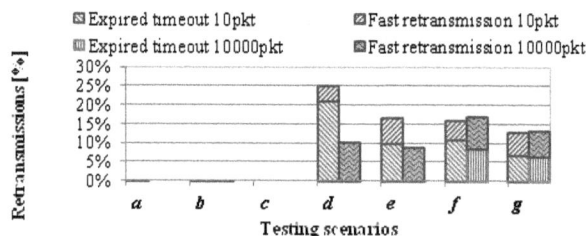

Figure 8. Frequency of TCP fast retransmits and timeout expirations (the SUT is acting as "sender").

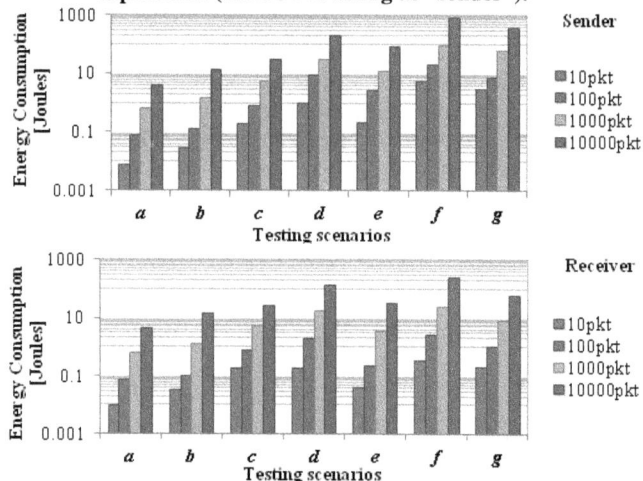

Figure 9. Energy Consumption of TCP connections at the SUT when acting as "sender" and "receiver".

increase is due to the C-state transition overheads, as discussed in section 4. In more detail, since the minimum packet inter-arrival time with the 10 Mbit bottleneck is approximately 1.2 ms, the CPU has enough time to wake up from the low power idle mode (i.e., the C_3 one) upon packet reception, to completely process the received packet, and to return to the C_3 state before the next packet arrives. So, a wake-up power peak (see Fig. 5) is spent for every packet.

Also the RTT plays a similar role. In fact, TCP connections in testing scenarios *c*, *f* and *g* exhibit larger energy requirements than in the cases *a*, *d* and *e*, respectively. The main cause of this behavior is due to the well-known bandwidth-delay product effect in the TCP: when this product is high (as in *c*, *f* and *g*) the TCP

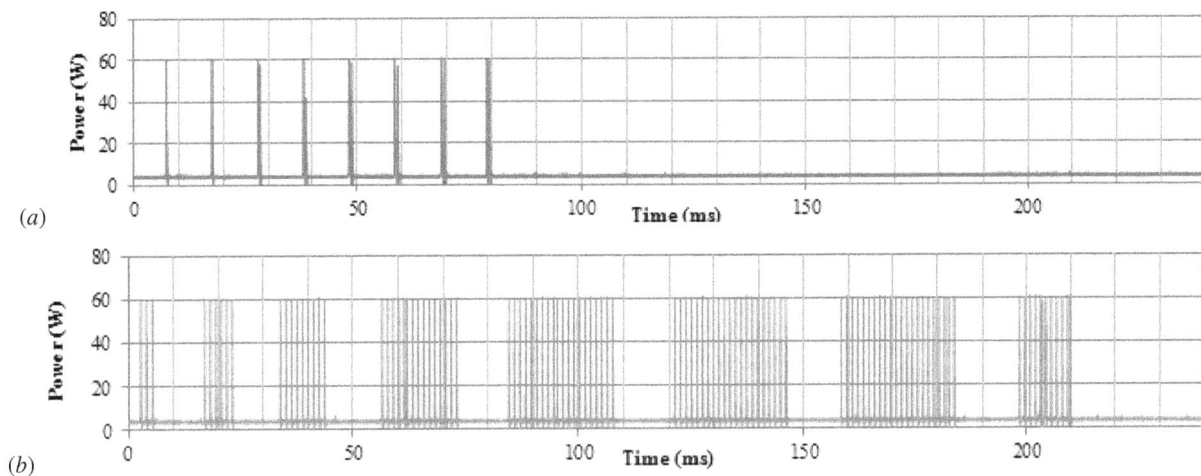

Figure 10. Energy consumption of the SUT CPU during a TCP connection in the testing scenarios *c* (*a*) and *f* (*b*). The SUT is acting as "receiver," and the connection carries 100 pkt.

throughput becomes more "bursty", in the sense that groups of segments are sent every RTT, and among these groups there are some "silence" periods. Every time a new group of segments arrives or is sent, the CPU SUT has to move from the C_3 state to the C_0 one, causing the wake-up energy spike of Fig. 5. When the product is low (e.g., scenario *a*), the silence periods, and then the C_3 - C_0 transitions, are rarer.

All these dynamics and the impact of both bottleneck bandwidth and the RTT are evident in the transitory measurements in Fig. Figure 10. As expected, in the testing scenario b, the average energy consumption is larger than in the case a: the loss of segments obviously causes the sender to wake-up an additional number of times in order to retransmit the lost packets, or to receive it separately from the original burst. Moreover, after the packet loss, also the congestion window is decreased, and, consequently, the TCP transmission may become more "bursty" (it may happen to have some additional silence periods between two groups of segments, given the decrease of congestion window after loss events).

Besides the considerations above, it is worth noting that the energy consumption for sending data through the TCP connection appears to be higher than the one needed to receive them in all the considered scenarios.

5.3 Profiling CPU SW activities

From the results in the previous section, it is clear that the energy consumption of the SUT mainly depends on two key aspects: (*i*) the number of transitions between CPU idle and active states, which can be caused by excessive inter-packet and inter-burst arrival times; and (*ii*) the CPU activity time. The latter certainly depends on the SW computational load, which directly impacts on the processing time of incoming and outgoing traffic.

In this respect, Fig. 11 shows the average CPU uptime values (i.e., the time the CPU has spent in the C_0 state, this parameter can be thought as an indirect estimation of SW computational complexity) as experimentally measured in all the testing scenarios for both SUT sender and receiver modes, but only for connection lengths equal to 1.4 and 1425.8 KB.

As expected, the CPU uptime appears to be much lower as the network scenario allows higher TCP performance levels: increasing the efficiency of the TCP transfer, also the CPU SW

obviously behaves in a more effective way. However, when the bottleneck bandwidth decreases, or the RTT increases, CPU uptimes may enlarge of almost two magnitude orders. Moreover, Fig. 11 clearly outlines how the TCP data reception operations appear to require more CPU uptime than the transmission ones.

In order to understand insights of the computational complexity of SUT SW components (and then where a part of the energy consumption arises), we used the Oprofile tool. As introduced in section 5.3, this tool allows estimating which source-code function of the Linux kernel or of an application is utilizing the CPU, and which is its time share of the overall uptime. A selection of the results obtained is shown in Figs. 12

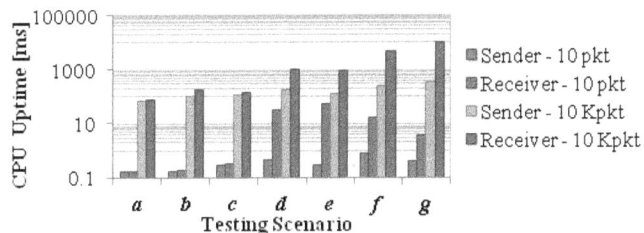

Figure 11. Average value of CPU uptime for 10 pkt and 10 Kpkt length TCP connections.

Table 2. Oprofile Function Categories for the Entire SUT SW

Name	Description
Scheduler	the operating system scheduler
ACPI	the ACPI and especially the functions related to idle-active transitions and vice versa
Socket	management of the TCP "socket" interface
Driver	the hardware driver of the NIC
Packet Handling	kernel processes related to the reception and transmission of traffic
Kernel User Interface	functions needed to copy data from the kernel to the user-space and vice versa
Tcptest	the TCP testing application
IRQ	management of HW interrupts, mainly due to the NIC.
IP	traffic processing at the IP layer
Memory	management of kernel memory
TCP	traffic processing at the TCP layer
Other	other spurious sources of CPU activity

and 13. In order to make these results as intuitive as possible, we grouped together the source code functions into 12 categories, each one representing a different functional block of the SUT SW architecture. The selected categories are introduced in Table 2.

Figs. 12 and 13 exhibit that the processing of TCP protocol account for 10% - 17% of the CPU uptime. This value tends to be higher in sender mode, given the additional TCP processing to retransmit the lost segments (see the next-subsection). The Tcptest application appears to not significantly affect the CPU uptime (approximately 1-2%). The OS scheduler generally accounts for a major share, since it ranges between 10% and 41%, which appears to be somehow dependent on the HW interrupts (2-8%). In fact, the OS is generally wakened up by HW interrupts generated by the NIC upon traffic reception. Once waken up, the OS scheduler decides for the next job to be executed (i.e., serving incoming traffic). The scheduler is also recalled when the CPU completes its job backlog, and then is ready to enter into the low power idle state (executing ACPI functions – ranging between 5% and 15%). For this reason, scheduler, IRQ, and ACPI functions tend to increase significantly in all the testing scenarios with high number of active-idle transitions. So, these three categories may be thought as sources of undesired CPU uptimes, and then of additional overhead in the energy consumption.

5.4 Breakdown of TCP functionalities

In order to analyze the TCP behavior under the different network scenarios in more detail, we decided to further refine the Oprofile results in the previous section. The functions related to the TCP were broken down into the new categories of Table 3, and the obtained results are in Figs. Figure 14 and Figure 15. From these figures, the difference between the sender and the receiver modes become manifest.

In the sender mode, the most time-consuming part is the Tx_data, which ranges 25% and 45%. This percentage depends on both the amount of packets sent and the network scenario. In the cases *b*, *d*, *e*, *f* and *g*, where the bottleneck speed is lower and/or the RTT is relatively high, we can note also a significant weight of retransmissions (1-16%) due to fast re-transmissions and timeout expirations (see Fig. 8).

The ACK management on the sender is much more consuming that at the sender (14-21%): the reason is that the sender must generate an ACK in each sent segment, while the receiver can exploit delayed ACKs in a more effective way. On this respect, we can also note that the Rx_data and Acknowledge categories are proportionally dependent, and they require almost the same percentage of TCP CPU cycles in the sender mode. The Open_Close and the State_Machine categories do not represent a significant part of the overall power consumption of the TCP, since they account for less than 7%. The relative weight of the Open_Close and the State_Machine categories is obviously larger in case of shorter TCP connections.

The management of timeout counters and the estimation of RTT appear to be light-weight operations (less than 4%). When multiple packet losses occour, and the SACK recovering is applied, it consumes up to 9% of the TCP CPU uptime share.

On the receiver side, the most consuming part of the TCP processing is the Rx_data; it spends between 40% and 60% of CPU cycles due to TCP elaboration. The ACK processing appears to be less consuming than on the sender, since this functionality accounts approximately for the same figure of

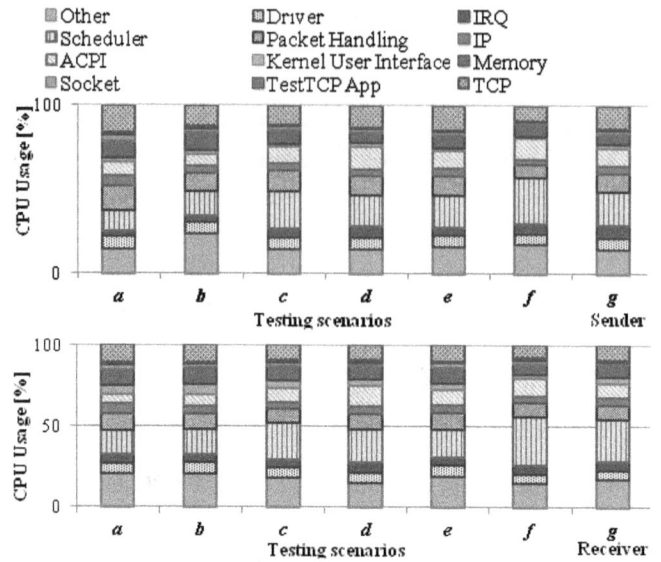

Figure 12. Breakdown of the CPU usage of a TCP connection of 10 pkt.

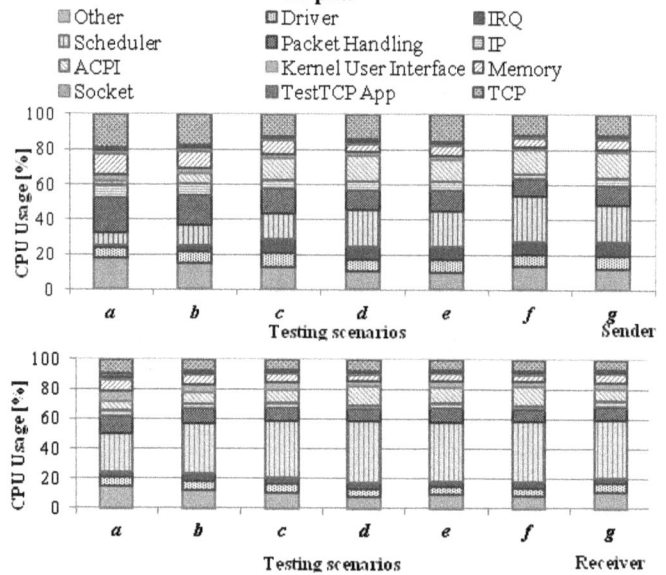

Figure 13. Breakdown of the CPU usage of a TCP connection of 10000 pkt.

Table 3 Oprofile Function Categories for the TCP Module

ID	Name	Description
A	Timer_RTT	Estimation of RTT and management of timeouts
B	Rx_data	Processing of received segments
C	Tx_data	Processing of segments to be sent
D	Options	Processing of TCP options (e.g., SACK)
E	Congestion_Control	Calculation of congestion control window
F	Flow_Control	Management of the flow control mechanism
G	Checksum	Checksumming of Rx and Tx segments
H	Acknowledge	Generation of the Acknowledgements
I	Open_Close	Management of connetion starting and ending phases
J	State_Machine	Management of TCP protocol state machine
K	Retransmission	Operations to retransmit lost packets

Tx_data, 11% of the TCP process. They both represent the elements with higher percentage of CPU use, after Rx_data. We can also underline how the Flow Control mechanism only accounts for 5% of CPU usage.

6. CONCLUSIONS

In this paper, we conducted an extensive experimental evaluation to study the impact of TCP traffic on a networked device under a number of realistic scenarios. From the numerical results, we revealed that the power consumption of SUT performing TCP data transfers is strongly dependent on the network conditions, being the most power consuming when low speed bottleneck are present (e.g., 10Mbps).

Analyzing into the detail the SUT behavior with a number of internal HW and SW probes, we noticed how this increased power consumption is due to (i) the arrival time of each packet, and how it may trigger the power saving mechanisms available in the SUT, and (ii) the packet loss rate, which may cause an additional TCP processing. Unfortunately, we also deduced that in network environments, similar to network residential accesses, for the reasons above, TCP connections may induce SUT consumption levels two orders of magnitude more than in high-speed network scenarios.

A complete SW profiling has been also reported with the aim of deeply understanding which SW modules and TCP functionalities are the main sources of energy consumption.

7. ACKNOWLEDGMENTS

This work has been supported by the ECONET (low Energy Consumption NETworks) project, co-funded by the European Commission under the 7th Framework Programme (FP7), Grant Agreement no. 258454, and by the TREND (Towards Real Energy-efficient Network Design) network of excellence, co-funded by the European Commission under the 7th Framework Programme (FP7), Grant Agreement no. 257740.

8. REFERENCES

[1] GESI, "Smart 2020," report, DOI=http://www.smart2020.org/_assets/files/02_Smart2020 Report.pdf

[2] Bolla R., Bruschi R., Carrega A., Davoli F., Suino D., Vassilakis C., Zafeiropoulos A., "Cutting the energy bills of Internet Service Providers and telecoms through power management: An impact analysis," Computer Networks, Elsevier, vol. 56, no. 10, pp. 2320-2342, July 2012.

[3] Bolla R., Bruschi R., Christensen. K., Cucchietti F. Davoli, F., Singh S., "The Potential Impact of Green Technologies in Next Generation Wireline Networks - Is There Room for Energy Savings Optimization?," IEEE Communications Magazine, vol. 49, no. 8, pp. 80-86, Aug. 2011.

[4] Cisco Visual Networking Index: Forecast and Methodology, 2010-2015.

[5] Bolla R., Bruschi R., Davoli F., Cucchietti F., "Energy Efficiency in the Future Internet: A Survey of Existing Approaches and Trends in Energy-Aware Fixed Network Infrastructures," IEEE Communications Surveys & Tutorials, vol. 13, no. 2, pp. 223-244, 2nd Qr. 2011.

[6] C. Panarello, M. Ajmone Marsan, A. Lombardo, M. Mellia, M. Meo, G. Schembra, "On the Intertwining between

Capacity Scaling and TCP Congestion Control" Proc. of the 3rd International Conference on Energy-Efficient Computing and Networking, e-Energy'12, Madrid, Spain, May 2012.

[7] A. Finamore, M. Mellia, M. Meo, M. M. Munafo, D. Rossi, "Experiences of Internet Traffic Monitoring with Tstat," IEEE Network, vol. 25, no. 3, pp. 8-14, May-June 2011.

[8] S. Nedevschi, L. Popa, G. Iannaccone, D. Wetherall, S. Ratnasamy, "Reducing Network Energy Consumption via Sleeping and Rate-Adaptation", Proc. of the 5th USENIX Symp. on Networked Systems Design and Impl. (NSDI), San Francisco, CA, 2008, pp. 323-336.

[9] B. Wang and S. Singh, "Computational Energy Cost of TCP," Proc. of the 23rd IEEE INFOCOM'04, Hong Kong, March 7-11, 2004, vol. 2, pp. 785-795.

[10] M. Zorzi, R. Rao, "Is TCP energy efficient?" Proc. of 1999 IEEE Mobile Multimedia Communications (MoMuC '99), San Diego, CA, USA, Nov. 1999.

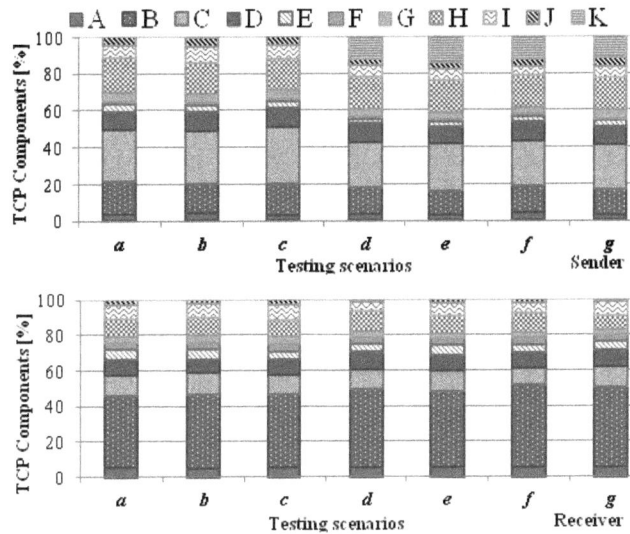

Figure 14. Breakdown of the TCP process in a TCP connection of 10 pkt.

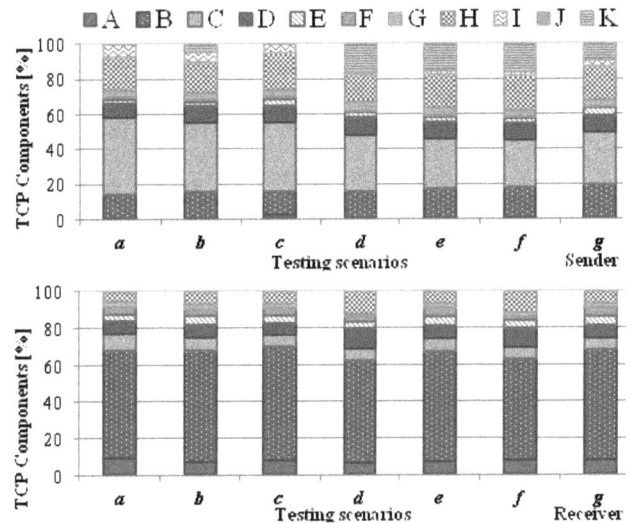

Figure 15. Breakdown of the TCP process in a TCP connection of 10000 pkt.

[11] V. Tsaoussidis, H. Badr, K. Pentikousis, X. Ge, "Enegy/Throughput Tradeoffs of TCP Error Control Strategies" Proc. of 5th IEEE international symposium on computers and communications (ISCC), Antibes-Juan, France, 2000.

[12] A. Sassu, C. Scarso, F. Cuomo, "TCP behavior over a greened network", Sustainable Internet and ICT for Sustainability (SustainIT), Pisa, Italy, Oct. 2012.

[13] Intel® Core™ i5 Processor Datasheet, DOI=http://download.intel.com/design/processor/datashts/322164.pdf

[14] S. Ha, I. Rhee, and L. Xu, "CUBIC: a new TCP-friendly high-speed TCP variant," ACM SIGOPS Oper. Syst. Rev., vol. 42, no. 5, July 2008, pp. 64-74.

[15] B. Wei, V.W.S. Wong,, V.C.M. Leung,, "A Model for Steady State Throughput of TCP CUBIC," Proc. of IEEE GLOBECOM 2010, pp.1-6, 6-10, Miami, FL, USA, Dec. 2010.

[16] The TCP Test application, source code and download at DOI=http://www.tnt.dist.unige.it/Tcptest/

[17] S. Hemminger, "Network Emulation with NetEm," Proc. of the 6th Australia's National Linux Conf. (LCA), Canberra, Australia Apr. 2005.

[18] Dominion PX-5297 Technical Specifications, DOI=http://www.raritan.com/ px-5000/px-5297/tech-specs/

[19] Adex Electronics, DDR3 Riser Card, DOI=http://www.adexelec.com/other. htm#DDR3.

[20] Adex Electronics, PEX16LX PCI-e x16 tall extender, DOI=http://www.adexelec.com/pciexp.htm#PEX16LX.

[21] The ATX specification 2.1, DOI=http://www.formfactors.org/developer/ specs/atx2_1.pdf

[22] Agilent U2356A 64-Channel 500kSa/s USB Modular Multifunction Data Acquisition, DOI=http://www.home.agilent.com/agilent/product.jspx? nid=-33627.384462.00

[23] Oprofile, DOI= http://oprofile.sourceforge.net/

[24] The Advanced Configuration and Power Interface (ACPI), DOI=http://www.acpi.info/

[25] "Energy-Efficient Platforms – Considerations for Application Software and Services," Intel Whitepaper March 2011, http://download.intel.com/ technology/pdf/Green_Hill_Software.pdf

Keynote Talk

A New Industrial Revolution
for a Sustainable Energy Future

Dr. Arun Majumdar

Google Inc.

How to Auto-Configure Your Smart Home? High-Resolution Power Measurements to the Rescue

Frank Englert[1], Till Schmitt[1], Sebastian Kößler[1], Andreas Reinhardt[2],

Ralf Steinmetz[1]

[1] Multimedia Communications Lab
Technische Universität Darmstadt
Darmstadt, Germany
{frank.englert, till.schmitt,
sebastian.koessler, ralf.steinmetz}
@kom.tu-darmstadt.de

[2] School of Computer Science and Engineering
The University of New South Wales
Sydney, Australia
andreasr@cse.unsw.edu.au

ABSTRACT

Most current home automation systems are confined to a timer-based control of light and heating in order to improve the user's comfort. Additionally, these systems can be used to achieve energy savings, e.g., by turning the appliances off during the user's absence. The configuration of such systems, however, represents a major hindrance to their widespread deployment, as each connected appliance must be individually configured and assigned an operation schedule. The detection of active appliances as well as their current operating mode represents an enabling technology on the way to truly smart buildings. Once appliance identities are known, the devices can be deactivated to save energy or automatically controlled to increase the user's comfort.

In this paper, we propose an approach to have buildings informed about the presence and activity of electric appliances. It relies on distributed high-frequency measurements of electrical voltage and current and a feature extraction process that distills the collected data into distinct features. We utilize a supervised machine learning algorithm to classify readings into the underlying device type as well as its operation mode, which achieves an accuracy of up to 99.8%.

Categories and Subject Descriptors

B.m [**Hardware**]: Miscellaneous; H.4.m [**Information Systems**]: Miscellaneous

Keywords

Smart Home, Device Classification, High Accuracy, Operating Mode Identification, Harmonic Spectrum Analysis

1. INTRODUCTION

The rise of home automation systems has greatly improved the comfort in modern homes. Current systems are, however, generally confined to controlling temperature [14] and lighting settings and they do not yet offer complete building automation. Besides increasing the user comfort, various further functionalities can be realized, e.g., detecting unexpected behavior to realize building security or deactivating appliances to achieve energy savings. In recent years, numerous researchers have addressed the latter issue. Example applications include to cut off standby loads [9], trim appliances to the most energy efficient setting [21], defer the usage of devices [13] until the grid-wide energy demand is low, or infer information about the user activities at home [1].

However, an information gap exists between the building automation system and the electric appliances, as the system is not aware of available devices, their location, and their state. In order to realize such functionalities without manual configuration of all appliances and their operation state, the building automation system needs to autonomously acquire this information. Only when detailed information about appliances and their mode of operation are known, can the system exert control over the electric appliances and actuate them to save energy and increase the user comfort.

In this paper, we present our solution to this problem, consisting of a hardware board for high resolution power measurements and a software framework for classifying the connected appliance and its state. The hardware board is carefully designed to achieve galvanic insulation between mains and the line-level output, as well as maintaining a low noise floor and a flat frequency response. In order to fulfill these properties, we have conducted an evaluation of different current transducers to select the best match for our use case. The resulting hardware board outputs line-level signals for both voltage and current. These signals can be simply converted to the digital domain by a computer system's audio interface or a dedicated analog-to-digital converter to process them further with our proposed software framework. This setup allows us to sample the power consumption of a connected device with up to 96 kHz sample rate.

Our software framework analyses the current and voltage waveform obtained from the connected appliance, and ex-

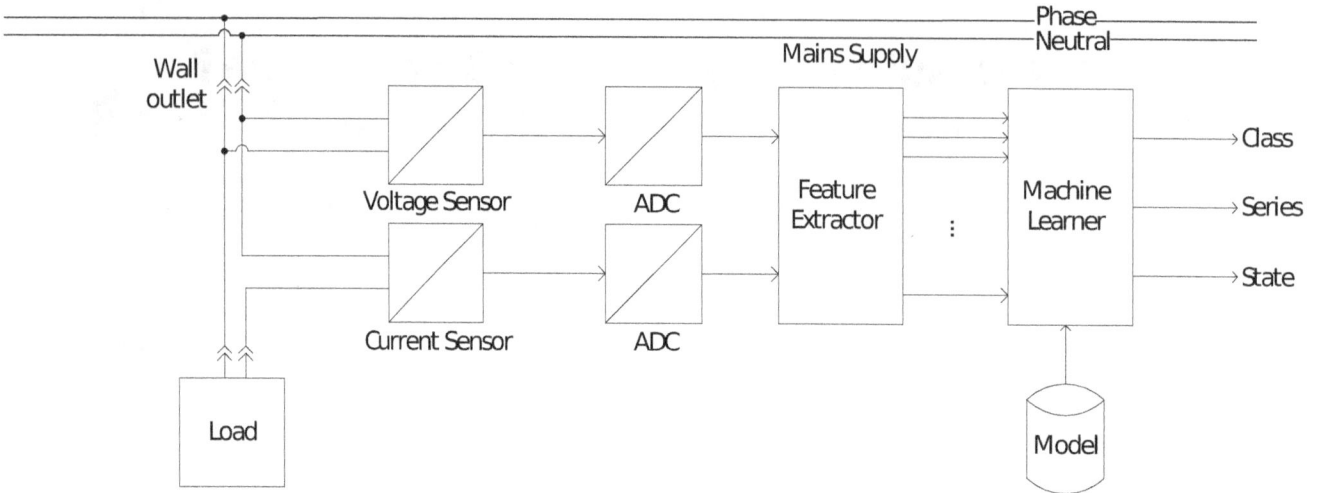

Figure 1: A schematic system overview: Our hardware board measures the voltage and the current of the load. The resulting signals are then converted to the digital domain and post processed by a machine learning toolkit.

tracts different features from the shape of the voltage and current consumption curves. Those features are used to build up two kinds of models, namely (1) a coarse-grained model to detect the appliance type attached as well as (2) a fine-grained model to differentiate between products of an appliance type and also to determine its mode of operation. Splitting the classification step into two stages allows to extend the list of supported devices dynamically. Our software framework is specifically tailored to cope with high data rates obtained from the hardware board. The system is designed for low latencies: Typically it takes at most one second for recording, feature extraction, and classification. Our approach requires no disaggregation algorithm because each device has its own metering unit.

This work closes the information gap between existing building automation systems and the electric appliances available in a household. By inferring type and state information, our system allows building automation systems to control electric appliances, e.g., in order to reduce their standby power consumptions. We make the following contributions:

- We present our hard- and software design for the device classification and operation mode determination system.

- We evaluate the classification accuracy values for the device type identification based on a comprehensive collection of traces.

- We assess the accuracy of the operation mode determination based on harmonic component fingerprints.

This paper is structured as follows. We discuss work related to the field of device classification and operation mode detection in Section 2, and subsequently introduce our general concept in Section 3. The hardware design for our power measurement device is detailed in Section 4, which also summarizes our comprehensive analysis of several current transducers in order to achieve a high spectral resolution and low noise levels. In the subsequent Section 5, we present our PISI software system, which extracts the features from the

collected traces and inserts them into a machine learning model. We evaluate the detection sensitivity and the overall classification accuracy in Section 6 where we also discuss the results. Finally, we conclude this paper in Section 7.

2. RELATED WORK

Numerous approaches exist to classify electric appliances based on their power consumption. These approaches can be divided into three different groups: Non-Intrusive approaches meter the current draw at a central position and then apply a disaggregation step to separate between different devices. In contrast Intrusive approaches provide a measuring unit for each household appliance. Finally integrated approaches require the household appliance to contain a metering and communication module. In this section we will describe these three different groups and discuss their advantages and disadvantages.

Centralized approaches meter a household's whole electricity consumption in order to detect attached appliances. The first who introduced such a solution was Hart in 1992. His Non Intrusive Appliance Monitoring paper [7] described a system consisting of a power meter for real and reactive power measurement and an attached state machine to track steps in the power consumption. Since then many researchers picked up his idea and improved it by sampling with a higher frequency [16, 17], better features [20, 5] and better disaggregation algorithms [2, 10, 15]. The main advantage of such a system is the requirement to place only one metering unit at a central position next to the fuse box. But as Zeifman summarizes in [24], centralized approaches are known to suffer from problems in detecting low-power appliances and devices with variable power consumption. Also the centralized approaches only allow the monitoring of currently running appliances and not controlling them from remote.

Alternatively distributed smart meters could be used to measure the power consumption at appliance level. One of the first researchers to explore this field was Ito [8]. He built up a system for classifying devices connected to a special wall outlet based on their power consumption. Many researchers improved his system by using better features [22, 10] or a wireless sensor network for metering devices [11, 4]. Kim even used different audio, light, temperature, and vibration sensors in combination with power meters to classify the currently running appliances [12]. As a main advantage, this solution does not rely on an error-prone load disaggregation method to split the power trace of a whole household into power traces for single devices. Furthermore, the distributed smart meters can control the attached appliances. But on the other hand each appliance requires its own metering unit. To the best of our knowledge, none of these systems can classify the state of an attached appliance.

There is a third field of related work namely the fact that an increasing number of business and household appliances will be equipped with networking capabilities in the future. If those appliances expose an interface to their functionality they become controllable from remote. A smartphone application to control home and office appliances was shown by Nichols [19]. Other researchers developed smart home servers to control such appliances from the Web [18]. Although those systems would make it possible to forward the internal state of appliances to a smart home system without the necessity of deploying additional measurement devices, we do not expect such functionalities to be present in low-end products in the coming years. As a result, this renders the integration of intelligent monitoring components into everyday appliances unlikely in the mid-term.

3. CONCEPT DESCRIPTION

In this section, we describe the general concept of our system, followed by the selected system architecture and a short description of its individual components.

In order to extract suitable features from voltage and current waveforms, the collection of these analog quantities is required, followed by their conversion to the digital domain and their appropriate processing. Hence, a hardware system with transducers for both physical parameters is needed, which needs to be capable of its interfacing to a data processing system. As a result, we have conducted both hard- and software design in order to address the problem at hand. We have especially designed our system to fulfill the following fundamentals:

- The system has to record power traces including associated current and voltage waveforms to enable the calculation of real, reactive, and apparent power, as well as the calculation of the phase shift.

- The collected power traces need to have a high resolution in both time and quantization. This also makes the usage of low-noise components necessary in order to minimize the impact of unwanted error signals.

- The provided current waveforms must be processed appropriately in order to enable machine learning tools to find and evaluate current and voltage waveform characteristics of appliances.

- The stored data has to be normalized with respect to the power value and grid frequency in order to cater for the repeatable nature of the conducted evaluations.

The resulting architecture of our system is shown in Figure 1. Its main components can be decomposed into three tiers: a hardware board, the software framework and the machine learner. The hardware board is plugged between the wall outlet and the electric appliance to easily measure the power consumption. It features voltage and current sensors to analyze device-specific high frequency characteristics of its energy consumption. It outputs two analog signals, one which is proportional to the grid voltage, whilst the second signal is directly proportional to the device's current draw. Those signals are then sampled by an analog-to-digital converter and further processed by our software framework. This framework is responsible for extracting features from the power recordings and to forward the extracted features to the widely used Weka data mining toolkit [6]. In order to maintain workplace safety, the hardware board furthermore ensures a galvanic decoupling from the mains voltage.

4. POWER MEASUREMENT

In order to collect the power consumption of electric appliances at a high resolution, distributed measurement units with high sample rates are required. Commercially available metering platforms, like Plugwise[1] or Wattson[2] however, often output mean power consumption values instead of allowing for direct access to the collected samples. Additionally, the phase shift is commonly not reported by these units, such that real power, but not reactive power components, can be measured.

As a result, these existing solutions are not suitable for the task at hand, and in consequence, we have designed a specific circuit board that fulfills the following set of criteria.

- High sample rate: In order to collect all possibly relevant characteristics from an appliance's current consumption, a sample rate as high as possible is desired. This requirement is based on the observations made in related work (e.g., [16]). At the same time, higher sampling frequencies allow for a better spectral resolution of the frequency domain plot, and thus better means for the subsequent signal analysis.

- Galvanic decoupling: The most straightforward approach to collect voltage and current measurements relies on the use of resistors (shunts for current measurement and resistor dividers to sample the mains voltage). Both approaches, however, suffer from the drawback of galvanic coupling to the mains voltage, which can destroy expensive measurement equipment and is not touch-safe. As a result, our prerequisite to the data collection unit was the galvanically decoupled transmission of both voltage and current to the sampling unit.

- Linear transducers: Especially when higher frequency components are analyzed in the signal, non-linear signal distortions can falsify the signal. As a result, the transducers used to galvanically decouple the signals from the mains voltage need to behave as linear as possible and add a minimum amount of noise onto the signals.

[1]Plugwise: Smart Wireless Solution for Energy Saving, Monitoring and Switching.
http://www.plugwise.com. Accessed: 10.01.2013
[2]Wattson: Energy Monitoring from Energeno.
http://www.diykyoto.com/uk/. Accessed: 10.01.2013

- Line-level output: In order to process the collected readings using small-signal processing components, e.g. studio recording equipment, the device needs to provide adjustable output voltages up to 2V RMS. In our case 775mV RMS for line-level was used.

4.1 Hardware Design

As a result of the commercial unavailability of galvanically decoupled high-resolution voltage and current measurement units in affordable price classes, we have developed our own printed circuit board that meets aforementioned design criteria. The resulting device is visualized in Figure 2, in which its four main functional sections are highlighted.

Figure 2: The PCB of the hardware board. Part 1 shows the capacitive power supply. Part 2 illustrates the small signal processing, Part 3 and 4 show the voltage and current measuring components.

4.1.1 Mains Voltage Components

The mains voltage is connected to the voltage and current transducers. In our practical realization, we have chosen a current transducer manufactured by LEM (a detailed discussion on its selection is given in Section 4.2). In order to enable our device to also collect voltage readings, an Avago Technologies HCPL-7840 optocoupler has been used, which caters for a galvanic decoupling of analog signals. While the LEM transducer does not require any power supply on the mains voltage side, the linear voltage optocoupler requires an input voltage of 5 volts DC to operate, which is not galvanically decoupled from mains. We have hence included a capacitive power supply on the mains side, which provides the linear optocoupler with its required operating voltage. The widely available switch mode power supplies have deliberately not been used in order to avoid switching noise on the captured signals.

4.1.2 Current and Voltage Sampling

The second part of our circuit board comprises the signal conditioning of the current signal. Instead of directly interfacing the transducer's output voltage to the signal processing system, however, we have included an operational amplifier in the signal path. The operational amplifier is used as voltage follower to increase the output driving strength.

Similar to the current conditioning part, the output of the linear optocoupler is interfaced to an operational amplifier. This amplifier setup adjusts the fully differential output signal of the optocoupler to different line levels and increases the output driving strength. Both voltage and current signals at the secondary side are decoupled using capacitors.

4.2 Sensor Selection

Many different sensor technologies are readily available for current sensing. Emerald [3] and Ziegler [25] give a good overview of the performance characteristics, advantages and disadvantages for the major sensor technologies. For the hardware board, we require a current sensor with a current range from 5mA up to 15A, which corresponds to a power draw ranging from 1.15W up to 3.5kW given the European grid voltage of 230V. The sensor should be capable of measuring frequencies from 50Hz up to 50kHz with a flat frequency response. Ideally it should have linear I/V correlation. Because we are mainly interested in the characteristics of different devices a precise calibration for current and voltage was not a major requirement.

To evaluate the sensors we generated sinusoidal, rectangular, and SMPS-like (Switched Mode Power Supply) signals using a HAMEG HMF 2525 frequency generator. Those signals were amplified using a BEAK BAA 1000 high current amplifier with a maximum output power of 1250VA and a maximum frequency of 50kHz. To avoid measuring errors caused by distortions of the amplifier a reference current was metered using a shunt resistor. To simulate a load we used high power resistors wrapped in two CPU coolers. To keep the temperature and thus the resistance of the load network nearly constant we heated the whole circuit until its temperature reached a steady state.

For each sensor we measured the I/V linearity, the frequency response using a sweep, the noise floor and the correlation with the reference signal for rectangular and sinusoidal signals and repeated all of those measurements with different currents.

We evaluated a set of different technologies current transducers, namely LTS 15-NP (hall effect), ACS712 (hall effect), CAS 15-NP (fluxgate), and ACS1050 (current transformer) called CT using the stated criteria. For each criterion we ranked the sensors and finally we calculated the average rank of each sensor. According to the criteria the CT and the CAS performed best. Both have the same average rank. The CT has the lowest non-linear distortions and a flat frequency response. On the other hand the signal obtained from the CAS has the highest correlation with the reference signal and also the lowest noise floor. As the frequency response could also be flatten with a post processing step we decided to build up the hardware board with a CAS current sensor.

4.3 Analog-to-Digital Conversion

To digitize the data a Realtek ACL 880 Intel HDA compatible sound card was connected to our hardware board via line-in. The sound card has a maximum sample rate of 96kHz. With respect to the Nyquist Theorem this results in a maximum usable sample rate of 48kHz. But due to an internal analog low pass filter, frequencies above 32kHz are attenuated and the maximum feasible frequency ranges up to 40kHz. Above this frequency, the signals are too weak to process them further. This low pass behavior is important to avoid aliasing effects if signal components above the Nyquist frequency are present in the input. The sound card has a sampling resolution of 20 Bit and an inbound voltage range of $1 V_{RMS}$. To obtain the readings from the sound card we use the ALSA sound system in combination with the octave package *aarecord*.

We used a soundcard for recording the current and voltage signals because of the wide availability of such Intel HDA compatible sound capturing devices. Nearly every laptop and desktop computer is equipped with such a hardware component. This should allow a flexible change of the data processing component, e.g. different PCs.

5. SOFTWARE FRAMEWORK

As we have described in Section 3, a software framework is responsible for recording and post processing the digitized sample stream. Our software framework called PISI is written in GNU Octave and performs a number of processing steps, including a segmentation of the recorded data, the feature extraction, and the export to the Attribute Relations File Format (ARFF), supported by the Weka toolkit. Next, we describe each of these steps in detail.

The first step in data processing is the generation of raw sample recordings. Those raw recordings contain many periods of the alternating voltage and current signals. As the recording starts at a random point in time, the alternating voltage signal has an undetermined phasing. To avoid any influence of this indeterminism, the software framework performs a segmentation of the recorded data. A segment starts on a zero-crossing of the voltage signal from a negative to a positive value and has a fixed length of N periods. The segmentation process is graphically shown in Figure 3. Shorter values of N are faster to obtain, whereas larger values of N are expected to yield better classification results due to the smaller impact of unexpected signal variations.

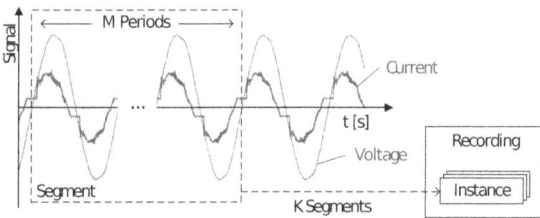

Figure 3: The segmentation process of the PISI .

In the second stage, our software framework extracts features from the segments which we present in more detail in Section 5.1. The resulting feature vector is stored and can be exported as a ARFF file. This is an important task to interface the Weka machine learning toolkit [6, 23] which is used for the classification. Numerous feature generation functions exist to extract the feature vector. Those functions take a segment as input and produce a set of features as output. PISI was designed with the requirement in mind to add new feature generation functions on demand. Therefore PISI organizes all recordings in a project environment. The environment keeps track of historic recordings with their corresponding class, the available feature generation functions and a list of available classes.

Last but not least the software framework is able to live predict the type and state of an electric appliance, which is attached to the hardware board. In order to do so, the software framework records the current and voltage signals for several seconds. Then it extracts segments from this recording and generates the features for each segment. The last step is the invocation of a Weka machine learner with a pre-built context model to predict the type of electric appliance. Then PISI loads a device related model which contains all possible states of the given device. This machine learning model is used to classify the state of the currently connected appliance.

5.1 Feature Extraction

One of the most important issues in the machine learning process is the feature extraction. Hence, a modular and flexible feature extraction solution was chosen to enable post processing the data prior to their forwarding to the data mining component. In general, we differentiate between two classes of features, namely *waveform* and *classical* features. In the context of this paper we use active power $P[W]$, phase shift $\varphi[°]$, and the current's crest factor C_i as the so called classical features. For continuous values the real power P_{cont} can be calculated from voltage and current waveforms over one or more periods T (cf. Equation 1). Its corresponding iterative calculation multiplies current and voltage samples over one or more periods N_P (cf. Equation 2).

$$P_{cont} = \frac{1}{T} \int_{t}^{t+T} v(t) \cdot i(t)\, \mathrm{d}t \qquad (1)$$

$$P = \frac{1}{N_P} \sum_{x=n}^{n+N_P} v_x \cdot i_x \qquad (2)$$

Equation 3 gives the shift from a current to an according voltage signal in degrees. In this context the voltage is supplied from grid, and thus considered to have constant amplitude during the short sampling window. Hence, only the current is considered to vary.

$$\varphi = \varphi_v - \varphi_i \qquad (3)$$

At this point we have to mention that the phase shift is calculated using the cross-correlation of v and i due to the diversity of current draw signals.

The crest factor measurement is the ratio from a waveforms amplitude peak to its RMS (Root Mean Square) value. For ideal sinusoidal signals it results in the factor of $\sqrt{2}$, which can be assumed for the grid voltage. On the other hand the current signal i varies and its crest factor C_i is specified according to Equation 4. The crest factor and its variance from $\sqrt{2}$ allow getting a first estimation of a signal's waveform characteristics.

$$C_i = \frac{i_{peak}}{i_{RMS}} \qquad (4)$$

To acquire specific information about the appliance current draw characteristics, a harmonic analysis is commonly used. In our implementation a discrete Fourier transform (DFT) is applied to the digitized current signal. The result is the representation of the signal in the frequency domain. Then the current signal harmonics (H_I) are selected by integer multiples of the fundamental frequency up to the Nyquist frequency which is defined as the half sampling frequency f_S. Since the current is directly dependent on the grid voltage, the first harmonic h_1 is allocated to the grid and thus the fundamental frequency f_{grid}. Hence the number of the highest harmonic h_N is defined by Equation 6.

$$H_I = [h_1, h_2, ..., h_N] \qquad (5)$$

$$\underline{with}: \quad N = \frac{f_S}{2}\frac{1}{f_{grid}} \qquad (6)$$

According to Parseval's theorem the energy in the time and frequency domain are equal. Because of this, the absolute amplitude of the harmonics depends on the power consumption of the metered appliance. To eliminate this coupling, we normalize the harmonics so that the first harmonic has a value of 0dB. This is leading to clearly separated features of signal strength (P) and signal waveform (H_I).

In theory, the even harmonics should be zero if the stimulating signal has a symmetric waveform and the stimulated system is linear time invariant (LTI). In practice, the even harmonics are not zero because the observed real-world systems were never totally linear. Therefore, we introduce two more features in addition to the harmonics for the evaluation of even and odd harmonic's influences (cf. Equations 7 and 8).

$$RMS_{h,odd} = \sqrt{\frac{2}{N}\sum_{x=1}^{N} h_{2x+1}^2} \qquad (7)$$

$$RMS_{h,even} = \sqrt{\frac{2}{N}\sum_{x=1}^{N} h_{2x}^2} \qquad (8)$$

Equation 9 shows the feature vector which is the base for this research in summary. The setup has a sampling frequency f_S of $96kHz$ which - together with the European grid frequency ($f_{grid} = 50Hz$) and Equation 6 - allows us to choose up to 960 harmonics to gain a wide range spectrum. Due to the low pass characteristics of the Metering Unit (c.f. Section 4.3), we have used only the first 800 harmonics in our implementation.

$$F = [P, \varphi, C_i, RMS_{h,odd}, RMS_{h,even}, h_1, h_2, ..., h_{800}] \quad (9)$$

5.2 Instance Extraction

The feature vector introduced in Equation 9 holds all features of one segment and represents one class instance for the machine learner. A set of K instances which are stored together are named recording. Obviously it requires less space to save the features of one segment as vector F. But this reduction of the size causes no substantial information loss because the information is then represented by classical features and harmonics. According to Figure 3 this allows us to adopt the resolution of the feature by changing the number of periods per segment (M). This relation can be observed in Figure 4 which shows the odd harmonics spectrum of a single instance and the standard deviation borders (SD+, SD-) of many instances.

The spectrum of each instance is noisy in between the characteristic standard deviation borders. This is caused by aforementioned sampling resolution. We were able to identify this noise as non-deterministic by means of Figure 4b, which contains the same base data and standard deviation borders as in Figure 4, but shows a flattened spectrum in between these.

The spectrum was calculated as arithmetic average of odd harmonics from all instances of one recording, and effectively shows the noise cancellation as compared to a single spectrum. Obviously the flattened spectrum is visibly better suited for classification than a single harmonic spectrum of an instance. Therefore we have used the averaged spectrum in further waveform-related analysis.

We have also observed that there is a trade-off between harmonic features and features derived from the power consumption of one instance. Classical features of one instance do not show the huge variance which is visible for the harmonics as shown in Figure 4. Thus, their features can be extracted from short segments in a sufficiently good quality and being used for classification. On the other hand, averaged harmonic features could help to distinguish between different appliances of one type or with similar power consumptions with the drawback of an increased demand for data, time, and processing. To exploit both benefits during the evaluation process (cf. Section 6), we have decided to split the machine learning models into two parts: One coarse-grained model with power related features to separate between different appliances and fine-grained models to distinguish between devices of one appliance type with nearly the same power consumption using average values. Our software enables the user to build training and testing data sets using feature vectors defined in Equation 9 as input. One can extract new features from this base value F and define a set with a subset of features easily. It is also possible to merge instances with the above mentioned arithmetic mean calculation. All these combinations are stored as ARFF files and processed in the data mining framework Weka.

6. ACCURACY EVALUATION

We have evaluated the system in different settings, which are described as follows.

6.1 Evaluation Setup

Based on the observations in Section 4 we used the current transducer CAS 15-NP for measuring different electric appliances. We have interfaced different electric appliances, as shown in Table 1.

For each appliance, we have measured 30 segments, where each segment covers voltage and current readings of 10 periods. Hence the measurement of one segment needs 200ms at 50Hz European grid frequency and one recording consists of 30 instances where each holds a complete feature vector F. The measurements of the devices and their operating modes were repeated several times under different conditions. We used independent recordings to create the training and test set for the machine learning process. More precisely: Both sets were recorded on different days with the attempt of different conditions for the measured appliance (e.g. temperature and running time of device). Using this approach we built one machine learning model for the device classification to detect the appliance (Section 6.2.1). A second model was built to differentiate between products of one appliance type (Section 6.2.2). In a third step we analyzed specified operating modes of selected devices (Section 6.3). The quality of the machine learning models was optimized through the feature and the classifier selection.

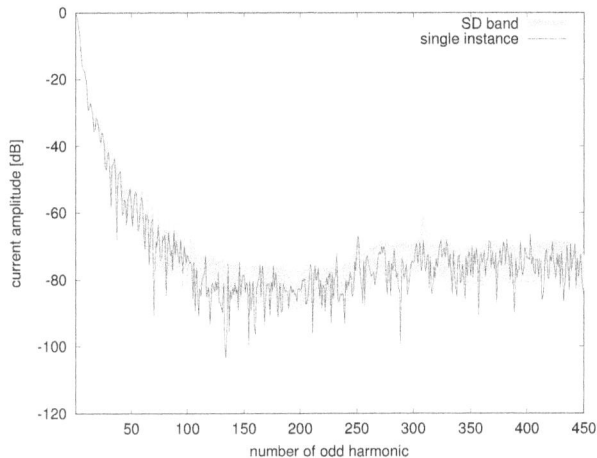

(a) The noisy spectrum of a single instance.

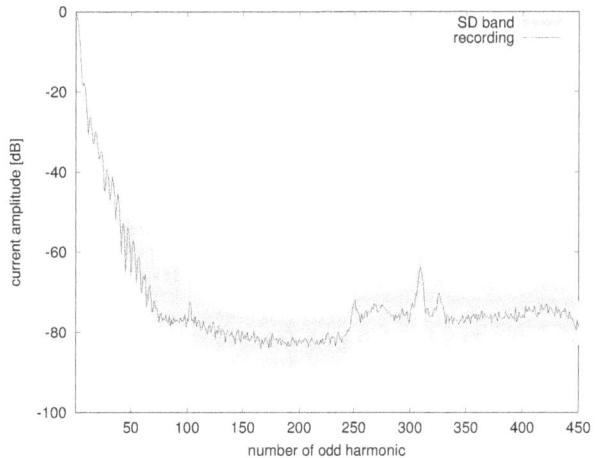

(b) The mean spectrum of multiple instances.

Figure 4: The harmonic spectrum for a laptop device. Both plots are based on the same data.

6.2 Device Identification

For the evaluation, we selected different appliances which are typical for a household or an office environment. Those appliances are listed in Table 1. We took care to select diverse devices: Some of them are movable - others are stationary, several appliances have the same power consumption - others differ in a spread spectrum and a few devices are even of the same production series. We recorded all of those appliances according to the setup described in Section 6.1 and extracted the full set of features described in Section 5.1 from their voltage and current readings.

6.2.1 Coarse-Grained Classification

In the first step we tried to classify the appliance type only with the classical features as described in Section 5.1. We define this as coarse-grained classification. The first column of Table 1 lists the different appliances to be classified in this step. Some of them were recorded in distinctive major operating modes. This results in a total of 14 appliance types. Each set consists of four recordings with 30 instances each. We used the Random Forest classifier to build our machine learning model. With this setup we obtained a accuracy of 99.6% for the device types. With the additional features that are calculated trough the RMS value of all even respectively all odd harmonics (see Equations 7 and 8) we can improve the result to 99.8%. This means that even if the device is in standby mode we can differentiate the device type. In general this is a hard task, because the standby power consumption is typically located in the range of 0.5-5 W. Since the classical feature values showed small variance, we were able to use single instances of recordings as classifier input. In our scenario an appliance type can hence be classified by a 200ms snapshot of its I/V characteristics.

6.2.2 Fine-Grained Classification

The classification of different series of devices within the same class using the coarse-grained classification was only partially satisfying. For example the six devices of the monitor type reached a maximum of 87% classification accuracy using the features described in the last section. This clearly shows that the coarse-grained classification would not scale

with respect to a broad training set. Again, the harmonic RMS features brought an accuracy improvement of 4%. This indicates that it might be useful to infer more information from the harmonic content. To do so, we introduced as second step, the so called fine-grained classification that considers also the device specific waveforms. This step was also motivated through the findings we made during the visualization of the current draw in the frequency domain (see Figure 5). Since all errors in the aforementioned classification appeared in between the products of the monitor type, we will use the fine-grained classification only for different monitors in standby and on status. Therefore the fine-grained classification is another escalation layer to separate between similar devices.

In a first step we processed a Greedy Stepwise feature ranking on a global set. This global set holds only one instance for each monitor which was calculated as the arithmetic average of all recordings. The aim of such a set is to provide a spectrum with product specific harmonics and thus to prevent the ranking of single instance's noise. We chose the top 20 ranked features, including 3 classical features (P, φ, $RMS_{h,even}$) and 17 harmonics: In total 3 even harmonics, 9 harmonics below 5kHz, and 7 harmonics up to 38.95kHz are present. These features build the new feature vector for the fine grained classification. As next step we created an independent training and test set with recordings from different days. To reduce the collected information and to concentrate on finding patterns we calculated the mean of 30 instances for each recording. Thus - and with a Random Forest classification model built from the train set - we gained a classification rate of 100%. This rate has to be interpreted as optimistic for a generalized case but it is valid for this specific scenario. It might decrease with the use of more products or with the use of test sets recorded under yet new conditions. To determine the influence of varying environmental conditions over the time, we investigated the long term behavior in Section 6.4.

6.3 Operating Mode Identification

In the last section, we have shown that it is possible to distinguish between different appliances. Furthermore the question is, if it is even possible to detect the operating mode of an appliance by analyzing its power consumption characteristics. To do so, we have selected different devices with several operating modes. Those devices have either operating modes with different power consumptions in each mode or their power consumption is nearly constant in all operating modes. To detect the operating mode, we trained a machine learning model for each appliance. Due to this design decision, it is possible to use different feature sets for each machine learning model. This flexibility is especially useful if some features do not differ between operating modes.

To evaluate the models, we performed the classification of a test set for each model. Those results are shown in Table 2. The model of a smartphone charger showed a high accuracy. It was always possible to detect, whether the charger was running idle, loading an empty smartphone battery or trickle loading an attached smartphone. No incorrect classification occurred by the use of only waveform features. We repeated the same procedure to classify the number of clients wired to an Ethernet switch and also if those clients were transferring data or not. Our system can classify the number of wired ports accurately but it was impossible to detect whether there is a network transfer happening or not. This can be seen in Table 2: The state c0 shows the switch running idle, c1 shows the switch connected with one client, c2 with two clients, c2d1 shows the results with two clients and a pending network transfer. As one can see in the table, it is impossible, to separate between the state c2 and c2d1. Last but not least we built a model with several operating modes of a TFT monitor. As a TFT monitor consists of millions of different pixels and a dimmable back light it has a nearly infinitely large set of different operating modes. From all of these operating modes we have chosen seven different

Table 1: Description of the devices tested during the evaluation.

Appliance [status]	Product	Power/W
Monitor [on, stby]	Fujitsu Siemens 24"	[77, 1]
	Fujitsu Siemens 19"	[34, 1]
	Fujitsu Siemens 17"	[34, 2]
	3 x Dell 20"	[48, 1]
LCD TV [on, stby]	Samsung 40"	[178, 1]
Laptop	Lenovo T430s	28
	Lenovo T420s	22
	Macbook 2.1	22
	Dell	29
Freezer [on, idle]	Liebherr FKS 3602	[130, 13]
Lamp	110W CCFL	105
	10W LED	6
	60W resistive	63
	100W resistive	98
Charger [stby]	Samsung Phone	0.5
	Apple Laptop	0.5
	Lenovo Laptop	0.5
Switch	Netgear 8 Port	6
Fan [level 1-3]	Tevion	[32-45]
USB-Hub	LogiLink 4 Port	0.5

Table 2: Results of the operating mode evaluation.

a	b	c	d	e	f	g	h	i	j	k	l	m	n	o	classified as
60	0	0													a charger idle
0	60	0													b charger loading
0	0	60													c charger done
			30	0	0	0									d switch c0
			0	30	0	0	0								e switch c1
			0	0	22	8	0								f switch c2
			0	0	16	14	0								g switch c2d1
			0	0	0	0	30								h switch c3
								149	30	0	0	0	0	0	i monitor black
								0	26	0	0	0	0	154	j monitor red
								0	0	175	0	0	2	0	k monitor white
								0	0	0	170	0	10	0	l monitor picture 1
								0	0	0	6	172	0	0	m monitor picture 2
								0	0	40	0	0	139	0	n monitor website 1
								0	1	0	0	1	0	177	o monitor website 2

100%   83%   81%

screen contents. Those operating modes include screens of the colors black, red, white, two pictures and two web sites. The initial machine learning model with the classical feature set performed poorly. By looking at the distribution of the features we noticed that only the power varies for different screen outputs and there are no noticeable changes in the frequency domain. Therefore this model considers only the power feature. In spite of the fact that we are merely able to classify the operating mode by the power we achieved an accuracy of 81%. Most remarkably, colors with high contrast like black and white can be separated successfully. An opposite effect is shown regarding the website one, which is the eEnergy 2013 website, and the color white. In this case the classification may be mixed up because of the large white areas of the eEnergy 2013 website.

It appears from the measurements that different colors do not affect the power consumption. But differences in the contrast are significantly measureable. Because of this fact, the classification of different screen outputs is possible. The variation of the power consumption ranges from 0.1W to 3W for the different operating modes. To separate between different states of the TFT monitor, there must be a difference of 0.5W. Differences below 0.5W might be caused by environmental noise and thus should not be used as feature.

6.4 Long Term Stability

We measured the energy consumption of particular devices at different points in time. But due to changes in the environment (varying grid voltage, grid frequency, temperature, SMPS switching frequency), it might be the case that the recordings of devices vary over time in a way that causes the machine learner to fail. To analyze if variable environment conditions could cause problems, we measured the energy consumption of one monitor over the period of one week. The Figure 5 shows the spectral variance of the obtained recordings. The blue line shows the mean of a recording which was obtained by averaging 30 continuously recorded segments. The gray area indicates the standard deviation around the mean for each harmonic. The thickness of the band shows the noise level in this harmonic spectrum. This noise of single instances could either be caused by the environment or by the appliance itself. Characteristic for this device is the oscillation around the 325th harmonic which is caused by the switching power supply (SMPS). This peak was present in all recordings of this particular device. Therefore frequencies in this region are good features for recognizing the device.

As the Figure 5 shows, the instances contain time vari-

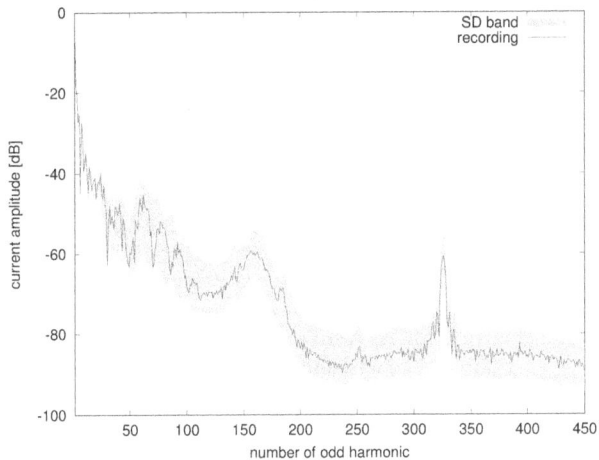

Figure 5: The variance of the spectrum of a monitor over a period of one week.

ant noise. The fine-grained device classification and also the operating mode classification have to deal with such distortions to avoid overfitting of the machine learning model to particular environment conditions. Two possible solutions exist to deal with those issues: One could either use a hand selected set of omnipresent features or one could repeat measurements over a period of time to face varying environmental conditions by averaging all measurements. Regardless of the time variant noise, the shape of the harmonics remains a characteristically feature for the classification process.

6.5 Discussion

As we have shown in Section 2, various related works perform a device classification based on the current draw of electrical appliances. But even recent works have diverse limitations: either they provide no contemporary information or the classification results are inaccurate for certain devices under certain circumstances. Moreover, to the best of our knowledge, no solution for operation mode estimation was presented yet.

Our work enhances the state of the art by carefully crafted methodologies for high accuracy device- and operating mode classification. Our two layered approach achieves a high precision of up to 99.8% together with low latencies. Furthermore, our evaluation shows, that both waveform and classical features carry valuable information with regard to the device classification.

These achieved improvements close the information gap between electric appliances and the smart home system, and thus enable the development of a new generation of smart home systems. These next generation smart home systems could improve the user comfort, reduce the energy consumption of the household or detect unexpected behavior and increase the safety. For example, one can build an energy saving module, which uses the provided information to cut off standby loads, trim appliances to the most energy efficient setting or defer devices until the time of use pricing is low.

In the next step, we will implement our approach on an embedded system. In this case, the amount of data to process is no serious problem: an adaptive sampling schema together with a decent Digital Signal Processor will easily handle the feature extraction step. A classification of the attached appliance is only required, if turn-on or switching transients happen. This classification task could be handled by a centralized processing unit. The price of such a solution might be very low, if mass production cuts the costs per unit.

7. CONCLUSION

Enhancing current building automation systems by the capabilities to identify electric appliances is a major step on the way to truly smart buildings. We have thus presented an approach to determine both the type as well as the operation mode of an electric appliance based on measuring its voltage and current waveforms at high resolution. To this end, we have designed a hardware system to collect and condition the readings from the physical environment, as well as the PISI framework to extract relevant features from the raw data and add them to the model of a machine learner. PISI has been shown to classify the type and the state of connected electric appliances with a precision of up to 99.8%.

Our solution bridges the information gap between the building automation system and the electric appliances, and thus enables the development of the next generation home automation systems. Based on the appliance identification, novel features can be realized, e.g., saving energy by automated deactivation of appliances while their operation is not required by the user. Similarly, improvements to the user's comfort and safety as well as support for Ambient Assisted Living can be made when the presented functionality is available to building automation systems.

Acknowledgements

This work was supported by the BMWi Project IP-KOM-OeV. Sincere thanks are given to Christian Hatzfeld and Jan Lotichius from the research team of the Institute of Electromechanical Design for supporting us with equipment and helpful discussions for the sensor selection.

8. REFERENCES

[1] C. Beckel, L. Sadamori, and S. Santini. Towards Automatic Classification of Private Households Using Electricity Consumption Data. In *4th Intl. Workshop on Embedded Sensing Systems for Energy-Efficiency in Buildings (BuildSys)*. ACM, 2012.

[2] H. Chang, C. Lin, and J. Lee. Load Identification in Nonintrusive Load Monitoring Using Steady-State and Turn-On Transient Energy Algorithms. In *14th Intl. Conference on Computer Supported Cooperative Work in Design (CSCWD)*, pages 27–32. IEEE, 2010.

[3] P. Emerald. Non Intrusive'Hall Effect Current Sensing Techniques Provide Safe, Reliable Detection and Protection for Power Electronics. In *Intl. Appliance Technical Conference (IATC)*, pages 1–2. IEEE, 1998.

[4] T. Ganu, D. Seetharam, V. Arya, R. Kunnath, J. Hazra, S. Husain, L. De Silva, and S. Kalyanaraman. nPlug: A Smart Plug for Alleviating Peak Loads. In *3rd Intl. Conference on Future Energy Systems: Where Energy, Computing and Communication Meet (e-Energy)*, pages 1–10. ACM, 2012.

[5] S. Gupta, M. S. Reynolds, and S. N. Patel. ElectriSense: Single-Point Sensing Using EMI for

Electrical Event Detection and Classification in the Home. In *12th Intl. Conference on Ubiquitous Computing (UbiComp)*, pages 139–148. ACM, 2010.

[6] M. Hall, E. Frank, G. Holmes, B. Pfahringer, P. Reutemann, and I. H. Witten. The WEKA Data Mining Software: An Update. *SIGKDD Explorations*, 11(1):10–18, 2009.

[7] G. Hart. Nonintrusive Appliance Load Monitoring. *Proceedings of the IEEE*, 80(12):1870–1891, 1992.

[8] M. Ito, R. Uda, S. Ichimura, K. Tago, T. Hoshi, and Y. Matsushita. A Method of Appliance Detection Based on Features of Power Waveform. In *4th Intl. Conference on Symposium on Applications and the Internet (SAINT)*, pages 291–294. IEEE, 2004.

[9] M. Jahn, M. Jentsch, C. Prause, F. Pramudianto, A. Al-Akkad, and R. Reiners. The Energy Aware Smart Home. In *5th Intl. Conference on Future Information Technology (FutureTech)*, pages 1–8. IEEE, 2010.

[10] L. Jiang, S. Luo, and J. Li. An Approach of Household Power Appliance Monitoring Based on Machine Learning. In *5th Intl. Conference on Intelligent Computation Technology and Automation (ICICTA)*, pages 577–580. IEEE, 2012.

[11] X. Jiang, S. Dawson-Haggerty, P. Dutta, and D. Culler. Design and Implementation of a High-Fidelity AC Metering Network. In *8th Intl. Conference on Information Processing in Sensor Networks (IPSN)*, pages 253–264. IEEE, 2009.

[12] Y. Kim, T. Schmid, Z. M. Charbiwala, and M. B. Srivastava. ViridiScope: Design and Implementation of a Fine Grained Power Monitoring System for Homes. In *11th Intl. Conference on Ubiquitous Computing (UbiComp)*, pages 245–254. ACM, 2009.

[13] D. Kirschen. Demand-Side View of Electricity Markets. *IEEE Transactions on Power Systems*, 18(2):520–527, 2003.

[14] W. Kleiminger, S. Santini, and M. Weiss. Opportunistic Sensing for Smart Heating Control in Private Households. In *2nd Intl. Workshop on Networks of Cooperating Objects (CONET)*, 2011.

[15] Kolter, Zico and Johnson, Matthew. REDD: A Public Data Set for Energy Disaggregation Research. In *1st Intl. Workshop on Data Mining Applicatins in Sustainability (SustKDD)*. ACM, 2011.

[16] C. Laughman, K. Lee, R. Cox, S. Shaw, S. Leeb, L. Norford, and P. Armstrong. Power Signature Analysis. *IEEE Power and Energy Magazine*, 1(2):56–63, 2003.

[17] K. Lee, S. Leeb, L. Norford, P. Armstrong, J. Holloway, and S. Shaw. Estimation of Variable-Speed-Drive Power Consumption from Harmonic Content. *IEEE Transactions on Energy Conversion*, 20(3):566–574, 2005.

[18] T. Mantoro, M. A. Ayu, and E. E. Elnour. Web-Enabled Smart Home Using Wireless Node Infrastructure. In *9th Intl. Conference on Advances in Mobile Computing and Multimedia (MoMM)*, pages 72–79. ACM, 2011.

[19] J. Nichols and B. Myers. Controlling Home and Office Appliances with Smart Phones. *Pervasive Computing*, 5(3):60–67, 2006.

[20] S. Patel, T. Robertson, J. Kientz, M. Reynolds, and G. Abowd. At the Flick of a Switch: Detecting and Classifying Unique Electrical Events on the Residential Power Line. In *9th Intl. Conference on Ubiquitous Computing (UbiComp)*, pages 271–288. ACM, 2007.

[21] J. Pierce, D. Schiano, and E. Paulos. Home, Habits, and Energy: Examining Domestic Interactions and Energy Consumption. In *28th Intl. Conference on Human Factors in Computing Systems (CHI)*, pages 1985–1994. ACM, 2010.

[22] A. Reinhardt, P. Baumann, D. Burgstahler, M. Hollick, H. Chonov, M. Werner, and R. Steinmetz. On the Accuracy of Appliance Identification Based on Distributed Load Metering Data. *Lamp*, 6:45, 2012.

[23] I. Witten, E. Frank, and M. Hall. *Data Mining: Practical Machine Learning Tools and Techniques*. Morgan Kaufmann, 2011.

[24] M. Zeifman and K. Roth. Nonintrusive Appliance Load Monitoring: Review and Outlook. *IEEE Transactions on Consumer Electronics*, 57(1):76–84, 2011.

[25] S. Ziegler, R. Woodward, H. Iu, and L. Borle. Current Sensing Techniques: A Review. *IEEE Sensors Journal*, 9(4):354–376, 2009.

Smart Air-Conditioning Control by Wireless Sensors: An Online Optimization Approach

Muhammad Aftab, Chi-Kin Chau, and Peter Armstrong
Masdar Institute of Science and Technology
Abu Dhabi, UAE
{muhaftab, ckchau, parmstrong}@masdar.ac.ae

ABSTRACT

One of the most prominent applications of smart technology for energy saving is in buildings, in particular, for optimizing heating, ventilation, and air-conditioning (HVAC) systems. Traditional HVAC systems rely on wired temperature regulators and thermostats installed at fixed locations, which are both inconvenient for deployment and ineffective to cope with dynamic changes in the thermal behavior of buildings. New generation of wireless sensors are increasingly becoming popular due to their convenience and versatility for sophisticated monitoring and control of smart buildings. However, there also emerge new challenges on how to effectively harness the potential of wireless sensors. First, wireless sensors are energy-constrained, because they are often powered by batteries. Extending the battery lifetime, therefore, is a paramount concern. The second challenge is to ensure that the wireless sensors can work in uncertain environments with minimal human supervision as they can be dynamically displaced in new environments. Therefore, in this paper, we study a fundamental problem of optimizing the trade-off between the battery lifetime and the effectiveness of HVAC remote control in the presence of uncertain (even adversarial) fluctuations in room temperature. We provide an effective offline algorithm for deciding the optimal control decisions of wireless sensors, and a 2-competitive online algorithm that is shown to attain performance close to offline optimal through extensive simulation studies. The implication of this work is to shed light on the fundamental trade-off optimization in wireless sensor controlling HVAC systems.

Categories and Subjects: [Computer systems organization]: *Embedded and cyber-physical systems: Sensors and actuators*

General Terms: Algorithms, Design, Management

Keywords: Wireless Sensors, Smart Buildings, HVAC, Air-Conditioning, Online Algorithms

1. INTRODUCTION

Buildings are among the largest consumers of energy, topping 40% of total energy usage in many countries [9]. A significant portion of energy use in buildings is attributed to the heating, ventilation, and air conditioning (HVAC) systems, which account for up to 50% of the total energy consumption in buildings [7]. Therefore, improving energy efficiency of buildings, in particular, optimizing HVAC system is critically important and will have a significant impact in reducing the overall energy consumption.

Usually, the air conditioning systems need to maintain room temperature within a certain desirable range. To detect the variations of temperature, traditional air conditioning systems rely on wired temperature regulators and thermostats installed at fixed locations. These classical controllers are both inconvenient for deployment and ineffective to cope with dynamic changes in the thermal behavior of buildings. In particular, the temperature distribution is not spatially uniform. Having sensors installed at fixed and limited locations cannot react to the rapidly varying room conditions due to transient and non-stationary human behavior.

New generation of wireless sensors are revolutionizing the design of HVAC systems. Wireless sensors, being not limited by wired installation, can be deployed strategically close to the fluctuating thermal sources in an ad hoc fashion (e.g., near to doors, windows and computers). With wireless sensors, demand responsive air-conditioning control can be developed that dynamically adjusts the room temperature according to intelligent monitoring and tracking of human behavior and room conditions. Furthermore, wireless sensors can be integrated with home security and infotainment systems, enabling more sophisticated smart home control systems.

Despite the promising potential, wireless sensors also introduce several new challenges:

1. **Battery Lifetime**: Wireless sensors are often battery-powered and typically have to operate for prolonged periods of time. Therefore, one of the primary goals is to maximize the battery lifetime of sensors. According to a survey of several commercial wireless sensors (see Appendix-C), the communication operations consume the most energy. Thus, an effective way to extend battery lifetime is to reduce the communication frequency, inducing limited communication among wireless sensors.

2. **Control Effectiveness**: Wireless sensors are also dis-

tributed autonomous computing devices. They can be programmed to intelligently optimize their energy consumption with respect to the effectiveness of their control operations. Intuitively, energy consumption is inversely proportional to the effectiveness (i.e., sleeping all the time can effectively reduce energy consumption, but is ineffective to satisfy the control requirement). The ability to balance the energy consumption and effectiveness is critical to the usefulness of these wireless sensors, particularly for smart home applications.

3. **Uncertain Deployment**: Wireless sensors are supposed to be deployed in an ad hoc fashion, without a-prior measurement or calibration. It is critical to ensure that wireless sensors operate robustly and reliably in the presence of uncertainty of new environments. They should be able to rapidly cope with dynamic displacements with minimal human supervision. An important question is to investigate the fundamental ability of wireless sensors to control room temperature without assuming any a-prior or stochastic knowledge of the temperature fluctuations.

In this paper, we study a fundamental problem of optimizing the trade-off between the lifetime of the wireless sensors and the effectiveness of HVAC remote control in the presence of uncertain (even adversarial) fluctuations in room temperature. The novelty of our work lies in the fact that unlike most intelligent HVAC control techniques (as summarized in the related work section), our approach is to solve the optimization problem in an online manner without stochastic modeling or machine learning methods. The key contributions of this work are summarized as follows.

1. We formulate a new online optimization problem of balancing the trade-off between communication frequency of wireless sensor and the effectiveness of HVAC remote control. Our goal is to simultaneously maintain thermal comfort and maximize the battery lifetime of the wireless sensor. In other words, we aim to maximize the sensor energy efficiency while meeting the required control performance. To the best of our knowledge, this specific problem has not been studied before.

2. We present an effective offline algorithm, which is based on dynamic programming, for determining the optimal control decisions by wireless sensors when all future temperature fluctuations are known in advance. The offline algorithm is useful to benchmark the online algorithm we propose.

3. We devise an online algorithm that optimizes the control decisions without the knowledge about future temperature fluctuations. We prove that our online algorithm is 2-competitive against offline optimal algorithm.

4. We experimentally evaluate the performance of our algorithm through simulations and show that our online algorithm can attain performance close to the offline optimal solution.

The rest of the paper is organized as follows. In Section 2, we present the background of online algorithmic approach, competitive analysis, and a related problem known as dynamic TCP acknowledgement problem. We present the models and formulations of ambient room temperature and wireless sensor network control in Section 3. In Section 4, we provide the offline and online algorithms and competitive analysis. In Section 5, we evaluate the performance of our algorithms through extensive simulations. In Section 6, we present a review of related work. Finally, we summarize and discuss several future extensions in Section 7.

2. BACKGROUND

In this section, we present the background information about online algorithms and a well-known online problem known as dynamic TCP acknowledgment problem, which is closely related to our problem.

2.1 Online Algorithms

Online algorithms have received considerable attention in the literature for their fundamental principles and practical applications. In an online problem, a sequence of input is revealed gradually over time. The algorithm needs to make certain decisions and generate output instantaneously over time, based on only the part of the input that has been seen so far, without knowing the rest of the input to be revealed in the future. There are many practical problems studied in the online algorithmic setting that require real-time and instantaneous decisions, such as real-time resource allocation in operating systems, data structuring, robotics or communication networks [1,8]. The performance of online algorithms is evaluated using competitive analysis. The *competitive ratio* of an online algorithm is defined as the worst-case ratio between the cost of the solution obtained by the online algorithm versus that of an offline optimal solution obtained by knowing the all input sequence in the future [19].

Online algorithms have several practical implications. First, they do not require a-prior or stochastic knowledge of the input sequence, which makes them robust in any uncertain (even adversarial) environments. Second, online algorithms uses often simple decision-making mechanisms, without being hampered by inaccurate or slow convergent machine learning techniques. Third, online algorithms can give a fundamental characterization without further assumptions of the problems, which are useful to benchmark other sophisticated and more complicated decision-making mechanisms. In this paper, we adopt the online algorithmic approach to study the fundamental problem of optimizing the trade-off between the battery lifetime and the effectiveness of HVAC remote control in the presence of uncertain fluctuations in room temperature.

2.2 Dynamic TCP Acknowledgment

A well-known example involving online algorithms is the dynamic TCP acknowledgment problem as described as follows. A stream of packets arrives at a destination. The packets must be acknowledged in order to notify the sender that the transmission was successful. However, it is possible to simultaneously acknowledge multiple packets using a single acknowledgments packet. The delayed acknowledgment mechanism reduces the frequency of the acknowledgments, but it might also add excessive latency to the TCP connection and interfere with the TCP's congestion control mechanisms [10]. The problem is to find an optimal trade-off between the total number of acknowledgments sent and the latency cost introduced due to delaying acknowledgment.

More specifically, Dooly et al. [6] formulated this trade-off as the dynamic TCP acknowledgement problem as follows.

In the dynamic TCP acknowledgement problem, a sequence of n packets $\sigma = (p_1, p_2, ..., p_n)$ arrive at a certain destination. An algorithm divides the received sequence σ into m subsequences $\sigma_1, \sigma_2, ..., \sigma_m$, where a single acknowledgment is sent at the end of each subsequence. All the packets contained in $\sigma_j (1 \leq j \leq m)$ are acknowledged together by the j-th acknowledgement at time t_j. The objective is to choose an optimal acknowledgment time sequence that minimizes the weighted sum of the cost for transmitting acknowledgements and the cost of the latency of delayed acknowledgements. The decision of transmitting an acknowledgment time is decided in an online fashion without knowing the future packet arrivals.

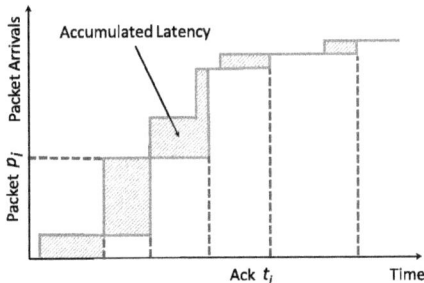

(a) Dynamic TCP acknowledgment

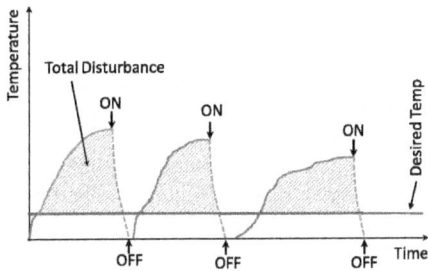

(b) Wireless sensor controlling AC system

Figure 1: A pictorial comparison between dynamic TCP acknowledgment and wireless sensor controlling AC system.

Comparison to Our Problem: Our problem is somewhat similar to the dynamic TCP acknowledgment problem. In TCP, random arrivals of packets are received, such that the receiver makes online decisions when to transmit acknowledgments considering the weighted total cost of number of acknowledgment and latency. In our problem, random fluctuations of temperature and external thermal sources are perceived by the wireless sensor, and the wireless sensor makes online decisions when to transmit control commands to remote air conditioning system considering the weighted total cost of transmissions and effectiveness (defined by the disturbance of temperature compared to a desirable temperature). A pictorial comparison between the two problems is provided in Fig. 1.

Despite the similarity, our results are not direct applications of the dynamic TCP acknowledgment problem. In particular, the dynamic TCP acknowledgment problem assumes latency as a linearly increasing function of time, whereas in our problem the total disturbance of temperature changes non-linearly with time. This requires a non-trivial extension of the original TCP acknowledgment problem to the new context of air-conditioning control. Furthermore, we present extensive simulation studies that are specific to the air-conditioning control setting for corroborating the usefulness of our online algorithms for this new problem.

3. MODEL AND FORMULATION

The goal of our study is to optimize the trade-off between the wireless sensor battery lifetime and the effectiveness of ambient room temperature control in the presence of uncertain fluctuations. In this section, we present the models of ambient room temperature and wireless sensor control. We note that a table of notations with explanations is provided in Table 5 in the Appendix. It is worth mentioning that we make several assumptions in order to improve the tractability of our models and for convenience of analysis.

3.1 Assumptions of Ambient Room Temperature

The thermal behavior of buildings is a complex system. The mathematical models in the literature typically involve several empirical constants, non-linear functions and uncertain factors such as heat flow and material properties [16]. Moreover, external factors, such as weather condition (e.g., temperature, humidity), soil temperature, radiation effects and other sources of energy (e.g., human activities, lighting and equipment), also play a critical role in determining the thermal behavior of buildings [16].

Tractable mathematical models of building thermal behavior are particularly useful for the design of intelligent controls and regulations of HVAC systems. Therefore, assumptions are often imposed to improve the tractability of the thermal models of buildings.

In this work, we employ a simple yet commonly used thermal model for a single room. This model considers several major factors, such as the outdoor environment, the thermal characteristics of the room, and the air-conditioning system. We mostly consider the setting of cooling, where the air-conditioning system is required to make continual adjustment to the room temperature for maintaining a (lower) desirable temperature level. We remark that our results can be applied to the setting of heating with minor modifications.

First, we list several common assumptions of the ambient room temperature in the literature [21] for improving the tractability:

- The air in the room is assumed to be fully mixed.
- The temperature distribution is assumed to be uniform and the dynamics can be expressed using a lump capacity model.
- The room behaves ideally, such that the effect of each wall is uniformly equivalent.
- The density of the air is constant and is not affected by the changes in temperature and humidity.

3.2 Dynamic Model of Ambient Room Temperature

Based on the above assumptions, a simple dynamic model of ambient room temperature can be formulated as follows.

We consider the setting of continuous time, and model the ambient room temperature at time t by a function $T(t)$, which depends on several major factors:

1. The *initial ambient room temperature* T_0 at time $t = 0$.

2. The *influence of outdoor temperature* $T_{\mathrm{od}}(t)$, which is a function of time affected by time-of-day and weather. A simple example is a sinusoidal function depending on the time-of-day. We assume that the variation of $T_{\mathrm{od}}(t)$ is relatively slow, as compared to the effect of air-conditioning system. Hence, we simply write $T_{\mathrm{od}}(t)$ as a constant T_{od}.

3. The *external thermal sources* entering into the room, for example, due to human body heat or human activities (e.g., computers). We model the arrivals of thermal sources by a function $W(t)$, such that there is a level of thermal intensity $W(t)$ (measured by degree Celsius) arriving at time t.

4. The *heat absorptivity and insulation properties* of the materials in a room (e.g., walls). Heat can be retained in a room for a longer period of time in a well-insulated room with sufficiently absorptive materials.

5. The *air-conditioning system output*. This is the control variable we seek to optimize in order to maintain the ambient room temperature within a desirable range.

a) Without external thermal sources: Throughout this paper, we rely on a widely-used model of dynamic ambient room temperature [5]. First, we assume that there is no external thermal sources entering into the room (i.e., $W(t) = 0$ for all t). In particular, we denote the ambient room temperature without external thermal sources as $\tilde{T}(t)$. Given the initial ambient room temperature T_0 and outdoor temperature T_{od}, the dynamic behavior of $\tilde{T}(t)$ can be described by the following differential equations

$$\frac{\mathrm{d}\tilde{T}(t)}{\mathrm{d}t} = \frac{1}{\mathsf{c} \cdot M_{\mathrm{air}}} \cdot \left(\frac{\mathrm{d}Q_{\mathrm{in}}(t)}{\mathrm{d}t} - \frac{\mathrm{d}Q_{\mathrm{ac}}(t)}{\mathrm{d}t} \right) \quad (1)$$

$$\frac{\mathrm{d}Q_{\mathrm{in}}(t)}{\mathrm{d}t} = \frac{T_{\mathrm{od}} - \tilde{T}(t)}{R_{\mathrm{eq}}} \quad (2)$$

$$\frac{\mathrm{d}Q_{\mathrm{ac}}(t)}{\mathrm{d}t} = \frac{\mathsf{c} \cdot M_{\mathrm{ac}} \cdot (\tilde{T}(t) - T_{\mathrm{ac}})}{E_{\mathrm{ac}}} \quad (3)$$

where T_{ac} is the temperature output by the air-conditioning system, $Q_{\mathrm{in}}(t)$ is the net heat transfer from outdoor, $Q_{\mathrm{ac}}(t)$ is the net heat chilled by the air-conditioning system, M_{air}, M_{ac}, E_{ac}, c, R_{eq} are constants that model the heat absorptivity and insulation properties in the room (see Appendix for full explanations). By substitution, one can solve the differential equations by the following lemma.

LEMMA 1. *In the above model, the solution to Eqns. (1)-(3) is given by*

$$\tilde{T}(t) = \frac{\mathsf{C}_1}{\mathsf{C}_2} - \left(\frac{\mathsf{C}_1}{\mathsf{C}_2} - \tilde{T}(0) \right) \cdot e^{-\mathsf{C}_2 \cdot t} \quad (4)$$

where

$$\mathsf{C}_1 = \frac{\mathsf{c} \cdot T_{\mathrm{ac}} \cdot M_{\mathrm{ac}} \cdot R_{\mathrm{eq}} + E_{\mathrm{ac}} \cdot T_{\mathrm{od}}}{\mathsf{c} \cdot M_{\mathrm{air}} \cdot R_{\mathrm{eq}} + E_{\mathrm{ac}}} \quad (5)$$

$$\mathsf{C}_2 = \frac{E_{\mathrm{ac}} + \mathsf{c} \cdot M_{\mathrm{ac}} \cdot R_{\mathrm{eq}}}{\mathsf{c} \cdot E_{\mathrm{ac}} \cdot M_{\mathrm{air}} \cdot R_{\mathrm{eq}}} \quad (6)$$

We provide the proof in Appendix-A.

b) With external thermal sources: Next, we consider the setting with external thermal sources. We consider $W(t)$ as a sequence of *impulsive thermal sources*, such that

$$W(t) = \sum_{i=1}^{m} w_i \cdot \delta(t - t_i) \quad (7)$$

where $\delta(t)$ is Dirac delta function, and w_i is the level of thermal intensity entering into the room at time t.

Impulsive thermal sources are a reasonable assumption for modeling short-lived thermal sources (e.g., temporarily opening a door). Further, any arbitrary $W(t)$ can be approximated by a sequence of appropriately placed impulsive thermal sources by taking $w_i = W(t_i)$ (see Fig. 2 for an illustration). Note that, in this paper, we do not assume any a-priori knowledge of the stochastic property of $W(t)$. We

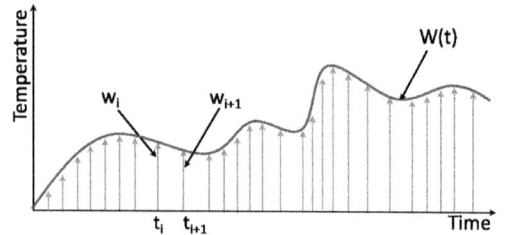

Figure 2: An illustration for using impulsive heat sources to approximate arbitrary $W(t)$.

denote $\mathbf{a} \triangleq ((w_i, t_i) : i = 1, ..., m)$ for a sequence of arrivals of impulsive thermal sources, where m is the total number of arrivals. Given $\mathbf{a}$, the ambient room temperature at time t can be obtained recursively as follows. For $i \in \{1, ..., m\}$, we note that there is no external thermal source during interval $t_{i-1} < t < t_i$. We denote the ambient room temperature during interval $t_{i-1} \leq t < t_i$ by $\tilde{T}_i(t)$. Thus, following by Lemma 1, we obtain

$$\tilde{T}_i(t) = \frac{\mathsf{C}_1}{\mathsf{C}_2} - \left(\frac{\mathsf{C}_1}{\mathsf{C}_2} - \tilde{T}_{i-1}(t_{i-1}) - w_{i-1} \right) \cdot e^{-\mathsf{C}_2 \cdot (t - t_{i-1})} \quad (8)$$

where $\tilde{T}_{i-1}(t_{i-1}) + w_{i-1}$ is the initial temperature at t_{i-1}.

For completeness, we let $t_0 = 0$, $w_0 = 0$ and $\tilde{T}_0(t_0) = T_0$. Hence, we obtain the ambient room temperature for given external thermal sources $\mathbf{a}$ and initial ambient room temperature T_0 as

$$T(t; \mathbf{a}, T_0) = \tilde{T}_i(t), \text{ if } t_{i-1} \leq t < t_i \quad (9)$$

3.3 Model of Wireless Sensor Control

To model wireless sensor control, we consider a wireless sensor deployed in the target zone for sensing the ambient temperature. The wireless sensor issues control commands to a remote air-conditioning system when the locally sensed ambient temperature exceeds a certain desirable temperature range. There are several issues considered in our sensor model.

a) Trade-off: Since wireless sensors are energy constrained and often powered by batteries, the wireless sensor is required to optimize the battery lifetime without affecting the thermal comfort. Although various operations are performed in wireless sensors (e.g., computations and sensing), the wireless communication operations typically consumes most of the energy in a wireless sensor (see Appendix-C). Hence, it is crucial to reduce the number of wireless communication operations for extending the battery lifetime.

There are two prominent conflicting factors that a wireless sensor needs to optimize:

1. The *update frequency* of control commands to remote air-conditioning system in the presence of random fluctuating thermal sources, which characterizes the effectiveness of ambient room temperature control.

2. The *communication operations* for transmitting the control commands, which critically governs the wireless sensor battery lifetime.

Note that increasing of the number of communication operations will reduce the battery lifetime. This naturally gives rise to an online decision problem, where the wireless sensor decides the update frequency in an online manner without a-prior information of random fluctuating arrivals of thermal sources.

b) Air-conditioning Operations: Let $T_{\mathrm{des}}^{\max}$ be the maximally desirable temperature (e.g., 25 degree Celsius), and $T_{\mathrm{des}}^{\min}$ be the minimally desirable temperature (e.g., 21 degree Celsius). The desirable ambient room temperature is aimed to be retained within $[T_{\mathrm{des}}^{\min}, T_{\mathrm{des}}^{\max}]$.

A simple setting of control command by wireless sensor is the "ON/OFF" or hysteresis control, such that when the ambient room temperature is sufficiently higher than $T_{\mathrm{des}}^{\max}$, an ON command is communicated to air-conditioning system, whereas when the sensed ambient room temperature is sufficiently lower than $T_{\mathrm{des}}^{\min}$, an OFF command is communicated to air-conditioning system[1]. This induces an ON/OFF cycle of air-conditioning operations (see Fig. 3 for an illustration), which is one of the most commonly used control strategy in today's air-conditioning systems [13].

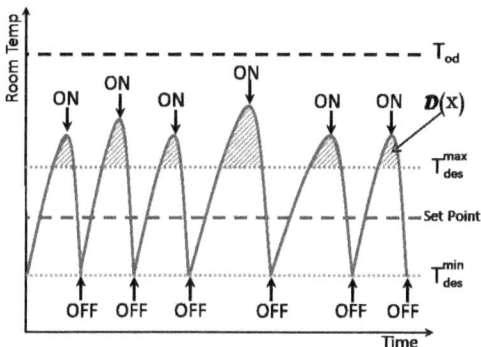

Figure 3: An illustration of the ON/OFF cycle of air-conditioning. Note that we may allow the ambient room temperature to exceed $T_{\mathrm{des}}^{\max}$ temporarily.

Furthermore, for the sake of tractability, we assume that an OFF command is automatically issued when the ambient room temperature drops below $T_{\mathrm{des}}^{\min}$, and the cooling process is rather efficient, i.e.,cooling can be achieved in a relatively short time. However, we may allow the ambient room temperature to exceed $T_{\mathrm{des}}^{\max}$ temporarily. Hence, our study is simplified to only optimize the ON command decisions in order to balance the trade-off between the wireless sensor

---

[1]In our ambient room temperature model, the air-conditioning system can be disabled by letting $M_{\mathrm{ac}} = 0$

battery lifetime and the effectiveness of ambient room temperature control, without considering the OFF commands.

We consider a finite time horizon for any $t \in [0, B]$. We define the *decision variables* as $\mathbf{x} = (x_k \in [0, B])_{k=1}^K$, where each x_k is the time that the k-th ON command is issued by the wireless sensor, while K is the total number of ON commands which the wireless sensor needs to optimize without affecting the thermal comfort.

c) Disturbance of Temperature: We characterize the thermal comfort by a metric defined as the total disturbance of ambient temperature exceeding the desirable temperature range.

For given time τ, we let $\mathbf{a}_\tau$ be the sub-sequence, such that

$$\Big((w_i, t_i - \tau), (w_{i+1}, t_{i+1} - \tau), (w_{i+2}, t_{i+2} - \tau), ...\Big) \quad (10)$$

where t_i is defined such that $t_{i-1} < \tau \le t_i$. Namely, $\mathbf{a}_\tau$ is a truncated sequence of $\mathbf{a}$ starting at τ.

We define $T_\tau(t)$ to be the temperature function $T(t; \mathbf{a}, T_0)$ starting at time τ with initial temperature $T_0 = T_{\mathrm{des}}^{\min}$ and sequence of thermal sources $\mathbf{a}_\tau$. That is, for any $t \ge \tau$,

$$T_\tau(t) \triangleq T\Big(t - \tau; \mathbf{a}_\tau, T_{\mathrm{des}}^{\min}\Big) \quad (11)$$

Hence, the total *disturbance* given decision variables $\mathbf{x}$ is defined by (also shown in Fig. 3)

$$\mathcal{D}(\mathbf{x}) \triangleq \sum_{k=1}^K \int_{t=x_k}^{x_{k+1}} [T_{x_k}(t) - T_{\mathrm{des}}^{\max}]^+ \mathrm{d}t \quad (12)$$

where $[x]^+ = \max(x, 0)$ and $T_{\mathrm{des}}^{\max}$ is the maximal desirable temperature threshold.

DEFINITION 1. *Formally, we define the decision problem for wireless sensor controlling air-conditioning (WSAC) as follows:*

WSAC problem:

$$\min_{\mathbf{x}} \mathrm{Cost}(\mathbf{x}) \triangleq \min_{\mathbf{x}} \eta \cdot K + (1 - \eta) \cdot \mathcal{D}(\mathbf{x}) \quad (13)$$

where $\eta \in [0, 1]$ is a weight assigned to balance the update frequency and the thermal comfort.

In the offline decision setting, $\mathbf{x}$ is decided given a-priori information of $\mathbf{a}$ and T_{od} without any restriction; whereas in the online decision setting, we require $\mathbf{x}$ to be decided such that x_k only considers the thermal sources before time x_k: $\{(w_i, t_i) \mid t_i \le x_k\}$.

Let $\mathbf{x}^*$ be the offline optimal solution to WSAC problem, while $\mathbf{x}_\mathcal{A}$ is the output solution given by an online algorithm $\mathcal{A}$. We define the competitive ratio as

$$\mathsf{CR}(\mathcal{A}) \triangleq \max_{\mathbf{a}, T_{\mathrm{od}}} \frac{\mathrm{Cost}(\mathbf{x}_\mathcal{A})}{\mathrm{Cost}(\mathbf{x}^*)} \quad (14)$$

In our problem, we seek to find an optimal online algorithm $\mathcal{A}$ to solve WSAC problem with the minimal $\mathsf{CR}(\mathcal{A})$.

4. RESULTS

In this section, we provide an effective offline algorithm to solve WSAC problem, and a 2-competitive online algorithm.

4.1 Offline Algorithm

While the rest of paper considers online algorithm, we first devise an effective offline algorithm to solve WSAC problem based on dynamic programming. The ramifications are that (1) the offline algorithm will enable us to compute the competitive ratio under diverse simulation settings; (2) the offline algorithm is useful in the setting with predictable $\mathbf{a}$. For example, based on the past history and statistics of $\mathbf{a}$, one can effectively solve WSAC problem by offline algorithm.

In the offline decision setting, we assume that all future temperature fluctuations are given in advance. We present our offline algorithm ($\mathcal{A}_{\text{OFL}}$) in Algorithm 1 that gives an optimal solution to WSAC problem.

Algorithm 1 Optimal Offline Algorithm $\mathcal{A}_{\text{OFL}}$, Input(a)

1: $\text{Cost}_{\min}[0] \leftarrow 0$
2: $\text{Cost}[1, 1] \leftarrow 1 \cdot \eta + (1 - \eta) \cdot \left[\int_{t=0}^{t_1} [T_{t_0}(t) - T_{\text{des}}^{\max}]^+ \mathrm{d}t \right]$
3: $\text{Cost}_{\min}[1] \leftarrow \text{Cost}[1, 1]$, $\text{idx}[1] \leftarrow 1$
4: **for** $i \in [2, m]$ **do**
5:     **for** $j \in [1, i]$ **do**
6:         $\text{Cost}[i, j] \leftarrow 1 \cdot \eta$
            $+ (1 - \eta) \cdot \left[\int_{t=t_{i-j}}^{t_i} [T_{t_{i-j}}(t) - T_{\text{des}}^{\max}]^+ \mathrm{d}t \right]$
            $+ \text{Cost}_{\min}[i - j]$
7:         **if** $\text{Cost}[i, j] < \text{Cost}_{\min}[i]$ **then**
8:             $\text{Cost}_{\min}[i] \leftarrow \text{Cost}[i, j]$
9:             $\text{idx}[i] \leftarrow j$
10:         **end if**
11:     **end for**
12: **end for**
13: $y_1 \leftarrow t_m$, $k' \leftarrow 1$, $r \leftarrow m$ ▷ *backtrack to find* $\mathbf{x}^*$
14: **while** $r > 1$ **do**
15:     $r \leftarrow r - \text{idx}[r]$, $k' \leftarrow k' + 1$
16:     $y_{k'} \leftarrow t_r$
17: **end while**
18: $K \leftarrow k'$
19: Output $(x_k = y_{K-k+1})_{k=1}^{K}$

The basic idea of $\mathcal{A}_{\text{OFL}}$ is based on dynamic programming, which relies on solving a sub-problem to decide when the previous ON command should be transmitted, assuming all the previous ON commands can be decided optimally.

Recall that t_i is the arrival time of the i-th external thermal source in sequence $\mathbf{a}$. Let $\text{Cost}[i, j]$ be the minimum cost when the last ON command is transmitted at time t_i and the second to last ON command is transmitted at time t_{i-j}, over all possible $\mathbf{x}$ with fixed $x_K = t_i$ and $x_{K-1} = t_{i-j}$. Also, let $\text{Cost}_{\min}[i]$ be the minimum cost when the last ON command is transmitted at time t_i. We note that $\text{Cost}[i, j]$ and $\text{Cost}_{\min}[i]$ can be computed recursively in Algorithm 1.

Once $\text{Cost}_{\min}[m]$ is found, the optimal decision $\mathbf{x}^*$ can be determined by backtracking. To enable backtracking, we maintain indices $\text{idx}[i]$ to record j when $\text{Cost}_{\min}[i] \leftarrow \text{Cost}[i, j]$.

THEOREM 1. *$\mathcal{A}_{\text{OFL}}$ in Algorithm 1 outputs an optimal solution to WSAC problem*

PROOF. The proof can be achieved in two steps.

    (i) WSAC problem exhibits the optimal sub-structure property;
    (ii) $\mathcal{A}_{\text{OFL}}$ explores all sub-problems and thus gives an optimal solution.

To prove **(i)**, we consider a subsequence of thermal sources

$$\Big((w_1, t_1), (w_2, t_2), ..., (w_i, t_i), \Big) \tag{15}$$

where the last ON command is transmitted at time $x_k = t_i$. Let us assume that we know that (perhaps told by an oracle) the second to last ON command is transmitted after the $(i - j)$-th arrival of thermal sources (i.e., $x_{k-1} = t_{i-j}$) is optimal, then we only need to optimize the subsequence $\Big((w_1, t_1), (w_2, t_2), ..., (w_{i-j}, t_{i-j}) \Big)$ in order to obtain the full optimal solution. Thus, the problem exhibits the optimal sub-structure property.

To prove **(ii)**, we need to examine the execution of $\mathcal{A}_{\text{OFL}}$. We note that there are two FOR-loops. For each iteration of the outer loop (i.e., upon arrival of each new thermal source), the inner loop is executed from start to i (i.e., all subsequences in $\Big((w_1, t_1), (w_2, t_2), ..., (w_i, t_i), \Big)$ are traversed). This process is repeated for each new thermal source until we reach the end of the sequence. By doing so, $\mathcal{A}_{\text{OFL}}$ is able to explore all subsequences and, therefore, all subproblems. □

4.2 Online Algorithm

In this section, we present a deterministic online algorithm that optimizes the trade-off between the frequency of ON commands and the thermal comfort. Our online algorithm achieves so by balancing the cost of transmitting the ON command immediately with the cost of delaying the ON command.

We assume that a wireless temperature sensor continuously tracks the change of temperature. Without the arrival of external thermal sources, the change in ambient temperature occurs smoothly as given by the differential equations Eqns. (1)-(3). However, when there is an arrival of external thermal source, the wireless sensor will be able to detect a sudden spike (because we assume impulsive thermal sources) in temperature, and hence, infer the arrival time of thermal source.

Recall that the j-th thermal source arrives at t_j. Let

$$\sigma_k \triangleq \{ i \in \{1, ..., m\} \mid x_{k-1} < t_i \leq x_k \} \tag{16}$$

Namely, σ_k is the set of thermal sources arrived between the $(k - 1)$-th and the k-th ON commands. Upon each new arrival of thermal source, our online algorithm sets a timer such that the total cost (i.e., sum of transmission and disturbance costs) for σ_k if an ON command is transmitted immediately is equal to the disturbance cost for σ_k if an ON command is transmitted after waiting for some time τ.

To be specific, suppose the last ON command is transmitted at time x_k. We decide the transmission time of the next ON command (x_{k+1}). The cost incurred if an ON command is transmitted immediately (i.e., at time t_j) is given by

$$\eta + (1 - \eta) \cdot \int_{t=x_k}^{t_j} [T_{x_k}(t) - T_{\text{des}}^{\max}]^+ \mathrm{d}t \tag{17}$$

On the other hand, the total cost if an ON command is transmitted after waiting for time τ (i.e., at $t_j + \tau$) is given by

$$(1 - \eta) \cdot \left[\int_{t=x_k}^{t_j} [T_{x_k}(t) - T_{\text{des}}^{\max}]^+ \mathrm{d}t + \int_{t=t_j}^{t_j + \tau} [T_{x_k}(t) - T_{\text{des}}^{\max}]^+ \mathrm{d}t \right] \tag{18}$$

Equating Eqn. (17) and Eqn. (18), we obtain τ as a solution to the following equation.

$$\frac{\eta}{(1-\eta)} = \int_{t=t_j}^{t_j+\tau} [T_{x_k}(t) - T_{\text{des}}^{\max}]^+ \mathrm{d}t \qquad (19)$$

However, if there is an arrival of a new thermal source (at t_{j+1}) before timer expires, then we have to reset the timer and obtain a new τ as a solution to the following equation.

$$\frac{\eta}{(1-\eta)} = \int_{t=t_j}^{t_{j+1}+\tau} [T_{x_k}(t) - T_{\text{des}}^{\max}]^+ \mathrm{d}t \qquad (20)$$

Thus, upon each new arrival, we increment the upper integration limit in Eqn. (20) and get a new τ. The complete algorithm is presented in Algorithm 2 ($\mathcal{A}_{\text{ONL}}$).

Algorithm 2 OnlineAlgorithm $\mathcal{A}_{\text{ONL}}$, Input(t_{now})

1: Global variables: τ, timer
2: Initialization: $\tau \leftarrow 0$, timer $\leftarrow 0$

3: **if** $t_{\text{now}} >$ timer **then**   ▷ *upon the beginning or after each OFF command*
4:    Find τ such that

$$\frac{\eta}{(1-\eta)} = \int_{t=t_{\text{now}}}^{t_{\text{now}}+\tau} [T_{x_k}(t) - T_{\text{des}}^{\max}]^+ \mathrm{d}t$$

5:    timer $\leftarrow t_{\text{now}} + \tau$
6: **end if**
7: **if** $t_{\text{now}} =$ timer **then**   ▷ *timer has expired*
8:    Transmit an ON command
9: **else if** $t_{\text{now}} <$ timer **then**   ▷ *timer has not expired yet*
10:    **if** j-th new thermal source is detected at t_{now} **then**
11:       Let t_j be the time after the last ON command
12:       Find τ such that   ▷ *decrease the timer due to new thermal source*

$$\frac{\eta}{(1-\eta)} = \int_{t=t_j}^{t_{\text{now}}+\tau} [T_{x_k}(t) - T_{\text{des}}^{\max}]^+ \mathrm{d}t$$

13:       timer $\leftarrow t_{\text{now}} + \tau$
14:    **else**
15:       Do not transmit   ▷ *wait for timer expiry*
16:    **end if**
17: **end if**
18: **if** Room Temperature $\leq T_{\text{des}}^{\min}$ **then**
19:    Transmit an OFF command
20: **end if**

Selecting the timer in such a manner will make $\mathcal{A}_{\text{ONL}}$ behave as follows. Upon the arrival of a each new temperature command, the algorithm sets a timer such that the expiry of timer will indicate that the comfort level threshold has reached and an ON command needs to be transmitted to the air-conditioning system. If an additional thermal source arrives before the timer expires, then a new smaller timer is set because the comfort level threshold will reach sooner due to the additional thermal source. In any case, whenever the timer expires, an ON command is transmitted and the current outstanding sequence is ended.

Example: We provide an example to illustrate the operations of offline optimal and online algorithms. In the example, the outdoor temperature is assumed to follow sinusoidal pattern. The input temperature sampled by the

wireless sensor as a result of thermal sources entering the room at at random intervals are given by Table. 1. For convenience, we restrict the example to 10 input samples (i.e., $m = 10$). The maximally desirable temperature $T_{\text{des}}^{\max}$ is 24 degree Celsius.

Table 1: Arrivals of impulsive thermal sources

t_1	t_2	t_3	t_4	t_5	t_6	t_7	t_8	t_9	t_{10}
4	12	15	21	26	30	34	35	40	43
w_1	w_2	w_3	w_4	w_5	w_6	w_7	w_8	w_9	w_{10}
24	23	24	25	23	29	24	27	25	28

For the arrivals shown in Table. 1, we execute $\mathcal{A}_{OFL}$. Table. 2 lists the entries Cost$[i,j]$, where the minimum costs (i.e.,Cost$_{\min}[i]$) are highlighted in yellow.

Table 2: Cost$_{\min}[i]$ **and** Cost$[i,j]$ **for offline optimal algorithm**

i,j	1	2	3	4	5	6	7	8	9	10
1	1.4									
2	4.1	4.5								
3	7.4	7.1	7.9							
4	11.5	12.0	12.6	14.4						
5	17.1	17.2	18.5	20.0	22.5					
6	23.9	23.8	24.6	26.6	28.7	31.8				
7	31.0	31.1	31.8	33.2	35.9	38.6	42.3			
8	37.4	36.6	36.7	37.4	38.8	41.5	44.2	47.9		
9	44.8	46.3	47.2	49.1	51.4	54.4	58.6	62.7	67.8	
10	54.0	53.2	54.6	55.5	57.4	59.7	62.7	66.9	71.0	76.1

After obtaining Cost$_{\min}[m]$, we use backtracking to determine the optimal decision variables $\mathbf{x}^*$ as

$$\mathbf{x}^* = (t_1, t_3, t_4, t_6, t_8, t_{10})$$

where each t_i is the time to transmit an ON command.

For the same arrivals, the online algorithm online algorithm $\mathcal{A}_{\text{ONL}}$ gives the following solution

$$\mathbf{x}_{\text{ONL}} = (t_6, t_{10})$$

The decision made by both algorithms are illustrated in Fig. 4.

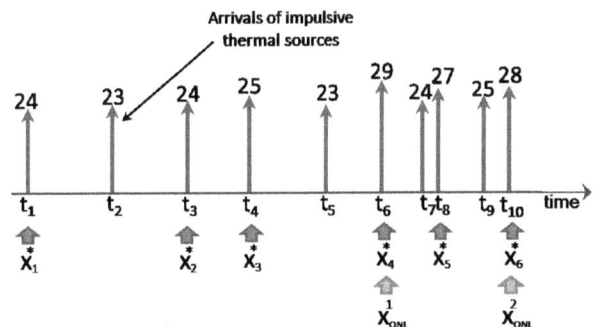

Figure 4: An illustration of the decisions by the offline optimal and online algorithms

Finally, the costs of both algorithms and the competitive ratio are computed as:

$$\text{Cost}(\mathbf{x}^*) = 53 \qquad \text{Cost}(\mathbf{x}_{\text{ONL}}) = 63 \qquad \text{CR}(\mathcal{A}_{\text{ONL}}) = 1.19$$

4.3 Competitive Analysis

Let $\mathbf{x}^*$ be the offline optimal solution, while $\mathbf{x}_{\text{ONL}}$ is the output solution given by online algorithm $\mathcal{A}_{\text{ONL}}$. We define the competitive ratio as

$$\mathsf{CR}(\mathcal{A}_{\text{ONL}}) \triangleq \max_{\mathbf{a}, T_{\text{od}}} \frac{\text{Cost}(\mathbf{x}_{\text{ONL}})}{\text{Cost}(\mathbf{x}^*)} \tag{21}$$

We show that the competitive ratio i.e., $\mathsf{CR}(\mathcal{A}_{\text{ONL}}) \leq 2$.

THEOREM 2. $\text{Cost}(\mathbf{x}_{\text{ONL}}) \leq 2 \cdot \text{Cost}(\mathbf{x}^*)$

PROOF. Assume that $\mathcal{A}_{\text{ONL}}$ sends a total of m ON commands for certain external thermal source arrivals, thus partitioning the sequence into m subsequences, where each subsequence ends with an ON command being transmitted to the air-conditioning system. The total cost by $\mathcal{A}_{\text{ONL}}$ for the input $\mathbf{a}$ is the sum of the cost for transmitting m ON commands and the extra latency cost for each subsequence, which can be calculated as follows. First, as shown previously, $\mathcal{A}_{\text{ONL}}$ sets τ, such that

$$\frac{\eta}{(1-\eta)} = \int_{t=t_j}^{t_j+\tau} [T_{x_k}(t) - T_{\text{des}}^{\text{max}}]^+ \mathrm{d}t \tag{22}$$

Note that $\int_{t=t_j}^{t_j+\tau} [T_{x_k}(t) - T_{\text{des}}^{\text{max}}]^+ \mathrm{d}t$ is a strictly increasing function in τ. Hence, the solution τ always exists and is uniquely defined. Also, it can be seen from Eqn. (22) that the timer is set in a manner that equalizes the total thermal disturbance of the subsequence to $\eta/(1-\eta)$. Thus, the disturbance cost for each subsequence is $\frac{\eta}{(1-\eta)} \cdot (1-\eta) = \eta$. The total cost incurred by $\mathcal{A}_{\text{ONL}}$, therefore, is

$$
\begin{aligned}
\text{Cost}(\mathbf{x}_{\text{ONL}}) &= \text{cost of } m \text{ ON commands} \\
&\quad + \text{disturbance cost for } m \text{ subsequences} \\
&= m\eta + m\eta = 2m\eta \tag{23}
\end{aligned}
$$

To calculate $\text{Cost}(\mathbf{x}^*)$, let m^* be the number of ON commands transmitted to the air-conditioning system in an optimal solution. When $m \leq m^*$, it immediately follows that $\text{Cost}(\mathbf{x}^*) \geq m^*\eta \geq m\eta$. Thus $\text{Cost}(\mathbf{x}_{\text{ONL}})/\text{Cost}(\mathbf{x}^*) \leq 2$.

We now consider the case when $m > m^*$. Since the m^* optimal ON commands are distributed over the m subsequences partitioned by $\mathcal{A}_{\text{ONL}}$. Thus, at least $m - m^*$ subsequences in online algorithm partition have no ON command at their end from the corresponding optimal solution. We claim that for each such a sequence, the disturbance cost is at least η in $\mathcal{A}_{\text{ONL}}$, because $\mathcal{A}_{\text{ONL}}$ decides ON command in such a way that the disturbance cost is equal to weighted cost of ON command (i.e., η). It is straightforward to see that disturbance cost of such a subsequence is at least η, because $\mathcal{A}_{\text{ONL}}$ resets the room temperature to $T_{\text{des}}^{\text{min}}$ at the beginning of each subsequence, whereas offline optimal algorithm does not. This induces a total disturbance cost of at least $(m - m^*)\eta$ to the optimal solution. The total cost of offline optimal algorithm is:

$$\text{Cost}(\mathbf{x}^*) \geq m^*\eta + (m - m^*)\eta = m\eta \tag{24}$$

Thus, $\text{Cost}(\mathbf{x}^*) \geq m\eta$, which is at least half of $\text{Cost}(\mathbf{x}_{\text{ONL}})$. □

5. SIMULATION STUDIES

In this section, we present the results of the simulations to experimentally evaluate the performance of our algorithms. We have used the classical ON/OFF algorithm as a baseline control model. In the classical ON/OFF technique (also known as bang-bang or hysteresis control), the wireless sensor sends an ON command to the air-conditioner whenever the room temperature reaches $T_{\text{des}}^{\text{max}}$ and OFF command when the temperature drops to $T_{\text{des}}^{\text{min}}$. First we compare the online solution against the baseline algorithm. We, then, provide a detailed cost comparison between the online and offline algorithms under different models of random thermal sources and different values of η.

In the first experiment, all three algorithms were run multiple times for different values of η to determine their relative performance against each other. Fig. 5 shows the results of the experiment. The input size during all experiments was 1000. As can be seen, the average cost ratio of the online algorithm against offline algorithm is always below 1.5 which is much better than the theoretical ratio of 2. We can also see that our algorithm always perform better than than classic ON/OFF control technique.

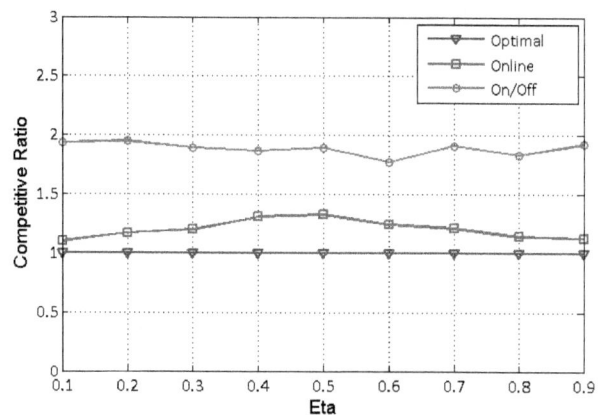

Figure 5: Simulation results showing the performance comparison between the online, the offline, and the classical ON/OFF algorithms.

We now compare the performance of the online algorithm with the optimal offline algorithm under different models of random thermal sources. For the next experiment, we draw the random thermal sources from Poisson distribution. Poisson distribution is with one parameter, where parameter, λ, is both the mean and the variance of the distribution. Thus, we can change the behaviour of random thermal source by changing λ. Poisson distribution is suitable in situations that involve counting the number of times a random event occurs in a given interval (e.g.,time, distance, area etc.). We ran the simulations for different models of random thermal sources generated by varying the parameter λ. Fig. 6 shows the simulations results for $\lambda \in \{10, 20, 30\}$. and $\eta \in \{0.1, 0.2, ..., 0.9\}$. The vertical axis gives the ratio of the cost of the online algorithm's solution to the cost of the optimal solution and the horizontal axis represents the relative cost weighting of sending a control signal to the air-conditioner. By looking at each line, it can be seen that the cost ratio gets closer to one when the value of η approaches either zero or one. *This means that the online algorithm performs better when the relative weighting of sending a control signal is either very low or very high. It can also be observed that the performance of the algorithm improves as we decrease λ (i.e.,reducing the random thermal disturbances).*

Similar results were observed when the experiment was re-

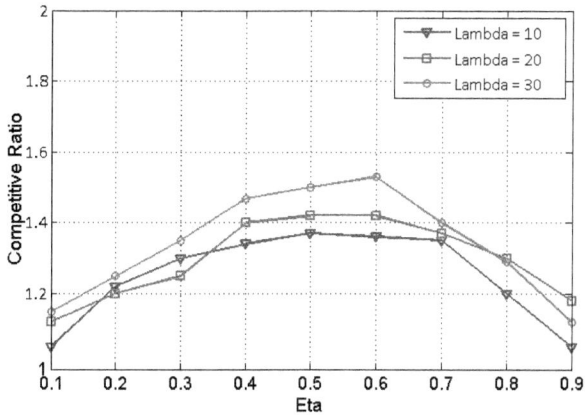

Figure 6: Competitive ratio of the online algorithm against the optimal algorithm when random thermal sources are drawn from Poisson distribution

peated, with random thermal sources drawn from Binomial distribution (see Fig. 7). Binomial distribution requires a parameter p, the probability of success. In our case, p is the probability of a random thermal source entering the room at a certain time. The results shown are for $p \in \{0.2, 0.5, 0.75\}$ and $\eta \in \{0.1, 0.2, ..., 0.9\}$. *Once again, as expected, the algorithm's performance improves as we reduce the value of p (i.e.,the probability of occurrence of thermal disturbances).*

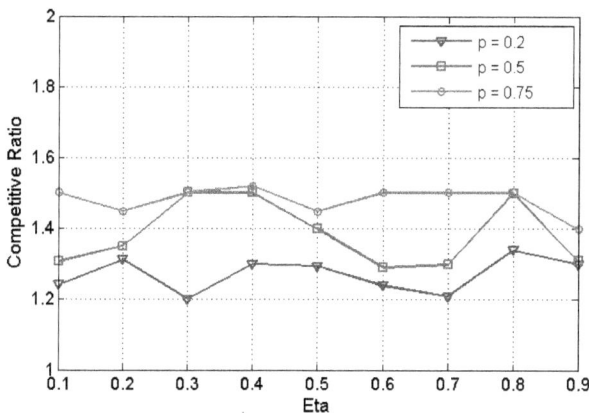

Figure 7: Competitive ratio of the online algorithm against the optimal algorithm when random thermal sources drawn from Binomial distribution.

6. RELATED WORK

Recently, many studies have explored the use of intelligent methods to control HVAC systems. These methods vary from simple manipulation of set-point temperatures to more sophisticated techniques such as fuzzy logic, neural networks, genetic algorithms etc. In this section, we first summarize a few papers that are relatively simple extensions to the classical HVAC control techniques, then we discuss several state-of-the-art intelligent control techniques employed in HVAC systems. We also present a brief survey of the recent works on HVAC control through WSN. We conclude this section by discussing a paper that is somewhat related to our work in that it also aims to optimize the wireless sensors cost while maintaining the control performance within an acceptable range.

Extensions of Classical Techniques: In [12], the authors proposed a relatively simple way of controlling the HVAC systems in which the set-point temperature of the regulator and thermostat is manipulated. They developed an adaptive module of classical regulator to control the peak consumption and provide thermal comfort. Their regulator is based on varying temperature set-point of the air conditioning in response to maximum permissible power. Similar approach has been used in [13], where an optimal control scheme for compressor ON/OFF cycling operations has been proposed.

Intelligent HVAC Control: The design of an intelligent comfort control system by using human learning strategy for an HVAC system was proposed in [14]. Based on a standard thermal comfort model, a human learning strategy was designed to tune the user's comfort zone by learning the specific user's comfort preference. The integration of comfort zone with the human learning strategy was applied for thermal comfort control. The authors in [22] proposed a multi-objective particle swarm optimization algorithm, embedded in a controller. The algorithm was used to determine the amount of energy dispatched to HVAC equipment based on utilizing swarm intelligence technique.

A method based on fuzzy logic controller dedicated to the control of HVAC systems has been proposed in [2]. They obtained the initial knowledge-base required by fuzzy logic controller from human experts and control engineering knowledge which they subsequently tuned by a genetic algorithm. In [17], a hierarchical structure for the control of an HVAC system using the Model Predictive Control (MPC) algorithms and fuzzy control algorithms has been proposed. The main task of the proposed hierarchical control system is to provide thermal comfort and minimize energy consumption. Their technique showed a good comparison between two conflicted objectives: thermal comfort and energy consumption. The authors of [3] used model-predictive control technique to learn and compensate for the amount of heat due to occupants and equipment. They used statistical methods together with a mathematical model of thermal dynamics of the room to estimate heating loads due to inhabitants and equipment and control the AC accordingly. *However, majority of the existing intelligent HVAC control techniques rely on stochastic knowledge about the input which makes them less robust in uncertain environments.*

WSN-based AC Control: In [11], an air-conditioning control system for a dynamical situation in wide public spaces has been proposed. They tracked people movement through multiple large scale scanners. Also, networked temperature sensors were deployed in the target space for temperature monitoring. The obtained temperature distribution was integrated with the results of people tracking in real-time to direct HVAC to locations with high population density and insufficient temperature. In [20], the authors presented the conceptual design of an adaptive multi zone HVAC control system that utilized WSN for predicting the occupancy pattern of people in a building. Their control strategy involved turning off the AC in unoccupied zones and manipulating the set-point temperature. A multi-sensor non-learning control strategy has been proposed in [18]. This paper evaluates the energy and comfort performance of three multi-sensor control strategies that use wireless temperature and humidity sensors and that can be applied to existing ON/OFF central HVAC system. The multi-sensor control strategies adjust the temperature set point of a thermostat to (i) control the average of all room temperatures using a temperature threshold logic, (ii) minimize aggregate discomfort of all rooms, or (iii) maximize the number of rooms within a comfort zone. The strategies were evaluated in a real occupied house and were found to outperform single-sensor control strategies.

In [15], the authors proposed somewhat similar approach to our work. They introduced a co-design methodology that optimizes the sensor network cost while maintaining the control performance within an acceptable range. They applied the developed co-design methodology to a distributed control for building lighting systems. They empirically compared the developed system for building lighting control with a baseline control method and reported significant reduction in energy use and saving in the network cost while maintaining the user comfort.

7. CONCLUSION AND FUTURE WORK

While intelligent systems for smart buildings have been a popular research topic, online optimization approach has been explored to a lesser extent. This paper investigates a new breed of research problems by applying online algorithms to wireless sensor based smart building control. We provide the first study of optimizing the trade-off between the battery lifetime of wireless sensor and the effectiveness of HVAC remote control in the presence of uncertain fluctuations in room temperature. We present both an effective offline optimal algorithm and a 2-competitive online algorithm.

There are plenty of research opportunities to extend the results of this work to a more general context. So far, we devised a deterministic online algorithm. It is well-known that randomized online algorithms can exhibit both improved theoretical competitive ratio and practical performance. For the on-going work, we will study randomized online algorithms for wireless sensor controlling air-conditioning systems, and evaluate their performance.

In this paper, we only consider a single sensor control setting. In a general setting, there may be multiple sensors and multiple air conditioning systems. The interaction among multi-input and multi-control systems in a networked setting will be a challenging yet important research problem.

Finally, we are implementing our control algorithms in real-world air-conditioning systems. More empirical studies will be presented in the extended version of this work.

8. REFERENCES

[1] Susanne Albers and Stefano Leonardi. On-line algorithms. *ACM Comput. Surv.*, 31(3es), September 1999.

[2] Rafael Alcalá, Jose M. Benítez, Jorge Casillas, Oscar Cordón, and Raúl Pérez. Fuzzy control of hvac systems optimized by genetic algorithms. *Applied Intelligence*, 18(2):155–177, March 2003.

[3] A. Aswani, N. Master, J. Taneja, D. Culler, and C. Tomlin. Reducing transient and steady state electricity consumption in hvac using learning-based model-predictive control. *Proceedings of the IEEE*, 100(1):240 –253, jan. 2012.

[4] Chi-Kin Chau, Fei Qin, S. Sayed, M.H. Wahab, and Yang Yang. Harnessing battery recovery effect in

wireless sensor networks: Experiments and analysis. *Selected Areas in Communications, IEEE Journal on*, 28(7):1222 –1232, september 2010.

[5] Bartkevics D. Power and control aspects in integrated building management systems with event-based control, 2011.

[6] Daniel Dooly, Sally A. Goldman, and Stephen D. Scott. Tcp dynamic acknowledgment delay: Theory and practice. In *In Proceedings of the Thirtieth Annual ACM Symposium on Theory of Computing*, pages 389–398. ACM Press, 1998.

[7] Varick L. Erickson, Yiqing Lin, Ankur Kamthe, Rohini Brahme, Amit Surana, Alberto E. Cerpa, Michael D. Sohn, and Satish Narayanan. Energy efficient building environment control strategies using real-time occupancy measurements. In *Proceedings of the First ACM Workshop on Embedded Sensing Systems for Energy-Efficiency in Buildings*, BuildSys '09, pages 19–24, New York, NY, USA, 2009. ACM.

[8] A. Fiat, Internationales Begegnungs und Forschungszentrum für Informatik Dagstuhl, and Dagstuhl Seminar on On-line Algorithms. 1996. *On-line Algorithms: 24.06. - 28.06.96 (9626); [Dagstuhl Seminar on On-line Algorithms]*. Dagstuhl-Seminar-Report. Geschäftsstelle Schloss Dagstuhl, Univ. des Saarlandes, 1996.

[9] P. Huovila, United Nations Environment Programme. Sustainable Consumption, and Production Branch. *Buildings and Climate Change: Summary for Decision-makers*. United Nations Environment Programme, Sustainable Consumption and Production Branch, 2009.

[10] Anna R. Karlin, Claire Kenyon, and Dana Randall. Dynamic tcp acknowledgement and other stories about e/(e-1. In *Algorithmica*, pages 502–509. Press, 2001.

[11] K. Katabira, Huijing Zhao, Y. Nakagawa, and R. Shibasaki. Real-time monitoring of people flows and indoor temperature distribution for advanced air-conditioning control. In *Intelligent Transportation Systems, 2008. ITSC 2008. 11th International IEEE Conference on*, pages 664 –668, oct. 2008.

[12] K. Le, T. Tran-Quoc, J.C. Sabonnadiere, C. Kieny, and N. Hadjsaid. Peak load reduction by using air-conditioning regulators. In *Electrotechnical Conference, 2008. MELECON 2008. The 14th IEEE Mediterranean*, pages 713 –718, may 2008.

[13] Bin Li and A.G. Alleyne. Optimal on-off control of an air conditioning and refrigeration system. In *American Control Conference (ACC), 2010*, pages 5892 –5897, 30 2010-july 2 2010.

[14] J. Liang and R. Du. Design of intelligent comfort control system with human learning and minimum power control strategies. *Energy Conversion and Management*, 49(4):517 – 528, 2008.

[15] Alie El-Din Mady, Gregory Provan, and Ning Wei. Designing cost-efficient wireless sensor/actuator networks for building control systems. In *Proceedings of the Fourth ACM Workshop on Embedded Sensing Systems for Energy-Efficiency in Buildings*, BuildSys '12, pages 138–144, New York, NY, USA, 2012. ACM.

[16] Mendes N, Oliveira G H C, AraÃžjo H X, and

Coelho L S. A matlab-based simulation tool for building thermal performance analysis, 2003.

[17] M. Nowak and A. Urbaniak. Utilization of intelligent control algorithms for thermal comfort optimization and energy saving. In *Carpathian Control Conference (ICCC), 2011 12th International*, pages 270 –274, may 2011.

[18] Nathan Ota, Ed Arens, and Paul Wright. Energy efficient residential thermal control with wireless sensor networks: A case study for air conditioning in california. *ASME Conference Proceedings*, 2008(48692):43–52, 2008.

[19] Daniel D. Sleator and Robert E. Tarjan. Amortized efficiency of list update and paging rules. *Commun. ACM*, 28(2):202–208, February 1985.

[20] Y. Tachwali, H. Refai, and J.E. Fagan. Minimizing hvac energy consumption using a wireless sensor network. In *Industrial Electronics Society, 2007. IECON 2007. 33rd Annual Conference of the IEEE*, pages 439 –444, nov. 2007.

[21] Karla Vega. Thermal modeling for buildings performance analysis, 2009.

[22] Rui Yang and Lingfeng Wang. Optimal control strategy for hvac system in building energy management. In *Transmission and Distribution Conference and Exposition (T D), 2012 IEEE PES*, pages 1 –8, may 2012.

APPENDIX

A. PROOF OF LEMMA 1

In this subsection, we prove Lemma 1 that we used in Section 3.2. The differential equations are again listed here:

$$\frac{d\tilde{T}(t)}{dt} = \frac{1}{c \cdot M_{air}} \cdot \left(\frac{dQ_{in}(t)}{dt} - \frac{dQ_{ac}(t)}{dt} \right) \quad (25)$$

$$\frac{dQ_{in}(t)}{dt} = \frac{T_{od} - \tilde{T}(t)}{R_{eq}} \quad (26)$$

$$\frac{dQ_{ac}(t)}{dt} = \frac{c \cdot M_{ac} \cdot (\tilde{T}(t) - T_{ac})}{E_{ac}} \quad (27)$$

where $\frac{dQ_{in}(t)}{dt}$ is the heat flowing into the room from outside environment and and $\frac{dQ_{ac}(t)}{dt}$ is the chilled air flowing from air conditioning system into the room. By substituting, Eqn. (26) and Eqn. (27) into Eqn. (25), we obtain

$$\frac{d\tilde{T}(t)}{dt} = \frac{1}{c \cdot M_{air}} \cdot \left(\frac{T_{od} - \tilde{T}(t)}{R_{eq}} - \frac{c \cdot M_{ac} \cdot (\tilde{T}(t) - T_{ac})}{E_{ac}} \right)$$

$$= \frac{E_{ac} \cdot T_{od} - E_{ac} \cdot \tilde{T}(t) - R_{eq} \cdot c \cdot M_{ac} \cdot (\tilde{T}(t) - T_{ac})}{c \cdot M_{air} \cdot R_{eq} \cdot E_{ac}}$$

$$= \frac{E_{ac} \cdot T_{od} + R_{eq} \cdot c \cdot M_{ac} \cdot T_{ac}}{c \cdot M_{air} \cdot R_{eq} \cdot E_{ac}}$$

$$- \frac{E_{ac} + R_{eq} \cdot c \cdot M_{ac}}{c \cdot M_{air} \cdot R_{eq} \cdot E_{ac}} \cdot \tilde{T}(t) \quad (28)$$

Let

$$C_1 = \frac{E_{ac} \cdot T_{od} + R_{eq} \cdot c \cdot M_{ac} \cdot T_{ac}}{c \cdot M_{air} \cdot R_{eq} \cdot E_{ac}}$$

$$C_2 = \frac{E_{ac} + R_{eq} \cdot c \cdot M_{ac}}{c \cdot M_{air} \cdot R_{eq} \cdot E_{ac}}$$

Then, Eqn. (28) can be written as:

$$\frac{d\tilde{T}(t)}{dt} = C_1 - C_2 \cdot \tilde{T}(t)$$

By rearrangement,

$$\frac{\frac{d\tilde{T}(t)}{dt}}{\frac{C_1}{C_2} - \tilde{T}(t)} = C_2 \cdot dt$$

Integrating both sides with respect to t,

$$-\log\left|\frac{C_1}{C_2} - \tilde{T}(t)\right| = C_2 \cdot t + C$$

By substituting $t = 0$ (i.e.,initial condition), we obtain

$$-\log\left|\frac{C_1}{C_2} - \tilde{T}(t)\right| = C_2 \cdot t - \log\left|\frac{C_1}{C_2} - \tilde{T}(0)\right|$$

$$e^{C_2 \cdot t} = \frac{\frac{C_1}{C_2} - \tilde{T}(0)}{\frac{C_1}{C_2} - \tilde{T}(t)}$$

$$\tilde{T}(t) = \frac{C_1}{C_2} - \left(\frac{C_1}{C_2} - \tilde{T}(0) \right) \cdot e^{-C_2 \cdot t} \quad (29)$$

This concluded the proof as Eqn. (29) is the same as Eqn. (4).

B. CALCULATION OF ROOM THERMAL RESISTANCE

The building thermal model used in this paper (both during the theoretical part and simulations) requires the total equivalent (also called lumped) thermal resistance, R_{eq}, of the entire room. Therefore, we include a simple example on how to calculate R_{eq} using the rooms dimensions, number and sizes of windows and the type of insulation used in walls. Table 3 shows the room geometry and insulation details used for calculation of R_{eq}.

Table 3: Room geometry and insulation details

Description	Value
Room length (Len_{room})	10 m
Room width (Wid_{room})	5 m
Room height (Ht_{room})	4 m
Roof pitch (Pit_{roof})	40
Number of windows ($Num_{windows}$)	4
Height of windows ($Ht_{windows}$)	1 m
Width of windows ($Wid_{windows}$)	1 m
Wall insulation having glass wool (L_{walls})	0.2 m
Window insulation ($L_{windows}$)	0.01 m
Thermal conductivity of walls (K_{walls})	0.038
Thermal conductivity of windows ($K_{windows}$)	0.78

From the values in Table 3, we can calculate the equivalent resistances of the walls as follows.

$$R_{Wall} = \frac{L_{Wall}}{k_{Wall} \times Wall_{area}} \quad (30)$$

Where,

$$Wall_{area} = (2 \cdot Len_{room} \cdot Ht_{room}) + (2 \cdot Wid_{room} \cdot Ht_{room})$$
$$+ [2 \cdot (1/\cos(Pit_{roof}/2)] \cdot (Wid_{room} \cdot Len_{room})$$
$$+ [(\tan(Pit_{roof}) \cdot Wid_{room})] - Window_{area}$$

Similarly, the equivalent resistance of windows is calculated as:

$$R_{Window} = \frac{L_{Window}}{k_{Window} \times Window_{area}} \quad (31)$$

Where,

$$Window_{area} = Num_{windows} \cdot Ht_{windows} \cdot Wid_{windows}$$

From Eqns. 31 and 30, R_{eq} is calculated as.

$$R_{eq} = \frac{R_{Wall} \times R_{Window}}{R_{Wall} + R_{Window}} \quad (32)$$

C. SENSOR POWER CONSUMPTION

In order to maximize the battery life-time of wireless sensors, it is important to understand the energy consumed by each component of a wireless sensor node. Therefore, we provide power consumption data for each unit (i.e.,transceiver, micro-controller, and sensor) in common wireless sensor nodes (see Tables 4-6). From the tables, it is evident that radio communication is most energy-intensive among the three operations (i.e.,sensing, processing, and communication). Specifically, the transceiver power consumption can get as high as 28 times compared the power consumption of micro-controller (see Table 4 and 5). The ratio becomes even higher when compared to the power consumption of the sensor modules. For these reasons, we aim to maximize the battery life-time of the wireless sensor by optimizing the update frequency of the control commands sent to the air-conditioner.

Table 4: Power Consumptions of Transceivers and in Common Wireless Sensors. [4]

Transceiver Model	Transmission (mA)	Reception (mA)	Sleep (mA)
TR1000	12	3.8	0.0007
CC1000	10.4	7.4	0.03
CC2500	21.6	12.8	0.0004
nRF2401A	10.5	18	0.0004
CC2420	17.4	18.8	0.4
RF230	14.5	15.5	0.00002
MC13192	30	37	0.5
JN5121	45	50	0.0004

Table 5: Power Consumptions of MCUs in Common Wireless Sensors. [4]

MCU Model	Active (mA)	Sleep (mA)
AT163	5	0.025
AT128	5.5	0.015
80c51	4.3	0.19
MSP430	1.8	0.00512
HCS08	4.3	0.0005

Table 6: Power Consumptions of Sensor Module in Common Wireless Sensors. [4]

Sensor Module	Function	Current (mA)
SHT15	Humidity, Temperature	0.55
TSL2561	Light	0.24
ADXL202	Accelerometer	0.6

Table 7: Key Notations in This Paper

Notation	Definition
$T(t)$	Ambient room temperature at time t (unit: degree Celsius)
T_0	Initial ambient room temperature at time $t = 0$
T_{od}	Outdoor temperature
T_{ac}	Temperature of the cold air from air conditioner
M_{air}	Total air mass inside the room
M_{ac}	Air mass flow through air conditioner (Kg/hr)
E_{ac}	Air conditioner efficiency
c	Heat capacity of the air at constant pressure
R_{eq}	Equivalent thermal resistance of the entire room
$W(t)$	Sequence of impulsive thermal sources
w_i	Level of thermal intensity entering the room at time t
a	Sequence of arrivals of impulsive thermal sources
T_{des}^{max}	Maximal desirable temperature
T_{des}^{min}	Minimal desirable temperature
$T_\tau(t)$	Temperature of thermal sources
X	Set of decision variables
x_k	Time that the k_{th} ON command is issued by the wireless sensor
$\mathcal{D}(\mathbf{x})$	Thermal disturbance given decision variable x
$[x]^+$	max(x, 0)
$T_{t_k}(t)$	Temperature of the room after k_{th} ON command
η	Weight assigned to balance the update frequency and the thermal comfort
$\mathcal{A}_{OFL}$	Offline Algorithm
$Cost[i,j]$	Minimum cost when the last and second to last ON command are transmitted at time t_i and t_{i-j} respectively
$Cost_{min}[i]$	Minimum cost when the last ON command is transmitted at time t_i
$idx[i]$	Array to record j when $Cost_{min}[i] \leftarrow Cost[i,j]$
σ_k	Set of thermal sources arrived between the $(k-1)$-th and the k-th ON commands
$\mathcal{A}_{ONL}$	Online Algorithm
$\mathcal{D}_{ij}(\tau)$	Total thermal disturbance accumulated from the start of the subsequence to the latest arrival
t_j	The time when the timer was first set after transmission of the last ON command
λ	Mean and variance of the Poission distribution
p	Success probability. A parameter required by Binomial distribution

SPOT: A Smart Personalized Office Thermal Control System

Peter Xiang Gao
School of Computer Science
University of Waterloo
Waterloo, Ontario, Canada

S. Keshav
School of Computer Science
University of Waterloo
Waterloo, Ontario, Canada

ABSTRACT

Heating, Ventilation, and Air Conditioning (HVAC) accounts for about half of the energy consumption in buildings. HVAC energy consumption can be reduced by changing the indoor air temperature setpoint, but changing the setpoint too aggressively can overly reduce user comfort. We have therefore designed and implemented SPOT: a Smart Personalized Office Thermal control system that balances energy conservation with personal thermal comfort in an office environment. SPOT relies on a new model for personal thermal comfort that we call the Predicted Personal Vote (PPV) model. This model quantitatively predicts human comfort based on a set of underlying measurable environmental and personal parameters. SPOT uses a set of sensors, including a Microsoft Kinect, to measure the parameters underlying the PPV model, then controls heating and cooling elements to dynamically adjust indoor temperature to maintain comfort. Based on a deployment of SPOT in a real office environment, we find that SPOT can accurately maintain personal comfort despite environmental fluctuations and allows a worker to balance personal comfort with energy use.

Categories and Subject Descriptors

H.4.0 [**Information Systems Applications**]: General

Keywords

HVAC, Personalization, Human Thermal Comfort

1. INTRODUCTION

About 30% to 50% of the residential and commercial energy consumption in most developed countries is used by Heating, Ventilation, and Air Conditioning (HVAC) systems [3, 5, 22, 25]. Increasing the efficiency of HVAC systems, therefore, can greatly reduce the overall energy footprint of a commercial building.

The focus of our work is thermal comfort in office environments. We assume that workers in offices have work areas that are relatively thermally isolated from each other, such as separate offices or cubicles with walls. Thus, heating and cooling within a personal work space would be for the benefit of a single worker.

We suggest that the overall building temperature level be set to a value lower than normal in winter and to a value higher than normal in summer. Then, a personal thermal controller in each work space could provide an offset to this base temperature. For instance, most commercial buildings today are heated to 23°C in winter. Instead, we suggest that the buildings be heated only to, say, 20°C, and that each work space have a small computer-controlled radiant heater that can heat the work space to a personalized higher level. In summer, symmetrically, a small fan can provide additional cooling below a building setpoint of, say, 26°C [28]. The role of the personal thermal control system, therefore, is to automatically control the per-workspace radiant heater or fan to maintain the comfort level of individual workers when they are actually present. In contrast, existing time-based or motion-based sensor control often suffers from irksome false positives and false negatives. Manual control, of course, would have no such errors, but this requires human effort, and office workers have no incentive to participate.

It has been found that user comfort is not just a function of room temperature. Two persons who are differently dressed would experience different levels of comfort for the same room temperature. Ideally, an HVAC control system should control room temperature not to achieve a temperature setpoint, but a particular human comfort level. This is the key idea that motivates the design of SPOT: a Smart Personalized Office Thermal control system.

SPOT uses an ensemble of sensors to measure the six parameters that have been found to contribute to human comfort: air temperature, radiant temperature, humidity, air speed, clothing level, and activity level. This lets it compute human comfort according to the ISO 7730 standard called the Predicted Mean Vote (PMV) model [4]. We have extended this model to allow per-user personalization; we call our personalized model the Predicted Personal Vote (PPV) model. SPOT uses the PPV model to maintain a desired comfort level despite environmental fluctuations. We have deployed SPOT and evaluated its performance in a realistic office environment. Our work makes it possible to trade off a decrease in human comfort for a reduction in energy usage.

The major contributions of our paper are:

- We extend the ISO 7730 standard [4] to define the PPV model for user comfort and use it to design SPOT, an HVAC control system that maintains user comfort, rather than merely air temperature

- We have implemented SPOT and deployed it in a realistic environment

- We find that SPOT can accurately maintain personal comfort despite environmental fluctuations and allows a user to balance personal comfort with energy use.

2. BACKGROUND

HVAC control systems traditionally put user comfort first, expending energy freely to achieve a given setpoint. 'Dumb' thermostats use the same setpoint all day, and smarter, programmable thermostats allow users to vary setpoints by time of day and day of week. Some thermostats allow remote control. For example, in Ontario, the PeakSaver [1] thermostat responds to an emergency broadcast radio signal and increases the cooling set point by up to two degrees, thereby reducing home electricity usage by up to 37%. Other 'smart' thermostats, such as the Nest [6], learn user occupancy patterns to intelligently control HVAC usage by means of proprietary algorithms. Nevertheless, none of these thermostats are aware of user comfort: they focus, instead, only on controlling room temperature.

The basis of our work is a quantitative model for human comfort called the PMV model that is defined in the ISO 7730 Standard [4]. The PMV model computes a numerical comfort level, called a *vote*, that describes the degree of comfort of a typical person in a moderate thermal environment. The PMV model predicts human comfort as a function of four environmental variables (air temperature, radiant temperature, air speed, and humidity) and two personal variables (clothing and physical activity). Given these variables, it predicts the mean value of a group of people's votes in a 7-point ASHRAE [2] thermal sensation scale.

Vote	Comfort Level
+3	Hot
+2	Warm
+1	Slightly Warm
0	Neutral
-1	Slightly Cool
-2	Cool
-3	Cold

Table 1: 7-point ASHRAE scale in PMV model

The PMV model was first proposed by Fanger [15] in 1970 and it is widely used for evaluating thermal comfort [7] [20]. Although the model is based on a theoretically well-grounded physical thermal balance model, it has been found to be problematic to use in practice [18]. Many variations of the PMV model have been developed to fix these problems. For example, De Dear et. al. [10] developed a model to capture the sociological and geographical factors that may affect human's thermal preference, such as people living in warmer areas preferring warmer indoor temperature than people living in cooler areas. Similarly, Nicol et. at. [24] have shown that people can use physiological and psychological adaptations to be comfortable in a wider range of temperatures than supposed by the PMV model; their model reflects this observation.

Although these newer models improve the accuracy of the PMV model, they all predict the average thermal comfort of a large group of people. However, in a micro-climate such as an office work area, comfort is usually relevant only for one person or a small number of people. This motivates the design of a *personalized* thermal comfort model. In our work, we extend the PMV model to the Predicted Personal Vote (PPV) model to capture individual thermal preference. We use PPV model to automatically adjust an HVAC control system's temperature setpoint so that a worker always feel comfortable.

2.1 Predicted Mean Vote

We now describe the PMV model [15] in greater detail. It assigns a numerical comfort value $pmv(\mathbf{x})$ based on a vector $\mathbf{x}$ with six elements

$$\mathbf{x} = \{t_a, \bar{t}_r, v_{ar}, p_a, M, I_{cl}\}^\top$$

- t_a is the air temperature
- $\bar{t}_r$ is the mean background radiant temperature
- v_{ar} is the air velocity
- p_a is the humidity level
- M is the metabolic rate of a person
- I_{cl} is the clothing insulation factor of a person

We can evaluate PMV using the function:

$$pmv = pmv(\mathbf{x}) \quad (1)$$

The details of the function can be evaluated in practise are in the appendix.

3. DESIGN

We now describe our design in more detail. Recall that SPOT's goal is to maintain a particular comfort level (PPV value) based on sensor measurements and its control over the operation of a small personal radiant heater or fan. We first discuss the PPV model and clothing level estimation, then SPOT's control strategy.

3.1 Predicted Personal Vote Model

The PMV model reflects the thermal comfort of a large group of people. However, individual workers may have their own thermal preference. We have, therefore, modified the PMV model to create a model we call the Predicted Personal Vote (PPV) model.

For each person, the Predicted Personal Vote function has two parts, the PMV part and the personal part:

$$ppv(\mathbf{x}) = pmv(\mathbf{x}) + personal(\mathbf{x}) \quad (2)$$

where $pmv(\mathbf{x})$ is the output of the PMV model and $personal(\mathbf{x})$ models how the current user is different from an average person. We model $personal(\mathbf{x})$ as a linear function:

$$personal(\mathbf{x}) = \mathbf{a}^\top \mathbf{x} + b \quad (3)$$

where $\mathbf{a}$ is a vector of size 6 that models the users sensitivity to each variable.

$$\mathbf{a} = \{a_{temp}, a_{radiant}, a_{velocity}, a_{humidity}, a_{metabolic}, a_{clothing}\}^\top \quad (4)$$

. For example, a person who is more sensitive to humidity than average will have a relatively large $a_{humidity}$ value. Variable b denotes the thermal preference of the user. A person prefers warmer temperatures will have negative b value and vice versa.

Using the PPV model requires a training phase. In the training phase, SPOT measures the environmental variables $\mathbf{x}$ and also records the worker's personal vote apv. Suppose we have a training set $\{(\mathbf{x}_k, apv_k)\}_{k=1}^K$ of size K, where apv_k is the k-th actual personal vote, and $\mathbf{x}_k$ is the vector of environmental and personal variables when the user gives the k-th vote. This allows us to estimate parameters $\mathbf{a}$ and b using straightforward linear regression. In the absence of a training set, SPOT simply reverts to the PMV model. Similarly, when there are not enough data points to do a linear regression for Equation 3, we train a simpler linear function $g(\cdot)$ to estimate PPV:

$$ppv(\mathbf{x}) = g(pmv(\mathbf{x})) \quad (5)$$

The function $g(\cdot)$ is trained by least square regression.

3.2 Clothing Level Estimation

Five out of the six underlying parameters of the PPV model can be measured in a relatively straightforward manner using appropriate sensors (this is discussed in more detail in §4.2). However, measuring the 'clothing level' parameter is non-trivial (see Table 12), and the focus of this subsection.

The key idea behind our approach to clothing level estimation is the fact that most humans have a relatively constant skin temperature of about 34°C. The greater the level of clothing worn, the greater the degree of insulation, and the lower the temperature of the outermost layer of clothes. Thus, the clothing level can be estimated by measuring the temperature of the clothing using an infrared sensor as discussed in §4.2.2.

Specifically, we build a linear regression model to estimate the clothing level I_{cl} as:

$$I_{cl} = f(t_{ir}) \qquad (6)$$

where $f(\cdot)$ is a linear function and t_{ir} is the infrared intensity of the clothing. We fit the function $f(\cdot)$ using least square linear regression. The model is trained using a data set of I_{cl} estimates from Table 12 and t_{ir} measured by the infrared sensor.

Note that this assumes that the worker's body temperature is in the normal range. In case the worker has a fever, estimation accuracy can be affected. We can solve this problem by using the infrared intensity of the worker's face as a reference; however, we have not currently implemented this refinement.

3.3 Control Strategy

SPOT maintains the PPV of a worker by controlling the air temperature, because this is the factor underlying the PPV that is the easiest to control. It uses a simple reactive control strategy rather than a complex model-based predictive approach, such as the one in [7]. This is possible because personal heating and cooling systems such as radiant heaters and fans affect human comfort almost immediately[1], unlike centralized HVAC systems, such as air conditioners and forced-air heaters, which can take tens of minutes to take effect.

Specifically, when the Kinect sensor indicates the presence of a worker in the work space during the prior five minutes, the system chooses an operative temperature setpoint such that $ppv(\mathbf{x}) = dc$, where dc is the desired comfort level (nominally 0). The five-minute window allows thermal comfort to be maintained despite brief absences. For example, in winter, when a worker is detected and $ppv(\mathbf{x}) < dc$, the heater is turned on to increase the room temperature. Otherwise, the heater is turned off.

This reactive control strategy takes into account both real-time occupancy and personal thermal comfort. An occupancy-aware reactive controller will always use less energy than an occupancy-unaware controller. On the other hand, a worker may choose a personal comfort value that is much higher than normal, causing the corresponding heater or fan to expend more energy than normal. However, we believe that this would be more than made up by the reduction of energy use in common areas that are heated or cooled to lower or higher than a nominal setpoint, respectively.

4. IMPLEMENTATION

We now describe the implementation of our system in greater detail. SPOT has three principal components: controller, sensors and actuators.

---

[1]We validate this observation in §5.3.

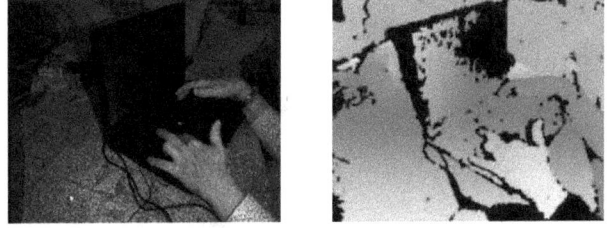

Figure 2: The image at left is from the camera and the image at right is the depth image. Lighter portions of the image are closer to the camera; the green portion of the image is too far to be measured, and there is no depth information available from black portions of the image.

4.1 Controller

The SPOT controller is a PC with an Intel i5-3450 processor and 8GB of RAM, running Windows 7 Enterprise edition. The entire project code is written in C#. All sensors and actuators are connected to this PC, and all control logic is implemented on this machine.

4.2 Sensors

SPOT uses multiple sensors to measure the environmental variables (air temperature, background radiant temperature, humidity and wind speed) and personal variables (users' clothing level and activity level) that underlie the PPV model. We describe these next.

4.2.1 Microsoft Kinect

A Microsoft Kinect sensor provides 3D information in real-time about the location of humans in a scene. The Kinect sensor was originally designed for the Microsoft Xbox as a natural user interface. By using a Kinect sensor, video game players can interact with Xboxes without actually touching the game controllers.

The Kinect has an RGB camera and an infrared camera. The RGB camera is similar to a normal webcam that captures image from the visible light spectrum. The infrared camera works together with an infrared projector, which emits infrared laser signal with a predefined pattern. The infrared camera collects the reflected laser beams and calculate the distance to the laser point by the time difference between sending and receiving the signal. The Kinect can generate 640×480 resolution depth images with a sensitivity of 1mm. Figure 2 shows the raw infrared image and the depth image generated by the Kinect sensor.

By using both the color images generated by the RGB camera and the depth images generated by the infrared camera, the Kinect can build a 3D motion model for the player. When a player enters the frame, the Kinect starts to track the skeleton points of the player, and report the locations of these skeleton points (such as head, hands, knees) as a *skeleton stream*. Gesture-based Xbox applications can use the skeleton stream as user input.

For our research prototype, we use Kinect for Windows, which is a special sensor designed for Windows developers. We implemented our system using Visual Studio 2010 and Kinect for Windows SDK v1.6.

SPOT uses the Kinect for three purposes:

1. It is used as an *occupancy sensor*. When a person is tracked in the skeleton frame, the work space is treated as occupied and the system starts to control the room temperature. We also use the Kinect skeleton tracking APIs to determine the worker's activity level.

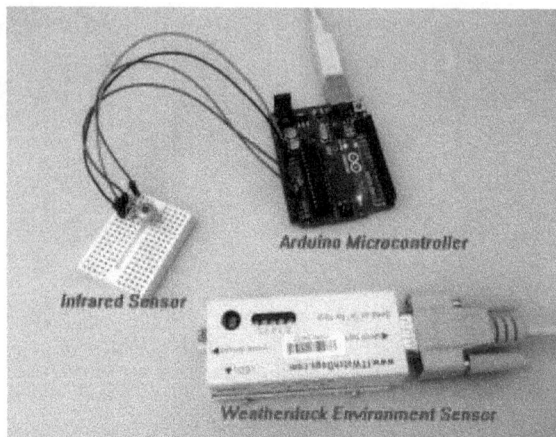

Figure 3: Infrared Thermometer and WeatherDuck Climate Monitor

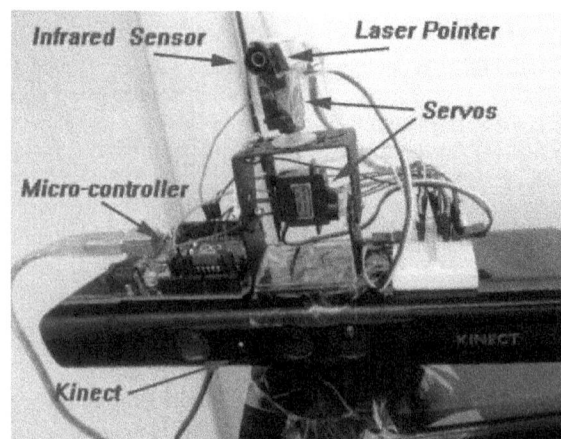

Figure 4: Kinect with infrared sensor mounted. The infrared thermal sensor and laser pointer are installed on top of the two servos, which can adjust the rotation angles of the infrared sensor and laser pointer. The micro-controller controls the laser pointer and servos. It also pulls infrared readings from the sensor.

2. It is used as a worker *location sensor*. We use the location information to point an infrared thermal sensor mounted on a tracking system at the worker to measure the worker's clothing surface temperature. This allows us to estimate the clothing level (see §3.2).

3. We also use the Kinect to allow workers to *customize their PPV model* parameters using simple gestures. To record a vote, a worker simply points to the Kinect and raises his or her hand in the air to indicate a particular comfort level (the selected comfort level is shown on a screen connected to the Kinect). The system then records one data point of the form $(\mathbf{x}_k, apv_k)$ (see §3.1).

4.2.2 Infrared Thermometer

A MLX90614 Infrared Thermometer (Figure 3 upper part) detects background radiant temperature between -40°C to +85°C with a resolution of 0.02°C. It is connected to an Arduino Uno[2] board, which reads the measured radiant temperature and sends the value to a PC via a USB cable every second.

To measure the surface clothing temperature, we mounted two servos[3] and an infrared sensor on top of the Kinect (Figure 4). The infrared sensor and a laser pointer are placed on the two servos such that they can face any direction. The laser pointer is used for calibration, and turned off during normal operation. The two servos, the infrared sensor, and the laser pointer are connected to an Arduino micro-controller. The micro-controller sends signals to control the angle of the servos and the on/off state of the laser pointer. The micro-controller is connected to the PC with a USB cable, and it communicates with the PC program using a virtual serial port.

When a worker enters the work space, the Kinect tracks the worker and sends a skeleton stream to the PC. The PC finds the location of the worker's body center and calculates the rotation angle of the servos. It then communicates with the micro-controller to adjust the angle of the two servos so that the infrared sensor is facing the body center. When the tracked worker is moving, the infrared sensor may not be actually facing towards the worker. Therefore, we introduce a 0.5 second measurement delay into the system. That

is, the infrared sensor starts collecting data only when the worker has been standing still for at least 0.5s. The system then estimates the clothing insulation by the clothing surface temperature as described in §3.2.

4.2.3 Environment Sensor

SPOT senses environmental variables using the WeatherDuck Climate Monitor [4] (Figure 3 lower part), a low-cost sensor that monitors air temperature, humidity and air flow. It can detect air temperature from -10°C to 85°C and relative humidity from 0% - 100%. Its air flow sensor can detect wind speed from 0 to 100 CFM. It also measures the light and sound level of the room as side channels for occupancy detection. The WeatherDuck Climate Monitor is connected to the PC via a serial to USB converter.

4.3 Actuator

SPOT controls a SunBeam SLP3300CN heater with a maximum power rating of approximately 1350W using a power plug that is controlled over a Z-Wave wireless network. Z-Wave is specially designed for reliable, low-latency communication of small data packets, which is desirable for home appliance control. Z-Wave devices use command classes to achieve different tasks. In our research prototype, we use a DSC06106 Smart Energy Switch to control and sense the energy consumption of the heater. The Smart Energy Switch is controlled wirelessly by a Silicon Labs CP201s Z-Wave controller, which supports command classes to control the on/off state of a device and measure the energy consumption of that device.

4.4 Occupancy Detection

SPOT detects room occupancy using the Microsoft Kinect. Recall that the Kinect APIs allows the controller to obtain near-real-time skeleton tracking. When there is a skeleton tracked by Kinect, SPOT considers the room as occupied. However, it does not turn the heater on immediately after it detects a worker to deal with transient occupancy of the work space. Instead, we have implemented a leaky-bucket based low-pass filter that turns on the heater only if

---

[2]http://arduino.cc

[3]A servo is similar to a stepper motor in that its degree of rotation can be precisely controlled.

---

[4]http://www.itwatchdogs.com

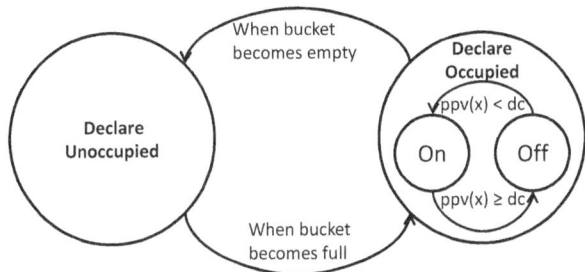

Figure 5: Heater control logic showing the leaky-bucket based low-pass filter. An office room will be declared as occupied only if it is occupied for a certain fraction of the past few minutes.

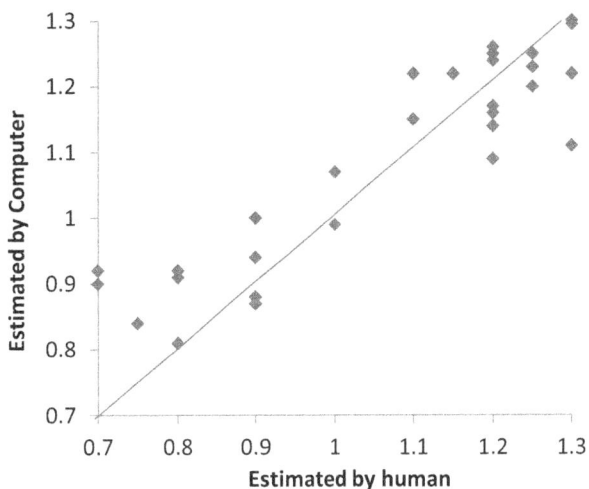

Figure 6: Clothing level estimated by human v.s. estimated by our algorithm. The RMSE of the estimation is 0.8466 and the Pearson correlation coefficient was 0.9201 indicating good linear correlation.

the work space has been occupied for a sufficiently long fraction of the prior few minutes.

Specifically, the system has a virtual leaky bucket of size 5 units. The bucket is initially empty. At the end of each minute, if the Kinect sensor reports that the work space was occupied in the past minute, one unit of "water" is added to the leaky bucket, up to a maximum bucket size of 5 units; otherwise, one unit is subtracted. When the "water" in the leaky bucket reaches 5 units, the work space is declared to be occupied. Conversely, when the "water" in the bucket reaches 0, the work space is declared to be unoccupied. When SPOT thinks that the work space is occupied, it evaluates the current $ppv(\mathbf{x})$ value and compares it to the desired comfort set point dc. At the beginning of each minute, if $ppv(\mathbf{x}) < dc$, the heater is turned on until the PPV value reaches dc; otherwise the heater is turned off. The detailed heater control logic is demonstrated in Figure 5.

5. EVALUATION

We discuss the evaluation of our system in this section. Since the evaluation period was in winter, we validated our design by implementing a heating control system. Our ideas, however, are applicable to a cooling system, where we would replace the heater with a small personal fan.

To evaluate the effectiveness of our system, we deployed the prototype to a $11.9m^2$ office room at the University of Waterloo. The office room is owned by a professor and it is usually occupied from 8:30 AM to 5:30 PM on weekdays. Note that the room is in a building that also has its own HVAC control system whose design goal is to maintain a constant temperature of $23°C$ throughout the day. Therefore, the PPV setpoint is chosen to correspond to a comfort level that is somewhat warmer than usual (corresponding to a worker who prefers warm working conditions), as a positive offset to this nominal base value.

5.1 Accuracy of Clothing Level Estimation

We first discuss the effectiveness of the clothing level estimation. Since the system is designed for indoor thermal control, we assume that the clothing level is between 0.7 (a shirt) and 1.3 (shirt, sweater and jacket), which are common in our office environment. In the training phase, 23 data points were collected as training data, with the clothing level ranging from 0.7 to 1.25. This allowed us to compute a linear regression to estimate the clothing level from the infrared sensor reading.

Subsequently, about 20 volunteers were selected to participate in a test of accuracy. For volunteers wearing a jacket, we first tested

the clothing level estimation algorithm when they were wearing their jackets. We then tested the clothing level after they took off their jackets. Therefore, we collected 35 testing data points in total.

To test our algorithm, the clothing level of each volunteer was first estimated by one of the authors using Table 12. It was then evaluated using the estimation algorithm. The results are shown as a scatter plot in Figure 6. The root mean square error (RMSE) of the prediction was 0.8466 and the Pearson correlation coefficient was 0.9201 indicating good linear correlation.

Note that the infrared sensor we are using has a 5 degree detection angle. We found that when a subject was more than 2 meters away from the sensor, the clothing estimation result was inaccurate because of noise from background infrared radiation. Therefore, in a real deployment, we need to install the IR sensor no more than 2 meters from the worker. If this is an issue, for example in a large office, we advocate using sensors with a smaller detection angle.

5.2 Accuracy of PPV Estimation

This subsection discusses the accuracy of comfort level estimation using the PPV model. The evaluation was done in an actual office at the University of Waterloo for several days.

On the first day, the office owner gave votes to the system on the thermal environment to train the PPV model. Over the training period, 12 votes were collected as training data. We then tested the PPV model by comparing the predicted votes with 8 actual votes on the following days. The results are plotted in Figure 7. We found that the RMSE of PPV estimation was 0.5377 and the Pearson correlation coefficient was 0.8182 indicating good linear correlation.

5.3 Responsiveness of the Work Space to Thermal Control

This section discusses the responsiveness of the experimental workspace to thermal control using the radiant heater. To test responsiveness, we turned on the radiant heater at 4:20PM and measured the PPV every minute. The result is plotted in Figure 8.

Before the heater was turned on at 4:20 PM, the room temperature was maintained by the central HVAC system and the PPV was

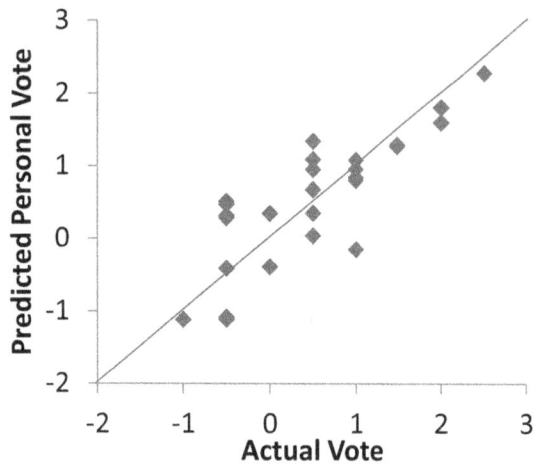

Figure 7: Actual personal vote v.s. predicted personal vote. The RMSE of the estimation is 0.5377 and the Pearson correlation coefficient was 0.8182 indicating good linear correlation.

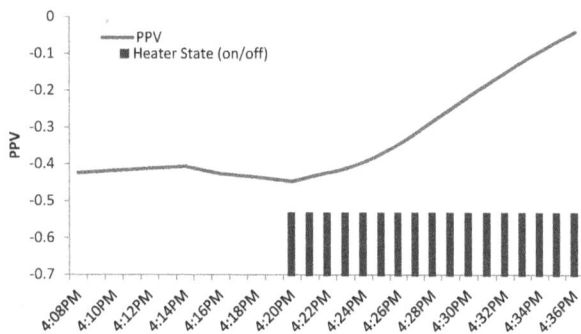

Figure 8: Room temperature and the heater state. The room temperature increased 1 degree Celsius after turning on the heater for 15 minutes.

between -0.5 to -0.4. When the heater was turned on at 4:20 PM, the room temperature as well as the PPV started to increase immediately, reaching the target PPV of 0 in 15 minutes at 4:35 PM.

We conclude that it is feasible to use reactive control for personal thermal comfort without significantly reducing human comfort.

5.4 Thermal Comfort Over a Day

We now discuss the performance of SPOT over the course of a typical day. Our experimental setup was the same as for the other experiments. However, we required SPOT to maintain a PPV of 0. Figure 9 shows the results of this experiment, depicting room occupancy, PPV, and room temperature over the day.

Note that both the room temperature and PPV are relatively low before the worker entered the office at 10 AM in the morning, with the PPV of -1.5 being the comfort level corresponding to the temperature setpoint chosen by the central HVAC system. SPOT turned on the heater within five minutes of the worker's arrival and both the temperature and the PPV increased steadily to 0 over the next 45 minutes.

The PPV of the office was always maintained around 0 when the worker was in the office, with small excursions above zero when the HVAC heating system turned on from time to time. The brief

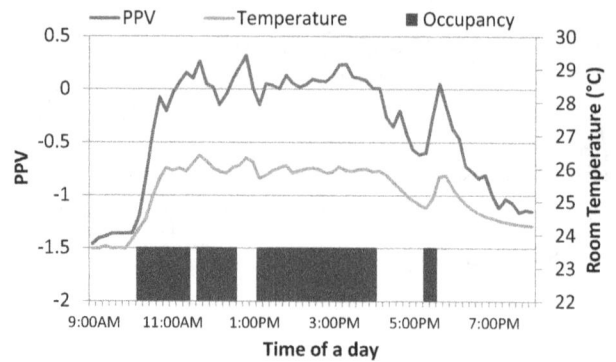

Figure 9: Room occupancy and PPV over a day. Each tick on the X axis is 10 minutes. For most of the time when the room is occupied, the PPV is maintained around 0, even through there are external disturbances from the central HVAC system.

change in occupancy just before 11 AM, of about 10 minutes, was too short to cause any appreciable change in PPV.

The worker left the office at 4 PM for an hour. During that time, SPOT turned off the heater to save energy. This reduced the PPV to -0.5, but the PPV returned to 0 soon after the worker returned at 5 PM. When the worker finally left at 5:30 PM, the PPV declines, eventually reaching -1.5.

This demonstrates that SPOT can maintain the PPV at a chosen comfort value over the course of a day, despite the periodic activation of the central HVAC heating system and changes in office occupancy.

Note that, in this instance, PPV tracks room temperature quite closely. This is because there was little change in other environmental and personal factors, such as humidity and clothing level. In other circumstances, such as when the worker may put on or take off a jacket, SPOT would be able to maintain the comfort level by appropriately reducing the room temperature.

5.5 Trade-off between PPV and Energy Consumption

SPOT allows a building's energy consumption to be decreased in three ways.

- It allows the common areas of the building to be heated or cooled to a lesser degree than the ASHRAE standard of 23°C.

- It only heats or cools a work space when the worker is actually present.

- It allows the worker to choose a comfort level that is lower than 0, thus saving energy.

Here, we focus on the third element above.

To evaluate the possible amount of energy saving by lowering the PPV value, we measured the relationship between PPV and heater energy consumption[5]. We did this by setting the heater power to different values and recording the PPVs when the room temperature had converged. When the heater was turned off, the room temperature was maintained by the centralized HVAC system at around 23 degrees, corresponding to PPV values between -0.5 and -1.24 depending on the central HVAC system's phase in its heating cycle. In contrast, when the radiant heater was set to its maximum power,

---

[5]This relationship is necessarily noisy because temperature is not the only determinant of PPV. Nevertheless, the trend is distinct.

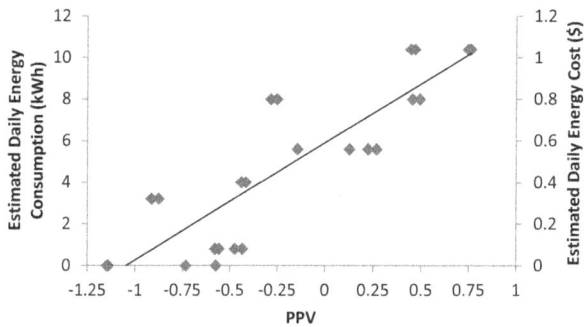

Figure 10: Relationship between daily energy consumption of the heater and the PPV. Maintaining a PPV of 0 consumes about 6 kWh electricity daily. By setting the target PPV to -0.25, we can save about 1.5 kWh electricity per day.

the PPV was about 0.75 with the estimated power consumption per day was about 10.5 kWh.

Figure 10 shows this trade-off between PPV and the heater energy consumption in a day. We see that a reduction in PPV of 0.1, which is hardly noticeable by a human, results in the reduction in usage of 0.6 kWh of electricity in a day. This allows us to quantitatively select the trade-off between personal thermal comfort and heating energy consumption. For instance, an energy-aware office worker can set the target PPV value dc (as mentioned in §4.4) to -0.5 in order to save energy.

6. RELATED WORK

6.1 PMV Model

The PMV model is widely used for evaluating the performance of building temperature control systems. Yang et. al. [28] used PMV to reduce the energy consumption of a building's HVAC system. Since radiant temperature is a significant factor in PMV in hot and humid areas, air temperature and humidity control is not enough for cooling in summer. In their system, they control the air velocity as well in order to maintain PMV at the comfortable range. Aswani [7] et. al. have also used the PMV model in their building temperature control system. They use Learning-Based Model Predictive Control (LBMPC) to control the building HVAC system such that different zones of the building maintain PMVs close to 0. Our approach, instead, uses the personalized PPV model to achieve personalized thermal control.

The Thermovote [12] system allows workers in a building to vote on the current temperature. Instead of using PMV to predict users' feeling, they use the actual vote of the workers to adjust their comfort. However, the system requires the users to vote frequently, which is onerous. We avoid this problem by building personalized models for each individual. SPOT only requires votes during the training phase to calibrate the PPV model. Subsequently, the thermal preference of the user is used to control the HVAC system and no more voting is required.

6.2 Occupancy-based HVAC Control

There has been a considerable amount of work on improving the energy-efficiency of HVAC systems. For example, Aswani et. al. [7] use Learning-Based Model Predictive Control (LBMPC) to model and control HVAC systems in a large university building. They were able to save an average of 1.5MWh of electricity per day

in their testbed. Fong et. al [16] used evolutionary programming (EP) to find the optimal HVAC setting, and apply this setting instead of the default one. They found that about 7% of energy could be saved by replacing the default HVAC settings by their optimized ones. Although sophisticated, these approaches control only the room temperature, rather than attempt to achieve a certain level of user comfort, as we do.

Turning off an HVAC system when no humans are present is an obvious technique to reduce energy use. It is also important to turn on heating in advance of human occupancy, because it can take tens of minutes to heat a cold building to tolerable levels. Most occupancy prediction methods use previously collected occupancy data. Lu et. al. [21] showed that by learning occupancy and sleep patterns, it is possible to save about 28% of energy use in a home environment. The PreHeat system [27] uses occupancy sensors to predict home occupancy patterns and automatically adjusts the HVAC temperature setpoint to save energy. If we consider the previous history of work space occupancy as a time series, and current time as a function of previous entries in the time series, we can learn this function as a Gaussian Process. This approach has been used in Erickson's and Rogers' papers [14, 26]. In another learning-based approach, home occupancy is modeled as a Markov chain and room occupancy is encoded as a state in the Markov model [13]. Mozer et. al. use a neural network and a lookup table to predict home occupancy in their Neuralthermostat project [23]. To build human interpretable model, Leephakpreeda [19] applied a grey model for occupancy prediction. Ardakanian et. al [8] uses sound and light level of a room to infer occupancy and applies POMDP for optimal HVAC control.

Most learning based models require relatively large amount of data in order to produce accurate predictions. Hence, to predict occupancy with limited historical occupancy data, there exists other approaches to employ some side channels to assist prediction. For example, Gupta et. al. [17] uses GPS sensors on mobile phones to estimate the arrival time of home owners and heat the house before they arrive.

HVAC control based on occupancy can be used in conjunction with our techniques to allow the HVAC controller to pre-heat or pre-cool a work space to achieve a target comfort level rather than a target temperature. We discuss this further in §7.3

7. DISCUSSION AND FUTURE WORK

7.1 Extreme Sensing

The SPOT system, with its plethora of sensors, can be viewed as a somewhat extremal point in the space of HVAC control systems. We are keenly aware that our approach is hardware and compute intensive, and has a price point that may put it out of reach of most offices. Nevertheless, we believe that our approach is interesting for at least two reasons.

First, with the proliferation of sensing and compute systems, even high-end sensors such as the Microsoft Kinect will be much cheaper in the near future. Second, even if maintaining per-worker comfort is too expensive in terms of sensing, per-worker temperature control, a far more achievable goal, is cheap, effective, and well within reach in existing offices. We believe, therefore, that SPOT establishes an interesting data point in the thermal control design space.

7.2 Predictive Control

Our system uses reactive control to adjust the temperature of the room. An alternative approach is to use Model Predictive Control (MPC) [9] to adjust the setpoint of the HVAC system. MPC is fundamentally different from reactive control because it builds a thermal model for the system. Since we can estimate the internal mechanics of the system, we can predict the future control outputs given the current control inputs and the states of the system. Model Predictive Control is a white box approach, hence it is usually easier to tune the model parameters to meet the optimal control objective. For example, by using MPC, we can build a thermal model of the room using the heater power as the input. With this model, we can easily find the optimal control strategy that minimizes the energy consumption (i.e., the integral of power). We decided not to use this more complex approach because simple reactive control appears to be adequate for a small work space with negligible thermal mass.

7.3 Incorporating Occupancy Prediction

We have already discussed many well-known approaches for occupancy prediction. Our system can take advantage of existing occupancy prediction algorithms to make control decisions *before* room occupancy changes. For example, in Figure 9, in the morning, it took about 45 minutes to get the PPV to 0. Instead, with occupancy prediction, as in Preheat [27], we could start to heat the room 1 hour before the estimated arrival time of the worker and maintain a PPV of -0.5. We could then heat the room to the target PPV of 0 only when the worker actually arrived.

Similarly, the thermal mass of a work space causes it to cool down over many tens of minutes. Therefore, we can stop heating the work space in advance of the worker leaving it by using an occupancy prediction algorithm [11]. We intend to explore this direction in future work.

Note that prediction accuracy critically affects the performance control system. False negative prediction will reduce worker comfort and false positive prediction will lead to a waste of energy. Therefore, when using occupancy prediction, we must evaluate prediction accuracy and the cost of false predictions.

7.4 Optimal Control

With accurate occupancy prediction, we can use an optimal control framework to further reduce energy consumption. For example, in Figure 9, when the heater was turned off at 5:30 PM, the room temperature was at 26° C. It took 2.5 hours for the office to cool down to 24° C. If the office had very good insulation, the cool down process would have been even slower. Therefore, we can save energy by turning off the heater earlier if we know that the office will be unoccupied in the near future. Conversely, if we can predict the arrival time of the worker, we can heat the room up to a comfortable temperature before the worker arrives. However, to take advantage of these approaches, we need to decide how much time in advance should we turn on or turn off the heater. An optimal control framework would allow us to decide the best timing for any control action over a planning horizon.

7.5 Human Factors in Automation

Our discussion so far has assumed that the worker has little role to play in thermal control. In fact, workers themselves can be active participants in a thermal control system if they receive and act on energy-saving tips. For example, SPOT could, instead of turning on a heater, suggest to workers that they put on a jacket. This integration of humans into the control loop can be viewed as being unnecessarily intrusive. Nevertheless, we believe that, if properly presented to humans, such control actions can be both energy saving and marginally intrusive. We intend to explore this in future work.

7.6 Limitations

Our work has several inherent limitations that we discuss next.

Thermal isolation We assume that each work space is relatively thermally isolated from other work spaces. This does not hold true, for example, in open-plan offices.

Personalized work spaces We assume that each worker has their own personal space, and that they do not move from space to space over time. This may not be a valid assumption for all workplaces.

Cost Each SPOT system costs about $1,000. This may be too high a cost to pay for modest increases in worker comfort. We expect this cost to rapidly decline over the next few years.

Calibration SPOT requires worker participation to calibrate personal comfort levels. This can take a day or so, and can be viewed as onerous by some workers.

Validation Because we do not have control over our building's temperature setpoint, we are unable to validate our research hypothesis that personalized thermal control can save energy overall. This is certainly quite plausible, in that measurements have shown that a two degree increase in the temperature setpoint in summer can reduce home energy use by 37% [1], but we have no way to validate this conclusion.

Environment SPOT is blind to windows that are open versus closed, to HVAC state, and user mobility. In our experiment, the office is controlled by a centralized heating system and the window is always closed. These factors may affect the effectiveness of SPOT in other environment.

8. CONCLUSION

We have presented the design and implementation of SPOT, a smart personal thermal comfort system for office work spaces. SPOT builds on three underlying ideas. First, we extend the PMV model to create the PPV model to quantitatively estimate personal comfort. Second, we use a set of sensors, including a Microsoft Kinect sensor, to measure PPV model parameters, so that we can estimate a worker's comfort level at any point in time. Third, we use occupancy-aware simple reactive control to maintain the PPV despite changes in the environment. We deployed SPOT in a real office environment and validated that it can maintain comfort over the course of a typical work day. Moreover, we have shown how SPOT allows a worker to trade off a reduction in comfort for saving energy. We believe that our work demonstrates an interesting case study of how to maintain human comfort using extreme sensing. Finally, a limited version of our system, that only maintains personalized temperature offsets from a building-wide base setpoint, is not only easy to deploy, but is also likely to reduce overall building energy use.

9. REFERENCES

[1] 2009 peaksaver Residential Air Conditioner Measurement and Verification Study. *Ontario Power Authority, May 2010.*

[2] ANSI/ASHRAE Standard 55-1992, Thermal Environmental Conditions for Human Occupancy. *Atlanta: American Society of Heating, Refrigerating and Air-Conditioning Engineers.*

[3] Energy Efficiency Fact Sheet: Heating, ventilation & Air Conditioning. *http://www.originenergy.com.au/ files/SMEfs_HeatingAirCon.pdf.*

[4] Ergonomics of the Thermal Environment - Analytical Determination and Interpretation of Thermal Comfort using Calculation of the PMV and PPD Indices and Local Thermal Comfort Criteria. *ISO 7730:2005*.

[5] HVAC & Energy Systems. *http://canmetenergy.nrcan.gc.ca/buildings-communities/hvac-energy/908*.

[6] Nest, the learning thermostat. http://www.nest.com/.

[7] Anil Aswani and Neal Master and Jay Taneja and Andrew Krioukov and David E. Culler and Claire Tomlin. Energy-Efficient Building HVAC Control Using Hybrid System LBMPC. *CoRR*, abs/1204.4717, 2012.

[8] O. Ardakanian and S. Keshav. Using Decision Making to Improve Energy Efficiency of Buildings. *ICAPS-10 POMDP Practitioners Workshop, May 2010*.

[9] E. Camacho, C. Bordons, E. Camacho, and C. Bordons. *Model Predictive Control*, volume 303. Springer Berlin, 1999.

[10] R. de Dear and G. S. Brager. Developing an Adaptive Model of Thermal Comfort and Preference. *UC Berkeley: Center for the Built Environment, 1998*.

[11] C. Ellis, J. Scott, M. Hazas, and J. Krumm. EarlyOff: Using House Cooling Rates to Save Energy. In *Proceedings of the Fourth ACM Workshop on Embedded Sensing Systems for Energy-Efficiency in Buildings*.

[12] V. Erickson and A. Cerpa. Thermovote: Participatory Sensing for Efficient Building HVAC Conditioning. 2012.

[13] V. L. Erickson and A. E. Cerpa. Occupancy based Demand Response HVAC Control Strategy. In *Proceedings of the 2nd ACM Workshop on Embedded Sensing Systems for Energy-Efficiency in Building*, BuildSys '10, pages 7–12, New York, NY, USA, 2010. ACM.

[14] V. L. Erickson, Y. Lin, A. Kamthe, R. Brahme, A. Surana, A. E. Cerpa, M. D. Sohn, and S. Narayanan. Energy efficient building environment control strategies using real-time occupancy measurements. In *Proceedings of the First ACM Workshop on Embedded Sensing Systems for Energy-Efficiency in Buildings*, BuildSys '09, pages 19–24, New York, NY, USA, 2009. ACM.

[15] P. O. Fanger. Thermal comfort. *Analysis and Applications in Environmental Engineering, Danish Technical Press, Copenhagen, Denmark, 1970*.

[16] K. Fong, V. Hanby, and T. Chow. HVAC System Optimization for Energy Management by Evolutionary Programming. *Energy and Buildings*, 38(3):220 – 231, 2006.

[17] M. Gupta, S. S. Intille, and K. Larson. Adding GPS-Control to Traditional Thermostats: An Exploration of Potential Energy Savings and Design Challenges. In *Proceedings of the 7th International Conference on Pervasive Computing*, Pervasive '09, pages 95–114, Berlin, Heidelberg, 2009. Springer-Verlag.

[18] B. W. Jones. Capabilities and Limitations of Thermal Models for Use in Thermal Comfort Standards. *Energy and Buildings*, 34(6):653 – 659, 2002. Special Issue on Thermal Comfort Standards.

[19] T. Leephakpreeda. Adaptive Occupancy-based Lighting Control via Grey Prediction. *Building and Environment*, 40(7):881 – 886, 2005.

[20] W. Liping and W. N. Hien. The Impacts of Ventilation Strategies and Facade on Indoor Thermal Environment for Naturally Ventilated Residential Buildings in Singapore. *Building and Environment*, 42(12):4006–4015, 2007.

[21] J. Lu, T. Sookoor, V. Srinivasan, G. Gao, B. Holben, J. Stankovic, E. Field, and K. Whitehouse. The Smart Thermostat: Using Occupancy Sensors to Save Energy in Homes. In *Proceedings of the 8th ACM Conference on Embedded Networked Sensor Systems*, SenSys '10, pages 211–224, New York, NY, USA, 2010. ACM.

[22] G. K. C. Mehdi Shahrestani, Runming Yao. Performance Characterisation of HVAC&R Systems for Building Energy Benchmark. *http://www.cibse.org/content/cibsesymposium2012/Poster026.pdf*.

[23] M. Mozer, L. Vidmar, R. Dodier, et al. The Neurothermostat: Predictive Optimal Control of Residential Heating Systems. *Advances in Neural Information Processing Systems*, pages 953–959, 1997.

[24] J. Nicol and M. Humphreys. Adaptive Thermal Comfort and Sustainable Thermal Standards for Buildings. *Energy and Buildings*, 34(6):563 – 572, 2002. Special Issue on Thermal Comfort Standards.

[25] L. Prez-Lombard, J. Ortiz, and C. Pout. A Review on Buildings Energy Consumption Information. *Energy and Buildings*, 40(3):394 – 398, 2008.

[26] A. Rogers, S. Maleki, S. Ghosh, and J. Nicholas R. Adaptive Home Heating Control Through Gaussian Process Prediction and Mathematical Programming. In *Second International Workshop on Agent Technology for Energy Systems (ATES 2011)*, pages 71–78, May 2011. Event Dates: May 2011.

[27] J. Scott, A. Bernheim Brush, J. Krumm, B. Meyers, M. Hazas, S. Hodges, and N. Villar. PreHeat: Controlling Home Heating Using Occupancy Prediction. In *Proceedings of the 13th international conference on Ubiquitous computing*, UbiComp '11, pages 281–290, New York, NY, USA, 2011. ACM.

[28] K. Yang and C. Su. An Approach to Building Energy Savings Using the PMV Index. *Building and Environment*, 32(1):25 – 30, 1997.

10. APPENDIX: DETAILS OF THE PMV MODEL

The PMV [15] is computed as.

$$pmv(\mathbf{x}) = (0.303 \cdot exp(-0.036 \cdot M) + 0.028) \cdot$$

$$\begin{Bmatrix} (M - W) - 3.05 \cdot 10^{-3} \cdot (5733 - 6.99 \cdot (M - W) - p_a) \\ -0.42 \cdot ((M - W) - 58.15) - 1.7 \cdot 10^{-5} \cdot M \cdot (5867 - p_a) \\ -0.0014 \cdot M \cdot (34 - t_a) - 3.96 \cdot 10^{-8} \cdot f_{cl} \cdot ((t_{cl} + 273)^4 \\ -(\bar{t}_r + 273)^4) - f_{cl} \cdot h_c \cdot (t_{cl} - t_a) \end{Bmatrix}$$

(7)

where t_{cl} is the clothing surface temperature, and W is the effective mechanical power which is 0 for most indoor activities.

Variable t_{cl} can be evaluated by:

$$t_{cl} = 35.7 - 0.028 \cdot (M - W) - I_{cl} \cdot (3.96 \cdot 10^{-8} \cdot f_{cl} \cdot$$

$$((t_{cl} + 273)^4 - (\bar{t}_r + 273)^4) + f_{cl} \cdot h_c \cdot (t_{cl} - t_a)) \quad (8)$$

Variable h_c is the convective heat transfer coefficient, which is derived as

$$h_c = \begin{cases} 2.38 \cdot |t_{cl} - t_a|^{0.25} & \text{if } 2.38 \cdot |t_{cl} - t_a|^{0.25} > 12.1 \cdot \sqrt{v_{ar}} \\ 12.1 \cdot \sqrt{v_{ar}} & \text{if } 2.38 \cdot |t_{cl} - t_a|^{0.25} < 12.1 \cdot \sqrt{v_{ar}} \end{cases}$$

(9)

Variable f_{cl} is the clothing surface area factor, which is derived as:

$$f_{cl} = \begin{cases} 1.00 + 1.290 I_{cl} & \text{if } I_{cl} \leq 0.078 m^2 \cdot K/W \\ 1.05 + 0.645 I_{cl} & \text{if } I_{cl} > 0.078 m^2 \cdot K/W \end{cases}$$

(10)

In practice, the metabolic rate and the clothing insulation are first estimated by Table 11 and Table 12. Given the clothing insulation I_{cl}, we calculate the clothing surface temperature t_{cl} and the convective heat transfer coefficient h_c by iteratively applying Equation 8 and 9. Finally, by using Equation 7 and 10, we can estimated the Predicted Mean Vote.

Activity	Metabolic Rate	
	W/m^2	met
Reclining	46	0.8
Seated, relaxed	58	1.0
Sedentary activity	70	1.2
Standing, medium activity	93	1.6

Table 11: Metabolic Rates

Daily Wear Clothing	Clothing Insulation (I_{cl})	
	clo	$m^2 \cdot K/W$
Panties, T-shirt, shorts, light socks, sandals	0.30	0.050
Underpants, shirt with short sleeves, light trousers, light socks, shoes	0.50	0.080
Panties, petticoat, stockings, dress, shoes	0.70	0.105
Underwear, shirt, trousers, socks, shoes	0.70	0.110
Panties, shirt, trousers, jacket, socks, shoes	1.00	0.155
Panties, stockings, blouse, long skirt, jacket, shoes	1.10	0.170
Underwear with long sleeves and legs, shirt, trousers, V-neck sweater, jacket, socks, shoes	1.30	0.200
Underwear with short sleeves and legs, shirt, trousers, vest, jacket, coat, socks, shoes	1.50	0.230

Table 12: Thermal Insulation for different clothing level

An Opportunistic Activity-sensing Approach to Save Energy in Office Buildings

Marija Milenkovic
ACTLab, Signal Processing Systems
Electrical Engineering, TU Eindhoven
P.O. Box 513, NL-5600 MB Eindhoven
m.milenkovic@tue.nl

Oliver Amft
ACTLab, Signal Processing Systems
Electrical Engineering, TU Eindhoven
P.O. Box 513, NL-5600 MB Eindhoven
o.amft@tue.nl

ABSTRACT

In this work, we recognised office worker activities that are relevant for energy-related control of appliances and building systems using sensors that are commonly installed in new or refurbished office buildings. We considered desk-related activities and people count in office rooms, structured into desk- and room-cells. Recognition was performed using finite state machines (FSMs) and probabilistic layered hidden Markov models (LHMMs).

We evaluated our approach in a real living-lab office, including three private and multi-person office rooms. As example devices, we used different ceiling-mounted PIR sensors based on the EnOcean platform and plug-in power meters. In at least five days of study data per office room, including reference sensor data and occupant annotations, we confirmed that activities can be recognised using these sensors. For computer and desk work, an overall recognition accuracy of 95% was achieved. People count was estimated at 87% and 78% for the best-performing two office rooms. We furthermore present building simulation results that compare different control strategies. Compared to modern BEMS, our results show that 21.9% and 19.5% of electrical energy can be saved for controls based on recognised desk activity and estimated people count, respectively. These results confirm the relevance of building energy management based on activity sensing.

Categories and Subject Descriptors

G.3 [**Probability and Statistics**]: *Probabilistic algorithms (including Monte Carlo), Statistical computing*; I.2.6 [**Artificial Intelligence**]: Learning—*Parameter learning*; I.5.1 [**Pattern Recognition**]: Models—*Deterministic, Statistical*; I.5.2 [**Pattern Recognition**]: Design Methodology—*Classifier design and evaluation, Feature evaluation and selection, Pattern analysis*; I.6.6 [**Simulation and Modeling**]: Applications

Keywords

Activity recognition; office buildings; green ICT; energy saving, BEMS

1. INTRODUCTION

Commercial buildings are among the largest energy consumers and CO_2 producers worldwide [7]. While efficiency improvements in installation and appliances can contribute to lower consumption and cost, a large additional potential for energy savings exists in actively controlling building spaces according to their actual dynamic usage. Modern Building (Energy) Management Systems (BMS, BEMS) typically operate by adjusting heating, ventilation and air-conditioning (HVAC) and lighting based on readings from various types of sensors distributed within a building. Besides room conditions (e.g., temperature, humidity, light level), motion sensors and energy meters are frequently installed sensing modalities in new or refurbished commercial buildings. Although most energy consumers in offices support rapid control cycles and different operating states, the data provided by building sensors is not fully exploited in current BEMS. Where BEMS systems do not adapt dynamically to occupant activities energy could be wasted. In particular, dynamic information on user activities is currently not considered for building control. For example, to properly ventilate a building space, the actual people count in the office space is key to adjust air supply and temperature in order to maintain user comfort, as suggested by the ANSI/ASHRAE standard [2]. According to Erickson and Cerpa [5] an occupancy-driven ventilation strategy alone could reduce total energy usage by 8.1%. Through a business day, office building users follow various activities and occupy different building spaces. Current buildings only use presence or motion detectors in offices to switch lights on or off according to occupancy. Standby power detectors are used to switch off devices that remain in low-power mode, while not being used.

More detailed information on office activities could save energy. For example, detecting whether users perform paper-based or computer-based work could enable BEMS to dim lights during screen-work and increase lighting levels during paper-based work as well as to control office appliances [12]. To date, most approaches to activity recognition in offices relied on a variety of specifically added sensors, such as video and ultrasound sensors, or required users to use wearable technology (see related work in Sec. 2). In those approaches, the set of commonly installed and networked sensors that are available for BEMS management was not considered.

Adding and maintaining further sensors increases cost and burden for building management, thus should be minimised. Using existing sensors could provide options for activity-based control, ideally by updating BEMS software only.

In this paper we investigate an opportunistic approach to office activity recognition by using a subset of sensors that can be frequently found readily installed in office buildings. Our aim is thus to derive a set of activities that is relevant for energy-related control of appliances and building systems (HVAC, lighting, other office appliances). We then evaluate activity recognition performance using the opportunistic, hence often available, sensor set. In particular, this paper provides the following contributions:

1. We present our activity recognition and energy saving approach using sensors frequently installed in buildings. We detail our recognition approach using finite and probabilistic state models that can be applied with building sensor data. We recognise activities that have potential to lead to energy savings, including desk work activities and people count per office.

2. We give an overview on the living-lab installation of three different office rooms, in which we investigated our opportunistic sensing approach in a continuous recording study during five or more days per room. While occupants followed their regular office activities, data from the various in-building sensors were acquired together with complementary reference sensors to subsequently analyze recognition performances.

3. We determine the relevance of the considered activity sets regarding their energy saving potential using simulations. Here we compare several commonly used control strategies in buildings with our activity-based approach. Along the installation of our living-lab building, we illustrate the potential energy saving using opportunistic activity-based sensing.

Relations of office activities and building energy requirements have been shown previously through simulations. However, the feasibility of using standard office building sensors to recognise activities, which are relevant for energy saving, yet has to been confirmed. In this paper, we focus on identifying relevant office activities and recognition approaches as a first, but essential step towards improving energy efficient building operations. For this purpose, our work considers real offices, where participants work regularly.

2. RELATED WORK

Passive infrared sensors (PIR) are widely used in activity recognition, but most commonly in presence and motion detection, to track people's paths, in order to switch on and off lights, HVAC systems, and appliances. In [1] PIR sensors and magnetic door switches were used to detect occupancy of certain building spaces and control the HVAC system accordingly. Using information about occupancy from a pilot testbed and building simulation, potential energy savings between 10% to 15% were estimated.

Besides HVAC system control, PIR sensors together with other sensor modalities are commonly used for controlling lighting and appliances. In a work of Marchiori and Qi [13], PIR sensors and door switches were used together with energy controllers to manage different office appliances according to occupancy. Authors in [4] used motion and light sensors to determine high energy consumption points that can be optimized to gain greater efficiency. Their energy consumption optimization was based on occupancy and the level of ambient and artificial light. Savings from 58.6% in open-plan offices to 70.9% in corridors were estimated. To learn occupancy patterns and movement behavior, authors in [6] and [5] used wireless cameras. Camera nodes were placed at the boundaries between different areas to detect transition between them. Besides tracking people, the approach was used to estimate the number of people in the room and to predict room usage. Based on known occupancy patterns, HVAC control strategies could be optimized to save 8.1% of energy needs.

In several previous works, occupancy information was considered for control purposes. Even though occupancy information has proven to be valuable for controlling lighting, HVAC, and different appliances, and previous studies showed substantial energy saving potential, those works did not consider actual user activity. Our present study focuses on occupant activity and behaviour as a key information for building adaptation and energy saving, thus not limited to occupancy only.

Wojek et al. [23] used cameras and omni-directional microphones for room-level people tracking in offices and laboratories. The authors recognized whether users were participating in meetings, involved in discussions, paper work, phone calls, or the office was empty. Oliver and Horwitz used USB cameras and binaural microphones for detecting office activities [17, 18]. Based on audio data they were able to differentiate human speech, music, silence, ambient noise, phone ringing and typing. From video data, the system detected whether people were present in the office. At the highest layer in their model, activities, such as phone conversation, ongoing presentation, distant conversation, nobody in the office, and user present and engaged in some other activity were modelled. Since cameras and microphones are often considered to reveal privacy-sensitive details, authors in [24] proposed solution for user activity recognition and tracking by using a network of PIR sensors only. They grouped sensors into clusters to represent e.g. entering, leaving, tuning, walking up and down. Superclusters were used to model visiting, chatting and meeting activities. Their approach required a dense installation of PIR sensors. In [16] authors used PIR, pressure sensors and microphones for recognizing five different activities: working with PC, working without PC, having a meeting, presence, and absence. Jahn et al. [11] proposed ubiquitous sensor-based system for tracking user actions relevant for sustainable behaviour. In particular, power consumption, presence, lighting, window movement, and heating temperature was monitored. The work showed that user behaviour and awareness is relevant for reducing energy waste.

Studies investigating office activity recognition showed that high performance could be reached for several sensor modalities and information sources. Although cameras and microphones would provide rich information about user activities, they are often considered privacy-intrusive by occupants, and thus could affect user behaviour and comfort. PIR sensors and plug-in power meters, as used for motion detection and power measurements of computer screens in our study, could provide sufficient information to recognise activity. Since both, PIRs and plug-in power meters are regularly used in offices already, acceptance by occupants is likely. Furthermore, previous works on office activity recognition did not consider activities related to energy saving potential, as it is targeted in our study.

Opportunistic sensing approaches have been investigated to combat the constraints in obtrusiveness, privacy, and cost associated to the previously mentioned concepts, which are key to our application too. In the context of smart homes, infrastructure-mediated sensing or 'home bus snooping' was investigated to recognize user activities by single-point sensing. Patel et al. [20] analyzed electricity line noise of different device in homes. By observing the device operation, the authors derived information about user location and activity. In the approach of Froehlich et al. [9] water fixtures were monitored, including sink, toilet, shower, bathtub, clothes washer and dishwasher. Another approach was proposed by Patel et al. [19] for detecting human movement by differential pressure sensing at the home-based HVAC system. According to differential pressure it was possible to determine location of pressure disturbances and determine when people were passing through doorways as well as to detect door opening and closing. Although these approaches are easy to deploy and very promising, they are mainly applicable to private houses.

Our approach focuses on identifying activities in office buildings, where a single-point sensing approaches are not feasible due to the larger variability in installed appliances, variety in occupant behavior patterns, and need for scalable sensing solutions. We thus consider an opportunistic sensing approach that relies on modalities often available in modern or refurbished buildings.

In [10] and [22] authors used user access badges, identification via Wi-Fi points, user calendars, Instant Messaging clients, and computer system activity, in order to track users and recognize their activities. With this rich information about users, these opportunistic approaches showed good recognition accuracies and potential for energy saving. In contrast, our approach considers motion as a binary signal and energy consumption of computer screens only. In this work, sensor data is not used to identify users, neither to track them.

3. RECOGNITION CONCEPT

Our opportunistic sensing and recognition approach builds on a subset of sensors that are often already installed in office buildings. This section details modalities considered and describes the office activity recognition algorithms developed to derive activity information from the sensors.

3.1 Opportunistic in-building activity sensing

Since energy conservation and cost represent major challenges for building operators and facility managers, various wired and wireless in-building sensing systems have been introduced to support BMS control. Among them, motion detectors are typically used to control lights in different zones. Lights are switched on and off according to user presence. Power meters are used to measure consumption of plug-in appliances. Figure 1 illustrates the modalities commonly available and considered for activity recognition in this work.

Depending on building type and primary occupant use, additional modalities could be available, such as window switches (where windows can be opened), temperature, humidity, and many further modalities. In our approach, we

Figure 1: Illustration of sensor modalities that are frequently installed in modern buildings and considered here for activity recognition: (1) per-desk motion detectors at the ceiling for lighting and HVAC control, and (2) power meters to control plug-in appliances.

focus on a subset of sensing options that can provide information on office activities relevant for building control. Hence we neglected some data sources that may not be commonly available or are otherwise not providing information that is directly coupled to user relevant activity. In Section 6 we confirmed the relevance of the selected activity sets.

3.2 Desk-cell and room-cell activities

We matched general office activities, their relevance for energy-related control and information provided by the modalities described above to a shortlist of activities that can be considered by our opportunistic approach. In order to maintain scalability, we partitioned the recognition problem in desk-cell and room-cell, as there can be a variable number of desks in one room or building space.

Desk-cell activities. For an individual desk, major states include *Away* and *Presence*, where the latter can take different forms according to the actual activity: *Computer work*, summarizing computer-based activities, and *Desk work*, summarizing other desk work not actively working in front of the computer. The different activities during *Presence* can be used to control appliances, such as computer screens when users are not working with their computers, or to adjust lighting conditions for paper-based work. Clearly, *Away* and *Presence* as indicators of occupancy are crucial for HVAC, lighting, and appliances control.

Room-cell activities and states. Room-cell activities and states include *People count*, which describes the number of occupants actually working in a room. Based on *People count*, HVAC systems could adjust the ventilation of fresh air and temperature within a room.

All considered activities are listed together with related sensor modalities and intended use of the recognition result in Table 2. Subsequently, we derived algorithms that could process the opportunistic sensor data continuously and recognize office activities.

3.3 Desk activity recognition

We chose graphical models to implement our recognition approach in desk- and room-cells. Two different methods were used: finite state machines (FSMs) and layered hidden Markov models (LHMMs). Finite state machines allowed us to describe state logic and sensor data fusion. However, FSMs are deterministic, requiring expert knowledge and manual design of states and conditions.

Unlike FSMs, HMMs are non-deterministic models, where transitions between states and observation emissions are modeled probabilistically. During the LHMM training phase, unlabeled sensor data was used. Since labeled training data was not required, LHMMs can be derived even without expert knowledge. Moreover, as we showed in our previous study [15], less than two days of training data were needed to achieve robust recognition of activities in two office rooms and LHMMs can be learned independent of a particular occupant desk. Table 1 lists all features considered in our models and Table 2 summarizes recognition goals, activity sets, features, and example applications considered in this work.

Table 1: Features derived from PIR sensors and plug-in power meters that where used in our recognition models.

Nr.	Feature description	Symbol
1	PIR state	s_{PIR}
2	Screen energy consumption above standby threshold	$v_{Energy_i} > \phi_{Energy}$
3	Screen energy consumption below standby threshold	$v_{Energy_i} \leq \phi_{Energy}$

Finite state machines (FSM). For per-desk PIR sensors in new building installations, the states *Presence* and *Away* are conveniently detectable by s_{PIR_i} for desk i. Our *Desk activity* state-model however includes sub-states for *Computer work* and *Desk work* while in state *Presence*. The sub-states are recognized based on consumed energy v_{Energy_i} of the screen at desk i, as shown in Figure 2. This approach assumes that the screen will enter standby mode if no user activity is detected. Consequently, if the consumed outlet energy is below a known threshold, the screen is assumed to be in standby. With this hierarchical state-model, we obtained a more specific representation of activities while working at a desk. When a user is present, but the computer has switched off the screen (hence $v_{Energy_i} \leq \phi_{Energy}$), *Desk work* is reported and *Computer work* otherwise.

Nevertheless, most PIR sensors used in office buildings are detecting motion, thus may miss to detect presence accurately. To prevent falsely reporting *Away* while a user is sitting motionless, we introduced an intermediate state *Temporary away* and prolonged *Presence* state for a predefined time $\Delta t_{Presence}$. We observed that $\Delta t_{Presence}$ can be configured depending on the PIR sensor model used (see Sec. 4 for details). Only if $s_{PIR_i} = 0$ and $v_{Energy_i} \leq \phi_{Energy}$, *Away* is reported.

Layered Hidden Markov Models (LHMMs). A hidden Markov model (HMM) is a Markov model in which the observation is a probabilistic function of hidden states. To specify an HMM, two model parameters are needed: N representing the number of individual model states, and M,

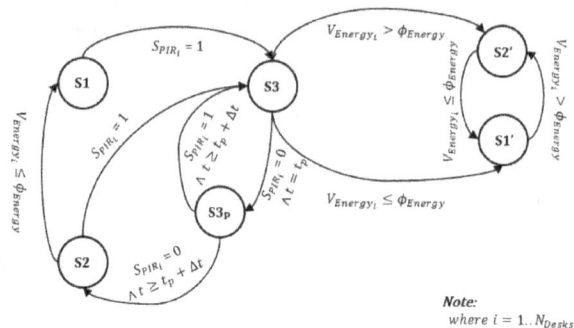

Figure 2: *Desk activity* state-model for classifying S_1-*Away*, S_2-*Temporary away* and S_3-*Presence*, S_1'-*Desk work* and S_2'-*Computer work* per office desk i. State S_{3p} represents the prolonged *Presence* state. See Section 3 for details. The energy consumption threshold ϕ_{Energy} and *Presence* state prolongation time $\Delta t_{Presence}$ were set empirically as detailed in Section 4.

describing the number of distinct observation symbols per state. The individual symbols are denoted as $V = \{v_1, v_2, \ldots, v_M\}$. Moreover, the HMM description requires specifying three sets of probability measures A, B and π, which are state-transition probability distribution, observation symbol probability distribution, and initial state distribution, respectively [21].

In order to derive HMM model parameters, unlabeled sequences of observations and states were used in a training step. The model parameter estimation, $\lambda = (A, B, \pi)$ was done using MATLAB [14].

Considering that *Computer work* and *Desk work* could exist only if the system is in a *Presence* state, we chose a layered HMM approach for recognizing activities. The classic LHMM approach used a bank of HMM classifiers to discriminate observation sequences [18]. The HMMs at a next level $L+1$ take outputs of the HMM at level L as their inputs. In our study, the first layer consisted of three nodes to model *Presence*, *Away* and *Temporary Away* states, as shown in Figure 3.

Unlike the classical LHMM approach, we used the Viterbi algorithm [21] to find most probable sequences of hidden states as a result of an observed event sequence. The result of the first layer was then used as an input for the second layer, which had two nodes representing *Computer work* and *Desk work* states.

3.4 People count estimation

To count people in an office room, we combined information from *Presence* states of all desk cells N_{Desks} in an office room. Our people count result is thus based on a combination of all distributed PIR sensor information. Upon a transition to or from the *Presence* state by our desk-cell recognition, people count was increased or decreased. To enable the system to perform rapid control decisions, we used the intermediate state *Temporary away* for updating the estimate, rather than *Away*. We applied the people count estimation based on results from both, FSM and LHMM models.

Table 2: Summary of our opportunistic sensing approach used for recognizing office activities. Activities were selected from a general set of office activities according to their relevance for energy-related control. Desk and room cells were considered to maintain scalability for office buildings of different sizes.

Recognition goal	Opportunistic sensor	Activities	Example energy saving application in offices
Desk-cell activities per desk $i = 1 \ldots N_{Desks}$			
"Desk activity"	Per-desk PIR (s_{PIR_i}) Per-desk PIR (s_{PIR_i}) Per-desk PIR (s_{PIR_i}), plug-in power meter (v_{Energy_i}) Per-desk PIR (s_{PIR_i}) plug-in power meter (v_{Energy_i})	Away Presence Computer work Desk work	Control appliances, e.g. shut-off computer screens when their users are away; decrease lighting for computer-based work.
Room-cell activities and states			
"People count"	Room PIR network ($s_{PIR_i} \forall i$)	People count	Room conditioning

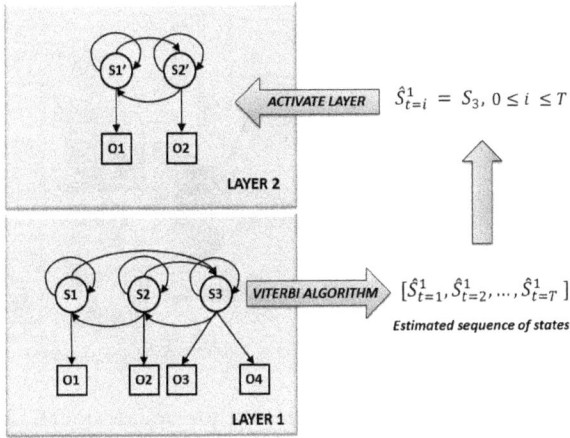

Figure 3: Layered representation of HMMs used in our recognition approach. Layer 1 consists of three states S_1-*Away*, S_2-*Temporary away* and S_3-*Presence*, where different PIR states and power measurements represent observations (O_1, O_2, O_3, O_4). The Viterbi algorithm was used to find optimal state sequences, linked to an observed event sequence. Our estimated state sequence represented the input for the second layer, where S_1'- *Desk work* and S_2'-*Computer work* states were activated during the S_3 (*Presence*) state.

4. IMPLEMENTATION AND STUDY DESIGN

This section details our living-lab implementation and study design used to analyze the opportunistic sensing approach. We describe the sensor selection to resemble commonly expectable performance. Study design considerations are described regarding opportunistic data acquisition and activity ground truth.

4.1 Living-lab implementation

In order to validate our approach, we set up a living-lab installation at multiple office rooms at the TU Eindhoven University campus. We investigated the scalability of our system and deployed installation in three offices with different structure: private (1-person office), a 3-persons office and a 4-persons offices. Each room cell was partitioned into several desk cells (N_{Desk}) according to number of users. Figure 4 shows a schematic representation of the office layout and sensor modalities used in the installation. Table 3 shows the type and quantity of the sensors used in our installations.

Table 3: Duration of the study, desk-cells configurations and type and quantity of sensors in private, 3-persons and 4-persons offices.

	Private office	3-persons office	4-persons office
Days of the study	5, including 2 weekend days	7, including 2 weekend days	5 working days
Number of desk-cells	1	3	4
Office sensors			
PIR Eltako FBH63AP	0	2	2
PIR Thermokon SR-MDS	1	1	2
Plug-in power meters Plugwise circles	1	3	4

In our living-lab installation, we used self-powered wireless PIR sensors based on the EnOcean wireless protocol[1] at 868 MHz with solar harvesting units: FBH63AP from Eltako[2] and SR-MDS from Thermokon[3]. All PIRs were ceiling-mounted, directly facing the desk areas. The office room height was ~3.2 m. Both models report measurements upon brightness changes of more than 10 lux, every 100 s, and directly when motion was detected. When no motion is detected, the Eltako model waits for 100 s before sending an off event, while the sensor from Thermokon waits for 1000 s. This duration was not adjustable for both models. Based on initial tests, we partly covered the PIR sensor lenses to narrow the sensor's field of view and to adjust focus to a specific desk. We set the state prolongation $\Delta t_{Presence}$ for the Eltako PIR to 240 s, and to 0 s for Thermokon.

For their solar harvesting operation, EnOcean sensors required sufficient lighting conditions. Nevertheless, our mount-

[1] www.enocean.com
[2] www.eltako.com
[3] www.thermokon.de

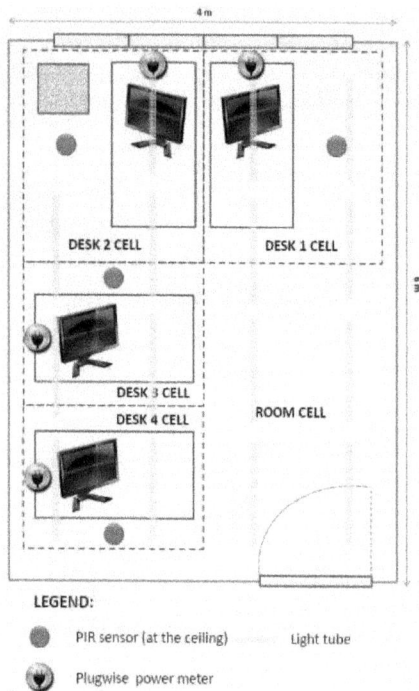

Figure 4: Schematic representation of a multi-user office room layout with four desk cells used in our investigation. Corresponding to the opportunistic sensor modalities introduced in Sec. 3, we used several sensor types and modalities. The room height is 3.2m.

ing locations did not hamper a continuous operation of the sensors.

To monitor energy consumption of desk screens we used Plugwise plug-in power meters[4] 'Circles' that communicate via ZigBee wireless protocol at 2.4 GHz. Plugwise provided instantaneous power consumption of the screens at a sampling frequency of ~1 min. Desk screens have a maximal power consumption rating of 41 Wh and a standby consumption of ~2 Wh. We set the energy consumption threshold to ϕ_{Energy}=2.2 Wh. The energy consumption threshold was chosen in accordance with Directive 2005/32/EC [8]. Screens standby time was configured to 2 min, based on the empirical observation in our living-lab that this setting is not considered uncomfortable for users. We used Context Recognition Network Toolbox (CRNT) [3] for recording and synchronization of the data streams via wireless USB interfaces.

4.2 Study design

Using the living-lab installation as detailed above, we implemented a multi-day study in the different offices with desks occupied by PhD students who were regular users of these desks. Participants were asked to maintain their working style as usual. No activities were scripted or prescribed in any form during the recordings. Recording durations and desk-cells configuration is shown in Table 3.

In order to obtain ground truth for the activities performed, including computer and desk work, participants were

[4]www.plugwise.com

asked to manually annotate their activities with a resolution of 1 min on a pencil-and-paper form. The form was designed such that regularly recurring activities could be rapidly filled in.

Since manual annotations can be inaccurate, we decided to use a pair of ultrasonic range finders (USR) as complementary sensor modalities to obtain reference data about the participants' presence. From USR data we derived presence as a threshold function of user's distance from the screen. The model SRF08 from Devantech was used and attached to the top of both sides of the screens except in the 4-persons office, where we used one sensor only. Sampling frequency was set to 1 s. The maximum range distance was set to 1 m to prevent unwanted reflections from the other objects. The actual installation setup for one desk cell is shown in Figure 5.

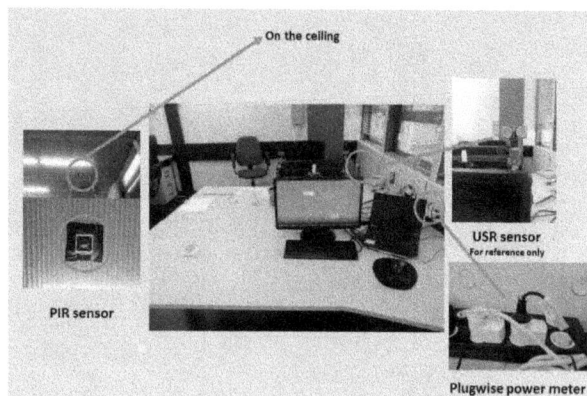

Figure 5: Living-lab installation for one desk cell, including ceiling-mounted PIR and plug-in power meter sensors. Ultrasound range sensors (USRs) were used as reference to derive ground truth for participant activities only.

4.3 Deriving ground truth

To complement the laborious manual annotations during the multi-day recordings, we used a USR-based activity detection of *Away* and *Presence*. In order to confirm the USRs' accuracy, we validated the classification performance in comparison to manual activity annotations of one user. The user was selected based on self-reported compliance in completing the manual annotation form.

The comparison between USR-based detection and user annotation resulted in a 94% agreement, except for the 4-persons office where only one USR was available. We considered this agreement as sufficient to further use the USR-based activity detection for evaluating the opportunistic sensor and recognition algorithm performances. To further obtain activity ground truth for the activities *Computer work* and *Desk work*, we merged USR-based activity detection with manual annotations of the users. In the 4-persons office dataset, manual annotations were used as they were found to be more accurate.

Subsequently, reference for people count was derived by considering the USR-based results from all desk cells. When the USR-based detection changed from *Away* to *Presence*, reference people count was increased, when passing to *Away*, reference people count was decreased.

252

5. RECOGNITION ANALYSIS

Figure 6 illustrates the sensor readings for one participant during a typical recording day in our study, including PIR sensor, screen power consumption, and the participant's activity annotation. As the waveforms illustrate, a general activity pattern is readily observable from sensor readings.

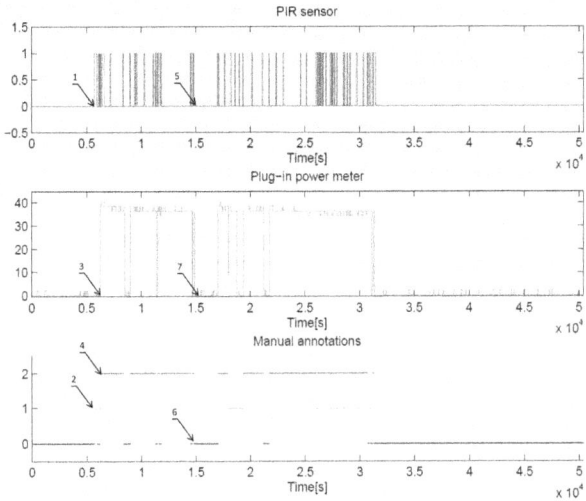

Figure 6: Example recording day from the study of one participant. Manual annotations of participant's activity were: 0=Away, 1=Desk work, 2=Computer work.

After entering the office room (1), the participant's initial activity was *Desk work* (2), which is confirmed by the screen's power consumption readings (3). Screen consumption alternated between 2 Wh and 41 Wh during the day according to the user's activity (4). When the participant left the desk (5), a delay was observable between the participant's activity transitions from *Computer work* to *Away* (6) and changes in the screen's consumption (7). This delay reflects the computer standby time of 2 min. We frequently observed that the PIR sensor falsely reported *Away* when a user was sitting still at the desk. It can be concluded that the PIR sensor was not sufficiently sensitive to recognize low amplitude motions. Our recognition could partly compensate this effect as the screen power consumption was incorporated too.

Desk activity recognition. Activity classification results for the *Desk activity* recognition approach are shown in Table 4. We derived the office-specific accuracy and reported durations of classified activities to represent the data amount for each condition. As the results show, FSMs and LHMMs yielded similar performance. Overall accuracy for *Presence* and *Away* was 82.4%. The results for *Presence* in the 3-persons and 4-persons office were lower than those for the private office, reflecting the different characteristics of PIR sensors used. This finding confirms observations made from the waveforms above and indicates that differences in hardware characteristics can affect accuracy.

Average performance for discriminating *Desk work* and *Computer work* was 95% for both, FSMs and LHMMs. The lower accuracy for *Desk work* was observed for the private office only. The private office occupant reported only very few minutes of *Desk work* throughout the study. We at-

Table 4: Classification results of the *Desk activity* recognition. Since ground truth was differently derived for *Presence* vs. *Away* and *Computer work* vs. *Desk work*, separate totals are shown.

Activity			Private office	3-persons office	4-persons office
Presence	FSM	Accuracy[%]	75.0	56.3	63.5
	LHMM	Accuracy[%]	75.0	56.3	63.5
		total time[h]	11.3	60.3	93.7
Away	FSM	Accuracy[%]	98.0	96.3	89.5
	LHMM	Accuracy[%]	98.0	96.3	89.5
		total time[h]	82.5	371.8	70.7
Total	FSM	Accuracy[%]	87.0	88.3	72.0
	LHMM	Accuracy[%]	87.0	88.3	72.0
		total time[h]	93.8	432.1	164.0
Desk work	FSM	Accuracy[%]	39.0	77.0	72.5
	LHMM	Accuracy[%]	43.0	77.7	73.0
		total time[h]	0.3	5.8	15.4
Computer work	FSM	Accuracy[%]	100.0	97.3	96.5
	LHMM	Accuracy[%]	100.0	98.3	94.5
		total time[h]	10.2	42.7	78.3
Total	FSM	Accuracy[%]	97.0	97.3	92.5
	LHMM	Accuracy[%]	98.0	98.0	91.8
		total time[h]	10.6	48.5	93.7

tributed the lower recognition performance to these short desk work activity interrupts of the occupant.

Figure 7 shows the average activity recognition performances for the *Desk activity* recognition. As the results confirm, the LHMM-based approach can obtain similar recognition performances compared to the manually designed FSMs.

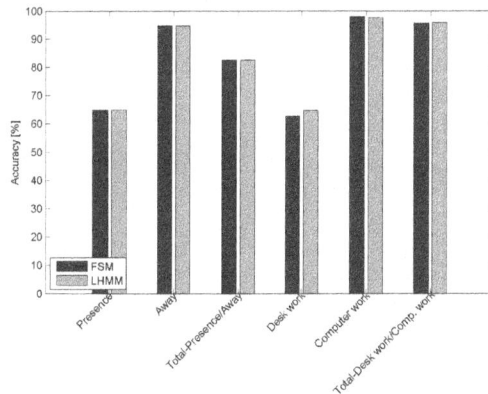

Figure 7: Average recognition performances for the *Desk activity* recognition. FSMs and LHMMs yielded similar results for all activities.

People count recognition. The people counting estimation performance and total people counts are shown in Table 5. Overall, an accuracy of 66.3% was obtained for FSMs and LHMMs. It could be noticed that accuracy of counting people for the 4-persons office was lower compared to the other two offices. This result can be explained by the low sensitivity of the PIR sensor model used in the 4-persons office, which resulted in a larger number of false negatives as discussed earlier. The accuracy per desk-cell represents the model's ability to recognize *Presence* and *Away* states, since the number of people will increase with activation of any PIR sensor and decrease with its deactivation. The relation of presence recognition and people count estimation is directly reflected in the accuracy of the private office setting.

Similar to the *Desk activity* recognition, FSMs and LH-MMs yielded identical estimation people count estimation results. In the LHMM-based approach, we learned data model without using labeled training data, while the FSM-based approach relied on a manually state design. The results confirmed that the unsupervised learning approach applied to derive the LHMMs is viable.

Table 5: People count estimation results. FSMs and LHMMs yielded identical estimation results due to a similar presence recognition performance. The PIR sensor model deployed in the 4-persons office showed insufficient sensitivity, resulting in reduced estimation performances.

	Activity	Private office	3-persons office	4-persons office
FSM	Accuracy [%]	87	78	34
LHMM	Accuracy [%]	87	78	34
	Total counts	468	2476	277

6. ENERGY CONSUMPTION ANALYSIS

We investigated whether the activities considered in our investigation are relevant for saving energy in office buildings. In this section, we compare the benefit of activity-based controls when using office-installed sensors, to several alternative control strategies without activity inference. These alternative controls can be frequently found in current BEMS.

6.1 Simulation approach

We used the EnergyPlus simulation software[5] for this analysis. EnergyPlus is an energy analysis and thermal load simulation tool developed by the US Department of Energy. It takes into account several parameters, including weather conditions, orientation and construction of the building, HVAC system, and occupancy. Key parameters of our simulation are shown in Table 6.

For all energy consumption simulations, we considered the 4-person multi-user office room shown in Figure 4 as an example. This room represents one thermal zone. We chose a fan coil unit as HVAC system and used yearly weather data for Amsterdam, the Netherlands, to simulate realistic environmental conditions. Space conditioning was determined according to setpoints that are typically applied in central Europe (see Tab. 6). The lighting system in this office room consisted of 15 dimmable fluorescent light tubes (length: 1200 mm), rated at 36 W per tube.

6.2 Desk activity recognition

Here we focused on lighting as one example to assess potential energy savings of activity-based control. To compare

[5]apps1.eere.energy.gov/buildings/energyplus

Table 6: Simulation parameters used in our energy consumption analysis to assess the benefit of activity sensing to save energy.

SIMULATION PARAMETERS	
HVAC system	Fan coil unit, 1 zone
Lighting system	15 dimmable fluorescent tubes, surface mounted
Area	24m^2
Location	Amsterdam, The Netherlands
Heating setpoints	21.1°C occupied, 12.8°C unoccupied
Cooling setpoints	23.9°C occupied, 40.0°C unoccupied (system off)

saving options, we considered several control types and used average activity patterns and sensor data from to our study to model user behaviours as described

Manual Control. Manual Control considers the situation where users would operate a lighting system. We assume here that the first person entering to the office room will switch lights on and the last person leaving the office will switch them off. Lights provided their maximal lux level, hence energy consumption would amount to 540 Wh. This situation can be considered as an extreme, uncontrolled condition, and serves as baseline for our comparison.

Presence-based Control. This control option considers that a PIR sensors would detect presence and consequently switch lights on. If no movement is detected, the PIR will switch lights off. The lights provide maximal lux level when activated. We derived PIR activations by combining measurements of all individual PIR sensors in our study.

Presence-per-desk-based Control. In the considered office room four light tubes are dedicated to each desk, except for desk 4, where there are only three tubes (see Figure 4). When $s_{PIR_i} = 1$ for desk i, lights dedicated to desk i will be activated and on maximal lux level. When $s_{PIR_i} = 0$ desk lights will be deactivated.

Activity-based Control. According to standard (BS EN 12464-1:2011), lower lighting levels (~500 lux) are sufficient for computer work, while desk-related work requires $\geq$750 lux. For activity-based control, we simulated a reduction of light level by ~30% during *Computer Work* and maintained lights maximal lux level during *Desk Work*. In our study, at least three light tubes were dedicated to each desk, providing total maximum of 7500 lm per desk. Per desk area (~5m^2), the available lux level was thus 1500 lux. If no presence was detected, lights would go off. Common lighting of the room was activated according to general room presence.

Analysis results.

In our simulations, we used actual data recorded from the 4-person office of our living-lab (see Table 3). The average power consumption per year of all control options is shown in Figure 8. For *Manual Control*, average power consumption on a yearly basis was 1199.48 kWh. For *Presence-based Control*, consumption decreased by 25.4%. When presence per desk was considered (*Presence-per-desk-based Control*), consumption decreased by 63.2% compared to *Presence-based Control*. By adding information on the participant's activity and adjusting lighting level accordingly in *Activity-based*

Control, average consumption per year was 257.5 kWh. Consumption in *Activity-based Control* was further 21.9% lower than the *Presence-per-desk-based Control*. Overall, the difference between *Manual Control* and *Activity-based Control* was 78.5%.

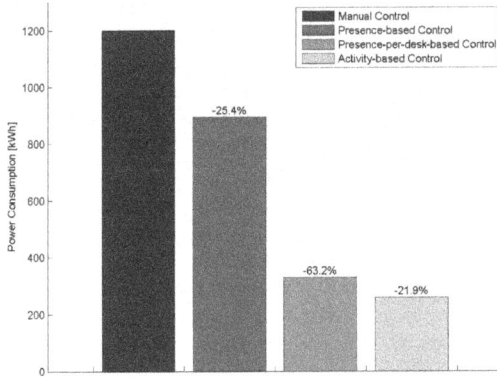

Figure 8: Lighting system average power consumption per year according to our energy consumption simulation. The results indicate savings for different control strategies, where *Activity-based Control* achieved an overall saving of 78.5% compared to *Manual Control*. This evaluation corresponds to the recognition of the 4-person office room in our living-lab.

6.3 People count estimation

People count in an office room has a large influence on the required ventilation rate. An increase in ventilation rate will increase power consumption of the fan coil unit. According to the ASHRAE Standard 62.1 [2] ventilation rates can be derived using:

$$V_{bz} = R_p P_z + R_a A_z, \qquad (1)$$

where R_p represents outdoor airflow rate per person, P_z is the population, R_a represents outdoor airflow rate per unit area and A_z is zone area.

We simulated people count in relation to required airflow, while considering a real fan coil unit. In this analysis, we used the UniTrane size 01[6] from Trane, as an example device. Based on simulated airflow, a relation of people count and HVAC energy consumption is illustrated in Table 7.

We combined real data recordings and recognition from our study and energy consumption estimations to assess the effect of actual people count recognition on HVAC energy consumption. In particular, we considered the following control strategies:

Time-based Control. In many buildings without presence detection, HVAC systems are operated according to fixed schedules. Facility managers set the HVAC to active before people would usually enter a building or office room, and shut it off when people left. This control procedure did not consider actual people count in a building. For our analysis, we assumed that the HVAC system will operate at

maximal speed and providing maximal air flow from 6:00 to 22:00.

Manual Control. This strategy assumes that users would operate the HVAC system. We considered that HVAC system would be activated when first person enters the office and deactivated when last person leaves the office.

Presence-based Control. Here, presence as reported by any PIR sensor in our example office room would control activation of the HVAC system. The HVAC would be operated at maximal speed.

People Count-based Control. In this strategy, we consider people count estimation and would control the HVAC system's fan coil unit accordingly. Thus, based to the detect number of occupants, power consumption of a fan coil unit will change. This approach resembles simulation results illustrated above.

Analysis results.

A comparison chart of the average power consumption derived in all control strategies is shown in Figure 9. Average power consumption per year for *Time-based Control* and *Manual Control* was 104 kWh and 55.5 kWh respectively, which is a difference of 46.6%. When using presence detection (*Presence-based Control*) at the office level, 25.4% was saved, compared to *Manual Control*. By considering the number of people in an office, fan speed was adjusted and with it air flow and power consumption. Here, energy consumption for *People Count-based Control* decreased by additionally 19.5%, resulting in a power consumption of 33.4 kWh on yearly basis. Nevertheless, here we estimated only the energy consumption related to fan operations, which does not consider energy needed for cooling and heating. Thus it can be expected that actual energy savings due to people count estimated could be larger.

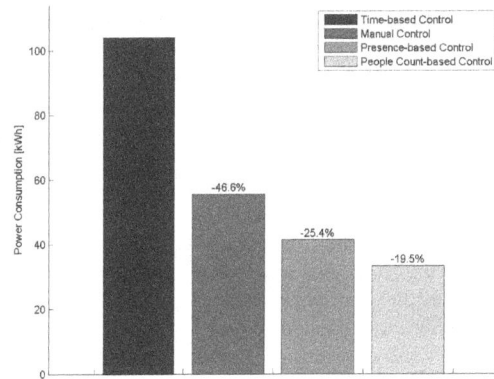

Figure 9: HVAC system average power consumption per year considering a fan coil unit operated according to different control strategies. Results show that actual people count estimates as obtained in the study can reduce energy consumption for 39.9% comparing to *Manual Control*.

Figure 10 illustrates the influence of different control strategies on lighting system power consumption per day: (a) *Manual Control*, (b) *Presence-based Control*, (c) *Presence-per-desk-based Control* and (d) *Activity-based Control*. In ad-

---

[6]www.engineer.trane.com

dition, activity annotations are shown for *Away*, DW-*Desk work*, and CW-*Computer work* of the four occupants. Occupants were assumed to be present if they reported their involvement in computer or desk-based activity. For this illustration, power consumptions and annotations were averaged in two minute windows. As the diagrams illustrate, power consumption decreases profoundly from *Manual Control* to *Presence-based Control*, and further reductions are noticeable for *Presence-per-desk-based Control*. During a period from 13:00 to 14:00, most of the occupants were involved in *Computer work*, and lighting system consumed less than 200 Wh, for *Activity-based Control*. This low consumption corresponds to the decrease of ∼30% compared to the same period in *Presence-per-desk-based control* (∼300 Wh).

Table 7: Influence of people count on airflow and power consumption of a fan coil unit. We considered here the HVAC unit UniTrane size 01 from Trane.

Number of people	Airflow $[m^3/h]$	Power [W]
0	0	0
1	36.36	17
2	72.72	23.84
3	109.44	25
4	145.80	25

7. DISCUSSION AND CONCLUSIONS

In this work, we investigated the potential of using a subset of commonly installed office building sensors for recognizing activities that are relevant for energy saving in appliances and building systems. While these sensors are currently installed in buildings to directly control appliances, lighting, and HVAC, we used them to extract more detailed office activities. We selected desk-related activities and people count in office rooms. FSM and LHMM recognition models were implemented to capture state logic and data fusion. As we showed here, the models and approach could be applied in private and multi-user offices.

Our living-lab study helped to confirm that standard office sensors are in principle sufficient for the recognition task. We consider this result relevant, as the considered sensor modalities and similar ones are frequently found in modern or recently refurbished office buildings and require minimal maintenance. While such installed sensors are intended for closed-loop operation, sensor are often integrated into wired or wireless building networks, thus information can be intercepted. Thus, our concept could be applied without additional sensor installation, as it was frequently assumed in previous investigations.

The activity recognition accuracy of our system could be compared to the study of Nguyen et al. [16], who used dedicated infrared, pressure and acoustic sensors. The authors focused on similar activities in an experimental office room occupied by a single user and recorded for five working days. Activities corresponding to *Away* and *Computer work* could be recognized at 96.5% and 100%, which is similar to the 98.0% and 100% obtained in our study, respectively. We observed lower accuracies for *Presence* (75.0%) and *Desk work* (43.0%), compared to 94.3% and ∼94.9% (*Working without PC*) in their study. We interpret that the reduced performances in our "Private office" setting could be explained by the small recording time for *Desk work* and by

using a PIR sensor and a power meter alone. Beside motion sensors, pressure sensors in chairs and microphones were used in the study of Nguyen et al. When compared to the overall performance across all offices in our study, we assume that the diversity in occupation with multiple users and natural working habits in our living-lab could explain differences. Unlike our study, where users were present for 165.34 h and away for 525 h, in their study only 4.42 h and 5.2 h were recorded, respectively.

Our energy consumption analysis showed energy saving potential for activity-based controls over using individual sensors only and manual control. Most of modern BMS and BEMS systems control lighting and HVAC systems based on occupancy only. Results of our study and simulations showed that by identifying user activities and people count per room, it is possible to save 21.9% and 19.5% of energy, respectively, over common BEMS systems. These results were based on our study data and recognition results in a 4-person office room. Although we consider the energy saving estimations as realistic for the considered office room, consumption may vary for other building setups that involve other users and include other installations. For example, fluorescent light tubes may not be very common in newly built commercial buildings. However, when considering modern LED lighting, e.g. the Philips Master LEDtube at 1200 mm length consuming 19 Wh, our activity-based control could still save ∼500 kWh per year for our example office room.

In our study, we divided office room into four zones where dedicated lights were operated according to activities. If an office room provides common lights for all zones as part of the overall lighting, per-room savings obtained by activity-based control will depend on the ratio between common lights and desk-related lights. While, additional common lights would decrease the saving potential, many office rooms already constrain common lighting to safety requirements.

We observed that sensor models from different manufacturers require specific recognition model settings to achieve optimal performance. To an extreme, sensitivity of some particular sensor models may not be adequate for all recognition tasks here. Furthermore, in our approach, we assumed that office computers were adequately configured and suspending the screen if no activity occurred. This may be acceptable in many environments, however could be a limitation if users work without a separately controllable screen. EnergyStar[7] states that 95% of deployed screens and 25% of office computers have power management features enabled, which supports our power metering approach.

While we recorded screen power in the study, our recognition approach does not require power metering. As alternative, a computer software could be used to track mouse and keyboard activities and thus identify computer work. More generally, office buildings include a variety of installations and appliances. We consider that our approach could be adapted to available sensor resources in a given building.

Since we modeled activities per desk-cell and used desk-specific sensors, information regarding the activity of individual occupants could be revealed. However, for our recognition approach it is not required to identify and track occupants.

Further elaboration of the recognition models and additional sensors, such as door and window switches, could in-

---

[7]www.energystar.gov

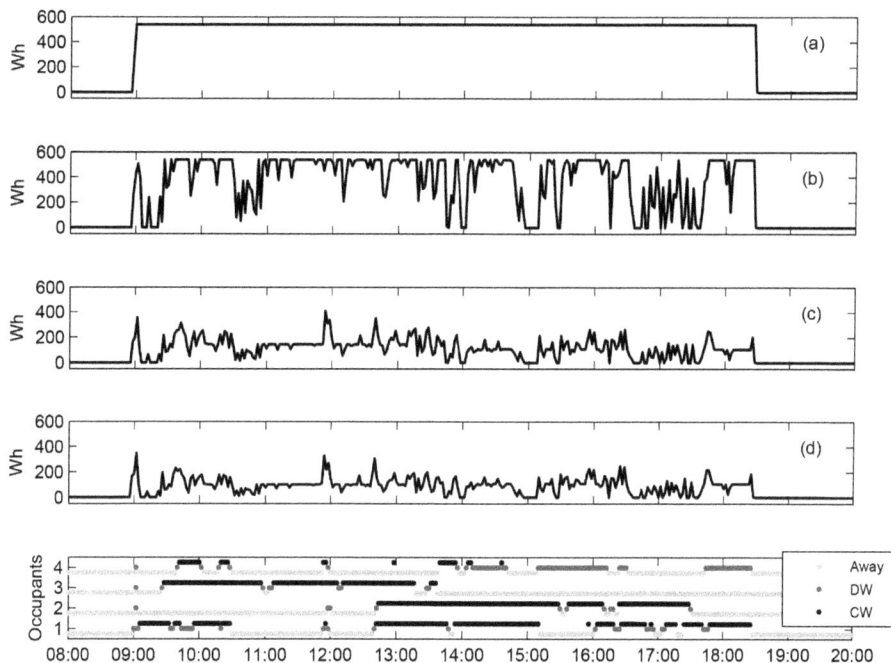

Figure 10: Illustration of the lighting system power consumption per day. The consumption traces were simulated based on the occupant activity annotations of four office users shown in the bottom plot (DW: *Desk work*, CW: *Computer work*). The following control strategies were considered: (a) *Manual Control*, (b) *Presence-based Control*, (c) *Presence-per-desk-based Control*, (d) *Activity-based Control*. Power consumptions and annotations were averaged using two minute windows.

crease the activity-based control options in buildings. The effort to configure model parameters during the deployment could be minimised through automatic teach-in procedures, similar to the commissioning of current building sensor networks.

Further studies, involving longer recording periods, more office rooms and participants should be conducted in order to confirm the recognition accuracy and potential energy savings in real office settings, as the living-lab used in this work.

8. ACKNOWLEDGMENTS

The work was kindly supported by the EU FP7 project GreenerBuildings, contract no. 258888, and the Netherlands Organisation for Scientific Research (NWO) project EnSO, contract no. 647.000.004.

9. REFERENCES

[1] Y. Agarwal, B. Balaji, R. Gupta, J. Lyles, M. Wei, and T. Weng. Occupancy-driven energy management for smart building automation. In *Proceedings of the 2nd ACM Workshop on Embedded Sensing Systems for Energy-Efficiency in Building*, BuildSys '10, pages 1–6, New York, NY, USA, 2010. ACM.

[2] American Society of Heating, Refrigerating and Air Conditioning Engineers, Inc. Ansi/ashrae standard 62.1-2010, ventilation for acceptable indoor air quality, 2010.

[3] D. Bannach, O. Amft, and P. Lukowicz. Rapid prototyping of activity recognition applications. *IEEE Perv Comput*, 7(2):22–31, April–June 2008.

[4] D. T. Delaney, G. M. P. O'Hare, and A. G. Ruzzelli. Evaluation of energy-efficiency in lighting systems using sensor networks. In *Proceedings of the First ACM Workshop on Embedded Sensing Systems for Energy-Efficiency in Buildings*, BuildSys '09, pages 61–66, New York, NY, USA, 2009. ACM.

[5] V. L. Erickson and A. E. Cerpa. Occupancy based demand response hvac control strategy. In *Proceedings of the 2nd ACM Workshop on Embedded Sensing Systems for Energy-Efficiency in Building*, BuildSys '10, pages 7–12, New York, NY, USA, 2010. ACM.

[6] V. L. Erickson, Y. Lin, A. Kamthe, B. Rohini, A. Surana, A. E. Cerpa, M. D. Sohn, and S. Narayanan. Energy efficient building environment control strategies using real-time occupancy measurements. In *Proceedings of the First ACM Workshop on Embedded Sensing Systems for Energy-Efficiency in Buildings*, BuildSys '09, pages 19–24, New York, NY, USA, 2009. ACM.

[7] European Commission. European union directive on the energy performance of buildings (EPBD). Technical Report 2002/91/EC, European Commission, 2002.

[8] European Commission. Guidelines accompanying commission regulation (ec) no 1275/2008, 2009. http://ec.europa.eu/energy/efficiency/ecodesign/doc/legislation/guidelines_for_smes_1275_2008_okt_09.pdf.

[9] J. E. Froehlich, E. Larson, T. Campbell, C. Haggerty, J. Fogarty, and S. N. Patel. Hydrosense: infrastructure-mediated single-point sensing of

whole-home water activity. In *Proceedings of the 11th international conference on Ubiquitous computing*, Ubicomp '09, pages 235–244, New York, NY, USA, 2009. ACM.

[10] S. K. Ghai, L. V. Thanayankizil, D. P. Seetharam, and D. Chakraborty. Occupancy detection in commercial buildings using opportunistic context sources. In *PerCom Workshops*, pages 463–466. IEEE, 2012.

[11] M. Jahn, T. Schwartz, J. Simon, and M. Jentsch. Energypulse: tracking sustainable behavior in office environments. In *Proceedings of the 2nd International Conference on Energy-Efficient Computing and Networking*, e-Energy '11, pages 87–96, New York, NY, USA, 2011. ACM.

[12] P. Jaramillo Garcia and O. Amft. Improving energy efficiency through activity-aware control of office appliances using proximity sensing - a real-life study. In *SEnAml 2013: Proceedings of the 5th International Workshop on Smart Environments and Ambient Intelligence*. IEEE, 2013. Accepted for publication.

[13] A. Marchiori and Q. Han. Distributed wireless control for building energy management? In *Proceedings of the 2nd ACM Workshop on Embedded Sensing Systems for Energy-Efficiency in Building*, BuildSys '10, pages 37–42, New York, NY, USA, 2010. ACM.

[14] MATLAB and Statistics Toolbox Release 2011b. *version 7.13.0 (R2011b)*. The MathWorks, Inc., Natick, Massachusetts, United States, 2011.

[15] M. Milenkovic and O. Amft. Recognizing energy-related activities using sensors commonly installed in office buildings. In *SEIT '13: Proceedings of the 3rd International Conference on Sustainable Energy Information Technology*. Procedia Computer Science, 2013. Accepted for publication.

[16] T. A. Nguyen and M. Aiello. Beyond indoor presence monitoring with simple sensors. In *Proceedings of the 2nd International Conference on Pervasive and Embedded Computing and Communication Systems*, 2012. To appear.

[17] A. G. Nuria Oliver, Eric Horvitz. Layered representation for human activity recognition. In *Fourth IEEE International Conference on Multimodal Interfaces*, pages 3 – 8, 2002.

[18] N. Oliver and E. Horvitz. A comparison of hmms and dynamic bayesian networks for recognizing office activities. In *UM'05 Proceedings of the 10th international conference on User Modeling*, pages 199–209. Springer, 2005.

[19] S. N. Patel, M. S. Reynolds, and G. D. Abowd. Detecting human movement by differential air pressure sensing in hvac system ductwork: An exploration in infrastructure mediated sensing. In *Proceedings of the 6th International Conference on Pervasive Computing*, Pervasive '08, pages 1–18, Berlin, Heidelberg, 2008. Springer-Verlag.

[20] S. N. Patel, T. Robertson, J. A. Kientz, M. S. Reynolds, and G. D. Abowd. At the flick of a switch: detecting and classifying unique electrical events on the residential power line. In *Proceedings of the 9th international conference on Ubiquitous computing*, UbiComp '07, pages 271–288, Berlin, Heidelberg, 2007. Springer-Verlag.

[21] L. Rabiner and B.-H. Juang. *Fundamentals of speech recognition*. Prentice-Hall, Inc., Upper Saddle River, NJ, USA, 1993.

[22] L. V. Thanayankizil, S. K. Ghai, D. Chakraborty, and D. P. Seetharam. Softgreen: Towards energy management of green office buildings with soft sensors. In *COMSNETS*, pages 1–6. IEEE, 2012.

[23] C. Wojek, K. Nickel, and R. Stiefelhagen. Activity recognition and room-level tracking in an office environment. In *Multisensor Fusion and Integration for Intelligent Systems, 2006 IEEE International Conference on*, page 25–30, 2006.

[24] C. R. Wren and E. M. Tapia. Toward scalable activity recognition for sensor networks. In *LoCA*, pages 168–185, 2006.

An ICT-based Energy Management System to Integrate Renewable Energy and Storage for Grid Balancing

Diego Arnone
Engineering I.I. S.p.A.
viale della Regione Siciliana N.O.
n.7275, 90146, Palermo, Italy
+390917511734

diego.arnone@eng.it

Massimo Bertoncini
Engineering I.I. S.p.A.
via Riccardo Morandi n.32,
00148, Roma, Italy
+390683074240

massimo.bertoncini@eng.it

Alessandro Rossi
Engineering I.I. S.p.A.
viale della Regione Siciliana N.O.
n.7275, 90146, Palermo, Italy
+390917511735

alessandro.rossi@eng.it

Fabrizio D'Errico
Mc Phy Energy
Z.A. Quartier Riétière
26190 La Motte-Fanjas, France
+33475711505

fabrizio.derrico@mcphy.com

Carlos García-Santiago
Tecnalia, Optima Unit
Parque Tecnológico de Álava
01510 Miñano, Spain
+34946430850

carlosalberto.garcia@tecnalia.com

Diana Moneta
Ricerca sul Sistema Energetico S.p.A
via Rubattino, 54
20134 Milano , Italy
+390239924662

diana.moneta@rse-web.it

Cristiano D'Orinzi
ENEL Distribuzione S.p.A.
Via Ombrone, 2
00195 Roma, Italy
+390683052699

cristiano.dorinzi@enel.com

ABSTRACT

Among the different renewable sources the "green" energy coming from wind or solar farms is often injected into the grid when it is not expected or there is no demand. The integration of photovoltaic and wind energy into the grid entails the need to provide proper solutions to a wide range of problems not entirely new to the network but still more critical. At the very heart of the INGRID European project, an ICT-based Energy Management System is being designed as the core of a self-adaptive & autonomic system in charge of dispatching the green energy among the smart grid, a hydrogen-based green-energy storage and an innovative urban mobility system.

Categories and Subject Descriptors

H.4 [**Information Systems Applications**]: Decision Support.

General Terms

Algorithms, Management, Measurement, Design.

Keywords

Energy Management Systems, Energy Storage, Electric Grid balancing, Renewable Sources integration.

e-Energy'13, May 21–24, 2013, Berkeley, California, USA.

ACM 978-1-4503-2052-8/13/05.

1. INTRODUCTION

The primary production of energy from renewable sources is rapidly expanding thanks to government directives and a growing environmental awareness. Among the different renewable sources (hydropower, wind power, solar power, tidal and wave power, geothermal power and power from biomass) the "green" energy coming from photovoltaic (PV) plants and wind farms is intrinsically characterized by *non-controllable variability*, *partial unpredictability* and *locational dependency* [1]. The integration of PV and wind energy into the grid entails the need to provide proper solutions to a wide range of problems not entirely new to the network but still more critical since, nowadays, a (smart) grid must compensate Renewable Energy (RE) output fluctuation, grid faults, conventional generation outages, load variation with more flexibility consisting in conventional generation flexibility, Demand Response, energy storage and Grid-friendly RE generation.

In this context, the INGRID European co-funded project is studying several solutions, whose this work, still in progress, is part of. The core innovation of the INGRID project will consist of combining solid-state high-density hydrogen storage systems with advanced ICT solutions for smart distribution grids which will monitor and control a large number of renewable energy sources in order to balance power supply and demand. In INGRID, the energy produced by RE sources can be injected into the grid or dispatched to the storage that, in its turn, can supply a fuel cell. The output of the fuel cell can be connected to the grid or can supply the Intelligent Dispenser (ID) of an innovative green urban mobility system [2]. Understanding how energy must be dispatched among grid, storage and urban mobility system is the main goal of the Energy Management System (EMS), which is the topic addressed by this work.

2. RELATED RESEARCH INITIATIVES

Several research projects and initiatives are focused on proposing new approaches and solutions in order to manage RE and/or balance energy supply and demand. Research projects like SEESGEN-ICT (http://seesgen-ict.rse-web.it/), INTEGRIS (http://fp7integris.eu/) or HiPerDNO (http://dea.brunel.ac.uk/hiperdno/), aim at designing and implementing new solutions for EMS, in particular through communication networks improvements. Other research works, like MIRABEL project (http://www.miracle-project.eu/), propose an approach on a conceptual and infrastructural level that allows energy distribution companies to balance the available supply of renewable energy sources and the current demand in ad-hoc fashion.

However, despite that a lot of research has been undertaken in ICT-based EMS, significant research challenges do exist in terms of improving sustainability, reliability and cost-efficiency of energy supply. So far, very partial and fragmented ICT-based solutions have been proposed. In general, current systems suffer from lack of a comprehensive yet fully integrated contextual data model and a suitable intelligent processing of the captured information; all of these will be fundamental for proactively predicting the energy production and accordingly fine tune the energy produced and delivered to the grid.

3. THE INGRID ICT-BASED EMS

The INGRID EMS is being developed on the concept of a closed-loop feedback control. An intelligent, self-learning and reasoning EMS is being designed to make the grid self-adaptive by implementing four key tasks: **collect**, **analyze**, **decide**, and **act** [3]. The feedback cycle starts with the *collection* and monitoring of relevant data from the relevant sources to feed a comprehensive contextual information model that will fully describe the whole status of the connected systems as well as the factors which affect the decisions for energy dispatching (i.e. weather forecast, historical energy demand, etc.). The analysis of the data coming from distributed and heterogeneous data sources will originally combine intelligent processing technologies (basic and advanced statistics, data mining, complex event processing, predictive analytics, etc) for predicting the RE production and adaptively match the RE farms output with the energy which could be injected into the grid, the storage

and the ID. Based on the results coming from the analysis, the EMS will take a *decision* about how to adapt the system in order to reach a desirable state, involving also external factors such as local policy for a risk analysis about reliability of the whole energy system. The decision will be properly implemented by an *action* through available effectors, for balancing energy flows and controlling the RES power quality. The EMS can follow either fully automatic or semi-automatic paradigm because, according to the requirements of reliability and/or risk policies, it could request the human intervention in order to achieve the desired goal. So, human operators will be able to question, and possibly over-ride, the decisions. This human knowledge will be stored in an automated Knowledge Base, together with both successful and unsuccessful past strategies. This knowledge will then be used to incrementally improve the system's performance.

4. CONCLUSIONS

This work-in-progress paper is an overview of an innovative ICT-based Energy Management System that is being developed at the very heart of the FP7 EC co-funded INGRID project. The EMS is in charge of dispatching the energy produced by PV and wind farms among an hydrogen high-capacity storage, an Intelligent Dispenser for electric car recharging and the grid. A concrete demonstrator that will be deployed in Italy, Apulia region (with more than 2.4 GW of already existing PV plants), will prove the effectiveness of the proposed EMS.

5. ACKNOWLEDGMENTS

Authors thank the members of the INGRID Consortium as well as the European Commission for supporting any project dissemination activities.

6. REFERENCES

[1] Grid integration of large-capacity Renewable Energy sources and use of large-capacity Electrical Energy Storage, white paper, IEC MSB October 2012.

[2] D'Errico, F., Screnci, A., and Romeo, M., A Green Urban Mobility System Solution from the EU Ingrid project, to be orally presented at REWAS 2013, San Antonio, Texas.

[3] Salehie, M., Pasquale, L., Omoronyia, I., and Nuseibeh, B., Towards self-protecting smart metering: investigating requirements for the MAPE loop, in 9th IEEE International Conference and Workshops on the Engineering of Autonomic and Autonomous Systems, 2012 (EASe'12).

Energy Eff cient Opportunistic Uplink Packet Forwarding in Hybrid Wireless Networks

Arash Asadi
arash.asadi@imdea.org

Vincenzo Mancuso
vincenzo.mancuso@imdea.org

Institute IMDEA Networks, and University Carlos III of Madrid
Madrid, Spain

ABSTRACT

Opportunistic schedulers have been primarily proposed to enhance capacity of cellular networks. However, little is known about opportunistic scheduling with fairness and energy efficiency constraints. In this work, we show that adapting opportunistic scheduling can dramatically ameliorate energy efficiency for uplink transmissions, while achieving near-optimal throughput and high fairness. To achieve this goal, we propose a novel two-tier uplink forwarding scheme in which users cooperate, in particular by forming clusters of dual-radio mobiles in hybrid wireless networks.

Keywords

LTE, opportunistic scheduling, clustering.

Categories and Subject Descriptors

C.2.1 [**Computer-communication networks**]: Network Architecture and Design—*Wireless communication*

1. INTRODUCTION

Designing a scheduler that maintains fairness, high spectral efficiency and low power costs in cellular networks is challenging. Although opportunistic schedulers exist and remedy bandwidth limitations [1,2], achieving joint fairness and low power consumption of smartphones is an open issue. For instance, the majority of smartphones are now equipped with at least two radio interfaces (i.e., LTE and WiFi) and use powerful processing hardware, which comes at the expense of potentially elevated power consumptions. Thereby, improving energy efficiency is of utmost importance.

In this paper, we propose an architecture which leverages cooperative communications and opportunistic scheduling to boost the energy efficiency of uplink communications (see Figure 1). Specifically, we exploit the secondary radio interface (e.g., WiFi) to form clusters among mobile devices. Cluster members use this secondary interface to forward their traffic to the *cluster head*, i.e., the cluster member with the best channel quality to the base station at that instant. Energy efficiency is substantially increased since only *cluster heads* talk to the base station, while using the secondary interface among cluster members requires low power.

Our proposed architecture benefits from opportunistic scheduling and cooperative communication techniques. Opportunistic schedulers have been extensively studied within past two decades (for a comprehensive survey see [3]). Clustering and relaying are common cooperative techniques used in

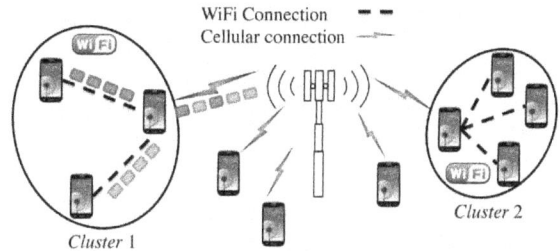

Figure 1: Our proposed clustered architecture.

wireless community, especially in sensor networks. An energy saving clustering approach for sensor networks is proposed in [4] where sensors volunteer to be the *cluster head*. Our approach is different in the sense that communications occur on distinct and heterogeneous radio interfaces and *cluster head* selection is opportunistic. The authors of [5,6] propose to use dual-radio mobiles/relay stations to improve the network performance. However, unlike our proposal there is no cluster formation among mobiles/relay stations. Moreover, the relay selection is not opportunistic that significantly impacts capacity. In [7], single antenna mobiles form cluster to create a virtual MIMO transmission. Unlike our proposal, in virtual MIMO all cluster members have to communicate with the base station.

2. CLUSTER-BASED SCHEDULERS

In this section, we illustrate the performance of our proposal by providing preliminary results obtained from numerical simulations run on Mathematica.

System model. We consider an LTE-like uplink communication scheme, using *20 MHz* bandwidth and operating in FDD mode. Uplink channel is assumed to follow a stationary Rayleigh fading model. For simplicity, we categorize the users into three predefined user channel quality classes (referred to as *user qualities*), namely, *poor*, *average*, and *good*. The mean achievable rates for *poor*, *average*, and *good* users are 20%, 50%, and 80% of the maximum transmission rate achievable in the system, respectively. In our evaluation scenario mobiles form three clusters with 6, 8, and 10 users, see Figure 1. Each mobile is equipped with LTE and WiFi interface and all queues are fully-backlogged. A cluster is simply a group of mobile users that communicate with each other over a WiFi network. We derive the power consumption of mobiles from the empirical power models proposed for LTE and WiFi in [8] and [9], respectively. The resulting power consumption of a device consists of: (*i*) a baseline power consumption of each of the two wireless interfaces; (*ii*) the power spent for LTE transmission by *cluster heads*; (*iii*) the power spent for WiFi reception by *cluster heads*; and (*iv*) the power spent for WiFi transmission by cluster members.

e-Energy'13, May 21–24, 2013, Berkeley, California, USA.
ACM 978-1-4503-2052-8/13/05.

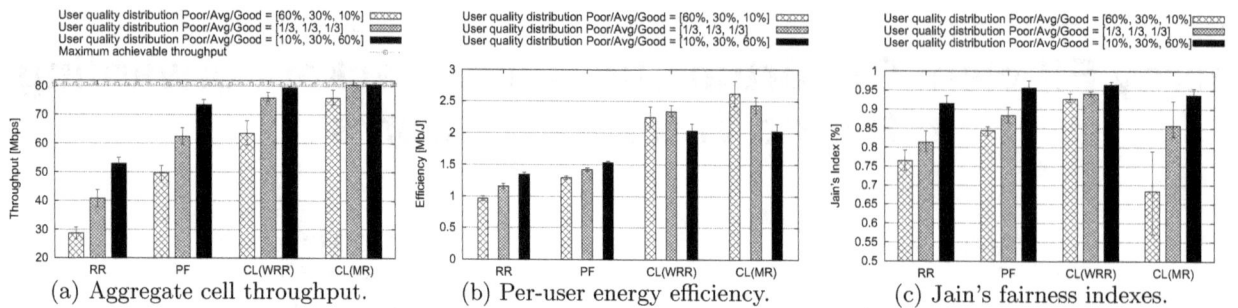

(a) Aggregate cell throughput. (b) Per-user energy efficiency. (c) Jain's fairness indexes.

Figure 2: aggregate throughput, energy efficiency and user fairness for single cell scenario.

Wireless interfaces are assumed to be in idle mode when no packet has to be transmitted or received

Cluster scheduling. Our scheduling architecture benefits from a two-tier system. The first-tier scheduler performs resource allocation among clusters by utilizing weighted round robin, namely CL(WRR), or MaxRate [1], namely CL(MR). CL(WRR) allocates resources to the clusters in a WRR manner with the weight being the cluster size. CL(MR) schedules the cluster whose *cluster head* is experiencing the best channel quality among all *cluster heads* in that instant. The second-tier scheduler is responsible for intra-cluster scheduling, and uses MaxRate to select the *cluster head*. We benchmark the proposal against Round Robin (RR) and Proportional Fair (PF) algorithms.

Evaluation. In addition to investigating different schedulers, we also explore the impact of user quality distribution on system performance. Hence, our evaluation includes the results for scenarios with: (*i*) equiprobable user quality distribution, (*ii*) statistically more *poor* users (i.e., the percentage of *poor*, *average*, and *good* users is {10%, 30%, 60%}, respectively), (*iii*) statistically more *good* users (i.e., 60% of *poor*, 30% of *average*, and 10% of *good* users).

Figure 2(a) confirms the superiority of cluster-based schedulers over RR and PF which is due to exploiting cooperative diversity and user channel diversity. Although PF also uses user channel diversity, it is still outperformed by cluster-based schedulers due to the cooperative architecture we propose. RR has the lowest throughput because it does not take advantage of cooperation and user channel diversities. Regarding the effect of user quality distributions, we can see that cluster-based schedulers almost achieve the maximum throughput with 33% *good* users in the network. Also the throughput differences among different schedulers shrink as the number of *good* users increases. This is expected because opportunistic gain relies on the channel diversity of users and increasing number of *good* users reduces this diversity.

In Figure 2(b) we see that cluster-based schedulers are more energy efficient than user-based schedulers. Since with cluster-based scheduling, the likelihood of transmitting under good channel quality is higher than PF and RR, mobiles can use high data rate with low packet loss, leading to higher energy efficiency. The evaluation results confirm that cluster-based schedulers provide a minimum 30% gain in energy efficiency with respect to PF, and the gain can reach up to 100% in presence of more *poor* users. The energy efficiency gain with respect to RR is even higher.

Figure 2(c) shows that fairness is lowest when there are more *poor* users in the system because of opportunistic schedulers bias toward serving *good* users. This also explains the reason why CL(MR) is outperformed by RR and PF. It is interesting to observe that CL(WRR) achieves the highest per-user fairness level in the system (even higher than PF).

In our clustering approach, the bandwidth is equally distributed among cluster members that smoothen the throughput differences among users. This improvement is the reason why CL(WRR) performs better than PF.

3. CONCLUSION

In this paper, we have shown that cluster-based schedulers can exploit the coexistence of LTE and WiFi interfaces in smartphones to enhance both system performance and device's battery life. Specifically, the impact of cluster-based schedulers is threefold: (*i*) they significantly improve the throughput of a cellular network, (*ii*) boost energy efficiency, and (*iii*) achieve fairness levels higher than PF.

4. ACKNOWLEDGEMENTS

This research was funded in part by the EU's FP7 program (ICT FLAVIA project, grant agreement n.257263).

5. REFERENCES

[1] R. Knopp and P. Humblet, "Information capacity and power control in single-cell multiuser communications," in *Proceedings of IEEE ICC*, 1995.

[2] P. Bender, P. Black, M. Grob, R. Padovani, N. Sindhushayana, and A. Viterbi, "CDMA/HDR: a bandwidth-efficient high-speed wireless data service for nomadic users," *IEEE Communications Magazine*, 2000.

[3] A. Asadi and V. Mancuso, "A survey on opportunistic scheduling in wireless communications," *IEEE Communications Surveys & Tutorials*, 2012.

[4] S. Bandyopadhyay and E. Coyle, "An energy efficient hierarchical clustering algorithm for wireless sensor networks," in *Proceedings of IEEE INFOCOM*, 2003.

[5] H. Wu, C. Qiao, S. De, and O. Tonguz, "Integrated cellular and ad hoc relaying systems: iCAR," *IEEE JSAC*, 2001.

[6] L. Le and E. Hossain, "Multihop cellular networks: Potential gains, research challenges, and a resource allocation framework," *IEEE Communications Magazine*, 2007.

[7] A. Sendonaris, E. Erkip, and B. Aazhang, "User cooperation diversity. Part I. System description," *IEEE Transactions on Communications*, 2003.

[8] J. Huang, F. Qian, A. Gerber, Z. Mao, S. Sen, and O. Spatscheck, "A close examination of performance and power characteristics of 4G LTE networks," in *Proceedings of ACM MobiSys*, 2012.

[9] A. Garcia-Saavedra, P. Serrano, A. Banchs, and G. Bianchi, "Energy consumption anatomy of 802.11 devices and its implication on modeling and design," in *Proceedings of ACM CoNEXT*, 2012.

Lessons Learned on Home Energy Monitoring and Management: Smartcity Málaga

Jaime Caffarel, Igor Gómez, Guillermo del Campo, Rocío Martínez and Carmen Lastres

CeDInt- Universidad Politécnica de Madrid

Campus de Montegancedo s/n, Pozuelo de Alarcón

28223, Madrid Spain

+34 914524900

jcaffarel@cedint.upm.es

ABSTRACT

The Smartcity Málaga project is one of Europe's largest eco-efficient city initiatives. The project has implemented a field trial in 50 households to study the effects of energy monitoring and management technologies on the residential electricity consumption. This poster presents some lessons learned on energy consumption trends, smart clamp's reliability and the suitability of power contracted by users, obtained after six months of data analysis.

Categories and Subject Descriptors

J.2 [**Physical Sciences and Engineering**]: Engineering

General Terms

Performance, Reliability, Experimentation

Keywords

Data Analysis, Data Reliability, Energy Consumption, Energy Monitoring.

1. THE SMARTCITY MÁLAGA PILOT

The Smartcity Málaga project aims to concentrate a wide range of sustainable technologies in the city: smart metering, smart grid, renewable energy generation as well as storage and e-vehicle recharging, among other innovations. Specifically, it evaluates the impact on reducing residential energy consumption by means of improving the energy consumption feedback. In Spain, customers usually receive information about their energy consumption (billing) every two months. Previous reviews [1] report saving up to 15%, while recent projects show effects around 5% [2].

An energy monitoring system consisting of a smart meter, a smart clamp and a set of on-off controllers and power monitoring plugs has been installed and configured in each participant's household. The end user has access to instant and average energy consumption and configuration information via smartphone and over the Internet.

Moreover, through an acquisition, storage and exploitation system (DASE), all the configuration values, together with hourly energy measurements, historic and current billing data, weather information and working calendar have been stored in a relational database.

Based on a segmentation process, an initial group of 50 participants were identified, according to their power consumption and higher technical knowledge. Finally, only 25 users that had more than 75% of right measurements (i.e. without long term disconnection of devices) were retained for analysis. The trial field phase started in December 2011 and ended in December 2012. In this poster, we present the structure of the DASE along with the results from the data analysis corresponding to the first semester of year 2012.

2. DATA ACQUISITION, STORAGE AND EXPLOITATION SYSTEM

As a first step, collecting and formatting all the relevant information (i.e. metering, billing, configuration and weather) has been necessary to perform a reliable analysis.

Each data source has a different format (XML, CSV and XLS), structure and time stamp. Using the DataStage tool (IBM), information is collected, transformed and stored in a database. A relational database was designed and implemented using the DB2 tool (IBM). Cognos tool (IBM) has been used to represent both numerically and graphically all the information. This visualization process has allowed the identification and correction of erroneous data.

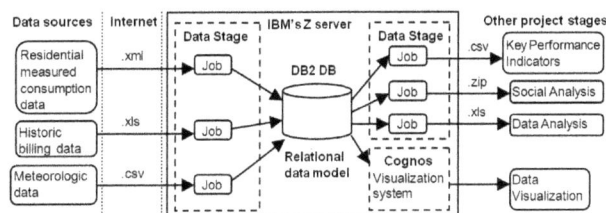

Figure 1. DASE system

In addition, error-free data is exported and sent daily to other systems for further analysis and exploitation (i.e. calculation of key performance indicators or statistics).

3. RESULTS AND DISCUSSION

Preliminary results obtained from the data analysis covering from 1st January 2012 to 30th June 2012 are presented in the following sections.

3.1 Consumption trend

Billing data has been compared to the mean of historic billing data (2008-2011 mean) to evaluate the consumption trend of the participants. 42% of participants have achieved a remarkable consumption reduction (above 10%) while 33% have kept their previous consumption level (within ±10%). The remaining 25% have increased their consumption more than 10% (see fig 2).

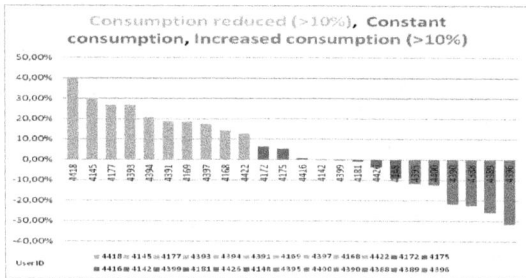

Figure 2. Consumption trend by participant (%)

Setting aside the behavior of each participant it is difficult to assure that the consumption reduction was motivated by the use of the installed devices. It could have been motivated by exogenous causes such as the economic crisis, variations in the number of occupants or the substitution of electric appliances for more efficient ones.

3.2 Contracted power

The maximum power to be contracted by a user is usually estimated by considering the maximum power of the appliances in the home and the patterns of use. The availability of real hourly energy consumption suggested the possibility of contrasting the power contracted by users with real energy measurements.

Considering the total amount of hourly consumption values, we have represented a histogram for each participant. The histogram intervals correspond to the increments for contracted power available in Spanish contracts (1.1kW). Assuming the maximum power demand is covered when 99% of measurements are below a certain level, 70% of participants would be able to reduce their contracted power in at least one step. However, there are some types of electrical appliances such as washing machines or HVAC systems whose peak consumption is not well represented by hourly measurements. Increasing the measurement frequency, (e.g. to 15 minutes) would help to better validate the suitability of this analysis.

3.3 Data reliability

When providing measurements for external use and analysis, data reliability is a must. Detecting errors and identifying the causes (e.g. measurement or communication failures, devices disconnection, etc.) has helped us to automate pre-storage and pre-exploitation processes. On the other hand, the detection of anomalous behaviors and malfunction of devices has been used to select the participants to be considered qualified for the project. The majority of the problems were caused by the disconnection of gateways, reflected by the system as "no data".

4. CONCLUSIONS

The pilot of the Smart City Project in Málaga has provided a good set of lessons learned for future applications in new pilot projects.

In the first place, we must note the difficulty of drawing conclusions on energy consumption variations related to the implementation of energy monitoring technologies. In a real-life pilot, many other factors can cause these changes: climate fluctuations in different years, economic and social circumstances of end users, occupancy of households, etc.

Another added value provided by detailed monitoring is the possibility of optimizing users' type of contract and utilities' sizing of the network by estimating the optimum contracted power. This estimation could be improved using 15 minutes rate measurements or peak power meters.

Finally, it is necessary to take into account that some smart clamps used for in-home monitoring do not have as much accuracy as the smart meters. Consequently, they are suitable to provide energy feedback to users, but not for billing purposes. Their suitability for specific demand-response services should be further investigated. However, the interviews with participants have yielded alternative uses of the devices, not only for energy monitoring, but also for remote management and supervision of their homes.

5. ACKNOWLEDGMENTS

Authors would like to thank Endesa for the support provided during the project and CDTI and FEDER for the funding of the SmartCity Málaga project.

6. REFERENCES

[1] C. Fischer, "Feedback on household electricity consumption: A tool for saving energy?" *Energy Efficiency 1*, 79-104, 2008.

[2] Schleich, J.; Klobasa, M.; Golz, S., "Does smart metering reduce residential electricity demand?," *European Energy Market (EEM), 2012 9th International Conference on the* , vol., no., pp.1-4, 10-12 May 2012 doi: 10.1109/EEM.2012.6254779

Minimizing Building Electricity Costs in a Dynamic Power Market: Algorithms and Impact on Energy Conservation

Dawei Pan[*][†]   Dan Wang[†]   Jiannong Cao[†]   Yu Peng[*]   Xiyuan Peng[*]

[†]The Hong Kong Polytechnic University,   [*]Harbin Institute of Technology

Categories and Subject Descriptors

J.7 [**Computer Applications**]: Computer in Other Systems

Keywords

Energy, Thermal Storage, Room Management, Smart Grid

1. INTRODUCTION

Energy is a global concern and the electricity bills nowadays are leading to unprecedented costs. Electricity price is market-based and dynamic. In this paper, we investigate how to cut the electricity bills of commercial buildings in a dynamic power market. The building thermal systems (e.g., air-conditioning), which dominate electricity bills, has a special property of thermal storage, i.e., the energy will not immediately dissipate from thermal air/water. Intuitively, with storage, the energy can be "stored" in the thermal system, making it possible to purchase electricity in low price and use it at appropriate time. Clearly, the building thermal supply and electricity purchasing depends on human activities such as classes and meetings that the building should support. In this paper, we develop a holistic planning of electricity purchasing schedule with thermal storage management, and appropriate room assignment schedules for classes/meetings usage, with the objective of minimizing electricity bills.

Such system is cyber-physical in natural. We develop wireless sensing systems to collect fine-grained data which are used to assist the cross-disciplinary physical thermal modeling. We then formulate an optimization problem, analyze its complexity and develop algorithms. Our primary focus is to minimize electricity bills, which matches the incentives of the commercial buildings. We show, however, that this does not coincide with energy conservation. We thus further investigate the relationship of minimization of electricity bills and minimization of energy consumption. Our evaluation shows that our algorithms can achieve a 40% cost reduction as compared to current scheduling.

There are more recent works on electricity cost, and we harvest from studies [1][2], etc, which take advantage of storage and real time pricing to save electricity bill. However, we focus on detailed human activities in commercial buildings in this paper.

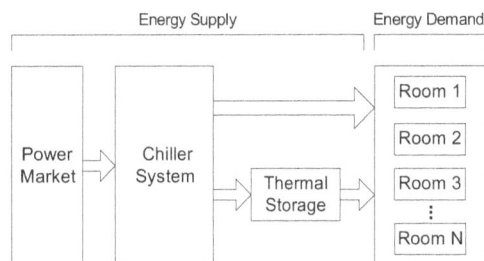

Figure 1: The diagram of the thermal energy flow.

2. BACKGROUND AND AN OVERVIEW

The energy supply and demand in a building can be abstracted as Fig. 1. The energy demands come from the rooms when scheduled to hold human activities, i.e., meetings. This meeting is meant to be general. In a campus context, this can be translated into class schedules and in a commercial building context, this can be translated into office planning and meeting schedules. The energy supplies come from the chiller system and the thermal storage. The chiller system is electrically charged to support the energy demands from rooms when the thermal storage is low. All of these finally are electrically supported by the power market.

Minimizing building electricity bills in such system falls into an optimization problem. Yet we face difficulties both in algorithm design and in physical thermal modeling. On the computing side, we need to develop two schedules: 1) meeting/class schedule and room assignment schedule for meetings; and 2) electricity purchase schedule from the dynamic power market. On the physical side, we need to model: 1) thermal storage capacity, and 2) energy (electricity) requirement for each room if they are assigned for meetings/classes.

The linkage between the computing side and the physical side is that computing schedules need inputs from the physical side. From a high level point of view, we develop equations that link the dynamics between the demands (air-conditioning), and the supplies (thermal storage and electricity charges for the chiller system).

Thermal computing falls into the expertise of Building and Service Engineering. They have sophisticated tools Energy-Plus. EnergyPlus is a complex model and there can have hundreds of parameters. Our physical computing is based on EnergyPlus. However, to model the rooms, there may be high complication if we need to find out the parameters for EnergyPlus room-by-room. We adopt a wireless sensor system assisted approach as follows. The major complexity of EnergyPlus comes from some compound parameters that are not easy to obtain directly. For example, a key parameter is thermal conductivity of a wall. It involves knowledge

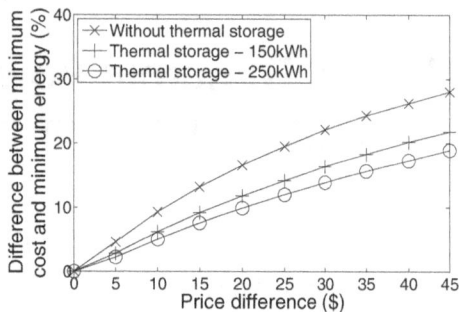

Figure 2: Difference between minimizing electricity cost and minimizing energy consumption.

of sub-parameters of materials etc. We found that these parameters are invariants, however; as it will not change subject to environments. We can inversely calibrate it if we can first collect a set of data on electricity usage, temperature of the room, etc. We develop wireless sensor networks for this. We thus can substantially reduce the number of parameters to be input to EnergyPlus. The details of such approach can be found in [3].

3. ELECTRICITY COST VS. ENERGY

Before we go into the problem and algorithms, we analyze the relationship between minimizing electricity cost and minimizing energy consumption. As said, these two minimizations do not coincide with each other.

Intrinsically, if 1) there is no price difference on the supply side, or 2) there is no difference on the demand side, saving cost can be achieved only by saving energy. Thus, the two minimizations become identical. We can generalize this as follows. Let $\mathcal{E}(r,t)$ be the energy consumption of room r at time t. Define *cost-energy in-conflict condition* as:

(1) Given $\mathcal{E}(r,t) > \mathcal{E}(r',t')$ and $P_t < P_{t'}$, $\forall r,t,r',t'$, $\mathcal{E}(r,t)P_t > \mathcal{E}(r',t')P_{t'}$; or (2) Given $\mathcal{E}(r,t) < \mathcal{E}(r',t')$ and $P_t > P_{t'}$, $\forall r,t,r',t'$, $\mathcal{E}(r,t)P_t < \mathcal{E}(r',t')P_{t'}$.

LEMMA 1. *If cost-energy in-conflict condition holds, minimizing electricity cost and minimizing energy consumption are identical.*

Next lemma shows the role that thermal storage can play: it can mask the difference between the two minimizations. Intuitively, thermal storage hides the impact of the electricity price difference at different times.

LEMMA 2. *Given that the thermal storage has infinite capacity, minimizing electricity cost and minimizing energy consumption are identical.*

To quantitatively understand the difference of the two minimization, we plot an illustration in Fig. 2. We see that when the price difference is $25, there can be a difference of around 20%. With a thermal storage capacity of 150kWh, the difference is 15%. The thermal storage capacity of 150kWh in our setting indicates that the storage can hold for all the building rooms in operation for one hour. This is reasonable practice in real world and $25 - $30 price differences are also conservative.

4. THE PROBLEM AND ALGORITHM

Our problem is to compute the meeting and room assignment schedules and the electricity purchasing schedules

from the power market, so as to minimize the cost. Our formal formulation of a Minimize Building Electricity Cost (MBEC) problem can be found in [3] and we have shown that MBEC is NP-complete.

Our philosophy in developing the heuristic for MBEC is as follows. We need to develop two schedules, 1) the meeting schedule and the room assignment schedule and 2) the electricity demand schedule. Accordingly, we develop two algorithms: 1) given the electricity demand schedule fixed, find the best meeting schedule and room assignment schedule; we call it best-Assignment() and 2) given the meeting schedule and room assignment schedule fixed, find the best electricity demand schedule. We solve the overall MBEC by a Lagrangian relaxation structure using the two algorithms as sub-routines. Details can be found in [3].

5. SIMULATION

We evaluate the impact of meeting schedules and thermal storage on electricity costs. We compare MBEC with 1) room scheduling algorithm that just satisfies the meeting time and room capacity requirements (denoted as just-fit) and 2) best-Assignment() only.

Fig. 3 shows the electricity cost as against to the meeting numbers. Specifically, we see that if there are 200 meetings, the total electricity cost needed by just-fit, best-Assignment() and MBEC() is $47.5, $34.5 and $30.0 showing that MBEC() has a saving of 36.8%. Fig. 4 shows the electricity cost as against to thermal storage capacity. This is not surprising as just-fit and best-Assignment() do not use the thermal storage for cost saving. When the thermal storage is 1000kWh (approximately support all rooms for 3 hours), it can introduce a saving of 22.3% as compared to best-Assignment().

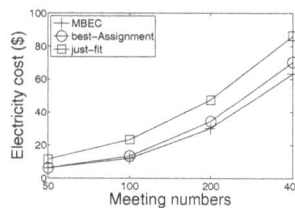

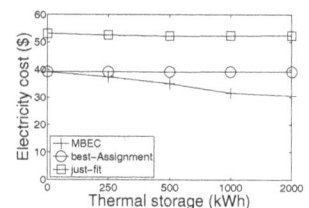

Figure 3: Electricity cost vs. number of meetings.

Figure 4: Electricity cost vs. thermal storage.

6. ACKNOWLEDGEMENT

Dawei Pan and Dan Wang are affiliated to The Hong Kong Polytechnic University Shenzhen Research Institute, Shenzhen 518057, China. The work described in this paper was supported in part by the National Natural Science Foundation of China (No. 61272464).

7. REFERENCES

[1] B. Daryanian, R. Bohn, and R. Tabors. An experiment in real time pricing for control of electric thermal storage systems. *IEEE Trans. Power Systems*, 6(4):1356–1365, 1991.
[2] A. Mishra, D. Irwin, P. Shenoy, J. Kurose, and T. Zhu. Smartcharge: cutting the electricity bill in smart homes with energy storage. In *Proc. e-Energy*, 2012.
[3] D. Pan, D. Wang, J. Cao, Y. Peng, and X. Peng. Minimizing building electricity costs in a dynamic power market: Algorithms and impact on energy conservation. Technical report, 2012. available at – http://www4.comp.polyu.edu.hk /~csdwang/Projects/wBACnet.htm.

Sustainable Performance in Energy Harvesting - Wireless Sensor Networks *

Xenofon Fafoutis
Denmark Technical University
Matematiktorvet
2800 Kgs Lyngby, Denmark
xefa@imm.dtu.dk

Alessio Di Mauro
Denmark Technical University
Matematiktorvet
2800 Kgs Lyngby, Denmark
adma@imm.dtu.dk

Nicola Dragoni
Denmark Technical University
Matematiktorvet
2800 Kgs Lyngby, Denmark
ndra@imm.dtu.dk

ABSTRACT

In this practical demo we illustrate the concept of "sustainable performance" in Energy-Harvesting Wireless Sensor Networks (EH-WSNs). In particular, for different classes of applications and under several energy harvesting scenarios, we show how it is possible to have sustainable performance when nodes in the network are powered by ambient energy.

Categories and Subject Descriptors

C.2 [**Computer-Communication Networks**]: Network Architecture and Design; C.2 [**Computer-Communication Networks**]: Network Protocols

Keywords

Wireless Sensor Networks, MAC Protocols

1. INTRODUCTION

Recent advancements in energy harvesting have led to the possibility of powering wireless embedded devices by small-scale ambient energy. Depending on the nature of the application, several environmental energy sources can be harvested, such as solar power or heat from radiators. In contrast to the limited amount of energy a battery can store, energy harvesting can potentially produce an infinite amount of energy. Therefore, the continuous operation of the system is only limited by hardware or software failures. As a result, the installation and maintenance costs of a network are significantly reduced. Furthermore, energy harvesting constitutes an environmental friendly means to power embedded devices, as it is an efficient driver to cut down wasted energy and battery wastes.

The system goal of Energy Harvesting - Wireless Sensor Networks (EH-WSNs) is fundamentally different from the one of traditional (battery-powered) WSNs (that is, to maximize network lifetime by minimizing energy consumption). Indeed, as long as the harvested energy is more than or equal to the energy consumed, energy does not constitute a limitation on the lifespan of the embedded device. Furthermore, any additional harvested energy can be used to improve the performance of the application. Thus, the system goal of EH-WSNs is twofold. Sustainable operation constitutes the primary goal, while application performance represents the secondary goal whenever the energy input allows it. In other words, we aim at achieving max sustainable performance.

In this demo we show some results on our ongoing research on EH-WSNs. In our previous work, we introduced ODMAC [3], a Medium Access Control (MAC) protocol specifically designed for EH-WSNs. The MAC scheme plays a central part in the design of energy-efficient WSNs since it controls the duty cycle of each node and the radio component (which dominates the energy consumption of a node). In particular, our analytic results [2] and simulations [3] shows that ODMAC supports sustainable performance. The contribution of this demo is to illustrate these findings on a real EH-WSN test bed. In particular, for different classes of WSN applications (i.e., applications having different performance requirements) as well as under several energy harvesting scenarios, we show how ODMAC supports sustainable operation of nodes powered by ambient energy.

2. EH-WSN TEST BED

Hardware. The system has been implemented on Texas Instruments' eZ430-rf2500 sensor nodes. Each node consists of an MSP430 microcontroller and a CC2500 radio, operating on the 2.4 GHz band. In addition to batteries, the nodes can be powered by external energy harvesting boards. In particular, we use Cymbet's CBC-EVAL-10 and CBC-EVAL-9 energy harvester boards (Figure 1). The boards harvest energy from different sources and store it into embedded batteries ($100 \mu Ah$ capacity). The energy harvester board can power the radio in active state for approximately $150 - 200$ ms every 10 seconds.

Figure 1: Ex. of Nodes in our EH-WSN Test-Bed

Firmware. Our aim is to have complete control of the system, mitigating the use of external libraries. As a result, the only external library we use is a Texas Instruments' Mini-

*Research partially supported by the IDEA4CPS project granted by the Danish National Research Foundation.

mal Radio Frequency Interface (*MRFI*), which provides an interface for the radio transceiver. More explicitly, it allows for the actual communication between the microcontroller and the radio components, it provides fundamental primitives (*receive, transmit with/without clear channel assessment*) and it gives control over the powering of the radio module itself (*sleep, listen*). On the other hand additional overhead and unnecessary information are added to the exchanged frames. For this reason, this layer will be replaced and rewritten in the future to better suit our needs.

On Demand MAC Scheme. On top of MRFI we implemented ODMAC [3], a receiver-initiated Medium Access Control protocol we specifically designed for EH-WSNs. The key feature of ODMAC is the possibility for each node in the network to dynamically and independently choose its own duty-cycle in order to accommodate its energy constraints. Furthermore, ODMAC supports anycast routing, namely *opportunistic forwarding*. The idea is that each node, instead of waiting for the optimal relay, sends messages to the recipient that wakes up first (that is, to the first recipient that makes itself available for reception). This recipient must belong to a set of possible forwarders provided by the routing protocol. The main consequence from the adoption of ODMAC is that the sensor network is characterized by some important features, such as autonomous load balancing, reduced delay and reduced idle listening. These characteristics, in particular the former and the latter, become crucial from an energy consumption perspective.

Embedded Security. A security suite inspired by TinySec [4] has been designed and implemented at the MAC layer. In order to support sustainable performance, the suite provides four security modes, namely *no security, authentication, encryption, both*. The specific mode can be chosen on a per-message basis. Authentication of beacons (control messages) has also been implemented. Both confidentiality and authentication are provided through the same encryption primitive. The adopted algorithm is *Skipjack* [1], and other implementations are underway (e.g. AES, PRESENT) for further security analyses. Encryption is always performed by using secure modes of operation, specifically *Cipher-block Chaining* with *Cyphertext Stealing* to keep the message size unmodified. Since the suite works at the MAC layer, Message Authentication Codes and encryptions are verified and reconstructed at each hop. This requires more CPU intensive work, but allows for forged or malformed packet to be identified and discarded right away, thus saving transmissions of useless data and assuring greater energy savings. Key managements schemes and adaptive security (that take into account the energy harvesting rate of the nodes) are work-in-progress extensions of the current security layer.

3. DEMO

The purpose of this demo is to show how our EH-WSN prototype can run under several energy harvesting scenarios (i.e. varying the amount of salvageable energy) always adapting itself in order to maintain a sustainable operational state. As WSN applications are mission-specific, we run our sensor network in three scenarios characterizing different classes of WSN applications. Each class will favor a different performance metric.

Scenario 1: Delay Sensitive Applications. The focus of this class of applications is to minimize the end-to-end delay which is defined as the amount of time elapsed since a message generation until its reception, while maintaining the sustainability of the system. This scenario is characterized by a sender and an energy harvesting powered receiver. The beaconing period of the receiver controls the level of the delay. The specific measurement that we will evaluate is the delay of the messages generated by the sender node. The energy harvesting node has to find the right compromise between survivability and low delay. Minimal delay is crucial for surveillance type of applications, where an anomaly should be reported as soon as possible.

Scenario 2: Datalogging Applications. The focus of this class is to gather as many messages as possible for long-term monitoring or offline analysis. The optimal working point here would be the one that allows for the highest number of successful transmission, independent of the time it takes a specific message to travel across the network, while maintaining sustainability. Here we use a basic two nodes single-hop topology, with the sink on one side and an energy harvesting powered node on the other.

Scenario 3: Security Sensitive Applications. This class of applications handles data having significant security requirements. In this scenario we focus on the impact of the possible security extensions over the energy consumption. Depending on the specific application, different messages may have different security requirements. We will use the same topology of scenario 2 to measure the network's throughput, while cycling through different security modes.

4. CONCLUSION

The use of energy harvesting technologies to power embedded wireless devices has changed the fundamental optimization goals of sensor networks. Instead of focusing on maximizing the limited lifetime of the nodes, energy harvesting can provide nodes with potentially infinite energy that can be used to prioritize the requirements dictated by the underlying application rather than merely prolonging the network's life. The key challenge is to optimize the performance of the application without compromising the sustainability of the system. In this practical demonstration, we show how ODMAC is able to support the aforementioned goal in three different scenarios, namely delay-sensitive, datalogging and security-sensitive WSN applications.

5. REFERENCES

[1] SKIPJACK and KEA Algorithm Specifications 2.0, National Institute of Standards and Technology. Technical report, May 1998.

[2] X. Fafoutis and N. Dragoni. Adaptive Media Access Control for Energy Harvesting - Wireless Sensor Networks. In *Proceedings of INSS'12, IEEE*.

[3] X. Fafoutis and N. Dragoni. ODMAC: An On-Demand MAC Protocol for Energy Harvesting - Wireless Sensor Networks. In *Proceedings of PE-WASUN'11, ACM*.

[4] C. Karlof, N. Sastry, and D. Wagner. TinySec: a Link Layer Security Architecture for Wireless Sensor Networks. In *Proceedings of SenSys'04, ACM*.

Using Clustering Mechanisms for Defining Consumer Energy Services

Frank Feather
Marina Thottan
Alcatel-Lucent, Bell Labs
600 Mountain Ave
Murray Hill, NJ 07974

{frank.feather & marina.thottan}
@alcatel-lucent.com

Dayu Huang
Dept of Electrical and Computer
Engineering
University of Illinois at Urbana-
Champaign
901 W Illinois St, Urbana, IL
dayuhuang@gmail.com

Katherine Farley
Strategic Research, EPB
10 W. ML King Blvd
Chattanooga, TN 37402
farleykl@epb.net

ABSTRACT

The ongoing smart grid transformation in utility networks is making available fine grained measurements of electricity consumption. To realize the full potential of the collected data we apply sophisticated data analytics and machine learning techniques to correlate consumption with other types of demographic data (household surveys and tax records) to place the collected consumption data within the right context. This context setting is achieved by a rigorous feature selection procedure, followed by clustering to group customers into peer groups. The statistical information gleaned from these peer groups are then used to identify outliers and define new services both for the utility (energy audits) and the end consumer (Home Energy Health Management systems). Analysis shows that outlier detection within clusters is better able to target customers than outlier detection without clustering: on average, half the outliers found in the clusters would not be outliers in the overall population.

Categories and Subject Descriptors

I.5.3 [**Computing Methodologies**]: Pattern Recognition–Clustering

Keywords

Smart Grid; Energy Services; Clustering; Feature Selection

1. METHODOLOGY

Home Energy Health Management (HEHM) has emerged as a key component for residential energy management in an overall Smart Grid architecture. But to enlist the cooperation of the consumer to achieve sustainability goals, the consumer will need information on his consumption levels along with recommendations of how to save energy and money. One area of HEHM is *innovative billing* or *comparative billing*, which compares a customer's energy usage to his or her neighbors. Comparative billing is the idea

that consumers can see if they consume a higher, similar, or lower amount of energy as compared to their neighbors, and thus be motivated to conserve (i.e., peer pressure) [1][4]. The biggest challenge when doing this comparative analysis is the selection of the neighbors, that is, finding other customers that have similar characteristics and then forming the appropriate comparison group [2][4][6].

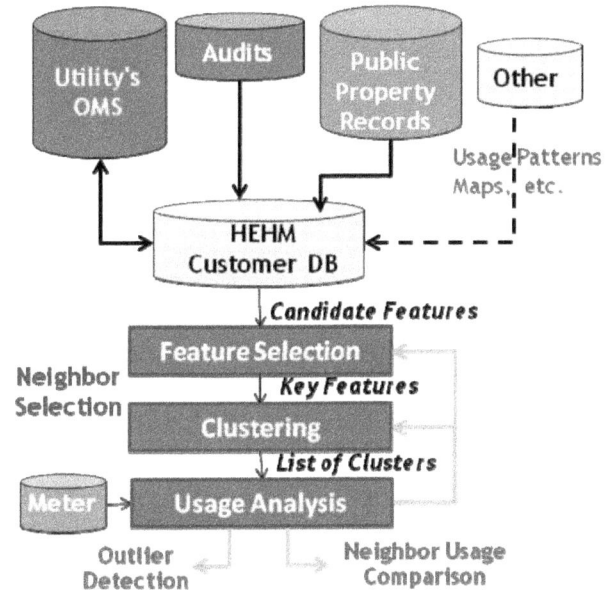

Figure 1: Method for forming comparison groups

Our study developed a method for feature selection and formation of comparison groups. The method will extract relevant features by utilizing information available from various sources. These features are analyzed and reduced to the set of features that most affect usage. These features are then used in cluster analysis, which will form comparison groups, i.e., houses with similar characteristics. Each group can be analyzed to provide comparative reporting for a customer. Analysis of these groups can also be used by the utility to identify outlier households with higher than normal usage who could be candidates for an energy audit, or participants in a demand response trial. Figure 1 above illustrates our approach.

2. RESULTS

For this work, we utilize data from an electric utility with over 170K customers. We use data from three sources shown in Figure 1, including detailed 5000 audits of homes. After correlation, the total number of customers for this study was 983. We analyzed usage for three seasons: spring (April), summer (July) and winter (December), but only present December, which is heavily dependent on electric heating. The data was analyzed using the method in Figure 1.

2.1 Feature Selection

For feature selection we start with over two dozen available attributes which we reduce it to 8 key attributes, and then use feature selection to reduce it to the most more pertinent features, removing redundant features. We use a rigorous selection process, based on mutual information process and a heuristic procedure in which we approximate the statistical dependence between multiple variables by Markov dependence using the Chow-Liu algorithm [3]. We found the five features that affect the monthly electricity usage for December are: Home Value, Area, Year Built, Occupancy, and Presence of Electric Heating.

2.2 Clustering

Using the above five features, we then use the k-means algorithm to create clusters. We run the algorithm with different number of clusters and then select the best number of clusters using various statistical analyses of the derived clusters. That analysis includes: within cluster sum of square (WCSS) test to determine the best value(s) of k; median differential entropy test to confirm that clustering is maintaining randomness; and Mann-Whitney-Wilcoxon test [5] to help determine the maximum value of k. The final analysis also looked at the distribution of usage within each cluster.

Based on our analysis we produced the following five clusters (the number in brackets is cluster size):

Cluster #1 [184]: Electric Heat; Older homes (1980); 2-6 occupants (3.8); Low value ($41K); Average size (1800 ft^2)

Cluster #2 [61]: Electric Heat; New homes (1998); 2-5 occupants (3.3); High value ($270K); Large size (3360 ft^2)

Cluster #3 [261]: Electric Heat; Older homes (1980) 1-2 occupants (1.7); Average value ($95K); Average size (1750 ft^2)

Cluster #4 [166]: Gas Heat; Newer homes (1991); 2-5 occupants (3.1); High value ($240K); Large size (3080 ft^2)

Cluster #5 [311]: Gas Heat; Older homes (1980); 1-4 occupants (2.4); Average value ($110K); Average size (1980 ft^2)

2.3 Usage Analysis

Our usage analysis is outlier detection for each of the clusters, and for the whole study population for comparison. An outlier is defined as being points above the second standard deviation. Outlier detection on the clusters found 45 outliers with most (29) being in clusters #3 & #5. In contrast, for the whole population there were 46 outliers, but most were with houses in the first three clusters, especially targeting the large expensive home (cluster #2). Only about half the outliers found in the clusters were outliers for the whole population. We conclude that *clustering delivers a more targeted set of energy outliers that is particular to the type of homes in the cluster.*

3. CONCLUSION

By correlating consumption data with other pertinent demographic data we were able to extract valuable information that could be used to design customized energy services. Our analysis shows that outlier detection within clusters is better able to target customers than outlier detection without clustering. On average, half the outliers found in the clusters would not be outliers in the overall population.

The methods outlined here could be used by the utility company to identify customers that use more electricity than expected for their household group, and thus recommend them for energy audit or participation in a Demand Response program. These techniques can benefit the consumer through innovative billing, which would allow a household to view their monthly usage compared to peers. Finally, the techniques employed could also be used to detect anomalous usage patterns that may be indicative of fraudulent use of electricity.

4. References

[1] Kevin Bengtson, *Can Better Utility Bills Save Energy?* Home Energy Magazine Online, May/June 1997

[2] Chicco, G., Napoli, R., and Piglione F., *Comparisons among clustering techniques for electricity customer classification*, IEEE Transactions on Power Systems, vol.21, no.2, May 2006, pp. 933- 940

[3] Chow, C.K., and Liu, C.N., *Approximating Discrete Probability Distributions with Dependence Trees*, IEEE Transactions on Information Theory 14, no 3 (1968): 462-467

[4] Iyer, M., Kempton, W., and Payne, C., *Comparison groups on bills: Automated, personalized energy information*, Elsevier B.V., Energy and Buildings 38 (2006) 988-996, www.sciencedirect.com

[5] Mann–Whitney–Wilcoxon test, http://en.wikipedia.org/wiki/Mann-Whitney_U

[6] Rasanen, T., Ruuskanen, J., and Kolehmainen, M., *Reducing energy consumption by using self-organizing maps to create more personalized electricity use information*, Elsevier, Applied Energy 85 (9), pp. 830-840, Sept 2008

Smart Street Light System with Energy Saving Function Based on the Sensor Network

Yusaku Fujii
Department of Electronic Engineering
Gunma University
1-5-1 Tenjin-cho
Kiryu, Gunma Japan
fujii@el.gunma-u.ac.jp

Noriaki Yoshiura
Department of Information and
Computer Sciences
Saitama University
255 Shimo-ookubo Sakura-ku
Saitama City, Japan
yoshiura@fmx.ics.saitama-u.ac.jp

Akihiro Takita
Department of Electronic Engineering
Gunma University
1-5-1 Tenjin-cho
Kiryu, Gunma Japan
takita@el.gunma-u.ac.jp

Naoya Ohta
Department of Computer Science
Gunma University
1-5-1 Tenjin-cho
Kiryu, Gunma Japan
ohta@cs.gunma-u.ac.jp

ABSTRACT

Our project for developing a smart street light system is reviewed. In this project, the street light system, in which lights on when needed and light-off when not needed.

Currently, in the whole world, enormous electric energy is consumed by the street lamps, which are automatically turn on when it becomes dark and automatically turn off when it becomes bright. This is the huge waste of energy in the whole world and should be changed.

Our smart street light system consists of a LED light, a brightness sensor, a motion sensor and a short-distance communication network. The lights turn on before pedestrians and vehicles come and turn off or reduce power when there is no one. It will be difficult for pedestrians and drivers of vehicles to distinguish our smart street lamps and the conventional street lights, since our street lamps all turn on before they come.

The present status and the future prospects of our smart start light project will be reviewed.

Categories and Subject Descriptors

H.1.0 [Information Systems]: Models and principles

Keywords

Street light, Energy saving, Sensor network

1. INTRODUCTION

Currently, in the whole world, enormous electric energy is consumed by the street lights, which are controlled by means of the embedded brightness sensors. They are automatically turn on when it becomes dark and automatically turn off when it becomes bright. This is the huge waste of energy in the whole world and should be changed.

There are some attempts, in which the energy wastes of the street lights are reduced. A sensor light, which is controlled by the brightness sensor and the motion sensor, is sometimes used [1]. It only turns on for w while when the motion is detected in front of the light and it is dark. However, it usually is too late to turn the light on when a person or a car comes in front of it. The light should turn on before a person or a car comes.

On the other hand, some companies and universities have developed centrally-controlled smart street light systems with the host computers [2,3]. They might be suitable for being applied to a large area or a newly developed area based on the total plan. However, they might not be suitable for being applied to a small area.

We propose an autonomous-distributed-controlled light system, in which the lights turn on before pedestrians come and turn off or reduce power when there is no one by means of a distributed-installed sensor network.

2. PROPOSED SYSTEM

Figure 1 shows the components, with which our smart street light system is realized.

(a) Lamp unit:

It consists of power-adjustable LED array, the brightness sensor, the motion sensor, the communication device, such as ZigBee module, and the controller. It turns on for several minutes under the conditions that a motion is detected in the defined area by the sensors including its own sensor. Then, it sends the message to other units. It turns off or reduced power under the condition that any motion is not detected in the defined area.

(b) Sensor unit:

It consists of the motion sensor, the communication device and the controller. It sends out the message to other units under the condition that motion is detected. This unit is placed to many locations, such as at electric poles, at house gates, at house fence and inside or outside of the door, to ensure that every street light turn on before pedestrians notice that. As for power supply, the solar battery can be a good option.

(c) Access point:

It consists of the communication device and the controller. It is used in the case that the distance between the lamp units and the sensor units are too large to communicate each other.

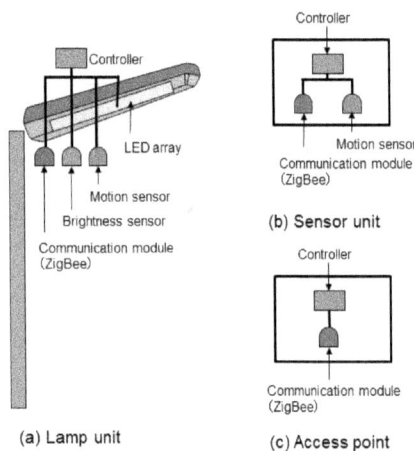

Figure 1. The components for the smart street light

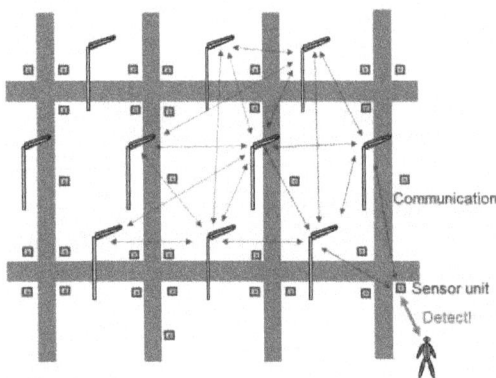

Figure 2. The components for the smart street light

As for communication devices, a power-saving short-distance device, such as ZigBee, is appropriate for our system. As for the position information, each controller has plural addresses, which correspond to the adjacent different networks.

Figure 2 shows an example of our smart street light system. The street lights turn on before the pedestrians come and turn off or reduce power when there is no one by means of a distributed-installed sensor network.

3. DISCUSSIONS

The targets of our development are as follows,

- Easy installation and extension: Each unit can be installed one by one to the network by setting the parameters. The system is autonomous-distributed controlled. No host computer is needed.

- Low cost: Only the parts of mass production are used.

- Easy update: The firmware of each unit can be updated easily. The control algorisms should be developed for the situations, such as a quiet residential area, a shopping street, a part, a main road and a mountain road.

- Self-diagnosis: The worst event is that the light does not turn on when the pedestrian come. Each unit records the failures, in which the motion is detected in front of it without the advanced notification from the other units.

4. ACKNOWLEDGMENTS

This study was supported by the Grant-in-Aid for Scientific Research (B) 24300246 (KAKENHI 24300246).

5. REFERENCES

[1] Velaga, R. and Kumar, A. 2012. Techno-economic evaluation of the feasibility of a smart street system: A case study of rural India. Procedia Social and Behavioral Sciences. 62, 1220-1224.

[2] Echelon Corp. https://www.echelon.com/applications/street-lighting/

[3] Bruno, A., Di Franco, F. and Rasconà, G. 2012. Smart street lighting. EE Times http://www.eetimes.com/design/smart-energy-design/4375167/Smart-street-lighting

[4] The e-JIKEI Network Promotion Institute, et al. Smart street light system with communication means. Published unexamined patent application in Japan P2011-165573A (in Japanese).

A Greedy Algorithm for the Unforecasted Energy Dispatch Problem with Storage in Smart Grids

Giorgos Georgiadis
Computer Science and Engineering
Chalmers University of Technology
Göteborg, Sweden
georgiog@chalmers.se

Marina Papatriantafilou
Computer Science and Engineering
Chalmers University of Technology
Göteborg, Sweden
ptrianta@chalmers.se

ABSTRACT

Integration of renewable and distributed energy sources is today possible through the use of the Smart Grid but these technologies bring benefits as well as challenges, such as their intermittent nature that leads to utilization problems for the grid. On the other hand, upcoming storage technologies, such as electrical cars, hold the potential to store and utilize this intermittent supply at a later time but bring challenges of their own, for example efficient storage utilization and intermittent energy demand.

In this work we propose a novel modeling of the problem of unforecasted energy dispatch with storage as a scheduling problem of tasks on machines and an associated greedy algorithm with a guaranteed performance, along with an efficient algorithm for the problem. Finally, we outline an extensive simulation study for a variety of scenarios based on data from a large network of customers.

Categories and Subject Descriptors

F.2.2 [**Nonnumerical Algorithms and Problems**]: Scheduling algorithms

Keywords

online scheduling; resource allocation; smart grid

1. INTRODUCTION

In recent years there has been an organized effort on an international level to modernize the power grid by adding resilience properties, precise accounting and new services through the use of information technologies, collectively leading to a new type of grid commonly called *smart grid*. It is expected that these changes will enable the incorporation of renewable (e.g. photovoltaic arrays and wind generator farms) and distributed (e.g. electric car fleets) energy sources on a large scale but these technologies bring benefits as well as challenges.

While established models (cf [2] and bibliography therein) are able to accurately predict energy demand and compensate with sufficient supply, the intermittent nature of renewable energy sources, such as wind generator farms, challenges the way we utilize energy when it is available, compensate for when it is not and rely on weather forecasts for the grids' daily operation. On the other hand, distributed

energy sources such as electric car fleets can act as storage options and balance the demand and supply of electrical energy but bring challenges of their own, for example efficient storage utilization and intermittent energy demand. In the present work we focus on solutions that adapt the demand and use storage capabilities to mitigate the intermittent effects of renewable and distributed energy sources.

By drawing on the problem of scheduling tasks to machines, we model the problem of energy dispatch from producers to consumers in the distribution level of the grid, using an extended *online load alancing ro lem*. We present a novel scheduling and resource dispatching algorithm that is able to cope with the inherent unpredictability of renewable energy sources without the use of forecasts, as well as take advantage of available storage options in the grid. We also show analytically that the proposed algorithm is within a logarithmic factor of the optimal solution for the specific problem. Finally we conduct an extensive simulation study for a variety of scenarios based on data from a large number of customers and show that the presented algorithm is highly competitive to methods that use forecasts and assume total knowledge about the incoming load demand requests for the same problem.

2. SYSTEM DEFINITION

We focus on the distribution system of the grid as a high level abstraction, by considering a number of *nodes* being connected through a simple topology, for example low voltage power lines in a radial feeder configuration [3]. These nodes issue *load demand re ests* on the feeder at irregular intervals, independently from each other, and energy is being dispatched from generation sites to satisfy the demand. For reasons of convention, we are assuming that requests are scheduled in hourly timeslots and are coming in a diurnal pattern, i.e. a demand refers to at most the next 24 hours. Note that we are considering both electrical and thermal services offered in two separate feeder lines on all nodes, an electrical and a thermal one respectively. In total, we identify three orthogonal axis to characterize load demand requests and we support demands of any of the eight types that these axis jointly define: **elastic/inelastic**, according to the ability to shift the demanded load over time or not[1], **electrical/thermal**, depending on whether the demanded load can be serviced using only electrical energy or both

---

[1]Elastic loads can be scheduled to be serviced within a set of timeslots, while inelastic loads must be serviced necessarily in a specific time slot.

electrical and thermal, and **storage/simple**, according to the ability to store the demanded load for future use or not.

The *unforecasted energy dispatch problem with storage* that is addressed here is the problem of dispatching generated electrical and thermal energy to end consumers in a way that minimizes peak energy consumption within a given time interval, without using forecasts and by taking into account any storage capabilities present.

3. MODELING AND ALGORITHMIC APPROACH

We model the unforecasted energy dispatch problem with storage to a problem of scheduling tasks to restricted machines[2] as follows. We regard load demand requests as an input sequence of tasks to be run on machines and the electrical and thermal feeder lines as the machines themselves. To capture the dimension of time, each feeder line corresponds to 24 machines, one for each hourly timeslot, leading to a total of 48 under this modeling. Each machine has the capability of storing energy if a storage task is run on it, and subsequent machines have a portion[3] of this storage also available, to reflect the fact that stored energy is available in later timeslots. In total, the elastic/inelastic characteristic of demand requests is expressed in the set of machines that the demand is allowed to run, the electrical/thermal characteristic in the machines themselves and the storage/simple characteristic in the accumulated storage on the machines.

We consider the case that encapsulates the above modeling, where machines are defined as previously and can accumulate storage and tasks can either be simple (i.e. only inducing load on a machine) or storage (i.e. inducing load and generating storage). We call the resulting problem from this transformation an *online load demand balancing problem with storage*, which is the problem of assigning the tasks to the machines while minimizing the maximum load on the machines.

Based on this modeling, we propose the STORAGEGREEDY algorithm, which assigns each incoming task to the allowed machine that has the minimum load-storage difference (breaking ties arbitrarily). In practice, this is expressed on a system level and for a load demand request with specific restrictions by scheduling the request to be executed on the allowed timeslot with the minimum load-storage difference. Note here that the scheduling is taking place on the node that issued the request (where a copy of the algorithm is running), without a centralized decision center.

Following a methodology similar to [1], we show that the solution cost of our algorithm lies within a factor of $\lceil \log n \rceil + 1$ when compared to the cost of the optimal algorithm, where n is the number of machines (according to our modeling, $n = 48$). Note that no assumptions are made regarding the dissipation of the stored energy and the results apply both for dissipating and non-dissipating storage.

4. EXPERIMENTAL STUDY

Here we outline a small part of the study that was conducted using the reported Swedish load demand mix for

---

[2]I.e. where a task is allowed to run only on selected machines.

[3]Depending on whether losses in storage are being considered.

households, using data collected from 400 households over the course of approximately 4 years, on behalf of the Swedish Energy Agency [4]. These data were extrapolated in order to model different households (customer profiles) and different types of loads (i.e. elastic/inelastic, storage etc). A number of scenarios were subsequently created, with a varying amount of houses and penetration of renewables and flexible loads, ranging from a mix of 65%-27%-8% inelastic electric, inelastic thermal and elastic electric loads respectively (business-as-usual scenario) to a mix of 45%-12%-12%-8%-3% inelastic electric, inelastic thermal, elastic electric, elastic thermal loads and elastic thermal, elastic electric storage respectively (smart house/neighborhood scenario).

During our experiments we aimed at minimizing the peak consumption and focused on the comparison between our STORAGEGREEDY algorithm and a simple, yet powerful algorithm for the same problem that uses forecasts, called the LPT algorithm: tasks are sorted globally by decreasing processing time and each task is assigned to the machine that has the least load (breaking ties arbitrarily) (see figure 1).

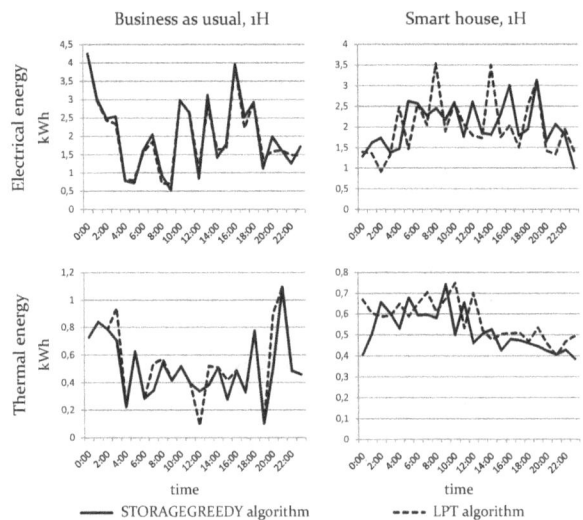

Figure 1: Comparison of load demand curves for the *Business as usual* and *Smart house* scenarios for one household

It is easy to see that for the business as usual scenario the two algorithms have comparable performance. However, under the smart house scenario the STORAGEGREEDY algorithm succeeds in lowering peak consumption by taking advantage of storage, even though the LPT algorithm has full access to forecasted demands.

5. REFERENCES

[1] A. Borodin and R. El-Yaniv. *Online Computation and Competitive Analysis*. Cambridge University Press, Apr. 1998.

[2] W. H. Kersting. *Distribution System Modeling and Analysis*. CRC Press, 1 edition, Aug. 2001.

[3] A. von Meier. *Electric Power Systems: A Conceptual Introduction*. Wiley-IEEE Press, 1 edition, July 2006.

[4] J. P. Zimmermann. End-use metering campaign in 400 households in sweden assessment of the potential electricity savings. *Energimyndigheten Sweden*, 2009.

Missing Data Handling for Meter Data Management System*

Ru-Sen Jeng[1], Chien-Yu Kuo[1], Yao-hua Ho[1], Ming-Feng Lee[2], Lin-Wen Tseng[2],
Chia-Lin Fu[2], Pei-Fang Liang[2], and Ling-Jyh Chen[3]

[1] Department of Computer Science and Information Engineering, National Taiwan Normal University
[2] Green Energy and Environment Research Lab, Industrial Technology Research Institute
[3] Institute of Information Science, Academia Sinica

ABSTRACT

We study the meter data management systems (MDMS) with a focus on missing data handling, and propose two approaches, called Lookback-N and Sandwich-N, based on the historical data. Using a realistic dataset, we demonstrate that Lookback-N is effective for online processing, and Sandwich-N outperforms the conventional offline approach, in terms of estimation accuracy, in all test cases. The proposed approaches are simple, effective, and show promise in handling missing data for emerging smart meter data management systems.

Categories and Subject Descriptors

H.4.2 [**Information Systems Applications**]: Types of Systems— *Decision support (e.g., MIS)*

Keywords

Missing data, Smart meter, MDMS

1. INTRODUCTION

In this study, we tackle the meter data management systems (MDMS) with a focus on missing data handling. The issue is challenging because 1) it has to be accurate as it is the basis for electricity billing, planning, and provisioning; and 2) it has to be simple in computation to support large-scale deployments. As a consequence, conventional missing data handling methods are not applicable to smart meter data management because 1) they are based on sophisticated statistical models that are computationally expensive [3, 5]; and 2) they require additional knowledge of environment and social factors [1, 4] that are infeasible in real deployments.

In the following sections, we propose two approaches to deal with missing data for meter data management systems, namely, the *Lookback-N* and *Sandwich-N* schemes. Using a realistic dataset, we evaluate the proposed schemes against the baseline scheme, which is based on linear interpolation and has been widely used by electric power companies. The results demonstrate that 1) the *Lookback-N* scheme can support *online processing* while keeping estimation error acceptable; and 2) the *Sandwich-N* scheme can yield a smaller estimation error than the baseline scheme, especially when missing data takes place in a burst.

*The authors wish to express their sincere gratitude to Bureau of Energy, Ministry of Economic Affairs, Taiwan, for financial support.

2. PROPOSED APPROACHES

We propose the *Lookback-N* scheme and the *Sandwich-N* scheme to deal with missing data for meter data management systems. Specifically, let M_t^i be the reading of the i-th meter on time t; and let δ_t^i be a status variable that is equal to 0 if M_t^i is missing and equal to 1 otherwise. Suppose the reading of the i-th meter is missing at time t_k (i.e., $\delta_{t_k}^i$ 0), and the last/next non-missing reading before/after $M_{t_k}^i$ is on time t_x and t_y respectively. Let $\widetilde{M}_{t_k}^i$ be the estimate of $M_{t_k}^i$, the conventional approach (i.e., the baseline scheme) implements the linear interpolation method to obtain $\widetilde{M}_{t_k}^i$, i.e., $\widetilde{M}_{t_k}^i$ $\frac{M_{t_x}^i + M_{t_y}^i}{2}$.

2.1 The Lookback-N Approach

The rationale of the *Lookback-N* scheme is that *'similar things come together'*. Let $f(i, t_u, j, t_v, n)$ be a matching function that returns 1 when the last n readings of the i-th meter since t_u are the same as the last n readings of the j-th meter since t_v; and it returns 0 otherwise. There are two cases in *Lookback-N* to estimate $M_{t_k}^i$:

1. When t_x $t_k - 1$, it looks up the historical data to identify the set of the 'similar' meter readings that have the same values in its preceding N readings as $M_{t_k}^i$ by

$$R_{t_k}^i = [M_{t_u}^j | \forall j \forall t_u \in [t_k - L, t_k) : f(i, t_k, j, t_u, N) = 1], \quad (1)$$

where L is the *lookback* factor that determines the length of the historical data to look up. Then, it estimates $\widetilde{M}_{t_k}^i$ based on $R_{t_k}^i$ using *roulette-wheel selection* [2].

2. When $t_x < t_k - 1$, it estimates $\widetilde{M}_{t_x+1}^i$ using the procedure used in the above case, and then it uses the estimated value to estimate $\widetilde{M}_{t_v}^i$, for t_v $t_x + 2, ..., t_k$.

2.2 The Sandwich-N Approach

The *Sandwich-N* scheme is similar to *Lookback-N*, except that it considers both the preceding and succeeding N readings of the missing one when identifying the set of 'similar' meter readings. We let $g(i, t_u, j, t_v, n)$ be a matching function that returns 1 when the preceding and succeeding n readings of the i-th meter since t_u are the same as those of the j-th meter since t_v, and it returns 0 otherwise.

When t_x $t_k - 1$, the scheme identifies the 'similar' set using Eq. 2, and it uses *roulette-wheel selection* to estimate $\widetilde{M}_{t_k}^i$ based on $S_{t_k}^i$. When $t_x < t_k - 1$, it estimates $\widetilde{M}_{t_x+1}^i$ first, and then it uses the estimated value to estimate $\widetilde{M}_{t_v}^i$, for t_v $t_x + 2, ..., t_k$, iteratively.

$$S_{t_k}^i = [M_{t_u}^j | \forall j \forall t_u \in [t_k - L, t_k) : g(i, t_k, j, t_u, N) = 1]. \quad (2)$$

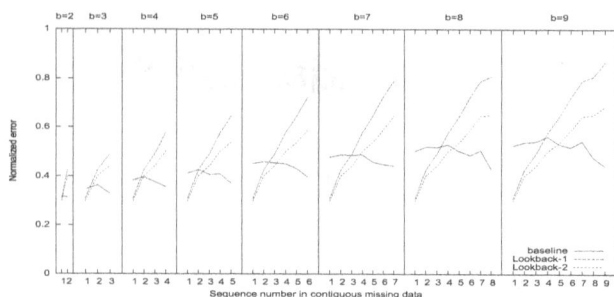

Figure 1: Comparison of the *Lookback-N* scheme and the baseline scheme under contiguous missing data

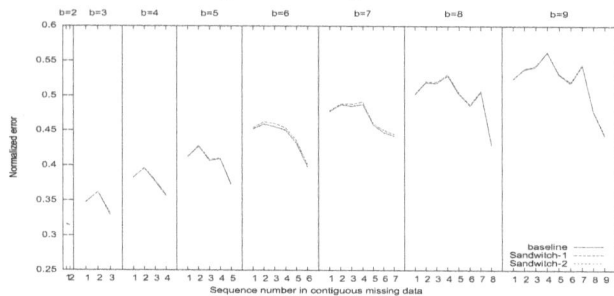

Figure 2: Comparison of the *Sandwich-N* scheme and the baseline scheme under contiguous missing data

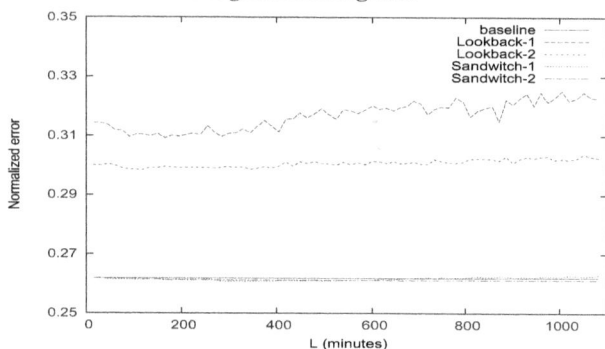

Figure 3: Comparison of the different schemes under different lengths of historical data to look up (i.e., L)

3. EVALUATION

We evaluate the proposed schemes using a real dataset from the Pilot Smart Meter Deployment in Taiwan. The dataset comprises eleven household electricity meters, from 2011/6/30 to 2011/9/30, in the same building in Taipei, Taiwan; and the sample rate of each meter is one sample every 15 minutes. There are no missing data in the dataset, and we suppose the energy consumption behavior of each household is unchanged in the measurement period.

We randomly generate different lengths of contiguous missing data in the dataset, and observe the normalized error (i.e., the estimation error over the original value) achieved by each scheme. Figures 1 and 2 show the average results based on 10,000 runs for each setting, and there are four observations:

1. The error achieved by *Lookback-N* is greater than that by the other schemes, but it remains affordable for *online processing* in meter data management systems.

2. When *Lookback-N* is used, the error accumulates with the

sequence number of missing data in a burst, and the error is smaller than that of the baseline scheme when estimating the first few missing data in a burst.

3. When *Sandwich-N* or the baseline scheme is used, the error increases with the sequence number of the contiguous missing data.

4. *Sandwich-N* is comparable to the baseline scheme when dealing with contiguous missing data. The error achieved by the two schemes is greater when estimating the middle few missing data, and it is smaller when estimating the last few missing data in a burst.

Figure 3 shows the normalized error achieved by different schemes under different lengths of historical data to look up (i.e., L). We observe that the two *Lookback-N* curves ($N = 1$ and 2) have greater normalized errors, and the error increases with L. The reason is that, when N is small or when L is large, *Lookback-N* is more likely to yield an estimate based on wrong energy consumption behavior models (e.g., it may estimates the missing data on working hours based on the non-working hours behavior model), thereby leading to a greater estimation error.

Moreover, we also find that the estimation error is smaller when a large N is used in *Sandwich-N*. As shown in Figure 3, the normalized error of *Sandwich-2* is consistently lower than *Sandwich-1* and the baseline scheme. However, we note that the larger the value of N used in *Sandwich-N*, the fewer 'similar' smart meters might be included in $S_{t_k}^i$, leading to bias in estimation. Moreover, the larger the value of L is, the higher computational complexity is resulted for both *Lookback-N* and *Sandwich-N*.

4. CONCLUSION

In this paper, we propose two missing data handling schemes, namely, *Lookback-N* and *Sandwich-N*, for emerging smart electricity meter systems. Using a realistic dataset, we show that *Sandwich-N* can achieve the lowest estimation error in all settings, and *Lookback-N* is effective for online missing data handling, as well as estimating the first few missing data in a burst. Work on considering additional factors when identifying 'similar' meter data is ongoing, and we plan to report the results in the near future.

5. REFERENCES

[1] J.-Y. Fan and J. D. McDonald. A Real-Time Implementation of Short-Term Load Forecasting for Distribution Power System. *IEEE Transactions on Power Systems*, 9(2):988–994, May 1994.

[2] D. E. Goldberg. *Genetic algorithms in search, optimisation, and machine learning*. Addison Wesley Longman, Inc., 1989.

[3] N. J. Horton and K. P. Kleinman. Much ado about nothing: A comparison of missing data methods and software to fit incomplete data regression models. *The American Statistician*, 61(1):79–90, February 2007.

[4] D. Matheson, C. Jing, and F. Monforte. Meter Data Management for the Electricity Market. In *International Conference on Probabilistic Methods Applied to Power Systems*, 2004.

[5] T. D. Pigott. A Review of Methods for Missing Data. *Educational Research and Evaluation*, 7(4):353–383, 2001.

Efficient Demand Assignment in Multi-Connected Microgrids

Kirill Kogan
University of Waterloo
kirill.kogan@gmail.com

Sergey Nikolenko
Steklov Mathematical Institute
sergey@logic.pdmi.ras.ru

Srinivasan Keshav
University of Waterloo
keshav@uwaterloo.ca

Alejandro Lopez-Ortiz
University of Waterloo
alopez-o@uwaterloo.ca

ABSTRACT

With the proliferation of distributed generation, an electrical load can be satisfied either by a centralized generator or by local/nearby distributed generators. Given a set of resource demands in a collection of geographically co-located microgrids that are connected to the central grid and also potentially to each other, each such demand characterized by a power level and a duration, we study algorithms that allocate generation resources to the set of demands by configuring switched paths from sources to loads.

Categories and Subject Descriptors

C.4 [**Measurement techniques; Performance attributes; Design studies**]: Miscellaneous

Keywords

switching; scheduling; worst-case analysis

1. INTRODUCTION AND MOTIVATION

In recent years, electricity generation has been rapidly becoming more diverse: power is generated today not only from large, capital-intensive plants but also from numerous smaller-capacity resources including solar panels, wind turbines, and diesel gensets. This proliferation has made it possible for electric demands to be met with local generation, reducing distribution losses and simultaneously increasing energy security. Increasingly, sets of loads can rely nearly entirely on local generation resources, forming a *microgrid*, with access to the central grid used only as a backup.

We anticipate that in the future geographically-close microgrids will opportunistically form connections with each other to increase reliability. This would allow, for instance, a set of apartment complexes to augment their own diesel gensets with shared solar generation from a nearby office complex on weekends. This is a natural recapitulation of the self-organizing process by which electricity grids were formed in the first place, before centralized generation essentially eliminated micro-generation a century ago.

The focus of our work is on efficient demand satisfaction in the context of multi-connected microgrids, where a demand can be met by different generation resources: local, nearby, or on a regional grid. Specifically, we are concerned with minimizing the delay in satisfying a set of resource demands (assuming that these demands

can be temporarily delayed but non-preemptive (*elastic* [1]). An additional concern is to minimize the number of switching operations needed to meet a particular set of demands because the wear and tear induced by each re-organization of switches eventually leads to equipment failure.

With some simplifying assumptions, we find that the abstract problem of meeting time-limited loads (i.e., each load requires a certain power for a certain time) from a set of generation resources using a set of distributed switches is similar to the problem of assigning packets of a certain length arriving at the input ports of a rearrangable optical switch to a set of output ports. Each packet corresponds to a demand, each input port to a generation resource, and each output port to a load Given a set of demands, the minimum make-span assignment of these demands to loads is also the assignment that minimizes total delay, while the minimum set of configurations also minimizes switch wear and tear. We therefore extend past work in demand assignments in rearrangable optical switches [2] to compute lower and upper bounds on the minimum number of rearrangements needed to meet a set of demands.

2. PROBLEM STATEMENT AND NOTATION

We make two simplifying assumptions in our work. First, we assume that all generators have the same cost of power production. Second, we assume that distribution losses are negligible.

We model a set of multi-connected microgrids with a switching system $(I, \mathcal{D})$ that consists of a set of inputs I (the generators) with port capacities c_i (the nominal power that they currently generate) and a set of demands $\mathcal{D}$ (elastic electrical loads) that are to be scheduled; a demand d is characterized by its length $l(d)$ (how long the demand lasts), width $w(d)$ (the power level of the demand), and a *load balancing vector* $v(d)$ that contains the set of input ports available to process d (i.e., the set of generators that can feasibly meet this demand).

Time is discrete; we denote by L and l respectively the longest and shortest length in time slots among all given demands. If a demand d is assigned to input i at time t, d uses $w(d)$ bandwidth of port i during the time interval $[t, t + l(d) - 1]$. A schedule P is a sequence of configurations, where each configuration is a partial mapping of the demands to the inputs that has to satisfy constraints imposed by port capacities and load balancing vectors. The length of a configuration C is defined by the longest demand that is scheduled during C. There is a non-negligible penalty, called *configuration overhead*, of V time slots between two consecutive configurations. Our goal is to satisfy loads in $\mathcal{D}$ as fast as possible. Note that the value of V can impact a scheduling decision. Therefore,

Algorithm 1 GREEDYSCHEDULINGPOLICY($\mathcal{D}, I$)

1: $D := \mathcal{D}, \mathcal{C} := \emptyset$.
2: **while** $D \neq \emptyset$ **do**
3:     start new configuration $C := \emptyset, I' := I$;
4:     **while** there are available ports and demands **do**
5:         $(i, d) := $ CHOOSEPORTDEMAND(D, I');
6:         $C := C \cup \{(i, d)\}, c'_i := c'_i - w(d), D := D \setminus \{d\}$;
7:     **end while**
8:     $\mathcal{C} := \mathcal{C} \cup \{C\}, D := D \setminus \{d \mid d \in C\}$.
9: **end while**
10: Return $\mathcal{C}$.

Algorithm 2 SG

1: **function** CHOOSEPORTDEMAND($\{\mathcal{D}_i\}_i, I$)
2:     **for** $i := 2$ **to** I **do**
3:         **if** $c_i > w(d)$ for some $d \in \mathcal{D}_i$ **then**
4:             return $(i, $ CHOOSEDEMAND$(\mathcal{D}_i, c_i))$;
5:         **end if**
6:     **end for**
7:     Return $(1, $ CHOOSEFIRST$(\{\mathcal{D}_i\}_i, I))$.
8: **end function**

Algorithm 3 SLD

1: **function** CHOOSEDEMAND($\mathcal{D}_i, c_i$)
2:     Return arg max $\{l(d) \mid d \in \mathcal{D}_i\}$.
3: **end function**
4: **function** CHOOSEFIRST($\mathcal{D} = \{\mathcal{D}_i\}_i, I$)
5:     $D := \{d \in \mathcal{D} \mid l(d) = \max_{d'} l(d')\}$.
6:     Return arg $\max_{d \in D} \{k(\mathcal{D}_i) \mid d \in \mathcal{D}_i\}$.
7: **end function**

Algorithm 4 SLP

1: **function** CHOOSEDEMAND($\mathcal{D}_i, c_i$)
2:     Return arg max $\{l(d) \mid d \in \mathcal{D}_i\}$.
3: **end function**
4: **function** CHOOSEFIRST($\mathcal{D} = \{\mathcal{D}_i\}_i, I$)
5:     $I' := \{i \mid k(\mathcal{D}_i) = \max_j k(\mathcal{D}_j)\}$.
6:     Return arg max $\{l(d) \mid d \in \mathcal{D}_i, i \in I'\}$.
7: **end function**

we consider an additional objective: to minimize the total number of configurations.

The practically interesting case is one where each demand can be met from exactly two input ports, and one of them is shared among all demands. This situation arises naturally if local distribution networks, each covering its own region, are supplemented by a central grid. In this case, the problem is to reuse the central grid input in the most efficient manner in order to optimize either makespan or the number of configurations.

3. ANALYTICAL STUDY

Formally speaking, we say that an algorithm A has approximation ratio α (is α-approximate) with respect to some objective function if for every input $(\mathcal{D}, I)$, A produces a schedule with objective function value at most α times greater than the optimal objective function value.

We concentrate our efforts on simple policies, amenable to efficient implementation since such policies can also scale well. The general algorithm describing such a policy is presented in Algorithm 1. Given a set of demands $\mathcal{D}$ and a set of input ports I with capacities $c_i, i \in I$, a greedy scheduling policy creates each consecutive configuration by greedily choosing the next demand to process. Once there are no more ports (i.e., generators) available to meet the existing demands, so that all relevant capacities have been exhausted, the current configuration is finalized and a new configuration begins.

The heart of Algorithm 1 is the CHOOSEPORTDEMAND procedure that takes current state (remaining demands and leftover capacities) as input and outputs the input-demand pair (i, d) for the next assignment. Various algorithms considered in this work differ from each other precisely in their CHOOSEPORTDEMAND procedures.

ALG	Unit capacities		Unit widths		General case	
	Lower	Upper	Lower	Upper	Lower	Upper
SG	1	3/2	5/3	2	-	4
SLD	$3/2 - 2^{-(I-1)}$	3/2	5/3	2	-	4
SLP	1	1	1	1	-	2

Table 1: Results summary for minimizing the number of configurations.

The obvious general algorithm for this case is SG, which stands for "Shared Greedy" (Algorithm 2): that first fills the capacities of every port except the first, then chooses demands for the first port.

Different algorithms may differ in choosing a demand for a single port (CHOOSEDEMAND procedure) and in choosing which demand to send to the first port for extra processing (CHOOSEFIRST procedure).

The basic tradeoff here is the balance between minimizing the number of configurations and minimizing their total length (duration). In this regard, we define two algorithms from the SG family: SLD ("Shared Longest Demand", Algorithm 3) and SLP ("Shared Longest Port", Algorithm 4). SLD chooses the longest available demand for the current configuration; for the CHOOSEDEMAND procedure it does not matter which one, for the CHOOSEFIRST procedure SLD splits ties with the largest port heuristic (maximal $k(\mathcal{D}_i)$). SLP, on the other hand, chooses for the CHOOSEFIRST procedure a demand from the port with maximal normalized load $k(\mathcal{D}_i)$; for splitting ties and CHOOSEDEMAND, it uses the longest demand heuristic.

The interplay of the following four parameters define the behaviour of a scheduling policy: (i) input port capacities, (ii) demand lengths, (iii) demand widths, and (iv) "normalized load". Clearly, all parameters that have an impact on the number of configurations also affect the schedule length objective. The notion of "normalized load" has significant impact on the both objectives. Observe that demand length has no impact on the number of configurations but can have significant influence on the total length of schedule as $\frac{L}{l}$ grows. The impact of input capacity constraint is interesting even for unit-sized demand widths. There is a correlation between the number of configurations and utilization of input capacities. During our study we carefully explore the impact of each one of the considered parameters on the performance of scheduling policies. The main contribution of this paper is an analysis of characteristics that should be implemented by an "ideal" policy. A short summary of our theoretical results is shown in Table 1. Our work provides the first steps towards establishing a strong theoretical foundation for the scheduling of demands in multi-connected microgrids.

4. REFERENCES

[1] S. Keshav and C. Rosenberg. On load elasticity. In *Proc. IEEE COMSOC MMTC E-Letter*, To Appear.

[2] A. Kesselman and K. Kogan. Nonpreemptive scheduling of optical switches. *IEEE Transactions on Communications*, 55(6):1212–1219, 2007.

'Just Enough' Sensing to ENLITEN

A preliminary demonstration of sensing strategy for the 'ENergy LIteracy Through an Intelligent Home ENergy Advisor' (ENLITEN) project

Tom Lovett
Department of Architecture
and Civil Engineering
University of Bath
Bath, UK
t.r.lovett@bath.ac.uk

Elizabeth Gabe-Thomas
Department of Psychology
University of Bath
Bath, UK
e.g.thomas@bath.ac.uk

Sukumar Natarajan
Department of Architecture
and Civil Engineering
University of Bath
Bath, UK
s.natarajan@bath.ac.uk

Eamonn O'Neill
Department of Computer
Science
University of Bath
Bath, UK
eamonn@cs.bath.ac.uk

Julian Padget
Department of Computer
Science
University of Bath
Bath, UK
j.a.padget@bath.ac.uk

ABSTRACT

The ENLITEN project aims to reduce carbon emissions attributable to energy use within buildings – particularly homes – by understanding and influencing occupants' habits and behaviours with respect to energy use. To achieve this we are developing a system based on a whole building energy model that, uniquely, integrates (i) a thermal model, (ii) a model of occupants' habits and requirements and (iii) a disaggregated model of energy use in the dwelling. A "minimal sensor set" is being identified for delivery of the live data feed to the whole building energy model. This will be complemented by collection and analysis of occupant data that will be used to construct a model of occupants' energy-related attitudes, behaviours and habits. The whole building energy model will contribute to an interactive in-building tool to help occupants identify and break poor energy habits, form better ones and reduce energy demand and carbon emissions. This system is to be deployed in 200 homes for a period of 2 years in the city of Exeter (UK). This poster and demo presents our prototype sensor deployment and the visualisation of live data collected from the trial home group, through which the minimal sensor set is being developed.

Categories and Subject Descriptors

H.4 [**Information Systems Applications**]: Miscellaneous

Keywords

ENLITEN; Energy use; sensing; intelligence; interaction

1. INTRODUCTION

Buildings are the single largest contributors to UK carbon emissions (greater than transport) [3] with a total footprint of about 34%. To meet the UK's national target of 80% emissions reductions by 2050 [1], energy use in buildings has to be reduced. It is clear that savings from efficiency improvements alone (e.g. insula-

tion, more efficient boilers, etc.), while substantial, will not achieve this target. Micro-generation and improvements to building design will be important, but none of these factors are likely to work unless we have a deeper understanding of how occupants interact with their internal environments, the buildings they occupy and the systems they use within the buildings. Furthermore, the occupants themselves require a better understanding of their energy use and the implications – both environmentally and financially – of their energy-related behaviour.

Current work into this problem has concentrated on various factors [8], including: (i) thermal properties of buildings, i.e. understanding thermal models of buildings and heat loss reduction through improved building and heating, ventilation and air conditioning (HVAC) system design [2] (ii) inference of energy use from appliances, e.g. using disaggregation of sensed electrical data [5, 7] and (iii) occupant behaviour, e.g. inferring occupants' energy-related activities from automatically sensed data [6], providing occupants with information on energy use and its implications [4], and attempting to intervene in energy-related behaviour.

This poster and demo describes the ENLITEN[1] project, which aims to understand these factors better and to use sensing technology to inform occupants both about their energy use and actively encourage them to change their energy-related behaviour. Unlike previous work, which explores sensing capabilities and occupancy inference, e.g. [5, 8], we concentrate on the trade-off between sensing capability and *affordability*, i.e. an appropriate sensor choice for energy-related sensing applications.

2. APPROACH

Figure 1 illustrates the conceptual architecture for the ENLITEN project. We are developing a building *energy model*, which comprises of a thermal model and occupant model, both of which actively configure and adapt themselves through the inference and learning of incoming data. The *thermal model* is concerned with heat generation, control and loss within each home. Based on real time modelling of buildings' thermal responses [2], the thermal model will capture and learn about environmental behaviours

---

[1]http://cs.bath.ac.uk/enliten

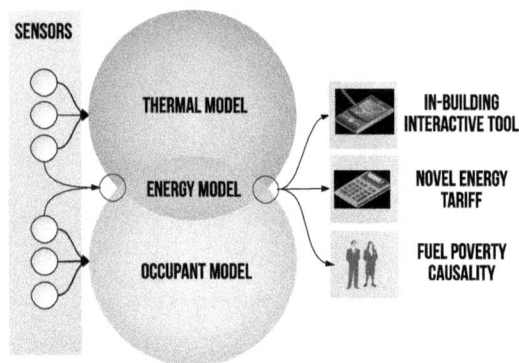

Figure 1: ENLITEN conceptual architecture

in order to inform occupants about areas of concern or potential improvement. The *occupant model* is concerned with occupants' energy-related activities within the home. Using a combination of electrical load monitoring [5, 7], gas sensing, occupancy detection and qualitative survey data, the model will build a representation of occupants' behaviour to auto-suggest changes to their energy-related behaviour.

3. CURRENT WORK

The initial goal for ENLITEN is to deploy an *affordable* sensor set in a sample of $\approx$ 200 homes (from a candidate set of 5000 homes) in the city of Exeter in the UK. The key challenge is the design of a sensor set that trades off data value – that is, the importance of each sensor to the set value – against the set cost – that is monetary and energy consumption costs.

Sensing equipment can be expensive, particularly for high-precision sensing [5], leading to a trade-off between data quality (and therefore cost) and scalability. If sensors are too expensive, it limits the number of deployments. Moreover, an expensive system will have little impact in the real world where consumer affordability is a major issue. Conversely, if the sensing equipment is of low quality, we risk compromising on data integrity and our understanding of energy consumption. Given the fundamental contribution of data sensing to the energy reduction problem and to ENLITEN, low integrity data is likely to have a negative impact on higher-level thermal and occupancy modelling.

Our current work, and the focus of this poster and demo, is a study of energy-related sensing requirements and the identification of a "minimal sensor set" to sense 'just enough' about each of the 200 buildings and their occupants. The first step in this process is comprehensive sensing in a small group of trial homes ($\approx$ 5), in order to identify superfluous sensors and proxy sensors to achieve both sufficient coverage and redundancy at low cost. Table 1 lists the sensors and the relevance of each to the 3 models described above. Following deployment of a large sensor set in each of the trial homes, occupants will undertake scripted events so that we can compare sensor data against ground truth and measure sensor subset information. This will be combined with a measure of subset cost in order to analyse the value-cost trade-off of each subset. The demo visualizes this sensor deployment and demonstrates how varying sensor subsets affects the trade-off between sensor value and cost using appropriate measures.

Sensor	Model relevance		
	Thermal	Energy	Occupant
"Environmental" sensing			
Temperature	X		X
Temperature (water pipe)	X	X	X
CO$_2$	X		X
Air pressure	X		X
Humidity	X		X
Internal light		X	X
Insolation	X		
Sound level			X
"Energy" sensing			
Gas usage		X	
Real power (appliance)		X	X
Reactive power (appliance)		X	X
Real power (circuit)		X	X
Reactive power (circuit)		X	X
Real power (household)		X	X
Reactive power (household)		X	X
Other sensing			
Passive Infra Red (PIR)			X
Window open/close	X		X
Water flow		X	X

Table 1: List of sensors and model relevance

4. SUMMARY

We have presented the ENLITEN project, outlining its high level architecture, key project components and planned outputs. In particular, we describe our current work on sensor set design, which will be presented in the demo. Full scale deployment is scheduled for 3Q2013.

5. REFERENCES

[1] Climate Change Act 2008. Retrieved from http://www.legislation.gov.uk/ukpga/2008, 20130122, 2008.

[2] D. Coley and J. Penman. Second order system identification in the thermal response of real buildings. Paper II: recursive formulation for on-line building energy management and control. *Building and Environment*, 27(3):269–277, 1992.

[3] Committee on Climate Change. Meeting Carbon Budgets – ensuring a low-carbon recovery. Retrieved from http://www.theccc.org.uk/reports, 20130122, June 2010.

[4] E. Costanza, S. Ramchurn, and N. Jennings. Understanding Domestic Energy Consumption through Interactive Visualisation: a Field Study. In *Proc. Ubicomp '12*, pages 216–225, 2012.

[5] J. Kolter and M. Johnson. REDD: A public data set for energy disaggregation research. In *Workshop on Data Mining Applications in Sustainability (SIGKDD), San Diego, CA*, 2011.

[6] S. Mamidi, Y. Chang, and R. Maheswaran. Improving building energy efficiency with a network of sensing, learning and prediction agents. In *Proc. AGENTS '12*, pages 45–52, 2012.

[7] O. Parson, S. Ghosh, M. Weal, and A. Rogers. Nonintrusive load monitoring using prior models of general appliance types. In *Proc. AAAI '12*, 2012.

[8] R. Zhang, K. Lam, Y. Chiou, and B. Dong. Information-theoretic environment features selection for occupancy detection in open office spaces. *Building Simulation*, 5(2):179–188, 2012.

BatNet: An Implementation of a 6LoWPAN Sensor and Actuator Network

Jorge Martín, Igor Gómez, Eduardo Montoya, Jie Song, Jorge Olloqui, Rocío Martínez

CeDInt-Universidad Politécnica de Madrid
Campus de Montegancedo s/n, Pozuelo de Alarcón
28223 Madrid (Spain)
+34 914524900 ext. 1749

jmartin@cedint.upm.es

ABSTRACT

This demo presents *BatNet*, a 6LoWPAN Wireless Transducer Network, in a Home Automation context. Its suitability for such application is shown by means of several performance and usability tests.

Categories and Subject Descriptors

J.2.2 [**Physical Science and Engineering**]: Engineering

General Terms

Measurement, Performance, Reliability, Experimentation, Verification.

Keywords

WSN; 802.15.4; 6LoWPAN; RPL; CoAP; Contiki OS; Home Automation; IoT.

1. INTRODUCTION

Improving energy efficiency in buildings is one of the goals of the Smart City initiatives and a challenge for the European Union [1]. The use of Wireless Transducer Networks to improve energy management in buildings has increased lately. In this demo, we show a 6LoWPAN Wireless Transducer Network (*BatNet*) as part of an open energy management system. This network has been designed to operate in buildings, to collect environmental information (temperature, humidity, illumination and presence) and electrical consumption in real time (voltage, current and power factor) as well as to control loads and systems such as HVAC (Heating, Ventilation and Air Conditioning), lighting or blinds.

2. SYSTEM DECRIPTION

BatNet system design [2] focuses on avoiding the traditional Home Automation systems (e.g.: LonWorks, KNX, X10) limitations in terms of cost, interoperability, power consumption and complexity. Its main characteristics are:

- *6LoWPAN-based network*: adaptation of IPv6 to Personal Area Networks. Physical and data link layers are based on IEEE 802.15.4 wireless protocol. RPL is used as routing protocol.

- *CoAP (Constrained Application Protocol):* (currently an IETF draft) specific web transfer protocol designed to be used with constrained nodes and networks.

- *Contiki OS*: open source operating system from SICS that enables multitasking and implements both IPv6 and IEEE 802.15.4 standards under low-capacitance hardware requirements.

BatNet is based on a modular functional architecture, which eases the development of the different transducers and the integration of external devices. Each node of the *BatNet* comprises a processing and communications module (*BatMote*) plus a transducer module.

The core of the *BatMote* is the ATmega 128RFA1 from Atmel

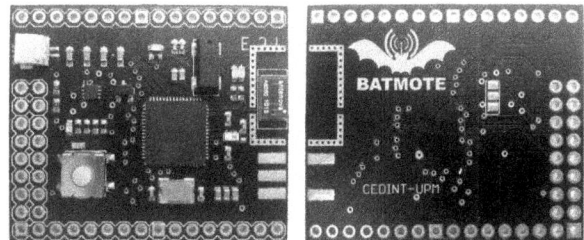

Figure 1. *BatMote* communications module

Corporation, which allows easy connection of transducer modules. *BatMote* can be powered by batteries or by other external power source (3.3-12V) and allows different low power operation modes[1].

In order to make information available remotely, several CoAP resources have been implemented for each mote, including those resources intended for the acquisition of parameter values, the system operating configuration or the display of system information.

---

[1] http://www.batmote.net

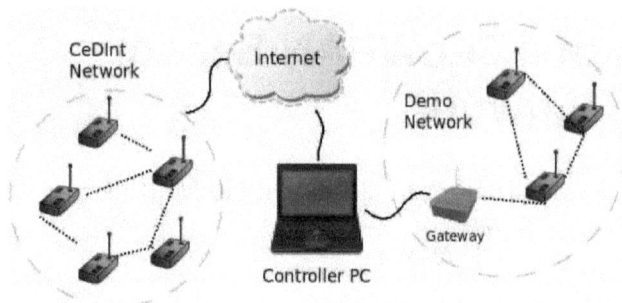

Figure 2. Demo deployment

The transducer modules that have been implemented are:

BatMeter: power meter sensor to be allocated in the distribution board, which calculates electrical power and energy consumption simultaneously.

BatSense: ambient multi-sensor module which includes temperature, humidity, illumination and presence sensors all together in a single device.

BatPlug: electric load controller and power meter based on TRIAC and relays to control the ON/OFF status. It also integrates a Hall Effect sensor to measure the electric current.

BatAmbientLight: RGB LED ambient light controller.

BatLink: Gateway device provided with both Ethernet and 802.15.4 interfaces.

3. DEMO SCENARIO

In this demo we show the proper functioning of the *BatNET* network in a Home Automation scenario. The demo will comprise the following elements:

- An ambient sensor mote, *BatSense*.

- A load controller and consumption meter, *BatPlug*.

- An ambient light controller, *BatAmbientLight*.

- An 802.15.4 to Ethernet gateway, *BatLink*.

- A PC connected to the Internet and to the 6LoWPAN network.

Besides, another Home Automation 6LoWPAN network will be set up in CeDInt building in Madrid, and accessed via Internet.

The existence of two networks, one local and one remote, allows us to simulate a real situation for an Internet of Things user regarding Home Automation, both when he or she is at home or away.

In the CeDInt building network other devices not present in the demo will be also accessible, such us a *BatMeter*, whose installation requires specific power grid conditions.

By the use of different software tools, attendees will be able to understand the following network related topics:

1. Network formation and evolution. IPv6 addressing. RSSI. Used software: Java based application, GNU/Linux IP commands.

2. Parameters visualization (real time and records) and control. Used software: Specific web-based interface.

3. Remote configuration of motes. Used software: Copper (Cu) CoAP user-agent for Firefox [3]

4. Ease of programming. Used software: Eclipse and GNU/Linux AVR tools.

More specific questions such as devices accuracy or development issues can be discussed during the demo.

4. CONCLUSIONS

In this demo we present an easy to deploy and affordable Home Automation network based on self-developed 6LoWPAN motes (*BatMotes*). Technical suitability is proved by different on-site and remote experiments.

Future work is required for achieving a higher delivery ratio for network packets, since the currently implemented path selection metric (ETX) does not optimize this factor [4].

5. REFERENCES

[1] Luis Pérez-Lombard, José Ortiz, Christine Pout. "A Review on buildings energy consumption information", Energy and Building, Elsevier, March 2007.

[2] G. Campo, E. Montoya, J. Martín, I. Gómez, A. Santamaría. "BatNet: a 6LoWPAN-based Sensors and Actuators Network", UCAmI, Vitoria, December 2012.

[3] Matthias Kovatsch. " Human–CoAP Interaction with Copper". Proceedings of the 7th IEEE International Conference on Distributed Computing in Sensor Systems (DCOSS 2011), Barcelona, Spain, June 2011.

[4] Bor-rong Chen, Kiran-Kumar Muniswamy-Reddy and Matt Welsh. "Lessons Learned from Implementing Ad-Hoc Multicast Routing in Sensor Networks". Harvard University Technical Report TR-22-05, November 2005.

Efficient Computation of Shapley Values for Demand Response Programs

Gearóid O'Brien
Department of Electrical
Engineering
Stanford University
Stanford, CA
gobrien@stanford.edu

Abbas El Gamal
Department of Electrical
Engineering
Stanford University
Stanford, CA
abbas@ee.stanford.edu

Ram Rajagopal
Department of Civil and
Environmental Engineering
Stanford University
Stanford, CA
ramr@stanford.edu

Categories and Subject Descriptors: G.3 [Probability and Statistics]: Probabilistic algorithms

General Terms: Algorithms, Economics, Theory

Keywords: Demand Response; Shapley Value; Game Theory; Cooperative Games

1. INTRODUCTION

Demand Response (DR) is currently a major area of research. We propose analyzing demand response schemes in a game theoretic setting and utilizing the Shapley Value for fairly compensating participants of such schemes. As exact computation of the Shapley Value is intractable in general, we propose a stratified sampling method that can dramatically reduce the variance of the Shapley Value estimate when compared to previous methods.

DR ([1] gives an overview of DR) can be achieved by a combination of three actions. Usage can be curtailed by reducing their consumption level, a process known as "load shedding." Local generation can be used to take loads "off-grid," and loads can be adjusted temporally so that the load consumes power at a later time [5].

We focus on the cases where penalties are submodular functions, and payments are supermodular functions of the set of participants. In this case, the Shapley Value concept provides a payment incentive that guarantees that every participant is better off joining a single grand coalition. Computing these optimal payments requires a combinatorial calculation. We propose a statistical approach to this challenge and show that the approach works well in practice. It generalizes an early idea of sampling coalitions [3], but furthers it by considering a careful statistical computation.

2. PROBLEM FORMULATION

A DR program consists of an operator and customers (agents) $\mathcal{X}$ who have agreed to participate. Each agent provides a load profile of power consumed which is not known in advance. The program manager requests a given level of DR at a certain time, and the agents must react as requested. If the coalition fails to meet the target, an ex-post penalty is charged which should be fairly shared by all participants in the program. As an example, agents receive a fixed payment c_i for participating and then subtracts a penalty. The total

paid out to the entire coalition is then $P(\mathcal{X}) = \sum_i c_i - f(\mathcal{X})$, where $f(\mathcal{X})$ is the total penalty assigned to the coalition and is known as the *characteristic function*. In many practical DR programs, the payment c_i has already been completed and so it is only necessary to fairly divide the total penalty $f(\mathcal{X})$ among the agents. If the game is structured so that the characteristic function is *submodular* (or conversely, if the payment function is supermodular), then no subset of players has any incentive to leave the coalition.

Shapley proposed a solution that is both unique and fair for dividing $f(\mathcal{X})$ among the agents. The Shapley value for agent $x \in \mathcal{X}$ is a weighted average of the marginal contributions of an agent to all possible coalitions and can be defined in the equivalent equations (1) and (2).

$$\phi_x(f) = \sum_{\mathcal{A} \subseteq \mathcal{X} \setminus \{x\}} \frac{|\mathcal{A}|!(|\mathcal{X}| - |\mathcal{A}| - 1)!}{|\mathcal{X}|!} [\rho_x(\mathcal{A})] \quad (1)$$

$$= \frac{1}{|\mathcal{X}|!} \sum_R \left[\rho_x(\mathcal{A}_x^R) \right] \quad (2)$$

$\rho_x(\mathcal{Y}) = f(\mathcal{Y} \cup \{x\}) - f(\mathcal{Y})$ is the marginal contribution of agent x to coalition $\mathcal{Y}$. In (2), R is an ordering of the agents in $\mathcal{X}$ and $\mathcal{A}_x^R$ is the set of agents that precede x in a given ordering R.

Aside from directly calculating (1), generating functions have been previously examined [2]. This approach requires large arrays in order to reduce complexity to polynomial time, although it does produce an exact solution for the Shapley Value. The intractability of calculating the Shapley value has given rise to a number of studies which attempt to estimate it via approximation methods [3]. In [4], the authors use the Shapley value to allocate transmission service costs among network users in energy markets.

The main contribution of this paper is proposing a new stratified sampling method for estimating the Shapley Value, thus reducing the variance of the estimated value when compared to other randomized sampling methods.

3. ESTIMATING THE SHAPLEY VALUE US-ING STRATIFIED SAMPLING

Stratified sampling is a method of dividing the population into strata and then sampling from these strata. Here, the population is the marginal contributions of a given participant to every possible coalition. We divide this population into strata where each stratum contains the marginal contribution of a given participant to coalitions with an equal number of agents.

It can be shown that the Shapley Value is equal to the mean of the average marginal contributon from each stratum. By sampling appropriately from the strata, the variance of the Shapley Value estimate can be reduced. We introduce the notation σ^2 to represent a vector of the strata variances and μ to represent a vector of the strata means, indexing both vectors by i. Further, N is the total number of samples taken and n is the number of strata.

The variance of the Shapley Value estimate calculated by stratified sampling is minimized by sampling from all strata in proportion to its standard deviation. The variance of this estimate is

$$\sigma_{\text{SD}}^2 = \frac{1}{N}\left[\frac{1}{n}\sum_{i=1}^{n}\sigma_i\right]^2.\tag{3}$$

Comparing this variance to that of random sampling, where we do not distinguish between strata the variance is

$$\sigma_{\text{RS}}^2 = \frac{1}{N}\left[\frac{1}{n}\sum_{i=1}^{n}\sigma_i^2 + \text{var}(\mu)\right].\tag{4}$$

As the strata standard deviations will not be available a priori in general, it is natural to ask what the *value* of this knowledge is. Applying the principle of indifference, we sample equally from each stratum. Here, the variance is

$$\sigma_{\text{EB}}^2 = \frac{1}{N}\left[\frac{1}{n}\sum_{i=1}^{n}\sigma_i^2\right].\tag{5}$$

It is possible to therefore show:

THEOREM 3.1. *It is always better to sample equally from each stratum than to sample randomly, and if the standard deviations of the strata are known, it is better again to sample in proportion to those, since:*

$$\sigma_{\text{SD}}^2 \leq \sigma_{\text{EB}}^2 \leq \sigma_{\text{RS}}^2.$$

Next, we propose a reinforcement learning algorithm to approximate standard deviation sampling.

3.1 Implementing SD Weighted Sampling

This is a typical reinforcement learning problem where we seek to *exploit* the information we know (regarding the standard deviations) in order to sample correctly, but must also *explore* in order to calculate the standard deviations with accuracy. Defining $\pi_i(t)$ as the probability of sampling from stratum i at sample number t, and $\hat{\sigma}_i$ as the estimated standard deviation of stratum i:

$$\pi_i(t) = \epsilon_t\frac{1}{n} + (1 - \epsilon_t)\frac{\hat{\sigma}_i}{\sum_j\hat{\sigma}_j},$$

A suitable function for ϵ_t would be similar to the sigmoid function $\epsilon_t = [1 + \exp(t - N)]^{-1}$, ensuring that the probability of sampling from each stratum starts equal and ends proportional to the standard deviation. For each sample, the reinforcement learning algorithm chooses a stratum at random, weighted according to $\pi_i(t)$. After sampling, the weights are updated according to ϵ_t and we sample again until N samples have been taken.

4. EXAMPLE

A simple example of a supermodular payment structure is the Load Shedding DR program where a limit is placed on the aggregate load. For every time step during which the aggregate load exceeds this limit, loads that consumed power will be penalized in proportion with their Shapley value. The characteristic function representing such a DR program is $f(\mathcal{X}) = \left[\sum_{x\in\mathcal{X}} x - M\right]_+$. To evaluate this DR program, we take a small sample set $\mathcal{X}$ containing 20 loads. We can calculate an exact Shapley value for a representative load and in the process compare how the three different sampling techniques would have fared.

Figure 1 shows the mean and standard deviation for each stratum when using the above characteristic function. Figure 2 shows the variance of the Shapley value estimate for the sampling techniques. SD weighted sampling significantly reduced variance of the Shapley Value.

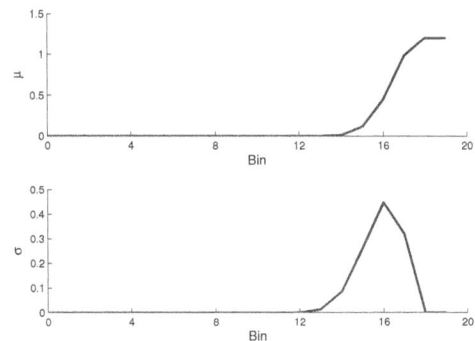

Figure 1: Strata Mean and Standard Deviation.

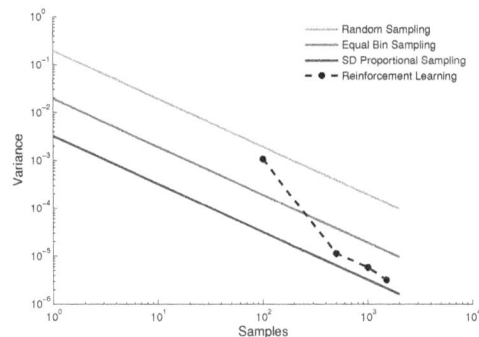

Figure 2: Change in variance with sample size.

5. REFERENCES

[1] M. Albadi and E. El-Saadany. Demand response in electricity markets: An overview. In *Power Engineering Society General Meeting, 2007. IEEE*.

[2] E. Algaba, J. Bilbao, J. F. Garcia, and J. Lopez. Computing power indices in weighted multiple majority games. *Mathematical Social Sciences*, 46(1):63 – 80, 2003.

[3] Y. Bachrach, E. Markakis, E. Resnick, A. D. Procaccia, J. S. Rosenschein, and A. Saberi. Approximating power indices: theoretical and empirical analysis. *Autonomous Agents and Multi-Agent Systems*, 20(2):105–122, Mar. 2010.

[4] M. Junqueira, L. Da Costa, L. Barroso, G. Oliveira, L. Thomé, and M. Pereira. An aumann-shapley approach to allocate transmission service cost among network users in electricity markets. *Power Systems, IEEE Transactions on*, 22(4):1532–1546, 2007.

[5] G. O'Brien and R. Rajagopal. A Method for Automatically Scheduling Notified Deferrable Loads. In *American Control Conference (ACC), 2013*. To appear.

Flexible Loads in Future Energy Networks

Jay Taneja, Ken Lutz, and David Culler
Computer Science Division
University of California, Berkeley
{taneja,culler}@cs.berkeley.edu, lutz@eecs.berkeley.edu

abstract
ABSTRACT

We develop a vignette of an information-rich energy network with flexible and responsive electrical loads in the form of a domestic refrigerator augmented with a thermal storage system and a supply-following controller that responds to the availability of fluctuating renewable sources. We fully characterize our prototype thermal storage-enhanced refrigerator. Using this, we investigate the behavior of a network of such loads at statewide scale in the context of a dynamic model of deep penetration of renewable sources. Our results show that with 10% penetration of thermal storage across refrigeration and freezer load on the California electricity grid, peak fossil fuel load can be reduced by nearly a gigawatt. The approach naturally extends to similar applications in thermal management of buildings and would operate in concert with smart vehicle charging.

Categories and Subject Descriptors: H.4 [Information Systems Applications]: Miscellaneous

General Terms: Design, Measurement

Keywords: Supply-Following, Renewables

1. FUTURE ENERGY NETWORKS

The nature of the electricity grid is evolving from a relatively small set of fully dispatchable generators powering a massively distributed set of oblivious loads to a more integrated, cooperative network of sources and loads with a far more extensive data overlay of control and management. Numerous trends across the energy sector – among them, renewables portfolio standards, the emergence of inexpensive wireless sensors and actuators for dense monitoring and management of energy systems, and the explosion in availability of competitive renewable generation – are pushing a transition where maintaining the match between supply and demand changes from a completely centralized task managed by electricity providers to a distributed one involving providers as well as consumers. These policy and technological trends point to a future grid where large supplies of non-dispatchable renewable generation dominate. However, deeper renewables penetration increases supply variability, exacerbating the challenge of supply-demand matching.

Some future reports have begun to document this challenge, though specific assumptions in models of future grids vary widely. Previous work introduced a model of a future

boilerplate
Copyright is held by the author/owner(s).
e-Energy'13, May 21–24, 2013, Berkeley, California, USA.
ACM 978-1-4503-2052-8/13/05.

instance of the California electricity grid with 60% of its electricity generated from renewable sources [1]. This grid presents a different set of challenges than today's grid does: instead of heavy fossil fuel use in the summer to accommodate a large air conditioning load, the scarcity in a grid with heavy renewables penetration occurs in the winter; this is the time the fossil fuel resource gets used. With an inability to control renewables generation, mismatches occur more often; flexible loads can respond to these mismatches. In this work, we extend the model to analyze the grid-scale ramifications of a massively distributed, agile energy storage resource represented by a network of energy loads.

2. FLEXIBLE LOADS WITH ENHANCED STORAGE

The class of flexible loads that incorporate energy storage are an underused and potentially valuable resource for electric grids. These loads, primarily refrigeration and freezer systems but also electric vehicles or rooftop ice storage systems, are already widely deployed – for example, over 99% of U.S. homes have a refrigerator, and more than half of refrigerators are replaced within seven years of purchase. Also, unlike batteries, thermal storage has high turn-around efficiency and nearly infinite charge cycles.

To evaluate the energy storage potential of this type of load, we developed a prototype system that couples this relatively simple mechanical system with IT and communications systems. We use a typical domestic refrigerator, a Whirlpool/KitchenAid unit with a 13.1 ft^3 refrigerator compartment and a 5.0 ft^3 freezer compartment. To better grasp physical system operation, we augmented it with a network of thermocouples and power sensors to monitor environmental and electrical operation. This network is connected via a 6LoWPAN-compliant IPv6 edge router.

To enhance device flexibility, we designed and installed a thermal storage system to augment a domestic refrigerator. The prototype system consists of three sealed tanks added to the freezer compartment and filled with a phase change material (19.7% ammonium chloride) for energy storage. In an aqueous solution, this nontoxic substance freezes at -15.4 ° C, just above the freezer operating range. To aid heat transfer, a low-freezing-point propylene glycol-based solution is circulated through the tanks to a heat exchanger in the refrigerator compartment by a small fluid transfer pump that is outside of the refrigerator. A comparison of the performance of the refrigerator with and without additional thermal storage is provided in Table 1. This is a prototype design; we believe that a commodity system would be integrated into the refrigerator and its internal control system, possibly providing insulation along with thermal storage capability while consuming minimal additional power.

Fridge Configuration	Unmodified	Enhanced with Energy Storage
Mean Power	76.69 W	86.55 W
Duty Cycle	53.32%	60.14%
Mean Compressor Duration	1558 s.	1762 s.
Mean Defrost Duration	11.4 mins.	11.4 mins.
Mean Heating Post Defrost	13.5 mins.	3.1 mins.
Defrost Frequency	15 hrs.	13.7 hrs.

Table 1: Comparison of power consumption and cycle behavior of two fridge configurations. Values are calculated over a 12-day, unperturbed period.

3. EVALUATION

In this section, we examine the value of a network of flexible refrigerators to a large electricity grid, first presenting a prototype implementation of a supply-following refrigerator before modeling its effect at grid-scale. Rather than the traditional model where loads obliviously consume energy, the supply-following refrigerator is able to use its communicative ability to monitor grid conditions, consuming electricity when most advantageous and avoiding it otherwise. We implement a control algorithm that observes a real-time renewables generation feed from the California ISO and selects from among three operating modes. It operates in either low-supply mode (<2250 MW), avoiding defrost cycles while maintaining a slightly warmer temperature; medium-supply mode (2250-2750 MW), operating conventionally; or high-supply mode (>2750 MW), using shorter, colder, more frequent cycles. Figure 1 shows operation of the fridge over 24 hours and provides mean power consumption in each mode.

Looking at the results, we notice how dynamic a supply-following refrigerator is compared to a conventional refrigerator. The fridge changes its setpoint and curtails a defrost cycle automatically to reduce power. Further, the fridge operates on precisely the opposite pattern from what today's grid operators favor – it consumes little energy at night when renewables are scarce, and consumes more during the day when solar energy is available. Also, note that medium-supply mode consumes nearly as much as high-supply mode – we believe this is an artifact of the particular day of operation and that in a longer deployment, medium-supply mode would approach its steady state consumption of 87 W, provided in Table 1. The stratification of power levels could be enhanced with forecasting of supply. Also, we acknowledge that temperature ranges in the compartments may be near the edges of their recommended ranges. We attribute this to the preliminary nature of the work. Nonetheless, we believe this is the first instance of a supply-following refrigerator, responding directly to real-time power availability.

The applications of such a technology are wide-ranging. A fleet of refrigerators could provide balancing to operate a renewable generation source as a baseload power plant. Instead, since utility expenses are often driven by high capital costs for underutilized facilities, flexible refrigerators could flatten duration curves of fossil fuel resources, relocating

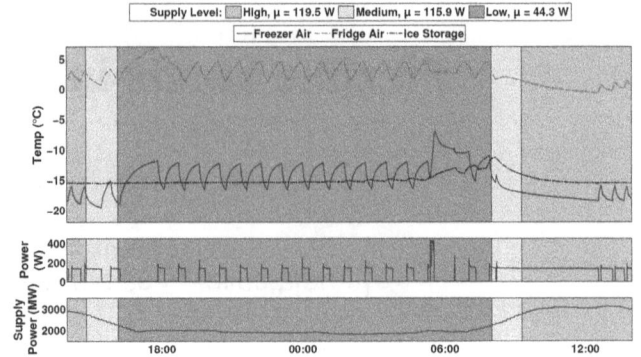

Figure 1: Operation of the thermal storage refrigerator as a supply-following load. The refrigerator operates in one of three power modes based on real-time grid availability of renewable power.

high-cost peak consumption to lower-cost times. Table 2 documents summaries of duration curves of natural gas generation in a 60% renewable grid with penetration of agile refrigerators ranging from 0% to the full capacity of refrigerators in California, 3 GW. Having even a fraction of the fridge capacity be agile provides extreme value; with 10% of the fridges agile, an entire gigawatt of supply can be obviated – nearly half the maximum possible peak natural gas reduction. Though very preliminary, these results highlight the potential of the emerging class of flexible electric loads.

	Fridge Penetration				
% of Hours	0% (GW)	10% (GW)	25% (GW)	50% (GW)	100% (GW)
Max	19.528	18.628	18.108	17.757	17.235
0.1	16.693	16.457	16.290	15.928	15.385
0.5	14.569	14.252	13.962	13.734	13.597
1.0	13.905	13.633	13.382	13.099	12.971
5.0	11.314	11.238	11.106	10.969	10.982
10.0	9.588	9.567	9.532	9.574	9.482
50.0	0.000	0.000	0.000	0.000	0.000
100.0	0.000	0.000	0.000	0.000	0.000

Table 2: Summaries of duration curves for natural gas generation sources in a 60% renewable grid. Columns represent different penetration levels of thermal storage refrigerators in California.

4. CONCLUSION

In this work, we present an inexpensive augmentation to a domestic refrigerator that enables a common household appliance to become a critically valuable electricity grid resource. By adding minimal equipment, improved sensing, and simple control functionality, we can reduce peak electricity demand and accommodate increased renewables penetration, creating a more sustainable electricity grid. We intend that these lessons can be assist in designing thermal and other energy storage systems and more advanced modeling and control processes for managing them.

Acknowledgments

This work is supported in part by the National Science Foundation under grants CPS-0932209 (LoCal), CPS-0931843 (ActionWebs), and CPS-1239552 (SDB). We thank Albert Goto and Stephen Dawson-Haggerty for deployment help.

5. REFERENCES

[1] J. Taneja, R. Katz, and D. Culler. Defining CPS Challenges in a Sustainable Electricity Grid. In *Proceedings of the Third ACM/IEEE Int'l Conf. on Cyber-Physical Systems*, 2012.

Towards Appliance Usage Prediction for Home Energy Management

Ngoc Cuong Truong, Long Tran–Thanh, Enrico Costanza and Sarvapali D. Ramchurn
Agent, Interaction and Complexity Group
Electronics and Computer Science
University of Southampton, UK
{nct1g10, ltt08r, ec, sdr}@ecs.soton.ac.uk

ABSTRACT

In this paper, we address the problem of predicting the usage of home appliances where a key challenge is to model the everyday routine of homeowners and the inter–dependency between the use of different appliances. To this end, we propose an agent based prediction algorithm that captures the everyday habits by exploiting their periodic features. We demonstrate that our approach outperforms existing methods by up to 40% in experiments based on real–world data from a prominent database of home energy usage.

Categories and Subject Descriptors

I.2.6 [**Artificial Intelligence**]: Learning

Keywords

Usage Prediction, Home Energy Management

1. INTRODUCTION

In the face of dwindling fossil fuels, an ageing electricity distribution infrastructure, and the adverse effects of high levels of green house gasses on climate change, the problem of generating affordable and clean electricity reliability is one of the greatest challenges of this century [1]. Now, to make the use of the electrical devices in the home more efficient, and thus, to reduce both carbon emissions and cost, a set of agent based demand side management techniques have recently been introduced to optimise the schedule of loads [4]. However, these techniques typically do not take into account the homeowner's preferences in their optimisation, and ignore inter-dependencies between the usage of different appliances.. Thus, the main challenge is to predict the energy consumption activities of homeowners, so that the agent can design optimal schedules by planning ahead the electricity usage that meets the human's preferences.

Against this background, we propose a novel approach to predicting the energy consumption of different home appliances, that takes into account both the human routine activities and the inter–dependency between appliances. To do so, we rely on the common assumption that human behaviour follows a certain cyclic pattern [2]. Based on this, we build a model that exploits this cyclic behaviour. To handle the inter–dependency between the appliances, we use the

e-Energy'13, May 21–24, 2013, Berkeley, California, USA.
ACM 978-1-4503-2052-8/13/05.

episode generation Hidden Markov model (EGH) [5] to efficiently identify the patterns that form the inter–dependency between the usage of the appliances. By putting the two models together, we demonstrate that our approach outperforms the state–of–the–art, that only focus on either human behaviour detection on inter–dependency pattern identification. We formalise our problem scenario in Section 2. Section 3 evaluates the algorithm and analyses the results.

2. PREDICTING THE USAGE ACTIVITIES OF APPLIANCES

We first describe the formalisation of our problem in Section 2.1, then introduce our algorithm in Section 2.2.

2.1 Model Description

We assume that we have a finite set of consumer activities, where different types of activities are distinguished by *labels* $l \in L$. An activity profile of label l $a_{l,t}$ is a tuple $\langle t, l, n \rangle$, composed of a time step t (measured in days), a label l and number of usage n, that denotes the number of occurrences of label l on day t. Let $x_t = \langle a_{1,t}, a_{2,t}, \ldots, a_{L,t} \rangle$ denote the usage profile of day t that contains the information about the usage of each label $l \in L$ on day t. The appliance usage history h_t of time slot t is the sequence $h_t = \{x_1, x_2, \ldots, x_t\}$. Our goal is to estimate x_{t+1} for any $t > 0$, given h_t.

2.2 The Prediction Algorithm

As mentioned earlier, the foundations of our prediction algorithm rely on the EGH method. However, as EGH is not designed for detecting human activities, we tailor the model to fit our settings by exploiting the periodic features of the human everyday routine. We build the inter–dependency model by relying on the EGH approach described by Srivatsa *et al.* [5]. Based on these models, we then construct a mixture model of the significant episodes (i.e. sets of possible inter–dependency rules) in order to calculate the probability of activities' occurrence in Section 2.2.2. Finally, Section 2.2.3 focuses on the prediction model in detail.

2.2.1 The Human Routine Model

We assume that human behaviour in home energy usage follows a weekly cycle. More formally, let K denote the number of occurrences of the target activity type l on the specific day d of the week in the activity usage history h_{t-1}. Thus, for each label l and the prediction day of the week d, from the original training dataset D, we extract a training set $D_{l,d} = \{X_i\}_{i=1}^{K}$, where $X_i = \langle x_{t_i-7}, \ldots, x_{t_i-1} \rangle$ is the weekly preceding window of activities from x that imme-

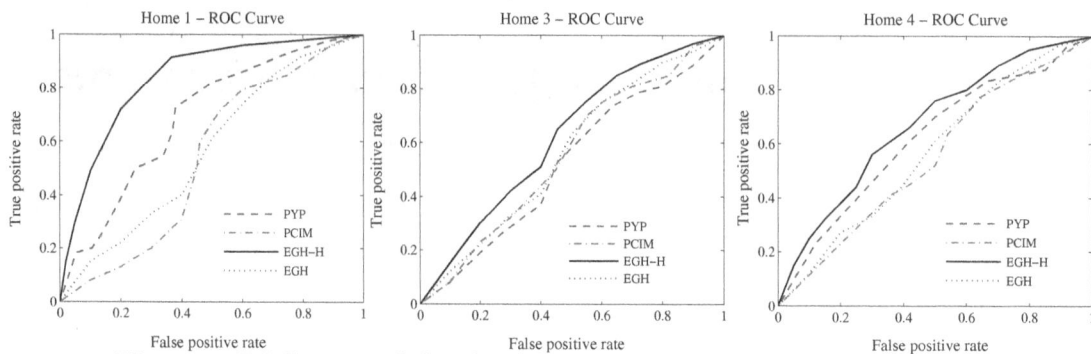

Figure 1: ROC curve of the algorithms run on three homes from REDD.

diately preceded the ith occurrence of l in x, and t_i is the time that the target activity type l occurred at the ith in the activity sequence. By doing so, we can reduce the computational costs and also improve the quality of prediction (as we will demonstrate later in Section 3).

2.2.2 *The Mixture Model*

Supppose that for a given training data set $D_{l,d} = \{X_i\}_{i=1}^{K}$, we have calculated a set of significant episodes, denoted as $F^s = \{\alpha_1, \ldots, \alpha_J\}$, and each HMM H_{α_j} of episode α_j. To model the effect of this joint influence, we compute a mixture model Λ_l (i.e. a combination of probabilistic processes) of the significant episodes' HMMs. The likelihood function of the training dataset D under a mixture model Λ_l is: $P[D|\Lambda_l] = \prod_{i=1}^{K} P[X_i|\Lambda_l] = \prod_{i=1}^{K} \left(\sum_{j=1}^{J} \theta_j P[X_i|H_{\alpha_j}] \right)$, where θ_j, $j = 1..J$ are the mixture coefficients of Λ_l (with $\theta_j \in [0,1]$ for all j, and $\sum_{j=1}^{J} \theta_j = 1$). We use the Expectation Maximisation (EM) algorithm to estimate the set of mixture coefficients of the mixture model Λ_l.

2.2.3 *The Prediction Model*

Let t denote the current time. For the set of target activity labels $l \in$, we want to predict their occurrences in the next day, $t + 1$. As we are mainly interested in occurrences of recent activities of the users, therefore, we construct a 7—length window of activities from the weekly period $[t - 7, t]$. We then estimate the likelihood of this recent activity sequence, given the mixture model, $\Lambda_l = \{(\alpha_j, \theta_j)\}_{j=1,\ldots,J}$, that is obtained from the training phase.

3. EMPIRICAL EVALUATION

Given the prediction model, we now demonstrate how our algorithm outperforms a set of benchmark algorithms, including Pitman–Yor Process (PYP), the piece–wise constant conditional intensity model (PCIM) [3]), and the original EGH method. Here, we perform the algorithms on the real–world REDD dataset [1] (see Section 3.1). We also compare the average running time of the algorithms in Section 3.2.

3.1 Performance on REDD Data

In overall, our method outperforms other state–of–the–art by up to 40%, based on the *F–score* measurement. In particular, it is better than PYP, EGH, and PCIM by approximately 73%, 40%, and 75% on average, respectively. From the figure of the receiver operating characteristic (ROC) curve (Figure 1), we can see that our algorithm dominates

[1] http://redd.csail.mit.edu/

all the others. In particular, the area under the curve (AUC) of EGH–H in home 1 is 0.84, while the AUC value for PYP, EGH, and PCIM is 0.68, 0.56, and 0.53, respectively. We can also observe that since data from homes 3 and 4 is less detailed, all the algorithms provides worse performance, compared to themselves in home 1. However, our algorithm still dominates the benchmark approaches.

3.2 Average Running Time of the Algorithms

We run the algorithms on an Intel(R) Xeon(R) computer (64–bit operating system) with 2.67 GHz and 12GB. We can observe that on average, our algorithm is 1504.78, 119.3, and 151.19 times faster than PYP, PCIM, and EGH on average.

4. CONCLUSIONS

We proposed EGH–H, the algorithm that addresses human behaviour prediction within the energy management domain. We also demonstrated through extensive evaluations, using real–world data taken from the REDD, that our algorithm outperforms state–of–the–art methods by up to 40% in prediction accuracy. As a result, our work could potentially form an efficient solution to real–world home energy management systems, where usage predictions are needed to optimally schedule the electrical consumption of the home. In addition, an improved version is described in [6].

5. REFERENCES

[1] Department of Energy & Climate Change. Smarter Grids : The Opportunity. (December), 2009.

[2] M. C. González, C. A. Hidalgo, and A. Barabási. Understanding individual human mobility patterns. *Nature*, 453(7196):779–82, June 2008.

[3] A. Gunawardana, C. Meek, and P. Xu. A Model for Temporal Dependencies in Event Streams. *NIPS*, 2011.

[4] S. D. Ramchurn, P. Vytelingum, A. Rogers, and N. Jennings. Agent-Based Control for Decentralised Demand Side Management in the Smart Grid. In *AAMAS*, pages 5 – 12, 2011.

[5] L. Srivatsan, P. Sastry, and K. Unnikrishnan. Discovering frequent episodes and learning hidden markov models: A formal connection. *IEEE Trans. on Knowledge and Data Eng.*, 17(11), 2005.

[6] N. C. Truong, J. McInerney, L. Tran-Thanh, E. Costanza, and S. Ramchurn. Forecasting multi-appliance usage for smart home energy management. *In Proc. of IJCAI*, 2013.

Mobile Location Sharing: An Energy Consumption Study

Ekhiotz Jon Vergara, Mihails Prihodko, Simin Nadjm-Tehrani
Department of Computer and Information Science
Linköping University, Sweden
ekhiotz.vergara@liu.se, mihpr362@student.liu.se, simin.nadjm-tehrani@liu.se

ABSTRACT

The use of a mobile device's battery for frequent transmissions of position data in a location sharing application can be more expensive than the location retrieval itself. This is in part due to energy-agnostic application development and in part dependent on choice of protocols. This paper studies the lightweight Message Queuing Telemetry Transport protocol (MQTT) as an application layer protocol on top of the third generation cellular communication. The energy efficiency and amount of data generated by the public/subscribe MQTT protocol is experimentally compared against the Hypertext Transfer Protocol (HTTP), which is currently used in typical location sharing applications.

The evaluation results indicate that MQTT is a good candidate as a protocol for location sharing. At comparable bandwidth and energy expenses MQTT offers better quality of user experience, since the subscribers are notified at once when the location of some interesting client has changed. Our measurements show that MQTT is more energy-efficient than HTTP in the idle state and when the number of other users with whom the client shares location is low. When the number of users increases beyond 3, HTTP becomes the preferred option in terms of energy efficiency at the cost of a higher notification delay.

Categories and Subject Descriptors

C.2.1 [**Computer Communication Networks**]: Wireless communication; C.4 [**Performance of Systems**]: Measurement techniques

Keywords

transmission energy; UMTS; location based services; MQTT; HTTP; mobile devices; Android;

1. INTRODUCTION

It is well-known that the ubiquitous access to location information, e.g., by exploiting geographic services such as the Global Positioning System, does not come cheaply in terms of energy resources of mobile devices. Location-based services are applications integrating geographic location with services, such as navigation systems, sports trackers or business information delivery. There are countless applications and services that support the user for sharing their location

data with other users (e.g., Foursquare, Google Latitude or Gowalla), also referred to Location-sharing services (LSS). It might therefore come as a surprise to the user of a LSS that the actual sharing of the location can be more expensive than the location retrieval [2]. The location data is often shared over cellular networks, which incurs an additional energy burden on the handset. In this work we isolate the energy overhead for *location sharing* which arises due to the cellular network data transmission regime. The data pattern of the application (i.e., packet size and inter-packet interval) highly influences the energy consumption of the mobile device. Our work focuses on the protocol impact by measuring the actual cost of sharing the location using 3G for two protocols.

We experimentally compare the energy cost for location sharing over a 3G network, at the device end, using MQTT and HTTP as application protocols in their standard usage mode. MQTT is a publish/subscribe messaging protocol designed for machine-to-machine or "Internet of things" contexts, e.g., constrained devices used in telemetry applications. Whenever a publisher generates a new location update, a message is sent to the subscribers (in a similar manner to server-push). The standard HTTP functions as request-response protocol using the client-server paradigm, the client periodically (every T) polls the server to retrieve other clients' location updates (similar to pull behaviour).

In order to evaluate the energy consumption and the traffic generated under realistic settings, we develop a location sharing application prototype for the Android platform able to share the location using both HTTP and MQTT. Carefully designed experiments emulate real user location traces, performed walking outdoors while carrying a smartphone with GPS signal over a predefined path at the university campus (10 minutes each) to collect the data transmission packet traces. The energy consumption is calculated using the EnergyBox, an accurate energy simulation tool that employs a detailed 3G radio resource state machine and uses the packet traces as input allowing repeatable experiments. EnergyBox has been evaluated against physical energy consumption measurements showing an accuracy of 98% [3].

2. EVALUATION RESULTS

Idle state: The mobile device may spend significant part of its time in idle state between aperiodic or infrequent location sharing updates (e.g., check-in mode or infrequent events). When there are no location updates by any of the users in the system, the standard HTTP technique requires to poll the server, which makes it to generate more traffic and consume more energy than the MQTT technique. The longer the update period T, the smaller the energy consumption becomes. However, the delay between the users' location update and the reception of it becomes greater.

Active state: When actual location data is transmitted, the low protocol overhead and the publish subscribe nature of MQTT creates lower amount of data traffic and makes MQTT more energy-efficient when the number of users in the system is low. When the load is increased, the HTTP technique takes advantage of aggregating all the location updates in a single burst being more energy-efficient when there are many updates. However, this comes at the cost of delaying the reception of the updates at the client side.

This is shown in Fig. 1, where the energy consumption of HTTP and MQTT is shown normalised to the greatest value of MQTT with 9 users (285 Joules). We vary the update period T and the number of simulated users that the client shares location data with, i.e., the load in the system. The case of 0 simulated users represents when only the client updates its own location. The energy consumption of HTTP is approximately the same for 3, 6 and 9 users, and therefore we only show the average value to compare with MQTT (shown as HTTP with users in Fig. 1).

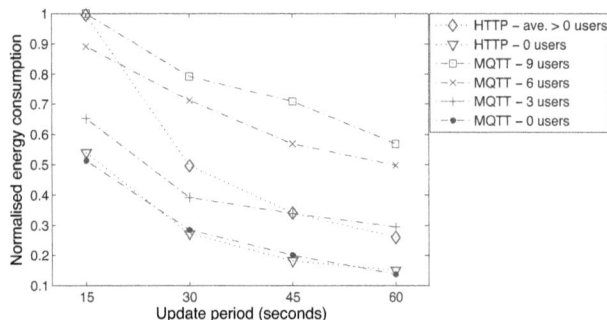

Figure 1: Normalised energy consumption for HTTP and MQTT in active state.

For 0 users, the energy consumption of HTTP and MQTT is similar. Fig. 1 shows that MQTT is more energy-efficient for shorter update periods when the number of users is 3. When the update period is increased, HTTP consumes less than MQTT since it updates the location of the users once every T. The HTTP protocol becomes more efficient when the number of users increases (6 and 9). The reason is that HTTP transmits all the data in a single burst (polling nature), at the cost of location sharing delay.

Protocol for check-in: Next we studied the energy footprint of the protocols when performing check-ins using data collected earlier [1]. Our study shows that MQTT is a more appropriate protocol for check-ins. We are interested in studying the time between updates (inter-update interval) for each user as an input to guide choosing between the two protocols. We calculate the inter-update interval for each user using the above dataset.

Fig. 2 (left) shows the empirical CDF of the inter-update interval of all the users in the dataset for short location update intervals in minutes. It shows that 15% of the general inter-update interval is longer, which causes the idle time to dominate the active time. The inter-update intervals shorter than 1 minute represent 5.6% of the total number of check-

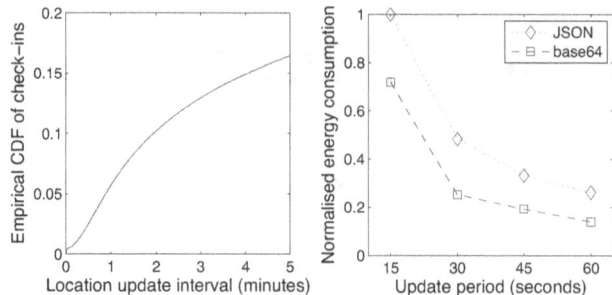

Figure 2: Empirical CDF of inter-update interval in minutes of the 22,387,930 check-ins from 224,804 users (left) and normalised energy consumption for the HTTP data encoding (right).

ins, and a single user rarely has more than 1 short consecutive updates per 1 minute (93% of the users do not have any updates in intervals shorter than 60 seconds). Therefore, MQTT appears to be the protocol of choice due to its instant location update delivery and lower energy consumption.

Data format: We next studied the contribution of data format to transmission energy footprint. Data format impacts the size of the payload and therefore the data pattern in terms of packet size. We show that applications using HTTP would benefit from using a compact data encoding format (Base64 Content-Transfer-Encoding format) over the standard verbose format (JavaScript Object Notation). Fig. 2 (right) shows the normalised energy compared to the update interval. It shows that the energy consumption can be reduced by 16 to 50% at the cost of losing human readability.

3. CONCLUSION

MQTT instantly delivers location data, creates less traffic and is more energy-efficient in the idle state or where number of sharing nodes is low. MQTT also appears as a suitable protocol option for check-in mode. When the load increases, aggregating location updates as the HTTP technique does become more competitive at the cost of delay. This suggests that an adaptive protocol that can switch the operation mode based on the number of users in the system to reduce the energy consumption might be viable.

4. REFERENCES

[1] Z. Cheng, J. Caverlee, K. Lee, and D. Z. Sui. Exploring Millions of Footprints in Location Sharing Services. In *Proceedings of the Fifth International Conference on Weblogs and Social Media*. AAAI, July 2011.

[2] M. Kjærgaard. Location-based services on mobile phones: minimizing power consumption. *IEEE Pervasive Computing*, 11(1):67 –73, 2012.

[3] E. J. Vergara and S. Nadjm-Tehrani. Energybox: A trace-driven tool for data transmission energy consumption studies. In *Proceedings of the International Conference on Energy Efficiency in Large Scale Distributed Systems (EE-LSDS 2013)*, Lecture Notes in Computer Science. Springer, April 2013.

A Multi-User Network Testbed for Wide-Area Monitoring and Control of Power Systems Using Distributed Synchrophasors

Matthew Weiss
North Carolina State University
mdweiss@ncsu.edu

Aranya Chakrabortty
North Carolina State University
achakra2@ncsu.edu

Yufeng Xin
Renaissance Computing Institute
yxin@renci.org

ABSTRACT

In this poster we describe an advanced hardware-in-loop simulation facility for real-time demonstration and validation of power system monitoring and control algorithms, currently under construction at NC State University. This facility integrates a real-time power system emulation lab with the GENI network and its associated cloud testbeds. The dynamic responses from the power system emulator are captured via real hardware Phasor Measurement Units (PMU) that are synchronized with the time-scale of the simulations via a common GPS reference. These responses are then sent to the computing and storage resource in GENI using the IEEE C37.118 protocol, running the smart grid control and management application simulations via QoS-guaranteed communications channels, all provisioned in a dynamic fashion.

Categories and Subject Descriptors

C.14 Distributed Architectures, C.4 Performance of Systems, D.4.6 Security and Protection

Author Keywords

Distributed computing, Synchrophasors, Wide-area monitoring and control, cyber-security.

1. INTRODUCTION

This poster describes a laboratory infrastructure developed at the FREEDM Systems Center in North Carolina State University, Raleigh, NC for hardware-in-loop simulations of power system networks using multiple Phasor Measurement Units (PMU) integrated with three racks of Real-time Digital Simulators. This facility is currently in the process of being extended to develop a multi-port, multi-user, and multi-vendor network of PMUs spread across the three campuses of NC State, Duke University and University of North Carolina Chapel Hill through an existing metro-scale fiber optic network called the Breakable Experimental Network (BEN), hosted by the Renaissance Computing Institute (RENCI) at UNC Chapel Hill. This PMU network will allow multiple users at various points to process, share, and communicate PMU data between each other, and collaboratively use them for critical Synchrophasor applications such as power oscillation monitoring, wide-area protection and damping control. The overall testbed will be a tremendously useful resource for testing, validation and demonstration of real-life power system operations using PMU data in a human-centric network environment. For example, local users in this PMU network can access artificial PMU data generated by simulations in real-time, apply their individual local algorithms on these data, and then communicate the results to neighboring users till the

loop reaches a global consensus over time. The network will also allow a detailed investigation of the sensitivity of these distributed algorithms on latencies, jitters, loss of GPS, etc.

2. Hardware-in-Loop PMU Testbed

In current state-of-art, using PMU data for research purposes is contingent on accessing the real data from specific utility companies that own the PMUs at the locations of interest. Gaining access to such data may not always be an easy task due to privacy and non-disclosure issues. More importantly, in many circumstances even if real PMU data are obtained they may not be sufficient for studying the detailed operation of the entire system because of their limited coverage. For example, in many studies on the dynamic analysis of the US west coast power grid, also referred to as the WECC (Western Electricity Coordinating Council) system, researchers have faced observability issues due to lack of PMU data at certain buses in the system that decide system connectivity. To circumvent this problem, over the past one year the first author and his students at FREEDM Systems Center at NC State University have started developing a hardware-in-loop simulation framework where high fidelity detailed models of large power systems can be simulated in Real-time Digital Simulators (RTDS), and the dynamic responses can be captured via real hardware Phasor Measurement Units that are synchronized via a common GPS reference. Two varieties of PMUs, namely SEL-421 and SEL-487 from Schweitzer Engineering Laboratories (SEL), are currently being used. Both have the capability of accepting an IRIG-B signal from the GPS clock, and producing data at a rate of at least 60 samples per second. They also have multiple current and voltage channels accessible, and are synchronized using a GPS clock that provides the IRIG-B signal. A Phasor Data Concentrator (PDC, SEL-3373) is used to collect measurements from multiple PMUs via Ethernet communications and EIA-232 protocol messages. The data is concentrated into one block and sent to the server via C37.118 communications protocol, or archived for later access. The user is able to view the phasor data on a computer screen using the SynchroWAVE PDC Assistant software that gathers real-time data from this PDC.

The second essential component of this set up is the Real Time Digital Simulator (RTDS), which is a power system simulator tool that allows the real-time simulation with a time-step of 50 microseconds. The RTDS comes with its own proprietary software known as RSCAD, which allows the user to develop detailed dynamic models of various components in prototype power systems. The RTDS also comes with digital cards that allow external hardware to interface with the simulation. The card of interest for our study is the Gigabit Transceiver Analog Output (GTAO) card, which will allow the user to view low level signals proportional to what voltages and currents lay on different buses of the system in real time. The GTAO card generates voltage and current waveforms, and communicates them to the PMUs. The PMUs measure these signals and send the resulting digitized phasor data calculations to the PDC. The PDC time-stamps and collects the data from all the PMUs, and sends them to the server for display and archival, when requested. The hardware and the software layers of this testbed are integrated with each other to create a substation-like environment within the confines of our research laboratory. The two layers symbiotically capture power system dynamic oscillations as if these measurements were made by real PMU data installed at the high voltage buses of a real transmission substation. So far we have implemented several prototype IEEE models such as the 9-bus Kundur System and IEEE 39-bus New England System, and carried out disturbance simulations on them to capture artificial PMU data. Currently these PMU data are being used for visualization and modal analysis. Integration of these models with Type 1 and Type 2 wind turbines isalso currently being tested. PMUs from multiple vendors are also being integrated into the system for the purpose of the testing phase angle differences due to differences in their internal filtering algorithms. Moreover, more realistic and meaningful models are being implemented in RTDS so that the use of PMUs for these models become more explicit. These models mainly represent some approximate and yet meaningful representations of realistic power system interconnections such as the US west coast grid (Western Electricity Coordinating Council, or WECC). The models can be tested by exciting them with different types of disturbances, and capturing their responses using the multi-vendor PMU set-up. Further tuning of the model parameters need to be done for validating the simulated responses.

3. Distributed Algorithms for WAMS via Exo-GENI

Besides the implementation of high-fidelity power system models in the RTDS and capturing their dynamic behavior using hardware PMUs, a parallel effort is also currently going on towards extending this facility to a much wider-scale phasor communication network using the BEN network of RENCI. The RTDS output ports are being connected to the BEN network at NC State campus, the necessary software codes in Matlab (or its open-source version depending on software license constraints, one option being Open PDC) are being developed to extract streaming PMU data from PMUs and PDCs, software images are being installed in all the virtual computers on the BEN PoPs, and finally, tests are being run to check if the data can be successfully accessed by all the nodes in sync with each other. The resulting network is referred to as the BEN-PMU network. This network will be an ideal resource to simulate, demonstrate and validate the fundamental challenges of distributed computation for any wide-area monitoring and control problem. The authors are currently planning to share only the artificial PMU data generated from the simulations using the RTDS-PMU facility. In future, however, this network can also be used for sharing actual PMU data from local utility companies. Moreover, as a first step all the hardware PMUs are being installed in the RTDS lab at NC State only. But the longer-term objective is to install more such PMUs at the remote nodes also, thereby creating a multi-sensing network. Database management, data archival and visualization strategies for this set up are also currently under development. BEN is a state-funded, metro-scale, multi-layered advanced dynamic optical network testbed operated by RENCI. It connects distributed cloud resources in local universities including NCSU and Duke University. BEN currently consists of four PoPs located at RENCI, UNC, NCSU and Duke, which use CISCO and Juniper routers above WDM/TDM bandwidth virtualization technology from Infinera. This allows dynamic multi-layer connections of up to 10 Gbps between the sites. Each PoP is also equipped with a Eucalyptus cloud cluster, and has abundant capacity to connect research labs in the BEN campus network. BEN also has 10 Gbps linkages to NLR and Internet 2 networks which allow us to set up dynamic high-speed connections to other universities in the nation. EXO-GENI uses control and management software called the Open Resource Control Architecture (ORCA) to orchestrate the networked cloud resource provisioning. ORCA platform was developed in earlier NSF-funded research and extended for use in the GENI initiative to enable experimenters to construct on demand private virtual networks spanning these research networks and GENI sites. In the end, ExoGENI will operate as a networked cloud infrastructure virtual laboratory for networking and computer science experiments that will help researchers advance the development of a faster, smarter and more reliable Internet.

We argue that the current design practice based on the centralized servers and IP-based Internet architecture is not an economical and efficient solution to satisfy the real-time requirement of processing large volume of Synchrophasor data. We envision a IaaS based solution would be more suitable. First of all, there exist different mathematical models and numerical analysis techniques that WAMS applications can use to estimate the dynamic state responses of the power grid. Different optimal control methods have also been developed to manage the stability and efficiency of the grid. These methods would produce different model accuracy with varying computation time for power systems of different topology and state conditions. EXO-GENI service allows dynamic provisioning of virtual machines of different CPU and memory capacities with customized software images. With this capability, the WAMS system can, therefore, automatically request for the right virtual machine to run the best real time algorithm for any given grid system. EXO-GENI also allows building QoS-guaranteed on-demand virtual topologies overlaid over the GENI and BEN physical network. With this capability, we can experiment various WAMS applications under real-time constraints, and answer critical questions such as: where to deploy the computing facilities (application servers and PDCs) to better facilitate the PMU data collection and processing, how to design better communication topologies, and what data transport protocols to use for more efficient control actions. Yet another major use of this testbed will be to emulate the effect of man-in-middle cyber attacks for closed-loop control using PMU measurements, either at the PDC level, or at the processing level.

4. REFERENCES

[1] A. G. Phadke, J. S. Thorp, and M. G. Adamiak, ``New Measurement Techniques for TrackingVoltage Phasors, Local System Frequency, and Rate of Change of Frequency," IEEE Transactions on Power Apparatus and Systems, vol. 102, pp. 1025-1038, May 1983.

[2] J. E. Dagle, ``Data Management Issues Associated with the August 14, 2003 Blackout Investigation," IEEE PES General Meeting, 2004.

[3] I. Baldine, Y. Xin, A. Mandal, and J. Chase, ``Exogeni: A multi-domain infrastructure-as-a-service testbed," 8th TrridentCom, March 2012.

Smart Metering using Application-Tailored Networks

Hans Wippel, Sören Finster
Institute of Telematics
Karlsruhe Institute of Technology (KIT)
Karlsruhe, Germany
{wippel, finster}@kit.edu

ABSTRACT

Smart Metering is one of the key components of upcoming Smart Grids. As with every new technology, there is no consensus on how it should be done and what dangers lie in its widespread application. Besides organization and management of data transfers and storage, the privacy of customers is also a major concern. A multitude of solutions exists, each covering specific requirements or use cases. We believe there is no single, generic solution that satisfies all requirements for all involved parties. Therefore, we see a necessity for a framework that provides flexibility in the choice of solutions and stability in deployment and design. We propose an architecture that enables stakeholders of the energy market to use application-tailored networks and protocols in order to solve their specific problems.

Categories and Subject Descriptors

C.2.1 [**Network Architecture and Design**]: Distributed Networks; C.2.4 [**Distributed Systems**]: Distributed Applications

General Terms

Design, Algorithms

Keywords

smart grid, smart metering, future internet, virtual networks

1. INTRODUCTION

The monitoring of energy consumption within a power grid using smart meters provides important data to many Smart Grid applications. But the operation of a Smart Metering service faces many challenges. It has to deal with heterogeneous deployments of smart meters and with different communication technologies. Transport of meter readings has to be done efficiently and securely. With large numbers of connected meters, the need for in-network preprocessing may arise. Depending on the application for which the metering data is intended, requirements for latency or resiliency have to be considered. And last but not least, concerns about customer privacy are raised. We believe, that there is no single solution that solves all challenges of Smart Metering at once. A Smart Metering implementation for

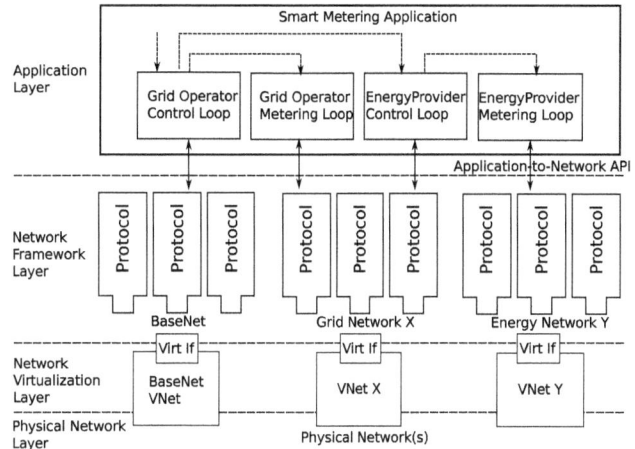

Figure 1: Smart meter overview.

a small electrical grid might look completely different from the implementation used for an electrical grid with millions of smart meters. The requirements of an implementation that is used to monitor an electrical grid in real-time are very different from a solution that simply collects long-term statistical data. Additionally, the need to support multiple concurrent Smart Metering implementations on a single smart meter at the same time may arise. For example, the smart meter could need to execute a grid operator's implementation to monitor the grid and the implementation of an energy provider to transmit important data for demand side management. A generic framework for Smart Metering solutions that copes with these extremely diverse requirements is desirable.

In this work, we propose a framework that builds on the concept of application-tailored networks. Using virtual networks, application-tailored protocols and the NENA middleware [1], we enable Smart Grid stakeholders to design and implement highly specialized Smart Metering solutions while reusing existing and proven building blocks. The deployment process uses virtual network infrastructure and enables even stakeholders with limited resources to deploy a Smart Metering solution simply, fast and without prior investment in new physical infrastructure. Additionally, smart meters can dynamically be attached to these networks and acquire all necessary protocols at runtime without modification of the Smart Metering application, giving customers flexibility to change energy providers easily.

2. PROPOSAL

Our proposal consists of a Smart Metering application, the NENA framework [1] as middleware and a set of virtual networks (see Figure 1). The Smart Metering application is pre-installed in conjunction with the NENA middleware on the smart meter. It accesses the sensors of the smart meter, e.g., to get metering data. Then, the Smart Metering application sends this metering data to a grid operator and an energy provider. Additionally, the application downloads control information from the grid operator's and energy provider's networks, in order to, for example, (re-)configure send intervals or read up-to-date energy costs. All these transfers are performed via the NENA middleware in the network framework layer. NENA selects the networks and protocols used for the respective transfers. The smart meter is attached to at least three virtual networks: the BaseNet, a grid operator network, and an energy provider network. The BaseNet hosts all the components required for the attachment process to other virtual networks. The grid operator network is used for the monitoring of an energy grid. The energy provider network is used to facilitate billing of energy consumption or demand side management.

When the smart meter is installed in a household, it is only connected to the BaseNet. The smart meter application is initially configured with a grid operator specific control URI. This URI is used to provide smart meters with a file that contains configuration parameters including a download URI for a list of available energy providers in the grid operator's energy network. The smart meter application retrieves this URI via the NENA framework. Since none of the networks NENA is currently connected with supports this URI, the attachment process [2] is started. During the attachment process, NENA establishes a virtual link to the network and acquires the protocols used in the network.

When the smart meter is connected to the grid'operator's network, the application downloads the list of available energy providers. This list contains each provider's meta information and control URI. This information is presented to the user via a web interface. At the first start of the smart meter or when the user wants to change energy providers, the user selects an energy provider and, thus, configures the respective control URI in the smart meter application. Then, the smart meter application retrieves control data from the newly configured URI. Since there is no connection to the energy provider's virtual network yet, NENA performs the attachment process [2] for the respective energy network. The control information is used to configure the metering URI and metering interval for the energy provider in the application. After these steps, the smart meter is attached to the virtual networks and the Smart Metering application performs its regular operation of sending meter data and retrieving control data.

For an example, we assume an energy provider that has the prospect of a large number of geographically scattered customers. The energy provider has currently no deployed infrastructure—especially not in the vicinity of his prospective customers. To provide the customers with time-of-use tariffs, the energy provider needs accurate monitoring of their energy consumption. The virtual network and protocol stack in Figure 2 are designed for the energy provider. In order to cope with the amount of data that is sent to the energy provider, a tree-shaped network with in-network aggregation of metering data is designed. Smart meters are

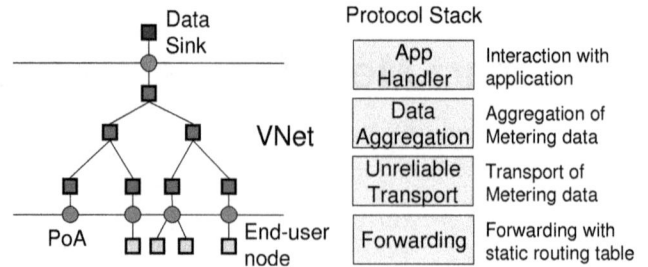

Figure 2: Smart Metering with data aggregation example.

attached to leaf-nodes of the tree that are geographically near. Metering data is sent from the smart meters to these leaf nodes, which aggregate the received metering data without removing customer specific information. On the path to the root of the tree, intermediate nodes keep aggregating the received data and forward them towards the root. After deployment of this virtual network, new customers of the energy provider automatically use the virtual network and the corresponding protocol stack. The energy provider receives accurate monitoring information and can scale his Smart Metering infrastructure with the influx of new customers by reconfiguring his virtual network.

3. CONCLUSION

In this work, we argue that the future Smart Grid is in need for highly specialized communication infrastructures and protocols. Smart Metering is one of the key components of the Smart Grid. A single infrastructure or protocol can not solve all challenges of Smart Metering at once. Therefore, we propose an architecture that utilizes application-tailored virtual networks to enable the coexistence of a multitude of specialized solutions. Our architecture enables stakeholders of the Smart Grid to design, implement and deploy their specialized infrastructures and protocols using virtual infrastructure. The architecture enables the coexistence of multiple virtual infrastructures while maintaining a consistent API to Smart Metering applications. The possibility of realizing complex tasks at the virtual network level allows for a generic Smart Metering application that is applicable for a multitude of Smart Metering solutions. For example, simple aggregation and complex peer-to-peer privacy solutions can be implemented at the virtual network level and use the same application on the smart meter. Additionally, the virtual networks and, thus, the functionality of the smart meter can be changed dynamically at runtime without modification of the Smart Metering application.

4. REFERENCES

[1] D. Martin, L. Völker, and M. Zitterbart. A Flexible Framework for Future Internet Design, Assessment, and Operation. *Computer Networks*, 55(4):910–918, Mar. 2011.

[2] H. Wippel and O. Hanka. End User Node Access to Application-Tailored Future Networks. In *Proceedings of the 21st International Conference on Computer Communication Networks (ICCCN 2012)*, Munich, Germany, Aug. 2012.

Author Index